Matrix Theory

From Generalized Inverses to Jordan Form

PURE AND APPLIED MATHEMATICS

A Program of Monographs, Textbooks, and Lecture Notes

MONOGRAPHS AND TEXTBOOKS IN PURE AND APPLIED MATHEMATICS

Recent Titles

W. J. Wickless, A First Graduate Course in Abstract Algebra (2004)

R. P. Agarwal, M. Bohner, and W-T Li, Nonoscillation and Oscillation Theory for Functional Differential Equations (2004)

J. Galambos and I. Simonelli, Products of Random Variables: Applications to Problems of Physics and to Arithmetical Functions (2004)

Walter Ferrer and Alvaro Rittatore, Actions and Invariants of Algebraic Groups (2005

Christof Eck, Jiri Jarusek, and Miroslav Krbec, Unilateral Contact Problems: Variationa Methods and Existence Theorems (2005)

M. M. Rao, Conditional Measures and Applications, Second Edition (2005)

A. B. Kharazishvili, Strange Functions in Real Analysis, Second Edition (2006)

Vincenzo Ancona and Bernard Gaveau, Differential Forms on Singular Varieties: De Rham and Hodge Theory Simplified (2005)

Santiago Alves Tavares, Generation of Multivariate Hermite Interpolating Polynomial (2005)

Sergio Macías, Topics on Continua (2005)

Mircea Sofonea, Weimin Han, and Meir Shillor, Analysis and Approximation of Contact Problems with Adhesion or Damage (2006)

Marwan Moubachir and Jean-Paul Zolésio, Moving Shape Analysis and Control: Applications to Fluid Structure Interactions (2006)

Alfred Geroldinger and Franz Halter-Koch, Non-Unique Factorizations: Algebraic, Combinatorial and Analytic Theory (2006)

Kevin J. Hastings, Introduction to the Mathematics of Operations Research with *Mathematica*®, Second Edition (2006)

Robert Carlson, A Concrete Introduction to Real Analysis (2006)

John Dauns and Yiqiang Zhou, Classes of Modules (2006)

N. K. Govil, H. N. Mhaskar, Ram N. Mohapatra, Zuhair Nashed, and J. Szabados, Frontiers in Interpolation and Approximation (2006)

Luca Lorenzi and Marcello Bertoldi, Analytical Methods for Markov Semigroups (2006)

M. A. Al-Gwaiz and S. A. Elsanousi, Elements of Real Analysis (2006)

Theodore G. Faticoni, Direct Sum Decompositions of Torsion-free Finite Rank Groups (2006)

R. Sivaramakrishnan, Certain Number-Theoretic Episodes in Algebra (2006)

Aderemi Kuku, Representation Theory and Higher Algebraic K-Theory (2006)

Robert Piziak and P. L. Odell, Matrix Theory: From Generalized Inverses to Jordan Form (2007)

Norman L. Johnson, Vikram Jha, and Mauro Biliotti, Handbook of Finite Translation Planes (2007)

Matrix Theory

From Generalized Inverses to Jordan Form

Robert Piziak

Baylor University

Texas, U.S.A.

P. L. Odell

Baylor University

Texas, U.S.A.

Chapman & Hall/CRC
Taylor & Francis Group
Boca Raton London New York

Chapman & Hall/CRC is an imprint of the
Taylor & Francis Group, an informa business

Chapman & Hall/CRC
Taylor & Francis Group
6000 Broken Sound Parkway NW, Suite 300
Boca Raton, FL 33487-2742

Printed in Canada on acid-free paper
10 9 8 7 6 5 4 3 2

International Standard Book Number-10: 1-58488-625-0 (Hardcover)
International Standard Book Number-13: 978-1-58488-625-9 (Hardcover)

Library of Congress Cataloging-in-Publication Data

Piziak, Robert.
Matrix theory : from generalized inverses to Jordan form / Robert Piziak and P.L. Odell.
p. cm. -- (Pure and applied mathematics)
Includes bibliographical references and index.
ISBN-13: 978-1-58488-625-9 (acid-free paper)
1. Matrices--Textbooks. 2. Algebras, Linerar--Textbooks. 3. Matrix inversion--Textbooks. I. Odell, Patrick L., 1930- II. Title. III. Series.

QA188.P59 2006
512.9'434--dc22 2006025707

Visit the Taylor & Francis Web site at
http://www.taylorandfrancis.com

and the CRC Press Web site at
http://www.crcpress.com

Dedication

Dedicated to the love and support of our spouses

Preface

This text is designed for a second course in matrix theory and linear algebra accessible to advanced undergraduates and beginning graduate students. Many concepts from an introductory linear algebra class are revisited and pursued to a deeper level. Also, material designed to prepare the student to read more advanced treatises and journals in this area is developed. A key feature of the book is the idea of "generalized inverse" of a matrix, especially the Moore-Penrose inverse. The concept of "full rank factorization" is used repeatedly throughout the book. The approach is always "constructive" in the mathematician's sense.

The important ideas needed to prepare the reader to tackle the literature in matrix theory included in this book are the Henderson and Searle formulas, Schur complements, the Sherman-Morrison-Woodbury formula, the LU factorization, the adjugate, the characteristic and minimal polynomial, the Frame algorithm and the Cayley-Hamilton theorem, Sylvester's rank formula, the fundamental subspaces of a matrix, direct sums and idempotents, index and the Core-Nilpotent factorization, nilpotent matrices, Hermite echelon form, full rank factorization, the Moore-Penrose inverse and other generalized inverses, norms, inner products and the QR factorization, orthogonal projections, the spectral theorem, Schur's triangularization theorem, the singular value decomposition, Jordan canonical form, Smith normal form, and tensor products.

This material has been class tested and has been successful with students of mathematics, undergraduate and graduate, as well as graduate students in statistics and physics. It can serve well as a "bridge course" to a more advanced study of abstract algebra and to reading more advanced texts in matrix theory.

Introduction

In 1990, a National Science Foundation (NSF) meeting on the undergraduate linear algebra curriculum recommended that at least one "second" course in linear algebra should be a high priority for every mathematics curriculum. This text is designed for a second course in linear algebra and matrix theory taught at the senior undergraduate and beginning graduate level. It has evolved from notes we have developed over the years teaching MTH 4316/5316 at Baylor University. This text and that course presuppose a semester of sophomore level introductory linear algebra. Even so, recognizing that certain basic ideas need review and reinforcement, we offer a number of appendixes that can be used in the classroom or assigned for independent reading. More and more schools are seeing the need for a second semester of linear algebra that goes more deeply into ideas introduced at that level and delving into topics not typically covered in a sophomore level class. One purpose we have for this course is to act as a bridge to our abstract algebra courses. Even so, the topics were chosen to appeal to a broad audience and we have attracted students in statistics and physics. There is more material in the text than can be covered in one semester. This gives the instructor some flexibility in the choice of topics. We require our students to write a paper on some topic in matrix theory as part of our course requirements, and the material we omit from this book is often a good starting point for such a project.

Our course begins by setting the stage for the central problem in linear algebra, solving systems of linear equations. In Chapter 1, we present three views of this problem: the geometric, the vector, and the matrix view. One of the main goals of this book is to develop the concept of "generalized inverses," especially the Moore-Penrose inverse. We therefore first present a careful treatment of ordinary invertible matrices, including a connection to the minimal polynomial. We develop the Henderson-Searle formulas for the inverse of a sum of matrices, the idea of a Schur complement and the Sherman-Morrison-Woodbury formula. In Chapter 2, we discuss the LU factorization after reviewing Gauss elimination. Next comes the adjugate of a matrix and the Frame algorithm for computing the coefficients of the characteristic polynomial and from which the Cayley-Hamilton theorem results.

In Chapter 3, we recall the subspaces one associates with a matrix and review the concept of rank and nullity. We derive Sylvester's rank formula and reap the many consequences that follow from this result. Next come direct sum decompositions and the connection with idempotent matrices. Then the idea of index of a matrix is introduced and the Core-Nilpotent factorization is proved. Nilpotent matrices are then characterized in what is probably the most challenging material in the book. Left and right inverses are introduced in part to prepare the way for generalized inverses.

In Chapter 4, after reviewing row reduced echelon form, matrix equivalence, and the Hermite echelon form, we introduce the all-important technique of full rank factorization. The Moore-Penrose inverse is then introduced using the concept of full rank factorization and is applied to the problem of solving a system of linear equations, giving a consistency condition and a description of all solutions when they exist. This naturally leads to the next chapter on other generalized inverses.

At this point, some choices need to be made. The instructor interested in pursuing more on generalized inverses can do so at the sacrifice of later material. Or, the material can be skipped and used as a source of projects. Some middle ground can also be chosen.

Chapter 6 concerns norms, vector and matrix. We cover this material in one day since it is really needed only to talk about minimum norm solutions and least squares solutions to systems of equations. The chapter on inner products comes next. The first new twist is that we are dealing with complex vector spaces so we need a Hermitian inner product. The concept of orthogonality is reviewed and the QR factorization is developed. Kung's approach to finding the QR factorization is presented. Now we are in a position to deal with minimum norm and least squares solutions and the beautiful connection with the Moore-Penrose inverse. The material on orthogonal projections in Chapter 8 is needed to relate the Moore-Penrose inverse to the orthogonal projections onto the fundamental subspaces of a matrix and to our treatment of the spectral theorem. Unfortunately, sections 2 and 5 are often skipped due to lack of time. In Chapter 9, we prove the all-important spectral theorem.

The highlights of Chapter 10 are the primary decomposition theorem and Schur's triangularization theorem. Then, of course, we also discuss singular value decomposition, on which you could spend an inordinate amount of time. The big finish to our semester is the Jordan canonical form theorem. To be honest, we have never had time to cover the last chapter on multilinear algebra, Chapter 12.

Now to summarize what we feel to be the most attractive features of the book:

1. The style is conversational and friendly but always mathematically correct. The book is meant to be read. Concrete examples are used to make abstract arguments more clear.

2. Routine proofs are left to the reader as exercises while we take the reader carefully through the more difficult arguments.

3. The book is flexible. The core of the course is complex matrix theory accessible to all students. Additional material is available at the discretion of the instructor, depending on the audience or the desire to assign individual projects.

4. The Moore-Penrose inverse is developed carefully and plays a central role, making this text excellent preparation for more advanced treatises in matrix theory, such as Horn and Johnson and Ben Israel and Greville.

5. The book contains an abundance of homework problems. They are not graded by level of difficulty since life does not present problems that way.

6. Appendixes are available for review of basic linear algebra.

7. Since MATLAB seems to be the language of choice for dealing with matrices, we present MATLAB examples and exercises at appropriate places in the text.

8. Most sections include suggested further readings at the end.

9. Our approach is "constructive" in the mathematician's sense, and we do not extensively treat numerical issues. However, some "Numerical Notes" are included to be sure the reader is aware of issues that arise when computers are used to do matrix calculations.

There are many debts to acknowledge. This writing project was made possible and was launched when the first author was granted a sabbatical leave by Baylor University in the year 2000. We are grateful to our colleague Ron Stanke for his help in getting the sabbatical application approved.

We are indebted to many students who put up with these notes and with us for a number of years. They have taken our courses, have written master's theses under our direction, and have been of immeasurable help without even knowing it. A special thanks goes to Dr. Richard Greechie and his Math 405 linear algebra class at LA TECH, who worked through an earlier version of our notes in the spring of 2004 and who suggested many improvements. We must also mention Curt Kunkel, who took our matrix theory class from this book and who went way beyond the call of duty in reading and rereading our notes, finding many misprints, and offering many good ideas. Finally, we appreciate the useful comments and suggestions of our colleague Manfred Dugas, who taught a matrix theory course from a draft version of our book.

Having taught out of a large number of textbooks over the years, we have collected a large number of examples and proofs for our classes. We have made some attempt to acknowledge these sources, but many have likely been lost in the fog of the past. We apologize ahead of time to any source we have failed to properly acknowledge. Certainly, several authors have had a noticeable impact on our work. G. Strang has developed a picture of "fundamental" subspaces that we have used many times. S. Axler has influenced us to be "determinant free" whenever possible, though we have no fervor against determinants and use them when they yield an easy proof. Of course, the wonderful book by C. D. Meyer, Jr., has changed our view on a number of sections and caused us to rewrite material to follow his insightful lead. The new edition of Ben-Israel and Greville clearly influences our treatment of generalized inverses.

We also wish to thank Roxie Ray and Margaret Salinas for their help in the typing of the manuscript. To these people and many others, we are truly grateful.

Robert Piziak
Patrick L. Odell

References

D. Carlson, E. R. Johnson, D. C. Lay, and A. D. Porter, The linear algebra curriculum study group recommendations for the first course in linear algebra, College Mathematics Journal, 24 (2003), 41–45.

J. G. Ianni, What's the best textbook?—Linear algebra, Focus, 24 (3) (March, 2004), 26.

An excellent source of historical references is available online at **http://www-history.mcs.st-andrews.ac.uk/** thanks to the folks at St. Andrews, Scotland.

Contents

Chapter 1

The Idea of Inverse

systems of linear equations, geometric view, vector view, matrix view

1.1 Solving Systems of Linear Equations

The central problem of linear algebra is the problem of solving a system of linear equations. References to solving simultaneous linear equations that derived from everyday practical problems can be traced back to Chiu Chang Suan Shu's book *Nine Chapters of the Mathematical Art,* about 200 B.C. [Smoller, 2005]. Such systems arise naturally in modern applications such as economics, engineering, genetics, physics, and statistics. For example, an electrical engineer using Kirchhoff's law might be faced with solving for unknown currents x, y, z in the system of equations:

$$\begin{cases} 1.95x + 2.03y + 4.75z = 10.02 \\ 3.45x + 6.43y - 5.02z = 12.13 \\ 2.53x + 7.01y + 3.61z = 19.46 \\ 3.01x + 5.71y + 4.02z = 10.52 \end{cases}$$

Here we have four linear equations (no squares or higher powers on the unknowns) in three unknowns x, y, and z. Generally, we can consider a system of m linear equations in n unknowns:

$$\begin{cases} a_{11}x_1 + a_{12}x_2 + \cdots + a_{1n}x_n = b_1 \\ a_{21}x_1 + a_{22}x_2 + \cdots + a_{2n}x_n = b_2 \\ a_{31}x_1 + a_{32}x_2 + \cdots + a_{3n}x_n = b_3 \\ \vdots \qquad\qquad\qquad\qquad \vdots \qquad \vdots \\ a_{m1}x_1 + a_{m2}x_2 + \cdots + a_{mn}x_n = b_m \end{cases} . \tag{1.1}$$

The coefficients a_{ij} and constants b_k are all complex numbers. We use the symbol $\mathbb{C}$ to denote the collection of complex numbers. Of course, real numbers, denoted $\mathbb{R}$, are just special kinds of complex numbers, so that case is automatically included in our discussion. If you have forgotten about complex numbers (remember $i^2 = -1$?) or never seen them, spend some time reviewing Appendix A, where we tell you all you need to know about complex numbers. Mathematicians allow the coefficients and constants in (1.1) to come from number domains more general than the complex numbers, but we need not be concerned about that at this point.

The concepts and theory derived to discuss solutions of a system of linear equations depend on at least three different points of view. First, we could view each linear equation individually as defining a *hyperplane* in the vector space of complex n-tuples $\mathbb{C}^n$. Recall that a hyperplane is just the translation of an $(n-1)$ dimensional subspace in $\mathbb{C}^n$. This view is the *geometric view* (or *row view*). If $n = 2$ and we draw pictures in the familiar Cartesian coordinate plane $\mathbb{R}^2$, a row represents a line; a row represents a plane in $\mathbb{R}^3$, etc. From this point of view, solutions to (1.1) can be visualized as the intersection of lines or planes. For example, in

$$\begin{cases} 3x - y = 7 \\ x + 2y = 7 \end{cases}$$

the first row represents the line (= hyperplane in $\mathbb{R}^2$) through the origin $y = 3x$ translated to go through the point $(0, -7)$ [see Figure 1.1], and the second row represents the line $y = -\frac{1}{2}x$ translated to go through $(0, \frac{7}{2})$. The solution to the system is the ordered pair of numbers $(3, 2)$ which, geometrically, is the intersection of the two lines, as is illustrated in Figure 1.1 and as the reader may verify.

Another view is to look "vertically," so to speak, and develop the *vector* (or *column*) *view*; that is, we view (1.1) written as

$$x_1 \begin{bmatrix} a_{11} \\ a_{21} \\ \vdots \\ a_{m1} \end{bmatrix} + x_2 \begin{bmatrix} a_{12} \\ a_{22} \\ \vdots \\ a_{m2} \end{bmatrix} + \cdots + x_n \begin{bmatrix} a_{1n} \\ a_{2n} \\ \vdots \\ a_{mn} \end{bmatrix} = \begin{bmatrix} b_1 \\ b_2 \\ \vdots \\ b_m \end{bmatrix} \tag{1.2}$$

recalling the usual way of adding column vectors and multiplying them by scalars. Note that the xs really should be on the right of the columns to produce the system (1.1). However, complex numbers satisfy the *commutative law of multiplication* ($ab = ba$), so it does not matter on which side the x is written. The problem has not changed. We are still obliged to find the unknown xs given the a_{ij}s and the b_ks. However, instead of many equations (the rows) we have just one vector equation. Regarding the columns as vectors (i.e., as n-tuples in $\mathbb{C}^n$), the problem becomes: find a linear combination of the columns on the left

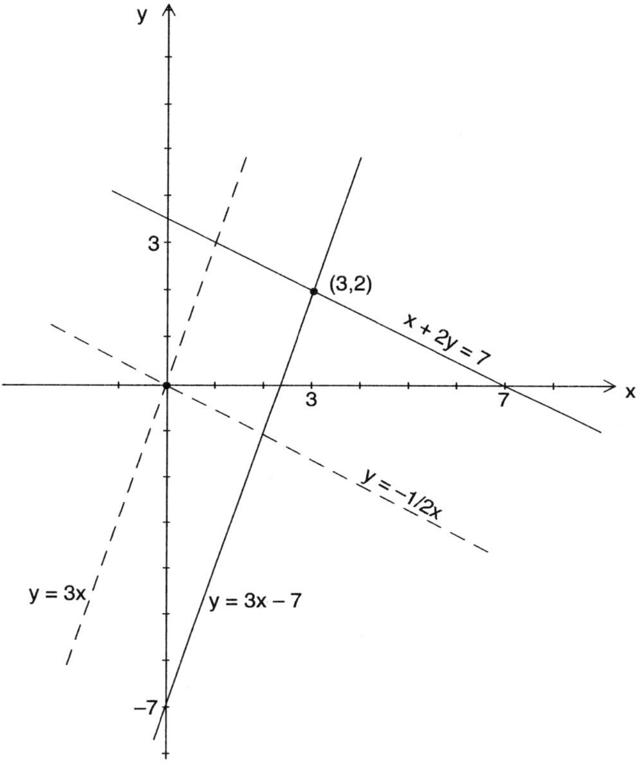

Figure 1.1: Geometric view.

to produce the vector on the right. That is, can you shrink or stretch the column vectors on the left in some way so they "resolve" to the vector on the right? And so the language of *vector spaces* quite naturally finds its way into the fundamental problem of linear algebra. We could phrase (1.2) by asking whether the vector $(b_1, b_2, \cdots, b_m)$ in $\mathbb{C}^m$ is in the subspace spanned by (i.e., generated by) the vectors $(a_{11}, a_{21}, \cdots, a_{m1})$, $(a_{12}, a_{22}, \cdots, a_{m2})$, $\cdots$, $(a_{1n}, a_{2n}, \cdots, a_{mn})$. Do you remember all those words? If not, see Appendix D.

Our simple example in the vector view is

$$x \begin{bmatrix} 3 \\ 1 \end{bmatrix} + y \begin{bmatrix} -1 \\ 2 \end{bmatrix} = \begin{bmatrix} 7 \\ 7 \end{bmatrix}.$$

Here the coefficients $x = 3$, $y = 2$ yield the solution. We can visualize the situation using arrows and adding vectors as scientists and engineers do, by the head-to-tail rule (see Figure 1.2).

The third view is the *matrix view*. For this view, we gather the coefficients a_{ij} of the system into an m-by-n matrix A, and put the xs and the bs into columns

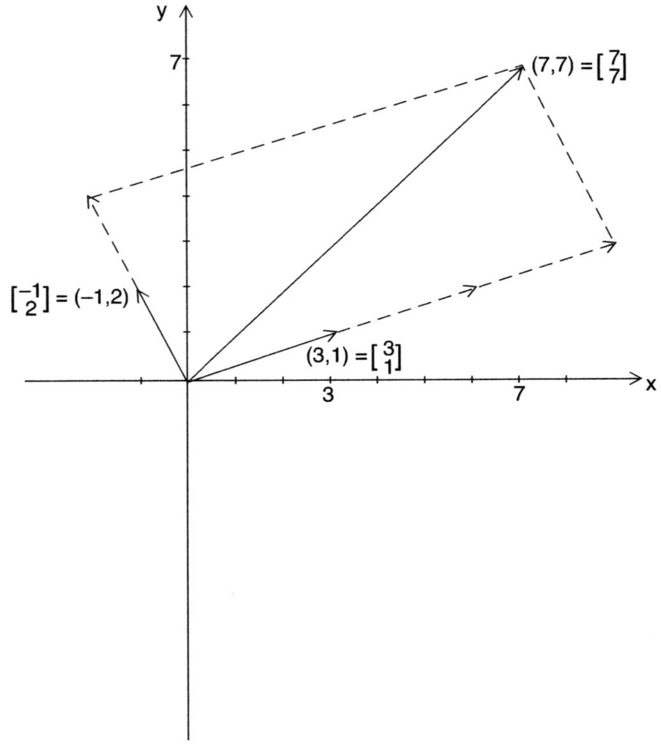

Figure 1.2: Vector view.

(n-by-1 and m-by-1, respectively). From the definition of matrix multiplication and equality, we can write (1.1) as

$$\begin{bmatrix} a_{11} & a_{12} & \cdots & a_{1n} \\ a_{21} & a_{22} & \cdots & a_{2n} \\ \vdots & \vdots & & \vdots \\ a_{m1} & a_{m2} & \cdots & a_{mn} \end{bmatrix} \begin{bmatrix} x_1 \\ x_2 \\ \vdots \\ x_n \end{bmatrix} = \begin{bmatrix} b_1 \\ b_2 \\ \vdots \\ b_m \end{bmatrix} \tag{1.3}$$

or, in the very convenient shorthand,

$$A\mathbf{x} = \mathbf{b} \tag{1.4}$$

where

$$A = \begin{bmatrix} a_{11} & a_{12} & \cdots & a_{1n} \\ a_{21} & a_{22} & \cdots & a_{2n} \\ \vdots & \vdots & & \vdots \\ a_{m1} & a_{m2} & \cdots & a_{mn} \end{bmatrix}, \quad \mathbf{x} = \begin{bmatrix} x_1 \\ x_2 \\ \vdots \\ x_n \end{bmatrix}, \quad \text{and} \quad \mathbf{b} = \begin{bmatrix} b_1 \\ b_2 \\ \vdots \\ b_m \end{bmatrix}.$$

This point of view leads to the rules of *matrix algebra* since we can view (1.1) as the problem of solving the matrix equation (1.4). The emphasis here is on "symbol pushing" according to certain rules. If you have forgotten the basics of manipulating matrices, adding them, multiplying them, transposing them, and so on, review Appendix B to refresh your memory. Our simple example can be expressed in matrix form as

$$\begin{bmatrix} 3 & -1 \\ 1 & 2 \end{bmatrix} \begin{bmatrix} x \\ y \end{bmatrix} = \begin{bmatrix} 7 \\ 7 \end{bmatrix}.$$

The matrix view can also be considered more abstractly as a *mapping* or *function view.* If we consider the function $\mathbf{x} \mapsto A\mathbf{x}$, we get a *linear transformation* from $\mathbb{C}^n$ to $\mathbb{C}^m$ if A is m-by-n. Then, asking if (1.1) has a solution is the same as asking if $\mathbf{b}$ lies in the range of this mapping. From (1.2), we see this is the same as asking if $\mathbf{b}$ lies in the column space of A, which we shall denote $Col(A)$. Recall that the *column space* of A is the subspace of $\mathbb{C}^m$ generated by the columns of A considered as vectors in $\mathbb{C}^m$. The connection between the vector view and the matrix view is a fundamental one. Concretely in the 3-by-3 case,

$$\begin{bmatrix} a & b & c \\ d & e & f \\ g & h & i \end{bmatrix} \begin{bmatrix} x \\ y \\ z \end{bmatrix} = \begin{bmatrix} a \\ d \\ g \end{bmatrix} x + \begin{bmatrix} b \\ e \\ h \end{bmatrix} y + \begin{bmatrix} c \\ f \\ i \end{bmatrix} z.$$

More abstractly,

$$\begin{aligned} A\mathbf{x} &= \begin{bmatrix} Col_1(A) & Col_2(A) & \cdots & Col_n(A) \end{bmatrix} \begin{bmatrix} x_1 \\ x_2 \\ \vdots \\ x_n \end{bmatrix} \\ &= x_1 Col_1(A) + x_2 Col_2(A) + \cdots + x_n Col_n(A). \end{aligned}$$

Now it is clear that $A\mathbf{x} = \mathbf{b}$ has a solution if $\mathbf{b}$ is expressible as a linear combination of the columns of A (i.e., $\mathbf{b}$ lies in the column space of A). A row version of this fundamental connection to matrix multiplication can also be useful on occasion

$$\begin{bmatrix} x & y & z \end{bmatrix} \begin{bmatrix} a & b & c \\ d & e & f \\ g & h & i \end{bmatrix} = x \begin{bmatrix} a & b & c \end{bmatrix} + y \begin{bmatrix} d & e & f \end{bmatrix} + z \begin{bmatrix} g & h & i \end{bmatrix}.$$

More abstractly,

$$\mathbf{x}^T A = \begin{bmatrix} x_1 & x_2 & \cdots & x_m \end{bmatrix} \begin{bmatrix} Row_1(A) \\ Row_2(A) \\ \vdots \\ Row_m(A) \end{bmatrix}$$
$$= x_1 Row_1(A) + x_2 Row_2(A) + \cdots + x_m Row_m(A).$$

You may remember (and easy pictures in $\mathbb{R}^2$ will remind you) that three cases can occur when you try to solve (1.1).

CASE 1.1
A unique solution to (1.1) exists. That is to say, if A and $\mathbf{b}$ are given, there is one and only one $\mathbf{x}$ such that $A\mathbf{x} = \mathbf{b}$. In this case, we would like to have an efficient algorithm for finding this unique solution.

CASE 1.2
More than one solution for (1.1) exists. In this case, infinite solutions are possible. Even so, we would like to have a meaningful way to describe all of them.

CASE 1.3
No solution to (1.1) exists. We are not content just to give up and turn our backs on this case. Indeed, "real life" situations demand "answers" to (1.1) even when a solution does not exist. So, we will seek a "best approximate solution" to (1.1), where we will make clear later what "best" means.

Thus, we shall now set some problems to solve. Our primary goal will be to give solutions to the following problems:

Problem 1.1
Let $A\mathbf{x} = \mathbf{b}$ where A and $\mathbf{b}$ are specified. Determine whether a solution for $\mathbf{x}$ exists (i.e., develop a "consistency condition") and, if so, describe the set of all solutions.

Problem 1.2
Let $A\mathbf{x} = \mathbf{b}$ where A and $\mathbf{b}$ are specified. Suppose it is determined that no solution for $\mathbf{x}$ exists. Then find the best approximate solution; that is, find $\mathbf{x}$ such that the vector $A\mathbf{x} - \mathbf{b}$ has minimal length.

Problem 1.3
Given a matrix A, determine the column space of A; that is, find all $\mathbf{b}$ such that $A\mathbf{x} = \mathbf{b}$ for some column vector $\mathbf{x}$.

To prepare for what is to come, we remind the reader that complex numbers can be conjugated. That is, if $z = a + bi$, then the *complex conjugate* of z is $\overline{z} = a - bi$. This leads to some new possibilities for operations on matrices that we did not have with real matrices. If $A = [a_{ij}]$, then $\overline{A} = [\overline{a}_{ij}]$ and A^* is the transpose of $\overline{A}$ (i.e., the *conjugate transpose*) of A. Thus if $A = \begin{bmatrix} 2+3i & 4+5i \\ 7-5i & 6-3i \end{bmatrix}$, then $\overline{A} = \begin{bmatrix} 2-3i & 4-5i \\ 7+5i & 6+3i \end{bmatrix}$ and $A^* = \begin{bmatrix} 2-3i & 7+5i \\ 4-5i & 6+3i \end{bmatrix}$. A matrix A is called *self-adjoint* or *Hermitian* if $A^* = A$. Also, $det(A)$ is our way of denoting the determinant of a square matrix A. (See Appendix C for a review on determinants.)

Exercise Set 1

1. Write and discuss the three views for the following systems of linear equations.

$$\begin{cases} 4x + 3y = 17 \\ 2x - \ y = 10 \end{cases}; \quad \begin{cases} 3x + 2y + z = 5 \\ 2x + 3y + z = 5 \\ x + y + 4z = 6 \end{cases}.$$

2. Give the vector and matrix view of

$$\begin{cases} (4+3i)z_1 + (7 - \pi i)z_2 = \sqrt{2} - \sqrt{3}i \\ (2 - i)z_1 + (5+13i)z_2 = 16 + 10i \end{cases}.$$

3. Solve

$$\begin{cases} (1+2i)z + (2-3i)w = 5+3i \\ (1-i)z + (4i)w = 10-4i \end{cases}.$$

4. Solve

$$\begin{cases} 3ix + 4iy + 5iz = 17i \\ 2iy + 7iz = 16i \\ 3iz = 6i \end{cases}.$$

5. Consider two systems of linear equations that are related as follows:

$$\begin{cases} a_{11}x_1 + a_{12}x_2 = k_1 \\ a_{21}x_1 + a_{22}x_2 = k_2 \end{cases} \quad \text{and} \quad \begin{cases} b_{11}y_1 + b_{12}y_2 = x_1 \\ b_{21}y_1 + b_{22}y_2 = x_2 \end{cases}.$$

Let A be the coefficient matrix of the first system and B be the coefficient matrix of the second. Substitute for the x's in the system using the second system and produce a new system of linear equations in the y's. How does the coefficient matrix of the new system relate to A and B?

6. Argue that the n-tuple $(s_1, s_2, \ldots, s_n)$ satisfies the two equations:

$$\begin{cases} a_{i1}x_1 + \cdots + a_{in}x_n = b_i \\ a_{j1}x_1 + \cdots + a_{jn}x_n = b_j \end{cases}$$

if and only if it satisfies the two equations:

$$\begin{cases} a_{i1}x_1 + \cdots + a_{in}x_n = b_i \\ (a_{j1} + ca_{i1})x_1 + \cdots + (a_{jn} + ca_{in})x_n = b_j + cb_i \end{cases}$$

for any constant c.

7. Consider the system of linear equations (1.1). Which of the following modifications of (1.1) will **not** disturb the solution set? Explain!
   (i) Multiply one equation through by a scalar.
   (ii) Multiply one equation through by a nonzero scalar.
   (iii) Swap two equations.
   (iv) Multiply the ith equation by a scalar and add the resulting equation to the jth equation, producing a new j^{th} equation.
   (v) Erase one of the equations.

8. Let $S = \{\mathbf{x} \mid A\mathbf{x} = \mathbf{b}\}$ be the solution set of a system of linear equations where A is m-by-n, $\mathbf{x}$ is n-by-1, and $\mathbf{b}$ is m-by-1. Argue that S could be empty (give a simple example); S could contain exactly one point but, if S contains two distinct points, it must contain infinitely many. In fact, the set S has a very special property. Prove that if $\mathbf{x}_1$ and $\mathbf{x}_2$ are in S, then $\lambda\mathbf{x}_1 + (1 - \lambda)\mathbf{x}_2$ is also in S for any choice of λ in $\mathbb{C}$. This says the set of solutions of a system of linear equations is an *affine subspace* of $\mathbb{C}^n$.

9. Argue that if $\mathbf{x}_1$ solves $A\mathbf{x} = \mathbf{b}$ and $\mathbf{x}_2$ solves $A\mathbf{x} = \mathbf{b}$, then $\mathbf{x}_1 - \mathbf{x}_2$ solves $A\mathbf{x} = \vec{0}$. Conversely, if $\mathbf{z}$ is any solution of $A\mathbf{x} = \vec{0}$ and $\mathbf{x}_p$ is a *particular solution* of $A\mathbf{x} = \mathbf{b}$, then argue that $\mathbf{x}_p + \mathbf{z}$ solves $A\mathbf{x} = \mathbf{b}$.

10. You can learn much about mathematics by making up your own examples instead of relying on textbooks to do it for you. Create a 4-by-3 matrix A

composed of zeros and ones where $A\mathbf{x} = \mathbf{b}$ has

(i) exactly one solution
(ii) has an infinite number of solutions
(iii) has no solution.

If you remember the concept of *rank* (we will get into it later in Section 3.1), what is the rank of A in your three examples?

11. Can a system of three linear equations in three unknowns over the real numbers have a complex nonreal solution? If not, explain why; if so, give an example of such a system.

Further Reading

[Atiyah, 2001] Michael Atiyah, Mathematics in the 20th Century, The American Mathematical Monthly, Vol. 108, No. 7, August-September, (2001), 654–666.

[D&H&H, 2005] Ian Doust, Michael D. Hirschhorn, and Jocelyn Ho, Trigonometric Identities, Linear Algebra, and Computer Algebra, The American Mathematical Monthly, Vol. 112, No. 2, February, (2005), 155–164.

[F-S, 1979] Desmond Fearnley-Sander, Hermann Grassmann and the Creation of Linear Algebra, The American Mathematical Monthly, Vol. 86, No. 10, December, (1979), 809–817.

[Forsythe, 1953] George Forsythe, Solving Linear Equations Can Be Interesting, Bulletin of the American Mathematical Society, Vol. 59, (1953), 299–329.

[Kolodner, 1964] Ignace I. Kolodner, A Note on Matrix Notation, The American Mathematical Monthly, Vol. 71, No. 9, November, (1964), 1031–1032.

[Rogers, 1997] Jack W. Rogers, Jr., Applications of Linear Algebra in Calculus, The American Mathematical Monthly, Vol. 104, No. 1, January, (1997), 20–26.

[Smoller, 2005] Laura Smoller, The History of Matrices, *http://www.ualr.edu/ lasmoller/matrices.html.*

[Wyzkoski, 1987] Joan Wyzkoski, An Application of Matrices to Space Shuttle Technology, The UMAP Journal Vol. 8, No. 3, (1987), 187–205.

1.1.1 Numerical Note

1.1.1.1 Floating Point Arithmetic

Computers, and even some modern handheld calculators, are very useful tools in reducing the drudgery that is inherent in doing numerical computations with matrices. However, calculations on these devices have limited precision, and blind dependence on machine calculations can lead to accepting nonsense answers. Of course, there are infinitely many real numbers but only a finite number of them can be represented on any given machine. The most common representation of a real number is *floating point representation.* If you have learned the scientific notation for representing a real number, the following description will not seem so strange.

A *(normalized) floating point number* is a real number x of the form

$$x = \pm .d_1 d_2 d_3 \cdots d_t \times b^e$$

where $d_1 \neq 0$, (that's the normalized part), b is called the *base*, e is the *exponent*, and d_i is an integer (digit) with $0 \leq d_i < b$. The number t is called the *precision*, and $\pm .d_1 d_2 d_3 \cdots d_t$ is called the *mantissa*. Humans typically prefer base $b = 10$; for computers, $b = 2$, but many other choices are available (e.g., $b = 8$, $b = 16$). Note 1.6×10^2 is scientific notation but $.16 \times 10^3$ is the normalized floating point way to represent 160. The exponent e has limits that usually depend on the machine; say $-k \leq e \leq K$. If $x = \pm .d_1 d_2 d_3 \cdots d_t \times b^e$ and $e > K$, we say x *overflows*; if $e < -k$, x is said to *underflow*. A number x that can be expressed as $x = \pm .d_1 d_2 d_3 \cdots d_t \times b^e$ given b, t, k, and K is called a *representable number*. Let $\mathcal{R}ep(b, t, k, K)$ denote the set of all representable numbers given the four positive integers b, t, k, and K. The important thing to realize is that $\mathcal{R}ep(b, t, k, K)$ is a finite set. Perhaps it is a large finite set, but it is still finite. Note zero, 0, is considered a special case and is always in $\mathcal{R}ep(b, t, k, K)$. Also note that there are bounds for the size of a number in $\mathcal{R}ep(b, t, k, K)$. If $x \neq 0$ in $\mathcal{R}ep(b, t, k, K)$, then $b^{k-1} \leq |x| \leq b^K(1 - b^{-t})$.

Before a machine can compute with a number, the number must be converted into a floating point representable number for that machine. We use the notation $fl(x) = \widetilde{x}$ to indicate that $\widetilde{x}$ is the floating point representation of x. That is

$$\widetilde{x} = fl(x) = x(1 + \delta),$$

where $\delta = \frac{\widetilde{x} - x}{x}$,

signifies that $\widetilde{x}$ is the floating point version of x. In particular, x is representable iff $fl(x) = x$. The difference between x and $\widetilde{x}$ is called *roundoff error*. The difference $\widetilde{x} - x$ is called the *absolute error*, and the ratio $\dfrac{|\widetilde{x} - x|}{|x|}$ is called the *relative error*. The maximum value for $|\delta|$ is called the *unit roundoff error* and is typically denoted by ϵ_0. For example, take $x = 51,022$. Say we wish to represent x in base 10, four-digit floating point representation. Then $fl(51022) = .5102 \times 10^5$. The absolute error is -2 and the relative error is about -3.9×10^{-5}.

There are two ways of converting a real number into a floating point number: *rounding* and *truncating* (or *chopping*). The rounded floating point version $\widetilde{x}_r$ of x is the t-digit number closest to x. The largest error in rounding occurs when a number is exactly halfway between two representable numbers. In the truncated version $\widetilde{x}_t$, all digits of the mantissa beyond the last to be kept are thrown away. The reader may verify the following inequalities:

$$\begin{aligned}
&\text{(i)}\ |\widetilde{x}_r - x| \le \frac{1}{2} b^{1-t}\, |x|\,, \\
&\text{(ii)}\qquad |\delta_r| \le \frac{1}{2} b^{1-t}, \\
&\text{(iii)}\ |\widetilde{x}_t - x| \le b^{1-t}\, |x|\,, \\
&\text{(iv)}\qquad |\delta_t| \le b^{1-t}.
\end{aligned}$$

The reader may also verify that the unit roundoff satisfies

$$\epsilon_0 = \begin{cases} \frac{1}{2} b^{1-t} & \text{for rounding} \\ b^{1-t} & \text{for truncating} \end{cases}.$$

1.1.1.2 Arithmetic Operations

Errors can occur when data are entered into a computer due to the floating point representation. Also, errors can occur when the usual arithmetic operations are applied to floating point numbers. For x and y floating point numbers, we can say

$$\begin{aligned}
\text{(i)}\ fl(x+y) &= (x+y)(1+\delta), \\
\text{(ii)}\quad fl(xy) &= xy(1+\delta), \\
\text{(iii)}\ fl(x-y) &= (x-y)(1+\delta), \\
\text{(iv)}\ fl(x \div y) &= (x \div y)(1+\delta).
\end{aligned}$$

Unfortunately, the usual laws of arithmetic fail. For example, there exists representable numbers x, y, and z, such that

$$\text{(v)}\ fl(fl(x+y)+z) \ne fl(x + fl(y+z)),$$

that is, the associative law fails. It may also happen that

$$\begin{aligned}(\text{vi})\ fl(x+y) &\neq fl(x)+fl(y),\\ (\text{vii})\quad fl(xy) &\neq fl(x)fl(y).\end{aligned}$$

For some good news, the commutative law still works.

1.1.1.3 Loss of Significance

Another phenomenon that can have a huge impact on the outcome of floating point operations is if small numbers are computed from big ones or if two nearly equal numbers are subtracted. This is called *cancellation error.* It can happen that the relative error in a difference can be many orders of magnitude larger than the relative errors in the individual numbers.

Exercises

1. How many representable numbers are in $\mathcal{R}ep(10, 1, 2, 1)$? Plot them on the real number line. Are they equally spaced?
2. Give an example of two representable numbers whose sum is not representable.
3. Write the four-digit floating point representation of 34248, and compute the absolute and relative error.

Further Reading

[Dwyer, 1951] P. S. Dwyer, *Linear Computations,* John Wiley & Sons, New York, (1951).

[F,M&M, 1977] G. E. Forsythe, M. A. Malcolm, and C. B. Moler, *Computer Methods for Mathematical Computations,* Prentice-Hall, Englewood Cliffs, NJ, (1977).

[G&vanL, 1996] Gene H. Golub and Charles F. Van Loan, *Matrix Computations,* 3rd edition, Johns Hopkins Press, Baltimore, MD, (1996).

[Higham, 1996] Nicholas J. Higham, *Accuracy and Stability of Numerical Algorithms,* SIAM, Philadelphia, (1996).

[Householder, 1964] Alston S. Householder, *The Theory of Matrices in Numerical Analysis,* Dover Publications, Inc., New York, (1964).

[Kahan, 2005] W. Kahan, How Futile are Mindless Assessments of Roundoff in Floating-Point Computation?, *http://www.cs.berkeley.edu/wkahan/Mindless.pdf.*

[Leon, 1998] Steven J. Leon, *Linear Algebra with Applications,* 5th edition, Prentice Hall, Upper Saddle River, NJ, (1998).

1.1.2 MATLAB Moment

1.1.2.1 Creating Matrices in MATLAB

MATLAB (short for MATrix LABoratory) is an interactive program first created by Cleve Moler in FORTRAN (1978) as a teaching tool for courses in linear algebra and matrix theory. Since then, the program has been written in C (1984) and improved over the years. It is now licensed by and is the registered trademark of The Math Works Inc., Cochituate Place, 24 Prime Park Way, Natick, MA. A student edition is available from Prentice Hall.

As its name implies, MATLAB is designed for the easy manipulation of matrices. Indeed, there is basically just one object it recognizes and that is a rectangular matrix. Even variable names are considered to be matrices. We assume the reader will learn how to access MATLAB on his or her local computer system. You will know you are ready to begin when you see the command line prompt: >>. The first command to learn is "help". MATLAB has quite an extensive internal library to assist the user. You can be more specific and ask for help on a given topic. Experiment! Closely related is the "lookfor" command. Type "lookfor" and a keyword to find functions relating to that keyword.

This first exercise helps you learn how to create matrices in MATLAB. There are a number of convenient ways to do this.

1.1.2.1.1 Explicit Entry Suppose you want to enter the matrix

$$\begin{bmatrix} 1 & 2 & 3 & 4 \\ 2 & 3 & 4 & 1 \\ 3 & 4 & 1 & 2 \end{bmatrix}.$$

There are a couple of ways to do this. First

$$>> A = [1\ 2\ 3\ 4; 2\ 3\ 4\ 1; 3\ 4\ 1\ 2]$$

will enter this matrix and give it the variable name A. Note we have used spaces to separate elements in a row (commas could be used too) and semicolons to delimit the rows. Another way to enter A is one row at a time.

$$
\begin{aligned}
>> A = [1\ 2\ 3\ 4 \\
2\ 3\ 4\ 1 \\
3\ 4\ 1\ 2]
\end{aligned}
$$

A row vector is just a matrix with one row:

$$>> a = [4\ 3\ 2\ 1]$$

A column vector can be created with the help of semicolons.

$$>> b = [1; 2; 3; 4]$$

The colon notation is a really cool way to generate some vectors. $m : n$ generates a row vector starting at m and ending at n. The implied step size is 1. Other step sizes can be used. For example, $m : s : n$ generates a row vector starting at m, going as far as n, in steps of size s.

For example

$$>> x = 1 : 5$$

returns

$$x = 1\ 2\ 3\ 4\ 5$$

and

$$>> y = 2.0 : -0.5 : 0$$

returns

$$y = 2.0000\ 1.5000\ 1.0000\ 0.5000\ 0$$

Now try to create some matrices and vectors of your own.

1.1.2.1.2 Built-in Matrices MATLAB comes equipped with many built-in matrices:

1. **The Zero Matrix:** the command $zeros(m, n)$ returns an m-by-n matrix filled with zeros. Also, $zeros(n)$ returns an n-by-n matrix filled with zeros.

2. **The Identity Matrix:** the command $eye(n)$ returns an n-by-n identity matrix (clever, aren't they?)

3. **The Ones Matrix:** the command $ones(m, n)$ returns an m-by-n matrix filled with ones. Of course, $ones(n)$ returns an n-by-n matrix filled with ones.

4. **Diagonal Matrices:** the function $diag$ creates diagonal matrices. For example, $diag([2\ 4\ 6])$ creates the matrix

$$\begin{bmatrix} 2 & 0 & 0 \\ 0 & 4 & 0 \\ 0 & 0 & 6 \end{bmatrix}$$

Now create some of the matrices described above. Try putting the matrix A above into *diag*. What happens? What is *diag(diag(A))*?

1.1.2.1.3 Randomly Generated Matrices MATLAB has two built-in random number generators. One is *rand,* which returns uniformly distributed numbers between zero and one, and the other is *randn*, which returns normally distributed numbers between zero and one. For example,

$$>> A = \text{rand}(3)$$

might create the 3-by-3 matrix

$$\begin{bmatrix} 0.9501 & 0.4860 & 0.4565 \\ 0.2311 & 0.8913 & 0.0185 \\ 0.6068 & 0.7621 & 0.8214 \end{bmatrix}.$$

We can generate a random 5-by-5 matrix with integers uniformly distributed between 0 and 10 by using

$$floor(11 * \text{rand}(5)).$$

A random 5-by-5 complex matrix with real and imaginary parts being integers normally distributed between 0 and 10 could be generated with

$$floor(11 * \text{rand}\, n(5)) + i * floor(11 * \text{rand}\, n(5)).$$

Experiment creating random matrices of various kinds.

1.1.2.1.4 Blocks A really neat way to build matrices is to build them up by blocks of smaller matrices to make a big one. The sizes, of course, must fit together.

For example,

$$\begin{aligned} >> A = [B\ 5 * ones(2) \\ eye(2)\ \ 3 * eye(2)] \end{aligned}$$

returns

$$\begin{matrix} 2 & 4 & 5 & 5 \\ 6 & 8 & 5 & 5 \\ 1 & 0 & 3 & 0 \\ 0 & 1 & 0 & 3 \end{matrix}$$

where $B = [2\ 4; 6\ 8]$ has been previously created. The matrix A could also have been created by

$$>> A = [B\ 5 * ones(2); eye(2)\ 3 * eye(2)]$$

Very useful for later in the text is the ability to create block diagonal matrices. For example,

$$>> A = blkdiag(7 * eye(3), 8 * eye(2))$$

returns the matrix

$$\begin{matrix} 7 & 0 & 0 & 0 & 0 \\ 0 & 7 & 0 & 0 & 0 \\ 0 & 0 & 7 & 0 & 0 \\ 0 & 0 & 0 & 8 & 0 \\ 0 & 0 & 0 & 0 & 8 \end{matrix}$$

Now create some matrices using blocks.

Further Reading

[H&H, 2000] Desmond J. Higham and Nicholas J. Higham, *MATLAB Guide,* SIAM, Philadelphia, (2000).

[L&H&F, 1996] Steven Leon, Eugene Herman, and Richard Faulkenberry, *ATLAST Computer Exercises for Linear Algebra*, Prentice Hall, Upper Saddle River, NJ, (1996).

[Sigmon, 1992] Kermit Sigmon, *MATLAB Primer*, 2nd edition, *http://math.ucsd.edu/~driver/21d-s99/matlab-primer.html,* (1992).

matrix inverse, uniqueness, basic facts, reversal law, Henderson and Searle formulas, Schur complement, Sherman-Morrison-Woodbury formula

1.2 The Special Case of "Square" Systems

In this section, we consider systems of linear equations (1.1) where the number of equations equals the number of unknowns; that is, $m = n$. In the form $A\mathbf{x} = \mathbf{b}$, A becomes a square (i.e., n-by-n) matrix while $\mathbf{x}$ and $\mathbf{b}$ are n-by-1 column vectors. Thinking back to the good old high school days in Algebra I, we learned the complete story concerning the equation

$$ax = b \tag{1.5}$$

over the real numbers $\mathbb{R}$.

If $a = 0$ and $b = 0$, any real number x provides a solution. If $a = 0$ but $b \neq 0$, then no solution x exists. If $a \neq 0$, then we have the unique solution $x = b/a$; that is, we divide both sides of (2.1) by a. Wouldn't it be great if we could solve the matrix equation

$$A\mathbf{x} = \mathbf{b} \tag{1.6}$$

the same way? However, as you well remember, dividing by a matrix was never an allowed operation, but matrix multiplication surely was. Another way to look at (1.5) is, rather than dividing by a, multiply both sides of (1.5) by the reciprocal (inverse) of a, $1/a$. Then $x = (1/a)b$ is the unique solution to (1.5). Now, the reciprocal of a in $\mathbb{R}$ is characterized as the unique number c such that $ac = 1 = ca$. Of course, we write c as $1/a$ or a^{-1} so we can write $x = a^{-1}b$ above. We can translate these ideas into the world of square matrices.

DEFINITION 1.1 *(matrix inverse)*
*An n-by-n matrix A is said to have an **inverse** if there exists another n-by-n matrix C with $AC = CA = I_n$, where I_n is the n-by-n identity matrix. Any matrix A that has an inverse is called **invertible** or **nonsingular**.*

It is important to note that under this definition, only "square" (i.e., n-by-n) matrices can have inverses. It is easy to give examples of square matrices that are not invertible. For instance $\begin{bmatrix} 0 & 1 \\ 0 & 2 \end{bmatrix}$ can not be invertible since multiplication by any 2-by-2 matrix yields

$$\begin{bmatrix} 0 & 1 \\ 0 & 2 \end{bmatrix}\begin{bmatrix} a & c \\ b & d \end{bmatrix} = \begin{bmatrix} b & d \\ 2b & 2d \end{bmatrix}.$$

To get the identity matrix $I_2 = \begin{bmatrix} b & d \\ 2b & 2d \end{bmatrix}$ as an answer, b would have to be 1 and $2b$ would have to be 0, a contradiction. On the other hand,

$$\begin{bmatrix} 0 & 1 \\ 1 & 1 \end{bmatrix}\begin{bmatrix} -1 & 1 \\ 1 & 0 \end{bmatrix} = \begin{bmatrix} 1 & 0 \\ 0 & 1 \end{bmatrix} = \begin{bmatrix} -1 & 1 \\ 1 & 0 \end{bmatrix}\begin{bmatrix} 0 & 1 \\ 1 & 1 \end{bmatrix}.$$

We see some square matrices have inverses and some do not. However, when a matrix is invertible, it can have only one inverse.

THEOREM 1.1 (uniqueness of inverse)

Suppose A is an n-by-n invertible matrix. Then only one matrix can act as an inverse matrix for A. As notation, we denote this unique matrix as A^{-1}. *Thus* $AA^{-1} = A^{-1}A = I_n$.

PROOF In typical mathematical fashion, to prove uniqueness, we assume there are two objects that satisfy the conditions and then show those two objects are equal. So let B and C be inverse matrices for A. Then $C = IC = (BA)C = B(AC) = BI = B$. You should, of course, be able to justify each equality in detail. We have used the neutral multiplication property of the identity, the definition of inverse, and the associative law of matrix multiplication. □

The uniqueness of the inverse matrix gives us a handy procedure for proving facts about inverses. If you can guess what the inverse should be, all you need to verify are the two equations in the definition. If they work out, you must have *the* inverse.

We next list some basic facts about inverses that, perhaps, you already know. It would be a good exercise to verify them.

THEOREM 1.2 (basic facts about inverses)

Suppose the matrices below are square and of the same size.

1. I_n *is invertible for any n and* $I_n^{-1} = I_n$.
2. *If* $k \neq 0$, kI_n *is invertible and* $(kI_n)^{-1} = \frac{1}{k}I_n$.
3. *If A is invertible, so is* A^{-1} *and* $(A^{-1})^{-1} = A$.
4. *If A is invertible, so is* A^n *for any natural number n and* $(A^n)^{-1} = (A^{-1})^n$.
5. *If A and B are invertible, so is AB and* $(AB)^{-1} = B^{-1}A^{-1}$. *(This is called the "reversal law for inverses" by some.)*
6. *If* $k \neq 0$ *and A is invertible, so is kA and* $(kA)^{-1} = \frac{1}{k}A^{-1}$.

7. *If A is invertible, so is $\overline{A}$, A^T, and A^*; moreover, $(A^T)^{-1} = (A^{-1})^T$, $\overline{(A^{-1})} = (\overline{A})^{-1}$, and $(A^*)^{-1} = (A^{-1})^*$.*

8. *If $A^2 = I_n$, then $A = A^{-1}$.*

9. *If A is invertible, so is AA^T, $A^T A$, AA^*, and A^*A.*

10. *If A and B are square matrices of the same size and the product AB is invertible, then A and B are both invertible.*

PROOF This is an exercise for the reader. □

There are many equivalent ways to say that a square matrix has an inverse. Again, this should be familiar material, so we omit the details. For the square matrix A, let the mapping from $\mathbb{C}^n$ to $\mathbb{C}^n$ given by $\mathbf{x} \mapsto A\mathbf{x}$ be denoted L_A.

THEOREM 1.3 (equivalents to being invertible)

For an n-by-n matrix A, the following statements are all equivalent (T.A.E.):

1. *A is invertible.*

2. *The equation $A\mathbf{x} = \vec{0}$ has only the trivial solution $\mathbf{x} = \vec{0}$.*

3. *$A\mathbf{x} = \mathbf{b}$ has a solution for every n-by-1 $\mathbf{b}$.*

4. *$A\mathbf{x} = \mathbf{b}$ has a unique solution for each n-by-1 $\mathbf{b}$.*

5. *$det(A) \neq 0$.*

6. *The columns of A are linearly independent.*

7. *The rows of A are linearly independent.*

8. *The columns of A span $\mathbb{C}^n$.*

9. *The rows of A span $\mathbb{C}^n$.*

10. *The columns of A form a basis of $\mathbb{C}^n$.*

11. *The rows of A form a basis of $\mathbb{C}^n$.*

12. *A has rank n (A has full rank).*

13. *A has nullity 0.*

14. *The mapping L_A is onto.*

15. *The mapping L_A is one-to-one.*

PROOF Again the proofs are relegated to the exercises. ☐

An important consequence of this theorem is the intimate connection between finding a basis of $\mathbb{C}^n$ and creating an invertible matrix by putting this basis as columns to form the matrix. We shall use this connection repeatedly. Now, returning to the question of solving a "square" system of linear equations, we get a very nice answer.

THEOREM 1.4 (solving a square system of full rank)
Suppose $A\mathbf{x} = \mathbf{b}$ *has an invertible coefficient matrix* A *of size* n*-by-*n*. Then the system has a unique solution, namely* $\mathbf{x} = A^{-1}\mathbf{b}$.

PROOF Suppose $A\mathbf{x} = \mathbf{b}$ has the invertible coefficient matrix A. We claim $\mathbf{x}_0 = A^{-1}\mathbf{b}$ is a solution. Compute $A\mathbf{x}_0 = A(A^{-1}\mathbf{b}) = (AA^{-1})\mathbf{b} = I\mathbf{b} = \mathbf{b}$. Next, if $\mathbf{x}_1$ is some other solution of $A\mathbf{x} = \mathbf{b}$, we must have $A\mathbf{x}_1 = \mathbf{b}$. But then, $A^{-1}(A\mathbf{x}_1) = A^{-1}\mathbf{b}$, and so $(A^{-1}A)\mathbf{x}_1 = A^{-1}\mathbf{b}$. Of course, $A^{-1}A = I$, so $I\mathbf{x}_1 = A^{-1}\mathbf{b}$, and hence, $\mathbf{x}_1 = A^{-1}\mathbf{b} = \mathbf{x}_0$, proving uniqueness. ☐

As an illustration, suppose we wish to solve

$$\begin{cases} 2x_1 + 7x_2 + x_3 = 2 \\ x_1 + 4x_2 - x_3 \ = 2 \\ x_1 + 3x_2 \qquad = 4 \end{cases}.$$

As a matrix equation, we have

$$\begin{bmatrix} 2 & 7 & 1 \\ 1 & 4 & -1 \\ 1 & 3 & 0 \end{bmatrix} \begin{bmatrix} x_1 \\ x_2 \\ x_3 \end{bmatrix} = \begin{bmatrix} 2 \\ 2 \\ 4 \end{bmatrix}.$$

The inverse of $\begin{bmatrix} 2 & 7 & 1 \\ 1 & 4 & -1 \\ 1 & 3 & 0 \end{bmatrix}$ is $\begin{bmatrix} -3/2 & -3/2 & 11/2 \\ 1/2 & 1/2 & -3/2 \\ 1/2 & -1/2 & -1/2 \end{bmatrix}$, so the solution to the system is $\begin{bmatrix} x_1 \\ x_2 \\ x_3 \end{bmatrix} = \begin{bmatrix} -3/2 & -3/2 & 11/2 \\ 1/2 & 1/2 & -3/2 \\ 1/2 & -1/2 & -1/2 \end{bmatrix} \begin{bmatrix} 2 \\ 2 \\ 4 \end{bmatrix} = \begin{bmatrix} 16 \\ -4 \\ -2 \end{bmatrix}$, as you may verify.

Theorem 1.4 is hardly the end of the story. What happens if A is *not* invertible? We still have a long way to go. However, we note that the argument given in

Theorem 1.4 did not need **x** and **b** to be n-by-1. Indeed, consider the matrix equation

$$AX = B$$

where A is n-by-n, X is n-by-p and, necessarily, B is n-by-p. By the same argument as in Theorem 1.4, we conclude that if A is invertible, then this matrix equation has the unique solution $X = A^{-1}B$. So, if

$$\begin{bmatrix} 1 & 2 \\ 3 & 4 \end{bmatrix} \begin{bmatrix} x & y \\ u & v \end{bmatrix} = \begin{bmatrix} 9 & 8 \\ 6 & 5 \end{bmatrix},$$

we get

$$\begin{bmatrix} x & y \\ u & v \end{bmatrix} = \begin{bmatrix} 1 & 2 \\ 3 & 4 \end{bmatrix}^{-1} \begin{bmatrix} 9 & 8 \\ 6 & 5 \end{bmatrix} = \begin{bmatrix} -2 & 1 \\ 1.5 & -.5 \end{bmatrix} \begin{bmatrix} 9 & 8 \\ 6 & 5 \end{bmatrix}$$
$$= \begin{bmatrix} -12 & -11 \\ 10.5 & 9.5 \end{bmatrix}.$$

Our definition of matrix inversion requires two equations to be satisfied, $AC = I$ and $CA = I$. Actually, one is enough (see exercise set 2, problem 21). This cuts down on our labor substantially and we will use this fact in the sequel.

1.2.1 The Henderson Searle Formulas

Dealing with the inverse of a product of invertible matrices is fairly easy in light of the "*reversal law*," $(AB)^{-1} = B^{-1}A^{-1}$. However, dealing with the inverse of a sum is more problematic. You may consider this surprising since matrix addition seems so much more straightforward than matrix multiplication. There are some significant facts about sums we now pursue. We present the formulas found in Henderson and Searle [H&S, 1981]. This notable paper gives some historical perspective on why inverting sums of matrices is of interest. Applications come from the need to invert partitioned matrices, inversion of a slightly modified matrix (rank one update), and statistics. Henderson and Searle give some general formulas from which many interesting special cases result. We present these general formulas in this section.

The following lemma is an important start to developing the Henderson-Searle formulas. Note that the matrices U and V below are included to increase the generality of the statements. They are quite arbitrary, but of appropriate size.

LEMMA 1.1

Suppose A is any n-by-n matrix with $I_n + A$ invertible. Then

1. $(I_n + A)^{-1} = I_n - A(I_n + A)^{-1} = I_n - (I_n + A)^{-1}A.$

In particular,

2. $A(I_n + A)^{-1} = (I_n + A)^{-1}A.$

PROOF Clearly $I_n = (I_n + A) - A$, so multiply by $(I_n + A)^{-1}$ from either side. ▯

Another important lemma is given next.

LEMMA 1.2
Suppose A, B, $(B + VA^{-1}U)$, $(A + UB^{-1}V)$ are all invertible with V s-by-n, A n-by-n, U n-by-s, and B s-by-s. Then $(B + VA^{-1}U)^{-1}VA^{-1} = B^{-1}V(A + UB^{-1}V)^{-1}$.

PROOF Note $VA^{-1}(A + UB^{-1}V) = V + VA^{-1}UB^{-1}V = (B + VA^{-1}U)B^{-1}V$ so $VA^{-1}(A + UB^{-1}V) = (B + VA^{-1}U)B^{-1}V$. Multiply both sides by $(A + UB^{-1}V)^{-1}$ from the right to get $VA^{-1} = (B + VA^{-1}U)B^{-1}V(A + UB^{-1}V)^{-1}$. Multiply both sides by $(B + VA^{-1}U)^{-1}$ from the left to get $(B + VA^{-1}U)^{-1}VA^{-1} = B^{-1}V(A + UB^{-1}V)^{-1}$. ▯

A quick corollary to Lemma 1.2 is given below.

COROLLARY 1.1
Suppose $(I_s + VU)$ and $(I_n + UV)$ are invertible. Then $(I_s + VU)^{-1}V = V(I_n + UV)^{-1}$.

PROOF Take $A = I_n$ and $B = I_s$ in Lemma 1.2. ▯

Finally, the Henderson-Searle formulas are given below.

THEOREM 1.5 (Henderson Searle formulas)
Suppose A is n-by-n invertible, U is n-by-p, B is p-by-q, and V is q-by-n. Suppose $(A + UBV)^{-1}$ exists. Then

1. $(A + UBV)^{-1} = A^{-1} - (I_n + A^{-1}UBV)^{-1}A^{-1}UBVA^{-1}$,
2. $(A + UBV)^{-1} = A^{-1} - A^{-1}(I_n + UBVA^{-1})^{-1}UBVA^{-1}$,
3. $(A + UBV)^{-1} = A^{-1} - A^{-1}U(I_p + BVA^{-1}U)^{-1}BVA^{-1}$,
4. $(A + UBV)^{-1} = A^{-1} - A^{-1}UB(I_q + VA^{-1}UB)^{-1}VA^{-1}$,

5. $(A+UBV)^{-1}=A^{-1}-A^{-1}UBV(I_n+A^{-1}UBV)^{-1}A^{-1}$,

6. $(A+UBV)^{-1}=A^{-1}-A^{-1}UBVA^{-1}(I_n+UBVA^{-1})^{-1}$.

PROOF First, note $I_n + A^{-1}UBV = A^{-1}(A+UBV)$, which is a product of invertible matrices (by hypothesis) and hence $I_n + A^{-1}UBV$ is invertible. For (1) then,

$$\begin{aligned}(A+UBV)^{-1} &= [A(I_n + A^{-1}UBV)]^{-1}\\ &= (I_n + A^{-1}UBV)^{-1}A^{-1}\\ &= ((I_n - (I_n + A^{-1}UBV)^{-1}A^{-1}UBV))A^{-1}\\ &= A^{-1} - (I_n + A^{-1}UBV)^{-1}A^{-1}UBVA^{-1}\end{aligned}$$

using Lemma 1.1. Using (2) of that lemma, we see $(I_n+A^{-1}UBV)^{-1}A^{-1}UBV = A^{-1}UBV(I_n + A^{-1}UBV)^{-1}$ so formula (5) follows.

Now for formula (2), note $(I_n + UBVA^{-1}) = (A+UBV)A^{-1}$, which is a product of invertible matrices and so $(I_n + UBVA^{-1})^{-1}$ exists. Using Lemma 1.1 again, we see $(A+UBV)^{-1} = [(I_n+UBVA^{-1})A]^{-1} = A^{-1}(I_n+UBVA^{-1})^{-1} = A^{-1}[I_n - (I_n + UBVA^{-1})^{-1}UBVA^{-1}] = A^{-1} - A^{-1}(I_n + UBVA^{-1})^{-1}UBVA^{-1}$. Again, using (2) of the lemma, formula (6) follows. Formula (3) follows from formula (2) and the corollary above if we knew $(I_p + BVA^{-1}U)^{-1}$ exists. But there is a determinant formula (see our appendix on determinants) that says $det(I_n + U(BVA^{-1})) = det(I_p + (BVA^{-1})U)$ so that follows. Now using the corollary, $A^{-1}[(I_n+(U)(BVA^{-1}))^{-1}U]BVA^{-1} = A^{-1}[U(I_p + (BVA^{-1})U)^{-1}]BVA^{-1}$, hence (3) follows from (2). Finally, (4) follows from (3) by a similar argument. □

The next corollary gives formulas for the inverse of a sum of two matrices.

COROLLARY 1.2
Suppose $A, B \in \mathbb{C}^{n\times n}$, A invertible, and $A + B$ invertible. Then

$$\begin{aligned}(A+B)^{-1} &= A^{-1} - (I_n + A^{-1}B)^{-1}A^{-1}BA^{-1}\\ &= A^{-1} - A^{-1}(I_n + BA^{-1})^{-1}BA^{-1}\\ &= A^{-1} - A^{-1}B(I_n + A^{-1}B)^{-1}A^{-1}\\ &= A^{-1} - A^{-1}BA^{-1}(I_n + BA^{-1})^{-1}.\end{aligned}$$

PROOF Take $U = V = I_n$ in Theorem 1.5. □

The next corollary is a form of the Sherman-Morrison-Woodbury formula we will develop soon using a different approach.

COROLLARY 1.3
Suppose A, B *and* $A + UBV$ *are invertible. Let* $D = -B^{-1}$ *so* $B = -D^{-1}$, *then* $(A - UD^{-1}V)^{-1} = A^{-1} + A^{-1}U(D - VA^{-1}U)^{-1}VA^{-1}$.

PROOF Using (3) from Theorem 1.5, $(A - UD^{-1}V)^{-1} =$
$A^{-1} - A^{-1}U[(I - D^{-1}VA^{-1}U)^{-1}(-D^{-1})]VA^{-1} =$
$A^{-1} - A^{-1}U[(D^{-1}D - D^{-1}VA^{-1}U)^{-1}(-D^{-1})]VA^{-1} =$
$A^{-1} + A^{-1}U(D - VA^{-1}U)^{-1}VA^{-1}$ using distributivity and the reversal law for inverses. □

We have a special case involving column vectors.

COROLLARY 1.4
Suppose A *is n-by-n invertible,* α *is a scalar,* $A + \alpha\mathbf{u}\mathbf{v}^*$ *is invertible,* $1 + \alpha\mathbf{v}^*A^{-1}\mathbf{u} \neq 0$, *and* $\mathbf{u}, \mathbf{v}$ *are n-by-1. Then*

$$(A + \alpha\mathbf{u}\mathbf{v}^*)^{-1} = A^{-1} - \frac{\alpha}{1 + \alpha\mathbf{v}^*A^{-1}\mathbf{u}}A^{-1}\mathbf{u}\mathbf{v}^*A^{-1}.$$

PROOF By (4) from Theorem 1.5 with $B = \alpha I$, $(A + \mathbf{u}B\mathbf{v}^*)^{-1} = (A + \mathbf{u}(\alpha I)\mathbf{v}^*)^{-1} = A^{-1} - A^{-1}\mathbf{u}(\alpha I)(I + \mathbf{v}^*A^{-1}\mathbf{u}(\alpha I))^{-1}\mathbf{v}^*A^{-1} =$
$$A^{-1} - \frac{\alpha A^{-1}\mathbf{u}\mathbf{v}^*A^{-1}}{1 + \alpha\mathbf{v}^*A^{-1}\mathbf{u}}.$$ □

We finish with a result for self-adjoint matrices.

COROLLARY 1.5
Suppose $A = A^*$, $B = B^*$, A *invertible, and* $A + UBU^*$ *invertible. Then* $(A + UBU^*)^{-1} = A^{-1} - A^{-1}UB(I + U^*A^{-1}UB)^{-1}U^*A^{-1}$.

PROOF Use (4) from Theorem 1.5 and take $V = U^*$. □

1.2.2 Schur Complements and the Sherman-Morrison-Woodbury Formula

We end this section by introducing the idea of a *Schur complement,* named after the German mathematician **Issai Schur** (10 January 1875–10 January 1941) and using it to prove the Sherman-Morrison-Woodbury formula. The Schur complement has to do with invertible portions of a partitioned matrix. To motivate the idea of a Schur complement, we consider the 2-by-2 matrix

$M = \begin{bmatrix} a & b \\ c & d \end{bmatrix} \in \mathbb{C}^{2\times 2}$. Suppose $a \neq 0$. Then

$$\begin{aligned} &\begin{bmatrix} 1 & 0 \\ -ca^{-1} & 1 \end{bmatrix}\begin{bmatrix} a & b \\ c & d \end{bmatrix}\begin{bmatrix} 1 & -a^{-1}b \\ 0 & 1 \end{bmatrix} \\ &= \begin{bmatrix} a & b \\ -ca^{-1}a + c & -ca^{-1}b + d \end{bmatrix}\begin{bmatrix} 1 & -a^{-1}b \\ 0 & 1 \end{bmatrix} \\ &= \begin{bmatrix} a & -aa^{-1}b + b \\ 0 & d - ca^{-1}b \end{bmatrix} = \begin{bmatrix} a & 0 \\ 0 & d - ca^{-1}b \end{bmatrix}. \end{aligned}$$

Since the matrices on either side of M are invertible, M is invertible iff $\begin{bmatrix} a & 0 \\ 0 & d - ca^{-1}b \end{bmatrix}$ is invertible iff $d - ca^{-1}b \neq 0$. Of course, all of this assumes that $a \neq 0$. Note that $det(M) = a(d - ca^{-1}b)$.

Similarly, if we assume $d \neq 0$ instead of $a \neq 0$, then $\begin{bmatrix} 0 & 1 \\ 1 & 0 \end{bmatrix}\begin{bmatrix} a & b \\ c & d \end{bmatrix} \times \begin{bmatrix} 0 & 1 \\ 1 & 0 \end{bmatrix} = \begin{bmatrix} d & c \\ b & a \end{bmatrix}$ is invertible iff M is. With different letters, we apply the above argument and conclude M is invertible iff $a - bd^{-1}c \neq 0$.

Again, there is the underlying assumption that $d \neq 0$. Now we extend these ideas to larger matrices. Let $M = \begin{bmatrix} A_{n\times n} & B_{n\times t} \\ C_{s\times n} & D_{s\times t} \end{bmatrix} \in \mathbb{C}^{(n+s)\times(n+t)}$. Assume A is nonsingular. Can we mimic what we did above? Let's try.

DEFINITION 1.2 *(Schur complements)*

Consider a matrix M partitioned as $M = \begin{bmatrix} A_{n\times n} & B_{n\times t} \\ C_{s\times n} & D_{s\times t} \end{bmatrix} \in \mathbb{C}^{(n+s)\times(n+t)}$. Define the ***Schur complement of A in M*** *by $M/A = D - CA^{-1}B \in \mathbb{C}^{s\times t}$ assuming, of course, that A is invertible. If M is partitioned $\begin{bmatrix} A_{s\times t} & B_{s\times n} \\ C_{n\times t} & D_{n\times n} \end{bmatrix} \in \mathbb{C}^{(n+s)\times(n+t)}$, where D is now invertible, define the* ***Schur complement of D in M*** *by $M//D = A - BD^{-1}C$.*

To illustrate, suppose $M = \begin{bmatrix} 1 & 5 & 9 & 13 \\ 2 & 3 & 4 & 5 \\ 6 & 7 & 8 & -7 \\ 5 & -4 & -3 & 2 \end{bmatrix}$, where A is the upper left 2-by-2 block $\begin{bmatrix} 1 & 5 \\ 2 & 3 \end{bmatrix}$. Then $M/A =$

$$\begin{bmatrix} 8 & -7 \\ -3 & 2 \end{bmatrix} - \begin{bmatrix} 6 & 7 \\ 5 & -4 \end{bmatrix}\begin{bmatrix} 1 & 5 \\ 2 & 3 \end{bmatrix}^{-1}\begin{bmatrix} 9 & 13 \\ 4 & 5 \end{bmatrix} = \begin{bmatrix} 0 & -16 \\ 10 & 24 \end{bmatrix}.$$

The next theorem is due to the astronomer/mathematician **Tadeusz Banachiewicz** (1882–1954).

THEOREM 1.6 (Banacheiwicz inversion formula, 1937)
Consider a matrix M of four blocks partitioned as
$M = \begin{bmatrix} A & B \\ C & D \end{bmatrix}$ *where A is r-by-r, B is r-by-s, C is s-by-r, and D is s-by-s.*

1. *if A^{-1} and $(M/A)^{-1}$exist, then the matrix $\begin{bmatrix} A & B \\ C & D \end{bmatrix}$ is invertible and*

$$\begin{bmatrix} A & B \\ C & D \end{bmatrix}^{-1} = \begin{bmatrix} A^{-1} + A^{-1}BS^{-1}CA^{-1} & -A^{-1}BS^{-1} \\ -S^{-1}CA^{-1} & S^{-1} \end{bmatrix}$$

where $S = M/A$.

2. *if D^{-1} and $(M//D)^{-1}$exist, then the matrix $\begin{bmatrix} A & B \\ C & D \end{bmatrix}$ is invertible and*

$$\begin{bmatrix} A & B \\ C & D \end{bmatrix}^{-1} = \begin{bmatrix} T^{-1} & -T^{-1}BD^{-1} \\ -D^{-1}CT^{-1} & D^{-1} + D^{-1}CT^{-1}BD^{-1} \end{bmatrix}$$

where $T = M//D$.

PROOF We will prove part (1) and leave part (2) as an exercise. Suppose A is invertible and $S = D - CA^{-1}B$ is invertible also. We verify the claimed inverse by direct computation, appealing to the uniqueness of the inverse. Compute

$$\begin{bmatrix} A & B \\ C & D \end{bmatrix} \begin{bmatrix} A^{-1} + A^{-1}BS^{-1}CA^{-1} & -A^{-1}BS^{-1} \\ -S^{-1}CA^{-1} & S^{-1} \end{bmatrix}$$

$$= \begin{bmatrix} AA^{-1} + AA^{-1}BS^{-1}CA^{-1} - BS^{-1}CA^{-1} & -AA^{-1}BS^{-1} + BS^{-1} \\ C[A^{-1} + A^{-1}BS^{-1}CA^{-1}] + D[-S^{-1}CA^{-1}] & -CA^{-1}BS^{-1} + DS^{-1} \end{bmatrix}$$

$$= \begin{bmatrix} I + IBS^{-1}CA^{-1} - BS^{-1}CA^{-1} & -IBS^{-1} + BS^{-1} \\ CA^{-1} + CA^{-1}BS^{-1}CA^{-1} - DS^{-1}CA^{-1} & -CA^{-1}BS^{-1} + DS^{-1} \end{bmatrix}.$$

A few things become clear. The first row is just what we hoped for. The first block is I_r and the second is $\mathbb{O}_{r\times s}$, so we are on the way to producing the identity matrix. Moreover, the lower right block

$$-CA^{-1}BS^{-1} + DS^{-1} = (-CA^{-1}B + D)S^{-1} = SS^{-1} = I_{s\times s}.$$

We are almost there! The last block is

$$\begin{aligned} &CA^{-1} + CA^{-1}BS^{-1}CA^{-1} - DS^{-1}CA^{-1} \\ &\quad = CA^{-1} + [CA^{-1}BS^{-1} - DS^{-1}]CA^{-1} \\ &\quad = CA^{-1} + [[CA^{-1}B - D]S^{-1}]CA^{-1} = CA^{-1} - SS^{-1}CA^{-1} = \mathbb{O} \end{aligned}$$

and we are done. ▯

It is of interest to deduce special cases of the above theorem where $B = \mathbb{O}$ or $C = \mathbb{O}$, or both B and C are $\mathbb{O}$. (See problem 30 of exercise set 2.) Note, in the latter case, we get

$$\begin{bmatrix} A & \mathbb{O} \\ \mathbb{O} & D \end{bmatrix}^{-1} = \begin{bmatrix} A^{-1} & \mathbb{O} \\ \mathbb{O} & D^{-1} \end{bmatrix}.$$

Statisticians have known an interesting result about the inverse of a certain sum of matrices since the late 1940s. A formula they discovered has uses in many areas. We look at a somewhat generalized version next.

THEOREM 1.7 (Sherman-Morrison-Woodbury formula)

Suppose A is n-by-n nonsingular and G is s-by-s nonsingular, where $s \leq n$. Suppose C and D are n-by-s and otherwise completely arbitrary. Then $A + CGD^*$ is invertible iff $G^{-1} + D^*A^{-1}C$ is invertible, in which case

$(A + CGD^*)^{-1} = A^{-1} - A^{-1}C(G^{-1} + D^*A^{-1}C)^{-1}D^*A^{-1}$.

PROOF We will use a Schur complement to prove this result. First, suppose that A and G are invertible and $S = G^{-1} + D^*A^{-1}C$ is invertible. We must show $A + CGD^*$ is invertible. We claim $\begin{bmatrix} A & C \\ -D^* & G^{-1} \end{bmatrix}$ is invertible and $\begin{bmatrix} A & C \\ -D^* & G^{-1} \end{bmatrix}^{-1} = \begin{bmatrix} A^{-1} - A^{-1}CS^{-1}D^*A^{-1} & -A^{-1}CS^{-1} \\ S^{-1}D^*A^{-1} & S^{-1} \end{bmatrix}$. Note the Schur complement of A is $G^{-1} - (-D^*)A^{-1}C = G^{-1} + D^*A^{-1}C$, which is precisely S. The theorem above applies; therefore

$$\begin{bmatrix} A & C \\ -D^* & G^{-1} \end{bmatrix}^{-1} = \begin{bmatrix} A^{-1} + A^{-1}CS^{-1}(-D^*)A^{-1} & -A^{-1}CS^{-1} \\ -S^{-1}(-D^*)A^{-1} & S^{-1} \end{bmatrix}$$

$$= \begin{bmatrix} A^{-1} - A^{-1}CS^{-1}D^*A^{-1} & -A^{-1}CS^{-1} \\ S^{-1}D^*A^{-1} & S^{-1} \end{bmatrix}.$$

The claim has been established. But there is more to show. The next claim is that $\begin{bmatrix} I & -CG \\ \mathbb{O} & I \end{bmatrix}$ is invertible. This is easy since $\begin{bmatrix} I & CG \\ \mathbb{O} & I \end{bmatrix}$ is its inverse, as you can easily check. Also, $\begin{bmatrix} I & \mathbb{O} \\ GD^* & I \end{bmatrix}$ is invertible since its inverse is $\begin{bmatrix} I & \mathbb{O} \\ -GD^* & I \end{bmatrix}$. Now consider

$$\begin{bmatrix} I & -CG \\ \mathbb{O} & I \end{bmatrix}\begin{bmatrix} A & C \\ -D^* & G^{-1} \end{bmatrix}\begin{bmatrix} I & \mathbb{O} \\ GD^* & I \end{bmatrix}$$
$$= \begin{bmatrix} I & -CG \\ \mathbb{O} & I \end{bmatrix}\begin{bmatrix} A+CGD^* & C \\ \mathbb{O} & G^{-1} \end{bmatrix}$$
$$= \begin{bmatrix} A+CGD^* & \mathbb{O} \\ \mathbb{O} & G^{-1} \end{bmatrix}.$$

Being the product of three invertible matrices, this matrix must be invertible. But then, the two nonzero diagonal block matrices must be invertible, so $A+CGD^*$ must be invertible. Moreover,

$$\begin{bmatrix} A+CGD^* & \mathbb{O} \\ \mathbb{O} & G^{-1} \end{bmatrix}^{-1} = \begin{bmatrix} (A+CGD^*)^{-1} & \mathbb{O} \\ \mathbb{O} & G \end{bmatrix}$$

but also equals

$$\begin{bmatrix} I & \mathbb{O} \\ GD^* & I \end{bmatrix}^{-1}\begin{bmatrix} A & C \\ -D^* & G^{-1} \end{bmatrix}^{-1}\begin{bmatrix} I & -CG \\ \mathbb{O} & I \end{bmatrix}^{-1}$$
$$= \begin{bmatrix} I & \mathbb{O} \\ -GD^* & I \end{bmatrix}\begin{bmatrix} A^{-1}-A^{-1}CS^{-1}D^*A^{-1} & -A^{-1}CS^{-1} \\ S^{-1}D^*A^{-1} & S^{-1} \end{bmatrix}\begin{bmatrix} I & CG \\ \mathbb{O} & I \end{bmatrix}$$
$$= \begin{bmatrix} A^{-1}-A^{-1}CS^{-1}D^*A^{-1} & \text{stuff} \\ \text{stuff} & \text{stuff} \end{bmatrix}.$$

Now compare the upper left blocks.

We leave the converse result as an easy exercise. □

The strength of a theorem is measured by the consequences it has. There are many useful corollaries to this theorem.

COROLLARY 1.6

For matrices of the appropriate size, we have

1. *If A, G, and A + G are invertible,* $(A+G)^{-1} = A^{-1} - A^{-1}(G^{-1} + A^{-1})^{-1}A^{-1}$.
2. *If A is invertible and n-by-n, C and D are n-by-s, and* $(I + D^*A^{-1}C)^{-1}$ *exists, then* $(A + CD^*)^{-1} = A^{-1} - A^{-1}C(I + D^*A^{-1}C)^{-1}D^*A^{-1}$.

3. *$(I + CD^*)^{-1} = I - C(I + D^*C)^{-1}D^*$, if $(I + D^*C)^{-1}$ exists.*

4. *Let $\mathbf{c}$ and $\mathbf{d}$ be n-by-1. Then $(A+\mathbf{cd}^*)^{-1} = A^{-1} - \dfrac{A^{-1}\mathbf{cd}^*A^{-1}}{1+\mathbf{d}^*A^{-1}\mathbf{c}}$, provided $\mathbf{d}^*A^{-1}\mathbf{c} \neq -1$ and A^{-1} exists.*

5. *Let $\mathbf{c}$ and $\mathbf{d}$ be n-by-1. Then $(I + \mathbf{cd}^*)^{-1} = I - \dfrac{\mathbf{cd}^*}{1+\mathbf{d}^*\mathbf{c}}$, if $1 + \mathbf{d}^*\mathbf{c} \neq 0$.*

6. *If $\mathbf{u}$ and $\mathbf{v}$ are n-by-1, then $(I - \mathbf{uv}^*)^{-1} = I - \dfrac{\mathbf{uv}^*}{\mathbf{v}^*\mathbf{u} - 1}$, if $\mathbf{v}^*\mathbf{u} \neq 1$.*

PROOF The proofs are left as exercises. □

We look at a special case of (2) above. Recall the "standard basis" vectors $\mathbf{e}_j = \begin{bmatrix} 0 \\ \vdots \\ 1 \\ \vdots \\ 0 \end{bmatrix}$ where the 1 appears in the jth position. Then we change the (i, j) entry of a matrix A by adding a specified amount α to this entry. If A is invertible and the perturbed matrix is still invertible, we have a formula for its inverse. Some people say we are inverting a "*rank one update*" of A.

$$(A + \alpha\mathbf{e}_i\mathbf{e}_j^*)^{-1} = A^{-1} - \alpha\frac{A^{-1}\mathbf{e}_i\mathbf{e}_j^*A^{-1}}{1 + \alpha\mathbf{e}_j^*A^{-1}\mathbf{e}_i} = A^{-1} - \alpha\frac{col_i(A^{-1})row_j(A^{-1})}{1 + \alpha ent_{ji}(A^{-1})}$$

We illustrate in the exercises (see exercise 16 in exercise set 2).

Exercise Set 2

1. If $A = \begin{bmatrix} i & 0 & 0 \\ 0 & i & 0 \\ 0 & 0 & i \end{bmatrix}$, what is A^{-1}? What if $A = \begin{bmatrix} 0 & 0 & i \\ 0 & i & 0 \\ i & 0 & 0 \end{bmatrix}$?

2. Find A^* and $(A^*)^{-1}$, if $A = \begin{bmatrix} 2+i & 3+i & 4+i \\ 5+3i & 6+3i & 6+2i \\ 8+2i & 8+3i & 9-3i \end{bmatrix}$.

3. If A is an m-by-n matrix, what is $A\mathbf{e}_j, \mathbf{e}_j^* A, \mathbf{e}_i^* A\mathbf{e}_j$, where $\mathbf{e}_i$ is the standard basis vector? What is $\mathbf{e}_j{}^*\mathbf{e}_i$? What is $\mathbf{e}_j\mathbf{e}_i{}^*$?

4. Suppose A and B are two m-by-n matrices and $A\mathbf{x} = B\mathbf{x}$ for all n-by-1 matrices $\mathbf{x}$. Is it true that $A = B$? Suppose $A\mathbf{x} = \overrightarrow{0}$ for all $\mathbf{x}$. What can you say about A?

5. Argue that $\begin{bmatrix} 1 & a & b \\ 0 & 1 & c \\ 0 & 0 & 1 \end{bmatrix}$ is always invertible regardless of what a, b, and c are and find a formula for its inverse. Do the same for $\begin{bmatrix} 1 & 0 & 0 \\ a & 1 & 0 \\ b & c & 1 \end{bmatrix}$.

6. Suppose A is n-by-n and $A^3 = I_n$. Must A be invertible? Prove it is or provide a counterexample.

7. Cancellation laws: Suppose B and C are m-by-n matrices. Prove
   (i) if A is m-by-m and invertible and $AB = AC$, then $B = C$.
   (ii) if A is n-by-n and invertible and $BA = CA$, then $B = C$.

8. Suppose A and B are n-by-n and invertible. Argue that $A^{-1} + B^{-1} = A^{-1}(A + B)B^{-1}$. If A and B were scalars, how would you have been led to this formula? (Hint: consider $(\frac{1}{a} + \frac{1}{b})$). If $(A + B)^{-1}$ exists, what is $(A^{-1} + B^{-1})^{-1}$ in view of your result above?

9. Suppose U is n-by-n and $U^2 = I_n$. Argue that $I + U$ is not invertible unless $U = I_n$.

10. Suppose A is n-by-n, $A^* = A$, and A is invertible. Prove that $(A^{-1})^* = A^{-1}$.

11. Suppose A is n-by-n and invertible. Argue that AA^* and A^*A are invertible.

12. Prove or disprove the following claim: Any square matrix can be written as the sum of two invertible matrices.

13. Verify the claims made in Theorem 1.2.

14. Complete the proof of Theorem 1.5.

15. Fill in the details of the proof of Corollary 1.1 and Corollary 1.6.

16. Let $A = \begin{bmatrix} 14 & 17 & 3 \\ 17 & 26 & 5 \\ 3 & 5 & 1 \end{bmatrix}$. Argue that A is invertible and $A^{-1} = \begin{bmatrix} 1 & -2 & 7 \\ -2 & 5 & -19 \\ 7 & -19 & 75 \end{bmatrix}$. Now consider the matrix $A_{new} = \begin{bmatrix} 14 & 17 & 3 \\ 17 & 24 & 5 \\ 3 & 5 & 1 \end{bmatrix}$. Use the rank one update formula above to find $(A_{new})^{-1}$.

17. Suppose A is invertible and suppose column j of A is replaced by a new column $\mathbf{c}$, so that the new matrix A_c is still invertible. Argue that $(A_c)^{-1} = A^{-1} - \dfrac{(A^{-1}\mathbf{c} - \mathbf{e}_j)row_j(A^{-1})}{row_j(A^{-1})\mathbf{c}}$. Continuing with the matrix in exercise 16, compute the inverse of $\begin{bmatrix} 1 & 17 & 3 \\ 1 & 26 & 5 \\ 1 & 5 & 1 \end{bmatrix}$.

18. Suppose we have a system of linear equations $A\mathbf{x} = \mathbf{b}$, where A is invertible. Suppose A is perturbed slightly to $A + \mathbf{cd}^*$. Consider the system of linear equations $(A + \mathbf{cd}^*)\mathbf{y} = \mathbf{b}$, where $(A + \mathbf{cd}^*)$ is still invertible. Argue that the solution of the perturbed system is $\mathbf{y} = A^{-1}\mathbf{b} - \dfrac{A^{-1}\mathbf{cd}^*A^{-1}\mathbf{b}}{1 + \mathbf{d}^*A^{-1}\mathbf{c}}$.

19. Argue that for ϵ small enough, $A + \epsilon\mathbf{e}_i\mathbf{e}_j^*$ remains nonsingular if A is.

20. Suppose $A \in \mathbb{C}^{m\times m}$ and $B \in \mathbb{C}^{n\times n}$ are invertible. What can you say about the matrix $\begin{bmatrix} A & \mathbb{O} \\ \mathbb{O} & B \end{bmatrix}$? (*Hint:* Use Theorem 1.6.) How about the converse?

21. In our definition of inverse, we required that the inverse matrix work on both sides; that is, $AC = I$ and $CA = I$. Actually, you can prove $AC = I$ implies $CA = I$ for square matrices. So, do it.

22. Prove the claims of Theorem 1.3. Feel free to consult your linear algebra book as a review.

23. If A is invertible and A commutes with B, then does A^{-1} commute with B as well? Recall we say "A commutes with B" when $AB = BA$.

24. If $I + AB$ is invertible, is $I + BA$ also invertible? If so, is there a formula for the inverse of $I + BA$?

25. Suppose $P = P^2$. Argue that $I - 2P$ is its own inverse. Generalize this to $(I - aP)^{-1}$ for any a.

26. Suppose A is not square but $AA^* = I$. Does $A^*A = I$ also?

27. Compute $det \begin{bmatrix} 1 & r & s \\ -r & 1 & t \\ -s & -t & 1 \end{bmatrix}$. What, if anything, can you conclude about the invertibility of this matrix?

28. Argue that A, a square matrix, can not be invertible if each row of A sums to 0.

29. Argue that $\begin{bmatrix} I & A \\ \mathbb{O} & I \end{bmatrix}$ is always an invertible matrix regardless of what A is. Exhibit the inverse matrix.

30. Investigate the special cases of the Schur complements theorem where $B = \mathbb{O}$, $C = \mathbb{O}$, or both are zero. Be clear what the hypotheses are.

31. For which values of λ is $\begin{bmatrix} 1-\lambda & 1 & 1 \\ 0 & 2-\lambda & 3 \\ 0 & -3 & 2-\lambda \end{bmatrix}$ an invertible matrix?

32. Is the matrix $\begin{bmatrix} a & b & c \\ d & e & 0 \\ f & 0 & 0 \end{bmatrix}$ invertible? Do you need any conditions on the entries?

33. Suppose $A\mathbf{x} = \lambda\mathbf{x}$, where λ is a nonzero scalar and $\mathbf{x}$ is a nonzero vector. Argue that if A is invertible, then $A^{-1}\mathbf{x} = \frac{1}{\lambda}\mathbf{x}$.

34. Can a skew symmetric matrix of odd order be invertible?

35. Let $A = nI_n + \begin{bmatrix} 1 & 1 & \cdots & 1 \\ 1 & 1 & \cdots & 1 \\ 1 & 1 & \cdots & 1 \\ 1 & 1 & \cdots & 1 \end{bmatrix} = nI_n + B$. Is A invertible? If so, what is A^{-1}? What is the sum of all entries in A^{-1}? (*Hint:* What is B^2?)

36. Suppose A and B are nonsingular. Argue that $[(AB)^{-1}]^T = [(AB)^T]^{-1}$.

37. Let $L_5(a) = \begin{bmatrix} 1 & 0 & 0 & 0 & 0 \\ a & 1 & 0 & 0 & 0 \\ 0 & a & 1 & 0 & 0 \\ 0 & 0 & a & 1 & 0 \\ 0 & 0 & 0 & a & 1 \end{bmatrix}$. What is $(L_5(a))^{-1}$? Can you generalize this example to $(L_n(a))^{-1}$?

38. Let A be a square and invertible matrix and let X be such that $A^{k+1}X = A^k$ for some k in $\mathbb{N}$. Argue that $X = A^{-1}$.

39. Find A^{-1} if $A = \begin{bmatrix} 1 & 0 \\ -2 & 1 \end{bmatrix}, \begin{bmatrix} 1 & 0 & 0 \\ -2 & 1 & 0 \\ 1 & -2 & 1 \end{bmatrix}, \begin{bmatrix} 1 & 0 & 0 & 0 \\ -2 & 1 & 0 & 0 \\ 1 & -2 & 0 & 0 \\ 0 & 1 & -2 & 1 \end{bmatrix},$ etc. Do you see a pattern?

40. Find A^{-1} if $A = \begin{bmatrix} 2 & -1 & 0 \\ -1 & 2 & -1 \\ 0 & -1 & 2 \end{bmatrix}, \begin{bmatrix} 2 & -1 & 0 & 0 \\ -1 & 2 & -1 & 0 \\ 0 & -1 & 2 & -1 \\ 0 & 0 & -1 & 2 \end{bmatrix}$, etc.

Do you see a pattern?

41. Let $A = \begin{bmatrix} A_{11} & A_{12} \\ A_{21} & A_{22} \end{bmatrix}$, where A_{11} is k-by-k and invertible. Argue that

$$A = \begin{bmatrix} I & \mathbb{O} \\ A_{21}A_{11}^{-1} & I \end{bmatrix} \begin{bmatrix} A_{11} & A_{12} \\ \mathbb{O} & A_{22} - A_{21}A_{11}^{-1}A_{12} \end{bmatrix}.$$

42. Consider $\begin{bmatrix} A & \mathbf{a} \\ \mathbf{d}^T & \alpha \end{bmatrix} \begin{bmatrix} \mathbf{x} \\ x_{n+1} \end{bmatrix} = \begin{bmatrix} \mathbf{b} \\ b_{n+1} \end{bmatrix}$, where A is n-by-n and invertible and $\mathbf{a}, \mathbf{b}, \mathbf{d} \in \mathbb{C}^n$. Argue that if $\alpha - \mathbf{d}^T A^{-1}\mathbf{a} \neq 0$, then $x_{n+1} = \dfrac{b_{n+1} - \mathbf{d}^T A^{-1}\mathbf{b}}{\alpha - \mathbf{d}^T A^{-1}\mathbf{a}}$ and $\mathbf{x} = A^{-1}\mathbf{b} - x_{n+1}A^{-1}\mathbf{a}$.

43. Argue that $det(A + \mathbf{c}\mathbf{d}^T) = det(A)(1 + \mathbf{d}^T A^{-1}\mathbf{c})$. In particular, deduce that $det(I + \mathbf{c}\mathbf{d}^T) = 1 + \mathbf{d}^T\mathbf{c}$.

44. Suppose $\mathbf{v}$ is an n-by-1, nonzero column vector. Argue that an invertible matrix exists whose first column is $\mathbf{v}$.

45. What is the inverse of $\begin{bmatrix} 1 & 1 & 1 & 1 \\ 1 & -i & -1 & i \\ 1 & -1 & 1 & -1 \\ 1 & i & -1 & -i \end{bmatrix}$? How about the inverse of $\begin{bmatrix} 0 & 1 & 1 & 1 \\ -1 & 0 & 1 & 1 \\ -1 & -1 & 0 & 1 \\ -1 & -1 & -1 & 0 \end{bmatrix}$? Do you notice anything interesting?

46. Under what conditions is $\begin{bmatrix} a & -b \\ b & a \end{bmatrix}$ invertible? Determine the inverses in these cases.

47. Argue that $\begin{bmatrix} A^{-1} & \mathbb{O} \\ -VA^{-1} & I \end{bmatrix}\begin{bmatrix} A & U \\ V & D \end{bmatrix} = \begin{bmatrix} I & A^{-1}U \\ \mathbb{O} & D - VA^{-1}U \end{bmatrix}$ where, of course, A is invertible. Deduce that $det(\begin{bmatrix} A & U \\ V & D \end{bmatrix})$ $= det(A)det(D - VA^{-1}U) = det(D)det(A - UD^{-1}V)$. As a corollary, conclude that $det(I + VU) = det(I + UV)$.

48. Suppose A, D, $A - BD^{-1}C$, and $A - BD^{-1}C$ are invertible. Prove that $\begin{bmatrix} A & B \\ C & D \end{bmatrix}^{-1} = \begin{bmatrix} (A - BD^{-1}C)^{-1} & -(A - BD^{-1}C)^{-1}BD^{-1} \\ -(D - CA^{-1}B)^{-1}CA & (D - CA^{-1}B)^{-1} \end{bmatrix}$. (*Hint:* See Theorem 1.6.)

49. Suppose A, B, C, and D are all invertible, as are $A - BD^{-1}C$, $C - DB^{-1}A$, $B - AC^{-1}D$, and $D - CA^{-1}B$. Argue that $\begin{bmatrix} A & B \\ C & D \end{bmatrix}^{-1}$ $= \begin{bmatrix} (A - BD^{-1}C)^{-1} & (C - DB^{-1}A)^{-1} \\ (B - AC^{-1}D)^{-1} & (D - CA^{-1}B)^{-1} \end{bmatrix}$.

50. Can you derive the Sherman-Morrison-Woodbury formula from the Henderson-Searle formulas?

51. Argue that $(D - CA^{-1}B)^{-1}CA^{-1} = D^{-1}C(A - BD^{-1}C)^{-1}$. (*Hint:* Show $CA^{-1}(A - BD^{-1}C) = (D - CA^{-1}B)D^{-1}C$.) From this, deduce as a corollary that $(I + AB)^{-1}A = A(I + BA)^{-1}$.

52. Argue that finding the inverse of an n-by-n matrix is tantamount to solving n systems of linear equations. Describe these systems explicitly.

53. Suppose $A + B$ is nonsingular. Prove that $A - A(A + B)^{-1}A = B - B(A + B)^{-1}B$.

54. Suppose A and B are invertible matrices of the same size. Prove that $A^{-1}+B^{-1}=A^{-1}(A+B)B^{-1}$.

55. Suppose A and B are invertible matrices of the same size. Suppose further that $A^{-1}+B^{-1}=(A+B)^{-1}$. Argue that $AB^{-1}A=BA^{-1}B$.

56. Suppose $det(I_m+AA^*)$ is not zero. Prove that $(I_m+AA^*)^{-1}=I-A(I_n+A^*A)^{-1}A^*$. Here A is assumed to be m-by-n.

57. Suppose $C=C^*$ is n-by-n and invertible and A and B are arbitrary and n-by-n. Prove that $(A-BC^{-1})C(A-BC^{-1})^*-BC^{-1}B^*=ACA^*-BA^*-(BA^*)^*$.

58. Refer to Theorem 1.6. Show that in case (1),
$$\begin{bmatrix} A & B \\ C & D \end{bmatrix}^{-1}=\begin{bmatrix} I & -A^{-1}B \\ \mathbb{O} & I \end{bmatrix}\begin{bmatrix} A^{-1} & \mathbb{O} \\ \mathbb{O} & S^{-1} \end{bmatrix}\begin{bmatrix} I & \mathbb{O} \\ -CA^{-1} & I \end{bmatrix}.$$

59. Continuing our extension of Theorem 1.6, show that $\begin{bmatrix} A & B \\ C & D \end{bmatrix}^{-1}$
$$=\begin{bmatrix} \left(A-BD^{-1}C\right)^{-1} & -A^{-1}B(D-CA^{-1}B)^{-1} \\ -D^{-1}C\left(A-BD^{-1}C\right)^{-1} & (D-CA^{-1}B)^{-1} \end{bmatrix}.$$

60. Use exercise 59 to derive the following identities:
    (a) $\left(A-BD^{-1}C\right)^{-1}BD^{-1}=A^{-1}B(D-CA^{-1}B)^{-1}$
    (b) $\left(A^{-1}-BD^{-1}C\right)^{-1}BD^{-1}=AB(D-CA^{-1}B)^{-1}$
    (c) $\left(A+BD^{-1}C\right)^{-1}BD^{-1}=A^{-1}B(D+CA^{-1}B)^{-1}$
    (d) $(D-CA^{-1}B)^{-1}=D^{-1}+D^{-1}C\left(A-BD^{-1}C\right)^{-1}BD^{-1}$
    (e) $(D-CA^{-1}B)^{-1}=D^{-1}-D^{-1}C\left(BD^{-1}C-A\right)^{-1}BD^{-1}$
    (f) $(D+CA^{-1}B)^{-1}=D^{-1}-D^{-1}C\left(BD^{-1}C+A\right)^{-1}BD^{-1}$
    (g) $(D^{-1}+CA^{-1}B)^{-1}=D-DC(BDC+A)^{-1}BD$
    (h) $(D-CAB)^{-1}=D^{-1}-D^{-1}C\left(BD^{-1}C-A^{-1}\right)^{-1}BD^{-1}$
    (i) $(D+CAB)^{-1}=D^{-1}-D^{-1}C\left(BD^{-1}C+A^{-1}\right)^{-1}BD$.

61. Apply Theorem 1.6 and the related problems above to the special case of the partitioned matrix $\begin{bmatrix} A & B \\ B^* & D \end{bmatrix}$. What formulas and identities can you deduce?

62. Prove the Schur determinant formulas. Let $M=\begin{bmatrix} A & B \\ C & D \end{bmatrix}$.
    (i) If $det(A)\neq 0$, then $det(M)=det(A)det(M/A)$.
    (ii) If $det(D)\neq 0$, then $det(M)=det(D)det(M//D)$.

Further Reading

[Agg&Lamo, 2002] Rita Aggarwala and Michael P. Lamoureux, Inverting the Pascal Matrix Plus One, The American Mathematical Monthly, Vol. 109, No. 4, April, (2002), 371–377.

[Banachiewicz, 1937] T. Banachiewicz, Sur Berechnung der Determinanten, wie auch der Inversen, und zur darauf basierten auflősung der Systeme Linearer Gleichungen, Acta Astronomica, Série C, 3 (1937), 41–67.

[B&R, 1986(2)] T. S. Blyth and E. F. Robertson, *Matrices and Vector Spaces*, Vol. 2, Chapman & Hall, New York, (1986).

[C&V 1993] G. S. Call and D. J. Velleman, Pascal's Matrices, The American Mathematical Monthly, Vol. 100, (1993), 372–376.

[Greenspan, 1955] Donald Greenspan, Methods of Matrix Inversion, The American Mathematical Monthly, Vol. 62, May, (1955), 303–318.

[Hager, 1989] W. W. Hager, Updating the Inverse of a Matrix, SIAM Rev., Vol. 31, (1989), 221–239.

[H&S, 1981] H. V. Henderson and S. R. Searle, On Deriving the Inverse of a Sum of Matrices, SIAM Rev., Vol. 23, (1981), 53–60.

[M&K, 2001] Tibor Mazùch and Jan Kozánek, New Recurrent Algorithm for a Matrix Inversion, Journal of Computational and Applied Mathematics, Vol. 136, No. 1–2, 1 November, (2001), 219–226.

[Meyer, 2000] Carl Meyer, *Matrix Analysis and Applied Linear Algebra*, SIAM, Philadelphia, (2000).

[vonN&G, 1947] John von Neumann and H. H. Goldstine, Numerical Inverting of Matrices of High Order, Bulletin of the American Mathematical Society, Vol. 53, (1947), 1021–1099.

[Zhang, 2005] F. Zhang, *The Schur Complement and Its Applications*, Springer, New York, (2005).

1.2.3 MATLAB Moment

1.2.3.1 Computing Inverse Matrices

If A is a square matrix, the command for returning the inverse, if the matrix is nonsingular, is

```
inv(A)
```

Of course, the answer is up to round off. Let's look at an example of a random 4-by-4 complex matrix.

```
>>format rat
>>A = fix(10 * rand(4)) + fix(10 * rand(4)) * i
A =
```

Columns 1 through 3

$$\begin{array}{ccc} 9+9i & 8 & 8+1i \\ 2+9i & 7+3i & 4+2i \\ 6+4i & 4+8i & 6+1i \\ 4+8i & 0 & 7+6i \end{array}$$

Column 4

$$\begin{array}{l} 9+2i \\ 7+1i \\ 1 \\ 4+7i \end{array}$$

```
>>inv(A)
ans =
```

Columns 1 through 3

$$\begin{array}{ccc} 1/2-1/6i & -959/2524+516/2381i & 291/4762+520/4451i \\ -1/10-2/15i & 1072/6651+139/1767i & 212/4359-235/2636i \\ -7/10+7/30i & 1033/2317-714/1969 & 71/23810-879/5098i \\ 1/10-11/30i & 129/1379+365/1057i & 31/11905+439/8181i \end{array}$$

Column 4

$$\begin{array}{l} -489/2381+407/2381i \\ 399/7241+331/3381i \\ 683/1844-268/1043i \\ -6/11905+81/601i \end{array}$$

We ask for the determinant of A which, of course, should not be zero

$$
\begin{aligned}
&>>\text{det(A)}\\
&\text{ans} =\\
&\quad -396 + 1248i\\
&>>\text{det(inv(A))}
\end{aligned}
$$

$$
\begin{aligned}
&\text{ans} =\\
&\quad -11/47620 - 26/35715i\\
&>>\text{det(A)} * \text{det(inv(A))}\\
&\text{ans} =\\
&\quad 1 - 1/1385722962267845i
\end{aligned}
$$

Note that, theoretically, det(A) and det(inv(A)) should be inverses to one another. But after all, look how small the imaginary part of the answer is in MATLAB. We can also check our answer by multiplying A and inv(A). Again, we must interpret the answer presented.

```
>> A*inv(A)
ans =
```

Columns 1 through 3

$$
\begin{matrix}
1 - * & * - * & * - * \\
* - * & 1 + * & * - * \\
* - * & * - * & 1 - * \\
* - * & * - * & * - *
\end{matrix}
$$

Column 4

$$
\begin{matrix}
0 + * \\
* + * \\
* + * \\
1 - *
\end{matrix}
$$

We need to recognize the identity matrix up to round off. To get a more satisfying result, try

```
>>round(A*inv(A)).
```

There may be occasions where you actually want a singular square matrix. Here is a way to construct a matrix that is not only singular but has a prescribed dependency relation among the columns. To illustrate let's say we want the third row of a matrix to be two times the first row plus three times the second.

```
>>A = [1111;2222;3333;4444]
A =
 1  1  1  1
 2  2  2  2
 3  3  3  3
 4  4  4  4
>>A(:, 3) = A(:, 1 : 2) * [2, 3]'
A =
 1  1   5  1
 2  2  10  2
 3  3  15  3
 4  4  20  4
>>det(A)
ans =
     0
>>
```

1.2.4 Numerical Note

1.2.4.1 Matrix Inversion

Inverting a matrix in floating point arithmetic has its pitfalls. Consider the simple example, $A = \begin{bmatrix} 1 & 1-\frac{1}{n} \\ 1 & 1 \end{bmatrix}$, where n is a positive integer. Exact arithmetic gives $A^{-1} = \begin{bmatrix} n & -n+1 \\ -n & n \end{bmatrix}$. However, when n is large, the machine may not distinguish $-n + 1$ from $-n$. Thus, the inverse returned would be $\begin{bmatrix} n & -n \\ -n & n \end{bmatrix}$, which has zero determinant and hence is singular. User beware!

1.2.4.2 Operation Counts

You probably learned in your linear algebra class how to compute the inverse of a small matrix using the algorithm $[A \mid I] \to [I \mid A^{-1}]$ by using Gauss-Jordan elimination. If the matrix A is n-by-n, the number of multiplications and the number of additions can be counted.

1. The number of additions is $n^3 - 2n^2 + n$.

2. The number of multiplications is n^3.

For a very large matrix, n^3 dominates and is used to approximate the number of additions and multiplications.

Chapter 2

Generating Invertible Matrices

Gauss elimination, back substitution, free parameters, Type I, Type II, and Type III operations; pivot

2.1 A Brief Review of Gauss Elimination with Back Substitution

This is a good time to go back to your linear algebra book and review the procedure of Gauss elimination with back substitution to obtain a solution to a system of linear equations. We will give only a brief refresher here. First, we recall how nice "upper triangular" systems are.

Suppose we need to solve

$$\begin{cases} 2x_1 + x_2 - x_3 = 5 \\ \qquad x_2 + x_3 = 3 \\ \qquad\quad 3x_3 = 6 \end{cases}.$$

There is hardly any challenge here. Obviously, the place to start is at the bottom. There is only one equation with one unknown; $3x_3 = 6$ so $x_3 = 2$. Now knowing what x_3 is, the second equation from the bottom has only one equation with one unknown. So, $x_2 + 2 = 3$, whence $x_2 = 1$. Finally, since we know x_2 and x_3, x_1 is easily deduced from the first equation; $2x_1 + 1 - 2 = 5$, so $x_1 = 3$. Thus the unique solution to this system of linear equations is (3, 1, 2). Do you see why this process is called *back substitution*? When it works, it is great! However, consider

$$\begin{cases} 2x_1 + x_2 - x_3 = 5 \\ \qquad\qquad x_3 = 2 \\ \qquad\quad 3x_3 = 6 \end{cases}.$$

We can start at the bottom as before, but there appears to be no way to obtain a value for x_2. To do this, we would need the diagonal coefficients to be nonzero.

Generally then, if we have a triangular system

$$\begin{cases} a_{11}x_1 + a_{12}x_2 + \ldots + a_{1n}x_n = b_1 \\ \qquad\quad a_{22}x_2 + \ldots + a_{2n}x_n = b_2 \\ \qquad\qquad\qquad\qquad \vdots \quad \vdots, \\ \qquad\qquad\qquad\quad a_{nn}x_n = b_n \end{cases}$$

and all the diagonal coefficients a_{ii} are nonzero, then back substitution will work recursively. That is, symbolically,

$$\begin{aligned} x_n &= b_n/a_{nn} \\ x_{n-1} &= \frac{1}{a_{n-1n-1}}(b_{n-1} - a_{n-1n}x_n) \\ &\vdots \\ x_i &= \frac{1}{a_{ii}}(b_i - a_{i,i+1}x_{i+1} - a_{i,i+2}x_{i+2} - \ldots - a_{in}x_n) \\ \text{for} \quad i &= n-1, n-2, \ldots, 2, 1. \end{aligned}$$

What happens if the system is not square? Consider $\begin{cases} x_1 + 2x_2 + 3x_3 + x_4 - x_5 = 2 \\ x_3 + x_4 + x_5 = -1 \\ x_4 - 2x_5 = 4 \end{cases}$. We can make it triangular, sort of:

$$\begin{cases} x_1 + 3x_3 + x_4 = 2 - 2x_2 + x_5 \\ x_3 + x_4 = -1 - x_5 \\ x_4 = 4 + 2x_5 \end{cases}.$$

Now back substitution gives

$$\begin{cases} x_1 = 13 - 2x_2 + 8x_5 \\ x_3 = -5 - 3x_5 \\ x_4 = 4 + 2x_5\,. \\ x_2 = x_2 \\ x_5 = x_5 \end{cases}$$

Clearly, x_2 and x_5 can be any number at all. These variables are called "*free*." So we introduce free parameters, say $x_2 = s$, $x_5 = t$. Then, all the solutions to this system can be described by the infinite set $\{(13 - 2s + 8t, s, -5 - 3t, 4 + 2t, t) \mid s, t \textit{ are arbitrary}\}$.

Wouldn't it be great if all linear systems were (upper) triangular? Well, they are not! The question is, if a system is not triangular, can you transform it into a triangular system without disturbing the set of solutions? The method we review next has ancient roots but was popularized by the German mathematician

Johann Carl Friedrich Gauss (30 April 1777 – 23 February 1855). It is known to us as *Gauss elimination.*

We call two linear systems *equivalent* if they have the same set of solutions. The basic method for solving a system of linear equations is to replace the given system with an equivalent system that is easier to solve. Three elementary operations can help us achieve a triangular system:

- *Type I*: Interchange two equations.
- *Type II*: Multiply an equation by a nonzero constant.
- *Type III*: Multiply an equation by any constant and add the result to another equation.

The strategy is to focus on the diagonal (*pivot*) position and use Type III operations to zero out (i.e., eliminate) all the elements below the pivot. If we find a zero in a pivot position, we use a Type I operation to swap a nonzero number below into the pivot position. Type II operations ensure that a pivot can always be made to equal 1. For example,

$$\left\{\begin{aligned} x_1 + 2x_2 + 3x_3 &= 6 \\ 2x_1 + 5x_2 + x_3 &= 9 \\ x_1 + 4x_2 - 6x_3 &= 1 \end{aligned}\right. \xrightarrow[a_{11}=1]{pivot} \left\{\begin{aligned} x_1 + 2x_2 + 3x_3 &= 6 \\ x_2 - 5x_3 &= -3 \\ 2x_2 - 9x_3 &= -5 \end{aligned}\right. \xrightarrow[a_{22}^{(1)}=1]{pivot}$$

$$\left\{\begin{aligned} x_1 + 2x_2 + 3x_3 &= 6 \\ x_2 - 5x_3 &= -3 \\ x_3 &= 1 \end{aligned}\right. .$$

Back substitution yields the unique solution $(-1, 2, 1)$. As another example, consider

$$\left\{\begin{aligned} x_1 + 2x_2 + x_3 + x_4 &= 4 \\ 2x_1 + 4x_2 - x_3 + 2x_4 &= 11 \\ x_1 + x_2 + 2x_3 + 3x_4 &= 1 \end{aligned}\right. \rightarrow \left\{\begin{aligned} x_1 + 2x_2 + x_3 + x_4 &= 4 \\ -3x_3 \qquad &= 3 \\ -x_2 + x_3 + 2x_4 &= -3 \end{aligned}\right. \rightarrow$$

$$\left\{\begin{aligned} x_1 + 2x_2 + x_3 + x_4 &= 4 \\ -x_2 + x_3 + 2x_4 &= -3 \\ -3x_3 \qquad &= 3 \end{aligned}\right. \rightarrow \left\{\begin{aligned} x_1 + 2x_2 + x_3 &= 4 - x_4 \\ -x_2 + x_3 &= -3 - 2x_4 \\ -3x_3 &= 3 \end{aligned}\right. .$$

We see x_4 is free, so the solution set is

$$\{(1 - 5t, 2 + 2t, -1, t) \mid t \text{ arbitrary}\}.$$

However, things do not always work out as we might hope. For example, if a zero pivot occurs, maybe we can fix the situation and maybe we cannot.

Consider

$$\left\{\begin{array}{r} x+y+z=4 \\ 2x+2y+5z=3 \\ 4x+4y+8z=10 \end{array}\right. \longrightarrow \left\{\begin{array}{r} x+y+z=4 \\ 3z=-5 \\ 4z=-6 \end{array}\right. .$$

There is no way these last two equations can be solved simultaneously. This system is evidently *inconsistent* — that is, no solution exists.

Exercise Set 3

1. Solve the following by Gauss elimination:

(a) $$\left\{\begin{array}{rcrcrcr} 2x_1 & + & x_2 & - & x_3 & = & 2 \\ x_1 & + & 3x_2 & + & 2x_3 & = & 1 \\ 6x_1 & - & 3x_2 & - & 3x_3 & = & 12 \\ 4x_1 & - & 4x_2 & + & 8x_3 & = & 20 \end{array}\right.$$

(b) $$\left\{\begin{array}{rcrcrcrcr} 2x_1 & + & 2x_2 & - & 6x_3 & - & 8x_4 & = & -2 \\ 4x_1 & + & 2x_2 & + & 10x_3 & + & 2x_4 & = & 10 \\ 9x_1 & + & 18x_2 & - & 6x_3 & + & 3x_4 & = & 24 \\ 10x_1 & + & 10x_2 & + & 10x_3 & - & 15x_4 & = & 10 \end{array}\right.$$

(c) $$\left\{\begin{array}{rcrcrcr} \sqrt{2}x_1 & + & 2x_2 & + & 3x_3 & = & 4 \\ 2\sqrt{2}x_1 & + & 3x_2 & + & x_3 & = & 8 \\ 3\sqrt{2}x_1 & + & x_2 & + & 2x_3 & = & 10 \end{array}\right.$$

(d) $$\left\{\begin{array}{rcrcrcrcrcr} 3ix_1 & + & 3ix_2 & - & 6ix_3 & + & 3ix_4 & + & 9ix_5 & = & 3i \\ 4x_1 & - & 2x_2 & + & 4x_3 & + & 4x_4 & + & 12x_5 & = & 4 \\ x_1 & + & \frac{2}{3}x_2 & - & \frac{4}{3}x_3 & - & x_4 & - & 3x_5 & = & 1 \end{array}\right.$$

(e) $$\left\{\begin{array}{rcrcrcrcr} x_1 & + & x_2 & + & x_3 & + & x_4 & = & 3 \\ 3x_1 & + & 4x_2 & - & 5x_3 & - & 6x_4 & = & 1 \\ 4x_1 & + & 5x_2 & - & 4x_3 & - & 6x_4 & = & 5 \end{array}\right.$$

(f) $$\left\{\begin{array}{rcrcr} 4x_1 & + & (2-2i)x_2 & = & i \\ (2-2i)x_1 & - & 2x_2 & = & i \end{array}\right.$$

(g) $\{x_1 + 2x_2 + 3x_3 = 4$

(h)
$$\begin{cases} x_1 + \frac{3}{2}x_2 - \frac{1}{2}x_3 - x_4 = 0 \\ x_2 - \frac{4}{7}x_3 - \frac{5}{7}x_4 = 0 \\ x_1 - \frac{1}{4}x_2 + \frac{1}{2}x_3 + \frac{1}{4}x_4 = 0 \\ x_1 - \frac{11}{2}x_2 + \frac{7}{2}x_3 + 4x_4 = 0 \end{cases}$$

(i)
$$\begin{cases} x_1 + x_2 + x_3 = 3 \\ 2x_1 + 3x_2 + x_3 = 6 \\ 5x_1 - x_2 + 3x_3 = 7 \end{cases}$$

(j)
$$\begin{cases} x_1 - \sqrt{2}x_2 + x_3 = \pi \\ 2x_1 + x_2 - \pi x_3 = \sqrt{2} \\ 3x_1 + 4x_2 - \sqrt{2}x_3 = 3\pi + 2\sqrt{2} \end{cases}$$

(k)
$$\begin{cases} 4x_1 + 6x_2 + (16+12i)x_3 = 26 \\ -4ix_1 + (2-6i)x_2 + (16-12i)x_3 = 8-26i \\ 3ix_2 + (-3+9i)x_3 = 3+12i \end{cases}.$$

2. Consider the nonlinear problem

$$\begin{cases} x^2 + 2xy + y^2 = 1 \\ 2x^2 + xy + 3y^2 = 29 \\ 3x^2 + 3xy + y^2 = 3 \end{cases}.$$

   Is there a way of making this a linear problem and finding x and y?

3. Suppose you need to solve a system of linear equations with three different right-hand sides but the same left-hand side — that is $A\mathbf{x} = \mathbf{b}_1$, $A\mathbf{x} = \mathbf{b}_2$, and $A\mathbf{x} = \mathbf{b}_3$. What would be an efficient way to do this?

4. Suppose you do Gauss elimination and back substitution on a system of three equations in three unknowns. Count the number of multiplications/divisions in the elimination part (11), in the back substitution part (6), and in the total. Count the number of additions/subtractions in the elimination part (8), in the back substitution part (3), and the total. (*Hint*: do not count creating zeros. The numerical people wisely put zeros where they belong and do not risk introducing roundoff error.)

5. If you are brave, repeat problem 4 for n equations in n unknowns. At the ith stage, you need $(n-i) + (n-i)(n-i+1)$ multiplications/divisions, so the total number of multiplications/divisions for the elimination is $(2n^3 + 3n^2 - 5n)/6$. At the ith stage, you need $(n-i)(n-i+1)$ additions/subtractions, so you need a total of $(n^3 - n)/3$ additions/subtractions. For the back substitution part, you need $(n^2+n)/2$

multiplications/divisions and $(n^2-n)/2$ additions/subtractions, so the total number of multiplications/divisions is $n^3/3+n^2-n/3$ and the total number of additions/subtractions is $n^3/3+n^2/2-5n/6$.

6. Create some examples of four equations in three unknowns (x, y, z) such that a) one solution exists, b) an infinite number of solutions exist, and c) no solutions exist. Recall our (zero, one)-advice from before.

7. Pietro finds 16 U.S. coins worth 89 cents in the Trevi Fountain in Rome. The coins are dimes, nickels, and pennies. How many of each coin did he find? Is your answer unique?

8. Suppose (x_1, y_1), (x_2, y_2), and (x_3, y_3) are distinct points in $\mathbb{R}^2$ and lie on the same parabola $y = Ax^2 + Bx + C$. Solve for A, B, and C in terms of the given data. Make up a concrete example and find the parabola.

9. A linear system is called *overdetermined* iff there are more equations than unknowns. Argue that the following overdetermined system is inconsistent: $\begin{cases} 2x+2y=1 \\ -4x+8y=-8 \\ 3x-3y=9 \end{cases}$. Draw a picture in $\mathbb{R}^2$ to see why.

10. A linear system of m linear equations in n unknowns is call *underdetermined* iff m < n (i.e., there are fewer equations than unknowns). Argue that it is not possible for an underdetermined linear system to have only one solution.

11. Green plants use sunlight to convert carbon dioxide and water to glucose and oxygen. We are the beneficiaries of the oxygen part. Chemists write

$$x_1CO_2 + x_2H_2O \to x_3O_2 + x_4C_6H_{12}O_6.$$

To balance, this equation must have the same number of each atom on each side. Set up a system of linear equations and balance the system. Is your solution unique?

Further Reading

[B&R, 1986(2)] T. S. Blyth and E. F. Robertson, *Matrices and Vector Spaces*, Vol. 2, Chapman & Hall, New York, (1986).

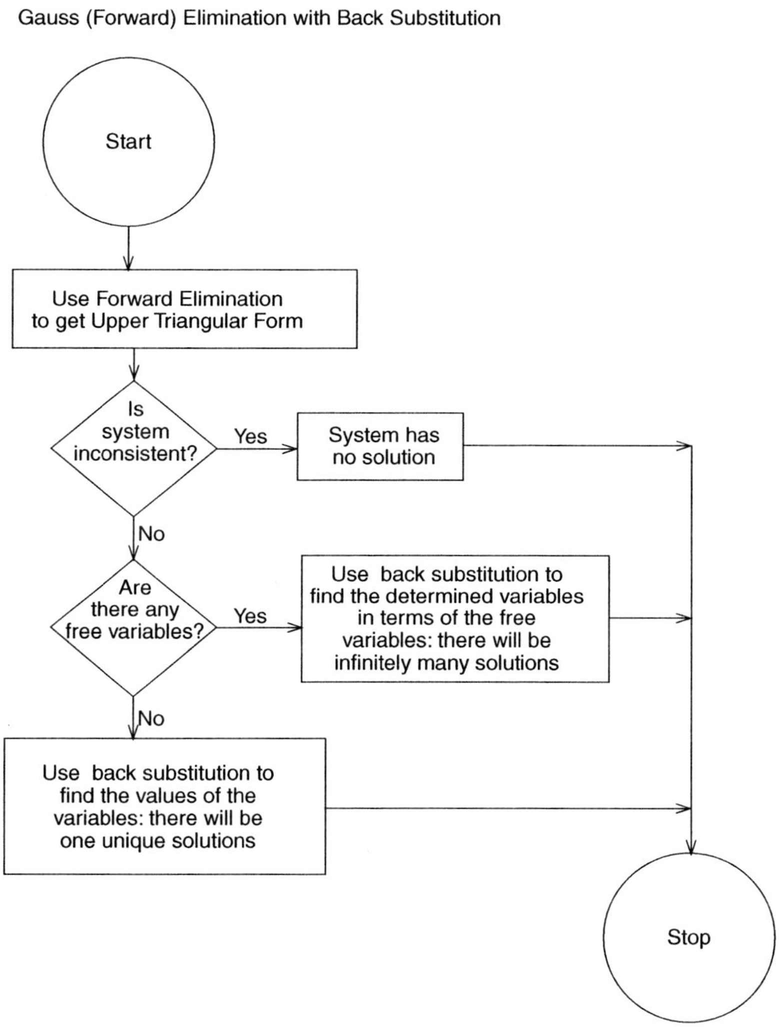

Figure 2.1: Gauss elimination flow chart.

2.1.1 MATLAB Moment

2.1.1.1 Solving Systems of Linear Equations

MATLAB uses the slash (/) and backslash (\) to return a solution to a system of linear equations. $X = A\backslash B$ returns a solution to the matrix equation $AX = B$, while $X = B/A$ returns a solution to $XA = B$. If A is nonsingular, then $A\backslash B$ returns the unique solution to $AX = B$. If A is m-by-n with $m > n$

(overdetermined), $A\backslash B$ returns a least squares solution. If A has rank n, there is a unique least squares solution. If A has rank less than n, then $A\backslash B$ is a basic solution. If the system is underdetermined ($m < n$), then $A\backslash B$ is a basic solution. If the system has no solution, the $A\backslash B$ is a least squares solution. Let's look at some examples. Let's solve the first homework problem of Exercise Set 3, 1a.

```
>>A=[2 1 -1;1 3 2;6 -3 -3;4 -4 8]
A =
   2    1   -1
   1    3    2
   6   -3   -3
   4   -4    8
>> b=[2;1;12;20]
b =
   2
   1
   12
   20
>> x=A\b
x =
    2.0000
   -1.0000
    1.0000
>> A*x
ans =
    2.0000
    1.0000
    12.0000
    1.0000
```

Next let's solve 1k since it has complex coefficients.

```
>> C=[4 6 16+12i;-4 2-6i 16-12i;0 3i -3+9i]
C =
     4.0000        6.0000             16.0000 + 12.0000i
    -4.0000     2.0000 - 6.0000i      16.0000 - 12.0000i
       0           0 + 3.0000i        -3.0000 + 9.0000i
>> d=[26;8-26i;3+12i]
d =
    26.0000
    8.0000 - 26.0000i
    3.0000 + 12.0000i
>> y=C\d
y =
    -0.1154 + 0.0769i
    1.1538 + .02308i
    0.7308 - 0.6538i
```

```
>> C*y
ans =
    26.0000
    8.0000 − 26.0000i
    3.0000 + 12.0000i
```

Now let's look at an overdetermined system.

```
>> E=[2 2;-4 8;3 -3]
E =
     2    2
    −4    8
     3   −3
>> f=[1; -8; 9]
    1
    −8
    9
>> z=E\f
z =
    1.6250
    −0.3750
>> E*z
ans =
    2.5000
    −9.5000
    6.0000
```

matrix units, transvections, dilations, permutation matrix, elementary row operations, elementary column operations, the minimal polynomial

---

2.2 Elementary Matrices

We have seen that when the coefficient matrix of a system of linear equations is invertible, we can immediately formulate the solution to the system in terms of the inverse of this matrix. Thus, in this section, we look at ways of generating invertible matrices. Several easy examples immediately come to mind. The n-by-n identity matrix I_n is always invertible since $(I_n)^{-1} = I_n$. So

$$\begin{bmatrix} 1 & 0 \\ 0 & 1 \end{bmatrix}^{-1} = \begin{bmatrix} 1 & 0 \\ 0 & 1 \end{bmatrix}, \begin{bmatrix} 1 & 0 & 0 \\ 0 & 1 & 0 \\ 0 & 0 & 1 \end{bmatrix}^{-1} = \begin{bmatrix} 1 & 0 & 0 \\ 0 & 1 & 0 \\ 0 & 0 & 1 \end{bmatrix},$$

and so on. Next, if we multiply I_n by a nonzero scalar λ, then λI_n is invertible by Theorem 1.2 of Chapter 1 page 18, with $(\lambda I_n)^{-1} = \frac{1}{\lambda} I_n$. Thus, for example, $\begin{bmatrix} 10 & 0 & 0 \\ 0 & 10 & 0 \\ 0 & 0 & 10 \end{bmatrix}^{-1} = \begin{bmatrix} 1/10 & 0 & 0 \\ 0 & 1/10 & 0 \\ 0 & 0 & 1/10 \end{bmatrix}$. More generally, when we take a diagonal matrix with nonzero entries on the diagonal, we have an invertible matrix with an easily computed inverse matrix. Symbolically, if $D = diag(d_1, d_2, \cdots, d_n)$ with all $d_i \neq 0$ for $i = 1, \cdots, n$, then $D^{-1} = diag(1/d_1, 1/d_2, \cdots, 1/d_n)$. For example, $\begin{bmatrix} 1 & 0 & 0 \\ 0 & 2 & 0 \\ 0 & 0 & 3 \end{bmatrix}^{-1} = \begin{bmatrix} 1 & 0 & 0 \\ 0 & 1/2 & 0 \\ 0 & 0 & 1/3 \end{bmatrix}$.

Now these examples are not going to impress our smart friends, so we need to be a bit more clever. Say we know an invertible matrix U. Then we can take a diagonal invertible matrix D and form $U^{-1}DU$, which has the inverse $U^{-1}D^{-1}U$. Thus, one nontrivial invertible matrix U gives us a way to form many nontrivial examples. For instance, take $D = \begin{bmatrix} 2 & 0 & 0 \\ 0 & 2 & 0 \\ 0 & 0 & 1 \end{bmatrix}$ and $U = \begin{bmatrix} 1 & 0 & 2 \\ 1 & 1 & 1 \\ -1 & 0 & -1 \end{bmatrix}$. Then $U^{-1} = \begin{bmatrix} -1 & 0 & -2 \\ 0 & 1 & 1 \\ 1 & 0 & 1 \end{bmatrix}$, as you may verify, and $\begin{bmatrix} -1 & 0 & -2 \\ 0 & 1 & 1 \\ 1 & 0 & 1 \end{bmatrix} \begin{bmatrix} 2 & 0 & 0 \\ 0 & 2 & 0 \\ 0 & 0 & 1 \end{bmatrix} \begin{bmatrix} 1 & 0 & 2 \\ 1 & 1 & 1 \\ -1 & 0 & -1 \end{bmatrix} = \begin{bmatrix} 0 & 0 & -2 \\ 1 & 2 & 1 \\ 1 & 0 & 3 \end{bmatrix}$ is an invertible matrix. Unfortunately, not all invertible matrices are obtainable this way (i.e., from a diagonal invertible matrix). We now turn to an approach that will generate all possible invertible matrices.

You probably recall that one of the first algorithms you learned in linear algebra class was Gauss elimination. We reviewed this algorithm in Section 2.1. This involved three kinds of "elementary" operations that had the nice property of not changing the solution set to a system of linear equations. We will now elaborate a matrix approach to these ideas.

There are three kinds of *elementary matrices,* all invertible. All are obtained from the identity matrix by perturbing it in special ways. We begin with some very simple matrices (not invertible at all) that can be viewed as the building blocks for all matrices. These are the so-called *matrix units* E_{ij}. Define E_{ij} to be the n-by-n matrix (actually, this makes sense for m-by-n matrices as well), with all entries zero except that the (i, j) entry has a one. In the 2-by-2 case, we can easily exhibit all of them:

$$E_{11} = \begin{bmatrix} 1 & 0 \\ 0 & 0 \end{bmatrix}, E_{12} = \begin{bmatrix} 0 & 1 \\ 0 & 0 \end{bmatrix}, E_{21} = \begin{bmatrix} 0 & 0 \\ 1 & 0 \end{bmatrix}, E_{22} = \begin{bmatrix} 0 & 0 \\ 0 & 1 \end{bmatrix}.$$

The essential facts about matrix units are collected in the next theorem, whose proof is left as an exercise.

THEOREM 2.1

1. $\sum_{i=1}^{n} E_{ii} = I_n$.
2. $E_{ij} = E_{pq}$ *iff* $i = p$ *and* $j = q$.
3. $E_{ij}E_{rs} = \begin{cases} \mathbb{O} & \textit{if } r \neq j \\ E_{is} & \textit{if } r = j \end{cases}.$
4. $E_{ii}^2 = E_{ii}$ *for all* i.
5. *Given any matrix* $A = [a_{ij}]$ *in* $\mathbb{C}^{n\times n}$, *we have* $A = \sum_{i=1}^{n} \sum_{j=1}^{n} a_{ij} E_{ij}$.
6. *The collection* $\{E_{ij} | i, j = 1, \cdots, n\}$ *is a basis for the vector space of matrices* $\mathbb{C}^{n\times n}$ *which, therefore, has dimension* n^2.

We will now use the matrix units to define the first two types of elementary matrices. The first type of elementary matrix (Type III) goes under a name derived from geometry. It is called a *transvection* and is defined as follows.

DEFINITION 2.1 *(transvection)*
For $i \neq j$, *define* $T_{ij}(c) = I_n + cE_{ij}$ *for any complex number* c.

Do not let the formula intimidate you. The idea is very simple. Just take the identity matrix and put the complex number c in the (i, j) position. In the 2-by-2 case, we have

$$T_{12}(c) = \begin{bmatrix} 1 & c \\ 0 & 1 \end{bmatrix} \text{ and } T_{21}(c) = \begin{bmatrix} 1 & 0 \\ c & 1 \end{bmatrix}.$$

The essential facts about transvections are collected in the next theorem.

THEOREM 2.2
Suppose c and d are complex numbers. Then

1. $T_{ij}(c)T_{ij}(d) = T_{ij}(d + c)$.
2. $T_{ij}(0) = I_n$.
3. *Each* $T_{ij}(c)$ *is invertible and* $T_{ij}(c)^{-1} = T_{ij}(-c)$.
4. $T_{ij}(c)T_{rs}(d) = T_{rs}(d)T_{ij}(c)$, *if* $j \neq r \neq s \neq i \neq j$.
5. $T_{ij}(cd) = T_{ir}(c)^{-1}T_{rj}(d)^{-1}T_{ir}(c)T_{rj}(d)$, *if* $r \neq i \neq j \neq r$.
6. $T_{ij}(c)^T = T_{ji}(c)$.
7. $T_{ij}(c)^* = T_{ji}(\overline{c})$.
8. $\det(T_{ij}(c)) = 1$.

The proofs are routine and left to the reader.

The second type of elementary matrix (Type II) also has a name derived from geometry. It is called a *dilation* and is defined as follows.

DEFINITION 2.2 (*dilation*)
For any $i = 1, \cdots, n$, *and any nonzero complex number* c, *define* $D_i(c) = I_n + (c - 1)E_{ii}$.

Once again, what we are doing is quite straightforward. To form $D_i(c)$, simply write down the identity matrix and replace the 1 in the diagonal (i, i) position with c. In the 2-by-2 case, we have

$$D_1(c) = \begin{bmatrix} c & 0 \\ 0 & 1 \end{bmatrix} \quad \text{and} \quad D_2(c) = \begin{bmatrix} 1 & 0 \\ 0 & c \end{bmatrix}.$$

The salient facts about dilations are collected in the next theorem.

THEOREM 2.3
Suppose c and d are nonzero complex numbers. Then

1. $D_i(c) = I - E_{ii} + cE_{ii} = cE_{ii} + \sum_{\substack{j=1 \\ j \neq i}}^{n} E_{jj}$.
2. $D_i(c)D_i(d) = D_i(cd)$.
3. $D_i(1) = I_n$.

4. *$D_i(c)$ is invertible and $D_i(c)^{-1} = D_i(c^{-1})$.*

5. *$det(D_i(c)) = c$.*

6. *$D_i(c)D_j(d) = D_j(d)D_i(c)$, if $i \neq j$.*

7. *$D_i(c)^T = D_i(c)$.*

8. *$D_i(c)^* = D_i(\overline{c})$.*

Once again, the easy proofs are left to the reader.

Finally, we have the third type of elementary matrix (Type I), which is called a *permutation matrix*. Let S_n denote the set of all permutations of the n element set $[n] := \{1, 2, \cdots, n\}$. Strictly speaking, a permutation is a one-to-one function from $[n]$ onto $[n]$. If σ is a permutation, we have σ change the identity matrix as follows.

DEFINITION 2.3 *(permutation matrix)*
Let σ be a permutation. Define the permutation matrix $P(\sigma) = [\delta_{i,\sigma(j)}]$ where $\delta_{i,\sigma(j)} = \begin{cases} 1 & \text{if } i = \sigma(j) \\ 0 & \text{if } i \neq \sigma(j) \end{cases}$. That is, $ent_{ij}(P(\sigma)) = \delta_{i,\sigma(j)}$.

Once again, a very simple idea is being expressed here. All we are doing is swapping rows of the identity matrix according to the permutation σ. Let's look at an example. Suppose σ, in cycle notation, is the permutation (123) of $[3] = \{1, 2, 3\}$. In other words, under σ, 1 goes to 2, 2 goes to 3, and 3 goes back to 1. Then,

$$P(123) = \begin{bmatrix} \delta_{12} & \delta_{13} & \delta_{11} \\ \delta_{22} & \delta_{23} & \delta_{21} \\ \delta_{32} & \delta_{33} & \delta_{31} \end{bmatrix} = \begin{bmatrix} 0 & 0 & 1 \\ 1 & 0 & 0 \\ 0 & 1 & 0 \end{bmatrix}.$$

Notice that this matrix is obtained from the identity matrix I_3 by sending the first row to the second, the second row to the third, and the third row to the first, just as σ indicated. Since every permutation is a product (i.e., composition) of transpositions (i.e., permutations that leave everything fixed except for swapping two elements), it suffices to deal with permutation matrices that are obtained by swapping only two rows of the identity matrix. For example, in $\mathbb{C}^{3\times 3}$,

$$P(12) = \begin{bmatrix} 0 & 1 & 0 \\ 1 & 0 & 0 \\ 0 & 0 & 1 \end{bmatrix} \quad \text{and} \quad P(23) = \begin{bmatrix} 1 & 0 & 0 \\ 0 & 0 & 1 \\ 0 & 1 & 0 \end{bmatrix}.$$

We collect the basic facts about permutation matrices next.

THEOREM 2.4

1. $P(\pi)P(\sigma) = P(\pi\sigma)$, *where* π *and* σ *are any permutations in* S_n.
2. $P(\iota) = I_n$, *where* ι *is the identity permutation that leaves every element of* $[n]$ *fixed.*
3. $P(\sigma^{-1}) = P(\sigma)^{-1} = P(\sigma)^T = P(\sigma)^*$.
4. $P(ij) = I - E_{ii} - E_{jj} + E_{ij} + E_{ji}$, *for any* i, j *with* $1 \leq i, j \leq n$.
5. $P(ij)^{-1} = P(ij)$.
6. $E_{jj} = P(ij)^{-1} E_{ii} P(ij)$.

In the 2-by-2 case, only two permutations exist: the identity ι and the transposition (12). Thus,

$$P(\iota) = \begin{bmatrix} 1 & 0 \\ 0 & 1 \end{bmatrix} \quad \text{and} \quad P(12) = \begin{bmatrix} 0 & 1 \\ 1 & 0 \end{bmatrix}.$$

So there you are. We have developed the three kinds of elementary matrices. Note that each is expressible in terms of matrix units and each is invertible with an inverse of the same type. Moreover, the inverses are very easy to compute! So why have we carefully developed these matrices? The answer is they do real work for us. The way they do work is given by the next two theorems.

THEOREM 2.5 (theorem on elementary row operations)
Let A be a matrix in $\mathbb{C}^{m \times n}$. *Then the matrix product*

1. $T_{ij}(c)A$ *amounts to adding the left c multiple of the* j*th row of A to the* i*th row of A.*
2. $D_i(c)A$ *amounts to multiplying the* i*th row of A on the left by c.*
3. $P(\sigma)A$ *amounts to moving the* i*th row of A into the position of the* $\sigma(i)$*th row for each i.*
4. $P(ij)A$ *amounts to swapping the* i*th and* j*th rows of A and leaving the other rows alone.*

We have similar results for columns.

THEOREM 2.6 (theorem on elementary column operations)
Let A be a matrix in $\mathbb{C}^{m\times n}$. Then the matrix product

1. *$AT_{ij}(c)$ amounts to adding the right c multiple of the ith column of A to the jth column of A.*
2. *$AD_i(c)$ amounts to multiplying the ith column of A on the right by c.*
3. *$AP(\sigma^{-1})$ amounts to moving the ith column of A into the position of the $\sigma(i)$th column.*
4. *$AP(ij)$ amounts to swapping the ith and jth columns of A and leaving the other columns alone.*

We illustrate in the 2-by-2 case. Let $A = \begin{bmatrix} a & b \\ c & d \end{bmatrix}$. Then $T_{12}(\alpha)A = \begin{bmatrix} 1 & \alpha \\ 0 & 1 \end{bmatrix}\begin{bmatrix} a & b \\ c & d \end{bmatrix} = \begin{bmatrix} a+\alpha c & b+\alpha d \\ c & d \end{bmatrix}$; $D_1(\alpha)A = \begin{bmatrix} \alpha a & \alpha b \\ c & d \end{bmatrix}$; $P(12)A = \begin{bmatrix} 0 & 1 \\ 1 & 0 \end{bmatrix}\begin{bmatrix} a & b \\ c & d \end{bmatrix} = \begin{bmatrix} c & d \\ a & b \end{bmatrix}$ and $AT_{12}(\alpha) = \begin{bmatrix} a & b \\ c & d \end{bmatrix}\begin{bmatrix} 1 & \alpha \\ 0 & 1 \end{bmatrix} = \begin{bmatrix} a & a\alpha+b \\ c & c\alpha+d \end{bmatrix}$.

Now that we have elementary matrices to work for us, we can establish some significant results. The first will be to determine all invertible matrices. However, we need a basic fact about multiplying an arbitrary matrix by an invertible matrix on the left. Dependency relationships among columns do not change.

THEOREM 2.7
Suppose A is an m-by-n matrix and R is an m-by-m invertible matrix. Select any columns $\mathbf{c}_1, \mathbf{c}_2, \cdots, \mathbf{c}_k$ from A. Then $\{\mathbf{c}_1, \mathbf{c}_2, \cdots, \mathbf{c}_k\}$ is independent if and only if the corresponding columns of RA are independent. Moreover, $\{\mathbf{c}_1, \mathbf{c}_2, \cdots, \mathbf{c}_k\}$ is dependent if and only if the corresponding columns of RA are dependent with the same scalars providing the dependency relations for both sets of vectors.

PROOF First note that the product RA is the same as $[R\mathbf{a}_1|R\mathbf{a}_2|\cdots|R\mathbf{a}_n]$, if A is partitioned into columns, $A = [\mathbf{a}_1|\mathbf{a}_2|\cdots|\mathbf{a}_n]$. Suppose $\{\mathbf{c}_1, \mathbf{c}_2, \cdots, \mathbf{c}_k\}$ is an independent set of columns. We claim the corresponding vectors $R\mathbf{c}_1$, $R\mathbf{c}_2, \cdots, R\mathbf{c}_k$ are also independent. For this we take the usual approach and assume we have a linear combination of these vectors that produces the zero vector, say

$$\alpha_1 R\mathbf{c}_1 + \alpha_2 R\mathbf{c}_2 + \cdots + \alpha_k R\mathbf{c}_k = \overrightarrow{0}.$$

But then, $R(\alpha_1\mathbf{c}_1 + \alpha_2\mathbf{c}_2 + \cdots + \alpha_k\mathbf{c}_k) = \overrightarrow{0}$. Since R is invertible, we conclude $\alpha_1\mathbf{c}_1 + \alpha_2\mathbf{c}_2 + \cdots + \alpha_k\mathbf{c}_k = \overrightarrow{0}$. But the **c**'s are independent, so we may conclude all the αs are zero, which is what we needed to show the vectors $R\mathbf{c}_1, R\mathbf{c}_2, \cdots, R\mathbf{c}_k$ to be independent. Conversely, suppose $R\mathbf{c}_1, R\mathbf{c}_2, \cdots, R\mathbf{c}_k$ are independent. We show $\mathbf{c}_1, \mathbf{c}_2, \cdots, \mathbf{c}_k$ are independent. Suppose $\alpha_1\mathbf{c}_1 + \alpha_2\mathbf{c}_2 + \cdots + \alpha_k\mathbf{c}_k = \overrightarrow{0}$. Then $R(\alpha_1\mathbf{c}_1 + \alpha_2\mathbf{c}_2 + \cdots + \alpha_k\mathbf{c}_k) = R\overrightarrow{0} = \overrightarrow{0}$. But then, $\alpha_1 R\mathbf{c}_1 + \alpha_2 R\mathbf{c}_2 + \cdots + \alpha_k R\mathbf{c}_k = \overrightarrow{0}$ and independence of the $R\mathbf{c}$s implies all the α's are zero, which completes this part of the proof. The rest of the proof is transparent and left to the reader. □

COROLLARY 2.1

With A and R as above, the dimension of the column space of A is equal to the dimension of the column space of RA. (Note: We did not say the column spaces are the same.)

We now come to our main result.

THEOREM 2.8

Let A be an n-by-n square matrix. Then A is invertible if and only if A can be written as a product of elementary matrices.

PROOF If A can be written as a product of elementary matrices, then A is a product of invertible matrices and hence itself must be invertible by Theorem 1.2 of Chapter 1 page 18.

Conversely, suppose A is invertible. Then the first column of A cannot consist entirely of zeros. Use a permutation matrix P, if necessary, to put a nonzero entry in the (1,1) position. Then, use a dilation D to make the (1,1) entry equal to 1. Now use transvections to "clean out" (i.e., zero) all the entries in the first column below the (1,1) position. Thus, a product of elementary matrices, some of which could be the identity, produces $TDPA = \begin{bmatrix} 1 & * & * & \cdots & * \\ 0 & * & * & & * \\ 0 & * & * & & * \\ \vdots & \vdots & \vdots & & \\ 0 & * & * & & * \end{bmatrix}$. Now there must be a nonzero entry at or below the (2,2) position of this matrix. Otherwise, the first two columns of this matrix would be dependent by Theorem 2.7. But this would contradict that A is invertible, so its columns must be independent. Use a permutation matrix, if necessary, to swap a nonzero entry into the (2,2) position and use a dilation, if necessary, to make it 1. Now use transvections to "clean out" above and below the (2,2) entry. Therefore, we achieve

$$T_1 D_1 P_1 TDPA = \begin{bmatrix} 1 & 0 & * & * & * & \cdots & * \\ 0 & 1 & * & & & & * \\ 0 & 0 & * & & & & * \\ \vdots & \vdots & \vdots & & & & \vdots \\ 0 & 0 & * & * & * & \cdots & * \end{bmatrix}.$$

Again, if all the entries of the third column at and below the (3,3) entry were zero, the third column would be a linear combination of the first two, again contradicting the invertibility of A. So, continuing this process, which must terminate after a finite number of steps, we get $E_1 E_2 E_3 \cdots E_p A = I_n$, where all the E_is are elementary matrices. Thus, $A = E_p^{-1} \cdots E_2^{-1} E_1^{-1}$, which again is a product of elementary matrices. This completes the proof. □

This theorem is the basis of an algorithm you may recall for computing by hand the inverse of, at least, small matrices. Begin with a square matrix A and augment it with the identity matrix I, forming $[A|I]$. Then apply elementary operations on the left attempting to turn A into I. If the process succeeds, you will have produced A^{-1}, where I was originally; in other words, $[I|A^{-1}]$ will be the result. If you keep track of the elementary matrices used, you will also be able to express A as a product of elementary matrices.

2.2.1 The Minimal Polynomial

There is a natural and useful connection between matrices and polynomials. We know the dimension of $\mathbb{C}^{n \times n}$ as a vector space is n^2 so, given a matrix A, we must have its powers $I, A, A^2, A^3, \cdots$ eventually produce a dependent set. Thus there must exist scalars, not all zero, such that $A^p + \alpha_{p-1} A^{p-1} + \cdots + \alpha_1 A + \alpha_0 I = \mathbb{O}$. This naturally associates to A the polynomial $p(x) = x^p + \alpha_{p-1} x^{p-1} + \cdots + \alpha_1 x + \alpha_0$, and we can think of A as being a "root" of p since replacing x by A yields $p(A) = \mathbb{O}$. Recall that a polynomial with leading coefficient 1 is called *monic.*

THEOREM 2.9

Every matrix in $\mathbb{C}^{n \times n}$ has a unique monic polynomial of least degree that it satisfies as a root. This unique polynomial is called the minimal (or minimum) polynomial *of the matrix and is denoted μ_A.*

PROOF Existence of such a polynomial is clear by the argument given above the theorem, so we address uniqueness. Suppose $f(x) = x^p + \alpha_{p-1}x^{p-1} + \cdots + \alpha_1 x + \alpha_0$ and $g(x) = x^p + \beta_{p-1}x^{p-1} + \cdots \beta_1 x + \beta_0$ are two polynomials of least degree satisfied by A. Then A would also satisfy $f(x) - g(x) = (\alpha_{p-1} - \beta_{p-1})x^{p-1} + \cdots + (\alpha_0 - \beta_0)$. If any coefficient of this polynomial were nonzero, we could produce a monic polynomial of degree less than p satisfied by A, a contradiction. Hence $\alpha_j = \beta_j$ for all j and so $p(x) = q(x)$. This completes the proof. ▯

You may be wondering why we brought up the minimal polynomial at this point. There is a nice connection to matrix inverses. Our definition of matrix inverse requires two verifications. The inverse must work on both sides. The next theorem saves us much work. It says you only have to check one equation.

THEOREM 2.10

Suppose A is a square matrix in $\mathbb{C}^{n\times n}$. If there exists a matrix B in $\mathbb{C}^{n\times n}$ such that $AB = I$, then $BA = I$ and so $B = A^{-1}$. Moreover, a square matrix A in $\mathbb{C}^{n\times n}$ has an inverse if and only if the constant term of its minimal polynomial is nonzero. If A^{-1} exists, it is expressible as a polynomial in A of degree $deg(\mu_A) - 1$. In particular, if A commutes with a matrix C, then A^{-1} commutes with C also.

PROOF Suppose that $AB = I$. We shall prove that $BA = I$ as well. Consider the minimal polynomial $\mu_A(x) = \beta_0 + \beta_1 x + \beta_2 x^2 + \cdots + x^m$. First we claim $\beta_0 \neq 0$. Otherwise, $\mu_A(x) = \beta_1 x + \beta_2 x^2 + \cdots + x^m = x(p(x))$, where $p(x) = \beta_1 + \beta_2 x + \cdots + x^{m-1}$. But then, $p(A) = \beta_1 + \beta_2 A + \cdots + A^{m-1} = \beta_1 I + \beta_2 AI + \cdots + A^{m-1}I = \beta_1 AB + \beta_2 AAB + \cdots + A^{m-1}AB = (\beta_1 A + \beta_2 A^2 + \cdots + A^m)B = \mu_A(A)B = \mathbb{O}$. But the degree of p is less than the degree of μ_A, and this is a contradiction. Therefore, $\beta_0 \neq 0$. This allows us to solve for the identity matrix in the minimal polynomial equation; $\beta_0 I = -\beta_1 A - \beta_2 A^2 - \cdots - A^m$ so that $I = -\dfrac{\beta_1}{\beta_0}A - \dfrac{\beta_2}{\beta_0}A^2 - \cdots - \dfrac{1}{\beta_0}A^m$. Multiplying through by B on the right we get $B = -\dfrac{\beta_1}{\beta_0}AB - \dfrac{\beta_2}{\beta_0}A^2B - \cdots - \dfrac{1}{\beta_0}A^mB = -\dfrac{\beta_1}{\beta_0}I - \dfrac{\beta_2}{\beta_0}A - \cdots - \dfrac{1}{\beta_0}A^{m-1}$. This expresses B as a polynomial in A of degree one less than that of the minimal polynomial and hence B commutes with A. Thus $I = AB = BA$. The remaining details are left to the reader. ▯

There is an algorithm that allows us to compute the minimum polynomial of matrices that are not too large. Suppose we are given an n-by-n matrix A. Start multiplying A by itself to form $A, A^2, \ldots, A^n$. Form a huge matrix B where the rows of B are $I, A, A^2, \ldots$ strung out as rows so that each row of B has n^2 elements and B is $(n+1)-by-n^2$. We must find a dependency among the first p rows of B where p is minimal. Append the identity matrix I_{n+1} to B and row reduce $[B \mid I_{n+1}]$. Look for the first row of zeros in the transformed B matrix. The corresponding coefficients in the transformed I_{n+1} matrix give the coefficients of a dependency. Let's illustrate with an example. Let $A = \begin{bmatrix} 1 & 0 & 3 \\ 2 & 1 & 1 \\ -1 & 0 & 3 \end{bmatrix}$.

Then $A^2 = \begin{bmatrix} -2 & 0 & 12 \\ 3 & 1 & 10 \\ -4 & 0 & 6 \end{bmatrix}$ and $A^3 = \begin{bmatrix} -14 & 0 & 30 \\ -5 & 1 & 40 \\ -10 & 0 & 6 \end{bmatrix}$. Then we form

$[B \mid I_4] =$

$$\begin{bmatrix} 1 & 0 & 0 & 0 & 1 & 0 & 0 & 0 & 1 & 1 & 0 & 0 & 0 \\ 1 & 0 & 3 & 2 & 1 & 1 & -1 & 0 & 3 & 0 & 1 & 0 & 0 \\ -2 & 0 & 12 & 3 & 1 & 10 & -4 & 0 & 6 & 0 & 0 & 1 & 0 \\ -14 & 0 & 30 & -5 & 1 & 40 & -10 & 0 & 6 & 0 & 0 & 0 & 1 \end{bmatrix}.$$

Now do you see what we mean about stringing out the powers of A into rows of B? Next row reduce $[B \mid I_4]$.

$$\begin{bmatrix} 1 & 0 & 0 & 0 & 1 & 0 & 0 & 0 & 1 & 0 & \frac{5}{3} & -\frac{5}{6} & \frac{1}{6} \\ 0 & 0 & 1 & 0 & \frac{2}{5} & \frac{17}{5} & -\frac{1}{3} & 0 & \frac{2}{3} & 0 & \frac{26}{45} & -\frac{23}{90} & \frac{7}{90} \\ 0 & 0 & 0 & 1 & 1 & -\frac{3}{5} & -\frac{6}{5} & 0 & 0 & 0 & -\frac{6}{5} & \frac{4}{5} & -\frac{1}{5} \\ 0 & 0 & 0 & 0 & 0 & 0 & 0 & 0 & 0 & 1 & -\frac{5}{3} & \frac{5}{6} & -\frac{1}{6} \end{bmatrix}.$$

The first full row of zeros in the transformed B part is the last row, so we read that an annihilating polynomial of A is $1-\frac{5}{3}x+\frac{5}{6}x^2-\frac{1}{6}x^3$. This is not a monic polynomial, however, but that is easy to fix by an appropriate scalar multiple. In this case, multiply through by -6. Then we conclude $\mu_A(x)=x^3-5x^2+10x-6$ is the minimal polynomial of A. The reader may verify this. A more efficient way to compute the minimal polynomial will be described later.

Exercise Set 4

1. Argue that $D_i(\frac{1}{a})T_{ji}(\frac{-c}{a}) = T_{ji}(-c)D_i(\frac{1}{a})$ $for\ a \neq 0$.

2. Fill in the details of the proofs of Theorems 2.1 through 2.3.

3. Fill in the details of Theorem 2.4.

4. Complete the proof of Theorem 2.5 and Theorem 2.6.

5. Argue that zero cannot be a root of μ_A if A is invertible.

6. Compute the inverse of $A = \begin{bmatrix} 1 & -1 & 2 & 5 \\ 2 & -1 & -1 & 4 \\ 1 & 3 & 2 & -1 \\ 2 & 1 & -1 & 2 \end{bmatrix}$ and express it as a polynomial in A and as a product of elementary matrices.

7. If $A = [a_{ij}]$, describe $P(\sigma)^{-1}AP(\sigma)$ for σ, a permutation in S_n.

8. Suppose $U = \begin{bmatrix} u_{11} & u_{12} & u_{13} \\ 0 & u_{22} & u_{23} \\ 0 & 0 & u_{33} \end{bmatrix}$. Find permutation matrices P_1 and P_2 such that $P_1 U P_2 = \begin{bmatrix} u_{33} & 0 & 0 \\ u_{23} & u_{22} & 0 \\ u_{13} & u_{12} & u_{11} \end{bmatrix}$.

9. Show that permutation matrices are, in a sense, redundant, since all you really need are transvections and dilations. (*Hint:* Show $P_{ij} = D_i(-1)T_{ji}(1)\ T_{ij}(-1)T_{ji}(1)$). Explain in words how this sequence of dilations and transvections accomplishes a row swap.

10. Consider the n-by-n matrix units E_{ij} and let A be any n-by-n matrix. Describe AE_{ij}, $E_{ij}A$, and $E_{ij}AE_{km}$. What is $(E_{ij}E_{jk})^T$?

11. Find the minimal polynomial of $A = \begin{bmatrix} 1 & 1 & 1 \\ 0 & 1 & 1 \\ 0 & 0 & 1 \end{bmatrix}$.

12. Make up a 4-by-4 matrix and compute its minimal polynomial by the algorithm described on pages 58–59.

13. Let $A = \begin{bmatrix} a & b \\ c & d \end{bmatrix}$ where $a \neq 0$. Argue that $A = \begin{bmatrix} 1 & 0 \\ x & 1 \end{bmatrix} \begin{bmatrix} u & v \\ 0 & w \end{bmatrix}$ for suitable choices of x, u, v, and w.

14. What can you say about the determinant of a permutation matrix?

15. Argue that multiplying a matrix by an elementary matrix does not change the order of its largest nonzero minor. Argue that nonzero minors go to nonzero minors and zero minors go to zero minors. Why does this mean that the rank of an m-by-n matrix is the order of its largest nonzero minor?

16. Let $p(x) = a_0 + a_1x + \cdots + a_nx^n$ be a polynomial with complex coefficients. We can create a matrix $p(A) = a_0I + a_1A + \cdots + a_nA^n$ if A is n-by-n. Argue that for any two polynomials $p(x)$ and $g(x)$, $p(A)$ commutes with $g(A)$.

17. Let $p(x) = a_0 + a_1x + \cdots + a_nx^n$ be a polynomial with complex coefficients. Let A be n-by-n and $\mathbf{v}$ be a nonzero column vector. There is a slick way to compute the vector $p(A)\mathbf{v}$. Of course, you could just compute all the necessary powers of A, combine them to form $p(A)$, and multiply this matrix by $\mathbf{v}$. But there is a better way. Form the *Krylov matrix* $\mathcal{K}_{n+1}(A, \mathbf{v}) = [\mathbf{v} \mid A\mathbf{v} \mid A^2\mathbf{v} \mid \cdots \mid A^n\mathbf{v}]$. Note you never have to compute the powers of A since each column of $\mathcal{K}_{n+1}(A, \mathbf{v})$ is just A times the previous column. Argue that $p(A)\mathbf{v} = \mathcal{K}_{n+1}(A, \mathbf{v})\begin{bmatrix} a_0 \\ a_1 \\ \vdots \\ a_n \end{bmatrix}$. Make up a 3-by-3 example to illustrate this fact.

18. Suppose $A^\wedge$ is obtained from A by swapping two columns. Suppose the same sequence of elementary row operations is performed on both A and $A^\wedge$, yielding B and $B^\wedge$. Argue that $B^\wedge$ is obtained from B by swapping the same two columns.

19. Find nice formulas for the powers of the elementary matrices, that is, for any positive integer m, $(T_{ij}(k))^m =$, $(D_i(c))^m =$, and $(P(\sigma))^m = ?$

20. Later, we will be very interested in taking a matrix A and forming $S^{-1}AS$ where S is, of course, invertible. Write a generic 3-by-3 matrix A and form $(T_{12}(\alpha))^{-1}AT_{12}(\alpha)$, $(D_2(\alpha))^{-1}AD_2(\alpha)$, and $\begin{bmatrix} 0 & 0 & 1 \\ 1 & 0 & 0 \\ 0 & 1 & 0 \end{bmatrix}^{-1}$ $A\begin{bmatrix} 0 & 0 & 1 \\ 1 & 0 & 0 \\ 0 & 1 & 0 \end{bmatrix}$. Now make a general statement about what happens in the n-by-n case.

21. Investigate how applying an elementary matrix affects the determinant of a matrix. For example, $det(T_{ij}(\alpha)A) = det(A)$.

22. What is the minimal polynomial of $A = \begin{bmatrix} a & 0 \\ 0 & a \end{bmatrix}$?

23. Suppose A is a square matrix and p is a polynomial with $p(A) = \mathbb{O}$. Argue that the minimal polynomial $\mu_A(x)$ divides $p(x)$. (*Hint:* Remember the division algorithm.)

24. Exhibit two 3-by-3 permutation matrices that do not commute.

Group Project

Elementary matrices can be generalized to work on blocks of a partitioned matrix instead of individual elements. Define a *Type I generalized elementary matrix* to be of the form $\begin{bmatrix} \mathbb{O} & I_m \\ I_n & \mathbb{O} \end{bmatrix}$, a *Type II generalized elementary matrix* multiplies a block from the left by a nonsingular matrix of appropriate size, and a *Type III generalized elementary matrix* multiplies a block by a matrix from the left and then adds the result to another row. So, for example, $\begin{bmatrix} \mathbb{O} & I_m \\ I_n & \mathbb{O} \end{bmatrix}\begin{bmatrix} A & B \\ C & D \end{bmatrix} = \begin{bmatrix} C & D \\ A & B \end{bmatrix}$, $\begin{bmatrix} S & \mathbb{O} \\ \mathbb{O} & I \end{bmatrix}\begin{bmatrix} SA & SB \\ C & D \end{bmatrix}$, and $\begin{bmatrix} I & \mathbb{O} \\ X & I \end{bmatrix}\begin{bmatrix} A & B \\ C & D \end{bmatrix} = \begin{bmatrix} A & B \\ XA + C & XB + D \end{bmatrix}$. The project is to develop a theory of generalized elementary matrices analogous to the theory developed in the text. For example, are the generalized elementary matrices all invertible with inverses of the same type? Can you write $\begin{bmatrix} A & B \\ \mathbb{O} & A^{-1} \end{bmatrix}$ as a product of generalized Type III matrices?

Further Reading

[B&R, 1986(2)] T. S. Blyth and E. F. Robertson, *Matrices and Vector Spaces*, Vol. 2, Chapman & Hall, New York, (1986).

[Lord, 1987] N. J. Lord, Matrices as Sums of Invertible Matrices, Mathematics Magazine, Vol. 60, No. 1, February, (1987), 33–35.

upper triangular, lower triangular, row echelon form, zero row, leading entry, LU factorization, elimination matrix, LDU factorization, full rank factorization

2.3 The LU and LDU Factorization

Our goal in this section is to show how to factor matrices into simpler ones. What this section boils down to is a fancy way of describing Gauss elimination. You might want to consult Appendix B for notation regarding entries, rows, and columns of a matrix.

The "simpler" matrices mentioned above are the *triangular matrices*. They come in two flavors: *upper* and *lower*. Recall that a matrix L is called *lower triangular* if $ent_{ij}(L) = 0$ for $i < j$, that is, all the entries of L above the main diagonal are zero. Similarly, a matrix U is called *upper triangular* if $ent_{ij}(U) = 0$ for $i > j$, that is, all the entries of U below the main diagonal are zero. Note that dilations, being diagonal matrices, are both upper and lower triangular, whereas transvections $T_{ij}(\alpha)$ are lower triangular if $i > j$ and upper triangular if $i < j$. We leave as exercises the basic facts that the product of lower (upper) triangular matrices is lower (upper) triangular, the diagonal of their products is the product of the diagonal elements, and the inverse of a lower (upper) triangular matrices, if it exists, is again lower (upper) triangular.

For example,

$$\begin{bmatrix} 2 & 0 & 0 \\ 3 & 4 & 0 \end{bmatrix} \begin{bmatrix} 1 & 0 & 0 \\ 5 & 2 & 0 \\ -1 & 3 & 4 \end{bmatrix} = \begin{bmatrix} 2 & 0 & 0 \\ 23 & 8 & 0 \end{bmatrix}$$

is the product of two lower triangular matrices, while

$$\begin{bmatrix} 1 & 0 & 0 \\ 5 & 2 & 0 \\ -1 & 3 & 4 \end{bmatrix}^{-1} = \begin{bmatrix} 1 & 0 & 0 \\ -5/2 & 1/2 & 0 \\ 17/8 & -3/8 & 1/4 \end{bmatrix}$$

is the inverse of the second matrix.

When we reviewed Gauss elimination in Section 2.1, we noted how easily we could solve a "triangular" system of linear equations. Now that we have developed elementary matrices, we can make that discussion more complete and precise. It is clear that the unknowns in a system of linear equations are convenient placeholders and we can more efficiently work just with the matrix of coefficients or with this matrix augmented by the right-hand side. Then elementary row operations (i.e., multiplying elementary matrices on the left) can be used to convert the augmented matrix to a very nice form called *row echelon form*. We make the precise definition of this form next.

Consider an m-by-n matrix A. A row of A is called a *zero row* if all the entries in that row are zero. Rows that are not zero rows will be termed (what else?) *nonzero rows*. A nonzero row of A has a first nonzero entry as you come from the left. This entry is called the *leading entry* of the row. An m-by-n matrix A is in *row echelon form* if it has three things going for it. First, all zero rows are below all nonzero rows. In other words, all the zero rows are at the bottom of the matrix. Second, all entries below a leading entry must be zero. Third, the leading entries occur farther to the right as you go down the nonzero rows of the matrix. In other words, the leading entry in any nonzero row appears in a column to the right of the column containing the leading entry of the row above it. In particular, then, a matrix in row echelon form is upper triangular! Do you see that the conditions force the nonzero entries to lie in a stair-step arrangement in the northeast corner of the matrix? Do you see how the word "echelon" came to be used? For example, the following matrix is in row echelon form:

$$\begin{bmatrix} 1 & 2 & 3 & 4 & 5 & 6 & 7 & 8 & 9 \\ 0 & 0 & 2 & 3 & 4 & 5 & 6 & 7 & 8 \\ 0 & 0 & 0 & 2 & 3 & 4 & 5 & 6 & 7 \\ 0 & 0 & 0 & 0 & 0 & 0 & 2 & 3 & 4 \\ 0 & 0 & 0 & 0 & 0 & 0 & 0 & 0 & 0 \end{bmatrix}.$$

We can now formalize Gauss elimination.

THEOREM 2.11
Every matrix $A \in \mathbb{C}^{m \times n}$ can be converted to a row echelon matrix by a finite number of elementary row operations.

PROOF First, use permutation matrices to swap all the zero rows below all the nonzero rows. Use more swaps to move $row_i(A)$ below $row_j(A)$ if the leading element of $row_i(A)$ occurs to the right of the leading element of $row_j(A)$. So far, all we have used are permutation matrices. If all these permutations have been performed and we have not achieved row echelon form, it has to be that there is a $row_s(A)$ above a $row_t(A)$ whose leading entry a_{sj} is in the same position as the leading entry a_{tj}. Use a transvection to zero out a_{tj}. That is, $T_{ts}(-a_{tj}/a_{sj})$ will put a zero where a_{tj} was. Continue in this manner until row echelon form is achieved. □

Notice that we did not use any dilations to achieve a row echelon matrix. That is because, at this point, we do not really care what the values of the leading entries are. This means we do not get a unique matrix in row echelon form starting with a matrix A and applying elementary row operations. We can multiply the rows by a variety of nonzero scalars and we still have a row echelon matrix associated with A. However, what is uniquely determined by A are the *positions* of leading entries. This fact makes a nice nontrivial exercise. Before we move on, we note a corollary to our theorem above.

COROLLARY 2.2

Given any matrix A in $\mathbb{C}^{m\times n}$, there exists an invertible matrix R such that RA is a row echelon matrix, that is, $RA = \begin{bmatrix} G \\ \cdots \\ \mathbb{O} \end{bmatrix}$ *and G has no zero rows.*

Let's motivate our next theorem with an example. Let

$A = \begin{bmatrix} 2 & 1 & 1 & 2 \\ 4 & -6 & 0 & 3 \\ -2 & 7 & 2 & 1 \end{bmatrix}$. Then, doing Gauss elimination,

$$T_{32}(1)T_{31}(1)T_{21}(-2)A = \begin{bmatrix} 2 & 1 & 1 & 2 \\ 0 & -8 & -2 & -1 \\ 0 & 0 & 1 & 2 \end{bmatrix} = U,$$

an upper triangular matrix. Thus, $A = T_{21}(-2)^{-1}T_{31}(1)^{-1}T_{32}(1)^{-1}U = T_{21}(2)T_{31}(-1)T_{32}(-1)U$

$= \begin{bmatrix} 1 & 0 & 0 \\ 2 & 1 & 0 \\ -1 & -1 & 1 \end{bmatrix}\begin{bmatrix} 2 & 1 & 1 & 2 \\ 0 & -8 & -2 & -1 \\ 0 & 0 & 1 & 2 \end{bmatrix} = LU$, the product of a lower and an upper triangular matrix. The entries of L are notable. They are the opposites of the multipliers used in Gauss elimination and the diagonal elements are all ones. What we have illustrated works great as long as you do not run into the necessity of doing a row swap.

THEOREM 2.12

Suppose $A \in \mathbb{C}^{m\times n}$ can be reduced to a row echelon matrix without needing any row exchanges. Then there is an m-by-m lower triangular matrix L with ones on the diagonal and an m-by-n upper triangular matrix U in row echelon form such that $A = LU$. Such a factorization is called an LU-factorization of A.

PROOF We know there exist elementary matrices $E_1, \dots, E_k$ such that $E_k \dots E_1 A = U$ is in row echelon form. Since no row swaps and no dilations were required, all the E_js are transvections $T_{ij}(\alpha)$ where $i > j$. These are lower triangular matrices with ones on their diagonals. The same is true for their inverses, so $L = E_1^{-1} \cdots E_k^{-1}$ is as required for the theorem. □

Instead of using elementary transvections one at a time, we can "speed up" the elimination process by clearing out a column in one blow. We now introduce some matrices we call *elimination matrices*. An elimination matrix E is a matrix of the form $E = I - \mathbf{u}\mathbf{v}^T = I_n - \begin{bmatrix} u_1 \\ u_2 \\ \vdots \\ u_n \end{bmatrix} [v_1 v_2 \dots v_n]$.

We know by Corollary 1.6 of Chapter 1 page 28 that E is invertible as long as $\mathbf{v}^T\mathbf{u} \neq 1$, and, in this case, $E^{-1} = I - \dfrac{\mathbf{u}\mathbf{v}^T}{\mathbf{v}^T\mathbf{u} - 1}$. We note that the elementary matrices already introduced are special cases of elimination matrices. Indeed, $T_{ij}(\alpha) = I + \alpha \mathbf{e}_i\mathbf{e}_j^T$, $D_k(\alpha) = I - (1-\alpha)\mathbf{e}_k\mathbf{e}_k^T$ and $P_{ij} = I - (\mathbf{e}_i - \mathbf{e}_j)(\mathbf{e}_i - \mathbf{e}_j)^T$.

Let $A = \begin{bmatrix} a_{11} & a_{12} & \dots & a_{1n} \\ a_{21} & a_{22} & \dots & a_{2n} \\ \vdots & \vdots & \vdots & \vdots \\ a_{m1} & a_{m2} & \dots & a_{mn} \end{bmatrix}$ with $a_{11} \neq 0$.

Let $\mathbf{u}_1 = \begin{bmatrix} 0 \\ a_{21}/a_{11} \\ \vdots \\ a_{m1}/a_{11} \end{bmatrix}$ and $\mathbf{v}_1 = \mathbf{e}_1 = \begin{bmatrix} 1 \\ 0 \\ \vdots \\ 0 \end{bmatrix}$. Let $\mu_{i1} = \dfrac{a_{i1}}{a_{11}}$ for $i = 2, 3, \dots, m$ and call μ_{i1} a "multiplier." Then $L_1 = I - \mathbf{u}_1\mathbf{v}_1^T =$
$\begin{bmatrix} 1 & 0 & \dots & \dots & 0 \\ -\mu_{21} & 1 & 0 & \dots & 0 \\ -\mu_{31} & 0 & 1 & \dots & 0 \\ \vdots & & & & \vdots \\ -\mu_{m1} & 0 & \dots & \dots & 1 \end{bmatrix}$ is lower triangular and

$$L_1 A = \begin{bmatrix} a_{11} & a_{12} & \dots & a_{1n} \\ 0 & a_{22}^{(2)} & \dots & a_{2n}^{(2)} \\ \vdots & \vdots & & \vdots \\ 0 & a_{m2}^{(2)} & \dots & a_{mn}^{(2)} \end{bmatrix} = A^{(2)},$$

where $a_{ij}^{(2)} = a_{ij} - \mu_{i1}a_{1j}$ for $i = 2, 3, \dots, n$, $j = 2, \dots, n$.

Let's look at an example. Suppose $A = \begin{bmatrix} 1 & 1 & 2 & 4 \\ 2 & 3 & 1 & -1 \\ 3 & -1 & 1 & 2 \end{bmatrix}$. Then $L_1 =$
$\begin{bmatrix} 1 & 0 & 0 \\ -2 & 1 & 0 \\ -3 & 0 & 1 \end{bmatrix}$ and $L_1 A = \begin{bmatrix} 1 & 1 & 2 & 4 \\ 0 & 1 & -3 & -9 \\ 0 & -4 & -5 & -10 \end{bmatrix}$. The idea is to repeat

the process. Let $L_2 = I - \mathbf{u}_2\mathbf{e}_2^T$, where $u_2 = \begin{bmatrix} 0 \\ 0 \\ \mu_{32} \\ \vdots \\ \mu_{m2} \end{bmatrix}$ and $\mathbf{e}_2 = \begin{bmatrix} 0 \\ 1 \\ 0 \\ \vdots \\ 0 \end{bmatrix}$. Here $\mu_{i2} = \dfrac{a_{i2}^{(2)}}{a_{22}^{(2)}}$ for $i = 3, \ldots, n$,

again assuming $a_{22}^{(2)} \neq 0$. With $(m-1)$ sweeps, we reduce our matrix A to row echelon form. For each k, $L_k = I - \mathbf{u}_k\mathbf{e}_k^T$ where $ent_i(\mathbf{u}_k) =$

$$\begin{cases} 0 & \text{if } i = 1, 2, \ldots, k \\ \mu_{ik} = \dfrac{a_{ik}^{(k)}}{a_{kk}^{(k)}} & \text{if } i = k+1, \ldots, m \end{cases}.$$

These multipliers are well defined as long as the elements $a_{kk}^{(k)}$ are nonzero, in which case $a_{ij}^{(k+1)} = a_{ij}^{(k)} - \mu_{ik}a_{kj}^{(k)}$, $i = k+1, \ldots, m$, $j = i, \ldots, n$. To continue our example, $L_2 = \begin{bmatrix} 1 & 0 & 0 \\ 0 & 1 & 0 \\ 0 & 4 & 1 \end{bmatrix}$. Then $L_2L_1A = \begin{bmatrix} 1 & 1 & 2 & 4 \\ 0 & 1 & -3 & -9 \\ 0 & 0 & -17 & -46 \end{bmatrix}$ $= U$. More generally we would find $A = LU$ where

$$L = \begin{bmatrix} 1 & 0 & 0 & \cdots & \cdots & 0 \\ \mu_{21} & 1 & 0 & \cdots & \cdots & 0 \\ \mu_{31} & \mu_{32} & 1 & 0 & \cdots & 0 \\ \vdots & & & & & \vdots \\ \mu_{m1} & \mu_{m2} & \cdots & \cdots & \cdots & 1 \end{bmatrix},$$

where $ent_{ik}(L) = \mu_{ik} = \dfrac{a_{ik}^{(k)}}{a_{kk}^{(k)}}$; $i = k+1, \ldots, m$ and $\mu_{kj} = a_{kj}^{(k)}$, $j = k, \ldots, n$.

Before you get too excited about the LU factorization, notice we have been assuming that nothing goes wrong along the way. However, the simple nonsingular matrix $\begin{bmatrix} 0 & 1 \\ 1 & 1 \end{bmatrix}$ does not have an LU factorization! (Why not?)

We do know a theorem that tells when a nonsingular matrix has an LU factorization. It involves the idea of *leading principal submatrices* of a matrix. They are the square submatrices you can form starting at the upper left-hand corner of A and working down. More precisely, if $A = [a_{ij}] \in \mathbb{C}^{n\times n}$, the leading principal submatrices of A are $A_1 = [a_{11}]$, $A_2 = \begin{bmatrix} a_{11} & a_{12} \\ a_{21} & a_{22} \end{bmatrix}, \dots,$

$$A_k = \begin{bmatrix} a_{11} & a_{12} & \dots & a_{1k} \\ a_{21} & a_{22} & \dots & a_{2k} \\ \vdots & \vdots & \dots & \vdots \\ a_{k1} & a_{k2} & \dots & a_{kk} \end{bmatrix}, \dots, A_n = A.$$

THEOREM 2.13

Let A be a nonsingular matrix in $\mathbb{C}^{n\times n}$. Then A has an LU factorization if and only if all the leading principal submatrices of A are nonsingular.

PROOF Suppose first that A has an LU factorization

$$A = LU = \begin{bmatrix} L_{11} & \mathbb{O} \\ L_{21} & L_{22} \end{bmatrix} \begin{bmatrix} U_{11} & U_{12} \\ \mathbb{O} & U_{22} \end{bmatrix} = \begin{bmatrix} L_{11}U_{11} & * \\ * & * \end{bmatrix},$$

where L_{11} and U_{11} are k-by-k. Being triangular with nonzero diagonal entries, L_{11} and U_{11} must be nonsingular, hence so is their product $L_{11}U_{11}$. This product is the leading principal submatrix A_k. This argument works for each $k = 1, \dots, n$.

Conversely, suppose the condition holds. We use induction to argue that each leading principal submatrix has an LU factorization and so A itself must have one. If $k = 1$, $A_1 = [a_{11}] = [1][a_{11}]$ is trivially an LU factorization and a_{11} cannot be zero since A_1 is invertible. Now, proceeding inductively, suppose $A_k = L_k U_k$ is an LU factorization. We must prove A_{k+1} has an LU factorization as well. Now $A_k^{-1} = U_k^{-1} L_k^{-1}$ so $A_{k+1} = \begin{bmatrix} A_k & \mathbf{u} \\ \mathbf{v}^T & \alpha_{k+1} \end{bmatrix} =$ $\begin{bmatrix} L_k & \mathbb{O} \\ \mathbf{v}^T U_k^{-1} & 1 \end{bmatrix} \begin{bmatrix} U_k & L_k^{-1}\mathbf{u} \\ \mathbb{O} & \alpha_{k+1} - \mathbf{v}^T A_k^{-1}\mathbf{u} \end{bmatrix}$, where $\mathbf{c}^T$ and $\mathbf{b}$ contain the first k components of $row_{k+1}(A_{k+1})$ and $col_{k+1}(A_{k+1})$. But $L_{k+1} = \begin{bmatrix} L_k & \mathbb{O} \\ \mathbf{v}^T U_k^{-1} & 1 \end{bmatrix}$ and $U_{k+1} = \begin{bmatrix} U_k & L_k^{-1}\mathbf{u} \\ \mathbb{O} & \alpha_{k+1} - \mathbf{v}^T A_k^{-1}\mathbf{u} \end{bmatrix}$ gives an LU factorization for A_{k+1}. The crucial fact is that $\alpha_{k+1} - \mathbf{v}^T A_k^{-1}\mathbf{u} \neq \overrightarrow{0}$. This is because $U_{k+1} = L_{k+1}^{-1} A_{k+1}$ is nonsingular. By induction, A_n has an LU factorization. □

There is good news on the uniqueness front.

THEOREM 2.14
If A is nonsingular and can be reduced to row echelon form without row swaps, then there exists a unique lower triangular matrix L with one's on the diagonal and a unique upper triangular matrix U such that $A = LU$.

PROOF We have existence from before, so the issue is uniqueness. Note $U = L^{-1}A$ is the product of invertible matrices, hence is itself invertible. In typical fashion, suppose we have two such LU factorizations, $A = L_1U_1 = L_2U_2$. Then $U_1U_2^{-1} = L_1^{-1}L_2$. However, $U_1U_2^{-1}$ is upper triangular and $L_1^{-1}L_2$ is lower triangular with one's down the diagonal. The only way that can be is if they equal the identity matrix. That is, $U_1U_2^{-1} = I = L_1^{-1}L_2$. Therefore $L_1 = L_2$ and $U_1 = U_2$, as was to be proved. □

Suppose we have a matrix A that does not have an LU factorization. All is not lost; you just have to use row swaps to reduce the matrix. In fact, a rather remarkable thing happens. Suppose we need permutation matrices in reducing A. We would have $L_{n-1}P_{n-1}L_{n-2}P_{n-2}\cdots L_2P_2L_1P_1A = U$ where $P_k = I$ if no row swap is required at the kth step. Remarkably, all we really need is one permutation matrix to take care of all the swapping at once. Let's see why. Suppose we have $L_2P_2L_1P_1A = U$. We know $P_k^TP_k = I$ so $L_2P_2L_1P_2^TP_2P_1A = L_2\widetilde{L}_1(P_2P_1)A$ where $\widetilde{L}_1$ is just L_1 reordered (i.e., $\widetilde{L}_1 = P_2L_1P_2^T$). To illustrate, suppose $L_1 = \begin{bmatrix} 1 & 0 & 0 \\ a & 1 & 0 \\ b & 0 & 1 \end{bmatrix}$ and $P = P(23)$.

Then

$$\widetilde{L}_1 = \begin{bmatrix} 1 & 0 & 0 \\ 0 & 0 & 1 \\ 0 & 1 & 0 \end{bmatrix} \begin{bmatrix} 1 & 0 & 0 \\ a & 1 & 0 \\ b & 0 & 1 \end{bmatrix} \begin{bmatrix} 1 & 0 & 0 \\ 0 & 0 & 1 \\ 0 & 1 & 0 \end{bmatrix} = \begin{bmatrix} 1 & 0 & 0 \\ b & 1 & 0 \\ a & 0 & 1 \end{bmatrix}.$$

More generally, $L_{n-1}P_{n-1}\cdots L_1P_1A = L_{n-1}\widetilde{L}_{n-2}\cdots\widetilde{L}_1(P_{n-1}\cdots P_1)A = \widetilde{L}PA = U$ where $\widetilde{L}_k = P_{n-1}\cdots P_{k+1}L_kP_{k+1}^T$ for $k = 1, 2, \ldots, n-2$ and $\widetilde{L} = L_{n-1}\widetilde{L}_{n-2}\cdots\widetilde{L}_1$ and $P = P_{n-1}\cdots P_1$. Then $PA = \widetilde{L}^{-1}U = LU$. We have argued the following theorem.

THEOREM 2.15
For any $A \in \mathbb{C}^{m\times n}$, there exists a permutation matrix P, an m-by-m lower triangular matrix L with ones down its diagonal, and an upper triangular m-by-n row echelon matrix U with $PA = LU$. If A is n-by-n invertible, L and U are unique.

For example, let $A = \begin{bmatrix} 0 & 2 & 16 & 2 \\ 2 & 1 & -1 & 3 \\ 4 & 3 & 2 & 7 \\ 8 & 5 & 15 & 24 \end{bmatrix}$.

Clearly there is no way to get Gauss elimination started with $a_{11} = 0$. A row swap is required, say

$P_{12}A = \begin{bmatrix} 2 & 1 & -1 & 3 \\ 0 & 2 & 16 & 2 \\ 4 & 3 & 2 & 7 \\ 8 & 5 & 15 & 24 \end{bmatrix}$. Then $T_{43}(\frac{11}{4})T_{42}(-\frac{1}{2})T_{32}(-\frac{1}{2})T_{41}(-4)$

$T_{31}(-2)P_{12}A = \begin{bmatrix} 2 & 1 & -1 & 3 \\ 0 & 2 & 16 & 2 \\ 0 & 0 & -4 & 0 \\ 0 & 0 & 0 & 11 \end{bmatrix}$.

The reader may verify

$$P_{12}A = \begin{bmatrix} 1 & 0 & 0 & 0 \\ 0 & 1 & 0 & 0 \\ 2 & \frac{1}{2} & 1 & 0 \\ 4 & \frac{1}{2} & -\frac{11}{4} & 1 \end{bmatrix} \begin{bmatrix} 2 & 1 & -1 & 3 \\ 0 & 2 & 16 & 2 \\ 0 & 0 & -4 & 0 \\ 0 & 0 & 0 & 11 \end{bmatrix}.$$

How does this LU business help us to solve systems of linear equations? Suppose you need to solve $A\mathbf{x} = \mathbf{b}$. Then, if $PA = LU$, we get $PA\mathbf{x} = P\mathbf{b}$ or $LU\mathbf{x} = P\mathbf{b} = \widetilde{\mathbf{b}}$. Now we solve $L\mathbf{y} = \widetilde{\mathbf{b}}$ by forward substitution and $U\mathbf{x} = \mathbf{y}$ by back substitution since we have triangular systems. So, for example,

$$\begin{cases} 2x_2 + 16x_3 + 2x_4 = 10 \\ 2x_1 + x_2 \quad - x_3 + 3x_4 = 17 \\ 4x_1 + 3x_2 + \ 2x_3 + 7x_4 = 4 \\ 8x_1 + 5x_2 + 15x_3 + 24x_4 = 5 \end{cases}$$ can be written

$$P_{12}A \begin{bmatrix} x_1 \\ x_2 \\ x_3 \\ x_4 \end{bmatrix} = P_{12} \begin{bmatrix} 10 \\ 17 \\ 4 \\ 5 \end{bmatrix} = \begin{bmatrix} 17 \\ 10 \\ 4 \\ 5 \end{bmatrix}.$$

Therefore,

$$L \begin{bmatrix} y_1 \\ y_2 \\ y_3 \\ y_4 \end{bmatrix} = \begin{bmatrix} 1 & 0 & 0 & 0 \\ 0 & 1 & 0 & 0 \\ 2 & \frac{1}{2} & 1 & 0 \\ 4 & \frac{1}{2} & -\frac{11}{4} & 1 \end{bmatrix} \begin{bmatrix} y_1 \\ y_2 \\ y_3 \\ y_4 \end{bmatrix} = \begin{bmatrix} 17 \\ 10 \\ 4 \\ 5 \end{bmatrix}$$

which yields

$$\begin{bmatrix} y_1 \\ y_2 \\ y_3 \\ y_4 \end{bmatrix} = \begin{bmatrix} 17 \\ 10 \\ -35 \\ -\frac{521}{4} \end{bmatrix}$$

and

$$\begin{bmatrix} 2 & 1 & -1 & 3 \\ 0 & 2 & 16 & 2 \\ 0 & 0 & -4 & 0 \\ 0 & 0 & 0 & 11 \end{bmatrix} \begin{bmatrix} x_1 \\ x_2 \\ x_3 \\ x_4 \end{bmatrix} = \begin{bmatrix} 17 \\ 10 \\ -35 \\ -\frac{521}{4} \end{bmatrix}.$$

It follows that

$$\begin{bmatrix} x_1 \\ x_2 \\ x_3 \\ x_4 \end{bmatrix} = \begin{bmatrix} \frac{3437}{44} \\ -\frac{4678}{44} \\ \frac{35}{4} \\ -\frac{521}{44} \end{bmatrix}.$$

It may seem a bit unfair that L gets to have ones down its diagonal but U does not. We can fix that (sometimes). Suppose A is nonsingular and $A = LU$. The trick is to pull the diagonal entries of U out and divide the rows of U by each diagonal entry. That is,

$$U = \begin{bmatrix} d_1 & 0 & \cdots & 0 \\ 0 & d_2 & \cdots & 0 \\ \vdots & \vdots & \ddots & \vdots \\ 0 & 0 & \cdots & d_n \end{bmatrix} \begin{bmatrix} 1 & \frac{\mu_{12}}{d_1} & \frac{\mu_{13}}{d_1} & \cdots & \frac{\mu_{1n}}{d_1} \\ 0 & 1 & \frac{\mu_{22}}{d_2} & \cdots & \frac{\mu_{2n}}{d_2} \\ \vdots & \vdots & \vdots & \ddots & \vdots \\ 0 & 0 & \cdots & 0 & 1 \end{bmatrix}.$$

Then we have $LDU_1 = A$ where D is a diagonal matrix and U_1 is upper triangular with ones on the diagonal as well. For example,

$$\begin{aligned} A &= \begin{bmatrix} -2 & 2 & 16 & 2 \\ 2 & 1 & -1 & 3 \\ 4 & 3 & 2 & 7 \\ 8 & 5 & 15 & 24 \end{bmatrix} \\ &= \begin{bmatrix} 1 & 0 & 0 & 0 \\ -1 & 1 & 0 & 0 \\ -2 & 7/3 & 1 & 0 \\ -4 & 13/3 & -14 & 1 \end{bmatrix} \begin{bmatrix} -2 & 2 & 16 & 2 \\ 0 & 3 & 15 & 5 \\ 0 & 0 & -1 & -2/3 \\ 0 & 0 & 0 & 1 \end{bmatrix} \\ &= \begin{bmatrix} 1 & 0 & 0 & 0 \\ -1 & 1 & 0 & 0 \\ -2 & 7/3 & 1 & 0 \\ -4 & 13/3 & -14 & 1 \end{bmatrix} \begin{bmatrix} -2 & 0 & 0 & 0 \\ 0 & 3 & 0 & 0 \\ 0 & 0 & -1 & 0 \\ 0 & 0 & 0 & 1 \end{bmatrix} \begin{bmatrix} 1 & -1 & -8 & -1 \\ 0 & 1 & 5 & 5/3 \\ 0 & 0 & 1 & 2/3 \\ 0 & 0 & 0 & 1 \end{bmatrix}. \end{aligned}$$

We summarize.

THEOREM 2.16 (LDU factorization theorem)
If A is nonsingular, there exists a permutation matrix P such that $PA = LDU$ where L is lower triangular with ones down the diagonal, D is a diagonal matrix, and U is upper triangular with ones down the diagonal. This factorization of PA is unique.

One application of the LDU factorization is for real symmetric matrices (i.e., matrices over $\mathbb{R}$ such that $A = A^T$). First, $A = LDU$ so $A^T = U^T D^T L^T = U^T D L^T$. But $A = A^T$ so $LDU = U^T D L^T$. By uniqueness $U = L^T$. Thus symmetric matrices A can be written LDL^T. Suppose $D = diag(d_1, d_2, \ldots, d_n)$ has all real positive entries. Then $D^{\frac{1}{2}} = diag(\sqrt{d_1}, \sqrt{d_2}, \ldots, \sqrt{d_n})$ makes sense and $(D^{\frac{1}{2}})^2 = D$. Let $S = D^{\frac{1}{2}} L^T$. Then $A = LDL^T = LD^{\frac{1}{2}} D^{\frac{1}{2}} L^T = S^T S$, where S is upper triangular with positive diagonal entries. Conversely, if $A = RR^T$ where R is lower triangular with positive diagonal elements, then $R = LD$ where L is lower triangular with ones on the diagonal and D is a diagonal matrix with all positive elements on the diagonal. Then $A = LD^2L^T$ is the LDU factorization of A.

Let's take a closer look at an LU factorization as a "preview of coming attractions." Suppose $A = LU = \begin{bmatrix} 1 & 0 & 0 & 0 \\ x & 1 & 0 & 0 \\ y & r & 1 & 0 \\ z & s & t & 1 \end{bmatrix} \begin{bmatrix} a & b & c & d \\ 0 & f & g & h \\ 0 & 0 & 0 & 0 \\ 0 & 0 & 0 & 0 \end{bmatrix}$

$$= \begin{bmatrix} a & b & c & d \\ xa & xb + f & xc + g & xd + h \\ ya & yb + rf & yc + rg & yd + rh \\ za & zb + sf & zc + sg & zd + sh \end{bmatrix}$$

$= \begin{bmatrix} 1 & 0 \\ x & 1 \\ y & r \\ z & s \end{bmatrix} \begin{bmatrix} a & b & c & d \\ 0 & f & g & h \end{bmatrix}$. The block of zeros in U makes part of L irrelevant. We could reconstruct the matrix A without ever knowing the value of t!

More specifically A

$$= \left[a \begin{bmatrix} 1 \\ x \\ y \\ z \end{bmatrix} \mid b \begin{bmatrix} 1 \\ x \\ y \\ z \end{bmatrix} + f \begin{bmatrix} 0 \\ 1 \\ r \\ x \end{bmatrix} \mid c \begin{bmatrix} 1 \\ x \\ y \\ z \end{bmatrix} + g \begin{bmatrix} 0 \\ 1 \\ r \\ x \end{bmatrix} \mid d \begin{bmatrix} 1 \\ x \\ y \\ z \end{bmatrix} + h \begin{bmatrix} 0 \\ 1 \\ r \\ s \end{bmatrix} \right]$$

so we have another factorization of A different from LU where only the crucial columns of L are retained so that all columns of A can be reconstructed using the nonzero rows of U as coefficients. Later, we shall refer to this as a *full rank factorization* of A. Clearly, the first two columns of L are independent, so Gauss elimination has led to a basis of the column space of A.

Exercise Set 5

1. Prove that the product of upper (lower) triangular matrices is upper (lower) triangular. What about the sums, differences, and scalar multiples of triangular matrices?

2. What can you say about the transpose and conjugate transpose of an upper (lower) triangular matrix?

3. Argue that every square matrix can be uniquely written as the sum of a strictly lower triangular matrix, a diagonal matrix, and a strictly upper triangular matrix.

4. Prove that the inverse of an upper (lower) triangular matrix, when it exists, is again upper (lower) triangular.

5. Prove that an upper (lower) triangular matrix is invertible iff the diagonal entries are all nonzero.

6. Argue that a symmetric upper (lower) triangular matrix is a diagonal matrix.

7. Prove that a matrix is diagonal iff it is both upper and lower triangular.

8. Prove the uniqueness of the LDU factorization.

9. Prove A is invertible iff A can be reduced by elementary row operations to the identity matrix.

10. Let $A = \begin{bmatrix} 2 & 1 & -1 & 3 \\ 4 & 2 & 2 & 7 \\ -2 & -1 & 16 & 2 \\ 8 & 4 & 15 & 24 \end{bmatrix}$.

Multiply $\begin{bmatrix} 1 & 0 & 0 & 0 \\ 2 & 1 & 0 & 0 \\ -1 & 3 & 1 & 0 \\ 4 & 1 & 5 & 1 \end{bmatrix} \begin{bmatrix} 2 & 1 & -1 & 3 \\ 0 & 0 & 4 & 1 \\ 0 & 0 & 3 & 2 \\ 0 & 0 & 0 & 1 \end{bmatrix}$ and

$\begin{bmatrix} 1 & 0 & 0 & 0 \\ 2 & 1 & 0 & 0 \\ -1 & \frac{15}{4} & 1 & 0 \\ 4 & \frac{19}{4} & \frac{29}{5} & 1 \end{bmatrix} \begin{bmatrix} 2 & 1 & -1 & 3 \\ 0 & 0 & 4 & 1 \\ 0 & 0 & 0 & 5/4 \\ 0 & 0 & 0 & 0 \end{bmatrix}$. What do you notice?
Does this contradict any of our theorems? Of course not, but the question is why not?

11. Let $A = \begin{bmatrix} 2 & 1 & -1 & 3 \\ 4 & 3 & 2 & 7 \\ -2 & 2 & 16 & 2 \\ 8 & 5 & 15 & 24 \end{bmatrix}$.
    (a) Find an LU factorization by multiplying A on the left by elementary transvections.
    (b) Find an LU factorization of A by "brute force." Set A
    $= \begin{bmatrix} 1 & 0 & 0 & 0 \\ x & 1 & 0 & 0 \\ y & r & 1 & 0 \\ z & s & t & 1 \end{bmatrix} \begin{bmatrix} 2 & 1 & -1 & 3 \\ 0 & a & b & c \\ 0 & 0 & d & e \\ 0 & 0 & 0 & f \end{bmatrix}$.
    Multiply out and solve. Did you get the same L and U ? Did you have to?

12. The leading elements in the nonzero rows of a matrix in row echelon form are called *pivots.* A *pivot,* or *basic column,* is a column that contains a pivot position. Argue that, while a matrix A can have many different pivots, the positions in which they occur are uniquely determined by A. This gives us one way to define a notion of "rank" we call *pivot rank.* The pivot rank of a matrix A is the number of pivot positions in any row echelon matrix obtained from A. Evidently, this is the same as the number of basic columns. The variables in a system of linear equations corresponding to the pivot or basic columns are called the *basic variables* of the system. All the other variables are called *free.* Argue that an m-by-n system of linear equations, $A\mathbf{x} = \mathbf{b}$ with variables $x_1, \ldots, x_n$, is consistent for all $\mathbf{b}$ iff A has m pivots.

13. Argue that an LU factorization cannot be unique if U has a row of zeros.

14. If A is any m-by-n matrix, there exists an invertible P so that $A = P^{-1}LU$.

Further Reading

[B&R, 1986(2)] T. S. Blyth and E. F. Robertson, *Matrices and Vector Spaces*, Vol. 2, Chapman & Hall, New York, (1986).

[E&S, 2004] Alan Edelman and Gilbert Strang, Pascal Matrices, The American Mathematical Monthly, Vol. 111, No. 3, March, (2004), 189–197.

[Johnson, 2003] Warren P. Johnson, An LDU Factorization in Elementary Number Theory, Mathematics Magazine, Vol. 76, No. 5, December, (2003), 392–394.

[Szabo, 2000] Fred Szabo, *Linear Algebra, An Introduction Using Mathematica*, Harcourt, Academic Press, New York, (2000).

2.3.1 MATLAB Moment

2.3.1.1 The LU Factorization

MATLAB computes LU factorizations. The command is

$$[L, U, P] = lu(A)$$

which returns a unit lower triangular matrix L, an upper triangular matrix U, and a permutation matrix P such that PA = LU. For example,

```
>>A=[0 2 16 2;2 1 -1 3;4 3 2 7;8 5 15 24]
A =
   0   2   16    2
   2   1   -1    3
   4   3    2    7
   8   5   15   24
>>[L,U,P]=lu(A)
L =
     1      0      0     0
     0      1      0     0
   1/2    1/4      1     0
   1/4   -1/8  11/38     1
```

```
U =
   8   5    15       24
   0   2    16        2
   0   0  -19/2   -22/19

P =
   0   0   0   1
   1   0   0   0
   0   0   1   0
   0   1   0   0
```

adjugate, submatrix, principal submatrices, principal minors, leading principal submatrices, leading principal minors, cofactors

---

2.4 The Adjugate of a Matrix

There is a matrix that can be associated to a square matrix and is closely related to the invertibility of that matrix. This is called the *adjugate matrix,* or *adjoint* matrix. We prefer to use the word "adjoint" in another context, so we go with the British and use "adjugate." Luckily, the first three letters are the same for both terms so the abbreviations will look the same.

Suppose A is an m-by-n matrix. If we erase $m - r$ rows and $n - c$ columns, what remains is called an *r-by-c submatrix* of A. For example, let

$$A = \begin{bmatrix} a_{11} & a_{12} & a_{13} & a_{14} & a_{15} & a_{16} & a_{17} \\ a_{21} & a_{22} & a_{23} & a_{24} & a_{25} & a_{26} & a_{27} \\ a_{31} & a_{32} & a_{33} & a_{34} & a_{35} & a_{36} & a_{37} \\ a_{41} & a_{42} & a_{43} & a_{44} & a_{45} & a_{46} & a_{47} \\ a_{51} & a_{52} & a_{53} & a_{54} & a_{55} & a_{56} & a_{57} \end{bmatrix}.$$

Suppose we strike out two rows, say the second and fifth, and three columns, say the second, fourth, and fifth, then the $(5 - 2)$-by-$(7 - 3)$ submatrix of A we obtain is

$$\begin{bmatrix} a_{11} & a_{13} & a_{16} & a_{17} \\ a_{31} & a_{33} & a_{36} & a_{37} \\ a_{41} & a_{43} & a_{46} & a_{47} \end{bmatrix} \in \mathbb{C}^{3\times 4}.$$

The $m-r$ rows to strike out can be chosen in $\binom{m}{m-r}$ ways and the $n-c$ columns can be chosen in $\binom{n}{n-c}$ ways, so there are $\binom{m}{m-r}\binom{n}{n-c}$ possible submatrices that can be formed from an m-by-n matrix. For example, the number of possible submatrices of size 3-by-4 from A above is $\binom{5}{3}\binom{7}{4} = 350$. You probably would not want to write them all down.

We are most interested in submatrices of square matrices. In fact, we are very interested in determinants of square submatrices of a square matrix.

The determinant of an r-by-r submatrix of $A \in \mathbb{C}^{n\times n}$ is called an r-by-r *minor* of A. There are $\binom{n}{r}^2$ such minors possible in A. We take the 0-by-0 minor of any matrix A to be 1 for convenience. Note that there are just as many minors of order r as of order n-r. For later, we note that the *principal submatrices* of A are obtained when the rows and columns deleted from A have the same indices. A submatrix so obtained is symmetrically located with respect to the main diagonal of A. The determinants of these principal submatrices are called *principal minors.* Even more special are the *leading principal submatrices* and their determinants, called the *leading principal minors.* If $A = \begin{bmatrix} a_{11} & a_{12} & a_{13} \\ a_{21} & a_{22} & a_{23} \\ a_{31} & a_{32} & a_{33} \end{bmatrix}$, the leading principal submatrices are $[a_{11}]$, $\begin{bmatrix} a_{11} & a_{12} \\ a_{21} & a_{22} \end{bmatrix}$, and $\begin{bmatrix} a_{11} & a_{12} & a_{13} \\ a_{21} & a_{22} & a_{23} \\ a_{31} & a_{32} & a_{33} \end{bmatrix}$.

Let $M_{ij}(A)$ be defined to be the $(n-1)$-by-$(n-1)$ submatrix of A obtained by striking out the ith row and jth column from A. For example, if $A = \begin{bmatrix} 1 & 2 & 3 \\ 4 & 5 & 6 \\ 7 & 8 & 9 \end{bmatrix}$, then $M_{12}(A) = \begin{bmatrix} 4 & 6 \\ 7 & 9 \end{bmatrix}$. The (i, j)-*cofactor* of A is defined by

$$cof_{ij}(A) = (-1)^{i+j} det(M_{ij}(A)).$$

For example, for A above, $cof_{12}(A) = (-1)^{1+2} det \begin{bmatrix} 4 & 6 \\ 7 & 9 \end{bmatrix} = (-1)(-6)$ $= 6$.

We now make a matrix of cofactors of A and take its transpose to create a new matrix called the *adjugate matrix* of A.

$$adj(A) := \left[cof_{ij}(A)\right]^T$$

For example,

$$adj\left(\begin{bmatrix} a & b \\ c & d \end{bmatrix}\right) = \begin{bmatrix} d & -b \\ -c & a \end{bmatrix}$$

and if $A = \begin{bmatrix} a_{11} & a_{12} & a_{13} \\ a_{21} & a_{22} & a_{23} \\ a_{31} & a_{32} & a_{33} \end{bmatrix}$,

$$adj(A) = \begin{bmatrix} det\begin{bmatrix} a_{22} & a_{23} \\ a_{32} & a_{33} \end{bmatrix} & -det\begin{bmatrix} a_{21} & a_{23} \\ a_{31} & a_{33} \end{bmatrix} & det\begin{bmatrix} a_{21} & a_{22} \\ a_{31} & a_{32} \end{bmatrix} \\ -det\begin{bmatrix} a_{12} & a_{13} \\ a_{32} & a_{33} \end{bmatrix} & det\begin{bmatrix} a_{11} & a_{13} \\ a_{31} & a_{33} \end{bmatrix} & -det\begin{bmatrix} a_{11} & a_{12} \\ a_{31} & a_{32} \end{bmatrix} \\ det\begin{bmatrix} a_{12} & a_{13} \\ a_{22} & a_{23} \end{bmatrix} & -det\begin{bmatrix} a_{11} & a_{13} \\ a_{21} & a_{23} \end{bmatrix} & det\begin{bmatrix} a_{11} & a_{12} \\ a_{21} & a_{22} \end{bmatrix} \end{bmatrix}^T.$$

You may be wondering why we took the transpose. That has to do with a formula we want that connects the adjugate matrix with the inverse of a matrix. Let's go after that connection.

Let's look at the 2-by-2 situation, always a good place to start. Let $A = \begin{bmatrix} a & b \\ c & d \end{bmatrix}$. Let's compute $A(adj(A)) = \begin{bmatrix} a & b \\ c & d \end{bmatrix}\begin{bmatrix} d & -b \\ -c & a \end{bmatrix} = \begin{bmatrix} ad+b(-c) & a(-b)+ba \\ cd+d(-c) & c(-b)+da \end{bmatrix} = \begin{bmatrix} ad-bc & 0 \\ 0 & ad-bc \end{bmatrix}$
$= \begin{bmatrix} det(A) & 0 \\ 0 & det(A) \end{bmatrix}$
$= det(A)\begin{bmatrix} 1 & 0 \\ 0 & 1 \end{bmatrix}$. That is a neat answer! Ah, but does it persist with larger matrices? In the 3-by-3 case, $A = \begin{bmatrix} a_{11} & a_{12} & a_{13} \\ a_{21} & a_{22} & a_{23} \\ a_{31} & a_{32} & a_{33} \end{bmatrix}$ and so

$$\begin{aligned} &A(adj(A)) \\ &= \begin{bmatrix} a_{11} & a_{12} & a_{13} \\ a_{21} & a_{22} & a_{23} \\ a_{31} & a_{32} & a_{33} \end{bmatrix} \\ &= \begin{bmatrix} a_{22}a_{33}-a_{23}a_{32} & -a_{12}a_{33}+a_{13}a_{32} & a_{12}a_{23}-a_{13}a_{22} \\ -a_{21}a_{33}+a_{31}a_{23} & a_{11}a_{33}-a_{13}a_{31} & -a_{11}a_{23}+a_{21}a_{13} \\ a_{21}a_{32}-a_{22}a_{31} & -a_{11}a_{32}+a_{12}a_{31} & a_{11}a_{22}-a_{12}a_{21} \end{bmatrix} \\ &= \begin{bmatrix} det(A) & 0 & 0 \\ 0 & det(A) & 0 \\ 0 & 0 & det(A) \end{bmatrix} = det(A)\begin{bmatrix} 1 & 0 & 0 \\ 0 & 1 & 0 \\ 0 & 0 & 1 \end{bmatrix}. \end{aligned}$$

Indeed, we have a theorem.

THEOREM 2.17

For any n-by-n matrix A, $A(adj(A)) = (det(A))I_n$. Thus, A is invertible iff $det(A) \neq 0$ and, in this case, $A^{-1} = \dfrac{1}{det(A)}adj(A)$.

PROOF The proof requires the Laplace expansion theorem (see Appendix C). We compute the (i, j)-entry: $ent_{ij}(A(adj(A)) = \sum_{k=1}^{n} ent_{ik}(A)ent_{kj}(adj(A)) = \sum_{k=1}^{n} a_{ik}(-1)^{j+k}det(M_{jk}(A)) = \begin{cases} det(A) & if i = j \\ 0 & if i \neq j \end{cases}$. □

While this theorem does not give an efficient way to compute the inverse of a matrix, it does have some nice theoretical consequences. If A were 10-by-10, just finding $adj(A)$ would require computing 100 determinants of 9-by-9 matrices! There must be a better way, even if you have a computer. We can, of course, illustrate with small examples.

Let $A = \begin{bmatrix} 6 & 1 & 4 \\ 3 & 0 & 2 \\ -1 & 2 & 2 \end{bmatrix}$. Then $adj(A) = \begin{bmatrix} -4 & 6 & 2 \\ -8 & 16 & 0 \\ 6 & -13 & -3 \end{bmatrix}$ and

$$Aadj(A) = \begin{bmatrix} 6 & 1 & 4 \\ 3 & 0 & 2 \\ -1 & 2 & 2 \end{bmatrix} \begin{bmatrix} -4 & 6 & 2 \\ -8 & 16 & 0 \\ 6 & -13 & -3 \end{bmatrix} = \begin{bmatrix} -8 & 0 & 0 \\ 0 & -8 & 0 \\ 0 & 0 & -8 \end{bmatrix},$$

so we see $det(A) = -8$ and $A^{-1} = -\frac{1}{8}\begin{bmatrix} -4 & 6 & 2 \\ -8 & 16 & 0 \\ 6 & -13 & -3 \end{bmatrix}$, as the reader may verify.

Exercise Set 6

1. Compute the adjugate and inverse of $\begin{bmatrix} 2 & 1 \\ 3 & 6 \end{bmatrix}$, $\begin{bmatrix} 2 & 4 & 3 \\ 0 & 2 & 4 \\ 0 & 0 & 0 \end{bmatrix}$, $\begin{bmatrix} 3 & 5 & 1 & 2 \\ -1 & 0 & 1 & 0 \\ 6 & 4 & 2 & 7 \\ 5 & 3 & 1 & 1 \end{bmatrix}$, if they exist, using Theorem 2.17 above.

2. Write down the generic 4-by-4 matrix A. Compute $M_{13}(A)$ and $M_{24}(M_{13}(A))$.

3. Establish the following properties of the adjugate where A and B are in $\mathbb{C}^{n \times n}$:

   (a) $adj(A^{-1}) = (adj(A))^{-1}$, provided A is invertible
   (b) $adj(cA) = c^{n-1}adj(A)$
   (c) if $A \in \mathbb{C}^{n \times n}$ with $n \geq 2$, $adj(adj(A)) = (det(A))^{n-2}A$
   (d) $adj(A^T) = (adj(A))^T$ so A is symmetric iff $adj(A)$ is
   (e) $adj(A^*) = (adj(A))^*$ so A is Hermitian iff $adj(A)$ is
   (f) $adj(AB) = adj(B)adj(A)$
   (g) $adj(adj(A)) = A$ provided $det(A) = 1$
   (h) $adj(\overline{A}) = \overline{adj(A)}$
   (i) the adjugate of a scalar matrix is a scalar matrix

(j) the adjugate of a diagonal matrix is a diagonal matrix
(k) the adjugate of a triangular matrix is a triangular matrix
(l) $adj(T_{ij}(-a)) = T_{ij}(a)$
(m) $adj(I) = I$ and $adj(\mathbb{O}) = \mathbb{O}$.

4. Find an example of a 3-by-3 nonzero matrix A with $adj(A) = \mathbb{O}$.

5. Argue that $det(adj(A))det(A) = (det(A))^n$ where A is n-by-n. So, if $det(A) \neq 0$, $det(adj(A)) = (det(A))^{n-1}$.

6. Argue that $det \begin{bmatrix} \mathbb{O} & I_n \\ I_m & \mathbb{O} \end{bmatrix} = (-1)^{nm}$ whenever $n \geq 1$ and $m \geq 1$.

7. Prove that $det \begin{bmatrix} A & \mathbf{u} \\ \mathbf{v}^* & \beta \end{bmatrix} = \beta det(A) - \mathbf{v}^*(adj(A))\mathbf{u} = det(A)(\beta - \mathbf{v}^* A^{-1}\mathbf{u})$ where β is a scalar and $\mathbf{u}$ and $\mathbf{v}$ are n-by-1. (*Hint:* Do a Laplace expansion by the last row and then more Laplace expansions by the last column.)

8. Argue that $adj(I - \mathbf{u}\mathbf{v}^*) = \mathbf{u}\mathbf{v}^* + (1 - \mathbf{v}^*\mathbf{u})I$ where $\mathbf{u}$ and $\mathbf{v}$ are n-by-1.

9. Prove that $det(adj(adj(A))) = (det(A))^{(n-1)^2}$.

10. If A is nonsingular, $adj(A) = det(A)A^{-1}$.

Further Reading

[Aitken, 1939] A. C. Aitken, *Determinants and Matrices,* 9th edition, Oliver and Boyd, Edinburgh and London, New York: Interscience Publishers, Inc., (1939).

[B&R, 1986(2)] T. S. Blyth and E. F. Robertson, *Matrices and Vector Spaces,* Vol. 2, Chapman & Hall, New York, (1986).

[Bress, 1999] David M. Bressoud, *Proofs and Confirmations: The Story of the Alternating Sign Matrix Conjecture,* Cambridge University Press, (1999).

[Bress&Propp, 1999] David Bressoud and James Propp, How the Alternating Sign Conjecture was Solved, Notices of the American Mathematical Society, Vol. 46, No. 6, June/July (1999), 637–646.

Group Project

Find out everything you can about the alternating sign conjecture and write a paper about it.

characteristic matrix, characteristic polynomial, Cayley-Hamilton theorem, Newton identities, Frame algorithm

2.5 The Frame Algorithm and the Cayley-Hamilton Theorem

In 1949, **J. Sutherland Frame** (24 December 1907 – 27 February 1997) published an abstract in the *Bulletin of the American Mathematical Society* indicating a recursive algorithm for computing the inverse of a matrix and, as a by-product, getting additional information, including the famous Cayley-Hamilton theorem. (Hamilton is the Irish mathematician **William Rowan Hamilton** (4 August 1805 – 2 September 1865), and Cayley is **Arthur Cayley** (16 August 1821 – 26 January 1895.) We have not been able to find an actual paper with a detailed account of these claims. Perhaps the author thought the abstract sufficient and went on with his work in group representations. Perhaps he was told this algorithm had been rediscovered many times (see [House, 1964, p. 72]). Whatever the case, in this section, we will expand on and expose the details of Frame's algorithm. Suppose $A \in \mathbb{C}^{n\times n}$. The *characteristic matrix* of A is $xI - A \in \mathbb{C}[x]^{n\times n}$, the collection of n-by-n matrices with polynomial entries. We must open our minds to accepting matrices with polynomial entries. For example, $\begin{bmatrix} x^2+1 & x-3 \\ 4x+2 & x^3-7 \end{bmatrix} \in \mathbb{C}[x]^{2\times 2}$. Determinants work just fine for these kinds of matrices. The determinant of $xI - A$, $det(xI - A) \in \mathbb{C}[x]$, the polynomials in x, and is what we call the *characteristic polynomial* of A:

$$\chi_A(x) = det(xI - A) = x^n + c_1x^{n-1} + \cdots + c_{n-1}x + c_n.$$

For example, if $A = \begin{bmatrix} 1 & 2 & 2 \\ 3 & 4 & 5 \\ 6 & 7 & 8 \end{bmatrix} \in \mathbb{C}^{3\times 3}$, then $xI_3 - A =$ $\begin{bmatrix} x-1 & -2 & -2 \\ -3 & x-4 & -5 \\ -6 & -7 & x-8 \end{bmatrix} \in \mathbb{C}[x]^{3\times 3}$. Thus $\chi_A(x) =$

$det\left(\begin{bmatrix} x-1 & -2 & -2 \\ -3 & x-4 & -5 \\ -6 & -7 & x-8 \end{bmatrix}\right) = x^3 - 13x^2 - 9x - 3$. This is computed using the usual familiar rules for expanding a determinant.

You may recall that the roots of the characteristic polynomial are quite important, being the *eigenvalues* of the matrix. We will return to this topic later. For now, we focus on the coefficients of the characteristic polynomial.

First, we consider the constant term c_n. You may already know the answer here, but let's make an argument. Now $det(A) = (-1)^n det(-A) = (-1)^n det(0I - A) = (-1)^n \chi_A(0) = (-1)^n c_n$. Therefore,

$$det(A) = (-1)^n c_n.$$

As a consequence, we see immediately that A is invertible iff $c_n \neq 0$, in which case

$$A^{-1} = \frac{(-1)^n}{c_n} adj(A),$$

where $adj(A)$ is the adjugate matrix of A introduced previously. Also recall the important relationship, $Badj(B) = det(B)I$. We conclude that

$$(xI - A)adj(xI - A) = \chi_A(x)I.$$

To illustrate with the example above, $(xI - A)adj(xI - A) =$

$$\begin{bmatrix} x-1 & -2 & -2 \\ -3 & x-4 & -5 \\ -6 & -7 & x-8 \end{bmatrix} \cdot \begin{bmatrix} x^2-12x-3 & 2x-2 & 2x+2 \\ 3x+6 & x^2-9x-4 & 5x+1 \\ 6x-3 & 7x+5 & x^2-5x-2 \end{bmatrix}$$

$$= \begin{bmatrix} x^3-13x^2-9x-3 & 0 & 0 \\ 0 & x^3-13x^2-9x-3 & 0 \\ 0 & 0 & x^3-13x^2-9x-3 \end{bmatrix}$$

$$= x^3 - 13x^2 - 9x - 3 \begin{bmatrix} 1 & 0 & 0 \\ 0 & 1 & 0 \\ 0 & 0 & 1 \end{bmatrix}.$$

Next, let $C(x) = adj(xI - A) \in \mathbb{C}[x]^{n \times n}$. We note that the elements of $adj(xI - A)$ are computed as (n-1)-by-(n-1) subdeterminants of $xI - A$, so the highest power that can occur in $C(x)$ is x^{n-1}. Also, note that we can identify $\mathbb{C}[x]^{n \times n}$, the n-by-n matrices with polynomial entries with $\mathbb{C}^{n \times n}[x]$, the polynomials with matrix coefficients, so we can view $C(x)$ as a polynomial in x with scalar matrices as coefficients. For example, $\begin{bmatrix} x^2+1 & x-3 \\ 4x+2 & x^3-7 \end{bmatrix} =$ $\begin{bmatrix} 0 & 0 \\ 0 & 1 \end{bmatrix} x^3 + \begin{bmatrix} 1 & 0 \\ 0 & 0 \end{bmatrix} x^2 + \begin{bmatrix} 0 & 1 \\ 4 & 0 \end{bmatrix} x + \begin{bmatrix} 1 & -3 \\ 2 & -7 \end{bmatrix}$. All you do is gather the coefficients of each power of x and make a matrix of scalars as the coefficient of that power of x. Note that what we have thusly created is an element of $\mathbb{C}^{n\times n}[x]$, the polynomials in x whose coefficients come from the n-by-n matrices over $\mathbb{C}$. Also note, $xB = Bx$ for all $B \in \mathbb{C}[x]^{n\times n}$, so it does not matter which side we put the x on. We now view $C(x)$ as such an expression in $\mathbb{C}^{n\times n}[x]$:

$$C(x) = B_0x^{n-1} + B_1x^{n-2} + \cdots + B_{n-2}x + B_{n-1}.$$

These coefficient matrices turn out to be of interest. For example, $adj(A) = (-1)^{n-1}adj(-A) = (-1)^{n-1}C(0) = (-1)^{n-1}B_{n-1}$, so

$$adj(A) = (-1)^{n-1}B_{n-1}.$$

Thus, if $c_n \neq 0$, A is invertible and we have

$$A^{-1} = \frac{-1}{c_n}B_{n-1}.$$

But now we compute

$$\begin{aligned}(xI - A)adj(xI - A) &= (xI - A)C(x) = (xI - A)(B_0x^{n-1} + B_1x^{n-2}\\ &\quad + \cdots + B_{n-2}x + B_{n-1})\\ &= x^nB_0 + x^{n-1}(B_1 - AB_0) + x^{n-2}(B_2 - AB_1) + \cdots\\ &\quad + x(B_{n-1} - AB_{n-2}) - AB_{n-1}\\ &= x^nI + x^{n-1}c_1I + \cdots + xc_{n-1}I + c_nI\end{aligned}$$

and we compare coefficients using the following table:

Compare Coefficients	**Multiply by**	**on the *Left***	**on the *Right***
$B_0 = I$	A^n	A^nB_0	A^n
$B_1 - AB_0 = c_1I$	A^{n-1}	$A^{n-1}B_1 - A^nB_0$	c_1A^{n-1}
$B_2 - AB_1 = c_2I$	A^{n-2}	$A^{n-2}B_2 - A^{n-1}B_1$	c_2A^{n-2}
$\vdots$			
$B_k - AB_{k-1} = c_kI$			
$\vdots$			
$B_{n-2} - AB_{n-3} = c_{n-2}I$	A^2	$A^2B_{n-2} - A^3B_{n-3}$	$c_{n-2}A^2$
$B_{n-1} - AB_{n-2} = c_{n-1}I$	A	$AB_{n-1} - A^2B_{n-2}$	$c_{n-1}A$
$-AB_{n-1} = c_nI$		$-AB_{n-1}$	c_nI
column	sum =	$\mathbb{O} =$	$\chi_A(A)$

So, the first consequence we get from these observations is that the Cayley-Hamilton theorem just falls out as an easy consequence. (Actually, Liebler [2003] reports that Cayley and Hamilton only established the result for matrices up to size 4-by-4. He says it was Frobenius (**Ferdinand Georg Frobenius** [26 October 1849 – 3 August 1917] who gave the first complete proof in 1878.)

THEOREM 2.18 (Cayley-Hamilton theorem)
For any n-by-n matrix A over $\mathbb{C}$, $\chi_A(A) = \mathbb{O}$.

What we are doing in the Cayley-Hamilton theorem is plugging a matrix into a polynomial. Plugging numbers into a polynomial seems reasonable, almost inevitable, but matrices? Given a polynomial $p(x) = a_0 + a_1 x + \cdots + a_k x^k \in \mathbb{C}[x]$, we can create a matrix $p(A) = a_0 I + a_1 A + \cdots + a_k A^k \in \mathbb{C}^{n \times n}$. For example, if $p(x) = 4 + 3x - 9x^3$, then $p(A) = 4I + 3A - 9A^3 = 4\begin{bmatrix} 1 & 0 & 0 \\ 0 & 1 & 0 \\ 0 & 0 & 1 \end{bmatrix} +$
$3\begin{bmatrix} 1 & 2 & 2 \\ 3 & 4 & 5 \\ 6 & 7 & 8 \end{bmatrix} - 9\begin{bmatrix} 1 & 2 & 2 \\ 3 & 4 & 5 \\ 6 & 7 & 8 \end{bmatrix}^3 = \begin{bmatrix} -2324 & -2964 & -3432 \\ -5499 & -7004 & -8112 \\ -9243 & -11\,778 & -13\,634 \end{bmatrix}$.
The Cayley-Hamilton theorem says that any square matrix is a "root" of its characteristic polynomial.

But there is much more information packed in those equations on the left of the table, so let's push a little harder. Notice we can rewrite these equations as

$$\begin{aligned} B_0 &= I \\ B_1 &= AB_0 + c_1 I \\ B_2 &= AB_1 + c_2 I \\ &\vdots \\ B_{n-1} &= AB_{n-2} + c_{n-1} I \\ \mathbb{O} &= AB_{n-1} + c_n I. \end{aligned}$$

By setting $B_n := \mathbb{O}$, we have the following recursive scheme clear from above: for $k = 1, 2, \ldots, n$,

$$\begin{aligned} B_0 &= I \\ B_k &= AB_{k-1} + c_k I. \end{aligned}$$

In other words, the matrix coefficients, the B_ks are given recursively in terms of the B_{k-1}s and c_ks. If we can get a formula for c_k in terms of B_{k-1}, we will get a complete set of recurrence formulas for the B_k and c_k. In particular, if we know B_{n-1} and c_n, we have A^{-1}, provided, of course, A^{-1} exists (i.e., provided

$c_n \neq 0$). For this, let's exploit the recursion given above:

$$\begin{aligned}
B_0 &= I \\
B_1 &= AB_0 + c_1 I = AI + c_1 I &&= A + c_1 I \\
B_2 &= AB_1 + c_2 I = A(A + c_1 I) + c_2 I &&= A^2 + c_1 A + c_2 I \\
B_3 &= &&= A^3 + c_1 A^2 + c_2 A + c_3 I \\
&\vdots
\end{aligned}.$$

Inductively, we see for $k = 1, 2, \ldots, n$,

$$B_k = A^k + c_1 A^{k-1} + \cdots + c_{k-1} A + c_k I.$$

Indeed, when $k = n$, this is just the Cayley-Hamilton theorem all over again. Now we have for $k = 2, 3, \ldots, n+1$,

$$B_{k-1} = A^{k-1} + c_1 A^{k-2} + \cdots + c_{k-2} A + c_{k-1} I.$$

If we multiply through by A, we get for $k = 2, 3, \ldots, n+1$,

$$AB_{k-1} = A^k + c_1 A^{k-1} + \cdots + c_{k-2} A^2 + c_{k-1} A.$$

Now we pull a trick out of the mathematician's hat. Take the trace of both sides of the equation using the linearity of the trace functional.

$$tr(AB_{k-1}) = tr(A^k) + c_1 tr(A^{k-1}) + \cdots + c_{k-2} tr(A^2) + c_{k-1} tr(A)$$
$$for\ k = 2, 3, \ldots, n+1.$$

Why would anybody think to do such a thing? Well, the appearance of the coefficients of the characteristic polynomial on the right is very suggestive. Those who know a little matrix theory realize that the trace of A^r is the sum of the rth powers of the roots of the characteristic polynomial and so Newton's identities leap to mind. Let s_r denote the sum of the rth powers of the roots of the characteristic polynomial. Thus, for $k = 2, 3, \ldots, n+1$,

$$tr(AB_{k-1}) = s_k + c_1 s_{k-1} + \cdots + c_{k-2} s_2 + c_{k-1} s_1.$$

2.5.1 Digression on Newton's Identities

Newton's identities go back aways. They relate the sums of powers of the roots of a polynomial recursively to the coefficients of the polynomial. Many proofs are available. Some involve the algebra of symmetric functions, but we do not want to take the time to go there. Instead, we will use a calculus-based argument following the ideas of [Eidswick 1968]. First, we need to recall some facts about polynomials. Let $p(x) = a_0 + a_1 x + \cdots + a_n x^n$. Then the coefficients of p

can be expressed in terms of the derivatives of p evaluated at zero (remember Taylor polynomials?):

$$p(x) = p(0) + p'(0)x + \frac{p''(0)}{2!}x^2 + \cdots + \frac{p^{(n)}(0)}{n!}x^n.$$

Now here is something really slick. Let's illustrate a general fact. Suppose $p(x) = (x-1)(x-2)(x-3) = -6 + 11x - 6x^2 + x^3$. Do a wild and crazy thing. Reverse the rolls of the coefficients and form the new reversed polynomial $q(x) = -6x^3 + 11x^2 - 6x + 1$. Clearly $q(1) = 0$ but, more amazingly, $q(\frac{1}{2}) = -\frac{6}{8} + \frac{11}{4} - \frac{6}{2} + 1 = \frac{-6+22-24+8}{8} = 0$. You can also check $q(\frac{1}{3}) = 0$. So the reversed polynomial has as roots the reciprocals of the roots of the original polynomial. Of course, the roots are not zero for this to work. This fact is generally true. Suppose $p(x) = a_0 + a_1x + \cdots + a_nx^n$ and the reversed polynomial is $q(x) = a_n + a_{n-1}x + \cdots + a_0x^n$. Note

$$q(0) = a_n,\ q'(0) = a_{n-1}, \ldots,\ \frac{q^{(n)}(0)}{n!} = a_0.$$

Then $r \neq 0$ is a root of p iff $\frac{1}{r}$ is a root of q.

Suppose $p(x) = a_0 + a_1x + \cdots + a_nx^n = a_n(x-r_1)(x-r_2)\cdots(x-r_n)$. The r_is are, of course, the roots of p, which we assume to be nonzero but not necessarily distinct. Then the reversed polynomial $q(x) = a_n + a_{n-1}x + \cdots + a_0x^n = a_0(x - \frac{1}{r_1})(x - \frac{1}{r_2})\cdots(x - \frac{1}{r_n})$. For the sake of illustration, suppose $n = 3$. Then form $f(x)$

$$= \frac{q'(x)}{q(x)} = \frac{(x-r_1^{-1})[(x-r_2^{-1})+(x-r_3^{-1})]+[(x-r_2^{-1})+(x-r_3^{-1})]}{(x-r_1)(x-r_2)(x-r_3)}$$

$$= \frac{1}{x-r_1^{-1}} + \frac{1}{x-r_2^{-1}} + \frac{1}{x-r_3^{-1}}.$$ Generally then,

$$f(x) = \sum_{k=1}^{n} \frac{1}{(x - r_k^{-1})}.$$

Let's introduce more notation. Let $s_m = \sum_{k=1}^{m} r_k^m$ for $m = 1, 2, 3, \ldots$. Thus, s_m is the sum of the mth powers of the roots of p. The derivatives of f are intimately related to the ss. Basic differentiation yields

$$\begin{array}{lllll} & & & f(0) & = -s_1 \\ f'(x) & = & \sum_{k=1}^{n} \frac{-1}{(x-r_k^{-1})^2} & f'(0) & = -s_2 \\ f''(x) & = & \sum_{k=1}^{n} \frac{-2}{(x-r_k^{-1})^3} & f''(0) & = -2s_3 \\ \vdots & & & & \\ f^{(k)}(x) & = & \sum_{k=1}^{n} \frac{-k!}{(x-r_k^{-1})^{k+1}} & f^{(k)}(0) & = -k!s_{k+1}. \end{array}$$

The last piece of the puzzle is the rule of taking the derivative of a product; this is the so-called *Leibnitz rule* for differentiating a product:

$$D^n(F(x)G(x)) = \sum_{j=0}^{n} \binom{n}{j} F^{(j)}(x)G^{(n-j)}(x).$$

All right, let's do the argument. We have $f(x) = \frac{q'(x)}{q(x)}$, so $q'(x) = f(x)q(x)$. Therefore, using the Leibnitz rule

$$q^{(m)}(x) = [f(x)q(x)]^{(m-1)} = \sum_{k=0}^{m-1} \binom{m-1}{k} f^{(k)}(x)q^{(m-1-k)}(x).$$

Plugging in zero, we get

$$\begin{aligned} q^{(m)}(0) &= \sum_{k=0}^{m-1} \binom{m-1}{k} f^{(k)}(0)q^{(m-1-k)}(0) \\ &= \sum_{k=0}^{m-1} \frac{(m-1)!}{k!(m-1-k)!}(-k!)s_{k+1}q^{(m-1-k)}(0). \end{aligned}$$

Therefore,

$$\frac{q^{(m)}(0)}{m!} = a_{n-m} = -\frac{1}{m}\sum_{k=0}^{m-1} \frac{q^{(m-1-k)}(0)}{(m-1-k)!}s_{k+1}.$$

One more substitution and we have the *Newton identities*

$$0 = ma_{n-m} + \sum_{k=0}^{m-1} a_{n-m+k+1}s_{k+1} \quad if\ 1 \leq m \leq n$$

$$0 = \sum_{k=m-n-1}^{m-1} a_{n-m+k+1}s_{k+1} \quad if\ m > n.$$

For example, suppose $n = 3$, $p(x) = a_0 + a_1x + a_2x^2 + a_3x^3$.

$$\begin{array}{ll} m = 1 & a_2 + s_1a_3 = 0 \\ m = 2 & 2a_1 + s_1a_2 + s_2a_3 = 0 \\ m = 3 & 3a_0 + s_1a_1 + a_2s_2 + a_3s_3 = 0 \\ m = 4 & s_1a_0 + s_2a_1 + s_3a_2 + s_4a_3 = 0 \\ m = 5 & a_0s_2 + a_1s_3 + a_2s_4 + a_3s_5 = 0 \\ m = 6 & a_0s_3 + a_1s_4 + a_2s_5 + a_3s_6 = 0 \\ \vdots & \vdots \end{array}$$

That ends our digression and now we go back to Frame's algorithm. We need to translate the notation a bit. Note $c_{n-k} = a_k$ in the notation above. So from

$$tr(AB_{k-1}) = s_k + c_1s_{k-1} + \cdots + c_{k-2}s_2 + c_{k-1}s_1 \quad for\ k = 2, 3, \ldots, n+1$$

we see

$$tr(AB_{k-1}) + kc_k = 0 \quad for\ k = 2, 3, \ldots, n.$$

In fact, you can check that this formula works when $k = 1$ (see exercise 1), so we have succeeded in getting a formula for the coefficients of the characteristic polynomial: for $k = 1, 2, \ldots, n$,

$$c_k = \frac{-1}{k} tr(AB_{k-1}).$$

Now we are in great shape. The recursion we want, taking $B_0 = I$, is given by the following: for $k = 1, 2, \ldots, n$,

$$c_k = \frac{-1}{k} tr(AB_{k-1})$$
$$B_k = AB_{k-1} + c_kI.$$

Note that the diagonal elements of AB_{n-1} are all equal (why?).

Let's illustrate this algorithm with a concrete example. Suppose $A = \begin{bmatrix} 2 & 3 & 5 & 2 \\ 4 & 7 & 10 & 3 \\ 0 & 3 & 1 & 4 \\ 2 & 5 & 1 & -27 \end{bmatrix}$. The algorithm goes as follows: first, find c_1.

$$c_1 = -tr(A) = -(-17) = 17.$$

Next find B_1:

$$B_1 = AB_0 + 17I = \begin{bmatrix} 19 & 3 & 5 & 2 \\ 4 & 24 & 10 & 3 \\ 0 & 3 & 18 & 4 \\ 2 & 5 & 1 & -10 \end{bmatrix}.$$

Then compute AB_1:

$$AB_1 = \begin{bmatrix} 54 & 103 & 132 & 13 \\ 110 & 225 & 273 & 39 \\ 20 & 95 & 52 & -27 \\ 4 & -6 & 51 & 293 \end{bmatrix}.$$

Now start the cycle again finding c_2:

$$c_2 = -\frac{1}{2}tr(AB_1) = -\frac{1}{2}624 = -312.$$

Next comes B_2:

$$B_2 = AB_1 + (-312)I = \begin{bmatrix} -258 & 103 & 132 & 13 \\ 110 & -87 & 273 & 39 \\ 20 & 95 & -260 & -27 \\ 4 & -6 & 51 & -19 \end{bmatrix}.$$

Now form AB_2:

$$AB_2 = \begin{bmatrix} -78 & 408 & -115 & -30 \\ -50 & 735 & -8 & -2 \\ 366 & -190 & 763 & 14 \\ -54 & 28 & -8 & 707 \end{bmatrix}.$$

Starting again, we find c_3:

$$c_3 = -\frac{1}{3}tr(AB_2) = -\frac{1}{3}(2127) = -709.$$

Then we form B_3:

$$B_3 = AB_2 - 709I = \begin{bmatrix} -787 & 408 & -115 & -30 \\ -50 & 26 & -8 & -2 \\ 366 & -190 & 54 & 14 \\ -54 & 28 & -8 & -2 \end{bmatrix}.$$

Now for the magic; form AB_3:

$$AB_3 = \begin{bmatrix} -2 & 0 & 0 & 0 \\ 0 & -2 & 0 & 0 \\ 0 & 0 & -2 & 0 \\ 0 & 0 & 0 & -2 \end{bmatrix}.$$

Next form c_4:

$$c_4 = -\frac{1}{4}tr(AB_3) = -\frac{1}{4}(-8) = 2.$$

Now we can clean up. First

$$det(A) = (-1)^4 c_4 = 2.$$

Indeed the characteristic polynomial of A is

$$\chi_A(x) = x^4 + 17x^3 - 312x^2 - 709x + 2.$$

Immediately we see A is invertible and

$$A^{-1} = \frac{(-1)^4}{c_4} adj(A) = \frac{1}{c_4}(-1)^3 B_3 = \frac{-1}{2}\begin{bmatrix} -787 & 408 & -115 & -30 \\ -50 & 26 & -8 & -2 \\ 366 & -190 & 54 & 14 \\ -54 & 28 & -8 & -2 \end{bmatrix}$$

$$= \begin{bmatrix} \frac{787}{2} & -204 & \frac{115}{2} & 15 \\ 25 & -13 & 4 & 1 \\ -183 & 95 & -27 & -7 \\ 27 & -14 & 4 & 1 \end{bmatrix}.$$

Moreover, we can express A^{-1} as a polynomial in A with the help of the characteristic polynomial

$$A^{-1} = \frac{709}{2} I + \frac{312}{2} A - \frac{17}{2} A^2 - \frac{1}{2} A^3.$$

2.5.2 The Characteristic Polynomial and the Minimal Polynomial

We end with some important results that connect the minimal and characteristic polynomials.

THEOREM 2.19
The minimal polynomial $\mu_A(x)$ divides any annihilating polynomial of the n-by-n matrix A. In particular, $\mu_A(x)$ divides the characteristic polynomial $\chi_A(x)$. However, $\chi_A(x)$ divides $(\mu_A(x))^n$.

PROOF The first part of the proof involves the division algorithm and is left as an exercise (see exercise 21). The last claim is a bit more challenging, so we offer a proof. Write $\mu_A(x) = \beta_r + \beta_{r-1}x + \cdots + \beta_1 x^{r-1} + x^r$. Let $B_0 = I_n$, and let $B_i = A^i + \beta_1 A^{i-1} + \cdots + \beta_i I_n$ for $i = 1$ to $r-1$. It is easy to see that $B_i - AB_{i-1} = \beta_i I_n$ for $i = 1$ to $r - 1$. Note that $AB_{r-1} = A^r + \beta_1 A^{r-1} + \cdots + \beta_{r-1}A = \mu_A(A) - \beta_r I_n = -\beta_r I_n$. Let $C = B_0 x^{r-1} + B_1 x^{r-2} + \cdots + B_{r-2}x + B_{r-1}$. Then $C \in \mathbb{C}[x]^{n\times n}$ and $(xI_n - A)C = (xI_n - A)(B_0 x^{r-1} + B_1 x^{r-2} + \cdots + B_{r-2}x + B_{r-1}) = B_0 x^r + (B_1 - AB0)x^{r-1} + \cdots + (B_{r-1} - AB_{r-2})x - AB_{r-1} =$

$\beta_r I_n + \beta_{r-1} x I_n + \cdots + \beta_1 x^{r-1} I_n + x^r I_n = \mu_A(x) I_n$. Now take the determinant of both sides and get $\chi_A(x) \det(C) = (\mu_A(x))^n$. The theorem now follows. ▯

A tremendous amount of information about a matrix is locked up in its minimal and characteristic polynomials. We will develop this in due time. For the moment, we content ourselves with one very interesting fact: In view of the corollary above, every root of the minimal polynomial must also be a root of the characteristic polynomial. What is remarkable is that the converse is also true. We give a somewhat slick proof of this fact.

THEOREM 2.20
The minimal polynomial and the characteristic polynomial have exactly the same set of roots.

PROOF Suppose r is a root of the characteristic polynomial χ_A. Then $det(rI - A) = 0$, so the matrix $rI - A$ is not invertible. This means there must be a dependency relation among the columns of $rI - A$. That means there is a nonzero column vector $\mathbf{v}$ with $(rI - A)\mathbf{v} = \vec{0}$ or what is the same, $A\mathbf{v} = r\mathbf{v}$. Given any polynomial $p(x) = a_0 + a_1 x + \cdots + a_k x^k$, we have $p(A)\mathbf{v} = a_0\mathbf{v} + a_1 A\mathbf{v} + \cdots + a_k A^k \mathbf{v} = a_0\mathbf{v} + a_1 r\mathbf{v} + \cdots + a_k r^k \mathbf{v} = p(r)\mathbf{v}$. This says $p(r)I - p(A)$ is not invertible, which in turn implies $det(p(r)I - p(A)) = 0$. Thus $p(r)$ is a root of $\chi_{p(A)}$. Now apply this when $p(x) = \mu_A(x)$. Then, for any root r of χ_A, $\mu_A(r)$ is a root of $\chi_{\mu_A(A)}(x) = \chi_{\mathbb{O}}(x) = det(xI - \mathbb{O}) = x^n$. The only zeros of x^n are 0 so we are forced to conclude $\mu_A(r) = 0$, which says r is a root of μ_A. ▯

As a consequence of this theorem, if we know the characteristic polynomial of a square matrix A factors as

$$\chi_A(x) = (x - r_1)^{d_1}(x - r_2)^{d_2} \cdots (x - r_k)^{d_k},$$

then the minimal polynomial must factor as

$$\mu_A(x) = (x - r_1)^{e_1}(x - r_2)^{e_2} \cdots (x - r_k)^{e_k}$$

where $e_i \leq d_i$ for all $i = 1, 2, \ldots, k$.

Exercise Set 7

1. Explain why the formula $tr(AB_{k-1}) + kc_k = 0$ for $k = 2, 3, 4, .., n$ also works for $k = 1$.

2. Explain why the diagonal elements of AB_{n-1} are all equal.

3. Using the characteristic polynomial, explain why A^{-1} is a polynomial in A when A is invertible.

4. Explain how to write a program on a handheld calculator to compute the coefficients of the characteristic polynomial.

5. Use the Newton identities and the fact that $s_k = tr(A^k)$ to find formulas for the coefficients of the characteristic polynomial in terms of the trace of powers and powers of traces of a matrix. (*Hint:* $c_2 = \frac{1}{2}[tr(A)^2 - tr(A^2)]$, $c_3 = -\frac{1}{6}tr(A)^3 + \frac{1}{2}tr(A)tr(A^2) - \frac{1}{3}tr(A^3), \ldots$.)

6. Consider the polynomial $p(x) = a_0 + a_1 x + \cdots + a_{n-1}x^{n-1} + x^n$. Find a matrix that has this polynomial as its characteristic polynomial. (*Hint:* Consider the matrix $\begin{bmatrix} 0 & 1 & 0 & \cdots & 0 \\ 0 & 0 & 1 & \cdots & 0 \\ \vdots & & & & \\ 0 & 0 & 0 & \cdots & 1 \\ -a_0 & -a_1 & -a_2 & \cdots & -a_{n-1} \end{bmatrix}$.)

7. Show how the Newton identities for $k > n$ follow from the Cayley-Hamilton theorem.

8. (D. W. Robinson) Suppose $A \in \mathbb{C}^{n \times n}$ and $B \in \mathbb{C}^{m \times m}$, where $m \leq n$. Argue that the characteristic polynomial of A is x^{n-m} times the characteristic polynomial of B if and only if $tr(A^k) = tr(B^k)$ for $k = 1, 2, \ldots, n$.

9. (H. Flanders, TAMM, Vol. 63, 1956) Suppose $A \in \mathbb{C}^{n \times n}$. Prove that A is nilpotent iff $tr(A^k) = 0$ for $k = 1, 2, \ldots, n$. Recall that nilpotent means $A^p = \mathbb{O}$ for some power p.

10. Suppose $A \in \mathbb{C}^{n \times m}$ and $B \in \mathbb{C}^{m \times n}$, where $m \leq n$. Prove that the characteristic polynomial of AB is x^{n-m} times the characteristic polynomial of BA.

11. What can you say about the characteristic polynomial of a sum of two matrices? A product of two matrices?

12. Suppose $tr(A^k) = tr(B^k)$ for $k = 1, 2, 3, \ldots$. Argue that $\chi_A(x) = \chi_B(x)$.

13. Suppose $A = \begin{bmatrix} B & \mathbb{O} \\ \mathbb{O} & C \end{bmatrix}$. What, if any, is the connection between the characteristic polynomial of A and those of B and C?

14. Verify the Cayley-Hamilton theorem for $A = \begin{bmatrix} 1 & 1 & 1 \\ 0 & 1 & 1 \\ 0 & 0 & 1 \end{bmatrix}$.

15. Let $A \in \mathbb{C}^{n \times n}$. Argue that the constant term of $\chi_A(x)$ is $(-1)^n$ $(product\ of\ the\ roots\ of\ \chi_A(x))$ and the coefficient of x^{n-1} is $-Tr(A) = -(sum\ of\ the\ roots\ of\ \chi_A(x))$. If you are brave, try to show that the coefficient of x^{n-j} is $(-1)^{n-j}$ times the sum of the j-by-j principal minors of A.

16. Is every monic polynomial of degree n in $\mathbb{C}[x]$ the characteristic polynomial of some n-by-n matrix in $\mathbb{C}^{n \times n}$?

17. Find a matrix whose characteristic polynomial is $p(x) = x^4 + 2x^3 - 3x^2 + 4x - 5$.

18. Explain how the Cayley-Hamilton theorem can be used to provide a method to compute powers of a matrix. Make up a 3-by-3 example and illustrate your approach.

19. Explain how the Cayley-Hamilton theorem can be used to simplify the calculations of matrix polynomials so that the problem of evaluating a polynomial expression of an n-by-n matrix can be reduced to the problem of evaluating a polynomial expression of degree less than n. Make up an example to illustrate your claim. (*Hint:* Divide the characteristic polynomial into the large degree polynomial.)

20. Explain how the Cayley-Hamilton theorem can be used to express the inverse of an invertible matrix as a polynomial in that matrix. Argue that a matrix is invertible iff the constant term of its characteristic polynomial is not zero.

21. Prove the first part of Theorem 2.19. (*Hint:* Recall the division algorithm for polynomials.)

22. Suppose U is invertible and $B = U^{-1}AU$. Argue that A and B have the same characteristic and minimal polynomial.

23. What is wrong with the following "easy" proof of the Cayley-Hamilton theorem: $\chi_A(x) = det(xI - A)$, so replacing x with A one gets $\chi_A(A) = det(AI - A) = det(\mathbb{O}) = 0$?

24. How can two polynomials be different and still have exactly the same set of roots? How can it be that one of these polynomials divides the other.

25. What are the minimal and characteristic polynomials of the n-by-n identity matrix I_n? How about the n-by-n zero matrix?

26. Is there any connection between the coefficients of the minimal polynomial for A and the coefficients of the minimal polynomial for A^{-1} for an invertible matrix A?

27. Suppose $A = \begin{bmatrix} B & \mathbb{O} \\ \mathbb{O} & C \end{bmatrix}$. What, if any, is the connection between the minimal polynomial of A and those of B and C?

28. Give a direct computational proof of the Cayley-Hamilton theorem for any 2-by-2 matrix.

Further Reading

[B&R, 1986(2)] T. S. Blyth and E. F. Robertson, *Matrices and Vector Spaces*, Vol. 2, Chapman & Hall, New York, (1986).

[B&R, 1986(4)] T. S. Blyth and E. F. Robertson, *Linear Algebra*, Vol. 4, Chapman & Hall, New York, (1986).

[Eidswick, 1968] J. A. Eidswick, A Proof of Newton's Power Sum Formulas, The American Mathematical Monthly, Vol. 75, No. 4, April, (1968), 396–397.

[Frame, 1949] J. S. Frame, A Simple Recursion Formula for Inverting a Matrix, Bulletin of the American Mathematical Society, Vol. 55, (1949), Abstracts.

[Gant, 1959] F. R. Gantmacher, *The Theory of Matrices*, Vol. 1, Chelsea Publishing Co., New York, (1959).

[H-W&V, 1993] Gilbert Helmberg, Peter Wagner, Gerhard Veltkamp, On Faddeev-Leverrier's Methods for the Computation of the Characteristic Polynomial of a Matrix and of Eigenvectors, Linear Algebra and Its Applications, (1993), 219–233.

[House, 1964] Alston S. Householder, *The Theory of Matrices in Numerical Analysis*, Dover Publications Inc., New York, (1964).

[Kalman, 2000] Dan Kalman, A Matrix Proof of Newton's Identities, Mathematics Magazine, Vol. 73, No. 4, October, (2000), 313–315.

[LeV, 1840] U. J. LeVerrier, Sur les Variations Séculaires des Élements Elliptiques des sept Planètes Principales, J. Math. Pures Appl. 5, (1840), 220–254.

[Liebler, 2003] Robert A. Liebler, *Basic Matrix Algebra with Algorithms and Applications*, Chapman & Hall/CRC Press, Boca Raton, FL, (2003).

[Mead, 1992] D. G. Mead, Newton's Identities, The American Mathematical Monthly, Vol. 99, (1992), 749–751.

[Pennisi, 1987] Louis L. Pennisi, Coefficients of the Characteristic Polynomial, Mathematics Magazine, Vol. 60, No. 1, February, (1987), 31–33.

[Robinson, 1961] D. W. Robinson, A Matrix Application of Newton's Identities, The American Mathematical Monthly, Vol. 68, (1961), 367–369.

2.5.3 Numerical Note

2.5.3.1 The Frame Algorithm

As beautiful as the Frame algorithm is, it does not provide a numerically stable means to find the coefficients of the characteristic polynomial or the inverse of a large matrix.

2.5.4 MATLAB Moment

2.5.4.1 Polynomials in MATLAB

Polynomials can be manipulated in MATLAB except that, without The Symbolic Toolbox, all you see are coefficients presented as a row vector. So if $f(x) = a_1x^n + a_2x^{n-1} + \cdots + a_nx + a_{n+1}$, the polynomial is represented as a row vector

$$[\mathrm{a}_1\ \mathrm{a}_2\ \mathrm{a}_3 \ldots \mathrm{a}_{n+1}].$$

For example, the polynomial $f(x) = 4x^3 + 2x^2 + 3x + 10$ is entered as

$$f = [4\ 2\ 3\ 10].$$

If $g(x)$ is another polynomial of the same degree given by g, the sum of $f(x)$ and $g(x)$ is just $f + g$; the difference is $f - g$. The product of any two polynomials f and g is conv(f,g). The roots of a polynomial can be estimated by the function roots( ). You can calculate the value of a polynomial f at a given number, say 5, by using polyval(f,5). MATLAB uses Horner's (nested) method to do the evaluation. You can even divide polynomials and get the quotient and remainder. That is, when a polynomial f is divided by a polynomial h there is a unique quotient q and remainder r where the degree of r is strictly less than the degree of h. The command is

$$[q, r] = deconv(f, h).$$

For example, suppose $f(x) = x^4 + 4x^2 - 3x + 2$ and $g(x) = x^2 + 2x - 5$. Let's enter them, multiply them, and divide them.

```
>> f=[1 0 4 -3 2]
f =
     1   0   4   −3   2
>> g =[1 2 -5]
g =
     1   2   −5
>> conv(f,g)
ans =
     1   2   −1   5   −24   19   −10
>> [q r] =deconv(f,g)
q=
     1   −2   13
r =
     0   0   0   −39   67
>> roots(g)
ans =
     −3.3395
      1.4495
>> polyval(f,0)
ans =
   2
```

Let's be sure we understand the output. We see

$$f(x)g(x) = x^6 + 2x^5 - x^4 + 5x^3 - 24x^2 + 19x - 10.$$

Next, we see

$$f(x) = g(x)(x^2 - 2x + 13) + (-39x + 67).$$

Finally, the roots of g are -3.3395 and 1.4495, while $f(0) = 2$.

One of the polynomials of interest to us is the *characteristic polynomial,* $\chi_A(x) = \det(xI - A)$. Of course, MATLAB has this built in as

$$poly(A).$$

All you see are the coefficients, so you have to remember the order in which they appear. For example, let's create a random 4-by-4 matrix and find its characteristic polynomial.

```
>>format rat
>>A=fix(11*rand(4))+i*fix(11*rand(4))
A =
  Columns 1 through 3
        10 + 10i      9      9 + 1i
         2 + 10i   8 + 3i    4 + 2i
         6 + 4i    5 + 8i    6 + 2i
         5 + 9i      0       8 + 6i
  Column 4
   10 + 2i
   8 + 2i
   1
   4 + 8i
>>p= poly(A)
p =
   Columns 1 through 3
        1   -28 - 23i   -4 + 196i
   Columns 4 through 5
        382 - 1210i   -176 + 766i
```

So we see the characteristic polynomial of this complex 4-by-4 matrix is

$$\chi_A(x) = x^4 - (28 + 23i)x^3 + (-4 + 196i)x^2 + (382 - 1210i)x + (-176 + 766i).$$

We can form polynomials in matrices using the polyvalm command. You can even check the Cayley-Hamilton theorem using this command. For example, using the matrix A above,

```
>> fix(polyvalm(p,A))
ans =
 0  0  0  0
 0  0  0  0
 0  0  0  0
 0  0  0  0
```

Finally, note how we are using the fix command to get rid of some messy decimals. If we did not use the rat format above, the following can happen.

>> A = fix(11*rand(4)) + i*fix(11*rand(4))
A =

$2.0000 + 3.0000i$	$1.0000 + 7.0000i$	$9.0000 + 4.0000i$	$9.0000 + 7.0000i$
$7.0000 + 3.0000i$	$7.0000 + 3.0000i$	$6.0000 + 7.0000i$	$7.0000 + 6.0000i$
$3.0000 + 3.0000i$	$4.0000 + 9.0000i$	$5.0000 + 6.0000i$	$8.0000 + 8.0000i$
$5.0000 + 5.0000i$	$9.0000 + 6.0000i$	$9.0000 + 4.0000i$	$7.0000 + 10.0000i$

>> p = poly(A)
p =
1.0e+003*
Columns 1 through 4
$0.0010 \quad -0.0210 - 0.0220i \quad -0.0470 - 0.1180i \quad -0.273 - 1.0450i$
Column 5
-2.087-1.3060i
>> polyvalm(p,A)
ans =
1.0e-009*

$0.1050 + 0.0273i$	$0.1398 + 0.0555i$	$0.1525 + 0.0346i$	$0.1892 + 0.0650i$
$0.1121 + 0.0132i$	$0.1560 + 0.0380i$	$0.1598 + 0.0102i$	$0.2055 + 0.0351i$
$0.1073 + 0.0421i$	$0.1414 + 0.0789i$	$0.1583 + 0.0539i$	$0.1937 + 0.0916i$
$0.1337 + 0.0255i$	$0.1837 + 0.0600i$	$0.1962 + 0.0277i$	$0.2447 + 0.0618i$

Notice the scalar factor of 1000 on the characteristic polynomial. That leading coefficient is after all supposed to be 1. Also, that last matrix is supposed to be the zero matrix but notice the scalar factor of 10^{-9} in front of the matrix. That makes the entries of the matrix teeny tiny, so we effectively are looking at the zero matrix.

Chapter 3

Subspaces Associated to Matrices

subspace, null space, nullity, column space, column rank, row space, row rank, rank-plus-nullity theorem, row rank=column rank

3.1 Fundamental Subspaces

In this section, we recall, in some detail, how to associate subspaces to a matrix A in $\mathbb{C}^{m\times n}$. First, we recall what a subspace is.

DEFINITION 3.1 *(subspace of $\mathbb{C}^n$)*
*A nonempty subset $M \subseteq \mathbb{C}^n$ is called a **subspace** of $\mathbb{C}^n$ iff M is closed under the formation of sums and scalar multiples. That is, if $\mathbf{u}, \mathbf{v} \in M$, then $\mathbf{u} + \mathbf{v} \in M$, and if $\mathbf{u} \in M$ and α is any scalar, then $\alpha\mathbf{u} \in M$.*

Trivial examples of subspaces are $(\overrightarrow{0})$, the set consisting only of the zero vector and $\mathbb{C}^n$ itself. Three subspaces are naturally associated to a matrix. We define these next.

DEFINITION 3.2 *(null space and nullity)*
*Let $A \in \mathbb{C}^{m\times n}$. We define the **null space** of A as $\mathcal{N}ull(A) = \{\mathbf{x} \in \mathbb{C}^n \,|\, A\mathbf{x} = \overrightarrow{0}\}$. The dimension of this subspace of $\mathbb{C}^n$ is called the **nullity** of A and is denoted $nlty(A)$.*

DEFINITION 3.3 *(column space and column rank)*
*Let $A \in \mathbb{C}^{m\times n}$. We define the **column space** of A as $\mathcal{C}ol(A) = \{A\mathbf{x} \mid \mathbf{x} \in \mathbb{C}^n\}$. The dimension of this subspace of $\mathbb{C}^m$ is called the **column rank** of A and is denoted $c\text{-}rank(A)$.*

DEFINITION 3.4 *(row space and row rank)*

Let $A \in \mathbb{C}^{m \times n}$. We define the **row space** *of A as the span of the rows of A in $\mathbb{C}^n$. In symbols, we write $\mathcal{R}ow(A)$ for the row space and the dimension of this subspace is the* **row rank** *of A and is denoted $r\text{-}rank(A)$.*

THEOREM 3.1

Let $A \in \mathbb{C}^{m \times n}$. Then

1. *$\mathcal{N}ull(A)$ is a subspace of $\mathbb{C}^n$ and $nlty(A) \leq n$.*
2. *$\mathcal{C}ol(A)$ is a subspace of $\mathbb{C}^m$ and $c\text{-}rank(A) \leq m$.*
3. *$\mathcal{C}ol(A)$ equals the span of the columns of A in $\mathbb{C}^m$.*
4. *$\mathcal{R}ow(A)$ is a subspace of $\mathbb{C}^n$ and $r\text{-}rank(A) \leq n$.*
5. *$\mathcal{R}ow(A) = \mathcal{C}ol(A^T)$ and $\mathcal{C}ol(A) = \mathcal{R}ow(A^T)$.*
6. *$\mathcal{R}ow(\bar{A}) = \mathcal{C}ol(A^*)$ and $\mathcal{C}ol(\bar{A}) = \mathcal{R}ow(A^*)$.*
7. *$\mathcal{N}ull(A^*A) = \mathcal{N}ull(A)$, so $nlty(A^*A) = nlty(A)$.*

PROOF We leave (1) through (6) as exercises. We choose to prove (7) to illustrate some points. To prove two sets are equal, we prove each is included in the other as a subset. First pick $\mathbf{x} \in \mathcal{N}ull(A)$. Then $A\mathbf{x} = \vec{0}$ so $A^*A\mathbf{x} = A^*\vec{0} = \vec{0}$ putting $\mathbf{x} \in \mathcal{N}ull(A^*A)$. To get the other inclusion, we need to note that if $\mathbf{x} = \begin{bmatrix} x_1 \\ x_2 \\ \vdots \\ x_m \end{bmatrix}$, then $\mathbf{x}^*\mathbf{x} = [\overline{x}_1\, \overline{x}_2 \cdots \overline{x}_m] \begin{bmatrix} x_1 \\ x_2 \\ \vdots \\ x_m \end{bmatrix} = |x_1|^2 + |x_2|^2 + \cdots + |x_m|^2$, using $\overline{z}z = |z|^2$ for a complex number z. Thus, if $\mathbf{x}^*\mathbf{x} = \vec{0}$, then $\sum_{j=1}^{m} |x_j|^2 = 0$, implying all x_js are zero, making $\mathbf{x} = \vec{0}$. Therefore, if $\mathbf{x} \in \mathcal{N}ull(A^*A)$, then $A^*A\mathbf{x} = \vec{0}$, so $\mathbf{x}^*A^*A\mathbf{x} = \mathbf{x}^*\vec{0} = \vec{0}$. Then $(A\mathbf{x})^*(A\mathbf{x}) = \vec{0}$. But $(A\mathbf{x})^*(A\mathbf{x}) = \vec{0}$ implies $A\mathbf{x} = \vec{0}$ by the above discussion. This puts $\mathbf{x} \in \mathcal{N}ull(A)$. Thus, we have proved $\mathcal{N}ull(A) \subseteq \mathcal{N}ull(A^*A)$ above and now $\mathcal{N}ull(A^*A) \subseteq \mathcal{N}ull(A)$. These together prove $\mathcal{N}ull(A) = \mathcal{N}ull(A^*A)$. ▯

Note that we strongly used a special property of real and complex numbers to get the result above. There are some other useful connections that will be needed later. We collect these next.

THEOREM 3.2
Let $A \in \mathbb{C}^{m\times n}$, $B \in \mathbb{C}^{q\times m}$. Then

1. $\mathcal{N}ull(A) \subseteq \mathcal{N}ull(BA)$
2. *If $U \in \mathbb{C}^{m\times m}$ is invertible, then $\mathcal{N}ull(UA) = \mathcal{N}ull(A)$ so $nlty(UA) = nlty(A)$.*

PROOF The proof is left as an exercise. ▯

We are often interested in factoring a matrix, that is, writing a matrix as a product of two other "nice" matrices. The next theorem gives us some insights on matrix factors.

THEOREM 3.3
Let $A \in \mathbb{C}^{m\times n}$, $B \in \mathbb{C}^{m\times p}$, $D \in \mathbb{C}^{q\times n}$. Then

1. *$\mathcal{C}ol(B) \subseteq \mathcal{C}ol(A)$ iff there exists a matrix C in $\mathbb{C}^{n\times p}$ such that $B = AC$.*
2. *$\mathcal{C}ol(AC) \subseteq \mathcal{C}ol(A)$ for any C in $\mathbb{C}^{n\times p}$, so c-$rank(AC) \leq$ c-$rank(A)$.*
3. *$\mathcal{R}ow(D) \subseteq \mathcal{R}ow(A)$ iff there exists a matrix K in $\mathbb{C}^{q\times m}$ such that $D = KA$.*
4. *$\mathcal{R}ow(KA) \subseteq \mathcal{R}ow(A)$ for any K in $\mathbb{C}^{q\times m}$, so r-$rank(KA) \leq$ r-$rank(A)$.*
5. *Let $S \in \mathbb{C}^{n\times k}$, $T \in \mathbb{C}^{n\times p}$. If $\mathcal{C}ol(S) \subseteq \mathcal{C}ol(T)$, then $\mathcal{C}ol(AS) \subseteq \mathcal{C}ol(AT)$. Moreover, if $\mathcal{C}ol(S) = \mathcal{C}ol(T)$, then $\mathcal{C}ol(AS) = \mathcal{C}ol(AT)$.*
6. *Let $S \in \mathbb{C}^{q\times m}$ and $T \in \mathbb{C}^{s\times m}$. Then, if $\mathcal{R}ow(S) \subseteq \mathcal{R}ow(T)$, then $\mathcal{R}ow(SA) \subseteq \mathcal{R}ow(TA)$. Moreover, if $\mathcal{R}ow(S) = \mathcal{R}ow(T)$, then $\mathcal{R}ow(SA) = \mathcal{R}ow(TA)$.*
7. *If $U \in \mathbb{C}^{m\times m}$ is invertible, then $\mathcal{R}ow(UA) = \mathcal{R}ow(A)$.*
8. *If $V \in \mathbb{C}^{n\times n}$ is invertible, then $\mathcal{C}ol(AV) = \mathcal{C}ol(A)$.*

PROOF The proofs are left as exercises. ▯

Next, we consider a very useful result whose proof is a bit abstract. It uses many good ideas from elementary linear algebra. It's a nice proof so we are going to give it.

THEOREM 3.4 (rank plus nullity theorem)

Let $A \in \mathbb{C}^{m\times n}$. Then the column rank of A plus the nullity of A equals the number of columns of A. That is, $c\text{-}rank(A) + nlty(A) = n$.

PROOF Take a basis $\{\mathbf{v}_1, \mathbf{v}_2, \cdots, \mathbf{v}_q\}$ of $\mathcal{N}ull(A)$. Extend this basis to a basis of all of $\mathbb{C}^n$, say $\{\mathbf{v}_1, \mathbf{v}_2, \cdots, \mathbf{v}_q, \mathbf{w}_1, \cdots, \mathbf{w}_r\}$ is the full basis. Note $n = q + r$. Now take $\mathbf{y}$ in $Col(A)$. Then, $\mathbf{y} = A\mathbf{x}$ for some $\mathbf{x}$ in $\mathbb{C}^n$. But then, $\mathbf{x}$ can be expressed in the basis (uniquely) as $\mathbf{x} = a_1\mathbf{v}_1 + a_2\mathbf{v}_2 + \cdots + a_q\mathbf{v}_q + b_1\mathbf{w}_1 + \cdots + b_r\mathbf{w}_r$. Thus $\mathbf{y} = A\mathbf{x} = a_1A\mathbf{v}_1 + a_2A\mathbf{v}_2 + \cdots + a_qA\mathbf{v}_q + b_1A\mathbf{w}_1 + \cdots + b_rA\mathbf{w}_r = \overrightarrow{0} + b_1A\mathbf{w}_1 + \cdots + b_rA\mathbf{w}_r$. This says the vectors $A\mathbf{w}_1, A\mathbf{w}_2, \cdots, A\mathbf{w}_r$ span $Col(A)$. Now the question is, are these vectors independent? To check, we set $c_1A\mathbf{w}_1 + c_2A\mathbf{w}_2 + \cdots + c_rA\mathbf{w}_r = \overrightarrow{0}$. Then $A(c_1\mathbf{w}_1 + c_2\mathbf{w}_2 + \cdots + c_r\mathbf{w}_r) = \overrightarrow{0}$. This puts the vector $c_1\mathbf{w}_1 + c_2\mathbf{w}_2 + \cdots + c_r\mathbf{w}_r$ in $\mathcal{N}ull(A)$. Therefore, this vector can be expressed in terms of the basis vectors $\mathbf{v}_1, \mathbf{v}_2, \cdots, \mathbf{v}_q$. Then, $c_1\mathbf{w}_1 + c_2\mathbf{w}_2 + \cdots + c_r\mathbf{w}_r = d_1\mathbf{v}_1 + d_2\mathbf{v}_2 + \cdots + d_q\mathbf{v}_q$, hence $d_1\mathbf{v}_1 + d_2\mathbf{v}_2 + \cdots + d_q\mathbf{v}_q - c_1\mathbf{w}_1 - c_2\mathbf{w}_2 - \cdots - c_r\mathbf{w}_r = \overrightarrow{0}$. But now, we are looking at the entire basis which is, of course, independent, so $d_1 = d_2 = \cdots = d_q = c_1 = c_2 = \cdots = c_r = 0$. Thus, the vectors $A\mathbf{w}_1, A\mathbf{w}_2, \cdots, A\mathbf{w}_r$ are independent and consequently form a basis for $Col(A)$. Moreover, $r = c\text{-}rank(A)$ and $q = nlty(A)$. This completes the proof. Isn't this a nice argument? ☐

COROLLARY 3.1

If $A \in \mathbb{C}^{m\times n}$ and $U \in \mathbb{C}^{m\times m}$ is invertible, then $c\text{-}rank(UA) = c\text{-}rank(A)$.

PROOF By Theorem 3.2(2) $\mathcal{N}ull(UA) = \mathcal{N}ull(A)$, so $nlty(UA) = nlty(A)$. By Theorem 3.4, $n = c\text{-}rank(A) + nlty(A) = c\text{-}rank(UA) + nlty(UA)$. Cancel and conclude $c\text{-}rank(A) = c\text{-}rank(UA)$. ☐

Notice that this corollary does not say $Col(A) = Col(UA)$. That is because this is not true! Consider $A = \begin{bmatrix} 1 & 4 \\ 2 & 8 \end{bmatrix}$. Now $U = \begin{bmatrix} 1 & 0 \\ -2 & 1 \end{bmatrix}$ is invertible and $UA = \begin{bmatrix} 1 & 4 \\ 0 & 0 \end{bmatrix}$. But $Col(A) = \{\alpha \begin{bmatrix} 1 \\ 2 \end{bmatrix} | \alpha \in \mathbb{C}\}$ and $Col(B) = \{\beta \begin{bmatrix} 1 \\ 0 \end{bmatrix} | \beta \in \mathbb{C}\}$ which are, evidently, different subspaces.

THEOREM 3.5

If U and V are invertible, then UAV has the same row rank and the same column rank as A.

PROOF We have $c\text{-}rank(A) = c\text{-}rank(UA)$ by Corollary 3.1 and $c\text{-}rank(UA) = c\text{-}rank(UAV)$ by Theorem 3.4(8). Thus $c\text{-}rank(A) = c\text{-}rank(UAV)$. Also, $\mathcal{R}ow(UAV) = \mathcal{C}ol((UAV)^T) = \mathcal{C}ol(V^T A^T U^T)$, so $r\text{-}rank(A) = c\text{-}rank(A^T) = r\text{-}rank(UAV) = c\text{-}rank(V^T A^T U^T)$. $\square$

We next look at a remarkable and fundamental result about matrices. You probably know it already. The row rank of a matrix is always equal to its column rank, even though A need not be square and $\mathcal{N}ull(A)$ and $\mathcal{C}ol(A)$ are contained in different vector spaces! Our next goal is to give an elementary proof of this fact.

THEOREM 3.6 (row rank equals column rank)
Let $A \in \mathbb{C}^{m\times n}$. Then $r\text{-}rank(A) = c\text{-}rank(A)$.

PROOF Let A be an m-by-n matrix of row rank r over $\mathbb{C}$. Let $\mathbf{r}_1 = row_1(A), \ldots, \mathbf{r}_m = row_m(A)$. Thus $\mathbf{r}_i = row_i(A) = [a_{i1}\ a_{i2} \cdots a_{in}]$ for $i = 1, 2, \ldots, m$. Choose a basis for the row space of A, $\mathcal{R}ow(A)$, say $\mathbf{b}_1, \mathbf{b}_2, \ldots, \mathbf{b}_r$. Suppose $\mathbf{b}_i = [b_{i1} b_{i2} \ldots b_{in}]$ for $i = 1, 2, \ldots, r$. It follows that each row of A is uniquely expressible as a linear combination of the basis vectors:

$$\begin{aligned} \mathbf{r}_1 &= c_{11}\mathbf{b}_1 + c_{12}\mathbf{b}_2 + \cdots + c_{1r}\mathbf{b}_r \\ \mathbf{r}_2 &= c_{21}\mathbf{b}_1 + c_{22}\mathbf{b}_2 + \cdots + c_{2r}\mathbf{b}_r \\ &\vdots \\ \mathbf{r}_m &= c_{m1}\mathbf{b}_1 + c_{m2}\mathbf{b}_2 + \cdots + c_{mr}\mathbf{b}_r. \end{aligned}$$

Now $[a_{11}a_{12}\ldots a_{1n}] = \mathbf{r}_1 = c_{11}[b_{11}b_{12}\ldots b_{1n}] + c_{12}[b_{21}b_{22}\ldots b_{2n}] + \cdots + c_{1r}[b_{r1}b_{r2}\ldots b_{rn}]$. A similar expression obtains for each row. By equating entries, we see for each j,

$$\begin{aligned} a_{1j} &= c_{11}b_{1j} + c_{12}b_{2j} + \cdots + c_{1r}b_{rj} \\ a_{2j} &= c_{21}b_{1j} + c_{22}b_{2j} + \cdots + c_{2r}b_{rj} \\ &\vdots \\ a_{mj} &= c_{m1}b_{1j} + c_{m2}b_{2j} + \cdots + c_{mr}b_{rj}. \end{aligned}$$

As a vector equation, we get

$$\begin{bmatrix} a_{1j} \\ a_{2j} \\ \vdots \\ a_{mj} \end{bmatrix} = b_{1j} \begin{bmatrix} c_{11} \\ c_{21} \\ \vdots \\ c_{m1} \end{bmatrix} + \cdots + b_{rj} \begin{bmatrix} c_{1r} \\ c_{2r} \\ \vdots \\ c_{mr} \end{bmatrix}, \quad j = 1, 2, \ldots, n.$$

This says that every column of A is a linear combination of the r vectors of cs. Hence the column space of A is generated by r vectors and so the dimension of the column space cannot exceed r. This says $c\text{-}rank(A) \leq r\text{-}rank(A)$. Applying the same argument to A^T, we conclude $c\text{-}rank(A^T) \leq r\text{-}rank(A^T)$. But then, we have $r\text{-}rank(A) \leq c\text{-}rank(A)$. Therefore, equality must hold. □

From now on, we will use the word *rank* to refer to either the row rank or the column rank of a matrix, whichever is more convenient, and we use the notation $rank(A)$ or $r(A)$. For some matrices, the rank is easy to ascertain. For example, $r(I_n) = n$. If $A = diag(d_1, d_2, \cdots, d_n)$, then $r(A)$ is the number of nonzero diagonal elements. For other matrices, especially for large matrices, the rank may not be so easily accessible.

We have seen that, given a matrix A, we can associate three subspaces and two dimensions to A. We have the null space $\mathcal{N}ull(A)$, the column space $\mathcal{C}ol(A)$, the row space $\mathcal{R}ow(A)$, the dimension $dim(\mathcal{N}ull(A))$, which is the nullity of A, and $dim(\mathcal{C}ol(A)) = dim(\mathcal{R}ow(A)) = r(A)$, the rank of A. But this is not the end of the story. That is because we can naturally associate other matrices to A. Namely, given A, we have the conjugate of A, $\overline{A}$; the transpose of A, A^T; and the conjugate transpose, $A^* = (\overline{A})^T$. This opens up a number of subspaces that can be associated with A:

$$\begin{array}{lll}
\mathcal{N}ull(A) & \mathcal{C}ol(A) & \mathcal{R}ow(A) \\
\mathcal{N}ull(\overline{A}) & \mathcal{C}ol(\overline{A}) & \mathcal{R}ow(\overline{A}) \\
\mathcal{N}ull(A^T) & \mathcal{C}ol(A^T) & \mathcal{R}ow(A^T) \\
\mathcal{N}ull(A^*) & \mathcal{C}ol(A^*) & \mathcal{R}ow(A^*).
\end{array}$$

Fortunately, not all 12 of these subspaces are distinct. We have $\mathcal{C}ol(A) = \mathcal{R}ow(A^T)$, $\mathcal{C}ol(\overline{A}) = \mathcal{R}ow(A^*)$, $\mathcal{C}ol(A^T) = \mathcal{R}ow(A)$, and $\mathcal{C}ol(A^*) = \mathcal{R}ow(\overline{A})$. Thus, there are actually eight subspaces to consider. If we wish, we can eliminate row spaces from consideration altogether and just deal with null spaces and column spaces. An important fact we use many times is that $rank(A) = rank(A^*)$ (see problem 15 of Exercise Set 8). This depends on the fact that if there is a dependency relation among vectors in $\mathbb{C}^n$, there is an equivalent dependency relation among the vectors obtained from these vectors by taking complex conjugates of their entries. In fact, all you have to do is take the complex conjugates of the scalars that effected the original dependency relationship.

We begin developing a heuristic picture of what is going on with the following diagram.

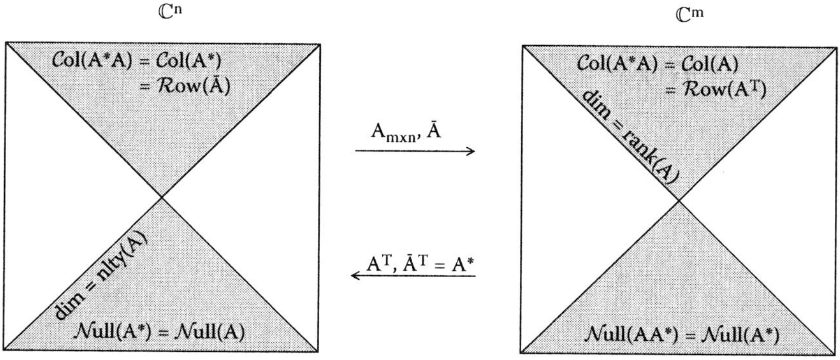

Figure 3.1: Fundamental subspaces.

Exercise Set 8

1. Give the rank and nullity of the following matrices:

$$\begin{bmatrix} 1 & 0 & 0 \\ 0 & 1 & 0 \\ 0 & 0 & 1 \end{bmatrix}, \begin{bmatrix} 2 & 0 & 0 \\ 0 & 3 & 0 \\ 0 & 0 & 0 \end{bmatrix}, \begin{bmatrix} 1 & i \\ 1 & i \end{bmatrix}, \begin{bmatrix} 1 & a & b & 0 \\ 0 & c & d & l \end{bmatrix}.$$

2. Let $A = \begin{bmatrix} 2 & 2+2i & 2-2i & 4 \\ 3+3i & 6i & 6 & 6+6i \end{bmatrix}$. Find the rank of A. Also, compute AA^* and A^*A and find their ranks.

3. If $A \in \mathbb{C}^{m \times n}$, argue that $r(A) \leq m$ and $r(A) \leq n$.

4. Fill in the proofs of Theorems 3.1, 3.2, and 3.3.

5. Argue that if A is square, A is invertible iff A^*A is invertible.

6. Prove that $Col(AA^*) = Col(A)$, so $c\text{-}rank(AA^*) = c\text{-}rank(A)$.

7. Prove that $c\text{-}rank(A) = c\text{-}rank(\overline{A})$ and $r\text{-}rank(A) = r\text{-}rank(\overline{A})$.

8. Argue that $\mathcal{N}ull(A) \cap Col(A^*) = (\vec{0})$ and $\mathcal{N}ull(A^*) \cap Col(A) = (\vec{0})$.

9. Let B be an m-by-n matrix. Argue that the rows of B are dependent in $\mathbb{C}^n$ iff there is a nonzero vector $\mathbf{w}$ with $\mathbf{w}B = \vec{0}$. Also the columns are dependent in $\mathbb{C}^m$ iff there is a nonzero vector $\mathbf{w}$ with $B\mathbf{w} = \vec{0}$.

10. Consider a system of linear equations $A\mathbf{x} = \mathbf{b}$ where $A \in \mathbb{C}^{m \times n}$, $\mathbf{x} \in \mathbb{C}^{n \times 1}$, and $\mathbf{b} \in \mathbb{C}^{m \times 1}$. Argue that this system is consistent (i.e., has a

solution) iff $rank(A) = rank([A \mid \mathbf{b}])$, where $[A \mid \mathbf{b}]$ is the augmented matrix (i.e., the matrix obtained from A by adding one more column, namely $\mathbf{b}$).

11. Define a map $J : \mathbb{C}^n \to \mathbb{C}^n$ by $J(x_1, x_2, \ldots, x_n) = (\overline{x}_1, \overline{x}_2, \ldots, \overline{x}_n)$. That is, J just takes the complex conjugate of each entry of a vector. Argue that $J^2 = I$. Is J a linear map? How close to a linear map is it? Is J one-to-one? Is J onto? If M is a subspace of $\mathbb{C}^n$, argue that $J(M)$ is a subspace and $dim(J(M)) = dim(M)$. If A is a matrix in $\mathbb{C}^{n \times n}$, what does J do to the fundamental subspaces of A?

12. Suppose A is an n-by-n matrix and $\mathbf{v}$ is a vector in $\mathbb{C}^n$. We defined the *Krylov matrix* $\mathcal{K}_s(A, \mathbf{v}) = [\mathbf{v} \mid A\mathbf{v} \mid A^2\mathbf{v} \mid \cdots \mid A^{s-1}\mathbf{v}]$. Then the *Krylov subspace* $\mathbf{K}_s(A, \mathbf{v}) = span\{\mathbf{v}, A\mathbf{v}, \ldots, A^{s-1}\mathbf{v}\} = \mathcal{C}ol(\mathcal{K}_s(A, \mathbf{v}))$. Suppose $A\mathbf{x} = \mathbf{b}$ is a linear system with A invertible. Suppose $deg(\mu_A) = m$. Argue that the solution to $A\mathbf{x} = \mathbf{b}$ lies in $\mathbf{K}_m(A, b)$. Note, every $\mathbf{x}$ in $\mathbf{K}_m(A, b)$ is of the form $p(A)\mathbf{b}$, where p is a polynomial of degree $m - 1$ or less.

13. Make up a concrete example to illustrate problem 12 above.

14. Argue that $\mathcal{N}ull(A) = (\overrightarrow{0})$ iff the columns of A are independent.

15. Prove that $r(A) = r(A^T) = r(A^*) = r(\bar{A})$.

16. Argue that $AB = \mathbb{O}$ iff $\mathcal{C}ol(B) \subseteq \mathcal{N}ull(A)$, where A and B are conformable matrices. Recall "conformable" means the matrices are of a size that can be multiplied.

17. Prove that $A^2 = \mathbb{O}$ iff $\mathcal{C}ol(A) \subseteq \mathcal{N}ull(A)$. Conclude that if $A^2 = \mathbb{O}$, then $rank(A) \leq \frac{n}{2}$ if A is n-by-n.

18. Let $Fix(A) = \{\mathbf{x} \mid A\mathbf{x} = \mathbf{x}\}$. Argue that $Fix(A)$ is a subspace of $\mathbb{C}^n$ when A is n-by-n. Find $Fix\left(\begin{bmatrix} 1 & 1 & 1 \\ 0 & 2 & 3 \\ 0 & 3 & 2 \end{bmatrix}\right)$.

19. Is $W = \{(z, \overline{z}, 0) \mid z \in \mathbb{C}\}$ a subspace of $\mathbb{C}^3$?

20. What is the rank of $\begin{bmatrix} 1 & 2 & 3 \\ 4 & 5 & 6 \\ 7 & 8 & 9 \end{bmatrix}$? What is the rank of $\begin{bmatrix} 1 & 2 & 3 & 4 \\ 5 & 6 & 7 & 8 \\ 9 & 10 & 11 & 12 \\ 13 & 41 & 15 & 16 \end{bmatrix}$? Do you see a pattern? Can you generalize?

21. Let $S = \{\begin{bmatrix} a & b \\ -b & c \end{bmatrix} \mid a, b, c \in \mathbb{C}\}$. Is S a subspace of $\mathbb{C}^{2\times 2}$?

22. Argue that $rank(AB) \leq \min(rank(A), rank(B))$ and $rank(B) = rank(-B)$.

23. Prove that if one factor of AB is nonsingular, then the rank of AB is the rank of the other factor.

24. Argue that if S and T are invertible, then $rank(SAT) = rank(A)$.

25. Prove that $rank(A + B) \leq rank(A) + rank(B)$ and $rank(A + B) \geq |rank(A) - rank(B)|$.

26. Argue that if $A \in \mathbb{C}^{m\times n}$ and $B \in \mathbb{C}^{n\times m}$ where $n < m$, then AB cannot be invertible.

27. For a block diagonal matrix $A = diag[A_{11}, A_{22}, \ldots, A_{kk}]$, argue that $rank(A) = \sum_{i=1}^{k} rank(A_{ii})$.

28. Suppose A is an n-by-n matrix and k is a positive integer. Argue that $\mathcal{N}ull(A^{k-1}) \subseteq \mathcal{N}ull(A^k) \subseteq \mathcal{N}ull(A^{k+1})$. Let $\{\mathbf{b}_1, \ldots, \mathbf{b}_r\}$ be a basis of $\mathcal{N}ull(A^{k-1})$; extend this basis to $\mathcal{N}ull(A^k)$ and get $\{\mathbf{b}_1, \ldots, \mathbf{b}_r, \mathbf{c}_1, \mathbf{c}_2, \ldots, \mathbf{c}_s\}$. Now extend this basis to get a basis $\{\mathbf{b}_1, \ldots, \mathbf{b}_r, \mathbf{c}_1, \mathbf{c}_2, \ldots, \mathbf{c}_s, \mathbf{d}_1, \ldots, \mathbf{d}_t\}$ of $\mathcal{N}ull(A^{k+1})$. Argue that $\{\mathbf{b}_1, \ldots, \mathbf{b}_r, A\mathbf{d}_1, \ldots, A\mathbf{d}_t\}$ is a linearly independent subset of $\mathcal{N}ull(A^k)$.

29. Argue that $\mathcal{C}ol(AB) = \mathcal{C}ol(A)$ iff $rank(AB) = rank(A)$.

30. Prove that $\mathcal{N}ull(AB) = \mathcal{N}ull(B)$ iff $rank(AB) = rank(B)$.

31. (Peter Hoffman) Suppose $A_1, A_2, \ldots, A_n$ are k-by-k matrices over $\mathbb{C}$ with $A_1 + A_2 + \cdots + A_n$ invertible. Argue that the block matrix
$$\begin{bmatrix} A_1 & A_2 & A_3 & \cdots & A_n & \mathbb{O} & \cdots & \mathbb{O} \\ \mathbb{O} & A_1 & A_2 & \cdots & A_{n-1} & A_n & \cdots & \mathbb{O} \\ \vdots & \vdots & \vdots & & & & & \vdots \\ \mathbb{O} & \mathbb{O} & \cdots & A_1 & A_2 & \cdots & \cdots & A_n \end{bmatrix}$$ has full rank.
What is this rank?

32. (Yongge Tian) Suppose A is an m-by-n matrix with real entries. What is the minimum rank of $A + iB$, where B can be any m-by-n matrix with real entries?

33. Prove that if $AB = \mathbb{O}$, then $rank(A) + rank(B) \leq n$, where A and $B \in \mathbb{C}^{n\times n}$.

34. Suppose A is nonsingular. Argue that the inverse of A is the unique matrix X such that $rank\begin{bmatrix} A & I \\ I & X \end{bmatrix} = rank(A)$.

35. If A is m-by-n of rank m, then an LU factorization of A is unique. In particular, if A is invertible, then LU is unique, if it exists.

36. Suppose A is n-by-n.
    (a) If $rank(A) < n - 1$, then prove $adj(A) = \mathbb{O}$.
    (b) If $rank(A) = n - 1$, then prove $rank(adj(A)) = 1$.
    (c) If $rank(A) = n$, then $rank(adj(A)) = n$.

37. Argue that $Col(A + B) \subseteq Col(A) + Col(B)$ and $Col(A + B) = Col(A) + Col(B)$ iff $Col(A) \subseteq Col(A + B)$. Also, $Col(A + B) = Col(A) + Col(B)$ iff $Col(B) \subseteq Col(A + B)$.

38. Prove that $Col([A \mid B]) = Col(A) + Col(B)$ so that if A is m-by-n and B is m-by-p, $rank([A \mid B]) \leq rank(A) + rank(B)$.

39. Suppose $A \in \mathbb{C}_r^{m\times n}$, $X \in \mathbb{C}^{n\times p}$, and $AX = \mathbb{O}$. Argue that the rank of X is less than or equal to $n - r$.

40. Suppose $A \in \mathbb{C}^{m\times n}$ and $m < n$. Prove that there exists $X \neq \mathbb{O}$ such that $AX = \mathbb{O}$.

41. Prove the Guttman rank additivity formula: suppose $M = \begin{bmatrix} A & B \\ C & D \end{bmatrix}$ with $det(A) \neq 0$. Then $rank(M) = rank(A) + rank(M/A)$.

Further Reading

[B&R, 1986(2)] T. S. Blyth and E. F. Robertson, *Matrices and Vector Spaces*, Vol. 2, Chapman & Hall, New York, (1986).

[Liebeck, 1966] H. Liebeck, A Proof of the Equality of Column and Row Rank of a Matrix, The American Mathematical Monthly, Vol. 73, (1966), 1114.

[Mackiw, 1995] G. Mackiw, A Note on the Equality of Column and Row Rank of a Matrix, Mathematics Magazine, Vol. 68, (1995), 285–286.

3.1.1 MATLAB Moment

3.1.1.1 The Fundamental Subspaces

MATLAB has a built-in function to compute the rank of a matrix. The command is

$$rank(A)$$

Unfortunately, MATLAB does not have a built-in command for the nullity of A. This gives us a good opportunity to define our own function by creating an M-file. To do this, we take advantage of the rank plus nullity theorem. Here is how it works.

Assuming you are on a Windows platform, go to the "File" menu, choose "New" and "M-file." A window comes up in which you can create your function as follows:

```
1   function nlty= nullity(A)
2-  [m,n]= size(A)
3-  nlty = n - rank(A).
```

Note that the "size" function returns the number of rows and the number of columns of A. Then do a "Save as" nullity.m, which is the suggested name. Now check your program on

$$A = \begin{bmatrix} 1 & -6 & 1 & 4 \\ -2 & 14 & 1 & -9 \\ 2 & -6 & 11 & 5 \end{bmatrix}$$

```
>> A=[1 -6 1 4;-2 14 1 -9;2 -6 11 5]

       1   -6   1    4
A =   -2   14   1   -9
       2   -6  11    5

>> rank(A)

ans =

    2

>> nullity(A)

ans =

  2
```

Now try finding the rank and nullity of

$$B = \begin{bmatrix} 1+i & 2+2i & 3+i \\ 2+2i & 4+4i & 9i \\ 3+3i & 6+6i & 8i \end{bmatrix}.$$

It is possible to get MATLAB to find a basis for the column space and a basis for the null space of a matrix. Again we write our own functions to do this. We use the rref command (row reduced echelon form), which we will review later. You probably remember it from linear algebra class. First, we find the column space. Create the M-file

```
function c = colspace(A)
[m,n] = size(A)
C = [A' eye(n)]
B = rref(C)'
c = B([1:m],[1:rank(A)]);
```

which we save as colspace.m. Next we create a similar function to produce a basis for the nullspace of A, which we call nullspace.m.

```
function N = nullspace(A)
[m,n] = size(A)
C = [A' eye(n)];
B = rref(C)'
N = B([m+1:m+n],[rank(A)]+1:n]);
```

Now try these out on matrix A above.

```
>> A=[1 -6 1 4;-2 14 1 -9;2 -6 11 5]
```

$$A = \begin{matrix} 1 & -6 & 1 & 4 \\ -2 & 14 & 1 & -9 \\ 2 & -6 & 11 & 5 \end{matrix}$$

```
>> colspace(A)
```

ans =

$$\begin{matrix} 1 & 0 \\ 0 & 1 \\ 8 & 3 \end{matrix}$$

```
>> format rat
>>nullspace(A)
```

ans =

$$\begin{matrix} 1 & 0 \\ 0 & 1 \\ -1/13 & -2/13 \\ -3/13 & 20/13 \end{matrix}$$

Now, determine the column space and nullspace of B above.

Finally we note that MATLAB does have a built-in command null(A), which returns an orthonormal basis for the nullspace, and orth(A), which returns an orthonormal basis for the column space. A discussion on this is for a later chapter.

By the way, the World Wide Web is a wonderful source of M-files. Just go out there and search.

Further Reading

[L&H&F, 1996] Steven Leon, Eugene Herman, Richard Faulkenberry, *ATLAST Computer Exercises for Linear Algebra*, Prentice Hall, Upper Saddle River, NJ, (1996).

Sylvester's rank formula, Sylvester's law of nullity, the Frobenius inequality

---

3.2 A Deeper Look at Rank

We have proved the fundamental result that row rank equals column rank. Thus, we can unambiguously use the word "rank" to signify either one of these numbers. Let $r(A)$ denote the rank of the matrix A. Then we know $r(A) = r(A^*) = r(A^T) = r(\overline{A})$. Also, the rank plus nullity theorem says $r(A) + nlty(A) =$ the number of columns of A. To get more results about rank, we develop a really neat formula which goes back to **James Joseph Sylvester** (3 September 1814 – 15 March 1897).

THEOREM 3.7 (Sylvester's rank formula)
Let $A \in \mathbb{C}^{m \times n}$ and $B \in \mathbb{C}^{n \times p}$. Then

$$r(AB) = r(B) - dim(\mathcal{N}ull(A) \cap \mathcal{C}ol(B)).$$

PROOF Choose a basis of $\mathcal{N}ull(A) \cap \mathcal{C}ol(B)$, say $\{\mathbf{b}_1, \mathbf{b}_2, \ldots, \mathbf{b}_s\}$, and extend this basis to a basis of $\mathcal{C}ol(B)$. Say $\mathcal{B} = \{\mathbf{b}_1, \mathbf{b}_2, \ldots, \mathbf{b}_s, \mathbf{c}_1, \mathbf{c}_2, \ldots, \mathbf{c}_t\}$ is this basis for $\mathcal{C}ol(B)$. We claim that $\{A\mathbf{c}_1, A\mathbf{c}_2, \ldots A\mathbf{c}_t\}$ is a basis for $\mathcal{C}ol(AB)$. As usual, there are two things to check. First, we check the linear independence of this set of vectors. We do this in the usual way. Suppose a linear combination $\alpha_1 A\mathbf{c}_1 + \alpha_2 A\mathbf{c}_2 + \cdots + \alpha_t A\mathbf{c}_t = \overrightarrow{0}$. Then, $A(\alpha_1\mathbf{c}_1 + \alpha_2\mathbf{c}_2 + \cdots + \alpha_t\mathbf{c}_t) = \overrightarrow{0}$. This puts $\alpha_1\mathbf{c}_1 + \alpha_2\mathbf{c}_2 + \cdots + \alpha_t\mathbf{c}_t$ in the null space of A. But the cs are in $\mathcal{C}ol(B)$, so this linear combination is in there also. Thus, $\alpha_1\mathbf{c}_1 + \alpha_2\mathbf{c}_2 + \cdots + \alpha_t\mathbf{c}_t \in \mathcal{N}ull(A) \cap \mathcal{C}ol(B)$. But we have a basis for this intersection, so there must exist scalars $\beta_1, \beta_2, \ldots, \beta_s$ so that $\alpha_1\mathbf{c}_1 + \alpha_2\mathbf{c}_2 + \cdots + \alpha_t\mathbf{c}_t = \beta_1\mathbf{b}_1 + \beta_2\mathbf{b}_2 + \cdots + \beta_s\mathbf{b}_s$. But then, $\alpha_1\mathbf{c}_1 + \alpha_2\mathbf{c}_2 + \cdots + \alpha_t\mathbf{c}_t - \beta_1\mathbf{b}_1 - \beta_2\mathbf{b}_2 - \cdots - \beta_s\mathbf{b}_s = \overrightarrow{0}$. Now the **c**s and the **b**s together make up an independent set so all the scalars, all the αs, and all the βs, must be zero. In particular, all the α's are zero, so this establishes the independence of $\{A\mathbf{c}_1, A\mathbf{c}_2, \ldots A\mathbf{c}_t\}$.

Is it clear that all these vectors are in $\mathcal{C}ol(AB)$? Yes, because each $\mathbf{c}_i$ is in $\mathcal{C}ol(B)$, so $\mathbf{c}_i = B\mathbf{x}_i$ for some $\mathbf{x}_i$; hence $A\mathbf{c}_i = AB\mathbf{x}_i \in \mathcal{C}ol(AB)$. Finally, we prove that our claimed basis actually does span $\mathcal{C}ol(AB)$. Let $\mathbf{y}$ be in $\mathcal{C}ol(AB)$. Then $\mathbf{y} = AB\mathbf{x}$ for some $\mathbf{x}$. But $B\mathbf{x}$ lies in $\mathcal{C}ol(B)$ so $B\mathbf{x} = \alpha_1\mathbf{b}_1 + \alpha_2\mathbf{b}_2 + \cdots + \alpha_s\mathbf{b}_s + \alpha_{s+1}\mathbf{c}_1 + \cdots + \alpha_{s+t}\mathbf{c}_t$. But then, $\mathbf{y} = AB\mathbf{x} = \alpha_1 A\mathbf{b}_1 + \alpha_2 A\mathbf{b}_2 + \cdots + \alpha_s A\mathbf{b}_s + \alpha_{s+1} A\mathbf{c}_1 + \cdots + \alpha_{s+t} A\mathbf{c}_t = \overrightarrow{0} + \alpha_{s+1} A\mathbf{c}_1 + \cdots + \alpha_{s+t} A\mathbf{c}_t$. Thus, we have established our claim. Now notice $t = dim(\mathcal{C}ol(AB)) = r(AB)$ and $r(B) = dim(\mathcal{C}ol(B)) = s + t = dim(\mathcal{N}ull(A) \cap \mathcal{C}ol(B)) + r(AB)$. The formula now follows. □

The test of a good theorem is all the consequences you can squeeze out of it. Let's now reap the harvest of this wonderful formula.

COROLLARY 3.2
For $A \in \mathbb{C}^{m\times n}$ and $B \in \mathbb{C}^{n\times p}$,

$$nlty(AB) = nlty(B) + dim(\mathcal{N}ull(A) \cap \mathcal{C}ol(B)).$$

PROOF This follows from the rank plus nullity theorem. □

COROLLARY 3.3
For $A \in \mathbb{C}^{m\times n}$ and $B \in \mathbb{C}^{n\times p}$, $r(AB) \le \min(r(A), r(B))$.

PROOF First, $r(AB) = r(B) - dim(\mathcal{N}ull(A) \cap \mathcal{C}ol(B)) \le r(B)$. Also $r(AB) = r((AB)^T) = r(B^T A^T) \le r(A^T) = r(A)$. □

COROLLARY 3.4
For $A \in \mathbb{C}^{m\times n}$ and $B \in \mathbb{C}^{n\times p}$, $r(A) + r(B) - n \le r(AB) \le \min(r(A), r(B))$.

PROOF First, $\mathcal{N}ull(A) \cap \mathcal{C}ol(B) \subseteq \mathcal{N}ull(A)$ so $dim\,(\mathcal{N}ull(A) \cap \mathcal{C}ol(B)) \leq dim(\mathcal{N}ull(A)) = nlty(A) = n - r(A)$ so $r(AB) = r(B) - dim(\mathcal{N}ull(A) \cap \mathcal{C}ol(B)) \geq r(B) - (n - r(A))$ so $r(AB) \geq r(B) + r(A) - n$. ▯

COROLLARY 3.5
Let $A \in \mathbb{C}^{m \times n}$; then $r(A^*A) = r(A)$.

PROOF It takes a special property of complex numbers to get this one. Let $\mathbf{x} \in \mathcal{N}ull(A^*) \cap \mathcal{C}ol(A)$. Then $A^*\mathbf{x} = \overrightarrow{0}$ and $\mathbf{x} = A\mathbf{y}$ for some $\mathbf{y}$. But then $\mathbf{x}^*\mathbf{x} = \mathbf{y}^*A^*\mathbf{x} = \overrightarrow{0}$ so $\sum |x_i|^2 = 0$. This implies all the components of $\mathbf{x}$ are zero, so $\mathbf{x}$ must be the zero vector. Therefore, $\mathcal{N}ull(A^*) \cap \mathcal{C}ol(A) = (\overrightarrow{0})$, and so $r(A^*A) = r(A) - dim(\mathcal{N}ull(A^*) \cap \mathcal{C}ol(A)) = r(A)$. ▯

COROLLARY 3.6
Let $A \in \mathbb{C}^{m \times n}$; then $r(AA^*) = r(A^*)$.

PROOF Replace A by A^* above. ▯

COROLLARY 3.7
Let $A \in \mathbb{C}^{m \times n}$; then $\mathcal{C}ol(A^*A) = \mathrm{Col}(A^*)$ and $\mathcal{N}ull(A^*A) = \mathcal{N}ull(A)$.

PROOF Clearly $\mathcal{C}ol(A^*A) \subseteq \mathcal{C}ol(A^*)$ and $\mathcal{N}ull(A) \subseteq \mathcal{N}ull(A^*A)$. But $dim(\mathcal{C}ol(A^*A)) = r(A^*A) = r(A) = r(A^*) = dim(\mathcal{C}ol(A^*))$. Also, $dim(\mathcal{N}ull(A)) = n - r(A) = n - r(A^*A) = dim(\mathcal{N}ull(A^*A))$ so $\mathcal{N}ull(A^*A) = \mathcal{N}ull(A)$. ▯

COROLLARY 3.8
Let $A \in \mathbb{C}^{m \times n}$; then $\mathcal{C}ol(AA^*) = \mathcal{C}ol(A)$ and $\mathcal{N}ull(AA^*) = \mathcal{N}ull(A^*)$.

PROOF Replace A by A^* above. ▯

Next, we get another classical result of Sylvester.

COROLLARY 3.9 (Sylvester's law of nullity [1884])
For square matrices A and B in $\mathbb{C}^{n \times n}$,

$$max(nlty(A), nlty(B)) \leq nlty(AB) \leq nlty(A) + nlty(B).$$

PROOF First $\mathcal{N}ull(B) \subseteq \mathcal{N}ull(AB)$ so $dim\,(\mathcal{N}ull(B)) \leq dim\,(\mathcal{N}ull(AB))$ and so $nlty(B) \leq nlty(AB)$. Also, $nlty(A) = n - r(A) \leq n - r(AB) = nlty(AB)$. Therefore, $max(nlty(A), nlty(B)) \leq nlty(AB)$. Also, by Corollary 3.4, $r(A) + r(B) - n \leq r(AB) = n - nlty(AB)$ so $n - nlty(A) + n - nlty(B) - n \leq n - nlty(AB)$. Canceling the ns gives the "minus" of the inequality we want. Thus, $nlty(AB) \leq nlty(A) + nlty(B)$. □

Another classical result goes back to F. G. Frobenius, whom we have previously mentioned.

COROLLARY 3.10 (the Frobenius inequality [1911])
Assume the product ABC exists. Then $r(AB) + r(BC) \leq r(B) + r(ABC)$.

PROOF Now $\mathcal{C}ol(BC) \cap \mathcal{N}ull(A) \subseteq \mathcal{C}ol(B) \cap \mathcal{N}ull(A)$ so $dim(\mathcal{C}ol(BC) \cap \mathcal{N}ull(A)) \leq dim(\mathcal{C}ol(B) \cap \mathcal{N}ull(A))$. But $dim(\mathcal{C}ol(BC) \cap \mathcal{N}ull(A)) = r(BC) - r(ABC)$ by Sylvester's formula. Also, $dim(\mathcal{C}ol(B) \cap \mathcal{N}ull(A)) = r(B) - r(AB)$ so $r(BC) - r(ABC) \leq r(B) - r(AB)$. Therefore, $r(BC) + r(AB) \leq r(B) + r(ABC)$. □

There is one more theorem we wish to present. For this we need some notation. The idea of augmented matrix is familiar. For A, B in $\mathbb{C}^{m \times n}$, define $[A \vdots B]$ as the m-by-$2n$ matrix formed by adjoining B to A on the right.

Similarly, define $\begin{bmatrix} A \\ \cdots \\ B \end{bmatrix}$ to be the $2m$-by-n matrix formed by putting B under A.

THEOREM 3.8

Let A and B be in $\mathbb{C}^{m \times n}$. Then $r(A+B) = r(A) + r(B) - dim(\mathcal{C}ol(\begin{bmatrix} A \\ \cdots \\ B \end{bmatrix}) \cap \mathcal{N}ull([I_m \vdots I_m])) - dim(\mathcal{C}ol(A^*) \cap \mathcal{C}ol(B^*))$. In particular, $r(A + B) \leq r(A) + r(B)$.

PROOF We note the $A + B = [I_m \vdots I_m] \begin{bmatrix} A \\ \cdots \\ B \end{bmatrix}$, so $r(A+B) = r(\begin{bmatrix} A \\ \cdots \\ B \end{bmatrix}) - dim(Col(\begin{bmatrix} A \\ \cdots \\ B \end{bmatrix}) \cap \mathcal{N}ull([I_m \vdots I_m]))$. But $r(\begin{bmatrix} A \\ \cdots \\ B \end{bmatrix}) = r(\begin{bmatrix} A \\ \cdots \\ B \end{bmatrix}^*) = r([A^* \vdots B^*]) = dim(Col([A^* \vdots B^*]) = dim(Col(A^*) + Col(B^*)) = dim(Col(A^*)) + dim(Col(B^*) - dim(Col(A^*) \cap Col(B^*)) = r(A^*) + r(B^*) - dim(Col(A^*) \cap Col(B^*)) = r(A) + r(B) - dim(Col(A^*) \cap Col(B^*))$ using the familiar dimension formula. ▯

Exercise Set 9

1. Can you discover any other consequences of Sylvester's wonderful formula?

2. Consider the set of linear equations $A\mathbf{x} = \mathbf{b}$. Then the set of equations $A^*A\mathbf{x} = A^*\mathbf{b}$ are called the *normal equations*.
   (a) Argue that the normal equations are always consistent.
   (b) If $A\mathbf{x} = \mathbf{b}$ is consistent, prove that $A\mathbf{x} = \mathbf{b}$ and the associated normal equations $A^*A\mathbf{x} = A^*\mathbf{b}$ have the same set of solutions.
   (c) If $\mathcal{N}ull(A) = (\overrightarrow{0})$, the unique solution to both systems is $(A^*A)^{-1} A^*\mathbf{b}$.

3. In this exercise, we develop another approach to rank. Suppose $A \in \mathbb{C}_r^{m \times n}$.
   (a) Suppose A is m-by-p and B is m-by-q. Argue that if the rows of A are linearly independent, then the rows of $[A \vdots B]$ are also linearly independent.
   (b) If the rows of $[A \vdots B]$ are dependent, argue that the rows of A are necessarily dependent.
   (c) Suppose $rank(A) = r$. Argue that all submatrices of order r + 1 are singular.
   (d) Suppose $rank(A) = r$. Argue that at least one nonsingular r-by-r submatrix of A exists.
   (e) If the order of the largest nonsingular submatrix of A is r-by-r, then argue A has rank r.

4. [Meyer, 2000] There are times when it is very handy to have a basis for $\mathcal{N}ull(A) \cap \mathcal{C}ol(B)$. Argue that the following steps will produce one.
   (a) Find a basis for $\mathcal{C}ol(B)$, say $\{\mathbf{x}_1, \mathbf{x}_2, \ldots, \mathbf{x}_r\}$.
   (b) Construct the matrix X in $\mathbb{C}^{n \times r}$; $X = [\mathbf{x}_1 \mid \mathbf{x}_2 \mid \ldots \mid \mathbf{x}_r]$.
   (c) Find a basis for $\mathcal{N}ull(AX)$, say $\{\mathbf{v}_1, \mathbf{v}_2, \ldots, \mathbf{v}_s\}$.
   (d) Argue that $\mathcal{B} = \{X\mathbf{v}_1, X\mathbf{v}_2, \ldots X\mathbf{v}_s\}$ is a basis for $\mathcal{N}ull(A) \cap \mathcal{C}ol(B)$. (*Hint:* Argue that $\mathcal{C}ol(X) = \mathcal{C}ol(B)$ and $\mathcal{N}ull(X) = (\vec{0})$; then use Sylvester's formula.)

5. Prove the $rank(\begin{bmatrix} A_{11} & A_{12} & \cdots & A_{1k} \\ \mathbb{O} & A_{22} & \cdots & A_{2k} \\ \cdots & \cdots & \cdots & \cdots \\ \mathbb{O} & \mathbb{O} & \cdots & A_{kk} \end{bmatrix}) \geq \sum_{i=1}^{k} rank(A_{ii})$. Compute the rank of $A = \begin{bmatrix} 0 & 1 & 0 & 0 \\ 0 & 0 & 1 & 0 \\ 0 & 0 & 0 & 0 \\ 0 & 0 & 0 & 1 \end{bmatrix}$ and compare it to the ranks of the 2-by-2 diagonal blocks. What does this tell you about the previous inequality?

6. Suppose A is n-by-n and invertible and D is square in $M = \begin{bmatrix} A & B \\ C & D \end{bmatrix}$. Argue that $rank(M) = n$ iff the Schur complement of A in M is zero.

7. If A is m-by-n and B is n-by-m and $m > n$, argue that $det(AB) = 0$.

8. Argue that the linear system $A\mathbf{x} = \mathbf{b}$ is consistent iff $rank[A \mid \mathbf{b}] = rank(A)$.

9. Suppose that $rank(CA) = rank(A)$. Argue that $A\mathbf{x} = \mathbf{b}$ and $CA\mathbf{x} = C\mathbf{b}$ have the same solution set.

10. Suppose A is m-by-n. Argue that $A\mathbf{x} = \vec{0}$ implies $\mathbf{x} = \vec{0}$ iff $m \geq n$ and A has full rank.

11. Prove that $A\mathbf{x} = \mathbf{b}$ has a unique solution iff $m \geq n$, the equation is consistent, and A has full rank.

12. Argue that the rank of a symmetric matrix (or skew-symmetric matrix) is equal to the order of its largest nonzero principal minor. In particular, deduce that the rank of a skew-symmetric matrix cannot be odd.

13. Let T be a linear map from $\mathbb{C}^n$ to $\mathbb{C}^m$, and let M be a subspace of $\mathbb{C}^n$. Argue that $dim(T(M)) = dim(M) - dim(M \cap Ker(T))$ so, in particular, $dim(T(M)) \leq dim(M)$.

14. Suppose T_1 and T_2 are linear maps from $\mathbb{C}^n$ to $\mathbb{C}^m$. Argue that
    (a) $Ker(T_1) \cap Ker(T_2) \subseteq Ker(T_1 + T_2)$.
    (b) $Im(T_1 + T_2) \subseteq Im(T_1) + Im(T_2)$.
    (c) $|rank(T_1) - rank(T_2)| \leq rank(T_1 + T_2) \leq rank(T_1) + rank(T_2)$.

15. Argue that $A\mathbf{x} = \mathbf{b}$ has no solution if $rank(A) \neq rank([A \mid \mathbf{b}])$. Otherwise, the general solution has $n - rank(A)$ free variables.

16. For A, B in $\mathbb{C}^{n\times n}$, argue that $r(AB - I) \leq r(A - I) + r(B - I)$.

17. For A, B in $\mathbb{C}^{m\times n}$, suppose $B^*A = \mathbb{O}$. Prove that $r(A + B) = r(A) + r(B) \leq n$.

18. For matrices of appropriate size, argue that
$$r(\left[\begin{array}{cc} A & B \\ C & D \end{array}\right]) \leq r(A) + r(B) + r(C) + r(D).$$

19. If A and $B \in \mathbb{C}^{n\times n}$, argue that
$$rank(\left[\begin{array}{cc} A & C \\ \mathbb{O} & B \end{array}\right]) \geq rank(A) + rank(B).$$

Further Reading

[Meyer, 2000] Carl Meyer, *Matrix Analysis and Applied Linear Algebra*, SIAM, Philadelphia, (2000).

[Zhang, 1999] Fuzhen Zhang, *Matrix Theory: Basic Results and Techniques*, Springer, New York, (1999).

complementary subspaces, direct sum decomposition, idempotent matrix, projector, parallel projection

---

3.3 Direct Sums and Idempotents

There is an intimate connection between direct sum decompositions of $\mathbb{C}^n$ and certain kinds of matrices. This correspondence plays an important role in our discussion of generalized inverses. First, let's recall what it means to have a direct sum decomposition. Let M and N be subspaces of $\mathbb{C}^n$. We say M and N are *disjoint* when their intersection is as small as it can be, namely, $M \cap N = (\overrightarrow{0})$. We can always form a new subspace $M + N = \{\mathbf{x} \mid \mathbf{x} = \mathbf{m} + \mathbf{n}$ where $\mathbf{m} \in M$ and $\mathbf{n} \in N\}$. Indeed, this is the smallest subspace of $\mathbb{C}^n$ containing both M and N. When M and N are disjoint and $M + N = \mathbb{C}^n$, we say M and N are *complementary subspaces* and that they give a *direct sum decomposition* of $\mathbb{C}^n$. The notation is $\mathbb{C}^n = M \oplus N$ for a direct sum decomposition. What is nice about a direct sum decomposition is that each vector in $\mathbb{C}^n$ has a unique representation as a vector from M plus a vector from N. For example, consider the subspace $M_1 = \{(z, 0) \mid z \in \mathbb{C}\}$ and $M_2 = \{(0, w) \mid w \in \mathbb{C}\}$. Clearly, any vector $\mathbf{v} = (z, w)$ in $\mathbb{C}^2$ can be written as $\mathbf{v} = (z, 0) + (0, w)$, so $\mathbb{C}^2 = M_1 + M_2$. Moreover, if $\mathbf{v} \in M_1 \cap M_2$, the second coordinate of $\mathbf{v}$ is 0 since $\mathbf{v} \in M_1$, and the first coordinate of $\mathbf{v}$ is 0 since $\mathbf{v} \in M_2$, so $\mathbf{v} = (0, 0) = \overrightarrow{0}$. Thus $\mathbb{C}^2 = M_1 \oplus M_2$.

This example is almost too easy. Let's try to be more imaginative. This time, let $M_1 = \{(x, y, z) \mid x + 2y + 3z = 0\}$ and $M_2 = \{(r, s, s) \mid r, s \in \mathbb{C}\}$. These are indeed subspaces of $\mathbb{C}^3$ and $(-5, 1, 1) \in M_1 \cap M_2$. Now any vector $\mathbf{v} = (x, y, z) \in \mathbb{C}^3$ can be written

$$\begin{aligned}(x, y, z) = &\left(x, -\frac{1}{5}x + \frac{3}{5}y - \frac{3}{5}z, -\frac{1}{5}x - \frac{2}{5}y + \frac{2}{5}z\right)\\ &+ \left(0, \frac{1}{5}(x + 2y + 3z), \frac{1}{5}(x + 2y + 3z)\right)\end{aligned}$$

so $\mathbb{C}^3 = M_1 + M_2$, but the sum is not direct.

Can we extend this idea of direct sum to more than two summands? What would we want to mean by $\mathbb{C}^n = M_1 \oplus M_2 \oplus M_3$? First, we would surely want any vector $\mathbf{v} \in \mathbb{C}^n$ to be expressible as $\mathbf{v} = \mathbf{v}_1 + \mathbf{v}_2 + \mathbf{v}_3$, where $\mathbf{v}_1 \in M_1$, $\mathbf{v}_2 \in M_2$, and $\mathbf{v}_3 \in M_3$. Then we would want this representation to be unique. What would it take to make it unique? Suppose $\mathbf{v} = \mathbf{v}_1 + \mathbf{v}_2 + \mathbf{v}_3 = \mathbf{w}_1 + \mathbf{w}_2 + \mathbf{w}_3$, where $\mathbf{v}_i, \mathbf{w}_i \in M_i$ for $i = 1, 2, 3$. Then $\mathbf{v}_1 - \mathbf{w}_1 = (\mathbf{w}_2 - \mathbf{v}_2) + (\mathbf{w}_3 - \mathbf{v}_3) \in M_2 + M_3$. Thus $\mathbf{v}_1 - \mathbf{w}_1 \in M_1 \cap (M_2 + M_3)$. To get $\mathbf{v}_1 - \mathbf{w}_1 = \overrightarrow{0}$, we need $M_1 \cap (M_2 + M_3) = (\overrightarrow{0})$. Similarly, we would need $M_2 \cap (M_1 + M_3) = (\overrightarrow{0})$ to get $\mathbf{v}_2 = \mathbf{w}_2$ and $M_3 \cap (M_1 + M_2) = (\overrightarrow{0})$ to get $\mathbf{v}_3 = \mathbf{w}_3$. Now that we have the idea, let's go for the most general case.

Suppose $M_1, M_2, \ldots, M_k$ are subspaces of $\mathbb{C}^n$. Then the sum of these subspaces, written $M_1 + M_2 + \cdots + M_k = \sum\limits_{i=1}^{k} M_i$, is defined to be the collection of all vectors of the form $\mathbf{v} = \mathbf{v}_1 + \mathbf{v}_2 + \cdots + \mathbf{v}_k$, where $\mathbf{v}_i \in M_i$ for $i = 1, 2, \ldots, k$.

THEOREM 3.9

Suppose $M_1, M_2, \ldots, M_k$ are subspaces of $\mathbb{C}^n$. Then $\sum\limits_{i=1}^{k} M_i$ is a subspace of $\mathbb{C}^n$. Moreover, it is the smallest subspace of $\mathbb{C}^n$ containing all the M_i, $i = 1, 2, \ldots, k$. Indeed, $\sum\limits_{i=1}^{k} M_i = span(\bigcup\limits_{i=1}^{k} M_i)$.

PROOF The proof is left as an exercise. □

Now for the idea of a direct sum.

THEOREM 3.10
Suppose $M_1, M_2, \ldots, M_k$ are subspaces of $\mathbb{C}^n$. Then T.A.E.:

1. *Every vector* $\mathbf{v}$ *in* $\sum_{i=1}^{k} M_i$ *can be written* uniquely $\mathbf{v} = \mathbf{v}_1 + \mathbf{v}_2 + \cdots + \mathbf{v}_k$ *where* $\mathbf{v}_i \in M_i$ *for* $i = 1, 2, \ldots, k$.
2. *If* $\sum_{i=1}^{k} \mathbf{v}_i = \overrightarrow{0}$ *with* $\mathbf{v}_i \in M_i$ *for* $i = 1, 2, \ldots, k$, *then each* $\mathbf{v}_i = \overrightarrow{0}$ *for* $i = 1, 2, \ldots, k$.
3. *For every* $i = 1, 2, \ldots, k$, $M_i \cap (\sum_{j \neq i} M_j) = (\overrightarrow{0})$.

PROOF The proof is left as an exercise. □

We write $\bigoplus_{i=1}^{k} M_i$ for $\sum_{i=1}^{k} M_i$, calling the sum a *direct sum* when any one (and hence all) of the conditions of the above theorem are satisfied. Note that condition (3) above is often expressed by saying that the subspaces $M_1, M_2, \ldots, M_k$ are *independent.*

COROLLARY 3.11
Suppose $M_1, M_2, \ldots, M_k$ are subspaces of $\mathbb{C}^n$. Then $\mathbb{C}^n = \bigoplus_{i=1}^{k} M_i$ iff $\mathbb{C}^n = \sum_{i=1}^{k} M_i$ and for each $i = 1, 2, \ldots, k$, $M_i \cap (\sum_{j \neq i} M_j) = (\overrightarrow{0})$.

Now suppose E is an n-by-n matrix in $\mathbb{C}^n$ (or a linear map on $\mathbb{C}^n$) with the property that $E^2 = E$. Such a matrix is called *idempotent.* Clearly, I_n and $\mathbb{O}_n$ are idempotent, but there are always many more. We claim that each idempotent matrix induces a direct sum decomposition of $\mathbb{C}^n$. First we note that, if E is idempotent, so is $I - E$. Next, we recall that $Fix(E) = \{\mathbf{v} \in \mathbb{C}^n \mid \mathbf{v} = E\mathbf{v}\}$. For an idempotent E, we have $Fix(E) = \mathcal{C}ol(E)$. Also $\mathcal{N}ull(E) = Fix(I - E)$ and $\mathcal{N}ull(I - E) = Fix(E)$. Now the big claim here is that $\mathbb{C}^n$ is the direct sum of $\mathcal{N}ull(E)$ and $\mathcal{C}ol(E)$. Indeed, it is trivial that any $\mathbf{x}$ in $\mathbb{C}^n$ can be written $\mathbf{x} = E\mathbf{x} + \mathbf{x} - E\mathbf{x} = E\mathbf{x} + (I - E)\mathbf{x}$. Evidently, $E\mathbf{x}$ is in $\mathcal{C}ol(E)$ and $E((I - E)\mathbf{x}) = E\mathbf{x} - EE\mathbf{x} = \overrightarrow{0}$, so $(I - E)\mathbf{x}$ is in $\mathcal{N}ull(E)$.

We claim these two subspaces are disjoint, for if $\mathbf{z} \in \mathcal{C}ol(E) \cap \mathcal{N}ull(E)$, then $\mathbf{z} = E\mathbf{x}$ for some $\mathbf{x}$ and $E\mathbf{z} = \overrightarrow{0}$. But then, $\overrightarrow{0} = E\mathbf{z} = EE\mathbf{x} = E\mathbf{x} = \mathbf{z}$. This establishes the direct sum decomposition. Moreover, if $\mathcal{B} = \{\mathbf{b}_1, \mathbf{b}_2, \ldots, \mathbf{b}_k\}$ is a basis for $\mathcal{C}ol(E) = Fix(E)$ and $\mathcal{C} = \{\mathbf{c}_1, \mathbf{c}_2, \ldots, \mathbf{c}_{n-k}\}$ is a basis of $\mathcal{N}ull(E)$, then $\mathcal{B} \cup \mathcal{C}$ is a basis for $\mathbb{C}^n$, so the matrix $S = [\mathbf{b}_1 \mid \cdots \mid \mathbf{b}_k \mid \mathbf{c}_1 \mid \cdots \mid \mathbf{c}_{n-k}]$ is invertible. Thus, $ES = [E\mathbf{b}_1 \mid \cdots \mid E\mathbf{b}_k \mid E\mathbf{c}_1 \mid \cdots \mid E\mathbf{c}_{n-k}] = [\mathbf{b}_1 \mid \cdots \mid \mathbf{b}_k \mid \overrightarrow{0} \mid \cdots \mid \overrightarrow{0}] = [B \mid \mathbb{O}]$, and so $E = [B \mid \mathbb{O}]S^{-1}$, where $B = [\mathbf{b}_1 \mid \cdots \mid \mathbf{b}_k]$. Moreover, if F is another idempotent and $\mathcal{N}ull(E) = \mathcal{N}ull(F)$ and $\mathcal{C}ol(E) = \mathcal{C}ol(F)$, then $FS = [B \mid \mathbb{O}]$, so $FS = ES$, whence $F = E$. Thus, there is only one idempotent that can give this particular direct sum decomposition.

Now the question is, suppose someone gives you a direct sum decomposition $\mathbb{C}^n = M \oplus N$. Is there a (necessarily unique) idempotent out there with $\mathcal{C}ol(E) = M$ and $\mathcal{N}ull(E) = N$? Yes, there is! As above, select a basis $\{\mathbf{b}_1, \mathbf{b}_2, \ldots, \mathbf{b}_k\}$ of M and a basis $\{\mathbf{c}_1, \mathbf{c}_2, \ldots, \mathbf{c}_{n-k}\}$ of N, and form the matrix $S = [B \mid C]$. Note $\{\mathbf{b}_1, \mathbf{b}_2, \ldots, \mathbf{b}_k, \mathbf{c}_1, \mathbf{c}_2, \ldots, \mathbf{c}_{n-k}\}$ is a basis of $\mathbb{C}^n$, $B \in \mathbb{C}^{n\times k}$ and $C \in \mathbb{C}^{n\times(n-k)}$. Then define $E = [B \mid \mathbb{O}_{n\times(n-k)}]S^{-1} = [B \mid C]\begin{bmatrix} I_k & \mathbb{O}_{n\times(n-k)} \\ \mathbb{O}_{(n-k)\times k} & \mathbb{O}_{(n-k)\times(n-k)} \end{bmatrix} S^{-1} = SJS^{-1}$. Then $E^2 = SJS^{-1}SJS^{-1} = SJJS^{-1} = SJS^{-1} = E$, so E is idempotent. Also we claim $E\mathbf{v} = \mathbf{v}$ iff $\mathbf{v} \in M$, and $E\mathbf{v} = \overrightarrow{0}$ iff $\mathbf{v} \in N$. First note $\mathbf{m} \in M$ iff $\mathbf{m} = B\mathbf{x}$ for some $\mathbf{x}$ in $\mathbb{C}^k$. Then $E\mathbf{m} = [B \mid C]J[B \mid C]^{-1}B\mathbf{x} = [B \mid C]J[B \mid C]^{-1}[B \mid C]\begin{bmatrix} \mathbf{x} \\ \overrightarrow{0} \end{bmatrix} = [B \mid C]J\begin{bmatrix} \mathbf{x} \\ \overrightarrow{0} \end{bmatrix} = [B \mid C]\begin{bmatrix} \mathbf{x} \\ \overrightarrow{0} \end{bmatrix} = B\mathbf{x} = \mathbf{m}$. Thus $M \subseteq Fix(E)$. Next if $\mathbf{n} \in N$, then $\mathbf{n} = C\mathbf{y}$ for some $\mathbf{y}$ in $\mathbb{C}^{n-k}$. Then $E\mathbf{n} = [B \mid C]J[B \mid C]^{-1}[B \mid C]\begin{bmatrix} \overrightarrow{0} \\ \mathbf{y} \end{bmatrix} = [B \mid C]J\begin{bmatrix} \overrightarrow{0} \\ \mathbf{y} \end{bmatrix} = [B \mid C]\begin{bmatrix} \overrightarrow{0} \\ \overrightarrow{0} \end{bmatrix} = \overrightarrow{0}$. Since E is idempotent, $\mathbb{C}^n = Fix(E) \oplus \mathcal{N}ull(E) = \mathcal{C}ol(E) \oplus \mathcal{N}ull(E)$. Now $M \subseteq Fix(E)$ and $k = dim(M) = rank(E) = dim(Fix(E))$ so $M = Fix(E)$. A dimension argument shows $N = \mathcal{N}ull(E)$ and we are done.

There is some good geometry here. The idempotent E is called *the projector of* $\mathbb{C}^n$ *onto M along N*. What is going on is *parallel projection*. To see this, let's look at an example. Let $E = \begin{bmatrix} \frac{1}{2} & \frac{1}{2} \\ \frac{1}{2} & \frac{1}{2} \end{bmatrix}$. Then $E = E^2$, so E must induce a direct sum decomposition of $\mathbb{C}^2$. We check that $\mathbf{v} = \begin{bmatrix} x \\ y \end{bmatrix}$ is fixed by E iff $x = y$ and $E\mathbf{v} = \overrightarrow{0}$ iff $y = -x$. Thus, $\mathbb{C}^2 = M \oplus N$, where $M = \{\begin{bmatrix} x \\ y \end{bmatrix} \mid x = y\}$ and $N = \{\begin{bmatrix} x \\ y \end{bmatrix} \mid y = -x\}$. The unique representation of an arbitrary vector is $\begin{bmatrix} x \\ y \end{bmatrix} = \begin{bmatrix} \frac{1}{2}x + \frac{1}{2}y \\ \frac{1}{2}x + \frac{1}{2}y \end{bmatrix} +$

$\begin{bmatrix} \frac{1}{2}x - \frac{1}{2}y \\ \frac{-1}{2}x + \frac{1}{2}y \end{bmatrix}$. We now draw a picture in $\mathbb{R}^2$ to show the geometry lurking in the background here.

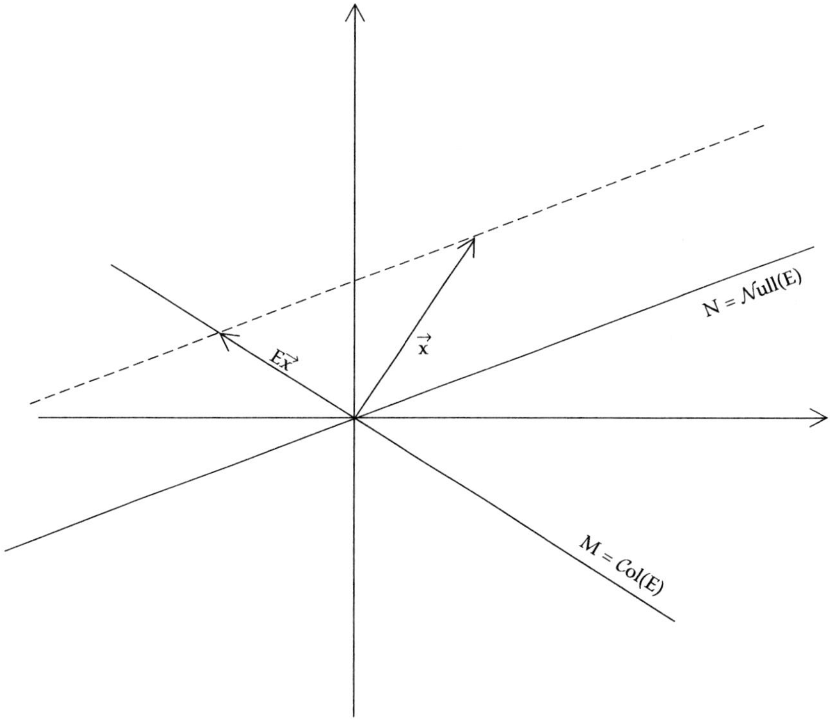

Figure 3.2: Parallel projection.

Of course, these ideas extend. Suppose $\mathbb{C}^n = M_1 \oplus M_2 \oplus M_3$. Select bases $\mathcal{B}_i = \{\mathbf{b}_{i1}, \mathbf{b}_{i2}, \ldots, \mathbf{b}_{ik_i}\}$ of M_i for $i = 1, 2, 3$. Then $\mathcal{B}_1 \cup \mathcal{B}_2 \cup \mathcal{B}_3$ is a basis for $\mathbb{C}^n$. Form the invertible matrix $S = [\mathbf{b}_{11}\mathbf{b}_{12}\cdots\mathbf{b}_{1k_1} \mid \mathbf{b}_{21}\mathbf{b}_{22}\cdots\mathbf{b}_{2k_2} \mid \mathbf{b}_{31}\mathbf{b}_{32}\cdots\mathbf{b}_{3k_3}]$. Form $E_1 = [\mathbf{b}_{11} \mid \mathbf{b}_{12} \mid \cdots \mid \mathbf{b}_{1k_1} \mid \mathbb{O}]S^{-1} = S\begin{bmatrix} I_{k_1} & \mathbb{O} \\ \mathbb{O} & \mathbb{O} \end{bmatrix}S^{-1}$,

$E_2 = [\mathbb{O} \mid \mathbf{b}_{21} \mid \mathbf{b}_{22} \mid \cdots \mid \mathbf{b}_{2k_2} \mid \mathbb{O}]S^{-1} = S\begin{bmatrix} \mathbb{O} & \mathbb{O} & \mathbb{O} \\ \mathbb{O} & I_{k_2} & \mathbb{O} \\ \mathbb{O} & \mathbb{O} & \mathbb{O} \end{bmatrix}S^{-1}$, and

$E_3 = [\mathbb{O} \mid \mathbf{b}_{31} \mid \mathbf{b}_{32} \mid \cdots \mid \mathbf{b}_{3k_3}] = S\begin{bmatrix} \mathbb{O} & \mathbb{O} \\ \mathbb{O} & I_{k_3} \end{bmatrix}S^{-1}$. Then $E_1 + E_2 +$

$E_3 = S\begin{bmatrix} I_{k_1} & \mathbb{O} \\ \mathbb{O} & \mathbb{O} \end{bmatrix}S^{-1} + S\begin{bmatrix} \mathbb{O} & \mathbb{O} & \mathbb{O} \\ \mathbb{O} & I_{k_2} & \mathbb{O} \\ \mathbb{O} & \mathbb{O} & \mathbb{O} \end{bmatrix}S^{-1} + S\begin{bmatrix} \mathbb{O} & \mathbb{O} \\ \mathbb{O} & I_{k_3} \end{bmatrix}S^{-1} =$

$SI_nS^{-1} = I_n$. Also, $E_1E_2 = S\begin{bmatrix} I_{k_1} & \mathbb{O} \\ \mathbb{O} & \mathbb{O} \end{bmatrix} S^{-1}S\begin{bmatrix} \mathbb{O} & \mathbb{O} & \mathbb{O} \\ \mathbb{O} & I_{k_2} & \mathbb{O} \\ \mathbb{O} & \mathbb{O} & \mathbb{O} \end{bmatrix} S^{-1} =$ $S\mathbb{O}S^{-1} = \mathbb{O} = E_2E_1$. Similarly, $E_1E_3 = \mathbb{O} = E_3E_1$ and $E_3E_2 = \mathbb{O} = E_2E_3$. Next, we see $E_1 = E_1I_n = E_1(E_1 + E_2 + E_3) = E_1E_1 + \mathbb{O} + \mathbb{O} = E_1^2$. Thus, E_1 is idempotent. A similar argument shows E_2 and E_3 are also idempotent. The language here is that idempotents E and F are called *orthogonal* iff $EF = \mathbb{O} = FE$. So the bottom line is that this direct sum decomposition has led to a set of pairwise orthogonal idempotents whose sum is the identity. Moreover, note that $Col(E_1) = Col([\mathbf{b}_{11} \mid \mathbf{b}_{12} \mid \cdots \mid \mathbf{b}_{1k_1} \mid \mathbb{O}]S^{-1}) =$ $Col([\mathbf{b}_{11} \mid \mathbf{b}_{12} \mid \cdots \mid \mathbf{b}_{1k_1} \mid \mathbb{O}]) = span(\{\mathbf{b}_{11}, \mathbf{b}_{12}, \cdots, \mathbf{b}_{1k_1}\}) = M_1$. Similarly, one shows $Col(E_2) = M_2$ and $Col(E_3) = M_3$. We finish by summarizing and extending this discussion with a theorem.

THEOREM 3.11

Suppose $M_1, M_2, \ldots, M_k$ are subspaces of $\mathbb{C}^n$ with $\mathbb{C}^n = M_1 \oplus M_2 \oplus \cdots \oplus M_k$. Then there exists a set of idempotents $\{E_1, E_2, \ldots, E_k\}$ such that

1. $E_iE_j = \mathbb{O}$ *if* $i \neq j$.
2. $E_1 + E_2 + \cdots + E_k = I$.
3. $Col(E_i) = M_i$ *for* $i = 1, 2, \ldots, k$.

Conversely, given idempotents $\{E_1, E_2, \ldots, E_k\}$ satisfying (1), (2), and (3) above, then $\mathbb{C}^n = M_1 \oplus M_2 \oplus \cdots \oplus M_k$.

PROOF In view of the discussion above, the reader should be able to fill in the details. ▯

To understand the action of a matrix $A \in \mathbb{C}^{n\times n}$, we often associate a direct sum decomposition of $\mathbb{C}^n$ consisting of subspaces of $\mathbb{C}^n$, which are invariant under A. A subspace M of $\mathbb{C}^n$ is *invariant under* A (i.e., A-*invariant*) iff $A(M) \subseteq M$. In other words, the vectors in M stay in M even after they are multiplied by A. What makes this useful is that if we take a basis of M, say $\mathbf{m}_1, \mathbf{m}_2, \ldots, \mathbf{m}_k$, and extend it to a basis of $\mathbb{C}^n$, $\mathcal{B} = \{\mathbf{m}_1, \mathbf{m}_2, \ldots, \mathbf{m}_k, \mathbf{b}_{k+1}, \ldots, \mathbf{b}_n\}$, we can form the invertible matrix S, whose columns are these basis vectors. Then the matrix $S^{-1}AS$ has a particularly nice form; it has a block of zeros in it. Let us illustrate. Suppose $M = span\{\mathbf{m}_1, \mathbf{m}_2, \mathbf{m}_3\}$, where $\mathbf{m}_1$, $\mathbf{m}_2$, and $\mathbf{m}_3$ are independent in $\mathbb{C}^5$. Now $A\mathbf{m}_1$ is a vector in M since M is A-invariant. Thus, $A\mathbf{m}_1$ is uniquely expressible as a linear combination of $\mathbf{m}_1$, $\mathbf{m}_2$, and $\mathbf{m}_3$. Say

$$A\mathbf{m}_1 = a_{11}\mathbf{m}_1 + a_{21}\mathbf{m}_2 + a_{31}\mathbf{m}_3.$$

Similarly,

$$\begin{aligned}
A\mathbf{m}_2 &= a_{12}\mathbf{m}_1 + a_{22}\mathbf{m}_2 + a_{32}\mathbf{m}_3 \\
A\mathbf{m}_3 &= a_{13}\mathbf{m}_1 + a_{23}\mathbf{m}_2 + a_{33}\mathbf{m}_3 \\
A\mathbf{b}_4 &= a_{14}\mathbf{m}_1 + a_{24}\mathbf{m}_2 + a_{34}\mathbf{m}_3 + \beta_{44}\mathbf{b}_4 + \beta_{54}\mathbf{b}_5 \\
A\mathbf{b}_5 &= a_{15}\mathbf{m}_1 + a_{25}\mathbf{m}_2 + a_{35}\mathbf{m}_3 + \beta_{45}\mathbf{b}_4 + \beta_{55}\mathbf{b}_5.
\end{aligned}$$

Since $A\mathbf{b}_4$ and $A\mathbf{b}_5$ could be anywhere in $\mathbb{C}^5$, we may need the entire basis to express them. Now form $AS = [A\mathbf{m}_1 \mid A\mathbf{m}_2 \mid A\mathbf{m}_3 \mid A\mathbf{b}_4 \mid A\mathbf{b}_5] =$

$$[a_{11}\mathbf{m}_1 + a_{12}\mathbf{m}_2 + a_{13}\mathbf{m}_3 \mid a_{21}\mathbf{m}_1 + a_{22}\mathbf{m}_2 + a_{23}\mathbf{m}_3 \mid a_{31}\mathbf{m}_1 + a_{32}\mathbf{m}_2 + a_{33}\mathbf{m}_3$$

$$\mid A\mathbf{b}_4 \mid A\mathbf{b}_5] = [\mathbf{m}_1 \mid \mathbf{m}_2 \mid \mathbf{m}_3 \mid \mathbf{b}_4 \mid \mathbf{b}_5] \begin{bmatrix} a_{11} & a_{12} & a_{13} & a_{14} & a_{15} \\ a_{21} & a_{22} & a_{23} & a_{24} & a_{25} \\ a_{31} & a_{32} & a_{33} & a_{34} & a_{35} \\ 0 & 0 & 0 & \beta_{44} & \beta_{45} \\ 0 & 0 & 0 & \beta_{54} & \beta_{55} \end{bmatrix} =$$

$$S \begin{bmatrix} a_{11} & a_{12} & a_{13} & a_{14} & a_{15} \\ a_{21} & a_{22} & a_{23} & a_{24} & a_{25} \\ a_{31} & a_{32} & a_{33} & a_{34} & a_{35} \\ 0 & 0 & 0 & \beta_{44} & \beta_{45} \\ 0 & 0 & 0 & \beta_{54} & \beta_{55} \end{bmatrix} = S \begin{bmatrix} A_1 & A_2 \\ \mathbb{O} & A_3 \end{bmatrix}.$$

It gets even better if $\mathbb{C}^n$ is the direct sum of two subspaces, both of which are A-invariant. Say $\mathbb{C}^n = M \oplus N$, where M and N are A-invariant. Take a basis of M, say $\mathcal{B} = \{\mathbf{b}_1, \ldots, \mathbf{b}_k\}$ and a basis of N, say $\mathcal{C} = \{\mathbf{c}_{k+1}, \ldots, \mathbf{c}_n\}$. Then $\mathcal{B} \cup \mathcal{C}$ is a basis of $\mathbb{C}^n$ and $S = [\mathbf{b}_1 \mid \ldots \mid \mathbf{b}_k \mid \mathbf{c}_{k+1} \mid \ldots \mid \mathbf{c}_n]$ is an invertible matrix with $S^{-1}AS = \begin{bmatrix} A_1 & \mathbb{O} \\ \mathbb{O} & A_2 \end{bmatrix}$. Let's illustrate this important idea in $\mathbb{C}^5$. Suppose $\mathbb{C}^5 = M_1 \oplus M_2$, where $A(M_1) \subseteq M_1$ and $A(M_2) \subseteq M_2$. In order to prepare for the proof of the next theorem, let's establish some notation. Let $\mathcal{B}_1 = \{\mathbf{b}_1^{(1)}, \mathbf{b}_2^{(1)}, \mathbf{b}_3^{(1)}\}$ be a basis for M_1 and $\mathcal{B}_2 = \{\mathbf{b}_1^{(2)}, \mathbf{b}_2^{(2)}\}$ be a basis for M_2. Form the matrices $B_1 = [\mathbf{b}_1^{(1)} \mid \mathbf{b}_2^{(1)} \mid \mathbf{b}_3^{(1)}] \in \mathbb{C}^{5\times 3}$ and $B_2 = [\mathbf{b}_1^{(2)} \mid \mathbf{b}_2^{(2)}] \in \mathbb{C}^{5\times 2}$. Then $AM_i = ACol(B_i) \subseteq Col(B_i)$ for $i = 1, 2$. Now A acts on a basis vector from $\mathcal{B}_1$ and produces another vector in M_1.

Therefore,

$$A\mathbf{b}_1^{(1)} = \alpha_1\mathbf{b}_1^{(1)} + \beta_1\mathbf{b}_2^{(1)} + \gamma_1\mathbf{b}_3^{(1)} = [\mathbf{b}_1^{(1)} \mid \mathbf{b}_2^{(1)} \mid \mathbf{b}_3^{(1)}]\begin{bmatrix}\alpha_1\\\beta_1\\\gamma_1\end{bmatrix} = B_1\begin{bmatrix}\alpha_1\\\beta_1\\\gamma_1\end{bmatrix}$$
$$A\mathbf{b}_2^{(1)} = \alpha_2\mathbf{b}_1^{(1)} + \beta_2\mathbf{b}_2^{(1)} + \gamma_2\mathbf{b}_3^{(1)} = [\mathbf{b}_1^{(1)} \mid \mathbf{b}_2^{(1)} \mid \mathbf{b}_3^{(1)}]\begin{bmatrix}\alpha_2\\\beta_2\\\gamma_2\end{bmatrix} = B_1\begin{bmatrix}\alpha_2\\\beta_2\\\gamma_2\end{bmatrix}.$$
$$A\mathbf{b}_3^{(1)} = \alpha_3\mathbf{b}_1^{(1)} + \beta_3\mathbf{b}_2^{(1)} + \gamma_3\mathbf{b}_3^{(1)} = [\mathbf{b}_1^{(1)} \mid \mathbf{b}_2^{(1)} \mid \mathbf{b}_3^{(1)}]\begin{bmatrix}\alpha_3\\\beta_3\\\gamma_3\end{bmatrix} = B_1\begin{bmatrix}\alpha_3\\\beta_3\\\gamma_3\end{bmatrix}$$

Form the matrix $A_1 = \begin{bmatrix}\alpha_1 & \alpha_2 & \alpha_3\\\beta_1 & \beta_2 & \beta_3\\\gamma_1 & \gamma_2 & \gamma_3\end{bmatrix}$. We compute that

$$AB_1 = [A\mathbf{b}_1^{(1)} \mid A\mathbf{b}_2^{(1)} \mid A\mathbf{b}_3^{(1)}] = \left[B_1\begin{bmatrix}\alpha_1\\\beta_1\\\gamma_1\end{bmatrix} \mid B_1\begin{bmatrix}\alpha_2\\\beta_2\\\gamma_2\end{bmatrix} \mid B_1\begin{bmatrix}\alpha_3\\\beta_3\\\gamma_3\end{bmatrix}\right] = B_1A_1.$$

A similar analysis applies to A acting on the basis vectors in $\mathcal{B}_2$;

$$A\mathbf{b}_1^{(2)} = \lambda_1\mathbf{b}_1^{(2)} + \mu_1\mathbf{b}_2^{(2)} = [\mathbf{b}_1^{(2)} \mid \mathbf{b}_2^{(2)}]\begin{bmatrix}\lambda_1\\\mu_1\end{bmatrix} = B_2\begin{bmatrix}\lambda_1\\\mu_1\end{bmatrix}$$
$$A\mathbf{b}_2^{(2)} = \lambda_2\mathbf{b}_1^{(2)} + \mu_2\mathbf{b}_2^{(2)} = [\mathbf{b}_1^{(2)} \mid \mathbf{b}_2^{(2)}]\begin{bmatrix}\lambda_2\\\mu_2\end{bmatrix} = B_2\begin{bmatrix}\lambda_2\\\mu_2\end{bmatrix}.$$

Form the matrix $A_2 = \begin{bmatrix}\lambda_1 & \lambda_2\\\mu_1 & \mu_2\end{bmatrix}$. Again compute that

$$AB_2 = A[\mathbf{b}_1^{(2)} \mid \mathbf{b}_2^{(2)}] = [A\mathbf{b}_1^{(2)} \mid A\mathbf{b}_2^{(2)}] = \left[B_2\begin{bmatrix}\lambda_1\\\mu_1\end{bmatrix} \mid B_2\begin{bmatrix}\lambda_2\\\mu_2\end{bmatrix}\right] = B_2A_2.$$

Finally, form the invertible matrix $S = [B_1 \mid B_2]$ and compute

$$AS = A[B_1 \mid B_2] = [AB_1 \mid AB_2] = [B_1A_1 \mid B_2A_2] = [B_1 \mid B_2]\begin{bmatrix}A_1 & \mathbb{O}\\\mathbb{O} & A_2\end{bmatrix} = S\begin{bmatrix}A_1 & \mathbb{O}\\\mathbb{O} & A_2\end{bmatrix}.$$

The more invariant direct summands you can find, the more blocks of zeros you can generate. We summarize with a theorem.

THEOREM 3.12

Suppose $A \in \mathbb{C}^{n\times n}$ *and* $\mathbb{C}^n = M_1 \oplus M_2 \oplus \cdots \oplus M_k$, *where each* M_i *is* A*-invariant for* $i = 1, 2, \ldots, k$. *Then there exists an invertible matrix* $S \in \mathbb{C}^{n\times n}$ *such that*

$$S^{-1}AS = BlockDiagonal[A_1, A_2, \ldots, A_k] = \begin{bmatrix} A_1 & \mathbb{O} & \mathbb{O} & \cdots & \mathbb{O} \\ \mathbb{O} & A_2 & \mathbb{O} & \cdots & \vdots \\ \vdots & \vdots & \ddots & & \mathbb{O} \\ \mathbb{O} & \mathbb{O} & \cdots & \cdots & A_k \end{bmatrix}.$$

PROOF In view of the discussion above, the details of the proof are safely left to the reader. We simple sketch it out. Suppose that M_i are invariant subspaces for A for $i = 1, 2, \ldots, k$. Form the matrices $B_i \in \mathbb{C}^{n\times d_i}$, whose columns come from a basis of M_i. Then $M_i = Col(B_i)$. Note $A\mathbf{b}_j^{(i)} = B_i\mathbf{y}_j^{(i)}$, where $\mathbf{y}_j^{(i)} \in \mathbb{C}^{d_i}$. Form the matrices $A_i = [\mathbf{y}_1^{(i)} \mid \mathbf{y}_2^{(i)} \mid \cdots \mid \mathbf{y}_{d_i}^{(i)}] \in \mathbb{C}^{d_i\times d_i}$. Compute that $AB_i = B_iA_i$, noting $n = d_1 + d_2 + \cdots d_k$. Form the invertible matrix $S = [B_1 \mid B_2 \mid \cdots \mid B_k]$. Then $S^{-1}AS = S^{-1}[AB_1 \mid AB_2 \mid \cdots \mid AB_k] = S^{-1}[B_1A_1 \mid B_2A_2 \mid \cdots \mid B_kA_k] = S^{-1}[B_1 \mid B_2 \mid \cdots \mid B_k]$

$$\begin{bmatrix} A_1 & \mathbb{O} & \mathbb{O} & \cdots & \mathbb{O} \\ \mathbb{O} & A_2 & \mathbb{O} & \cdots & \vdots \\ \vdots & \vdots & \ddots & & \mathbb{O} \\ \mathbb{O} & \mathbb{O} & \cdots & \cdots & A_k \end{bmatrix} = \begin{bmatrix} A_1 & \mathbb{O} & \mathbb{O} & \cdots & \mathbb{O} \\ \mathbb{O} & A_2 & \mathbb{O} & \cdots & \vdots \\ \vdots & \vdots & \ddots & & \mathbb{O} \\ \mathbb{O} & \mathbb{O} & \cdots & \cdots & A_k \end{bmatrix}.$$

□

Exercise Set 10

1. If $M \oplus N = \mathbb{C}^n$, argue that $dim(M) + dim(N) = n$.

2. If $M \oplus N = \mathbb{C}^n$, $\mathcal{B}$ is a basis for M, and $\mathcal{C}$ is a basis for N, then prove $\mathcal{B} \cup \mathcal{C}$ is a basis for $\mathbb{C}^n$.

3. Argue the uniqueness claim when $\mathbb{C}^n$ is the direct sum of M and N.

4. Prove that $M + N$ is the smallest subspace of $\mathbb{C}^n$ that contains M and N.

5. Argue that if E is idempotent, so is $I - E$. Moreover, $Col(I - E) = \mathcal{N}ull(E)$, and $\mathcal{N}ull(I - E) = Col(E)$.

6. Prove that if E is idempotent, the $Fix(E) = Col(E)$.

7. Verify that the constructed E in the text has the properties claimed for it.

8. Suppose E is idempotent and invertible. What can you say about E?

9. Prove that $E = E^2$ iff $\mathcal{C}ol(E) \cap \mathcal{C}ol(I - E) = (\overrightarrow{0})$.

10. Consider $\begin{bmatrix} 0 & 0 \\ 3 & 1 \end{bmatrix}, \begin{bmatrix} 1 & 0 \\ 1 & 0 \end{bmatrix}$, and $\begin{bmatrix} \frac{1}{5} & \frac{2}{5} \\ \frac{2}{5} & \frac{4}{5} \end{bmatrix}$. Show that these matrices are all idempotent. What direct sum decompositions do they induce on $\mathbb{C}^2$?

11. Show that if E is idempotent, then so is $Q = E + AE - EAE$ and $Q = E - AE + EAE$ for any matrix A of appropriate size.

12. Let $E = \begin{bmatrix} \frac{1}{2} & \frac{1}{2} \\ \frac{1}{2} & \frac{1}{2} \end{bmatrix} = \frac{1}{2}\begin{bmatrix} 1 & 1 \\ 1 & 1 \end{bmatrix}$. Argue that E is idempotent and symmetric. What is the rank of E? What direct sum decomposition does it induce on a typical vector? Let $E = \begin{bmatrix} \frac{1}{3} & \frac{1}{3} & \frac{1}{3} \\ \frac{1}{3} & \frac{1}{3} & \frac{1}{3} \\ \frac{1}{3} & \frac{1}{3} & \frac{1}{3} \end{bmatrix}$. Discuss the same issues for E. Do you see how to generalize to E n-by-n? Do it.

13. Recall from calculus that a function $f : \mathbb{R} \to \mathbb{R}$ is called *even* iff $f(-x) = f(x)$. The graph of such an f is symmetric with respect to the y-axis. Recall $f(x) = x^2$, $f(x) = cos(x)$, and so on are even functions. Also, $f : \mathbb{R} \to \mathbb{R}$ is called *odd* iff $f(-x) = -f(x)$. The graph of an odd function is symmetric with respect to the origin. Recall $f(x) = x^3$, $f(x) = sin(x)$, and so on are odd functions Let $V = \mathcal{F}(\mathbb{R})$ be the vector space of all functions on $\mathbb{R}$. Argue that the even (resp., odd) functions form a subspace of V. Argue that V is the direct sum of the subspace of even and the subspace of odd functions. (*Hint:* If $f : \mathbb{R} \to \mathbb{R}$, $f^e(x) = \frac{1}{2}[f(x) + f(-x)]$ is even and $f^o(x) = \frac{1}{2}[f(x) - f(-x)]$ is odd.)

14. Let $V = \mathbb{C}^{n \times n}$, the n-by-n matrices considered as a vector space. Recall a matrix A is *symmetric* iff $A = A^T$ and *skew-symmetric* iff $A^T = -A$. Argue that the symmetric (resp., skew-symmetric) matrices form a subspace of V and V is the direct sum of these two subspaces.

15. Suppose $M_1, M_2, \ldots, M_k$ are nontrivial subspaces of $\mathbb{C}^n$ that are independent. Suppose $\mathcal{B}_i$ is a basis of M_i for each $i = 1, 2, \ldots, k$. Argue that $\bigcup_{i=1}^{k} \mathcal{B}_i$ is a basis of $\bigoplus_{i=1}^{k} M_i$.

16. Suppose $M_1, M_2, \ldots, M_k$ are subspaces of $\mathbb{C}^n$ that are independent. Argue that $dim(M_1 \oplus M_2 \oplus \cdots \oplus M_k) = dim(M_1) + dim(M_2) + \cdots + dim(M_k)$.

17. Suppose $E^2 = E$ and $F = S^{-1}ES$ for some invertible matrix S. Argue that $F^2 = F$.

18. Suppose $M_1, M_2, \ldots, M_k$ are subspaces of $\mathbb{C}^n$. Argue that these subspaces are independent iff for each $j, 2 \leq j \leq k$, $M_j \cap (M_1 + \cdots + M_{j-1}) = (\vec{0})$. Is independence equivalent to the condition "pairwise disjoint," that is, $M_i \cap M_j = (\vec{0})$ whenever $i \neq j$?

19. Suppose E is idempotent. Is $I + E$ invertible? How about $I + \alpha E$? Are there conditions on α? Suppose E and F are both idempotent. Find invertible matrices S and T such that $S(E + F)T = \alpha E + \beta F$. Are there conditions on α and β?

20. Suppose A is a matrix and P is the projector on M along N. Argue that $PA = A$ iff $Col(A) \subseteq M$ and $AP = A$ iff $\mathcal{N}ull(A) \supseteq N$.

21. Suppose P is an idempotent and A is a matrix. Argue that $A(Col(P)) \subseteq Col(P)$ iff $PAP = AP$. Also prove that $\mathcal{N}ull(P)$ is A-invariant iff $PAP = PA$. Now argue that $A(Col(P)) \subseteq Col(P)$ and $A(\mathcal{N}ull(P)) \subseteq \mathcal{N}ull(P)$ iff $PA = AP$.

22. Prove the dimension formula: for M_1, M_2 subspaces of $\mathbb{C}^n$, $dim(M_1 + M_2) = dim(M_1) + dim(M_2) - dim(M_1 \cap M_2)$. How would this formula read if you had three subspaces?

23. Suppose M_1 and M_2 are subspaces of $\mathbb{C}^n$ and $dim(M_1 + M_2) = dim(M_1 \cap M_2) + 1$. Argue that either $M_1 \subseteq M_2$ or $M_2 \subseteq M_1$.

24. Suppose M_1, M_2, and M_3 are subspaces of $\mathbb{C}^n$. Argue that $dim(M_1 \cap M_2 \cap M_3) + 2n \geq dim(M_1) + dim(M_2) + dim(M_3)$.

25. Suppose $T{:}\mathbb{C}^n \rightarrow \mathbb{C}^n$ is a linear transformation. Argue that $dim(T(M)) + dim(Ker(T) \cap M) = dim(M)$, where M is any subspace of $\mathbb{C}^n$.

26. Prove Theorem 3.10 and Theorem 3.11.

27. Fill the details of the proofs of Theorem 3.13 and Theorem 3.14.

28. (Yongge Tian) Suppose P and Q are idempotents. Argue that (1) $rank(P + Q) \geq rank(P - Q)$ and (2) $rank(PQ + QP) \geq rank(PQ - QP)$.

29. Argue that every subspace of $\mathbb{C}^n$ has a complementary subspace.

30. Prove that if $M_i \cap \sum_{i \neq j} M_j = (\vec{0})$, then $M_i \cap M_j = (\vec{0})$ for $i \neq j$, where the M_i are subspaces of $\mathbb{C}^n$.

31. Suppose p is a polynomial in $\mathbb{C}[x]$. Prove that $\mathcal{C}ol(p(A))$ and $\mathcal{N}ull(p(A))$ are both A-invariant for any square matrix A.

32. Argue that P is an idempotent iff P^* is an idempotent.

33. Suppose P is an idempotent. Prove that $(I+P)^n = I + (2^n - 1)P$. What can you say about $(I - P)$?

34. Determine all n-by-n diagonal matrices that are idempotent. How many are there?

Further Reading

[A&M, 2005] J. Araújo and J. D. Mitchell, An Elementary Proof that Every Singular Matrix is a Product of Idempotent Matrices, The American Mathematical Monthly, Vol. 112, No. 7, August-September, (2005), 641–645.

[Tian, 2005] Yongge Tian, Rank Equalities Related to Generalized Inverses of Matrices and Their Applications, ArXiv:math.RA/0003224.

index, core-nilpotent factorization, Weyr sequence, Segre sequence, conjugate partition, Ferrer's diagram, standard nilpotent matrix

---

3.4 The Index of a Square Matrix

Suppose we have a matrix A in $\mathbb{C}^{n\times n}$. The rank-plus-nullity theorem tells us that

$$dim(\mathcal{C}ol(A)) + dim(\mathcal{N}ull\,(A)) = n.$$

It would be reasonable to suspect that

$$\mathbb{C}^n = \mathcal{N}ull(A) \oplus \mathcal{C}ol(A).$$

Unfortunately, this is not always the case. Of course, if A is invertible, then this is trivially true (why?). However, if $A = \begin{bmatrix} 0 & 2 \\ 0 & 0 \end{bmatrix}$, then $\begin{bmatrix} 2 \\ 0 \end{bmatrix} \in \mathcal{N}ull(A) \cap \mathcal{C}ol(A)$. We can say something important if we look at the powers of A. First, it is clear that $\mathcal{N}ull(A) \subseteq \mathcal{N}ull(A^2)$, for if $A\mathbf{x} = \vec{0}$, surely $A^2\mathbf{x} = A(A\mathbf{x}) = A\vec{0} = \vec{0}$. More generally, $\mathcal{N}ull(A^p) \subseteq \mathcal{N}ull(A^{p+1})$. Also it is clear that $\mathcal{C}ol(A^2) \subseteq \mathcal{C}ol(A)$ since, if $\mathbf{y} \in \mathcal{C}ol(A^2)$, then $\mathbf{y} = A^2\mathbf{x} = A(A\mathbf{x}) \in \mathcal{C}ol(A)$. These observations set up two chains of subspaces in $\mathbb{C}^n$ for a given matrix A:

$$(\vec{0}) = \mathcal{N}ull(A^0) \subseteq \mathcal{N}ull(A) \subseteq \mathcal{N}ull(A^2) \subseteq \cdots \subseteq \mathcal{N}ull(A^p) \\ \subseteq \mathcal{N}ull(A^{p+1}) \subseteq \cdots$$

and

$$\mathbb{C}^n = \mathcal{C}ol(A^0) \supseteq \mathcal{C}ol(A) \supseteq \mathcal{C}ol(A^2) \supseteq \cdots \supseteq \mathcal{C}ol(A^p) \supseteq \mathcal{C}ol(A^{p+1}) \supseteq \cdots .$$

Since we are in finite dimensions, neither of these chains can continue to strictly ascend or descend forever. That is, equality must eventually occur in these chains of subspaces. Actually, a bit more is true: once equality occurs, it persists.

THEOREM 3.13

Let $A \in \mathbb{C}^{n\times n}$. Then there exists an integer q with $0 < q \leq n$ such that

$$\mathbb{C}^n = \mathcal{N}ull(A^q) \oplus \mathcal{C}ol(A^q).$$

PROOF To ease the burden of writing, let $N^k = \mathcal{N}ull(A^k)$. Then we have $(\vec{0}) = N^0 \subseteq N^1 \subseteq \cdots \subseteq \mathbb{C}^n$. At some point by the dimension argument above, equality must occur. Otherwise $dim(N^0) < dim(N^1) < \ldots$ so $n < dim(N^{n+1})$, which is contradictory. Let q be the least positive integer where equality first happens. That is, $N^{q-1} \subsetneq N^q = N^{q+1}$. We claim there is nothing but equal signs in the chain from this point forward. In other words, we claim $N^q = N^{q+j}$ for all $j = 1, 2, \ldots$. We prove this by induction. Evidently the claim is true for $j = 1$. So, suppose the claim is true for $j = k - 1$. We argue that the claim is still true for $j = k$. That is, we show $N^{q+(k-1)} = N^{q+k}$. Now $A^{q+k}\mathbf{x} = \vec{0}$ implies $A^{q+1}(A^{k-1}\mathbf{x}) = \vec{0}$. Thus, $A^{k-1}\mathbf{x} \in N^{q+1} = N^q$ so $A^q A^{k-1}\mathbf{x} = \vec{0}$, which says $A^{q+(k-1)}\mathbf{x} = \vec{0}$, putting $\mathbf{x}$ in $N^{q+(k-1)}$. Thus, $N^{q+k} \subseteq N^{q+(k-1)} \subseteq N^{q+k}$. Therefore, $N^{q+(k-1)} = N^{q+k}$ and our induction proof is complete. By the rank-plus-nullity theorem, $n = dim(A^k) + dim(\mathcal{C}ol(A^k))$ so

$dim(Col(A^q)) = dim(Col(A^{q+1})) = \dots$. That is, the null space chain stops growing at exactly the same place the column space chain stops shrinking.

Our next claim is that $N^q \cap Col(A^q) = (\vec{0})$. Let $\mathbf{v} \in N^q \cap Col(A^q)$. Then $\mathbf{v} = A^q\mathbf{w}$ for some $\mathbf{w}$. Then $A^{2q}\mathbf{w} = A^q A^q \mathbf{w} = A^q\mathbf{v} = \vec{0}$. Thus $\mathbf{w} \in N^{2q} = N^{q+q} = N^{q+(q-1)} = \cdots = N^q$. Thus, $\vec{0} = A^q\mathbf{w} = \mathbf{v}$. This gives us the direct sum decomposition of $\mathbb{C}^n$ since $dim(N^q + Col(A^q)) = dim(N^q) + dim(Col(A^q)) - dim(N^q \cap Col(A^q)) = dim(N^q) + dim(Col(A^q)) = n$, applying the rank plus nullity theorem to A^q.

Finally, it is evident that $q \leq n$ since, if $q \geq n+1$, we would have a properly increasing chain of subspaces $N^1 \subsetneq N^2 \subsetneq \dots \subsetneq N^{n+1} \subsetneq \dots \subsetneq N^q$, which gives at least $n+1$ dimensions, an impossibility in $\mathbb{C}^n$. Thus, $\mathbb{C}^n = \mathcal{N}ull(A^q) \oplus Col(A^q)$. □

DEFINITION 3.5 *(index)*

For an n-by-n matrix A, the integer q in the above theorem is called the ***index*** *of A, which we will denote $index(A)$ or Ind(A).*

We now have many useful ways to describe the index of a matrix.

COROLLARY 3.12

Let $A \in \mathbb{C}^{n\times n}$. The index q of A is the least nonnegative integer k such that

1. $\mathcal{N}ull(A^k) = \mathcal{N}ull(A^{k+1})$
2. $Col(A^k) = Col(A^{k+1})$
3. $rank(A^k) = rank(A^{k+1})$
4. $Col(A^k) \cap \mathcal{N}ull(A^k) = (\vec{0})$
5. $\mathbb{C}^n = \mathcal{N}ull(A^k) \oplus Col(A^k)$.

Note that even if you do not know the index of a matrix A, you can always take $\mathbb{C}^n = \mathcal{N}ull(A^n) \oplus Col(A^n)$. Recall that a *nilpotent matrix* N is one where $N^k = \mathbb{O}$ for some positive integer k.

COROLLARY 3.13

The index q of a nilpotent matrix N is the least positive integer k with $N^k = \mathbb{O}$.

PROOF Suppose p is the least positive integer with $N^p = \mathbb{O}$. Then $N^{p-1} \neq \mathbb{O}$ so

$$\mathcal{N}ull(N) \subsetneq \mathcal{N}ull(N^2) \subsetneq \cdots \subsetneq \mathcal{N}ull(N^{p-1}) \subsetneq \mathcal{N}ull(N^p) = \mathcal{N}ull(N^{p+1}) = \cdots = \mathbb{C}^n.$$

Thus, $p = q$. □

Actually, the direct sum decomposition of $\mathbb{C}^n$ determined by the index of a matrix is quite nice since the subspaces are invariant under A; recall W is invariant under A if $\mathbf{v} \in W$ implies $A\mathbf{v} \in W$.

COROLLARY 3.14
Let $A \in \mathbb{C}^{n\times n}$ have index q. (Actually, q could be any nonnegative integer.) Then

1. $A(\mathcal{N}ull(A^q)) \subseteq \mathcal{N}ull(A^q)$
2. $A(\mathcal{C}ol(A^q)) \subseteq \mathcal{C}ol(A^q)$.

The index of a matrix leads to a particularly nice factorization of the matrix. It is called the *core-nilpotent factorization* in Meyer, [2000]. You might call the next theorem a "poor man's Jordan canonical form." You will see why later.

THEOREM 3.14 (core-nilpotent factorization)
Suppose A is in $\mathbb{C}^{n\times n}$, with index q. Suppose that $rank(A^q) = r$. Then there exists an invertible matrix S such that $A = S\begin{bmatrix} C_{r\times r} & \mathbb{O} \\ \mathbb{O} & N \end{bmatrix} S^{-1}$ where C is invertible and N is nilpotent of index q. Moreover, $\mu_A(x) = x^q\mu_C(x)$.

PROOF Recall from the definition of index that $\mathbb{C}^n = \mathcal{C}ol(A^q) \oplus \mathcal{N}ull(A^q)$. Thus, we can select a basis of $\mathbb{C}^n$ consisting of r independent vectors from $\mathcal{C}ol(A^q)$ and $n - r$ independent vectors from $\mathcal{N}ull(A^q)$. Say $\mathbf{x}_1, \mathbf{x}_2, \ldots, \mathbf{x}_r$ is a basis for $\mathcal{C}ol(A^q)$ and $\mathbf{y}_1, \mathbf{y}_2, \ldots, \mathbf{y}_{n-r}$ is a basis of $\mathcal{N}ull(A^q)$. Form a matrix S by using these vectors as columns, $S = [\mathbf{x}_1 \mid \mathbf{x}_2 \mid \cdots \mid \mathbf{x}_r \mid \mathbf{y}_1 \mid \cdots \mid \mathbf{y}_{n-r}] = [S_1 \mid S_2]$. Evidently, S is invertible. Now recall that $\mathcal{C}ol(A^q)$ and $\mathcal{N}ull(A^q)$ are invariant subspaces for A, and therefore, $S^{-1}AS = \begin{bmatrix} C & \mathbb{O} \\ \mathbb{O} & N \end{bmatrix}$. Raising both sides to the q power yields with appropriate partitioning $(S^{-1}AS)^q = S^{-1}A^qS = \begin{bmatrix} C^q & \mathbb{O} \\ \mathbb{O} & N^q \end{bmatrix} = \begin{bmatrix} T_1 \\ \cdots \\ T_2 \end{bmatrix} A^q \begin{bmatrix} S_1 \vdots S_2 \end{bmatrix}$ $= \begin{bmatrix} T_1 \\ \cdots \\ T_2 \end{bmatrix} \begin{bmatrix} A^qS_1 \vdots A^qS_2 \end{bmatrix} = \begin{bmatrix} T_1 \\ \cdots \\ T_2 \end{bmatrix} \begin{bmatrix} A^qS_1 \vdots \mathbb{O} \end{bmatrix} = \begin{bmatrix} T_1A^qS_1 & \mathbb{O} \\ T_2A^qS_1 & \mathbb{O} \end{bmatrix}$. Comparing the lower right corners, we see $N^q = \mathbb{O}$, so we have established that N is nilpotent.

Next we note that C^q is r-by-r and $rank(A^q) = r = rank(S^{-1}A^qS) = rank \begin{bmatrix} C^q & \mathbb{O} \\ \mathbb{O} & \mathbb{O} \end{bmatrix} = rank[C^q]$. Thus C^q has full rank and hence C is invertible. Finally, we need to determine the index of nilpotency of N. But if $index(N) < q$, then N^{q-1} would already be $\mathbb{O}$. This would imply $rank(A^{q-1})$ $= rank(S^{-1}A^{q-1}S) = rank(\begin{bmatrix} C^{q-1} & \mathbb{O} \\ \mathbb{O} & N^{q-1} \end{bmatrix}) = rank(\begin{bmatrix} C^{q-1} & \mathbb{O} \\ \mathbb{O} & \mathbb{O} \end{bmatrix}) =$ $rank\left(C^{q-1}\right) = r = rank(A^q)$, contradicting the definition of q. The rest is left to the reader. □

Now we wish to take a deeper look at this concept of index. We are about to deal with the most challenging ideas yet presented. What we prove is the crucial part of the Jordan canonical form theorem, which we will meet later. We determine the structure of nilpotent matrices. Suppose A in $\mathbb{C}^{n\times n}$ has index q. Then

$$\begin{aligned}(\vec{0}) &= \mathcal{N}ull(A^0) \subsetneq \mathcal{N}ull(A) \subsetneq \mathcal{N}ull(A^2) \subsetneq \cdots \subsetneq \mathcal{N}ull(A^q) \\ &= \mathcal{N}ull(A^{q+1}) = \cdots \subseteq \mathbb{C}^n.\end{aligned}$$

We associate a sequence of numbers with these null spaces.

DEFINITION 3.6 *(Weyr Sequence)*

Suppose A in $\mathbb{C}^{n\times n}$ has index q. The ***Weyr sequence*** *of A, $Weyr(A) = (\omega_1, \omega_2, \ldots, \omega_q)$, where $\omega_1 = nlty(A)$, $\omega_2 = nlty(A^2) - nlty(A)$, ... Generally, $\omega_i = nlty(A^i) - nlty(A^{i-1})$ for $i = 1, 2, \ldots, q$. Note $\omega_k = 0$ for $k \geq q$.*

In view of the rank-plus-nullity theorem, we can also express ω_i as

$$\omega_i = rank(A^{i-1}) - rank(A^i).$$

Note that $\omega_1 + \omega_2 + \cdots + \omega_i = dim(\mathcal{N}ull(A^i)) = n - dim(\mathcal{C}ol(A^i))$, $n \geq \omega_1 + \omega_2 + \cdots + \omega_q$ and $rank(A) \geq \omega_2 + \omega_3 + \cdots + \omega_q$. We may visualize what is going on here pictorially in terms of chains of subspaces.

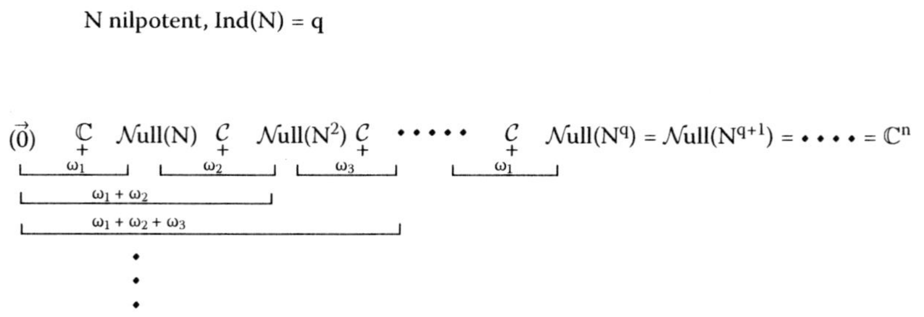

Figure 3.3: The Weyr sequence.

We may also view the situation graphically.

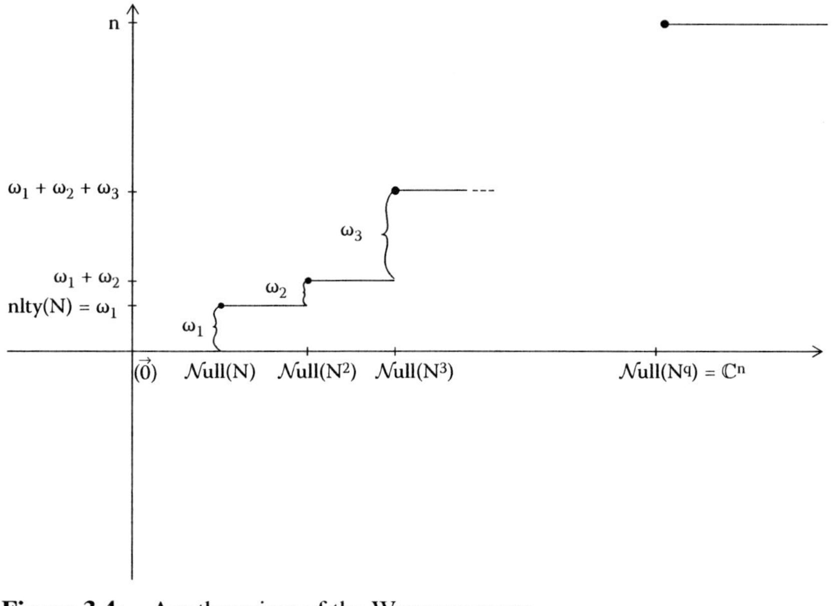

Figure 3.4: Another view of the Weyr sequence.

Let's look at some examples. We will deal with nilpotent matrices since these are the most important for our applications. First, let's define the "standard" k-by-k nilpotent matrix by

$$Nilp[k] = \begin{bmatrix} 0 & 1 & 0 & 0 & \cdots & 0 & 0 \\ 0 & 0 & 1 & 0 & \cdots & 0 & 0 \\ \vdots & \vdots & \vdots & \vdots & \cdots & 0 & 1 \\ 0 & 0 & 0 & 0 & \cdots & 0 & 0 \end{bmatrix} \in \mathbb{C}^{k \times k}.$$

Thus we have zeros everywhere except the superdiagonal, which is all ones. It is straightforward to see that $rank(Nilp[k]) = k - 1$, $index(Nilp[k]) = k$, and the minimal polynomial $\mu_{Nilp[k]}(x) = x^k$. Note $Nilp[1] = [0]$.

Let $N = Nilp[4] = \begin{bmatrix} 0 & 1 & 0 & 0 \\ 0 & 0 & 1 & 0 \\ 0 & 0 & 0 & 1 \\ 0 & 0 & 0 & 0 \end{bmatrix}$. Clearly $rank(N) = 3$, $nullity(N) = 1$; $N^2 = \begin{bmatrix} 0 & 0 & 1 & 0 \\ 0 & 0 & 0 & 1 \\ 0 & 0 & 0 & 0 \\ 0 & 0 & 0 & 0 \end{bmatrix}$, $rank(N^2) = 2$, $nullity(N^2) = 2$; $N^3 = \begin{bmatrix} 0 & 0 & 0 & 1 \\ 0 & 0 & 0 & 0 \\ 0 & 0 & 0 & 0 \\ 0 & 0 & 0 & 0 \end{bmatrix}$, $rank(N^3) = 1$, $nullity(N^3) = 3$; $N^4 = \mathbb{O}$, $rank(N^4) = 0$, $nullity(N^4) = 4$. Thus, we see $Weyr(N) = (1, 1, 1, 1)$ and $index(N) = 4$. Did you notice how the ones migrate up the superdiagonals as you raise N to higher powers? In general, we see that

$$Weyr(Nilp[k]) = (\underbrace{1, 1, 1, 1, \ldots, 1}_{k\ times}).$$

The next question is, can we come up with a matrix, again nilpotent, where the Weyr sequence is not all ones? Watch this. Let $N = \begin{bmatrix} 0 & 1 & 0 & 0 & & \\ 0 & 0 & 1 & 0 & & \\ 0 & 0 & 0 & 1 & & \\ 0 & 0 & 0 & 0 & & \\ & & & & 0 & 1 \\ & & & & 0 & 0 \end{bmatrix}$, where the blank spaces are understood to be all zeros. Now N is rank 4, nullity 2. Now look at N^2. $N^2 = \begin{bmatrix} 0 & 0 & 1 & 0 & & \\ 0 & 0 & 0 & 1 & & \\ 0 & 0 & 0 & 0 & & \\ 0 & 0 & 0 & 0 & & \\ & & & & 0 & 0 \\ & & & & 0 & 0 \end{bmatrix}$. We see $rank(N^2) = 2$, $nlty(N^2) = 4$. Next $N^3 = \begin{bmatrix} 0 & 0 & 0 & 1 & & \\ 0 & 0 & 0 & 0 & & \\ 0 & 0 & 0 & 0 & & \\ 0 & 0 & 0 & 0 & & \\ & & & & 0 & 0 \\ & & & & 0 & 0 \end{bmatrix}$. Clearly, $rank(N^3) = 1$, $nullity(N^3) = 5$ and finally $N^4 = \mathbb{O}$. Thus, $Weyr(N) = (2, 2, 1, 1)$.

Now the question is, can we exhibit a matrix with prescribed Weyr sequence? The secret was revealed above in putting standard nilpotent matrices in a block diagonal matrix. Note that we used $Nilp[4]$ and $Nilp[2]$ to create the Weyr sequence of (2, 2, 1, 1). We call the sequence (4,2) the *Segre sequence* of N above and write $Segre(N) = (4, 2)$ under these conditions. It may not be obvious to you that there is a connection between these two sequences, but there is!

DEFINITION 3.7 *(Segre sequence)*
Let $N = BlockDiagonal[Nilp[p_1], Nilp[p_2], \ldots, Nilp[p_k]]$. *We write* $Segre(N) = (p_1, p_2, \ldots, p_k)$. *Let's agree to arrange the blocks so that* $p_1 \geq p_2 \geq \cdots \geq p_k$. *Note* $N \in \mathbb{C}^{t \times t}$ *where* $t = \sum_{i=1}^{k} p_i$.

First, we need a lesson in counting. Suppose n is a positive integer. How many ways can we write n as a sum of positive integers? A *partition of n* is an m-tuple of nonincreasing positive integers whose sum is n. So, for example, (4, 2, 2, 1, 1) is a partition of $10 = 4 + 2 + 2 + 1 + 1$. There are many interesting counting problems associated with the partitions of an integer but, at the moment, we shall concern ourselves with counting how many there are for a given n. It turns out, there is no easy formula for this, but there is a recursive way to get as many as you could want.

Let p(n) count the number of partitions of n. Then

$p(1) = 1$ $\quad$ 1
$p(2) = 2$ $\quad$ $2, 1 + 1$
$p(3) = 3$ $\quad$ $3, 2 + 1, 1 + 1 + 1$
$p(4) = 5$ $\quad$ $4, 3 + 1, 2 + 2, 2 + 1 + 1, 1 + 1 + 1 + 1$
$p(5) = 7$ $\quad$ $5, 4 + 1, 3 + 2, 3 + 1 + 1, 2 + 2 + 1, 2 + 1 + 1 + 1, 1 + 1 + 1 + 1 + 1$
$p(6) = 11$
$p(7) = 15$
$p(8) = 22$

There is a formula you can find in a book on combinatorics or discrete mathematics that says

$$p(n) = p(n - 1) + p(n - 2) - p(n - 5) - p(n - 7) + \cdots.$$

For example, p(8) = p(7)+ p(6)− p(3)− p(1) = 15+11−3−1 = 22. Now, in the theory of partitions, there is the notion of a *conjugate partition* of a given partition and a *Ferrer's diagram* to help find it. Suppose $\alpha = (m_1, m_2, \ldots, m_k)$ is a partition of n. Define the *conjugate partition* of α to be $\alpha^* = (r_1^*, r_2^*, \ldots, r_s^*)$, where $r_i{}^*$ counts the number of m_js larger than or equal to i for $i = 1, 2, 3, \ldots$. So, for example, $(5, 3, 3, 1)^* = (4, 3, 3, 1, 1)$. Do not panic, there is an easy way to construct conjugate partitions from a Ferrer's diagram. Given a partition $\alpha = (m_1, m_2, \ldots, m_k)$ of n, its Ferrer's diagram is made up of dots (some people use squares), where each summand is represented by a horizontal row of dots as follows. Consider the partition of 12, (5, 3, 3, 1). Its Ferrer's diagram is

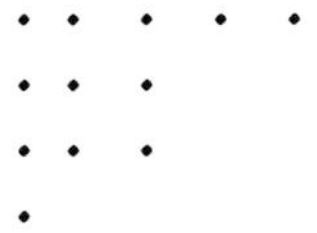

Figure 3.5: Ferrer's diagram.

To read off the conjugate partition, just read down instead of across. Do you see where (4, 3, 3, 1, 1) came from? Do you see that the number of dots in row i of α^* is the number of rows of α with i or more dots? Note that $\alpha^{**} = \alpha$, so the mapping on partitions of n given by $\alpha \mapsto \alpha^*$ is one-to-one and onto.

THEOREM 3.15

Suppose $N = BlockDiagonal[Nilp[p_1], Nilp[p_2], \ldots, Nilp[p_k]]$.

Then $Weyr(N)$ *is the conjugate partition of* $Segre(N) = (p_1, p_2, \ldots, p_k)$.

PROOF To make the notation a little easier, let $N = BlockDiagonal [N_1, N_2, \ldots, N_k]$, where $N_i = Nilp[p_i]$, which is size p_i-by-p_i for $i = 1, 2, 3, \ldots, k$. Now $N^2 = BlockDiagonal[N_1^2, N_2^2, \ldots, N_k^2]$, and generally $N^j = BlockDiagonal[N_1^j, N_2^j, \ldots, N_k^j]$ for $j = 1, 2, 3, \ldots$. Next note that, for each i, $rank(N_i) = p_i - 1$, $rank(N_i^2) = p_i - 2, \ldots, rank(N_i^j) = p_i - j$ as long as $j < p_i$; however, $rank(N_i^j) = 0$ if $j \geq p_i$. Thus, we conclude $rank(N_i^{j-1}) - rank(N_i^j) = \begin{cases} 1 \; if \; j \leq p_i \\ 0 \; if \; j > p_i \end{cases}$. The one acts like a counter signifying the presence of a block of at least size p_i. Next, we have $rank(N^j) = \sum_{i=1}^{k} rank(N_i^j)$, so $\omega_j = rank(N^{j-1}) - rank(N^j) = \sum_{i=1}^{k} rank(N_i^{j-1}) - \sum_{i=1}^{k} rank(N_i^j)$ $= \sum_{i=1}^{k} (rank(N_i^{j-1}) - rank(N_i^j))$. First, $\omega_1 = rank(N^0) - rank(N) = n - rank(N) = nullity(N) =$ the number of blocks in N since each N_i has nullity 1 (note the single row of all zeros in each N_i). Next, $\omega_2 = rank(N) - rank(N^2) = \sum_{i=1}^{k} (rank(N_i) - rank(N_i^2)) = (rank(N_1) - rank(N_1^2)) + (rank(N_2) - rank(N_2^2))$

$+\cdots+(rank(N_k)-rank(N_k^2))$. Each of these summands is either 0 or 1. A 1 occurs when the block N_i is large enough so its square does not exceed the size of N_i. Thus, ω_2 counts the number of N_i of size *at least* 2-by-2. Next, $\omega_3 = rank(N^2) - rank(N^3) = \sum_{i=1}^{k}(rank(N_i^2) - rank(N_i^3)) = (rank(N_1^2) - rank(N_1^3)) + (rank(N_2^2) - rank(N_2^3)) + \cdots + (rank(N_k^2) - rank(N_k^3))$. Again, a 1 occurs in a summand that has not yet reached its index of nilpotency, so a 1 records the fact that N_i is of size at least 3-by-3. Thus ω_3 counts the number of blocks N_i of size at least 3-by-3. Generally then, ω_j counts the number of blocks N_i of size at least j-by-j. And so, the theorem follows. □

COROLLARY 3.15

With N as above, the number of size j-by-j blocks in N is exactly $\omega_j - \omega_{j+1} =$
$rank(N^{j-1}) - 2rank(N^j) + rank(N^{j+1}) =$
$= [nlty(N^j) - nlty(N^{j-1})] - [nlty(N^{j+1}) - nlty(N^j)]$.

PROOF With the notation of the theorem above, $(rank(N_i^{j-1}) - (rank(N_i^j)) - (rank(N_i^j) - rank(N_i^{j+1})) = \begin{cases} 1 \; if \; j = p_i \\ 0 \; otherwise \end{cases}$. Thus, a 1 occurs exactly when a block of size p_i-by-p_i occurs. Thus the sum of these terms count the exact number of blocks of size j-by-j. □

Let's look at an example to be sure the ideas above are crystal clear. Consider

$$N = \begin{bmatrix} 0 & 1 & 0 & 0 & & & & & \\ 0 & 0 & 1 & 0 & & & & & \\ 0 & 0 & 0 & 1 & & & & & \\ 0 & 0 & 0 & 0 & & & & & \\ & & & & 0 & 1 & & & \\ & & & & 0 & 0 & & & \\ & & & & & & 0 & 1 & \\ & & & & & & 0 & 0 & \\ & & & & & & & & 0 \end{bmatrix}$$

. Clearly, $\omega_1 = 9 - rank(N) = 9 - 5 = 4$, so the theorem predicts 4 blocks in N of size at least 1-by-1. That is, it predicts (quite accurately) exactly 4 blocks in N. Next $\omega_2 = rank(N) - rank(N^2) = (rank(N_1) - rank(N_1^2)) + (rank(N_2) - rank(N_2^2)) - (rank(N_3) - rank(N_3^2)) + (rank(N_4) - rank(N_4^2)) = 1 + 1 + 1 + 0 = 3$ blocks of size at least 2-by-2, $\omega_3 = rank(N^2) - rank(N^3) = 1 + 0 + 0 + 0 = 1$ block of size at least 3-by-3, $\omega_4 = rank(N^3) - rank(N^4) = 1 + 0 + 0 + 0 = 1$ block of size at least 4-by-4, and $\omega_5 = rank(N^4) - rank(N^5) = 0$, which says there are no blocks of size 5 or more. Moreover, our formulas

$$\text{confirm}\begin{cases} \omega_1 - \omega_2 = 4 - 3 = 1 & \text{block of size 1-by-1} \\ \omega_2 - \omega_3 = 3 - 1 = 2 & \text{blocks of size 2-by-2} \\ \omega_3 - \omega_4 = 1 - 1 = 0 & \text{blocks of size 3-by-3} \\ \omega_4 - \omega_5 = 1 - 0 = 1 & \text{block of size 4-by-4} \end{cases}.$$

Next let's illustrate our theorem. Suppose we wish to exhibit a nilpotent matrix N with $Weyr(N) = (4, 3, 2, 1, 1)$. We form the Ferrer's diagram

Figure 3.6: Ferrer's diagram for (4, 3, 2, 1, 1).

and read off the conjugate partition (5, 3, 2, 1). Next we form the matrix $N = BlockDiagonal[Nilp[5], Nilp[3], Nilp[2], Nilp[1]] =$

$$\begin{bmatrix} 0&1&0&0&0&&&&&&\\ 0&0&1&0&0&&&&&&\\ 0&0&0&1&0&&&&&&\\ 0&0&0&0&1&&&&&&\\ 0&0&0&0&0&&&&&&\\ &&&&&0&1&0&&&\\ &&&&&0&0&1&&&\\ &&&&&0&0&0&&&\\ &&&&&&&&0&1&\\ &&&&&&&&0&0&\\ &&&&&&&&&&0 \end{bmatrix}.$$

Then $Weyr(N) = (4, 3, 2, 1, 1)$.

The block diagonal matrices we have been manipulating above look rather special but in a sense they are not. If we choose the right basis, we can make any nilpotent matrix look like these block diagonal nilpotent matrices. This is the essence of a rather deep theorem about the structure of nilpotent matrices. Let's look at an example to motivate what we need to achieve.

Suppose we have a nilpotent matrix N and we want N to "look like"

$$\begin{bmatrix} 0&1&0&0&0&0&0&0\\ 0&0&1&0&0&0&0&0\\ 0&0&0&1&0&0&0&0\\ 0&0&0&0&0&0&0&0\\ 0&0&0&0&0&1&0&0\\ 0&0&0&0&0&0&0&0\\ 0&0&0&0&0&0&0&1\\ 0&0&0&0&0&0&0&0 \end{bmatrix}$$

. That is, we seek an invertible matrix S

such that $S^{-1}NS$ will be this 8-by-8 matrix. This is the same as seeking a special basis of $\mathbb{C}^8$. If this is to work, we must have $NS = S$

$$\begin{bmatrix} 0 & 1 & 0 & 0 & 0 & 0 & 0 & 0 \\ 0 & 0 & 1 & 0 & 0 & 0 & 0 & 0 \\ 0 & 0 & 0 & 1 & 0 & 0 & 0 & 0 \\ 0 & 0 & 0 & 0 & 0 & 0 & 0 & 0 \\ 0 & 0 & 0 & 0 & 0 & 1 & 0 & 0 \\ 0 & 0 & 0 & 0 & 0 & 0 & 0 & 0 \\ 0 & 0 & 0 & 0 & 0 & 0 & 0 & 1 \\ 0 & 0 & 0 & 0 & 0 & 0 & 0 & 0 \end{bmatrix} =$$

$[\vec{0} \mid S\mathbf{e}_1 \mid S\mathbf{e}_2 \mid S\mathbf{e}_3 \mid \vec{0} \mid S\mathbf{e}_5 \mid \vec{0} \mid S\mathbf{e}_7] =$
$[\vec{0} \mid \mathbf{s}_1 \mid \mathbf{s}_2 \mid \mathbf{s}_3 \mid \vec{0} \mid \mathbf{s}_5 \mid \vec{0} \mid \mathbf{s}_7]$,
where $\mathbf{s}_i$ is the ith column of the matrix S. Equating columns we see

$$\begin{array}{lll} N\mathbf{s}_1 = \vec{0} & N\mathbf{s}_5 = \vec{0} & N\mathbf{s}_7 = \vec{0} \\ N\mathbf{s}_2 = \mathbf{s}_1 & N\mathbf{s}_6 = \mathbf{s}_5 & N\mathbf{s}_8 = \mathbf{s}_7 \\ N\mathbf{s}_3 = \mathbf{s}_2 & & \\ N\mathbf{s}_4 = \mathbf{s}_3 & & \end{array}.$$

In particular, we see that $\mathbf{s}_1$, $\mathbf{s}_5$, and $\mathbf{s}_7$ must form a basis for the null space of N. Note N is rank 5 so has nullity 3. Also note

$$\begin{array}{lll} N\mathbf{s}_1 = \vec{0} & N\mathbf{s}_5 = \vec{0} & N\mathbf{s}_7 = \vec{0} \\ N^2\mathbf{s}_2 = \vec{0} & N^2\mathbf{s}_6 = \vec{0} & N^2\mathbf{s}_8 = \vec{0} \\ N^3\mathbf{s}_3 = \vec{0} & & \\ N^4\mathbf{s}_4 = \vec{0} & & \end{array}$$

and so

$$\begin{array}{lll} \mathcal{N}ull(N) & = & sp\{\mathbf{s}_1, \mathbf{s}_5, \mathbf{s}_7\} \\ \mathcal{N}ull(N^2) & = & sp\{\mathbf{s}_1, \mathbf{s}_5, \mathbf{s}_7, \mathbf{s}_2, \mathbf{s}_6, \mathbf{s}_8\} \\ \mathcal{N}ull(N^3) & = & sp\{\mathbf{s}_1, \mathbf{s}_5, \mathbf{s}_7, \mathbf{s}_2, \mathbf{s}_6, \mathbf{s}_8, \mathbf{s}_3\} \\ \mathcal{N}ull(N^4) & = & sp\{\mathbf{s}_1, \mathbf{s}_5, \mathbf{s}_7, \mathbf{s}_2, \mathbf{s}_6, \mathbf{s}_8, \mathbf{s}_3, \mathbf{s}_4\} = \mathbb{C}^8 \end{array}.$$

Finally, note that

$$\begin{array}{lll} \mathbf{s}_4 & \mathbf{s}_6 & \mathbf{s}_8 \\ N\mathbf{s}_4 = \mathbf{s}_3 & N\mathbf{s}_6 = \mathbf{s}_5 & N\mathbf{s}_8 = \mathbf{s}_7 \\ N^2\mathbf{s}_4 = \mathbf{s}_2 & & \\ N^3\mathbf{s}_4 = \mathbf{s}_1 & & \end{array}.$$

The situation is a bit complicated but some clues are beginning to emerge. There seem to be certain "Krylov strings" of vectors that are playing an important role. We begin with a helper lemma.

LEMMA 3.1

For any square matrix A and positive integer k, $\mathcal{N}ull(A^{k-1}) \subseteq \mathcal{N}ull(A^k) \subseteq \mathcal{N}ull(A^{k+1})$. Let $\mathcal{B}_{k-1} = \{\mathbf{b}_1, \mathbf{b}_2, , \dots, \mathbf{b}_r\}$ be a basis for $\mathcal{N}ull(A^{k-1})$. Extend $\mathcal{B}_{k-1}$ to a basis of $\mathcal{N}ull(A^k)$; say $\mathcal{B}_k = \{\mathbf{b}_1, \mathbf{b}_2, , \dots, \mathbf{b}_r, \mathbf{c}_1, \mathbf{c}_2, \dots, \mathbf{c}_s\}$. Finally, extend $\mathcal{B}_k$ to a basis of $\mathcal{N}ull(A^{k+1})$; say $\mathcal{B}_{k+1} = \{\mathbf{b}_1, \mathbf{b}_2, , \dots, \mathbf{b}_r, \mathbf{c}_1, \mathbf{c}_2, \dots, \mathbf{c}_s, \mathbf{d}_1, \mathbf{d}_2, \dots, \mathbf{d}_t\}$. Then $\mathcal{T} = \{\mathbf{b}_1, \mathbf{b}_2, , \dots, \mathbf{b}_r, A\mathbf{d}_1, A\mathbf{d}_2, \dots, A\mathbf{d}_t\}$ is an independent subset of $\mathcal{N}ull(A^k)$ though not necessarily a basis thereof.

PROOF First we argue that $\mathcal{T} \subseteq \mathcal{N}ull(A^k)$. Evidently all the $\mathbf{b}_i$s are in $\mathcal{N}ull(A^{k-1})$ so $A^k\mathbf{b}_i = AA^{k-1}\mathbf{b}_i = A\overrightarrow{0} = \overrightarrow{0}$, putting the $\mathbf{b}_i$s in $\mathcal{N}ull(A^k)$. Next, the $\mathbf{d}_i$s are in $\mathcal{N}ull(A^{k+1})$ so $\overrightarrow{0} = A^{k+1}\mathbf{d}_i = A^k(A\mathbf{d}_i)$, which puts the $A\mathbf{d}_i$s in $\mathcal{N}ull(A^k)$ as well. Next, we argue the independence in the usual way. Suppose $\beta_1\mathbf{b}_1 + \cdots + \beta_r\mathbf{b}_r + \alpha_1 A\mathbf{d}_1 + \cdots + \alpha_t A\mathbf{d}_t = \overrightarrow{0}$. Then $A(\alpha_1\mathbf{d}_1 + \cdots + \alpha_t\mathbf{d}_t) = -(\beta_1\mathbf{b}_1 + \cdots + \beta_r\mathbf{b}_r) \in \mathcal{N}ull(A^{k-1})$. Thus, $A^{k-1}A(\alpha_1\mathbf{d}_1 + \cdots + \alpha_t\mathbf{d}_t) = \overrightarrow{0}$ putting $\alpha_1\mathbf{d}_1 + \cdots + \alpha_t\mathbf{d}_t \in \mathcal{N}ull(A^k)$. Therefore, $\alpha_1\mathbf{d}_1 + \cdots + \alpha_t\mathbf{d}_t = \gamma_1\mathbf{b}_1 + \cdots + \gamma_r\mathbf{b}_r + \delta_1\mathbf{c}_1 + \cdots + \delta_s\mathbf{c}_s$. But then $\alpha_1\mathbf{d}_1 + \cdots + \alpha_t\mathbf{d}_t - \gamma_1\mathbf{b}_1 - \cdots - \gamma_r\mathbf{b}_r - \delta_1\mathbf{c}_1 - \cdots - \delta_s\mathbf{c}_s = \overrightarrow{0}$. By independence, all the αs, γs, and δs must be zero. Thus $\beta_1\mathbf{b}_1 + \cdots + \beta_r\mathbf{b}_r = \overrightarrow{0}$ so all the βs must be zero as well by independence. We have proved $\mathcal{T}$ is an independent subset of $\mathcal{N}ull(A^k)$. □

Now let's go back to our example and see if we could construct a basis of that special form we need for S.

You could start from the bottom of the chain and work your way up, or you could start at the top and work your way down. We choose to build from the bottom (i.e., from the null space of N). First take a basis $\mathcal{B}_1$ for $\mathcal{N}ull(N)$; say $\mathcal{B}_1 = \{\mathbf{b}_1, \mathbf{b}_2, \mathbf{b}_3\}$. Then extend to a basis of $\mathcal{N}ull(N^2)$; say $\mathcal{B}_2 = \{\mathbf{b}_1, \mathbf{b}_2, \mathbf{b}_3, \mathbf{b}_4, \mathbf{b}_5, \mathbf{b}_6\}$. Extend again to a basis of $\mathcal{N}ull(N^3)$; say $\mathcal{B}_3 = \{\mathbf{b}_1, \mathbf{b}_2, \mathbf{b}_3, \mathbf{b}_4, \mathbf{b}_5, \mathbf{b}_6, \mathbf{b}_7\}$. Finally, extend to a basis $\mathcal{B}_4$ of $\mathcal{N}ull(N^4) = \mathbb{C}^8$, say $\mathcal{B}_4 = \{\mathbf{b}_1, \mathbf{b}_2, \mathbf{b}_3, \mathbf{b}_4, \mathbf{b}_5, \mathbf{b}_6, \mathbf{b}_7, \mathbf{b}_8\}$. The chances this basis has the special properties we seek are pretty slim. We must make it happen. Here is where things get a bit tricky. As we said, we will start from the full basis and work our way down. Let $\mathcal{T}_4 = \{\mathbf{b}_8\}$; form $\mathcal{B}_2 \cup \{N\mathbf{b}_8\} = \{\mathbf{b}_1, \mathbf{b}_2, \mathbf{b}_3, \mathbf{b}_4, \mathbf{b}_5, \mathbf{b}_6, N\mathbf{b}_8\}$. This is an independent subset of $\mathcal{N}ull(N^3)$ by our lemma. In this case, it must be a basis of $\mathcal{N}ull(N^3)$. If it was not, we would have had to extend it to a basis at this point. Next, let $\mathcal{T}_3 = \{N\mathbf{b}_8\}$; form $\mathcal{B}_1 \cup \{N^2\mathbf{b}_8\} = \{\mathbf{b}_1, \mathbf{b}_2, \mathbf{b}_3, N^2\mathbf{b}_8\} \subseteq \mathcal{N}ull(N^2)$. Again this is an independent set, but here we must extend to get a basis of $\mathcal{N}ull(N^2)$; say $\{\mathbf{b}_1, \mathbf{b}_2, \mathbf{b}_3, N^2\mathbf{b}_8, \mathbf{z}_1, \mathbf{z}_2\}$. Next, let $\mathcal{T}_2 = \{N^3\mathbf{b}_8\} \cup \{N\mathbf{z}_1, N\mathbf{z}_2\} = \{N^3\mathbf{b}_8, N\mathbf{z}_1, N\mathbf{z}_2\} \subseteq \mathcal{N}ull(N)$. This is a basis for $\mathcal{N}ull(N)$. So here is the modified basis we have constructed from our original one: $\{N^3\mathbf{b}_8, N\mathbf{z}_1, N\mathbf{z}_2, N^2\mathbf{b}_8, \mathbf{z}_1, \mathbf{z}_2, N\mathbf{b}_8, \mathbf{b}_8\}$. Next, we stack it carefully:

$$\begin{cases} \mathbf{b}_8 & & & in\ \mathcal{N}ull(N^4) \\ N\mathbf{b}_8 & & & in\ \mathcal{N}ull(N^3) \\ N^2\mathbf{b}_8 & \mathbf{z}_1 & \mathbf{z}_2 & in\ \mathcal{N}ull(N^2) \\ N^3\mathbf{b}_8 & N\mathbf{z}_1 & N\mathbf{z}_2 & in\ \mathcal{N}ull(N) \end{cases}$$

Now simply label from the bottom up starting at the left.

$$\begin{cases} \mathbf{b}_8 = \mathbf{s}_4 & & \\ N\mathbf{b}_8 = \mathbf{s}_3 & & \\ N^2\mathbf{b}_8 = \mathbf{s}_2 & \mathbf{z}_1 = \mathbf{s}_6 & \mathbf{z}_2 = \mathbf{s}_8 \\ N^3\mathbf{b}_8 = \mathbf{s}_1 & N\mathbf{z}_1 = \mathbf{s}_5 & N\mathbf{z}_2 = \mathbf{s}_7 \end{cases}$$

The matrix $S = [N^3\mathbf{b}_8 \mid N^2\mathbf{b}_8 \mid N\mathbf{b}_8 \mid \mathbf{b}_8 \mid N\mathbf{z}_1 \mid \mathbf{z}_1 \mid N\mathbf{z}_2 \mid \mathbf{z}_2]$ has the desired property; namely, $NS = [\overrightarrow{0} \mid N^3\mathbf{b}_8 \mid N^2\mathbf{b}_8 \mid N\mathbf{b}_8 \mid \overrightarrow{0} \mid N\mathbf{z}_1 \mid \overrightarrow{0} \mid N\mathbf{z}_2] = [\overrightarrow{0} \mid \mathbf{s}_1 \mid \mathbf{s}_2 \mid \mathbf{s}_3 \mid \overrightarrow{0} \mid \mathbf{s}_5 \mid \overrightarrow{0} \mid \mathbf{s}_7]$, which is exactly what we wanted. Now let's see if we can make the argument above general.

THEOREM 3.16

Suppose N is an n-by-n nilpotent matrix of index q. Then there exists an invertible matrix S with $S^{-1}NS = BlockDiagonal[Nilp[p_1], Nilp[p_2], \ldots, Nilp[p_k]]$. *The largest block is q-by-q and we may assume* $p_1 \geq p_2 \geq \cdots \geq p_k$. *Of course,* $p_1 + p_2 + \cdots + p_k = n$. *The total number of blocks is the nullity of N, and the number of j-by-j blocks is* $rank(N^{j-1}) - 2rank(N^j) + rank(N^{j+1})$.

PROOF Suppose N is nilpotent of index q so we have the proper chain of subspaces

$$(0) \subsetneq \mathcal{N}ull(N) \subsetneq \mathcal{N}ull(N^2) \subsetneq \cdots \subsetneq \mathcal{N}ull(N^{q-1}) \subsetneq \mathcal{N}ull(N^q) = \mathbb{C}^n.$$

As above, we construct a basis of $\mathbb{C}^n$ starting at the bottom of the chain. Then we adjust it working from the top down. Choose a basis $\mathcal{B}_1$ of $\mathcal{N}ull(N)$; say $\mathcal{B}_1 = \{\mathbf{b}_1, \ldots, \mathbf{b}_{k_1}\}$. Extend to a basis of $\mathcal{N}ull(N^2)$; say $\mathcal{B}_2 = \{\mathbf{b}_1, \ldots, \mathbf{b}_{k_1}, \mathbf{b}_{k_1+1}, \ldots, \mathbf{b}_{k_2}\}$. Continue this process until we successively produce a basis of $\mathbb{C}^n$, $\mathcal{B}_q = \{\mathbf{b}_1, \ldots, \mathbf{b}_{k_1}, \mathbf{b}_{k_1+1}, \ldots, \mathbf{b}_{k_2}, \ldots, \mathbf{b}_{k_{q-2}+1}, \ldots, \mathbf{b}_{k_{q-1}}, \mathbf{b}_{k_{q-1}+1}, \ldots, \mathbf{b}_{k_q}\}$. As we did in our motivating example, we begin at the top and work our way down the chain. Let $T_q = \{\mathbf{b}_{k_{q-1}+1}, \ldots, \mathbf{b}_{k_q}\}$; form $\mathcal{B}_{q-2} \cup \{N\mathbf{b}_{k_{q-1}+1}, \ldots, N\mathbf{b}_{k_q}\}$. This is an independent subset of $\mathcal{N}ull(N^{q-1})$ by our helper lemma. Extend this independent subset to a basis of $\mathcal{N}ull(N^{q-1})$; say $\mathcal{B}_{q-2} \cup \{N\mathbf{b}_{k_{q-1}+1}, \ldots, N\mathbf{b}_{k_q}\} \cup \{\mathbf{c}_{11}, \ldots, \mathbf{c}_{1t_1}\}$. Of course, we may not need any **c**s at all as our example showed. Next take $\mathcal{T}_{q-1} = \{N\mathbf{b}_{k_{q-1}+1}, \dots, N\mathbf{b}_{k_q}\} \cup \{\mathbf{c}_{11}, \dots, \mathbf{c}_{1t_1}\}$ and form $\mathcal{B}_{q-3} \cup \{N^2\mathbf{b}_{k_{q-1}+1}, \dots, N^2\mathbf{b}_{k_q}\} \cup \{N\mathbf{c}_{11}, \dots, N\mathbf{c}_{1t_1}\}$. This is an independent subset of $\mathcal{N}ull(N^{q-2})$. Extend this independent subset to a basis of $\mathcal{N}ull(N^{q-2})$; say $\mathcal{B}_{q-3} \cup \{N^2\mathbf{b}_{k_{q-1}+1}, \dots, N^2\mathbf{b}_{k_q}\} \cup \{N\mathbf{c}_{11}, \dots, N\mathbf{c}_{1t_1}\} \cup \{\mathbf{c}_{21}, \dots, \mathbf{c}_{2t_2}\}$. Continue this process until we produce a basis for $\mathcal{N}ull(N)$. Now stack the basis vectors we have thusly constructed.

$$\begin{array}{llllll}
\mathbf{b}_{k_{q-1}+1} & \cdots & \mathbf{b}_{k_q} & & & \\
N\left(\mathbf{b}_{k_{q-1}+1}\right) & \cdots & N\mathbf{b}_{k_q} & \mathbf{c}_{11} & \cdots & \mathbf{c}_{1t_1} \\
N^2\left(\mathbf{b}_{k_{q-1}+1}\right) & \cdots & N^2\left(\mathbf{b}_{k_q}\right) & N\mathbf{c}_{11} & \cdots & N\mathbf{c}_{1t_1} \\
\vdots & \vdots & \vdots & \vdots & \vdots & \vdots \\
N^{q-1}(\mathbf{b}_{k_{q-1}+1}) & \cdots & N^{q-1}\left(\mathbf{b}_{k_q}\right) & N^{q-2}(\mathbf{c}_{11}) & \cdots & N^{q-2}\left(\mathbf{c}_{1t_1}\right)
\end{array}$$

$$\begin{array}{llllll}
\mathbf{c}_{21} & \cdots & \mathbf{c}_{2t_2} & & & \\
\vdots & \vdots & \vdots & & & \\
N^{q-3}\left(\mathbf{c}_{21}\right) & \cdots & N^{q-3}\left(\mathbf{c}_{2t_2}\right) & \cdots & \mathbf{c}_{q-11} \cdots \mathbf{c}_{q-1t_{q-1}}
\end{array}$$

Note that every column except the last is obtained by repeated application of N to the top vector in each stack. Also note that each row of vectors belong to the same null space. As in our example, label this array of vectors starting at the bottom of the first row and continue up each column. Then $S =$

$$\begin{aligned}
[N^{q-1}(\mathbf{b}_{k_{q-1}+1}) \;&|\; N^{q-2}(\mathbf{b}_{k_{q-1}+1}) \mid \cdots \mid \mathbf{b}_{k_{q-1}+1} \mid N^{q-1}\left(\mathbf{b}_{k_q}\right) \mid \cdots \\
&|\; \cdots \mid \mathbf{c}_{q-11} \mid \cdots \mid \mathbf{c}_{q-1t_{q-1}}].
\end{aligned}$$

This matrix is invertible and brings N into the block diagonal form as advertised in the theorem. The remaining details are left to the reader. □

This theorem is a challenge so we present another way of looking at the proof due to **Manfred Dugas** (11 February, 1952). We illustrate with a nilpotent matrix of index 3. Let $N \in \mathbb{C}^{n\times n}$ with $N^3 = \mathbb{O} \neq N^2$. Now $(\overrightarrow{0}) \subsetneq \mathcal{N}ull(N) \subsetneq \mathcal{N}ull(N^2) \subsetneq \mathcal{N}ull(N^3) = \mathbb{C}^n$. Since $\mathcal{N}ull(N^2)$ is a proper subspace of $\mathbb{C}^n$, it has a complementary subspace M_3. Thus, $\mathbb{C}^n = \mathcal{N}ull(N^2) \oplus M_3$. Clearly, $N(M_3) \subseteq \mathcal{N}ull(N^2)$ and $N^2(M_3) \subseteq \mathcal{N}ull(N)$. Moreover, we claim that $N(M_3) \cap \mathcal{N}ull(N) = (\overrightarrow{0})$, for if $\mathbf{v} \in N(M_3) \cap \mathcal{N}ull(N)$, then $\mathbf{v} = N\mathbf{m}_3$ for some $\mathbf{m}_3$ in M_3 and $\overrightarrow{0} = N\mathbf{v} = N^2\mathbf{m}_3$. But this puts $\mathbf{m}_3$ in $\mathcal{N}ull(N^2) \cap M_3 = (\overrightarrow{0})$. Therefore, $\mathbf{m}_3 = \overrightarrow{0}$ so $\mathbf{v} = N\mathbf{m}_3 = \overrightarrow{0}$. Now $\mathcal{N}ull(N) \oplus N(M_3) \subseteq \mathcal{N}ull(N^2)$ and could actually equal $\mathcal{N}ull(N^2)$. In any case, we can find a supplement M_2 so that $\mathcal{N}ull(N^2) = \mathcal{N}ull(N) \oplus N(M_3) \oplus M_2$. Note $N(M_2) \subseteq \mathcal{N}ull(N)$. Even more, we claim $N(M_2) \cap N^2(M_3) = (\overrightarrow{0})$, for if $\mathbf{v} \in N(M_2) \cap N^2(M_3)$, then $\mathbf{v} = N\mathbf{m}_2 = N^2\mathbf{m}_3$ for some $\mathbf{m}_2$ in

M_2 and $\mathbf{m}_3$ in M_3. But then, $N(\mathbf{m}_2 - N\mathbf{m}_3) = (\overrightarrow{0})$ putting $\mathbf{m}_2 - N\mathbf{m}_3$ in $\mathcal{N}ull(N) \cap (N(M_3) \oplus M_2) = (\overrightarrow{0})$. Thus, $\mathbf{m}_2 = N\mathbf{m}_3 \in M_2 \cap N(M_3) = (\overrightarrow{0})$, making $\mathbf{v} = N\mathbf{m}_2 = \overrightarrow{0}$. Now $N(M_2) \oplus N^2(M_3) \subseteq \mathcal{N}ull(N)$, so there is a supplementary subspace M_1 with $\mathcal{N}ull(N) = N(M_2) \oplus N^2(M_3) \oplus M_1$. Thus, collecting all the pieces, we have $\mathbb{C}^n = M_3 \oplus NM_3 \oplus M_2 \oplus NM_2 \oplus N^2M_3 \oplus M_1$. We can reorganize this sum as

$$\mathbb{C}^n = [M_3 \oplus NM_3 \oplus N^2M_3] \oplus [M_2 \oplus NM_2] \oplus M_1 = L_3 \oplus L_2 \oplus L_1,$$

where the three blocks are N-invariant. To create the similarity transformation S, we begin with a basis $\mathcal{B}_3 = \{\mathbf{b}_1^{(3)}, \mathbf{b}_2^{(3)}, \dots, \mathbf{b}_{d_3}^{(3)}\}$ of M_3. We claim $\{N^2\mathbf{b}_1^{(3)}, N^2\mathbf{b}_2^{(3)}, \dots, N^2\mathbf{b}_{d_3}^{(3)}\}$ is an independent set. As usual, set $\overrightarrow{0} = \sum_{j=1}^{d_3} \alpha_j N^2\mathbf{b}_j^{(3)}$, which implies $\sum_{j=1}^{d_3} \alpha_j \mathbf{b}_j^{(3)} \in \mathcal{N}ull(N^2) \cap M_3 = (\overrightarrow{0})$. This means all the α_js are 0, hence the independence. Now if $0 = \sum_{j=1}^{d_3} \alpha_j N\mathbf{b}_j^{(3)}$, then $\sum_{j=1}^{d_3} \alpha_j N^2\mathbf{b}_j^{(3)} = 0$ so, as above, all the α_js are 0. Thus we have

$$\begin{aligned}
L_3 &= M_3 \oplus NM_3 \oplus N^2M_3 = span\{\mathbf{b}_1^{(3)}, \mathbf{b}_2^{(3)}, \dots, \mathbf{b}_{d_3}^{(3)}, N\mathbf{b}_1^{(3)}, N\mathbf{b}_2^{(3)}, \dots, \\
&\qquad N\mathbf{b}_{d_3}^{(3)}, N^2\mathbf{b}_1^{(3)}, N^2\mathbf{b}_2^{(3)}, \dots, N^2\mathbf{b}_{d_3}^{(3)}\} \\
L_2 &= M_2 \oplus NM_2 = span\{\mathbf{b}_1^{(2)}, \mathbf{b}_2^{(2)}, \dots, \mathbf{b}_{d_2}^{(2)}, N^2\mathbf{b}_1^{(2)}, N\mathbf{b}_2^{(2)}, \dots, N\mathbf{b}_{d_2}^{(2)}\} \\
L_3 &= M_1 span\{\mathbf{b}_1^{(1)}, \mathbf{b}_2^{(1)}, \dots, \mathbf{b}_{d_1}^{(1)}\},
\end{aligned}$$

again noting these spans are all N-invariant. Next we begin to construct S. Let $S_j^{(3)} = [N^2b_j^{(3)} \mid Nb_j^{(3)} \mid b_j^{(3)}] \in \mathbb{C}^{n\times 3}$ for $j = 1, \dots, d_3$. We compute that $NS_j^{(3)} = [\overrightarrow{0} \mid N^2b_j^{(3)} \mid Nb_j^{(3)}] = [N^2b_j^{(3)} \mid Nb_j^{(3)} \mid b_j^{(3)}] \begin{bmatrix} 0 & 1 & 0 \\ 0 & 0 & 1 \\ 0 & 0 & 0 \end{bmatrix} = S_j^{(3)}Nilp[3]$. Similarly, set $S_j^{(2)} = [N\mathbf{b}_j^{(2)} \mid \mathbf{b}_j^{(2)}]$ and find $NS_j^{(2)} = [\overrightarrow{0} \mid N\mathbf{b}_j^{(2)}] = [N\mathbf{b}_j^{(2)} \mid \mathbf{b}_j^{(2)}] \begin{bmatrix} 0 & 1 \\ 0 & 0 \end{bmatrix} = S_j^{(2)}Nilp[2]$. Finally, set $S_j^{(1)} = [\mathbf{b}_j^{(1)}]$. Compute that $NS_j^{(1)} = \begin{bmatrix} 0 \\ \vdots \\ 0 \end{bmatrix} = [\mathbf{b}_j^{(1)}][0]$. We are ready to take $S = [S_1^{(3)} \mid \cdots \mid S_{d_3}^{(3)} \mid S_1^{(2)} \mid \cdots \mid S_{d_2}^{(2)} \mid S_1^{(1)} \mid \cdots \mid S_{d_1}^{(1)}]$. Then

$$NS = S\begin{bmatrix} Nilp[3] & & & & & & & & \\ & \ddots & & & & & & & \\ & & Nilp[3] & & & & & & \\ & & & Nilp[2] & & & & & \\ & & & & \ddots & & & & \\ & & & & & Nilp[2] & & & \\ & & & & & & 0 & & \\ & & & & & & & \ddots & \\ & & & & & & & & 0 \end{bmatrix},$$

where there are d_3 copies of $Nilp[3]$, d_2 copies of $Nilp[2]$, and d_1 copies of $Nilp[1]$. So we have another view of the proof of this rather deep result about nilpotent matrices.

This theorem can be considered as giving us a "canonical form" for any nilpotent matrix under the equivalence relation of "similarity." Now it makes sense to assign a Segre sequence and a Weyr sequence to any nilpotent matrix. We shall return to this characterization of nilpotent matrices when we discuss the Jordan canonical form.

Exercise Set 11

1. Compute the index of $\begin{bmatrix} 1 & 3 & 1 & 0 \\ 0 & 1 & 0 & 1 \\ 1 & 4 & 1 & 1 \\ 0 & 0 & 2 & 1 \end{bmatrix}$.

2. Argue that $N = \begin{bmatrix} s-t & -2s & s+t \\ -t & 0 & t \\ -s-t & 2s & -s+t \end{bmatrix}$ is nilpotent. Can you say anything about its index?

3. Fill in the details of the proof of Corollary 3.12.

4. Fill in the details of the proof of Corollary 3.13.

5. What is the index of $\mathbb{O}$? What is the index of I_n? What is the index of an invertible matrix?

6. Suppose $P = P^2$ but $P \neq I$ or $\mathbb{O}$. What is the index of P?

7. Let $M = R \begin{bmatrix} 1 & 2 & 0 & 0 & 0 \\ 3 & 4 & 0 & 0 & 0 \\ 0 & 0 & 0 & 1 & 0 \\ 0 & 0 & 0 & 0 & 1 \\ 0 & 0 & 0 & 0 & 0 \end{bmatrix} R^{-1}$. What is the index of M?

8. Let $L = \begin{bmatrix} M & \mathbb{O} \\ \mathbb{O} & \mathbb{O} \end{bmatrix}$, where M is from above. What is the index of L?

9. Find the conjugate partition of (7, 6, 5, 3, 2).

10. Construct a matrix N with $Weyr(N) = (5, 3, 2, 2, 1)$.

11. Argue that the number of partitions of n into k summands equals the number of partitions of n into summands the largest of which is k.

12. Argue that if N is nilpotent of index q, then so is $S^{-1}NS$ for any invertible square matrix S.

13. Argue that if $B = S^{-1}AS$, then $Weyr(B) = Weyr(A)$.

14. Consider all possible chains of subspaces between $(\vec{0})$ and $\mathbb{C}^4$. How many types are there and what is the sequence of dimensions of each type? Produce a concrete 4-by-4 matrix whose powers generate each type of chain via their null spaces.

15. Consider the differentiation operator D on the vector space $V = \mathbb{C}[x]^{\leq n}$ of polynomials of degree less or equal to n. Argue that D is nilpotent. What is the index of nilpotency? What is the matrix of D relative to the standard basis $\{1, x, x^2, \ldots, x^n\}$?

16. An upper(lower) triangular matrix is called *strictly upper(lower) triangular* iff the diagonal elements consist entirely of zeros. Prove that strictly upper(lower) triangular matrices are all nilpotent. Let M be the matrix $\begin{bmatrix} \lambda & m_{12} & \cdots & m_{1n} \\ 0 & \lambda & \cdots & m_{2n} \\ \cdots & & \ddots & \vdots \\ 0 & 0 & \cdots & \lambda \end{bmatrix}$. Argue that $M - \lambda I$ is nilpotent.

17. Can you have a nonsingular nilpotent matrix? Suppose N is n-by-n nilpotent of index q. Prove that $I + N$ is invertible. Exhibit the inverse.

18. Compute p(9), p(10), p(11), and p(12).

19. Is the product of nilpotent matrices always nilpotent?

20. Suppose N is nilpotent of index q and α is a nonzero scalar. Argue that αN is nilpotent of index q. Indeed, if $p(x)$ is any polynomial with constant term zero, argue that $p(N)$ is nilpotent. Why is it important that $p(x)$ have a zero constant term?

21. Suppose N_1 and N_2 are nilpotent matrices of the same size. Is the sum nilpotent? What can you say about the index of the sum?

22. Suppose N is nilpotent of index q. Let $\mathbf{v} \in \mathbb{C}^n$ with $N^{q-1}\mathbf{v} \neq 0$. Then $q \leq n$ and the vectors $\mathbf{v}, N\mathbf{v}, N^2\mathbf{v}, \ldots, N^{q-1}\mathbf{v}$ are independent.

23. Suppose N is nilpotent of index q. Prove that $\mathcal{C}ol(N^{q-1}) \subseteq \mathcal{N}ull(N)$.

24. (M. A. Khan, CMJ, May 2003). Exponential functions such as $f(x) = 2^x$ have the functional property that $f(x + y) = f(x) + f(y)$ for all x and y. Are there functions $M{:}\mathbb{C}[x]^{n\times n} \to \mathbb{C}[x]^{n\times n}$ that satisfy the same functional equation (i.e., $M(x + y) = M(x) + M(y)$)? If there is such an M, argue that $I = M(0) = M(x)M(-x)$, so $M(x)$ must be invertible with $M(-x) = (M(x))^{-1}$. Also argue that $(M(x))^r = M(rx)$, so that $M(\frac{x}{r})^r = M(x)$. Thus, the rth root of $M(x)$ is easily found by replacing x by $\frac{x}{r}$. Suppose N is a nilpotent matrix. Then argue that $M(x) = I + Nx + \frac{N^2x^2}{2!} + \frac{N^3x^3}{3!} + \cdots$ is a matrix with polynomial entries that satisfies the functional equation $M(x + y) = M(x) + M(y)$. Verify Khan's example. Let $N = \begin{pmatrix} 7 & -10 & 7 & -4 \\ 4 & -1 & -8 & 5 \\ 1 & -4 & 1 & 2 \\ 4 & -13 & 16 & -7 \end{pmatrix}$. Argue that N is nilpotent and find $M(x)$ explicitly and verify the functional equation.

25. Suppose N is nilpotent of index q. What is the minimal polynomial of N? What is the characteristic polynomial of $Nilp[k]$?

26. Find a matrix of index 3 and rank 4. Can you generalize this to any rank and any index?

27. Suppose A is a matrix of index q. What can you say about the minimal polynomial of A? (*Hint:* Look at the core-nilpotent factorization of A.)

28. Draw a graphical representation of the Weyr sequence of a matrix using column spaces instead of null spaces, as we did in the text in Figure 3.4.

29. What is the trace of any nilpotent matrix?

30. Show that every 2-by-2 nilpotent matrix looks like $\begin{bmatrix} \alpha\beta & \beta^2 \\ -\alpha^2 & -\alpha\beta \end{bmatrix}$.

31. Suppose $N = Nilp[n]$ and A is n-by-n. Describe NA, AN, N^TA, AN^T, N^TAN, NAN^T, and N^rAN^s.

32. Suppose A is a matrix with invariant subspaces $(\overrightarrow{0})$ and $\mathbb{C}^n$. Prove that either A is nilpotent or A is invertible.

Further Reading

[Andr, 1976] George E. Andrews, *The Theory of Partitions*, Addison-Wesley, Reading, MA, (1976). Reprinted by Cambridge University Press, Cambridge, (1998).

[B&R, 1986(4)] T. S. Blyth and E. F. Robertson, *Linear Algebra*, Vol. 4, Chapman & Hall, New York, (1986).

[G&N, 2004] Kenneth Glass and Chi-Keung Ng, A Simple Proof of the Hook Length Formula, The American Mathematical Monthly, Vol. 111, No. 8, October, (2004), 700–704.

[Hohn, 1964] E. Hohn, *Elementary Matrix Theory*, 2nd Edition, The Macmillan Company, New York, (1958, 1964).

[Shapiro, 1999] Helene Shapiro, The Weyr Characteristic, The American Mathematical Monthly, Vol. 106, No. 10, December, (1999), 919–929.

3.4.1 MATLAB Moment

3.4.1.1 The Standard Nilpotent Matrix

We can easily create a function in MATLAB to construct the standard nilpotent matrix nilp[n]. Create the following M-file:

```
function N = nilp(n)
if n == 0.N = [].else,
N = diag(ones(n - 1), 1); end
```

This is an easy use of the logical format "if ... then ... else". Note, if n = 0, the empty matrix is returned. Try out your new function with a few examples.

There is a function to test if a matrix is empty. It is

$$isempty(A).$$

How could you disguise the standard nilpotent matrix to still be nilpotent but not standard? (*Hint:* If N is nilpotent, so is SNS^{-1} for any invertible S.)

3.5 Left and Right Inverses

As we said at the very beginning, the central problem of linear algebra is the problem of solving a system of linear equations. If $A\mathbf{x} = \mathbf{b}$ and A is square and invertible, we have a complete answer: $\mathbf{x} = A^{-1}\mathbf{b}$ is the unique solution. However, if A is not square or does not have full rank, inverses make no sense. This is a motivation for the need for "generalized inverses." Now we face up to the fact that it is very unlikely that in real-life problems our systems of linear equations will have square coefficient matrices. So consider $A\mathbf{x} = \mathbf{b}$, where A is m-by-n, $\mathbf{x}$ is n-by-1, and $\mathbf{b}$ is m-by-1. If we could find a matrix C n-by-m with $CA = I_n$, then we would have a solution to our system, namely $\mathbf{x} = C\mathbf{b}$. This leads us to consider one-sided inverses of a rectangular matrix, which is a first step in understanding generalized inverses.

DEFINITION 3.8 *(left, right inverses)*

*Suppose A is an m-by-n matrix. We say B in $\mathbb{C}^{n\times m}$ is a **left inverse** for A iff $BA = I_n$. Similarly we call C in $\mathbb{C}^{n\times m}$ a **right inverse** for A if $AC = I_m$.*

The first thing we notice is a loss of uniqueness. For example, let $A = \begin{bmatrix} 1 & 0 \\ 0 & 1 \\ 0 & 0 \end{bmatrix}$. Then any matrix $B = \begin{bmatrix} 1 & 0 & x \\ 0 & 1 & y \end{bmatrix}$ is a left inverse for any choice of x and y. Next we consider existence. Having a one-sided inverse makes a matrix rather special.

THEOREM 3.17

Suppose A is in $\mathbb{C}^{m\times n}$. Then

1. *A has a right inverse iff A has full row rank m.*

2. *A has a left inverse iff A has full column rank n.*

PROOF (1) Suppose A has a right inverse C. Then $AC = I_m$. But partitioning C into columns, $AC = [A\mathbf{c}_1|A\mathbf{c}_2|\cdots \mid A\mathbf{c}_m] = I_m = [\mathbf{e}_1 \mid \mathbf{e}_2 \mid \cdots \mid \mathbf{e}_m]$ so each $A\mathbf{c}_i$ is the standard basis vector $\mathbf{e}_i$. Thus, $\{A\mathbf{c}_1, A\mathbf{c}_2, \cdots, A\mathbf{c}_m\}$ is a basis of $\mathbb{C}^m$. In particular, the column space of A must equal $\mathbb{C}^m$. Hence the column rank of A is m. Therefore the row rank of A is m, as was to be proved.

Conversely, suppose A has full row rank m. Then its column rank is also m, so among the columns of A, there must be a basis of the column space, which is $\mathbb{C}^m$. Call these columns $\mathbf{d}_1, \mathbf{d}_2, \cdots, \mathbf{d}_m$. Now the standard basis vectors $\mathbf{e}_1, \mathbf{e}_2, \cdots, \mathbf{e}_m$ belong to $\mathbb{C}^m$ so are uniquely expressible in terms of the $\mathbf{d}$s: say

$$\begin{aligned}\mathbf{e}_1 &= \alpha_{11}\mathbf{d}_1 + \alpha_{12}\mathbf{d}_2 + \cdots + \alpha_{1m}\mathbf{d}_m\\ \mathbf{e}_2 &= \alpha_{21}\mathbf{d}_1 + \alpha_{22}\mathbf{d}_2 + \cdots + \alpha_{2m}\mathbf{d}_m\\ &\vdots\\ \mathbf{e}_m &= \alpha_{m1}\mathbf{d}_1 + \alpha_{m2}\mathbf{d}_2 + \cdots + \alpha_{mm}\mathbf{d}_m.\end{aligned}$$

Now we will describe how to construct a right inverse C with the help of these α_{ij}s. Put $\alpha_{11}, \alpha_{21}, \cdots, \alpha_{m1}$ in the row corresponding to the column of $\mathbf{d}_1$ in A. Put $\alpha_{12}, \alpha_{22}, \cdots, \alpha_{m2}$ in the row corresponding to the column $\mathbf{d}_2$ in A. Keep going in this manner and then fill in all the other rows of C with zeros. Then $AC = [\mathbf{e}_1|\mathbf{e}_2|\cdots|\mathbf{e}_m] = I_m$. Let's illustrate in a concrete example what just happened. Suppose $A = \begin{bmatrix} a & c & e & g \\ b & d & f & h \end{bmatrix}$ has rank 2. Then there must be two columns that form a basis of the column space $\mathbb{C}^2$, say $\mathbf{d}_1 = \begin{bmatrix} c \\ d \end{bmatrix}$, $\mathbf{d}_2 = \begin{bmatrix} g \\ h \end{bmatrix}$. Then $\mathbf{e}_1 = \begin{bmatrix} 1 \\ 0 \end{bmatrix} = \alpha_{11}\begin{bmatrix} c \\ d \end{bmatrix} + \alpha_{12}\begin{bmatrix} g \\ h \end{bmatrix}$ and $\mathbf{e}_2 = \begin{bmatrix} 0 \\ 1 \end{bmatrix} = \alpha_{21}\begin{bmatrix} c \\ d \end{bmatrix} + \alpha_{22}\begin{bmatrix} g \\ h \end{bmatrix}$. Thus $AC = \begin{bmatrix} a & c & e & g \\ b & d & f & h \end{bmatrix}\begin{bmatrix} 0 & 0 \\ \alpha_{11} & \alpha_{21} \\ 0 & 0 \\ \alpha_{12} & \alpha_{22} \end{bmatrix} = \begin{bmatrix} 1 & 0 \\ 0 & 1 \end{bmatrix} = I_2$.

(2) A similar argument can be used as above or we can be more clever. A has full column rank n iff A^* has full row rank n iff A^* has a right inverse iff A has a left inverse. □

Now we have necessary and sufficient conditions that show how special you have to be to have a left inverse. The nonuniqueness is not totally out of control in view of the next theorem.

THEOREM 3.18

If A in $\mathbb{C}^{m\times n}$ has a left inverse B, then all the left inverses of A can be written as $B + K$, where $KA = \mathbb{O}$. A similar statement applies to right inverses.

PROOF Suppose $BA = I$ and $B_1A = I$. Set $K_1 = B_1 - B$. Then $K_1A = (B_1 - B)A = B_1A - BA = I - I = \mathbb{O}$. Moreover, $B_1 = B + K_1$. □

Okay, that was a pretty trivial argument, so the theorem may not be that helpful. But we can do better.

THEOREM 3.19
Let A be in $\mathbb{C}^{m\times n}$.

1. *Suppose A has full column rank n. Then A^*A is invertible and $(A^*A)^{-1}A^*$ is a left inverse of A. Thus all left inverses of A are of the form $(A^*A)^{-1}A^* + K$, where $KA = \mathbb{O}$. Indeed, we can write $K = W[I_m - A(A^*A)^{-1}A^*]$, where W is arbitrary of appropriate size. Hence all left inverses of A look like*

$$(A^*A)^{-1}A^* + W[I_m - A(A^*A)^{-1}A^*].$$

2. *Suppose A has full row rank m. Then AA^* is invertible and $A^*(AA^*)^{-1}$ is a right inverse of A. Thus all right inverses of A are of the form $A^*(AA^*)^{-1} + K$, where $AK = \mathbb{O}$. Indeed $K = [I - A^*(AA^*)^{-1}A]V$, where V is arbitrary of appropriate size. Hence all right inverses of A look like*

$$A^*(AA^*)^{-1} + [I - A^*(AA^*)^{-1}A]V.$$

PROOF Suppose A has full column rank n. Then A^*A is n-by-n and $r(A^*A) = r(A) = n$. Thus A^*A has full rank and is thus invertible. Then $[(A^*A)^{-1}A^*]A = (A^*A)^{-1}(A^*A) = I_n$. The remaining details are left as exercises.

A similar proof applies. □

As an example, suppose we wish to find all the left inverses of the matrix $A = \begin{bmatrix} 1 & 2 \\ 2 & 1 \\ 3 & 1 \end{bmatrix}$. We find $A^*A = \begin{bmatrix} 14 & 7 \\ 7 & 6 \end{bmatrix}$ and $(A^*A)^{-1} = \frac{1}{35}\begin{bmatrix} 6 & -7 \\ -7 & 14 \end{bmatrix}$. Now we can construct one left inverse of A, namely $(A^*A)^{-1}A^* = \frac{1}{35}\begin{bmatrix} -8 & 5 & 11 \\ 21 & 0 & -7 \end{bmatrix}$. The reader may verify by direct computation that this matrix is indeed a left inverse for A. To get all left inverses we use

$$C = \left(A^*A\right)^{-1}A^* + W[I_m - A\left(A^*A\right)^{-1}A^*].$$

Let $W = \begin{bmatrix} a & b & c \\ d & e & f \end{bmatrix}$ be a parameter matrix. Now $A\,(A^*A)^{-1}\,A^* = \frac{1}{35}\begin{bmatrix} 34 & 5 & -3 \\ 5 & 10 & 15 \\ -3 & 15 & 26 \end{bmatrix}$. Then

$$C = \frac{1}{35}\begin{bmatrix} -8 & 5 & 11 \\ 21 & 0 & -7 \end{bmatrix} + \begin{bmatrix} a & b & c \\ d & e & f \end{bmatrix}\begin{bmatrix} \frac{1}{35} & -\frac{5}{35} & \frac{3}{35} \\ -\frac{5}{35} & \frac{25}{35} & -\frac{15}{35} \\ \frac{3}{35} & -\frac{15}{35} & \frac{9}{35} \end{bmatrix} =$$

$$\begin{bmatrix} -\frac{8}{35} + a - 5b + 3c & \frac{5}{35} - 5a + 25b - 15c & \frac{11}{35} + 3a - 15b + 9c \\ \frac{21}{35} + d - 5e + 3f & -5d + 25e - 15f & -\frac{7}{35} + 3d - 15e + 9f \end{bmatrix}$$

$$= \begin{bmatrix} -\frac{8}{35} + t & \frac{5}{35} - 5t & \frac{11}{35} + 3t \\ \frac{21}{35} + s & -5s & -\frac{7}{35} + 3s \end{bmatrix},$$ where $t = a - 5b + 3c$ and $s = d - 5e + 3f$.

The reader may again verify we have produced a left inverse of A.

So where do we stand now in terms of solving systems of linear equations? We have the following theorem.

THEOREM 3.20

Suppose we have $A\mathbf{x} = \mathbf{b}$*, where* A *is m-by-n of rank n,* $\mathbf{x}$ *is n-by-1, and* $\mathbf{b}$*, necessarily m-by-1. This system has a solution if and only if* $A(A^*A)^{-1}A^*\mathbf{b} = \mathbf{b}$ *(consistency condition). If the system has a solution, it is (uniquely)* $\mathbf{x} = (A^*A)^{-1}A^*\mathbf{b}$.

PROOF Note $n = rank(A) = rank(A^*A)$, which guarantees the existence of $(A^*A)^{-1}$, which is n-by-n and hence of full rank. First suppose the condition $A\,(A^*A)^{-1}\,A^*\mathbf{b} = \mathbf{b}$. Then evidently $\mathbf{x}_0 = (A^*A)^{-1}A^*\mathbf{b}$ is a solution for $A\mathbf{x} = \mathbf{b}$. Conversely, suppose $A\mathbf{x} = \mathbf{b}$ has a solution $\mathbf{x}_1$. Then $A^*A\mathbf{x}_1 = A^*\mathbf{b}$ so $(A^*A)^{-1}(A^*A)\mathbf{x}_1 = (A^*A)^{-1}A^*\mathbf{b}$ so $\mathbf{x}_1 = (A^*A)^{-1}A^*\mathbf{b}$ whence $\mathbf{b} = A\mathbf{x}_1 = A(A^*A)^{-1}A^*\mathbf{b}$.

Now suppose $A\mathbf{x} = \mathbf{b}$ has a solution. A has a left inverse C so $\mathbf{x} = C\mathbf{b}$ is a solution. But C can be written as $C = (A^*A)^{-1}\,A^* + W[I_m - A\,(A^*A)^{-1}\,A^*]$. So, $C\mathbf{b} = (A^*A)^{-1}\,A^*\mathbf{b} + W[I_m - A\,(A^*A)^{-1}\,A^*]\mathbf{b} = (A^*A)^{-1}\,A^*\mathbf{b} + \mathbb{O}$ using the consistency condition. Thus, $\mathbf{x} = (A^*A)^{-1}\,A^*\mathbf{b}$. ▯

For example, consider the system

$$\begin{cases} x + 2y = 1 \\ 2x + y = 1\,. \\ 3x + y = 1 \end{cases}$$

Then $\begin{bmatrix} 1 & 2 \\ 2 & 1 \\ 3 & 1 \end{bmatrix} \begin{bmatrix} x \\ y \end{bmatrix} = \begin{bmatrix} 1 \\ 1 \\ 1 \end{bmatrix}$. We check consistency with $A(A^*A)^{-1}A^* \begin{bmatrix} 1 \\ 1 \\ 1 \end{bmatrix} = \frac{1}{35} \begin{bmatrix} 34 & 5 & -3 \\ 5 & 10 & 15 \\ -3 & 15 & 26 \end{bmatrix} \begin{bmatrix} 1 \\ 1 \\ 1 \end{bmatrix} \neq \begin{bmatrix} 1 \\ 1 \\ 1 \end{bmatrix}$ so we conclude this system is inconsistent (i.e., has no solution). However, for

$$\begin{cases} x + 2y = 3 \\ 2x + y = 3, \\ 3x + y = 4 \end{cases}$$

$\frac{1}{35} \begin{bmatrix} 34 & 5 & -3 \\ 5 & 10 & 15 \\ -3 & 15 & 26 \end{bmatrix} \begin{bmatrix} 3 \\ 3 \\ 4 \end{bmatrix} = \begin{bmatrix} 3 \\ 3 \\ 4 \end{bmatrix}$ so the system is consistent and has the unique solution $\frac{1}{35} \begin{bmatrix} -8 & 5 & 11 \\ 21 & 0 & -7 \end{bmatrix} \begin{bmatrix} 3 \\ 3 \\ 4 \end{bmatrix} = \begin{bmatrix} 1 \\ 1 \end{bmatrix}$.

Exercise Set 12

1. (a) Suppose A has a left inverse and $AB = AC$. Prove that $B = C$.
   (b) Suppose A has a right inverse and suppose $BA = CA$. Prove that $B = C$.

2. Prove that $rank(SAT) = rank(A)$ if S has full column rank and T has full row rank.

3. Argue that $KA = \mathbb{O}$ iff there exists W such that $K = W[I - A(A^*A)^{-1}A^*]$. What can you say about the situation when $AK = \mathbb{O}$?

4. Argue that A has a left inverse iff $\mathcal{N}ull(A) = (\vec{0})$.

5. Suppose $A = LK$ where L has a left inverse and K has a right inverse. Argue that $r(A) = r(L) = r(K)$.

6. Construct all left inverses of $A = \begin{bmatrix} 1 & 0 \\ 0 & 1 \\ 1 & 1 \end{bmatrix}$.

7. Argue that A has a left inverse iff A^T has a right inverse.

8. Argue that a square singular matrix has neither a left nor a right inverse.

9. Let A be an m-by-n matrix of rank m. Let $B = A^*(AA^*)^{-1}$. Show B is a right inverse for A and $A = (B^*B)^{-1}B^*$. Moreover, B is the only right inverse for A such that B^* has the same row space as A.

10. If A is m-by-n and the columns of A span $\mathbb{C}^m$, then A has a right inverse and conversely.

11. Find all right inverses of $A = \begin{bmatrix} 1 & 0 & 0 & 2 & 1 \\ 0 & 1 & 1 & 0 & 1 \\ 1 & 0 & 1 & 2 & 1 \end{bmatrix}$.

12. Give an example of a matrix that has neither a left inverse nor a right inverse.

13. Suppose A is m-by-n, B is n-by-r, and $C = AB$. Argue that if A and B both have linearly independent columns, then C has linearly independent columns. Next argue that if the columns of B are linearly dependent, then the columns of C must be linearly dependent.

14. Let $T : \mathbb{C}^n \to \mathbb{C}^m$ be a linear map. Argue that the following statements are all equivalent:
   (a) T is left invertible.
   (b) $Ker(T) = (\overrightarrow{0})$.
   (c) T:$\mathbb{C}^n \to Im(T)$ is one-to-one and onto.
   (d) $n \leq m$ and $rank(A) = n$.
   (e) The matrix of T, $Mat(T;\mathcal{B},\mathcal{C})$ in $\mathbb{C}^{m\times n}$ has $n \leq m$ and has full rank.

15. Let $T : \mathbb{C}^n \to \mathbb{C}^m$ be a linear map. Argue that the following statements are all equivalent:
   (a) T is right invertible.
   (b) $rank(T) = m$.
   (c) $T : M \to \mathbb{C}^m$ is one-to-one and onto where $M \oplus Ker(T) = \mathbb{C}^n$.
   (d) $n \geq m$ and $nlty(T) = n - m$.
   (e) The matrix of T, $Mat(T;\mathcal{B},\mathcal{C})$ in $\mathbb{C}^{m\times n}$ has $n \geq m$ and has full rank.

16. Suppose A is m-by-n and $A = FG$. Suppose F is invertible. Argue that A has a right inverse iff G has a right inverse.

17. Suppose A has a left inverse C and the linear system $A\mathbf{x} = \mathbf{b}$ has a solution. Argue that this solution is unique and must equal $C\mathbf{b}$.

18. Suppose A has a right inverse B. Argue that the linear system $A\mathbf{x} = \mathbf{b}$ has at least one solution.

19. Suppose A is an m-by-n matrix of rank r. Discuss the existence and uniqueness of left and right inverses of A in the following cases: $r = m < n, r < m < n, r = m = n, r < m = n$, and $r = n < m, r < n < m$.

20. Argue that A has a left inverse iff A^* has a right inverse.

Further Reading

[Noble, 1969] Ben Noble, *Applied Linear Algebra,* Prentice Hall, Inc., Englewood Cliffs, NJ, (1969).

[Perlis, 1952] Sam Perlis, *Theory of Matrices,* Dover Publications Inc., New York, (1952).

Chapter 4

The Moore-Penrose Inverse

RREF, leading coefficient, pivot column, matrix equivalence, modified RREF, rank normal form, row equivalence, column equivalence

4.1 Row Reduced Echelon Form and Matrix Equivalence

Though we have avoided it so far, one of the most useful reductions of a matrix is to bring it into row reduced echelon form (RREF). This is the fundamental result used in elementary linear algebra to accomplish so many tasks. Even so, it often goes unproved. We have seen that a matrix A can be reduced to many matrices in row echelon form. To get uniqueness, we need to add some requirements. First, a little language. Given a matrix A, the *leading coefficient* of a row of A is the first nonzero entry in that row (if there is one). Evidently, every row not consisting entirely of zeros has a unique leading coefficient. A column of A that contains the leading coefficient of at least one row is called a *pivot column*.

DEFINITION 4.1 *(row reduced echelon form)*
A matrix A in $\mathbb{C}^{m\times n}$ is said to be in ***row reduced echelon form*** *iff*

1. *For some integer $r \geq 0$, the first r rows are nontrivial (not totally filled with zeros) and all the remaining rows (if there are any) are totally filled with zeros.*

2. *Row 1, row 2, $\cdots$ up to and including row r has its first nonzero entry a 1 (called a* leading one*).*

3. *Suppose the leading ones occur in columns $c_1, c_2, \cdots, c_r$. Then $c_1 < c_2 < \cdots < c_r$.*

4. *In any column with a leading one, all the other entries in that column are zero.*

For example,

$$\begin{bmatrix} 0 & 0 & 1 & 4 & 6 & 0 & 0 & 7 & 5 \\ 0 & 0 & 0 & 0 & 0 & 1 & 0 & 6 & 3 \\ 0 & 0 & 0 & 0 & 0 & 0 & 1 & 4 & 8 \\ 0 & 0 & 0 & 0 & 0 & 0 & 0 & 0 & 0 \\ 0 & 0 & 0 & 0 & 0 & 0 & 0 & 0 & 0 \end{bmatrix} \text{ is in row reduced echelon form.}$$

In other words, we have RREF iff each leading coefficient is one, any zero row occurs at the bottom, in each pair of successive rows that are not totally zero, the leading coefficient of the first row occurs in an earlier column than the leading coefficient of the later row and each pivot column has only one nonzero entry, namely a leading one. In particular, all entries below and to the left of a leading one are zero. Do you see better now why the word "echelon" is used? Notice, if we do not demand condition (4), we simply say the matrix is in row echelon form (REF) so that it may not be "reduced." In the next theorem, we shall prove that each matrix A in $\mathbb{C}^{m\times n}$ can be reduced by elementary row operations to a unique matrix in RREF. This is not so if condition (4) is not required.

THEOREM 4.1

Let A be in $\mathbb{C}^{m\times n}$. Then there exists a finite sequence of elementary matrices $E_1, E_2, \cdots, E_k$ such that $E_k E_{k-1} \cdots E_2 E_1 A$ is in RREF. Moreover, this matrix is unique and we denote it $RREF(A)$, though the sequence of elementary matrices that produce it is not. Moreover, if r is the rank of A, then $RA = \begin{bmatrix} G_{r\times n} \\ \cdots \\ \mathbb{O}_{(m-r)\times n} \end{bmatrix}$ where $R = E_k E_{k-1} \cdots E_1$ is in $\mathbb{C}^{m\times m}$ and is invertible. In fact, the rank of G is r where G is r-by-n. In particular, $A = R^{-1}\begin{bmatrix} G \\ \cdots \\ \mathbb{O} \end{bmatrix}$. Moreover, $\mathcal{R}ow(A) = \mathcal{R}ow(G)$.

PROOF If $A = \mathbb{O}$, then A is in RREF and $RREF(A) = \mathbb{O}$. So suppose $A \neq \mathbb{O}$. Then A must have a column with nonzero entries. Let c_1 be the number of the first such column. If the $(1, c_1)$ entry is zero, use a permutation matrix P to swap a nonzero entry into the $(1, c_1)$ position. Use a dilation, if necessary, to make this element 1. If any element below this 1 is nonzero, say α in the (j, c_1) position, use the transvection $T_{j1}(-\alpha)$ to make it zero. In this way, all the entries of column c_1 except the $(1, c_1)$ entry can be made zero.Thus far we have

$$A \rightarrow T_1 D_1 P_1 A = \begin{bmatrix} & 1 & * & \cdots & * \\ \mathbb{O} & 0 & * & & * \\ & 0 & * & & * \\ & \vdots & \vdots & & \vdots \\ & 0 & * & & * \end{bmatrix}$$. Now, if all the rows below the first row consist entirely of zeros, we have achieved RREF and we are done. If not, there will be a first column in the matrix above that has a nonzero entry, β say, below row 1. Suppose the number of this column is c_2. Evidently $c_1 < c_2$. By a permutation matrix (if necessary), we can swap rows and move β into row 2, keeping β in column c_2. By a dilation, we can make β be 1 and we can use transvections to "zero out"the entries above and below the $(2, c_2)$ position. Notice column c_1 is unaffected due to all the zeros produced earlier in that column. Continuing in this manner, we must eventually terminate when we get to a column c_r, which will be made to have a 1 in the (r, c_r) position with zeros above and below it. Either there are no more rows below row r or, if there are, they consist entirely of zeros.

Now we argue the uniqueness of the row reduced echelon form of a matrix. As above, let A be an m-by-n matrix. We follow the ideas of Yuster [1984] and make an inductive argument. For other arguments see Hoffman and Kunze [1971, p. 56] or Meyer [2000, p. 134]. Fix m. We proceed by induction on the number of columns of A. For $n = 1$, the result is clear (isn't it always?), so assume n > 1. Now suppose uniqueness of RREF holds for matrices of size m-by-(n-1). We shall show uniqueness also holds for matrices of size m-by-n and our result will follow by induction.

Suppose we produce from A a matrix in row reduced echelon form in two different ways by using elementary row operations. Then there exist R_1 and R_2 invertible with $R_1 A = Ech_1$ and $R_2 A = Ech_2$, where Ech_1 and Ech_2 are matrices in RREF. We would like to conclude $Ech_1 = Ech_2$. Note that if we partition A by isolating its last column, $A = [A' \mid col_n(A)]$, then $R_1 A = [R_1 A' \mid R_1 col_n(A)] = Ech_1$ and $R_2 A = [R_2 A' \mid R_2 col_n(A)] = Ech_2$. Here is a key point of the argument: any sequence of elementary row operations that yields RREF for A also puts A' into row reduced echelon form. Hence, by the induction hypothesis, we have $R_1 A' = R_2 A' = A''$ since A' is m-by-(n-1). We distinguish two cases.

CASE 4.1

Every row of A'' has a leading 1. This means there are no totally zero rows in this matrix. Then, by Theorem 2.7 of Chapter 2, the columns of A corresponding to the columns with the leading one's in A'' form an independent set and the last column of A is a linear combination of corresponding columns in A with the coefficients coming from the last column of Ech_1. But the same is true about the last column of Ech_2. By independence, these scalars are uniquely determined so the last column of Ech_1 must equal the last column of Ech_2. But the last column is the only place where Ech_1 could differ from Ech_2, so we conclude in this case, $Ech_1 = Ech_2$.

CASE 4.2

A'' has at least one row of zeros. Let's assume $Ech_1 \neq Ech_2$ and seek a contradiction. Now again, the only place Ech_1 can differ from Ech_2 is in the last column, so there must exist a j with $b_{jn} \neq c_{jn}$, where $b_{jn} = ent_{jn}(Ech_1)$ and $c_{jn} = ent_{jn}(Ech_2)$. Recall from Theorem 3.2 of Chapter 3 that $\mathcal{N}ull(Ech_1) = \mathcal{N}ull(A) = \mathcal{N}ull(Ech_2)$. Let's compare some null spaces. Suppose $\mathbf{x} = \begin{bmatrix} x_1 \\ \vdots \\ x_n \end{bmatrix} \in \mathcal{N}ull(A)$. Then $Ech_1\mathbf{x} = Ech_2\mathbf{x} = \overrightarrow{0}$ so $(Ech_1 - Ech_2)\mathbf{x} = \overrightarrow{0}$. But the first $n-1$ columns of $Ech_1 - Ech_2$ are zero so $\left[\mathbb{O} \mid \begin{matrix} b_{1n} - c_{1n} \\ \vdots \\ b_{mn} - c_{mn} \end{matrix}\right] \begin{bmatrix} x_1 \\ \vdots \\ x_n \end{bmatrix} = \begin{bmatrix} 0 \\ \vdots \\ 0 \end{bmatrix}$, which implies $(b_{in} - c_{in})x_n = 0$ for all i—in particular, when $i = j$, $(b_{jn} - c_{jn})x_n = 0$. By assumption, $b_{jn} - c_{jn} \neq 0$, so this forces $x_n = 0$, a rather specific value. Now if $\mathbf{u} = \begin{bmatrix} u_1 \\ \vdots \\ u_n \end{bmatrix} \in \mathcal{N}ull(Ech_1)$, then $\overrightarrow{0} = Ech_1\mathbf{u} = \left[A'' \mid \begin{matrix} b_{1n} \\ \vdots \\ b_{mn} \end{matrix}\right] \begin{bmatrix} u_1 \\ \vdots \\ u_n \end{bmatrix} = \begin{bmatrix} * \\ * \\ \vdots \\ b_{(k+1)n}u_n \\ b_{(k+2)n}u_n \\ \vdots \\ b_{mn}u_n \end{bmatrix}$ where k + 1 is the first full row of zeros in A''. If $b_{(k+1)n}, \ldots, b_{mn}$ all equal zero, then u_n can be any number and we can construct vectors $\mathbf{u}$ in $\mathcal{N}ull(Ech_1)$ without a zero last entry. This contradicts what we deduced above. Thus, some b in that list must be nonzero. If $b_{(k+1)n}$ were zero, this would contradict that Ech_1 is in RREF. So, $b_{(k+1)n}$ must be nonzero. Again quoting row reduced echelon form, $b_{(k+1)n}$ must be a leading one hence all the other bs other than $b_{(k+1)n}$ must be zero. But exactly the same argument applies to Ech_2 so $c_{(k+1)n}$ must be one and all the other cs zero. But then the last column of Ech_1 is identical to the last column of Ech_2, so once again we conclude, $Ech_1 = Ech_2$. This is our ultimate contradiction that establishes the theorem. □

For example,

$$RREF\begin{bmatrix} i & 2-4i & 3 & 4i & 0 \\ 3 & 2-7i & 1 & 2i & 0 \\ 3+2i & 6-15i & 7 & 10i & 0 \\ 0 & 0 & 0 & 0 & 0 \end{bmatrix}$$

$$= T_{12}(4+2i)D_2 114 - iT_{32}(-1)D_1(-i)T_{31}(-2+3i)$$

$$T_{21}(3i)\begin{bmatrix} i & 2-4i & 3 & 4i & 0 \\ 3 & 2-7i & 1 & 2i & 0 \\ 3+2i & 6-15i & 7 & 10i & 0 \\ 0 & 0 & 0 & 0 & 0 \end{bmatrix}$$

$$= \begin{bmatrix} \frac{-96}{197} - \frac{35}{197}i & \frac{54}{197} + \frac{32}{197}i & 0 & 0 \\ \frac{-3}{197} + \frac{42}{197}i & \frac{14}{197} + \frac{1}{197}i & 0 & 0 \\ -2 & -1 & 1 & 0 \\ 0 & 0 & 0 & 1 \end{bmatrix}\begin{bmatrix} i & 2-4i & 3 & 4i & 0 \\ 3 & 2-7i & 1 & 2i & 0 \\ 3+2i & 6-15i & 7 & 10i & 0 \\ 0 & 0 & 0 & 0 & 0 \end{bmatrix} =$$

$$\begin{bmatrix} 1 & 0 & \frac{-234}{197} - \frac{73}{197}i & \frac{76}{197} - \frac{276}{197}i & 0 \\ 0 & 1 & \frac{5}{197} + \frac{127}{197}i & \frac{-170}{197} + \frac{16}{197}i & 0 \\ 0 & 0 & 0 & 0 & 0 \\ 0 & 0 & 0 & 0 & 0 \end{bmatrix}.$$

Now, do you think we used a computer on this example or what?

The reader will no doubt recall that the process of Gauss elimination gives an algorithm for producing the row reduced echelon form of a matrix, and this gave a way of easily solving a system of linear equations. However, we have a different use of RREF in mind. But let's quickly sketch that algorithm for using RREF to solve a system of linear equations:

1. Write the augmented matrix of the system $A\mathbf{x} = \mathbf{b}$, namely $[A \mid \mathbf{b}]$.
2. Calculate $RREF([A \mid \mathbf{b}])$.
3. Write the general solution, introducing free variables for each nonpivot column.

Note that if every column of the RREF coefficient matrix has a leading one, the system can have at most one solution. If the final column also has a leading one, the system is inconsistent (i.e., has no solution). Recall that the number of leading ones in $RREF(A)$ is called the *pivot rank* ($p\text{-}rank(A)$) of A. Finally, recall that a system $A\mathbf{x} = \mathbf{b}$ with n unknowns has no solution if $p\text{-}rank(A) \neq p\text{-}rank([A \mid \mathbf{b}])$. Otherwise the general solution has $n - (p\text{-}rank(A))$ free variables.

4.1.1 Matrix Equivalence

Next, we consider *matrix equivalence*. Recall that we say that matrix $A \in \mathbb{C}^{m\times n}$ is *equivalent* to matrix $B \in \mathbb{C}^{m\times n}$ if B can be obtained from A by applying both elementary row and elementary column operations to A. In other words, A is equivalent to B if there exist invertible matrices $S \in \mathbb{C}^{m\times m}$ and T in $\mathbb{C}^{n\times n}$ such that $SAT = B$. Recall that invertible matrices are the same as products of elementary matrices so we have not said anything different.

DEFINITION 4.2 *(matrix equivalence)*
Let A and B be m-by-n matrices. We say A is equivalent to B and write in symbols $A \approx B$ iff there exist invertible matrices S in $\mathbb{C}^{m\times m}$ and T in $\mathbb{C}^{n\times n}$ such that $B = SAT$.

This is our first example of what mathematicians call an *equivalence relation* on matrices. There are in fact many such relations on matrices. But they all share the following crucial properties:

- *Reflexive Law*: Every matrix is related to itself.
- *Symmetric Law*: If matrix A is related to matrix B, then matrix B must also be related to matrix A.
- *Transitive Law*: If matrix A is related to matrix B and matrix B is related to matrix C, then matrix A is related to matrix C.

Clearly, *equality* is such a relation (i.e., $A = B$). Do the names of the Laws above make any sense to you? Let's make a theorem.

THEOREM 4.2
Matrix equivalence is an equivalence relation; that is,

1. *$A \approx A$ for all $A \in \mathbb{C}^{m\times n}$.*
2. *If $A \approx B$, then $B \approx A$ where A and B are in $\mathbb{C}^{m\times n}$.*
3. *If $A \approx B$ and $B \approx C$ then, $A \approx C$ where A, B and C are in $\mathbb{C}^{m\times n}$.*

PROOF The proof is left as an exercise. □

The nice thing about equivalence relations is that they partition the set of matrices $\mathbb{C}^{m\times n}$ into disjoint classes such that any two matrices that share a class are equivalent and any two matrices in different classes are not equivalent. Mathematicians like to ask, is there an easy way to check if two matrices are in the same class? They also want to know if each class is represented by a particularly nice matrix. In the meantime, we want to extend the notion of RREF.

You may have noticed that the leading ones of RREF do not always line up nicely to form a block identity matrix in the RREF. However, all it takes is some column swaps to get an identity matrix to show up. The problem is that this introduces column operations, whereas RREF was accomplished solely with row operations. However, the new matrix is still equivalent to the one we started with. Thus, we will speak of a *modified RREF* of a matrix when we use a permutation matrix on the right to get the leading ones of RREF to have an identity matrix block.

THEOREM 4.3

Suppose $A \in \mathbb{C}_r^{m\times n}$. Then there exists an invertible matrix R and a permutation matrix P such that $RAP = \begin{bmatrix} I_r & C \\ \mathbb{O} & \mathbb{O} \end{bmatrix}$. Moreover, the first r columns of AP form a basis for the column space of A, the columns of $P \begin{bmatrix} -C \\ \cdots \\ I_{n-r} \end{bmatrix}$ form a basis for the null space of A, and the columns of $P \begin{bmatrix} I_r \\ \cdots \\ C^* \end{bmatrix}$ form a basis of the column space of A^*.

PROOF We know there is an invertible matrix R such that $RREF(A) = RA = \begin{bmatrix} G \\ \cdots \\ \mathbb{O} \end{bmatrix}$. Suppose the pivot columns occur at $c_1, c_2, \ldots, c_r$ in RA. Consider the permutation σ that takes c_1 to 1, c_2 to 2, and so on, and leaves every thing else fixed. Form the permutation matrix $P(\sigma)$. In other words, column j of $P(\sigma)$ is $\mathbf{e}_{c_j}$ and the other columns agree with the columns of the identity matrix. Then $RAP = \begin{bmatrix} I_r & C \\ \mathbb{O} & \mathbb{O} \end{bmatrix}$ and the first r columns of AP form a basis for the column space of A. The remaining details are left to the reader. $\square$

For example, consider $A = \begin{bmatrix} 0 & 0 & 1 & 4 & 6 & 0 & 0 & 7 & 5 \\ 0 & 0 & 0 & 0 & 0 & 1 & 0 & 6 & 3 \\ 0 & 0 & 0 & 0 & 0 & 0 & 1 & 4 & 8 \\ 0 & 0 & 0 & 0 & 0 & 0 & 0 & 0 & 0 \\ 0 & 0 & 0 & 0 & 0 & 0 & 0 & 0 & 0 \end{bmatrix}$, which is already in RREF. The swaps we need to make are clear: $1 \leftrightarrow 3, 2 \leftrightarrow 6$, and $3 \leftrightarrow 7$. Thus, the permutation $\sigma = (37)(26)(13) = (173)(26)$ and the permutation matrix is $P(\sigma) = \begin{bmatrix} 0&0&0&0&0&0&1&0&0 \\ 0&0&0&0&0&1&0&0&0 \\ 1&0&0&0&0&0&0&0&0 \\ 0&0&0&1&0&0&0&0&0 \\ 0&0&0&0&1&0&0&0&0 \\ 0&1&0&0&0&0&0&0&0 \\ 0&0&1&0&0&0&0&0&0 \\ 0&0&0&0&0&0&0&1&0 \\ 0&0&0&0&0&0&0&0&1 \end{bmatrix}$. Finally,

$$AP(\sigma) = \begin{bmatrix} 1&0&0&4&6&0&0&7&5 \\ 0&1&0&0&0&0&0&6&3 \\ 0&0&1&0&0&0&0&4&8 \\ 0&0&0&0&0&0&0&0&0 \\ 0&0&0&0&0&0&0&0&0 \end{bmatrix}.$$

You may be wondering, why stop at a permutation matrix at the right of A. Once you have that identity matrix block, you can continue with transvections (i.e., column operations pivoting off the ones) and "zero out" the matrix C. This is, in fact, the case and leads to a very important result and normal form.

THEOREM 4.4 (rank normal form)

Any matrix $A \in \mathbb{C}_r^{m\times n}$ is equivalent to a unique matrix of the form I_r if $m = n = r$, $\begin{bmatrix} I_r & \mathbb{O} \\ \mathbb{O} & \mathbb{O} \end{bmatrix}$ if $m > r$, $n > r$, $[I_r \vdots \mathbb{O}]$ if $m = r < n$ or $\begin{bmatrix} I_r \\ \cdots \\ \mathbb{O} \end{bmatrix}$ if $m > r = n$, called the rank normal form *of A and denoted $RNF(A)$. The zero matrix is in a class by itself.*

PROOF Let $A \in \mathbb{C}_r^{m\times n}$. If we understand that there can be empty blocks of zeros, we argue that $A \approx \begin{bmatrix} I_r & \mathbb{O} \\ \mathbb{O} & \mathbb{O} \end{bmatrix}$. First apply row operations (i.e., elementary matrices on the left) to produce $RREF(A)$. Say the column numbers of the leading ones are $c_1, c_2, \cdots, c_r$. Then use column swaps (i.e., permutation matrices on the right) to produce $\begin{bmatrix} I_r & B \\ \mathbb{O} & \mathbb{O} \end{bmatrix}$. Finally, use transvections with the help of the ones in I_r to zero-out B row by row. You have now achieved $SAT = \begin{bmatrix} I_r & \mathbb{O} \\ \mathbb{O} & \mathbb{O} \end{bmatrix}$. ▯

COROLLARY 4.1

Let A, B belong to $\mathbb{C}^{m\times n}$. Then A is equivalent to B iff A and B have the same rank.

PROOF The proof is left as an exercise. □

These results give a very nice way to determine the classes under matrix equivalence. For example, $\mathbb{C}^{2\times 4}$ has three classes; $\{\mathbb{O}_{2\times 4}\}$, {all matrices of rank 1}, {all matrices of rank 2}. Actually, we can do a bit better. We can pick a *canonical representative* of each class that is very nice. So the three classes are $\{\mathbb{O}_{2\times 4}\}$, {all matrices equivalent to $\begin{bmatrix} 1 & 0 & 0 & 0 \\ 0 & 0 & 0 & 0 \end{bmatrix}$}, and {all matrices equivalent to $\begin{bmatrix} 1 & 0 & 0 & 0 \\ 0 & 1 & 0 & 0 \end{bmatrix}$}.

There is an algorithm for producing matrices S and T that put A into its rank normal form. First, adjoin the identity matrix to A, $[A \mid I]$. Next, row reduce this augmented matrix: $[A \mid I] \to [RREF(A) \mid S]$. Then, S is an invertible matrix with $SA = RREF(A)$. Now, form the augmented matrix $\left[\frac{RREF(A)}{I}\right]$ and column reduce it: $\left[\frac{RREF(A)}{I}\right] \to \left[\frac{RNF(A)}{T}\right]$. Then T is invertible and $SAT = RNF(A)$.

We can refine the notion of matrix equivalence to *row equivalence* and *column equivalence* by acting on just one side of the matrix with elementary operations.

DEFINITION 4.3 *(row equivalence, column equivalence)*

1. *We say the matrices A and B in $\mathbb{C}^{m\times n}$ are* row equivalent *and write $A \backsim_R B$ iff there exists an invertible matrix S with $SA = B$.*

2. *We say A and B are* column equivalent *and write $A \backsim_C B$ iff there is a nonsingular matrix T with $AT = B$.*

In other words, A is row equivalent to B if we can obtain B from A by performing a finite sequence of elementary row operations, and A is column equivalent to B iff B can be obtained from A by performing a finite sequence of elementary column operations on B.

THEOREM 4.5

Let A and B be in $\mathbb{C}^{m\times n}$. Then the following statements are all equivalent:

1. $A \backsim_R B$.

2. $RREF(A) = RREF(B)$.

3. $Col(A^T) = Col(B^T)$.

4. $\mathcal{N}ull(A) = \mathcal{N}ull(B)$.

PROOF The proof is left as an exercise. □

A similar theorem holds for column equivalence.

THEOREM 4.6

Let A and B be in $\mathbb{C}^{m\times n}$. Then the following statements are all equivalent:

1. $A \backsim_C B$.
2. $RREF(A^T) = RREF(B^T)$.
3. $\mathcal{C}ol(A) = \mathcal{C}ol(B)$.
4. $\mathcal{N}ull(A^T) = \mathcal{N}ull(B^T)$.

PROOF As usual, the proof is left to the reader. □

Exercise Set 13

1. Prove that matrix equivalence is indeed an equivalence relation. (This is Theorem 4.2.)

2. Prove that matrices A and B are equivalent if and only if they have the same rank. (This is Corollary 4.1.)

3. Describe the equivalence classes in $\mathbb{C}^{3\times 3}$ under $\approx$.

4. Reduce $A = \begin{bmatrix} 1240 & 862 & 1593 & 2278 \\ 2300 & 2130 & 1245 & 2620 \\ 2404 & 2200 & 1386 & 2818 \\ 488 & 438 & 309 & 598 \end{bmatrix}$ to its canonical form under $\approx$.

5. You may have noticed that your calculator has two operations that do row reductions, ref and rref. Note that rref is what we talked about above, RREF. This is unique. There is a weaker version of row reduction that is not unique, ref–row echelon form. Here you demand (1) all totally zero rows are at the bottom and (2) if the first nonzero entry in row i is at position k, then all the entries below the ith position in all previous columns are zero. Argue that the positions of the pivots are uniquely determined even though the row echelon form need not be unique. Argue that the number of pivots is the rank of A, which is the same as the number of nonzero rows in any row echelon form. If you call a column of A *basic*

if it contains a pivot position, argue the rank of A is the same as the number of basic columns.

6. Is there a notion of column reduced echelon form, $CREF(A)$? If so, formulate it.

7. Suppose A is square and nonsingular. Is $A \approx A^{-1}$? Is $A \backsim_R A^{-1}$? Is $A \backsim_C A^{-1}$?

8. Argue that $A \approx B$ iff $A^T \approx B^T$.

9. Prove that $A \backsim_R B$ iff $A^T \backsim_C B^T$.

10. Argue that $A \sim_C B$ or $A \backsim_R B$ implies $A \approx B$.

11. We say A and B are *simultaneously diagonable* with respect to equivalence if there exist nonsingular matrices S and T such that $SAT = D_1$ and $SBT = D_2$ where D_1 and D_2 are diagonal. Create a pair of 2-by-3 matrices that are simultaneously diagonable with respect to equivalence.

12. Prove Theorem 4.5.

13. Prove Theorem 4.6.

14. Argue that the linear relationships that exist among the columns of *RREF*(A), which are easy to see, are exactly the same as the linear relationships that exist among the columns of A. (*Hint:* Recall Theorem 2.7 of Chapter 2.)

15. Many people call the columns of A corresponding to the leading one columns of $RREF(A)$ the *basic columns* of A. Of course, the other columns of A are called *nonbasic columns*. Argue that the basic columns of A form a basis of the column space of A. Indeed, only the basic columns occurring to the left of a given nonbasic column are needed to express this nonbasic column as a linear combination of basic ones.

16. Argue that if A is row equivalent to B, then any linear relationship among the columns of A must also exist among the same columns of B with the same coefficients.

17. In view of exercise 16, what can you say if A is column equivalent to B?

18. Explain why the algorithm for producing the RNF of a matrix given in the text above works.

19. Here is another proof that row rank equals column rank: We use RREF in this argument. First let $R = RREF(A) = SA$, where S is m-by-m

invertible and A is m-by-n. Conclude that $\mathcal{R}ow(R) = \mathcal{R}ow(A)$. Write $A = [\mathbf{a}_1 \mid \mathbf{a}_2 \mid \cdots \mid \mathbf{a}_n]$ so that $R = SA = [S\mathbf{a}_1 \mid S\mathbf{a}_2 \mid \cdots \mid S\mathbf{a}_n]$. Let $\mathcal{B} = \{S\mathbf{a}_{j_1}, S\mathbf{a}_{j_2}, \ldots, S\mathbf{a}_{j_r}\}$ be the columns of R with the leading ones in them. Argue that $\mathcal{B}$ is a basis for $\mathcal{C}ol(R)$. Since S is invertible, argue that $\{\mathbf{a}_{j_1}, \mathbf{a}_{j_2}, \ldots, \mathbf{a}_{j_r}\}$ is an independent set. If $\mathbf{c}_j$ is any column of A, $S\mathbf{c}_j$ is a linear combination of the columns of $\mathcal{B}$. Therefore, conclude $\mathbf{c}_j$ is a linear combination of $\mathbf{c}_{j_1}, \mathbf{c}_{j_2}, \ldots, \mathbf{c}_{j_r}$. Conclude that $dim(\mathcal{R}ow(A)) = r = dim(\mathcal{C}ol(A))$.

20. If A is nonsingular n-by-n and B is n-by-r, argue that $RREF([A \mid B]) = [I \mid A^{-1}B]$.

21. If two square matrices are equivalent, argue that they are either both invertible or both singular.

22. Suppose T is a linear map from $\mathbb{C}^n$ to $\mathbb{C}^m$. Suppose $\mathcal{B}$ and $\mathcal{B}_1$ are bases of $\mathbb{C}^n$ and $\mathcal{C}$ and $\mathcal{C}_1$ are bases of $\mathbb{C}^m$. Argue that $Mat(T;\mathcal{B}_1,\mathcal{C}_1) = PMat(T;\mathcal{B},\mathcal{C})Q^{-1}$, where P is the transition matrix for $\mathcal{C}$ to $\mathcal{C}_1$ and Q is the transition matrix from $\mathcal{B}$ to $\mathcal{B}_1$. Deduce that any two matrix representations of T are matrix equivalent.

23. Tell whether the following matrices are in REF, RREF, or neither:
$$\begin{bmatrix} 2 & 4 & 6 & 8 \\ 0 & 1 & 3 & 5 \\ 0 & 0 & 7 & 9 \end{bmatrix}, \begin{bmatrix} 5 & 5 & 0 & 0 & 0 \\ 0 & 0 & 7 & 3 & 0 \\ 0 & 0 & 0 & 0 & 2 \end{bmatrix}, \begin{bmatrix} 1 & 0 & 0 \\ 0 & 1 & 2 \end{bmatrix},$$
$$\begin{bmatrix} 1 & 2 & 3 & 4 \\ 1 & 0 & 1 & 0 \end{bmatrix}, \begin{bmatrix} 1 & 0 & 0 & 0 & 7 \\ 0 & 0 & 1 & 0 & 5 \\ 0 & 1 & 0 & 2 & 3 \end{bmatrix}.$$

24. Make up an example to find two different REFs for one matrix.

25. If you are brave, do you think you can find the RREF of a matrix with polynomial entries? Try $\begin{bmatrix} x-4 & 3 & 3 \\ 2 & x-1 & 1 \\ -3 & -3 & x-9 \end{bmatrix}$.

26. Suppose two people try to solve $A\mathbf{x} = \mathbf{b}$ but choose different orders for listing the unknowns. Will they still necessarily get the same free variables?

27. Use rank normal form of a matrix A of rank r to prove that the largest number of columns (or rows) of A that are linearly independent is r. Argue that this is equivalent to saying A contains an r-by-r nonsingular submatrix and every $r+1$-by-$r+1$ submatrix is singular.

28. Fill in the details of the proof of Theorem 4.3.

Further Reading

[H&K, 1971] Kenneth Hoffman and Ray Kunze, *Linear Algebra*, 2nd Edition, Prentice Hall Inc., Englewood Cliffs, NJ, (1971).

[L&S, 2000] Steven L. Lee and Gilbert Strang, Row Reduction of a Matrix and $A = CaB$, The American Mathematical Monthly, Vol. 107, No. 8, October, (2000), 681–688.

[Yuster, 1984] Thomas Yuster, The Reduced Row Echelon Form of a Matrix Is Unique: A Simple Proof, The American Mathematical Monthly, Vol. 57, No. 2, March, (1984), 93–94.

4.1.2 MATLAB Moment

4.1.2.1 Row Reduced Echelon Form

MATLAB has a built in command to produce the RREF of a matrix A. The command is

$$rref(A)$$

Let's look at some examples.

```
>> B=round(10*rand(3,4))+round(10(3,4))*i

B =

  Columns 1 through 3

   1.0000 + 8.0000i   6.0000 + 7.0000i        0 + 7.0000i
   2.0000 + 5.000i    3.0000 + 8.0000i   7.0000 + 4.0000i
   2.0000 + 2.0000i        2.0000        4.0000 + 8.0000i

  Column 4

   9.0000 + 5.0000i
   5.0000 + 7.0000i
   4.0000 + 4.0000i

>>rref(B)

 Columns 1 through 3

   1.0000        0         0
        0   1.0000         0
        0        0    1.0000
```

Column 4

$-2.2615 - 1.2923i$
$1.5538 + 1.0308i$
$1.0462 + 0.1692i$

There is actually more you can do here. The command

$$[R,\ jb] = \mathit{rref}(A)$$

returns the RREF R and a vector jb so that jb lists the basic variables in the linear system Ax=b, r=length(jb) estimates the rank of A, A(:,jb) gives a basis for the column space of A. Continuing our example above,

```
>>[R, jb]=rref(B)

R=

 Columns 1 through 3

    1.0000       0         0
      0       1.0000       0
      0          0      1.0000

  Column 4

    -2.2615 - 1.2923i
    1.5538 + 1.0308i
    1.0462 + 0.1692i

jb =
   1  2  3
```

Let's get a basis for the column space of B.

```
>>B(:,jb)

ans =

   1.0000 + 8.0000i   6.0000 + 7.0000i        0 + 7.0000i
   2.0000 + 5.000i    3.0000 + 8.0000i   7.0000 + 4.0000i
   2.0000 + 2.0000i        2.0000        4.0000 + 8.0000i
```

Of course, this answer is not surprising. You might try experimenting with the matrix C=[1 2 3;2 4 5;3 6 9].

There is a really cool command called

$$\mathit{rrefmovie}(A)$$

This steps you through the process element by element as RREF is achieved for the given matrix.

4.1.3 Numerical Note

4.1.3.1 Pivoting Strategies

In theory, Gauss elimination proceeds just fine as long as you do not run into a zero diagonal entry at any step. However, pivots that are very small (i.e., near zero) can cause trouble in finite-precision arithmetic. If the pivot is small, the multipliers derived from it will be large. A smaller multiplier means that earlier errors are multiplied by a smaller number and so have less effect being carried forward. Equations (rows) can be *scaled* (multiplied by a nonzero constant), so we should choose as pivot an element that is relatively larger in absolute value than the other elements in its row. This is called *partial pivoting*. This will make the multipliers less than 1 in absolute value. One approach is to standardize each row by dividing row i by $\sum_j |a_{ij}|$. Or we can just choose the largest magnitude coefficient a_{kj} to eliminate the other x_k coefficients.

An easy example will illustrate what is going on. Consider the system

$$\begin{aligned} \frac{1}{n}x + y &= a \\ x + y &= b \end{aligned}$$

where n is very large compared to a and b. Using elementary operations, we get

$$\begin{aligned} \frac{1}{n}x + y &= a \\ (1-n)y &= b - na. \end{aligned}$$

Thus

$$\begin{aligned} y &= \frac{b-na}{1-n} \\ x &= (a-y)n. \end{aligned}$$

When n is very large, the computer will see $1-n$ as $-n$ and $b-na$ as $-na$ so the answer for y will be a, and hence x will be zero. In effect, b and 1 are overwhelmed by the size of n so as to disappear. On the other hand, if we simply swap the two equations,

$$\begin{aligned} x + y &= b \\ \frac{1}{n}x + y &= a \end{aligned}$$

and eliminate as usual, we get

$$\begin{aligned} x + y &= b \\ \left(1 - \frac{1}{n}\right) y &= a - \frac{b}{n} \end{aligned}$$

so

$$y = \frac{a - \frac{b}{n}}{1 - \frac{1}{n}}$$
$$x = b - y.$$

In summary then, the idea on partial pivoting is to look below the current pivot and locate the element in that column with the largest absolute value. Then do a row swap to get that element into the diagonal position. This ensures the multipliers will be less than or equal to 1 in absolute value. There is another strategy called *complete pivoting*. Here one searches not just below the current pivot, but in all remaining rows and columns. Then row and column swaps are necessary to get the element with largest absolute value into the pivot position. The problem is you have to keep track of row and column swaps. The column swaps do disturb the solution space so you have to keep up with changing the variables. Also, all this searching can use up lots of computer time.

4.1.3.2 Operation Counts

Operation counts give us a rough idea as to the efficiency of an algorithm. We count the number of additions/subtractions and multiplications/divisions. Suppose A is an n-by-n matrix and we wish to solve $A\mathbf{x} = \mathbf{b}$ (n equations in n unknowns).

Algorithm

1.	Gauss Elimination with back substitution	**Additions/Subtractions** $\frac{1}{3}n^3 + \frac{1}{2}n^2 - \frac{5}{6}n$ **Multiplications/Divisions** $\frac{1}{3}n^3 + n^2 - \frac{1}{3}n$
2.	Gauss-Jordan elimination (RREF)	**Additions/Subtractions** $\frac{1}{3}n^3 + \frac{1}{2}n^2 - \frac{5}{6}n$ **Multiplications/Divisions** $\frac{1}{3}n^3 + n^2 - \frac{1}{3}n$
3.	Cramer's rule	**Additions/Subtractions** $\frac{1}{3}n^4 - \frac{1}{6}n^3 - n^2 + \frac{1}{6}n$ **Multiplications/Divisions** $\frac{1}{3}n^4 + \frac{1}{3}n^3 + \frac{2}{3}n^2 + \frac{2}{3}n - 1$
4.	$\mathbf{x} = A^{-1}\mathbf{b}$ if A is invertible	**Additions/Subtractions** $n^3 - n^2$ **Multiplications/Divisions** $n^3 + n^2$

Interestingly, 1 and 2 have the same number of counts. To understand why, note that both methods reduce the augmented matrix to a REF. We leave it as an exercise to see that the number of operations to do back substitutions is the same as continuing to RREF.

Further Reading

[Anton, 1994] Howard Anton, *Elementary Linear Algebra*, 7th Edition, John Wiley & Sons, New York, (1994).

[C&deB 1980] Samuel D. Conte and Carl de Boor, *Elementary Numerical Analysis*, 3rd Edition, McGraw-Hill Book Company, New York, (1980).

[Foster, 1994] L. V. Foster, Gaussian Elimination with Partial Pivoting Can Fail in Practice, SIAM J. Matrix Anal. Appl., 15, (1994), 1354–1362.

4.2 The Hermite Echelon Form

There is another useful way to reduce a matrix, named in honor of the French mathematician **Charles Hermite** (24 December 1822 – 14 January 1901), that is very close to the RREF. However, it is only defined for square matrices. Statisticians have known about this for some time.

DEFINITION 4.4 *(Hermite echelon form)*
A matrix H in $\mathbb{C}^{n\times n}$ is in (upper) Hermite echelon form iff

1. *H is upper triangular ($h_{ij} = ent_{ij}(H) = 0$ if $i > j$).*
2. *The diagonal of H consists only of zeros and ones.*
3. *If a row has a zero on the diagonal, then every element of that row is zero; if $h_{ii} = 0$, then $h_{ik} = 0$ for all $k = 1, 2, \dots, n$.*
4. *If a row has a 1 on the diagonal, then every other element in the column containing that 1 is zero; if $h_{ii} = 1$, then $h_{ji} = 0$ for all $j = 1, 2, \dots, n$ except $j = i$.*

The first interesting fact to note is that a matrix in Hermite echelon form must be idempotent.

THEOREM 4.7
Let $H \in \mathbb{C}^{n\times n}$ be in Hermite echelon form. Then $H^2 = H$.

PROOF Let b_{ik} be the (i,k) entry of H^2. Then the definition of matrix multiplication gives $b_{ik} = \sum_{j=1}^{n} h_{ij}h_{jk} = \sum_{j=1}^{i-1} h_{ij}h_{jk} + h_{ii}h_{ik} + \sum_{j=i+1}^{n} h_{ij}h_{jk}$. If $i > k$, then $b_{ik} = 0$ since this is just a sum of zeros. Thus H^2 is upper triangular. If $i \leq k$, then $b_{ik} = \sum_{j=i}^{n} h_{ij}h_{jk}$. We consider cases. If $h_{ii} = 0$, then by (3), $h_{ij} = 0$ for all j = 1,2, ... , n so $b_{ik} = 0 = h_{ik}$. If $h_{ii} \neq 0$, then h_{ii} must equal 1, so $b_{ik} = h_{ik} + \sum_{j=i+1}^{n} h_{ij}h_{jk}$. Now whenever $h_{ij} \neq 0$ for $i + 1 \leq j \leq n$, we have by (4) that $h_{jj} = 0$ so from (3), $h_{jm} = 0$ for all m. This is so, in particular, for $m = k$. Thus, in any case, $b_{ik} = h_{ik}$ so $H^2 = H$. □

THEOREM 4.8
Every matrix A in $\mathbb{C}^{n\times n}$ can be brought into Hermite echelon form by using elementary row operations.

PROOF First we use elementary row operations to produce *RREF*(A). Then permute the rows of *RREF*(A) until each first nonzero element of each nonzero row is a diagonal element. The resulting matrix is in Hermite echelon form. □

For example, $RREF\left(\begin{bmatrix} 3 & 6 & 9 \\ 1 & 2 & 5 \\ 2 & 4 & 10 \end{bmatrix}\right) = \begin{bmatrix} 1 & 2 & 0 \\ 0 & 0 & 1 \\ 0 & 0 & 0 \end{bmatrix}$. Indeed $\begin{bmatrix} \frac{5}{6} & \frac{-3}{2} & 0 \\ \frac{-1}{6} & \frac{1}{2} & 0 \\ 0 & \frac{-1}{2} & \frac{1}{4} \end{bmatrix}\begin{bmatrix} 3 & 6 & 9 \\ 1 & 2 & 5 \\ 2 & 4 & 10 \end{bmatrix} = \begin{bmatrix} 1 & 2 & 0 \\ 0 & 0 & 1 \\ 0 & 0 & 0 \end{bmatrix}$. To get the Hermite echelon form, which we shall denote *HEF*(A), simply permute the second and third rows. Then $\begin{bmatrix} \frac{5}{6} & \frac{-3}{2} & 0 \\ 0 & \frac{-1}{2} & \frac{1}{4} \\ \frac{-1}{6} & \frac{1}{2} & 0 \end{bmatrix}\begin{bmatrix} 3 & 6 & 9 \\ 1 & 2 & 5 \\ 2 & 4 & 10 \end{bmatrix} = \begin{bmatrix} 1 & 2 & 0 \\ 0 & 0 & 0 \\ 0 & 0 & 1 \end{bmatrix} = H$. The reader may verify $H^2 = H$. Thus, our algorithm for finding *HEF*(A) is described as follows: use elementary row operations to produce $[A \mid I] \rightarrow$ $[HEF(A) \mid S]$. Then $SA = HEF(A)$.

COROLLARY 4.2
For any $A \in \mathbb{C}^{n\times n}$, there exists a nonsingular matrix S such that SA is in Hermite echelon form. Moreover, $ASA = A$.

For example, for $A = \begin{bmatrix} 1 & 2 & 1 \\ 2 & 3 & 1 \\ 1 & 1 & 0 \end{bmatrix}$,

$$D_2(-1)\,T_{32}(-1)T_{31}(-1)T_{21}(-2)\begin{bmatrix} 1 & 2 & 1 \\ 2 & 3 & 1 \\ 1 & 1 & 0 \end{bmatrix} = \begin{bmatrix} 1 & 0 & -1 \\ 0 & 1 & 1 \\ 0 & 0 & 0 \end{bmatrix} = H$$

or

$$\begin{bmatrix} -3 & 2 & 0 \\ 2 & -1 & 0 \\ 1 & -1 & 1 \end{bmatrix}\begin{bmatrix} 1 & 2 & 1 \\ 2 & 3 & 1 \\ 1 & 1 & 0 \end{bmatrix} = \begin{bmatrix} 1 & 0 & -1 \\ 0 & 1 & 1 \\ 0 & 0 & 0 \end{bmatrix}.$$

THEOREM 4.9
The Hermite echelon form of a matrix is unique, justifying the notation HEF(A) *for $A \in \mathbb{C}^{n\times n}$.*

PROOF (*Hint:* Use the uniqueness of *RREF*(A).)
The proof is left as exercise. ▯

Note that the sequence of elementary operations used to produce *HEF*(A) is far from unique. For the example above:

$$\begin{bmatrix} 0 & -1 & 3 \\ 0 & 1 & -2 \\ 1 & -1 & 1 \end{bmatrix}\begin{bmatrix} 1 & 2 & 1 \\ 2 & 3 & 1 \\ 1 & 1 & 0 \end{bmatrix} = \begin{bmatrix} 1 & 0 & -1 \\ 0 & 1 & 1 \\ 0 & 0 & 0 \end{bmatrix}.$$

The fact that $H = HEF(A)$ is idempotent means that there is a direct sum decomposition lurking in the background. The next result helps to indicate what that is.

COROLLARY 4.3
For any $A \in \mathbb{C}^{n\times n}$, $\mathcal{N}ull(A) = \mathcal{N}ull(HEF(A)) = \mathcal{C}ol(I - HEF(A))$. Moreover, the nonzero columns of $I - HEF(A)$ yield a basis for the null space of A. Also $rank(A) = rank(HEF(A)) = trace(HEF(A))$.

THEOREM 4.10
Let A and $B \in \mathbb{C}^{n\times n}$. Then $\mathrm{HEF}(A) = \mathrm{HEF}(B)$ *iff* $Col(A^*) = Col(B^*)$

PROOF First, suppose $Col(A^*) = Col(B^*)$. Then there exists an invertible matrix S with $A^*S^* = B^*$. This says that $SA = B$. Now there exists T nonsingular with $TB = HEF(B)$. Then $HEF(B) = TB = TSA = (TS)A$ is a matrix in Hermite echelon form. By uniqueness, $HEF(B) = HEF(A)$. Conversely, suppose $H = HEF(A) = HEF(B)$. Then, there are nonsingular matrices S and T with $SA = H = TB$. Then $A = S^{-1}TB$ so $A^* = B^*(S^{-1}T)^*$. But $(S^{-1}T)$ is nonsingular so $Col(A^*) = Col(B^*)$. □

COROLLARY 4.4

1. *For any $A \in \mathbb{C}^{n\times n}$,* $\mathrm{HEF}(A^*A) = \mathrm{HEF}(A)$.
2. *For $A \in \mathbb{C}^{n\times n}$ and $S \in \mathbb{C}^{n\times n}$ nonsingular,* $\mathrm{HEF}(SA) = \mathrm{HEF}(A)$.

In other words, row equivalent matrices have the same Hermite echelon form.

THEOREM 4.11
Let $H =$ $\mathrm{HEF}(A)$ *for some A in $\mathbb{C}^{n\times n}$. Suppose that the diagonal ones of H occurs in columns numbered $c_1, c_2, \ldots, c_k$. Then the corresponding columns of A are linearly independent.*

PROOF The proof is left as an exercise. □

COROLLARY 4.5
Consider the ith column of A, $\mathbf{a}_i$. This column is a linear combination of the set of linearly independent columns of A as described in the theorem above. The coefficients of the linear combinations are the nonzero elements of the ith column of $\mathrm{HEF}(A)$.

Exercise Set 14

1. Fill in the arguments for those theorems and corollaries given above.
2. Prove that $H = HEF(A)$ is nonsingular iff $H = I$.

3. Argue that if A is nonsingular, then $HEF(A) = I$.

4. Let S be a nonsingular matrix with $SA = HEF(A) = H$. Argue that $AH = A$.

5. Let $A \in \mathbb{C}^{n\times n}$, $S \in \mathbb{C}^{n\times n}$ nonsingular with $SA = HEF(A) = H$. Then prove that $AH = A$.

6. Let $H = HEF(A)$. Argue that A is idempotent iff $HA = H$.

7. Define $A \sim B$ iff $Col(A^*) = Col(B^*)$. Is $\sim$ an equivalence relation? Is HEF considered as a function on matrices constant on the equivalence classes?

8. Create an example of a matrix A with nonzero entries such that $H = HEF(A) = \begin{bmatrix} 1 & 0 & 1 \\ 0 & 1 & 1 \\ 0 & 0 & 0 \end{bmatrix}, \begin{bmatrix} 1 & 1 & 0 \\ 0 & 0 & 0 \\ 0 & 0 & 1 \end{bmatrix}, \begin{bmatrix} 1 & 1 & 1 \\ 0 & 0 & 0 \\ 0 & 0 & 0 \end{bmatrix}$.

9. Check to see that $AH = H$ in the examples you created above.

10. Spell out the direct sum decomposition induced by $H = HEF(A)$.

11. Fill in a proof for Corollary 4.2.

12. Make a proof for Theorem 4.9.

13. Fill in a proof for Corollary 4.10.

14. Fill in a proof for Corollary 4.4.

15. Make a proof for Theorem 4.11.

16. Fill in a proof for Corollary 4.5.

Further Reading

[Graybill, 1969] Franklin A. Graybill, *Introduction to Matrices with Applications in Statistics,* Wadsworth Publishing Co., Inc., Belmont, CA, (1969).

4.3 Full Rank Factorization

There are many ways to write a matrix as the product of others. You will recall the LU factorization we discussed in Chapter 3. There are others. In this section, we consider a factorization based on rank. It will be a major theme of our approach to matrix theory.

DEFINITION 4.5 *(full rank factorization)*
*Let A be a matrix in $\mathbb{C}_r^{m\times n}$ with $r > 0$. If there exists F in $\mathbb{C}_r^{m\times r}$ and G in $\mathbb{C}_r^{r\times n}$ such that $A = FG$, then we say we have a **full rank factorization** of A.*

There are the usual questions of existence and uniqueness. Existence can be argued in several ways. One approach is to take F to be any matrix whose columns form a basis for the column space of A. These could be chosen from among the columns of A or not. Then, since each column of A is uniquely expressible as a linear combination of the columns of F, the coefficients in the linear combinations determine a unique G in $\mathbb{C}^{r\times n}$ with $A = FG$. Moreover, $r = r(A) = r(FG) \leq r(G) \leq r$. Thus G is in $\mathbb{C}_r^{r\times n}$.

Another approach is to apply elementary matrices on the left of A to produce the unique RREF of A. That is, we produce an invertible matrix R in $\mathbb{C}_m^{m\times m}$ with $RA = \begin{bmatrix} G_{r\times n} \\ \cdots \\ \mathbb{O}_{(m-r)\times n} \end{bmatrix}$, where $r = r(A) = r(G)$ and $\mathbb{O}_{(m-r)\times n}$ is the $(m-r)$-by-n zero matrix. Then, $A = R^{-1}\begin{bmatrix} G_{r\times n} \\ \cdots \\ \mathbb{O}_{(m-r)\times n} \end{bmatrix}$. With a suitable partitioning of R^{-1}, say $R^{-1} = \begin{bmatrix} R_1 \vdots R_2 \end{bmatrix}$, where R_1 is m-by-r and R_2 is m-by-$(m-r)$, $A = \begin{bmatrix} R_1 \vdots R_2 \end{bmatrix}\begin{bmatrix} G \\ \cdots \\ \mathbb{O} \end{bmatrix} = R_1G + R_2\mathbb{O} = R_1G$. Take F to be R_1. Since R^{-1} is invertible, its columns are linearly independent so F has r independent columns and hence has full column rank. We summarize our discussion with a theorem.

THEOREM 4.12
Every matrix A in $\mathbb{C}_r^{m\times n}$ with $r > 0$ has a full rank factorization.

Even better, we will now describe a procedure, that is, an algorithm, for computing a full rank factorization of a given matrix A that works reasonably well for hand calculations on small matrices. It appears in [C&M, 1979]. Let A be in $\mathbb{C}_r^{m\times n}$.

Step 1. Use elementary row operations to reduce A to *RREF*(A).

Step 2. Construct a matrix F by choosing the columns of A that correspond to the columns with the leading ones in *RREF*(A) placing them in F in the same order they appear in A.

Step 3. Construct a matrix G by taking the nonzero rows of *RREF*(A) and placing them as the rows of G in the same order they appear in *RREF*(A).

Then, $A = FG$ is a full rank factorization of A.

Now for the bad news. As you may have guessed by now, not only do full rank factorizations exist, they abound. After all, in our first construction described above, there are many choices for bases of the column space of A, hence many choices for F. Indeed, if $A = FG$ is one full rank factorization of A in $\mathbb{C}_r^{m\times n}$ with $r > 0$, choose any invertible matrix R in $\mathbb{C}_r^{r\times r}$. Let $F_R = FR$ and $G_R = R^{-1}G$. Then clearly $A = F_R G_R$ is also a full rank factorization of A. Actually, this will turn out to be good news later since we will be able to select an R to produce very nice full rank factorizations. Again, we summarize with a theorem.

THEOREM 4.13

Every matrix A in $\mathbb{C}_r^{m\times n}$ with $r > 0$ has infinitely many full rank factorizations.

Example 4.1

Let $A = \begin{bmatrix} 3 & 6 & 13 \\ 2 & 4 & 9 \\ 1 & 2 & 3 \end{bmatrix}$. Then *RREF*$(A) = \begin{bmatrix} 1 & 2 & 0 \\ 0 & 0 & 1 \\ 0 & 0 & 0 \end{bmatrix}$ so we take $G = \begin{bmatrix} 1 & 2 & 0 \\ 0 & 0 & 1 \end{bmatrix}$ and $F = \begin{bmatrix} 3 & 13 \\ 2 & 9 \\ 1 & 3 \end{bmatrix}$. The reader may verify that $A = FG$ is indeed a full rank factorization of A.

Exercise Set 15

1. Compute full rank factorizations for the following matrices:
$$\begin{bmatrix} 1 & 1 & 0 \\ 1 & 0 & 1 \\ 1 & 1 & 0 \\ 1 & 0 & 1 \end{bmatrix}, \begin{bmatrix} 1 & 1 & 2 \\ 1 & 0 & 1 \\ 1 & 1 & 2 \\ 1 & 0 & 1 \end{bmatrix}, \begin{bmatrix} 1 & 2 & 3 \\ 1 & 2 & 3 \\ 1 & 2 & 3 \\ 1 & 2 & 3 \end{bmatrix}, \begin{bmatrix} 1 & 3 & 0 \\ 1 & 2 & 1 \\ 1 & 3 & 0 \\ 1 & 2 & 1 \end{bmatrix},$$
$$\begin{bmatrix} 1 & 1 & 1 & 1 \\ 1 & 0 & 1 & 0 \\ 0 & 1 & 0 & 1 \end{bmatrix}, \begin{bmatrix} 1 & 1 & 1 & 1 \\ 1 & 0 & 1 & 0 \\ 2 & 1 & 2 & 1 \end{bmatrix}.$$

2. Argue that any A in $\mathbb{C}_r^{m \times n}$ can be written $A = LK$, where L has a left inverse and K has a right inverse.

3. Suppose $A = FG$ is a full rank factorization of A and $\mathbb{C}^n = \mathcal{N}ull(A) \oplus \mathcal{C}ol(A)$. Argue that $(GF)^{-1}$ exists and $E = F(GF)^{-1}G$ is the projector of $\mathbb{C}^n$ onto $\mathcal{C}ol(A)$ along $\mathcal{N}ull(A)$.

4. Use full rank factorizations to compute the index of the following matrices:
$$\begin{bmatrix} 1 & 1 & 0 \\ 1 & 0 & 1 \\ 1 & 1 & 0 \\ 1 & 0 & 1 \end{bmatrix}, \begin{bmatrix} 1 & 1 & 2 \\ 1 & 0 & 1 \\ 1 & 1 & 2 \\ 1 & 0 & 1 \end{bmatrix}, \begin{bmatrix} 1 & 2 & 3 \\ 1 & 2 & 3 \\ 1 & 2 & 3 \\ 1 & 2 & 3 \end{bmatrix}, \begin{bmatrix} 1 & 3 & 0 \\ 1 & 2 & 1 \\ 1 & 3 & 0 \\ 1 & 2 & 1 \end{bmatrix},$$
$$\begin{bmatrix} 1 & 1 & 1 & 1 \\ 1 & 0 & 1 & 0 \\ 0 & 1 & 0 & 1 \end{bmatrix}, \begin{bmatrix} 1 & 1 & 1 & 1 \\ 1 & 0 & 1 & 0 \\ 2 & 1 & 2 & 1 \end{bmatrix}.$$

5. Suppose A has index q. Then $\mathbb{C}^n = \mathcal{C}ol(A^q) \oplus \mathcal{N}ull(A^q)$. We can use a full rank factorization of A^q to compute the projector of $\mathbb{C}^n$ onto $\mathcal{C}ol(A^q)$ along $\mathcal{N}ull(A^q)$. Indeed, let $A^q = FG$ be a full rank factorization. Argue that $F(GF)^{-1}G$ is the projector of $\mathbb{C}^n$ onto $\mathcal{C}ol(A^q)$ along $\mathcal{N}ull(A^q)$.

6. Suppose $A = FG$ is a full rank factorization. Argue that $A = A^2$ iff $GF = I$.

7. Argue that a full rank factorization of a matrix A can be obtained by first selecting a matrix G whose rows form a basis for $\mathcal{R}ow(A)$. Then F must be uniquely determined.

8. Explain how to produce a full rank factorization from the modified RREF of a matrix.

9. Suppose $A = A^2$. Prove the the rank of A is the trace of A.

10. (G. Trenkler) Suppose A and B are n-by-n idempotent matrices with $A + B + AB + BA = \mathbb{O}$. What can you conclude about A and B?

Further Reading

[C&M, 1979] S. L. Campbell and C. D. Meyer, Jr., *Generalized Inverses of Linear Transformations*, Dover Publications, Inc., New York, (1979).

4.3.1 MATLAB Moment

4.3.1.1 Full Rank Factorization

We can create an M-file to compute a full rank factorization of a matrix. By now, you should have this down.

```
function FRF=frf(A)
[R,jp] = rref(A)
r = rank(A)
for i = 1:r
G(I,:) = R(i,:)
end
F = A(:,jp)
G
```

Experiment with this routine on some matrices of your own creation.

4.4 The Moore-Penrose Inverse

In this section, we develop a key concept, the *Moore-Penrose inverse* (MP-inverse), also known as the *pseudoinverse*. What is so great about this inverse is that *every matrix has one,* square or not, full rank or not. Our approach to the pseudoinverse is to use the idea of full rank factorization; we build up from the factors of a full rank factorization. The idea of a generalized inverse of a singular matrix goes back to **E. H. Moore** [26 January, 1862 to 30 December, 1932] in a paper published in 1920. He investigated the idea of a "general reciprocal" of a matrix again in a paper in 1935. Independently, **R. Penrose** [8 August, 1931] rediscovered Moore's idea in 1955. We present the Penrose approach.

DEFINITION 4.6 *(Moore-Penrose inverse)*
Let A be any matrix in $\mathbb{C}^{m\times n}$. We say A has a ***Moore-Penrose inverse*** *(or just* ***pseudoinverse*** *for short) iff there is a matrix X in $\mathbb{C}^{n\times m}$ such that*

(MP1) $AXA = A$
(MP2) $XAX = X$
(MP3) $(AX)^* = AX$
(MP4) $(XA)^* = XA.$

These four equations are called the *Moore-Penrose equations* and the order in which they are written is crucial for our subsequent development. Indeed, later we will distinguish matrices that solve only a subset of the four Moore-Penrose equations. For example, a *1-inverse* of A would be a matrix X that is required to solve only MP1. A *{1,2}-inverse* of A would be required to solve only MP1 and MP2. Now, we settle the issue of uniqueness.

THEOREM 4.14 (the uniqueness theorem)
If A in $\mathbb{C}^{m\times n}$ has a pseudoinverse at all, it must be unique. That is, there can be only one simultaneous solution to the four MP-equations.

PROOF Suppose X and Y in $\mathbb{C}^{n\times m}$ both satisfy the four Moore-Penrose equations. Then $X = X(AX) = X(AX)^* = XX^*A^* = XX^*(AYA)^* = XX^*A^*(AY)^* = X(AX)^*(AY)^* = XAXAY = XAY = XAYAY = (XA)^*(YA)^*Y = A^*X^*A^*Y^*Y = (AXA)^*Y^*Y = A^*Y^*Y = (YA)^*Y = YAY = Y$.

The reader should be sure to justify each of the equalities above and note that all four Moore-Penrose equations were actually used. □

In view of the uniqueness theorem for pseudoinverses, we use the notation A^+ for the unique solution of the four MP-equations (when the solution exists, of course, which is yet to be established). Since this idea of a pseudoinverse may be quite new to you, the idea that all matrices have inverses may be surprising. We now spend some time on a few concrete examples.

Example 4.2

1. Clearly $I_n^+ = I_n$ for any n and $\mathbb{O}_{m\times n}^+ = \mathbb{O}_{n\times m}$. (Does this give us a chance to divide by zero?)

2. Suppose A is square and invertible. Then $A^+ = A^{-1}$. This is, of course, how it should be if the pseudoinverse is to generalize the idea of ordinary inverse. Let's just quickly verify the four MP-equations. $AA^{-1}A = AI = A$, $A^{-1}AA^{-1} = A^{-1}I = A^{-1}$, $(AA^{-1})^* = I^* = I = AA^{-1}$, and $(A^{-1}A)^* = I^* = I = A^{-1}A$. Yes, they all check.

3. Suppose P is a matrix such that $P = P^* = P^2$. Later we shall call such a matrix a *projection* (also known as a *Hermitian idempotent*). We claim for such a matrix, $P = P^+$. Again a quick check reveals, $PP^+P = PPP = PP = P$, $P^+PP^+ = PPP = P = P^+$, $(PP^+)^* = (PP)^* = P^* = P = PP = PP^+$, and $(P^+P)^* = (PP)^* = P^* = P = PP = P^+P$.

   Once again, we are golden. So, for example, $\begin{bmatrix} 1 & 0 & 0 \\ 0 & 1 & 0 \\ 0 & 0 & 0 \end{bmatrix}^+ = \begin{bmatrix} 1 & 0 & 0 \\ 0 & 1 & 0 \\ 0 & 0 & 0 \end{bmatrix}$, and $\begin{bmatrix} \frac{4}{5} & \frac{-2}{5} & 0 \\ \frac{-2}{5} & \frac{1}{5} & 0 \\ 0 & 0 & 0 \end{bmatrix}^+ = \begin{bmatrix} \frac{4}{5} & \frac{-2}{5} & 0 \\ \frac{-2}{5} & \frac{1}{5} & 0 \\ 0 & 0 & 0 \end{bmatrix}$.

4. Let's agree that for a scalar λ, $\lambda^+ = \frac{1}{\lambda}$ if $\lambda \neq 0$ and $\lambda^+ = 0$ if $\lambda = 0$. Let D be a diagonal matrix, say $D = diag(d_1, d_2, \ldots, d_n)$. We claim $D^+ = diag(d_1^+, d_2^+, \ldots, d_n^+)$. We leave the details as an exercise. In particular then, $\begin{bmatrix} 1 & 0 & 0 \\ 0 & 2 & 0 \\ 0 & 0 & 0 \end{bmatrix}^+ = \begin{bmatrix} 1 & 0 & 0 \\ 0 & \frac{1}{2} & 0 \\ 0 & 0 & 0 \end{bmatrix}$.

5. What can we say about an n-by-1 matrix? In other words we are just looking at one column that could be viewed as a vector. Say $\mathbf{b} = \begin{bmatrix} b_1 \\ b_2 \\ \vdots \\ b_n \end{bmatrix}$.

   If $\mathbf{b} = \overrightarrow{0}$, we know what $\mathbf{b}^+$ is so suppose $\mathbf{b} \neq \overrightarrow{0}$. A little trial and error leads us to $\mathbf{b}^+ = \frac{1}{\mathbf{b}^*\mathbf{b}}\mathbf{b}^*$. Remember, $\mathbf{b}^*\mathbf{b}$ is a scalar. In fact, it is the Euclidean length squared of $\mathbf{b}$ considered as a vector. We prefer to illustrate with an example and leave the formal proof as an exercise. We claim $\begin{bmatrix} 1 \\ 2 \\ 3i \end{bmatrix}^+ = \begin{bmatrix} \frac{1}{14} & \frac{2}{14} & \frac{-3i}{14} \end{bmatrix}$, where $\mathbf{b} = \begin{bmatrix} 1 \\ 2 \\ 3i \end{bmatrix}$. First note that $\mathbf{b}^*\mathbf{b} = 14$. Then $\begin{bmatrix} 1 \\ 2 \\ 3i \end{bmatrix} \begin{bmatrix} \frac{1}{14} & \frac{2}{14} & \frac{-3i}{14} \end{bmatrix} \begin{bmatrix} 1 \\ 2 \\ 3i \end{bmatrix} = \begin{bmatrix} 1 \\ 2 \\ 3i \end{bmatrix} [1] = \begin{bmatrix} 1 \\ 2 \\ 3i \end{bmatrix}$,

$$\left[\frac{1}{14}\,\frac{2}{14}\,\frac{-3i}{14}\right]\begin{bmatrix}1\\2\\3i\end{bmatrix}\left[\frac{1}{14}\,\frac{2}{14}\,\frac{-3i}{14}\right]=[1]\left[\frac{1}{14}\,\frac{2}{14}\,\frac{-3i}{14}\right]=\left[\frac{1}{14}\,\frac{2}{14}\,\frac{-3i}{14}\right],$$

$$\mathbf{b}^+\mathbf{b}=\left[\frac{1}{14}\,\frac{2}{14}\,\frac{-3i}{14}\right]\begin{bmatrix}1\\2\\3i\end{bmatrix}=[1] \text{ and } \mathbf{b}\mathbf{b}^+=\begin{bmatrix}1\\2\\3i\end{bmatrix}\left[\frac{1}{14}\,\frac{2}{14}\,\frac{-3i}{14}\right]$$

$$=\begin{bmatrix}\frac{1}{14} & \frac{2}{14} & \frac{-3i}{14}\\ \frac{2}{14} & \frac{4}{14} & \frac{-6i}{14}\\ \frac{3i}{14} & \frac{6i}{14} & \frac{9}{14}\end{bmatrix}.$$

Next, we come to crucial cases in which we can identify the pseudoinverse of a matrix.

THEOREM 4.15

1. *Suppose $F \in \mathbb{C}_r^{m\times r}$ — that is, F has full column rank. Then $F^+ = (F^*F)^{-1}F^*$.*

2. *Suppose $G \in \mathbb{C}_r^{r\times n}$ — that is, G has full row rank. Then $G^+ = G^*(GG^*)^{-1}$.*

PROOF

(1) We verify the four MP-equations. First, $FF^+F = F((F^*F)^{-1}F^*)F = F(F^*F)^{-1}(F^*F) = FI = F$. Next, $F^+FF^+ = ((F^*F)^{-1}F^*)F((F^*F)^{-1}F^*) = (F^*F)^{-1}(F^*F)(F^*F)^{-1}F^* = I((F^*F)^{-1}F^*) = F^+$. Now $F^+F = ((F^*F)^{-1}F^*)F = (F^*F)^{-1}(F^*F) = I_r$ so surely $(F^+F)^* = F^+F$. Finally, $(FF^+)^* = (F(F^*F)^{-1}F^*)^* = F^{**}(F^*F)^{-1}F^* = F(F^*F)^{*-1}F^* = F(F^*F)^{-1}F^* = FF^+$.

(2) This proof is similar to the one above and is left as an exercise. □

So we see that for matrices of full row or column rank, the pseudoinverse picks out a specific left(right) inverse of the matrix. From above, $F^+F = I_r$ and $GG^+ = I_r$. Now, for an arbitrary matrix A in $\mathbb{C}_r^{m\times n}$ with $r > 0$, we shall show how to construct the pseudoinverse.

DEFINITION 4.7 *(pseudoinverse)*

Let A be a matrix in $\mathbb{C}_r^{m\times n}$. Take any full rank factorization of $A = FG$. Then F^+ and G^+ exist by the theorem above. Define A^+ in $\mathbb{C}^{n\times m}$ by $A^+ := G^+F^+$. In other words, $A^+ = G^*(GG^*)^{-1}(F^*F)^{-1}F^*$.

THEOREM 4.16

For an arbitrary matrix A in $\mathbb{C}_r^{m\times n}$ with $r > 0$, A^+ defined above satisfies the four MP-equations and, hence, must be the unique pseudoinverse of A.

Moreover, $AA^+ = FF^+$ *and* $A^+A = G^+G$ *where* $A = FG$ *is any full rank factorization of* A.

PROOF Suppose the notation of (4.7). Then $AA^+A = AG^+F^+FG = AG^+G = FGG^+G = FG = A$. Next, $A^+AA^+ = G^+F^+AA^+ = G^+F^+FGA = G^+GA^+ = G^+GG^+F^+ = G^+F^+ = A^+$. Also, $AA^+ = FGG^+F^+ = FF^+$ and we know $(FF^+)^* = FF^+$. Finally, $A^+A = G^+F^+FG = G^+G$ and we know $(G^+G)^* = G^+G$. □

We now have established the uniqueness and existence of A^+ for any matrix A in $\mathbb{C}^{m\times n}$. The approach we used here goes back to Greville [1960], who credits A. S. Householder with suggesting the idea. There are some properties of pseudoinverses that are easy to establish. We collect some of these in the next theorem.

THEOREM 4.17
Let $A \in \mathbb{C}^{m\times n}$. *Then*

1. $(AA^+)^2 = AA^+ = (AA^+)^*$.
2. $(I_m - AA^+)^2 = (I_m - AA^+) = (I_m - AA^+)^*$.
3. $(A^+A)^2 = A^+A = (A^+A)^*$.
4. $(I_n - A^+A)^2 = (I_n - A^+A) = (I_n - A^+A)^*$.
5. $(I_m - AA^+)A = \mathbb{O}_{m\times n}$.
6. $(I_n - A^+A)A^+ = \mathbb{O}_{n\times m}$.
7. $A^{++} = A$.
8. $(A^*)^+ = (A^+)^*$.
9. $(A^*A)^+ = A^+A^{*+}$.
10. $A^* = A^*AA^+ = A^+AA^*$.
11. $A^+ = (A^*A)^+A^* = A^*(AA^*)^+$.
12. $(\lambda A)^+ = \lambda^+A^+$.

PROOF The proofs are left as exercises. □

Let's look at an example.

Example 4.3

We continue with the example from above: $A = \begin{bmatrix} 3 & 6 & 13 \\ 2 & 4 & 9 \\ 1 & 2 & 3 \end{bmatrix} = \begin{bmatrix} 3 & 13 \\ 2 & 9 \\ 1 & 3 \end{bmatrix} \begin{bmatrix} 1 & 2 & 0 \\ 0 & 0 & 1 \end{bmatrix}$, where $F = \begin{bmatrix} 3 & 13 \\ 2 & 9 \\ 1 & 3 \end{bmatrix}$ and $G = \begin{bmatrix} 1 & 2 & 0 \\ 0 & 0 & 1 \end{bmatrix}$ gives a full rank factorization of A. Then direct computation from the formulas in (4.15.1) and (4.15.2) yields $G^+ = \begin{bmatrix} \frac{1}{5} & 0 \\ \frac{2}{5} & 0 \\ 0 & 1 \end{bmatrix}$ and $F^+ = \begin{bmatrix} \frac{-3}{26} & \frac{-22}{26} & \frac{79}{26} \\ \frac{2}{26} & \frac{6}{26} & \frac{-18}{26} \end{bmatrix}$, and so $A^+ = \begin{bmatrix} \frac{-3}{130} & \frac{-11}{65} & \frac{79}{130} \\ \frac{-3}{65} & \frac{-22}{65} & \frac{79}{65} \\ \frac{1}{13} & \frac{3}{13} & \frac{-9}{13} \end{bmatrix} = \begin{bmatrix} \frac{-3}{130} & \frac{-22}{130} & \frac{79}{130} \\ \frac{-6}{130} & \frac{-44}{130} & \frac{158}{130} \\ \frac{10}{130} & \frac{30}{130} & \frac{-90}{130} \end{bmatrix}$. There is something of interest to note here. If you recall the formula for the inverse of a matrix in terms of the adjugate, we see $G^+(GG^*)^{-1} = \frac{1}{det(GG^*)} G^* adj(GG^*)$ and $F^+ = (F^*F)^{-1}F^* = \frac{1}{det(F^*F)} adj(F^*F)F^*$ so $A^+ = \frac{1}{det(F^*F)det(GG^*)} G^* adj(GG^*) adj(F^*F)F^*$. In particular, if the entries of A consist only of integers, the entries of A^+ will be rational numbers with the common denominator $det(F^*F)det(GG^*)$. In our example, $det(GG^*) = 26$ while $det(F^*F) = 5$, hence the common denominator of 130.

Before we finish this section, we need to tie up a loose end. We have already noted that a matrix A in $\mathbb{C}_r^{m \times n}$ with $r > 0$ has infinitely many full rank factorizations. We even showed how to produce an infinite collection using invertible matrices. We show next that this is the only way to get full rank factorizations.

THEOREM 4.18

Every matrix $A \in \mathbb{C}_r^{m \times n}$ with $r > 0$ has infinitely many full rank factorizations. However, if $A = FG = F_1G_1$ are two full rank factorizations of A, then there exists an invertible matrix R in $\mathbb{C}^{r \times r}$ such that $F_1 = FR$ and $G_1 = R^{-1}G$. Moreover, $(R^{-1}G)^+ = G^+R$ and $(FR)^+ = R^{-1}F^+$.

PROOF The first claim has already been established so suppose $A = FG = F_1G_1$ are two full rank factorizations of A. Then $F_1G_1 = FG$ so $F_1^+F_1G_1 = F_1^+FG$ so $G_1 = (F_1^+F)G$ since $F_1^+F_1 = I_r$. Note that F_1^+F is r-by-r and $r = r(G_1) = r((F_1^+F)G) \le r(F_1^+F) \le r$, so $r(F_1^+F) = r$ and so F_1^+F is invertible. Call $F_1^+F = S$. Similarly, $F_1G_1 = FG$ implies $F_1G_1G_1^+ = FGG_1^+ = F_1$ since $G_1G_1^+ = I_r$. Again note GG_1^+ is r-by-r of rank r so GG_1^+ is invertible. Name $R = GG_1^+$. Then $SR = F_1^+FGG_1^+ = F_1^+AG_1^+ = F_1^+F_1G_1G_1^+ = I_r$. Thus $S = R^{-1}$. Now we can see $G_1 = SG =$

$R^{-1}G$ and $F_1 = FGG_1^+ = FR$. To complete the proof, compute $(FR)^+ = ((FR)^*(FR))^{-1}(FR)^* = (R^*F^*FR)^{-1}R^*F^* = R^{-1}(F^*F)^{-1}R^{*-1}R^*F^* = R^{-1}(F^*F)^{-1}F^* = R^{-1}F^+$ and $(R^{-1}G)^+ = (R^{-1}G)^*((R^{-1}G)(R^{-1}G)^*)^{-1} = G^*(R^{-1})^*(R^{-1}GG^*R^{-1*})^{-1} = G^*(R^{-1})^*(R^{-1})^{*-1}(GG^*)^{-1}R = G^*(GG^*)^{-1}R = G^+R$. □

We end this section with a table summarizing our work on pseudoinverses so far.

TABLE 4.4.1: Summary Table

Dimension	Rank	Pseudoinverse
$n = m$	n	$A^+ = A^{-1}$
m-by-n	n	$A^+ = (A^*A)^{-1}A^*$
m-by-n	m	$A^+ = A^*(AA^*)^{-1}$
m-by-n	r	$A^+ = G^+F^+$ where A = FG is any full rank factorization of A

Exercise Set 16

1. Let $A = FG$ be a full rank factorization of A. Argue that $F^+A = G$, $FF^+A = A$, $AG^+ = F$, and $AG^+G = A$.

2. Suppose A^L is a left inverse of A — that is, $A^LA = I$. Is $A^L = A^+$ necessarily true? Suppose $A^*A = I$. What can you say about A^+?

3. Suppose $A^2 = A$ in $\mathbb{C}^{n\times n}$. Use a full rank factorization of A to prove that $rank(A) = trace(A)$ (i.e., the rank of A is just the trace of A when A is an idempotent matrix).

4. Justify each of the steps in the proof of the uniqueness theorem.

5. Determine the pseudoinverse of any diagonal matrix.

6. Go through the computations in detail of the numerical example in the text.

7. Prove (2) of Theorem 4.15.

8. Prove the 12 claims of Theorem 4.17.

9. Prove the following: $AA^+A^{+*} = A^{+*}$, $A^{+*}A^+A = A^{+*}$, $A^{*+}A^*A = A$, $AA^*A^{*+} = A$, $A^*A^{+*}A^+ = A^+$, $A^+A^{+*}A^* = A^+$.

10. Argue that the row space of A^+ is equal to the row space of A^*.

11. Argue that the column space of A^+ is equal to the column space of A^* and the column space of A^+A.

12. Argue that A, A^*, A^+, and A^{+*} all have the same rank.

13. Prove $(AA^*)^+ = A^{+*}A^+$, $(A^*A)^+ = A^+A^{+*} = A^+A^{*+}$ and $(AA^*)^+(AA^*) = AA^+$.

14. Prove $A = AA^*(A^+)^* = (A^+)^*A^*A = AA^*(AA^*)^+A$ and $A^* = A^*AA^+ = A^+AA^*$.

15. Prove $A^+ = A^+(A^+)^*A^* = A^*(A^+)^*A^+$.

16. Prove $A^+ = (A^*A)^+A^* = A^*(AA^*)^+$ so that $AA^+ = A(A^*A)^+A^*$.

17. Show that if $A = \sum A_i$ where $A_i^*A_j = \mathbb{O}$ whenever $i \neq j$, then $A^+ = \sum A_i^+$.

18. Argue that all the following matrices have the same rank: A, A^+, AA^+, A^+A, AA^+A, and A^+AA^+. The rank is $Tr(AA^+)$.

19. Argue that $(-A)^+ = -A^+$.

20. Suppose A is n-by-m and S is m-by-m invertible. Argue that $(AS)(AS)^+ = AA^+$.

21. Suppose $A^*A = AA^*$. Prove that $A^+A = AA^+$ and for any natural number n, $(A^n)^+ = (A^+)^n$. What can you say if $A = A^*$?

22. Prove that $A^+ = A^*$ if and only if A^*A is idempotent.

23. If $A = \begin{bmatrix} B & \mathbb{O} \\ \mathbb{O} & C \end{bmatrix}$, find a formula for A^+.

24. Suppose A is a matrix and X is a matrix such that $AXA = A$, $XAX = X$ and $AX = XA$. Argue that if X exists, it must be unique.

25. Why does $(F^*F)^{-1}$ exist in Theorem 4.15?

26. (MacDuffee) Let $A = FG$ be a full rank factorization of A in $\mathbb{C}_r^{m\times n}$. Argue that F^*AG^* is invertible and $A^+ = G^*(F^*AG^*)^{-1}F^*$. (*Hint:* First argue that F^*AG^* is in fact invertible.) Note $F^*AG^* = (F^*F)(GG^*)$ and these two matrices are r-by-r or rank r hence invertible. Then $(F^*AG^*)^{-1} = (GG^*)^{-1}(F^*F)^{-1}$.

27. Let $\mathbf{x} = \begin{bmatrix} x_1 \\ \vdots \\ x_n \end{bmatrix}$ and $\mathbf{y} = \begin{bmatrix} y_1 \\ \vdots \\ y_n \end{bmatrix}$. Show $(\mathbf{xy}^*)^+ = (\mathbf{x}^*\mathbf{x})^+(\mathbf{y}^*\mathbf{y})^+\mathbf{yx}^*$.

28. Find the MP inverse of a 2-by-2 matrix $\begin{bmatrix} a & b \\ c & d \end{bmatrix}$.

29. Find examples of matrices A and B with $(AB)^+ = B^+A^+$ and A and B with $(AB)^+ \neq B^+A^+$. Then argue Greville's [1966] result that $(AB)^+ = B^+A^+$ iff A^+A and BB^+ commute.

30. Find $\begin{bmatrix} 0 & 1 & 0 \\ 0 & 0 & 1 \\ 0 & 0 & 0 \end{bmatrix}^+$.

31. Remember the matrix units E_{ij}? What is E_{ij}^+?

32. Find necessary and sufficient conditions for $A^+ = A$.

33. (Y. Tian) Prove that the following statements are equivalent for m-by-n matrices A and B:
    (i) $Col\begin{bmatrix} A \\ A^+A \end{bmatrix} = Col\begin{bmatrix} B \\ B^+B \end{bmatrix}$
    (ii) $Col\begin{bmatrix} A \\ A^*A \end{bmatrix} = Col\begin{bmatrix} B \\ B^*B \end{bmatrix}$
    (iii) $A = B$.

34. In this exercise, we introduce the idea of a *circulant matrix.* An n-by-n matrix A is called a circulant matrix if its first row is arbitrary but its subsequent rows are cyclical permutations of the previous row. So, if the first row is $(a_1a_2a_3\cdots a_n)$, the second row is $(a_na_1a_2\cdots a_{n-1})$, and the last row is $(a_2a_3a_4 \quad a_na_1)$. There are entire books written on these kinds of matrices (see page 169). Evidently, if you know the first row, you know the matrix. Write a typical 3-by-3 circulant matrix. Is the identity matrix a circulant matrix? Let C be the circulant matrix whose first row is $(0100\cdots0)$. Argue that all powers of C are also circulant matrices. Moreover, argue that if A is any circulant matrix with first row $(a_1a_2a_3\cdots a_n)$, then $A = a_1I + a_2C + a_3C^2 + \cdots + a_nC^{n-1}$.

35. Continuing the problem above, prove that A is a circulant matrix iff $AC = CA$.

36. Suppose A is a circulant matrix. Argue that A^+ is also circulant and A^+ commutes with A.

37. (Cline 1964) If AB is defined, argue that $(AB)^+ = B_1^+ A_1^+$ where $AB = A_1 B_1$, $B_1 = A^+AB$ and $A_1 = AB_1B_1^+$.

38. If $rank(A) = 1$, prove that $A^+ = (tr(AA^*)^{-1})A^*$.

39. Prove that $AB = \mathbb{O}$ implies $B^+A^+ = \mathbb{O}$.

40. Prove that $A^*B = \mathbb{O}$ iff $A^+B = \mathbb{O}$.

41. Suppose $A^*AB = A^*C$. Prove that $AB = AA^+C$.

42. Suppose BB^* is invertible. Prove that $(AB)(AB)^+ = AA^+$.

43. Suppose that $AB^* = \mathbb{O}$. Prove that $(A + B)^+ = A^+ + (I_n - A^+B)[C^+ + (I - C^+C)MB^*(A^+)^*A^+(I - BC^+)$ where $C = (I_m - AA^+)B$ and $M = [I_n + (I_n - C^+C)B^*(A^+)A^+B(I_n - C^+C)]^{-1}$.

44. Prove that $\begin{bmatrix} A \\ \cdots \\ B \end{bmatrix}^+ = [A^+ - TBA^+ \mid T]$ where $T = E^+ + (I_n - E^+B)A^+(A^+)^*B^*K(I_p - EE^+)$ with $E = B(I_n - A^+A)$ and $K = [I_p + (I_p - EE^+)BA^+(A^+)^*B^*(I - EE^+)]^{-1}$.

45. Prove that $[A \vdots B]^+ = \begin{bmatrix} A^+ - A^+B(C^+ + D) \\ C^+ + D \end{bmatrix}$ where $C = (I_m - AA^+)B$ and $D =$
$(I_p - C^+C)[I_p + (I_p - C^+C)B^*(A^+)^*A^+B(I_p - C^+C)]^{-1}B^*(A^+)^*A^+(I_m - BC^+)$.

46. Argue Greville's [1966] results: $(AB)^+ = B^+A^+$ iff any one of the following hold true:
    (a) $A^+ABB^*A^* = BB^*A^*$ and $BB^+A^*AB = A^*AB$.
    (b) A^+ABB^* and A^*ABB^+ are self adjoint.
    (c) $A^+ABB^*A^*ABB^+ = BB^*A^*A$.
    (d) $A^+AB = B(AB)^+AB$ and $BB^+A^* = A^*AB(AB)^+$.

47. Suppose A is m-by-n and B is n-by-p and $rank(A) = rank(B) = n$. Prove that $(AB)^+ = B^+A^+$.

Further Reading

[Cline, 1964] R. E. Cline, Note on the Generalized Inverse of a Product of Matrices, SIAM Review, Vol. 6, January, (1964), 57–58.

[Davis, 1979] Philip J. Davis, *Circulant Matrices*, John Wiley & Sons, New York, (1979).

[Greville, 1966] T. N. E. Greville, Note on the Generalized Inverse of a Matrix Product, SIAM Review, Vol. 8, (1966), 518–524.

[Greville, 1960] T. N. E. Greville, Some Applications of the Pseudo Inverse of a Matrix, SIAM Review, Vol. 2, (1960), 15–22.

[H&M, 1977] Ching-Hsiand Hung and Thomas L. Markham, The Moore-Penrose Inverse of a Sum of Matrices, J. Australian Mathematical Society, Vol. 24, (Series A), (1977), 385–392.

[L&O, 1971] T. O. Lewis and P. L. Odell, *Estimation in Linear Models,* Prentice Hall, Englewood Cliffs, NJ, (1971).

[Lütkepohl, 1996] H. Lütkepohl, *Handbook of Matrices*, John Wiley & Sons, New York, (1996).

[Mitra, 1968] S. K. Mitra, On a Generalized Inverse of a Matrix and Applications, Sankhya, Series A, XXX:1, (1968), 107–114.

[M&O, 1968] G. L. Morris and P. L. Odell, A Characterization for Generalized Inverses of Matrices, SIAM Review, Vol. 10, (1968), 208–211.

[Penrose, 1955] R. Penrose, A Generalized Inverse for Matrices, Proc. Camb. Phil. Soc., Vol. 51, (1955), 406–413.

[R&M, 1971] C. R. Rao and S. K. Mitra, *Generalized Inverses of Matrices and its Applications*, John Wiley & Sons, New York, (1971).

[Rohde 1966] C. A. Rohde, Some Results on Generalized Inverses, SIAM Review, VIII:2, (1966), 201–205.

[Wong, 1981] Edward T. Wong, Polygons, Circulant Matrices, and Moore-Penrose Inverses, The American Mathematical Monthly, Vol. 88, No. 7, August/September, (1981), 509–515.

4.4.1 MATLAB Moment

4.4.1.1 The Moore-Penrose Inverse

MATLAB has a built-in command to compute the pseudoinverse of an m-by-n matrix A. The command is

$$pinv(A).$$

For example,

```
>> A=[1 2 3;4 5 6;7 8 9]
ans =
  -0.6389   -0.1667    0.3056
  -0.0556    0.0000    0.0556
   0.5278    0.1667   -0.1944
>> format rat
>> pinv(A)
ans =
  -23/36   -1/6    11/36
   -1/18    *       1/18
   19/36    1/6    -7/36.
```

For fun, find pinv(B) where B = ones(3), B = ones(4). Do you see a pattern?

4.5 Solving Systems of Linear Equations

Now, with the MP-inverse in hand, we consider an arbitrary system of linear equations $A\mathbf{x} = \mathbf{b}$ where A is m-by-n, $\mathbf{x}$ is n-by-1, and $\mathbf{b}$ is m-by-1.

THEOREM 4.19

$A\mathbf{x} = \mathbf{b}$ *has a solution if and only if* $AA^+\mathbf{b} = \mathbf{b}$. *If a solution exists at all, every solution is of the form* $\mathbf{x} = A^+\mathbf{b} + (I - A^+A)\mathbf{w}$, *where* $\mathbf{w}$ *is an arbitrary parameter matrix. Indeed, a consistent system always has* $A^+\mathbf{b}$ *as a particular solution.*

PROOF First, we verify the *consistency condition*. If $AA^+\mathbf{b} = \mathbf{b}$, then evidently $\mathbf{x}_o = A^+\mathbf{b}$ is a solution to the system. Conversely suppose a solution $\mathbf{y}$ exists. Then $A\mathbf{y} = \mathbf{b}$, so $A^+A\mathbf{y} = A^+\mathbf{b}$ whence $AA^+A\mathbf{y} = AA^+\mathbf{b}$. But $AA^+A = A$, so $\mathbf{b} = A\mathbf{y} = AA^+\mathbf{b}$, and we have the condition. Now, suppose the system has a solution $\mathbf{x}$ and let $\mathbf{x}_o = A^+\mathbf{b}$ as above. Then, if $\mathbf{w} = \mathbf{x} - \mathbf{x}_o$, $A\mathbf{w} = A(\mathbf{x} - \mathbf{x}_o) = A\mathbf{x} - A\mathbf{x}_o = \mathbf{b} - AA^+\mathbf{b} = \mathbf{b} - \mathbf{b} = \overrightarrow{0}$. Now clearly,

$\mathbf{x} = \mathbf{x}_o + \mathbf{w}$. But $A\mathbf{w} = \overrightarrow{0}$ implies $A^+A\mathbf{w} = \overrightarrow{0}$, so $\mathbf{w} = (I - A^+A)\mathbf{w}$. Thus we see $\mathbf{x} = \mathbf{x}_o + \mathbf{w} = A^+\mathbf{b} + (I - A^+A)\mathbf{w}$, and the proof is complete. □

So, while our previous concepts of inverse were highly restrictive, the MP-inverse handles arbitrary systems of linear equations, giving us a way to judge whether a solution exists and giving us a way to write down all solutions when they exist. We illustrate with an example.

Example 4.4

Let $A = \begin{bmatrix} 1 & 1 & 0 \\ 1 & 0 & 1 \\ 1 & 1 & 0 \\ 1 & 0 & 1 \end{bmatrix}$ in $\mathbb{C}^{4\times 3}$ and consider the system of linear equations $A\mathbf{x} = \mathbf{b}$, where $\mathbf{b} = \begin{bmatrix} 2 \\ 2 \\ 2 \\ 2 \end{bmatrix}$. First we compute the pseudoinverse of A. We write $A = FG = \begin{bmatrix} 1 & 1 \\ 1 & 0 \\ 1 & 1 \\ 1 & 0 \end{bmatrix} \begin{bmatrix} 1 & 0 & 1 \\ 0 & 1 & -1 \end{bmatrix}$, a full rank factorization of A. Then $A^+ = G^+F^+ = G^*(GG^*)^{-1}(F^*F)^{-1}F^* = \begin{bmatrix} \frac{1}{6} & \frac{1}{6} & \frac{1}{6} & \frac{1}{6} \\ \frac{2}{6} & \frac{-1}{6} & \frac{2}{6} & \frac{-1}{6} \\ \frac{-1}{6} & \frac{2}{6} & \frac{-1}{6} & \frac{2}{6} \end{bmatrix}$ and $AA^+ = \begin{bmatrix} \frac{1}{2} & 0 & \frac{1}{2} & 0 \\ 0 & \frac{1}{2} & 0 & \frac{1}{2} \\ \frac{1}{2} & 0 & \frac{1}{2} & 0 \\ 0 & \frac{1}{2} & 0 & \frac{1}{2} \end{bmatrix}$. Since AA^+ is idempotent, its rank is its trace, which is 2. Also $I - A^+A = \begin{bmatrix} \frac{1}{3} & \frac{-1}{3} & \frac{-1}{3} \\ \frac{-1}{3} & \frac{1}{3} & \frac{1}{3} \\ \frac{-1}{3} & \frac{1}{3} & \frac{1}{3} \end{bmatrix}$. We compute $AA^+\mathbf{b}$ to see if we have any solutions at all: $AA^+\mathbf{b} = \begin{bmatrix} \frac{1}{2} & 0 & \frac{1}{2} & 0 \\ 0 & \frac{1}{2} & 0 & \frac{1}{2} \\ \frac{1}{2} & 0 & \frac{1}{2} & 0 \\ 0 & \frac{1}{2} & 0 & \frac{1}{2} \end{bmatrix} \begin{bmatrix} 2 \\ 2 \\ 2 \\ 2 \end{bmatrix} = \begin{bmatrix} 2 \\ 2 \\ 2 \\ 2 \end{bmatrix}$. Yes, the system is consistent. To give all possible solutions we form $\mathbf{x} = A^+\mathbf{b} + [I - A^+A]\mathbf{w}$, where $\mathbf{w}$ is a column of free parameters: $\mathbf{x} =$

$$\begin{bmatrix} \frac{1}{6} & \frac{1}{6} & \frac{1}{6} & \frac{1}{6} \\ \frac{2}{6} & \frac{-1}{6} & \frac{2}{6} & \frac{-1}{6} \\ \frac{-1}{6} & \frac{2}{6} & \frac{-1}{6} & \frac{1}{6} \end{bmatrix} \begin{bmatrix} 2 \\ 2 \\ 2 \\ 2 \end{bmatrix} + \begin{bmatrix} \frac{1}{3} & \frac{-1}{3} & \frac{-1}{3} \\ \frac{-1}{3} & \frac{1}{3} & \frac{1}{3} \\ \frac{-1}{3} & \frac{1}{3} & \frac{1}{3} \end{bmatrix} \begin{bmatrix} w_1 \\ w_2 \\ w_3 \end{bmatrix} =$$

$$\begin{bmatrix} \frac{8}{6} \\ \frac{4}{6} \\ \frac{2}{6} \end{bmatrix} + \begin{bmatrix} \frac{1}{3}(w_1 - w_2 - w_3) \\ \frac{1}{3}(-w_1 + w_2 + w_3) \\ \frac{1}{3}(-w_1 + w_2 + w_3) \end{bmatrix} = \begin{bmatrix} \frac{4}{3} \\ \frac{2}{3} \\ \frac{1}{3} \end{bmatrix} + t \begin{bmatrix} \frac{1}{3} \\ \frac{-1}{3} \\ \frac{-1}{3} \end{bmatrix},$$ where $t = w_1 - w_2 - w_3$. This also tells us that the nullity of A is 1 and the null space is $\mathcal{N}ull(A) = span\left(\begin{bmatrix} \frac{1}{3} \\ \frac{-1}{3} \\ \frac{-1}{3} \end{bmatrix}\right)$. Next consider $A\mathbf{x} = \mathbf{b} = \begin{bmatrix} 1 \\ 2 \\ 1 \\ 0 \end{bmatrix}$. Again we check for consistency by computing $AA^+ \begin{bmatrix} 1 \\ 2 \\ 1 \\ 0 \end{bmatrix} = \begin{bmatrix} \frac{1}{2} & 0 & \frac{1}{2} & 0 \\ 0 & \frac{1}{2} & 0 & \frac{1}{2} \\ \frac{1}{2} & 0 & \frac{1}{2} & 0 \\ 0 & \frac{1}{2} & 0 & \frac{1}{2} \end{bmatrix} \begin{bmatrix} 1 \\ 2 \\ 1 \\ 0 \end{bmatrix}$

$= \begin{bmatrix} 1 \\ 1 \\ 1 \\ 1 \end{bmatrix} \neq \begin{bmatrix} 1 \\ 2 \\ 1 \\ 0 \end{bmatrix}$. Evidently, there is no solution to this system.

The astute reader has no doubt noticed that the full force of the pseudoinverse was not needed to settle the problem of solving systems of linear equations. Indeed, only equation (MP1) was used. This observation leads us into the next chapter.

Exercise Set 17

1. Use the pseudoinverse to determine whether the following systems of linear equations have a solution. If they do, determine the most general solution.

(a) $\begin{cases} 2x - 10y + 16z = 10 \\ 3x + y - 5z = 1 \\ 2x - y + z = 4 \\ 2x - 2y + z = 3 \end{cases}$

(b) $\begin{cases} 2x - 4y + 3z - 5w = -3 \\ 6x - 2y + 4z - 5w = 1 \\ 4x - 2y + z = 4 \end{cases}$

2. Create examples of three equations in two unknowns that

   (a) have unique solutions
   (b) have an infinite number of solutions
   (c) have no solutions.

   Are your examples compatible with the theory worked out above?

3. This exercise refers back to the Hermite echelon form. Suppose we desire the solutions of $A\mathbf{x} = \mathbf{b}$ where A is square but not necessarily invertible. We have showed how to use the pseudoinverse to describe all solutions to $A\mathbf{x} = \mathbf{b}$ if any exist. In this exercise, we consider a different approach. First form the augmented matrix $[A \mid \mathbf{b}]$. There is an invertible matrix S such that $SA = H = HEF(A)$ so form $[A \mid \mathbf{b}] \rightarrow [SA \mid S\mathbf{b}] = [H \mid S\mathbf{b}]$.

   (a) Argue that $rank(A) = rank(H) =$ the number of ones on the diagonal of H.
   (b) Prove that $A\mathbf{x} = \mathbf{b}$ is consistent iff $S\mathbf{b}$ has nonzero components only in the rows where H has ones.
   (c) If $A\mathbf{x} = \mathbf{b}$ is consistent, argue $S\mathbf{b}$ is a particular solution to the system.
   (d) Argue that if H has r ones down its diagonal, then $I - SA$ has exactly $n - r$ nonzero columns and these nonzero columns span $\mathcal{N}ull(A)$ and hence form a basis for $\mathcal{N}ull(A)$.
   (e) Argue that all solutions of $A\mathbf{x} = \mathbf{b}$ are described by $\mathbf{x} = S\mathbf{b} + (I - SA)D$, where D is a diagonal matrix containing $n - r$ free parameters.

Further Reading

[Greville, 1959] T. N. E. Greville, The Pseudoinverse of a Rectangular or Singular Matrix and its Application to the Solution of Systems of Linear Equations, SIAM Review, Vol. 1, (1959), 38–43.

[K&X, 1995] Robert Kalaba and Rong Xu, On the Generalized Inverse Form of the Equations of Constrained Motion, The American Mathematical Monthly, Vol. 102, No. 9, November, (1995), 821–825.

MP-Schur complement, the rank theorem, Sylvester's determinant formula, the quotient formula, Reidel's formula, parallel sum

---

4.6 Schur Complements Again (optional)

Now, we generalize the notion of a Schur complement to matrices that may not be invertible or even square. Let $M = \begin{bmatrix} A_{m\times n} & B_{m\times t} \\ C_{s\times n} & D_{s\times t}. \end{bmatrix}$. We define the *MP-Schur complement of A in M* to be

$$M/A = D - CA^{+}B.$$

Obviously, if A is square and nonsingular, $A^{-1} = A^{+}$ and we recapture the previous notion of the Schur complement. Similarly, we define

$$M//D = A - BD^{+}C.$$

Next, we investigate how far we can generalize the results of Chapter 1. Things get a bit more complicated when dealing with generalized inverses. We follow the treatment in Carlson, Haynsworth, and Markham [C&H&M, 1974].

THEOREM 4.20 (the rank theorem)

Let $M = \begin{bmatrix} A & B \\ C & D \end{bmatrix}$. *Then* $rank(M) \geq rank(A) + rank(M/A)$. *Moreover, equality holds iff*

1. $\mathcal{N}ull(M/A) \subseteq \mathcal{N}ull((I - A^{+}A)B)$
2. $\mathcal{N}ull((M/A)^{*}) \subseteq \mathcal{N}ull((I - A^{+}A)C^{*})$
3. $(I - AA^{+})B(M/A)^{+}C(I - A^{+}A) = \mathbb{O}$.

PROOF Let $P = \begin{bmatrix} I & \mathbb{O} \\ -CA^{+} & I \end{bmatrix}$ *and* $Q = \begin{bmatrix} I & -A^{+}B \\ \mathbb{O} & I \end{bmatrix}$. Note P and Q are invertible. Then $PMQ = \begin{bmatrix} I & \mathbb{O} \\ -CA^{+} & I \end{bmatrix} \begin{bmatrix} A & B \\ C & D \end{bmatrix} \begin{bmatrix} I & -A^{+}B \\ \mathbb{O} & I \end{bmatrix}$

$$= \begin{bmatrix} A & B \\ -CA^{+}A + C & -CA^{+}B + D \end{bmatrix} \begin{bmatrix} I & -A^{+}B \\ \mathbb{O} & I \end{bmatrix}$$

$$= \begin{bmatrix} A & -AA^{+}B + B \\ -CA^{+}A + C & CA^{+}AA^{+}B - CA^{+}B + -CA^{+}B + D \end{bmatrix}$$

$$= \begin{bmatrix} A & -AA^+B + B \\ -CA^+A + C & CA^+B - CA^+B + -CA^+B + D \end{bmatrix}$$

$$= \begin{bmatrix} A & -AA^+B + B \\ -CA^+A + C & D - CA^+B \end{bmatrix}.$$ *Let* $P =$

$$\begin{bmatrix} I & -(I - AA^+)B(M/A)^+ \\ \mathbb{O} & I \end{bmatrix}, \; M = \begin{bmatrix} A_{m\times n} & B_{m\times t} \\ C_{s\times n} & D_{s\times t} \end{bmatrix}.$$

Then $(M/A)_{s\times t} = D_{s\times t} - C_{s\times n}A^+_{n\times m}B_{m\times t} \in \mathbb{C}^{(m+s)\times(n+t)}$. □

The following theorem is *Sylvester's determinant formula* for noninvertible matrices.

THEOREM 4.21 (Sylvester's determinant formula)

Let $M = \begin{bmatrix} A_{k\times k} & B \\ C & D \end{bmatrix}$. *Let either* $\mathcal{N}ull(A) \subseteq \mathcal{N}ull(C)$ *or* $\mathcal{N}ull(A^*) \subseteq \mathcal{N}ull(B^*)$. *Let* $P = \begin{bmatrix} p_{ij} \end{bmatrix}$ *and*

$p_{ij} = det \begin{bmatrix} A & Col_j(B) \\ Row_i(C) & d_{ij} \end{bmatrix}$. *Then* $P = det(A)(M/A)$, *and if* M *is* n*-by-*n, $det(P) = (det(A))^{n-k-l} det\,(M)$.

PROOF Since either $\mathcal{N}ull(A) \subseteq \mathcal{N}ull(C)$ or $\mathcal{N}ull(A^*) \subseteq \mathcal{N}ull(B^*)$ holds, the determinant formula $det\,(M) = det\,(A) det(M/A)$ can be applied to the elements $p_{ij} : p_{ij} = det(A)(d_{ij} - Row_i(C)A^+Col_j(B)) = (det(A))\,(M/A)_{ij}$. If M is n-by-n, then by equation above $det\,(P) = (det(A))^{n-k} det(M/A) = (det(A))^{n-k-l} det(M)$. □

The *quotient formula* for the noninvertible case is stated and proved below.

THEOREM 4.22 (the quotient formula)

Let $A = \begin{bmatrix} B & G \\ H & K \end{bmatrix}$, *where* $B = \begin{bmatrix} C & D \\ E & F \end{bmatrix}$. *Let* $\mathcal{N}ull(A) \subseteq \mathcal{N}ull(C)$, $\mathcal{N}ull(A^*) \subseteq \mathcal{N}ull(B^*)$. *Then* $(A/B) = (A/C)/(B/C)$.

PROOF Since $\mathcal{N}ull(A) \subseteq \mathcal{N}ull(C)$ and $\mathcal{N}ull(A^*) \subseteq \mathcal{N}ull(B^*)$, we may write $A = \begin{bmatrix} B & BP \\ QP & K \end{bmatrix}$, $B = \begin{bmatrix} C & CR \\ SC & F \end{bmatrix}$. So, we have $A/B = K - QPBB^+BP = K - QBP$, $B/C = F - SCR$. Partition P, Q as $P = \begin{bmatrix} P_1 \\ P_2 \end{bmatrix}$, $Q = \begin{bmatrix} Q_1 & Q_2 \end{bmatrix}$. Then $A/C =$

$\begin{bmatrix} (B/C) & (B/C)P_2 \\ Q_2(B/C) & K-(Q_1+Q_2)C(P_1+RP_2) \end{bmatrix}$. Hence A/C satisfies $\mathcal{N}ull(A) \subseteq \mathcal{N}ull(C)$ and $\mathcal{N}ull(A^*) \subseteq \mathcal{N}ull(B^*)$. Since $B/C = F - SCR$, $(A/C)/(B/C) = K - Q_1CP_1 - Q_2SCP_1 - Q_2FP_2 = A/B$. □

The following theorem is an example using the MP-inverse to get a Sherman-Morrison-Woodbury type theorem. The theorem stated below is due to Reidel [1992].

THEOREM 4.23 (Reidel's formula)

Let $A_{l\times l}$ have rank l_1, where $l_1 < l$, V_1, V_2, W_1, W_2 be l-by-k and G be a k-by-k nonsingular matrix. Let the columns of V_1 belong to $\mathcal{C}ol(A)$ and the columns of W_1 be orthogonal to $\mathcal{C}ol(A)$. Let the columns of V_2 belong to $\mathcal{C}ol(A^*)$ and the columns of W_2 be orthogonal to $\mathcal{C}ol(A^*)$. Let $B = W_i^*W_i$ have rank k. Suppose $\mathcal{C}ol(W_1) = \mathcal{C}ol(W_2)$. Then the matrix $\pi = A + (V_1+W_1)G(V_2+W_2)^*$ has the MP-inverse

$$\pi^+ = A^+ - W_2(W_2^*W_2)^{-1}V_2^*A^+ - A^+V_1(W_1(W_1^*W_1)^{-1})^* + W_2(W_2^*W_2)^{-1}(G^+ + V_2^*A^+V_1)(W_1(W_1^*W_1)^{-1})^*.$$

The proof is computational but lengthy, so we leave it as a challenge to the reader. The corollary given below addresses some consequences of this theorem.

COROLLARY 4.6

The following are true under the assumptions of the previous theorem:

1. *If $G = I$, then* $(A+(V_1+W_1)(V_2+W_2)^*)^+$
$= A^+ - W_2(W_2^*W_2)^{-1}V_2^*A^+ - A^+V_1(W_1(W_1^*W_1)^{-1})^*$
$+ W_2(W_2^*W_2)^{-1}V_2^*A^+V_1(W_1(W_1^*W_1)^{-1})^*$.

2. *If $G = I$, and $A = I$, then* $(I+(V_1+W_1)(V_2+W_2)^*)^+$
$= I - W_2(W_2^*W_2)^{-1}V_2^* - V_1(W_1(W_1^*W_1)^{-1})^*$
$+ W_2(W_2^*W_2)^{-1}V_2^*V_1(W_1(W_1^*W_1)^{-1})^*$.

3. *If $V_1 = V_2 = \mathbb{O}$,*
then $((A+W_1W_2)^*)^+ = A^+ + W_2(W_2^*W_2)^{-1}G^+(W_1(W_1^*W_1)^{-1})^*$.

Fill and Fishkind [$F\&F$, 1999] noticed that the assumption $\mathcal{C}ol(W_1) = \mathcal{C}ol(W_2)$ is not used in the proof of Reidel's theorem, and they used this theorem to get a somewhat clean formula for the MP-inverse of a sum. They also noted that a rank additivity assumption cannot be avoided in Reidel's theorem.

THEOREM 4.24
Suppose $A, B \in \mathbb{C}^{n\times n}$ *with* $rank(A+B) = rank(A) + rank(B)$*. Then*

$$(A+B)^{+} = (I-S)A^{+}(I-T) + SB^{+}T$$

where $S = (B^{+}B(I-A^{+}A))^{+}$ *and* $T = ((I-AA^{+})BB^{+})^{+}$.

PROOF The proof uses Reidel's theorem and is again rather lengthy so we omit it. ∎

The following simple example is offered on page 630 of Fill and Fishkind [1999].

Example 4.5
Let A = B = [1]. Then rank(A+B) = rank(1+1) = rank(2) = 1≠ 2 = $rank[1]+ rank[1] = rank(A) + rank(B)$. Now S = $[[1][0]]^{+}$ = [0] and T = $[[0][1]]^{+}$ = [0]. Therefore, (I-S)A^{+}(I-T)+SB^{+}T = $[1][1]^{+}[1] + [0][1]^{+}[0] = [1]$ and $(A+B)^{+} = [\frac{1}{2}]$ and the theorem fails.

There is a nice application of the previous theorem to the parallel sum of two matrices. Given $A, B \in \mathbb{C}^{n\times n}$, we define the *parallel sum and A and B* to be

$$A \parallel B = (A^{+} + B^{+})^{+}.$$

COROLLARY 4.7
Suppose $A, B \in \mathbb{C}^{n\times n}$ *with* $rank(A \parallel B) = rank(A) + rank(B)$*. Then* $A \parallel B = (I-R)A(I-W) + RBW$*, where* $R = (BB^{+}(I-AA^{+})^{+}$ *and* $W = ((I-A^{+}A)B^{+}B)^{+}$.

Exercise Set 18

1. Let $M = \begin{bmatrix} A & B \\ C & D \end{bmatrix}$. Argue that $det(M) = det(A)det(M/A)$ if A is invertible. Moreover, if $AC = CA$, then $det(M) = AD - BC$.

Further Reading

[C&H&M, 1974] David Carlson, Emilie Haynsworth, and Thomas Markham, A Generalization of the Schur Complement by Means of the Moore Penrose Inverse, SIAM J. Appl. Math., Vol. 26, No. 1, (1974), 169–175.

[C&H, 1969] D. E. Crabtree and E. V. Haynsworth, An Identity for the Schur Complement of a Matrix, Proceedings of the American Mathematical Society, Vol. 22, (1969), 364–366.

[F&F, 1999] James Allen Fill and Donnielle E. Fishkind, The Moore-Penrose Generalized Inverse For Sums of Matrices, SIAM J. Matrix Anal., Vol. 21, No. 2, (1999), 629–635.

[Gant, 1959] F. R. Gantmacher, *The Theory of Matrices*, Vol. 1, Chelsea Publishing Company, New York, (1959).

[Riedel, 1992] K. S. Riedel, A Sherman-Morrison-Woodbury Identity for Rank Augmenting Matrices with Application to Centering, SIAM J. Matrix Anal., Vol. 13, No. 2, (1992), 659–662.

Chapter 5

Generalized Inverses

5.1 The $\{1\}$-Inverse

The Moore-Penrose inverse (MP-inverse) of a matrix A in $\mathbb{C}^{m \times n}$ can be considered as the unique solution X of a system of simultaneous matrix equations; namely,

$$\text{(MP1) } AXA = A$$
$$\text{(MP2) } XAX = X$$
$$\text{(MP3) } (AX)^* = AX$$
$$\text{(MP4) } (XA)^* = XA.$$

We have already noted that only MP1 was needed when seeking solutions of a system of linear equations. This leads us to the idea of looking for "inverses" of A that satisfy only some of the MP-equations. We introduce some notation. Let $A\{1\} = \{G \mid AGA = A\}$, $A\{2\} = \{H \mid HAH = H\}$, and so forth. For example, $A\{1,2\} = A\{1\} \cap A\{2\}$. That is, a $\{1,2\}$-inverse of A is a matrix that satisfies MP1 and MP2. We have established previously that $A\{1,2,3,4\}$ has just one element in it, namely A^+. Evidently we have the inclusions $A\{1,2,3,4\} \subseteq A\{1,2,3\} \subseteq A\{1,2\} \subseteq A\{1\}$. Of course, many other chains are also possible. You might try to discover them all.

In this section, we devote our attention to $A\{1\}$, the set of all 1-inverses of A. The idea of a 1-inverse can be found in the book by Baer [1952]. Baer's idea was later developed by Sheffield [1958] in a paper. Let's make it official.

DEFINITION 5.1 *(1-inverse)*

*Let $A \in \mathbb{C}^{m \times n}$. A **1-inverse** of A is a matrix G in $\mathbb{C}^{n \times m}$ such that $AGA = A$; that is, a matrix that satisfies MP1 is called a 1-inverse of A. Let $A\{1\}$ denote the collection of all possible 1-inverses of A.*

Our goal in this section is to describe $A\{1\}$. Of course, $A^+ \in A\{1\}$ so this collection is never empty and there is always a ready example of an element in $A\{1\}$. Another goal is to prove a fundamental result about 1-inverses. We begin by looking at a general matrix equation $AXB = C$ where $A \in \mathbb{C}^{m\times n}$, $X \in \mathbb{C}^{n\times p}$, $B \in \mathbb{C}^{p\times q}$ and, necessarily, $C \in \mathbb{C}^{m\times q}$. Here A, B, C are given and we are to solve for X.

THEOREM 5.1

Let $AXB = C$ be as above. Then this equation has a solution if and only if there exists an A^g in $A\{1\}$ and a B^g in $B\{1\}$ such that $AA^gCB^gB = C$. If solutions exist, they are all of the form $X = A^gCB^g + W - A^gAWBB^g$, where W is arbitrary in $\mathbb{C}^{n\times p}$.

PROOF Suppose the consistency condition $AA^gCB^gB = C$ holds for some A^g in $A\{1\}$ and some B^g in $B\{1\}$. Then clearly, $X_o = A^gCB^g$ solves the matrix equation. Conversely, suppose $AXB = C$ has a solution X_1. Then $AX_1B = C$ so $A^+AX_1BB^+ = A^+CB^+$. Thus, $AA^+AX_1BB^+B = AA^+CB^+B$ and also equals AX_1B, which is C. Therefore, $AA^+CB^+B = C$ and note that $A^+ \in A\{1\}$ and $B^+ \in B\{1\}$.

Now suppose $AXB = C$ has a solution, say Y. Let $K = Y - X_o$, where $X_o = A^gCB^g$ as above. Then $AKB = A(Y - X_o)B = AYB - AX_oB = C - AA^gCB^gB = \mathbb{O}$. Now, $AKB = \mathbb{O}$ implies $A^gAKBB^g = \mathbb{O}$, so $K = K - A^gAKBB^g$, so $Y = X_o + K = A^gCB^g + K - A^gAKBB^g$. On the other hand, if $X = A^gCB^g + W - A^gAWBB^g$ for some W, then $AXB = A(A^gCB^g)B + A(W - A^gAWBB^g)B = AA^gCB^gB + AWB - AA^gAWBB^gB = C$ using the consistency condition and the fact that A^g and B^g are 1-inverses. This completes the proof. □

The test of a good theorem is all the results that follow from it. We now reap the harvest of this theorem.

COROLLARY 5.1

Consider the special case $AXA = A$. This equation always has solutions since the consistency condition $AA^gAA^gA = A$ holds true for any A^g in $A\{1\}$, including A^+. Moreover, $A\{1\} = \{X \mid X = A^gAA^g + W - A^gAWAA^g\}$, where W is arbitrary and A^g is any 1-inverse of A. In particular, $A\{1\} = \{X \mid X = A^+ + W - A^+AWAA^+\}$, where W is arbitrary.

COROLLARY 5.2

Consider the matrix equation $AX = C$, A in $\mathbb{C}^{m\times n}$, X in $\mathbb{C}^{n\times p}$, and C in $\mathbb{C}^{m\times p}$. This is solvable iff $AA^gC = C$ for some 1-inverse A^g of A and the general solution is $X = A^gC + (I - A^gA)W$, where W is arbitrary.

COROLLARY 5.3
Consider the matrix equation $XB = C$, where X is n-by-p, $B \in \mathbb{C}^{p\times q}$, and $C \in \mathbb{C}^{n\times q}$. This equation is solvable iff $CB^gB = C$ for some 1-inverse B^g of B and then the general solution is $X = CB^g + W(I - BB^g)$, where W is arbitrary.

COROLLARY 5.4
Consider the matrix equation $AX = \mathbb{O}$, where $A \in \mathbb{C}^{m\times n}$ and X is n-by-p. Then this equation always has solutions since the consistency condition $AA^g\mathbb{O}I^gI = \mathbb{O}$ evidently holds for any 1-inverse of A. The solutions are of the form $X = (I - A^gA)W$, where W is arbitrary.

COROLLARY 5.5
Consider the matrix equation $XB = \mathbb{O}$. This equation always has solutions since the consistency condition evidently holds. The solutions are of the form $X = W(I - BB^g)$, where W is arbitrary and B^g is some 1-inverse of B.

COROLLARY 5.6
Consider a system of linear equations $A\mathbf{x} = \mathbf{c}$, where $A \in \mathbb{C}^{m\times n}$, $\mathbf{x}$ is n-by-1 and $\mathbf{c}$ is m-by-1. Then this system is solvable iff $AA^g\mathbf{c} = \mathbf{c}$ for any A^g in $A\{1\}$ and the solutions are all of the form $\mathbf{x} = A^g\mathbf{c} + (I - A^gA)\mathbf{w}$, where $\mathbf{w}$ is arbitrary of size n-by-1.

We recall that for a system of linear equations $A\mathbf{x} = \mathbf{b}$, A^{-1}, if it exists, has the property that $A^{-1}\mathbf{b}$ is a solution for every choice of $\mathbf{b}$. It turns out that the 1-inverses of A generalize this property for arbitrary A.

THEOREM 5.2
Consider the system of linear equations $A\mathbf{x} = \mathbf{b}$, where $A \in \mathbb{C}^{m\times n}$. Then $G\mathbf{b}$ is a solution of this system for every $\mathbf{b} \in Col(A)$ iff $G \in A\{1\}$.

PROOF Suppose first $G\mathbf{b}$ is a solution for every $\mathbf{b}$ in the column space of A. Then $AG\mathbf{b} = \mathbf{b}$ in $Col(A)$. Now $\mathbf{b} = A\mathbf{y}$ for some $\mathbf{y}$ so $AGA\mathbf{y} = A\mathbf{y}$. But $\mathbf{y}$ could be anything, so $AGA = A$. Conversely, suppose $G \in A\{1\}$. Take any $\mathbf{b}$ in $Col(A)$. Then $\mathbf{b} = A\mathbf{y}$ for some $\mathbf{y}$. Then $AG\mathbf{b} = AGA\mathbf{y} = A\mathbf{y} = \mathbf{b}$. ▯

As Campbell and Meyer [C&M, 1979] put it, the "equation solving" generalized inverses of A are exactly the 1-inverses of A. Next, we give a way of generating 1-inverses of a given matrix. First, we need the following.

THEOREM 5.3
If S and T are invertible matrices and G is a 1-inverse of A, then $T^{-1}GS^{-1}$ is a 1-inverse of $B = SAT$. Moreover, every 1-inverse of B is of this form.

PROOF First $B(T^{-1}GS^{-1})B = SAT(T^{-1}GS^{-1})(SAT) = SAIGIAT = SAGAT = SAT = B$, proving $T^{-1}GS^{-1}$ is a 1-inverse of B. Next, let K be any 1-inverse of B, that is $BKB = B$. Note also that $S^{-1}BT^{-1}(TKS)S^{-1}BT^{-1} = S^{-1}BKBT^{-1} = S^{-1}BT^{-1}$. But $S^{-1}BT^{-1} = A$ so if we take $G = SKT$, then $AGA = A(SKT)A = A$. That is, G is a 1-inverse of A and $K = T^{-1}GS^{-1}$ as claimed. ▯

THEOREM 5.4 (Bose)
If $A \in \mathbb{C}_r^{m\times n}$, there exist invertible matrices S and T with $SAT = \begin{bmatrix} I_r & \mathbb{O} \\ \mathbb{O} & \mathbb{O} \end{bmatrix}$. A matrix G is a 1-inverse of A iff $G = TN^{\#}S$, where $N^{\#} = \begin{bmatrix} I_r & Y \\ X & W \end{bmatrix}$ where X, Y, and W are arbitrary matrices of appropriate size.

PROOF Suppose $SAT = \begin{bmatrix} I_r & \mathbb{O} \\ \mathbb{O} & \mathbb{O} \end{bmatrix}$, where S and T are invertible. Let $G = T\begin{bmatrix} I_r & Y \\ X & W \end{bmatrix}S$, where X, Y, and W are arbitrary matrices of appropriate size. Then, $AGA = AT\begin{bmatrix} I_r & Y \\ X & W \end{bmatrix}SA =$

$$S^{-1}\begin{bmatrix} I_r & \mathbb{O} \\ \mathbb{O} & \mathbb{O} \end{bmatrix}T^{-1}T\begin{bmatrix} I_r & Y \\ X & W \end{bmatrix}SS^{-1}\begin{bmatrix} I_r & \mathbb{O} \\ \mathbb{O} & \mathbb{O} \end{bmatrix}T^{-1} =$$

$$S^{-1}\begin{bmatrix} I_r & \mathbb{O} \\ \mathbb{O} & \mathbb{O} \end{bmatrix}\begin{bmatrix} I_r & Y \\ X & W \end{bmatrix}\begin{bmatrix} I_r & \mathbb{O} \\ \mathbb{O} & \mathbb{O} \end{bmatrix}T^{-1} = S^{-1}\begin{bmatrix} I_r & Y \\ \mathbb{O} & \mathbb{O} \end{bmatrix}\begin{bmatrix} I_r & \mathbb{O} \\ \mathbb{O} & \mathbb{O} \end{bmatrix}T^{-1}$$

$= S^{-1}\begin{bmatrix} I_r & \mathbb{O} \\ \mathbb{O} & \mathbb{O} \end{bmatrix}T^{-1} = A$. Conversely, if G is a 1-inverse of A, then $T^{-1}GS^{-1}$ is a 1-inverse of $S^{-1}AT^{-1}$ by the previous theorem. That is, $N = T^{-1}GS^{-1}$ is a 1-inverse of $\begin{bmatrix} I_r & \mathbb{O} \\ \mathbb{O} & \mathbb{O} \end{bmatrix}$. Partition $N = \begin{bmatrix} M & Y \\ X & W \end{bmatrix}$, where M is r-by-r. But $\begin{bmatrix} I_r & \mathbb{O} \\ \mathbb{O} & \mathbb{O} \end{bmatrix}\begin{bmatrix} M & Y \\ X & W \end{bmatrix}\begin{bmatrix} I_r & \mathbb{O} \\ \mathbb{O} & \mathbb{O} \end{bmatrix} = \begin{bmatrix} M & \mathbb{O} \\ \mathbb{O} & \mathbb{O} \end{bmatrix}$ and also equals $\begin{bmatrix} I_r & \mathbb{O} \\ \mathbb{O} & \mathbb{O} \end{bmatrix}$. Therefore $M = I_r$. Thus $G = TNS = T\begin{bmatrix} I_r & Y \\ X & W \end{bmatrix}S$. ▯

Next, we consider an example.

Example 5.1

Let $A = \begin{bmatrix} 1 & 0 & 0 \\ 0 & 1 & 0 \\ 0 & 0 & 0 \end{bmatrix}$. Then $A = A^+$. To find all the 1-inverses of A, we compute $A^+ + W - A^+AWAA^+$ for an arbitrary W:

$$\begin{bmatrix} 1 & 0 & 0 \\ 0 & 1 & 0 \\ 0 & 0 & 0 \end{bmatrix} + \begin{bmatrix} x & y & z \\ u & v & w \\ r & s & t \end{bmatrix} - \begin{bmatrix} 1 & 0 & 0 \\ 0 & 1 & 0 \\ 0 & 0 & 0 \end{bmatrix} \begin{bmatrix} x & y & z \\ u & v & w \\ r & s & t \end{bmatrix} \begin{bmatrix} 1 & 0 & 0 \\ 0 & 1 & 0 \\ 0 & 0 & 0 \end{bmatrix} =$$

$\begin{bmatrix} 1 & 0 & z \\ 0 & 1 & w \\ r & s & t \end{bmatrix}$. Thus $A\{1\} = \{ \begin{bmatrix} 1 & 0 & z \\ 0 & 1 & w \\ r & s & t \end{bmatrix} \mid z, w, t, r, s$ are arbitrary$\}$.

We find that $AA^g = \begin{bmatrix} 1 & 0 & z \\ 0 & 1 & w \\ 0 & 0 & 0 \end{bmatrix}$, $A^gA = \begin{bmatrix} 1 & 0 & 0 \\ 0 & 1 & 0 \\ r & s & 0 \end{bmatrix}$, $I - A^gA = \begin{bmatrix} 0 & 0 & 0 \\ 0 & 0 & 0 \\ -r & -s & 1 \end{bmatrix}$, and $I - AA^g = \begin{bmatrix} 0 & 0 & -z \\ 0 & 1 & -w \\ 0 & 0 & 1 \end{bmatrix}$.

We introduced you to the Hermite echelon form for a reason. It helps to generate 1-inverses, indeed, nonsingular ones.

THEOREM 5.5

Suppose A is n-by-n and S is nonsingular with $SA = HEF(A) := H$. Then S is a 1-inverse of A.

PROOF The proof is left as an exercise. □

THEOREM 5.6 [Bose, 1959]

Let $H \in (A^*A)\{1\}$. Then $AHA^* = A(A^*A)^+ = AA^+$. In other words, AHA^* is the unique Hermitian idempotent matrix with the same rank as A.

PROOF We know $A^*AHA^*A = A^*A$ so $AHA^*A = A$ and $A^*AHA^* = A^*$. Let G_1 and G_2 be in $A^*A\{1\}$. Then $AG_1A^*A = AG_2A^*A$ so that $AG_1A^* = AG_2A^*$. This proves the uniqueness. The rest is left to the reader. □

In 1975, Robert E. Hartwig published a paper where he proposed a formula for the 1-inverse of a product of two matrices. We present this next.

THEOREM 5.7 [Hartwig, 1975]

For conformable matrices A and B, $(AB)^g = B^g A^g - B^g(I - A^g A)[(I - BB^g)(I - A^g A)]^g(I - BB^g)A^g$.

It would be reasonable to guess the same formula works for the MP-inverse; that is,

$$(AB)^+ = B^+A^+ - B^+(I - A^+A)[(I - BB^+)(I - A^+A)]^+(I - BB^+)A^+.$$

If we check it with our previous example $A = \begin{bmatrix} 1 & 0 & 1 \\ 1 & 1 & 0 \end{bmatrix}$ and $B = \begin{bmatrix} 1 & 0 \\ 0 & 1 \\ 1 & 1 \end{bmatrix}$, we find $(AB)^+ \neq B^+A^+$. However, the right-hand side computes to $\begin{bmatrix} \frac{5}{9} & \frac{-1}{9} \\ \frac{-1}{9} & \frac{2}{9} \end{bmatrix} -$
$\begin{bmatrix} \frac{2}{3} & \frac{-1}{3} & \frac{1}{3} \\ \frac{-1}{3} & \frac{2}{3} & \frac{1}{3} \end{bmatrix} \begin{bmatrix} \frac{1}{3} & \frac{-1}{3} & \frac{-1}{3} \\ \frac{-1}{3} & \frac{1}{3} & \frac{1}{3} \\ \frac{-1}{3} & \frac{1}{3} & \frac{1}{3} \end{bmatrix} \begin{bmatrix} 1 & 1 & -1 \\ -1 & -1 & 1 \\ -1 & -1 & 1 \end{bmatrix} \begin{bmatrix} \frac{1}{3} & \frac{1}{3} & \frac{-1}{3} \\ \frac{1}{3} & \frac{1}{3} & \frac{-1}{3} \\ \frac{-1}{3} & \frac{-1}{3} & \frac{1}{3} \end{bmatrix}$
$\begin{bmatrix} \frac{1}{3} & \frac{1}{3} \\ \frac{-1}{3} & \frac{2}{3} \\ \frac{2}{3} & \frac{-1}{3} \end{bmatrix} = \begin{bmatrix} 1 & -1 \\ -1 & 2 \end{bmatrix} = (AB)^+$. So the formula works! Before you get too excited, let $A = \begin{bmatrix} 1 & 0 \\ 0 & 1 \\ 1 & 1 \end{bmatrix}$ and $B = \begin{bmatrix} 1 \\ 0 \end{bmatrix}$. Now A has full columns, so $A^+A = I$, $A^*A = \begin{bmatrix} 2 & 1 \\ 1 & 2 \end{bmatrix}$, $(A^*A)^{-1} = \frac{1}{3}\begin{bmatrix} 2 & -1 \\ -1 & 2 \end{bmatrix}$, and
$A^+ = (A^*A)^{-1}A^* =$
$\frac{1}{3}\begin{bmatrix} 2 & 1 \\ 1 & 2 \end{bmatrix}\begin{bmatrix} 1 & 0 & 1 \\ 0 & 1 & 1 \end{bmatrix} = \frac{1}{3}\begin{bmatrix} 2 & -1 & 1 \\ -1 & 2 & 1 \end{bmatrix}$. Also $B^*B = [1]$, $B^+ = (B^*B)^{-1} = [10]$ and $AB = \begin{bmatrix} 1 \\ 0 \\ 1 \end{bmatrix}$.

$(AB)^*(AB) = [2]$, and $(AB)^+ = [(AB)^*(AB)]^{-1}(AB)^* = \frac{1}{2}[101]$. However, $B^+A^+ = [10]\frac{1}{3}\begin{bmatrix} 2 & -1 & 1 \\ -1 & 2 & 1 \end{bmatrix}$
$= \frac{1}{3}[2\ -1 1] \neq (AB)^+$. It would be nice to have necessary and sufficient conditions for the formula to hold, or better yet, to find a formula that always works.

Exercise Set 19

1. Verify that $G = \begin{bmatrix} 4 & -1 & 0 & 0 \\ -3 & 1 & 0 & 0 \\ 0 & 0 & 0 & 0 \\ 0 & 0 & 0 & 0 \end{bmatrix}$ is a 1-inverse of $A = \begin{bmatrix} 1 & 1 & 0 & 1 \\ 3 & 4 & 2 & 0 \\ 3 & 5 & 4 & -3 \\ 1 & -1 & -4 & 7 \end{bmatrix}$. Can you find some others? Can you find them all?

2. Suppose $A \in \mathbb{C}_r^{m \times n}$ and $A = \begin{bmatrix} A_{11} & A_{12} \\ A_{21} & A_{21}A_{11}^{-1}A_{12} \end{bmatrix}$, where A_{11} is r-by-r invertible. Show that $G = \begin{bmatrix} A_{11}^{-1} & \mathbb{O} \\ \mathbb{O} & \mathbb{O} \end{bmatrix}$ is a 1-inverse of A.

3. Suppose G_1 and G_2 are in $A\{1\}$ for some matrix A. Show that $\frac{1}{2}(G_1 + G_2)$ is also in $A\{1\}$.

4. Suppose G_1 and G_2 are in $A\{1\}$ for some matrix A. Show that $\lambda G_1 + (1 - \lambda)G_2$ is also in $A\{1\}$. In other words, prove that $A\{1\}$ is an affine set.

5. Suppose $G_1, G_2, \ldots, G_k$ are in $A\{1\}$ and $\lambda_1, \lambda_2, \ldots, \lambda_k$ are scalars that sum to 1. Argue that $\lambda_1 G_1 + \lambda_2 G_2 + \cdots + \lambda_k G_k$ is in $A\{1\}$.

6. Is it true that any linear combination of 1-inverses of a matrix A is again a 1-inverse of A?

7. Let S and T be invertible matrices. Then show $T^{-1}A^g S^{-1}$ is a 1-inverse of SAT for any A^g in $A\{1\}$.

8. Argue that B is contained in the column space of A if and only if $AA^g B = B$ for any A^g in $A\{1\}$.

9. Argue that $A\{1\} = \{A^+ + [I - A^+A]W_{21} + W_{12}[I - AA^+] \mid W_{21}, W_{12}$ arbitrary$\}$.

10. If $G \in A\{1\}$, then prove that GA and AG are both idempotent matrices.

11. If $G \in A\{1\}$, then prove $rank(A) = rank(AG) = rank(GA) = trace(AG) \leq rank(G)$. Show $rank(I - AG) = m - rank(A)$ and $rank(I - GA) = n - r(A)$.

12. Suppose A is n-by-n and $H = HEF(A)$. Prove A is idempotent if and only if H is a 1-inverse of A.

13. If G is a 1-inverse of A, argue that GAG is also a 1-inverse of A and has the same rank as A.

14. Show that it is always possible to construct a 1-inverse of $A \in \mathbb{C}_r^{m \times n}$ that has rank = min{m, n}. In particular, argue that every square matrix has an invertible 1-inverse.

15. Make clear how Corollary 5.1 to Corollary 5.6 follow from the main theorem.

16. Prove Theorem 5.5.

17. Prove Theorem 5.6.

18. Suppose $AGA = A$. Argue that AG and GA are idempotents. What direct sum decompositions do they generate?

19. What is $\mathbb{O}_{n \times n}\{1\}$? What is $I_n\{1\}$? Remember the matrix units $E_{ij} \in \mathbb{C}^{m \times n}$? What is $E_{ij}\{1\}$?

20. Is $A\{2, 3, 4\}$ ever empty for some weird matrix A?

21. Find a 1-inverse for $\begin{bmatrix} 1 & 0 & 2 & 3 \\ 0 & 1 & 4 & 5 \\ 0 & 0 & 0 & 0 \\ 0 & 0 & 0 & 0 \end{bmatrix}$.

22. Show that $A \in \mathbb{C}_r^{m \times n}$ can have a $\{1\}$-inverse of any rank between r and $\min\{m, n\}$. (*Hint:* rank$\begin{bmatrix} I_r & \mathbb{O} \\ \mathbb{O} & D \end{bmatrix} = r + rank(D)$.)

23. Argue that any square nonsingular matrix has a unique 1-inverse.

24. Suppose $A \in \mathbb{C}^{m \times n}$ and $G \in A\{1\}$. Argue that $G^* \in A^*\{1\}$.

25. Suppose $A \in \mathbb{C}^{m \times n}$ and $G \in A\{1\}$ and $\lambda \in \mathbb{C}$. Argue that $\lambda^+ G \in (\lambda A)\{1\}$ where, recall, $\lambda^+ = \begin{cases} 0 & if\ \lambda = 0 \\ \lambda^{-1} & if\ \lambda \neq 0 \end{cases}$.

26. Suppose $G \in A\{1\}$. Argue that GA and AG are idempotent matrices that have the same rank as A.

27. Suppose $G \in A\{1\}$. Argue that $rank(G) \geq rank(A)$.

28. Suppose $G \in A\{1\}$. Argue that if A is invertible, $G = A^{-1}$.

29. Suppose $G \in A\{1\}$ with S and T invertible. Argue that, $T^{-1}GS^{-1} \in (SAT)\{1\}$.

30. Suppose $G \in A\{1\}$. Argue that $\mathcal{C}ol(AG) = \mathcal{C}ol(A)$, $\mathcal{N}ull(GA) = \mathcal{N}ull(A)$, and $\mathcal{C}ol((GA)^*) = \mathcal{C}ol(A^*)$.

31. Suppose $G \in A\{1\}$, where $A \in \mathbb{C}_r^{m\times n}$. Argue that $GA = I$ iff $r = n$ iff G is a left inverse of A.

32. Suppose $G \in A\{1\}$, where $A \in \mathbb{C}_r^{m\times n}$. Argue that $AG = I$ iff $r = m$ iff G is a right inverse of A.

33. Suppose $G \in A\{1\}$ and $\mathbf{v} \in \mathcal{N}ull(A)$. Argue that $G_1 = [\mathbf{g}_1 \mid \cdots \mid \mathbf{g}_j + \mathbf{v} \mid \cdots \mid \mathbf{g}_m] \in A\{1\}$.

34. Let $A \in \mathbb{C}_r^{m\times n}$. Argue that $G \in A\{1\}$ iff $G = S \begin{bmatrix} I_r & \mathbb{O} \\ \mathbb{O} & D \end{bmatrix} T$ for some invertible S and T.

35. Let $G \in A\{1\}$. Argue that $H = G + (W - GAWAG)$ is also a 1-inverse and all 1-inverses of A look like H.

36. (Penrose) Prove that $AX = B$ and $XC = D$ have a common solution X iff each separately has a solution and $AD = BC$.

37. Suppose $G \in A\{1\}$. Prove that $G + B - GABAG \in A\{1\}$ and $G + B(I_m - AG) = (I_n - GA)C \in A\{1\}$.

38. Suppose $G \in A\{1\}$. Prove that $rank(A) = rank(GA) = rank(AG)$ and $rank(G) = rank(A)$ iff $GAG = G$.

39. Suppose $G \in A^*A\{1\}$. Prove that $AGA^*A = A$ and $A^*AGA^* = A^*$.

40. Suppose $BAA^* = CAA^*$. Prove that $BA = CA$.

41. Suppose $G \in AA^*\{1\}$ and $H \in A^*A\{1\}$. Prove that $A = AA^*GA$, $A = AHA^*A$, $A = AA^*G^*A$, and $A = AH^*A^*A$.

42. Suppose A is n-by-m, B is p-by-m, and C is n-by-q with $\mathcal{C}ol(B^*) \subseteq \mathcal{C}ol(A^*)$ and $\mathcal{C}ol(C) \subseteq \mathcal{C}ol(A)$. Argue that for all G in $A\{1\}$, $BGC = BA^+C$.

43. Suppose A is n-by-n, $\mathbf{c}$ is n-by-1, and $\mathbf{c} \in \mathcal{C}ol(A) \cap \mathcal{C}ol(A^*)$. Argue that for all G in $A\{1\}$, $\mathbf{c}^*G\mathbf{c} = \mathbf{c}^*A^+\mathbf{c}$. How does this result read if $A = A^*$?

44. Find matrices G and A such that $G \in A\{1\}$ but $A \notin G\{1\}$.

Further Reading

[Baer, 1952] Reinhold Baer, *Linear Algebra and Projective Geometry*, Academic Press, Inc., New York, (1952).

[B-I&G, 2003] Adi Ben-Israel and Thomas N. E. Greville, *Generalized Inverses: Theory and Applications*, 2nd Edition, Springer, New York, (2003).

[B&O, 1971] Thomas L. Boullion and Patrick L. Odell, *Generalized Inverse Matrices*, Wiley-Interscience, New York, (1971).

[C&M, 1979] S. L. Campbell and C. D. Meyer, Jr., *Generalized Inverses of Linear Transformations*, Dover Publications, Inc., New York, (1979).

[Sheffield, 1958] R. D. Sheffield, A General Theory For Linear Systems, The American Mathematical Monthly, February, (1958), 109–111.

[Wong, 1979] Edward T. Wong, Generalized Inverses as Linear Transformations, Mathematics Gazette, Vol. 63, No. 425, October, (1979), 176–181.

5.2 {1,2}-Inverses

C. R. Rao in 1955 made use of a generalized inverse that satisfied MP1 and MP2. This type of inverse is sometimes called a *reflexive generalized inverse*. We can describe the general form of {1,2}-inverses as we did with 1-inverses. It is interesting to see the extra ingredient that is needed. We take the constructive approach as usual.

THEOREM 5.8

Let $A \in \mathbb{C}_r^{m\times n}$. *There are matrices S and T with* $SAT = \begin{bmatrix} I_r & \mathbb{O} \\ \mathbb{O} & \mathbb{O} \end{bmatrix}$. *A matrix G is a {1,2}-inverse of A if and only if* $G = TN^{\#}S$, *where* $N^{\#} = \begin{bmatrix} I_r & Y \\ X & XY \end{bmatrix}$, *where X and Y are arbitrary of appropriate size.*

PROOF First suppose $G = TN^{\#}S$ as given in the theorem. Then G is a 1-inverse by Theorem 5.3. To show it is also a 2-inverse we compute $GAG = GS^{-1}\begin{bmatrix} I_r & \mathbb{O} \\ \mathbb{O} & \mathbb{O} \end{bmatrix} T^{-1}G = TN^{\#}\begin{bmatrix} I_r & \mathbb{O} \\ \mathbb{O} & \mathbb{O} \end{bmatrix} N^{\#}S =$ $T\begin{bmatrix} I_r & Y \\ X & XY \end{bmatrix}\begin{bmatrix} I_r & \mathbb{O} \\ \mathbb{O} & \mathbb{O} \end{bmatrix}\begin{bmatrix} I_r & Y \\ X & XY \end{bmatrix} S = T\begin{bmatrix} I_r & \mathbb{O} \\ X & \mathbb{O} \end{bmatrix}\begin{bmatrix} I_r & Y \\ X & XY \end{bmatrix}$ $S = T\begin{bmatrix} I_r & Y \\ X & XY \end{bmatrix} S = G$. Conversely, suppose G is a {1,2}-inverse of A. Then, being a 1-inverse, we know $G = TN^{\#}S$, where $N^{\#} = \begin{bmatrix} I_r & Y \\ X & W \end{bmatrix}$. But to be a 2-inverse, $G = GAG =$ $T\begin{bmatrix} I_r & Y \\ X & W \end{bmatrix} SS^{-1}\begin{bmatrix} I_r & \mathbb{O} \\ \mathbb{O} & \mathbb{O} \end{bmatrix} T^{-1}T\begin{bmatrix} I_r & Y \\ X & W \end{bmatrix} S = T\begin{bmatrix} I_r & \mathbb{O} \\ X & \mathbb{O} \end{bmatrix}$ $\begin{bmatrix} I_r & Y \\ X & W \end{bmatrix} S = T\begin{bmatrix} I_r & Y \\ X & XY \end{bmatrix} S$. Comparing matrices, we see $W = XY$. ▯

Exercise Set 20

1. Suppose S and T are invertible matrices and G is a $\{1,2\}$-inverse of A. Then $T^{-1}GS^{-1}$ is a $\{1,2\}$-inverse of $B = SAT$. Moreover, every $\{1,2\}$-inverse of B is of this form.

2. (Bjerhammar) Suppose G is a 1-inverse of A. Argue that G is a $\{1,2\}$-inverse of A iff $rank(G) = rank(A)$.

3. Argue that $G \in A\{1, 2\}$ iff $G = G_1AG_2$, where $G_1, G_2 \in A\{1\}$.

4. Argue that $G = E(HAE)^{-1}H$ belongs to $A\{2\}$, where H and E are selected judiciously so that HAE is nonsingular.

5. Argue that $G = E(HAE)^{+}H$ belongs to $A\{2\}$, where H and E are chosen of appropriate size.

6. Suppose A and B are $\{1,2\}$-inverses of each other. Argue that AB is the projector onto $\mathcal{C}ol(A)$ along $\mathcal{N}ull(B)$ and BA is the projector of $\mathcal{C}ol(B)$ along $\mathcal{N}ull(A)$.

7. Argue that $G \in A\{1, 2\}$ iff there exist S and T invertible with $G = S\begin{bmatrix} I_r & O \\ O & O \end{bmatrix} T$ and $SAT = \begin{bmatrix} I_r & B \\ O & O \end{bmatrix}$.

5.3 Constructing Other Generalized Inverses

In this section, we take a constructive approach to building a variety of examples of generalized inverses of a given matrix. The approach we adopt goes back to the fundamental idea of reducing a matrix to row echelon form. Suppose $A \in \mathbb{C}_r^{m \times n}$. Then there exists an invertible matrix R with $RA = \begin{bmatrix} G \\ \cdots \\ \mathbb{O} \end{bmatrix}$, where G has $r = rank(A) = rank(G)$. Now G has full row rank and, therefore, we know $G^+ = G^*(GG^*)^{-1}$. Indeed, it is clear that $GG^+ = I_r$.

Now let's define $A^{g_1} = [G^+ + (I - G^+G)X \mid V]R$, where X and V are arbitrary matrices of appropriate size. We compute

$$AA^{g_1}A = R^{-1}\begin{bmatrix} G \\ \cdots \\ \mathbb{O} \end{bmatrix}[G^+ + (I - G^+G)X \vdots V]RR^{-1}\begin{bmatrix} G \\ \cdots \\ \mathbb{O} \end{bmatrix} =$$

$$= R^{-1}\begin{bmatrix} G \\ \cdots \\ \mathbb{O} \end{bmatrix}[G^+G + (I - G^+G)XG + \mathbb{O}] =$$

$$R^{-1}\begin{bmatrix} G^+G + (I - G^+G)XG + \mathbb{O} \\ \cdots \\ \mathbb{O} \end{bmatrix} = R^{-1}\begin{bmatrix} G \\ \cdots \\ \mathbb{O} \end{bmatrix} = A.$$ We see A^{g_1} is a $\{1\}$-inverse of A. For example,

let $A = \begin{bmatrix} 1 & 2 \\ 3 & 6 \end{bmatrix}$. Then $\begin{bmatrix} 1 & 0 \\ -3 & 1 \end{bmatrix}\begin{bmatrix} 1 & 2 \\ 3 & 6 \end{bmatrix} = \begin{bmatrix} 1 & 2 \\ 0 & 0 \end{bmatrix}$ so

$G = \begin{bmatrix} 1 & 2 \end{bmatrix}$, $GG^* = [5]$, $G^+ = \begin{bmatrix} 1/5 \\ 2/5 \end{bmatrix}$, $G^+G =$
$\begin{bmatrix} 1/5 & 2/5 \\ 2/5 & 4/5 \end{bmatrix}$, $I - G^+G = \begin{bmatrix} 4/5 & -2/5 \\ -2/5 & 1/5 \end{bmatrix}$. Then

$$A^{g_1} = \left[\begin{matrix} 1/5 \\ \\ \\ 2/5 \end{matrix} + \begin{bmatrix} 4/5 & -2/5 \\ -2/5 & 1/5 \end{bmatrix}\begin{bmatrix} r \\ u \end{bmatrix} \begin{matrix} \vdots & x \\ \\ \\ \vdots & y \end{matrix}\right]\begin{bmatrix} 1 & 0 \\ -3 & 1 \end{bmatrix} =$$

$$\begin{bmatrix} \frac{1}{5} + \frac{4r}{5} - \frac{2}{5}u \vdots x \\ \\ \frac{2}{5} - \frac{2r}{5} + \frac{u}{5} \vdots y \end{bmatrix}\begin{bmatrix} 1 & 0 \\ -3 & 1 \end{bmatrix}$$

$= \begin{bmatrix} \frac{1}{5} + \frac{4r}{5} - \frac{2}{5}u - 3x & x \\ \frac{2}{5} - \frac{2r}{5} + \frac{u}{5} - 3y & y \end{bmatrix}$, where r, u, x, and y can be chosen arbitrarily.

The skeptical reader can compute $AA^{g_1}A$ directly and watch the magic of just the right cancellations in just the right place at just the right time. Suppose in our example $u = r = 1$ and $x = y = \frac{1}{5}$. Then $A^{g_1} = \begin{bmatrix} 0 & 1/5 \\ -2/5 & 1/5 \end{bmatrix}$. However,

$$AA^{g_1} = \begin{bmatrix} 1 & 2 \\ 3 & 6 \end{bmatrix} \begin{bmatrix} 0 & 1/5 \\ -2/5 & 1/5 \end{bmatrix} = \begin{bmatrix} -4/5 & 3/5 \\ -12/5 & 9/5 \end{bmatrix} \text{ and}$$
$$A^{g_1}A = \begin{bmatrix} 0 & 1/5 \\ -2/5 & 1/5 \end{bmatrix} \begin{bmatrix} 1 & 2 \\ 3 & 6 \end{bmatrix} = \begin{bmatrix} 3/5 & 6/5 \\ 1/5 & 2/5 \end{bmatrix},$$

neither of which is symmetric. Also $A^{g_1}AA^{g_1} =$

$$\begin{bmatrix} 0 & 1/5 \\ -2/5 & 1/5 \end{bmatrix} \begin{bmatrix} -4/5 & 3/5 \\ -12/5 & 9/5 \end{bmatrix} = \begin{bmatrix} \frac{-12}{25} & \frac{9}{25} \\ -4/25 & 3/25 \end{bmatrix} \neq A^{g_1}.$$

So this choice of u, r, x, and y produces a $\{1\}$-inverse of A that satisfies none of the other MP-equations. In order to gain additional equations, we make some special choices.

As a next step, we choose $X = \mathbb{O}$ in our formula for A^{g_1} to produce $A^{g_{14}} = \begin{bmatrix} G^+ & \vdots & V \end{bmatrix} R$. As you may have already guessed from the notation, we claim now to have added some symmetry, namely equation MP-4. Surely MP-1 is still satisfied, and now

$A^{g_{14}}A = \begin{bmatrix} G^+ & \vdots & V \end{bmatrix} RR^{-1} \begin{bmatrix} G \\ \cdots \\ \mathbb{O} \end{bmatrix} = G^+G + V\mathbb{O} = G^+G$, which is evidently a projection. From our example above,

$$A^{g_{14}} = \begin{bmatrix} 1/5 - 3x & x \\ 2/5 - 3y & y \end{bmatrix}.$$

Continuing with our choice of $x = y = 1/5$, we have $A^{g_{14}} = \begin{bmatrix} -2/5 & 1/5 \\ -1/5 & 1/5 \end{bmatrix}$ is in $A\{1, 4\}$. However $AA^{g_{14}} = \begin{bmatrix} 1 & 2 \\ 3 & 6 \end{bmatrix} \begin{bmatrix} -2/5 & 1/5 \\ -1/5 & 1/5 \end{bmatrix} = \begin{bmatrix} -4/5 & 3/5 \\ -12/5 & 9/5 \end{bmatrix}$ is not symmetric and $A^{g_{14}}AA^{g_{14}} = \begin{bmatrix} -2/5 & 1/5 \\ -1/5 & 1/5 \end{bmatrix} \begin{bmatrix} -4/5 & 3/5 \\ -12/5 & 9/5 \end{bmatrix}$

$$= \begin{bmatrix} -4/25 & 3/25 \\ 8/25 & 6/25 \end{bmatrix} \neq A^{g_{14}}.$$

So $A^{g_{14}}$ satisfies MP-1 and MP-4 but not MP-2 or MP-3.

All right, suppose we want to get MP-2 to be satisfied. What more can we demand? Let's let mathematics be our guide. We want

$$A^{g_{14}} = A^{g_{14}}AA^{g_{14}} = \begin{bmatrix} G^+ & \vdots & V \end{bmatrix} RR^{-1} \begin{bmatrix} G \\ \cdots \\ \mathbb{O} \end{bmatrix} \begin{bmatrix} G^+ & \vdots & V \end{bmatrix} R$$

$$= \begin{bmatrix} G^+ & \vdots & V \end{bmatrix} \begin{bmatrix} G \\ \cdots \\ \mathbb{O} \end{bmatrix} \begin{bmatrix} G^+ & \vdots & V \end{bmatrix} R = [G^+G + \mathbb{O}] \begin{bmatrix} G^+ & \vdots & V \end{bmatrix} R$$

$= \begin{bmatrix} G^+GG^+ & \vdots & G^+GV \end{bmatrix} R = \begin{bmatrix} G^+ & \vdots & G^+GV \end{bmatrix} R$. We are close but we badly need $G^+GV = V$. So we select $V = G^+W$, where W is arbitrary of appropriate size. Then $G^+GV = G^+GG^+W = G^+W = V$ as we desired.

Now $A^{g_{124}} = \left[G^+ \vdots G^+W\right] R$ will be in $A\{1,2,4\}$ as the reader may verify. Continuing with our example,

$G^+W = \begin{bmatrix} 1/5 \\ 2/5 \end{bmatrix} [a] = \begin{bmatrix} a/5 \\ 2a/5 \end{bmatrix}$ so

$A^{g_{124}} = \begin{bmatrix} 1/5 & a/5 \\ 2/5 & 2a/5 \end{bmatrix} \begin{bmatrix} 1 & 0 \\ -3 & 1 \end{bmatrix} = \begin{bmatrix} 1/5 - \frac{3a}{5} & a/5 \\ 2/5 - \frac{6a}{5} & 2a/5 \end{bmatrix}$. Choosing $a = 1$, we get $A^{g_{124}} = \begin{bmatrix} -2/5 & 1/5 \\ -4/5 & 2/5 \end{bmatrix} \in A\{1,2,4\}$. However, $AA^{g_{124}} = \begin{bmatrix} -10/5 & 5/5 \\ -30/5 & 15/5 \end{bmatrix} = \begin{bmatrix} -2 & 1 \\ -6 & 3 \end{bmatrix}$ is not symmetric, so $A^{g_{124}} \notin A\{3\}$.

We are now very close to $A^+ = A^{g_{1234}}$. We simply need the $AA^{g_{124}}$ to be symmetric. But $AA^{g_{124}} = \begin{bmatrix} 1 & 2 \\ 3 & 6 \end{bmatrix} \begin{bmatrix} 1/5 & -3a/5 & a/5 \\ 2/5 & -6a/5 & 2a/5 \end{bmatrix} = \begin{bmatrix} 1 & -3a & a \\ 3 & -9a & 3a \end{bmatrix}$. We need $3 - 9a = a$ or $10a = 3$, so $a = 3/10$. The reader may verify that $A^+ = \begin{bmatrix} 1/50 & 3/50 \\ 2/50 & 6/50 \end{bmatrix}$. It is not at all clear how to get A^+ by making another special choice for W, say. We have $AA^{g_{124}} = R^{-1} \begin{bmatrix} G \\ \cdots \\ \mathbb{O} \end{bmatrix} \left[G^+ \vdots G^+W\right] R = R^{-1} \begin{bmatrix} I_r & W \\ \mathbb{O} & \mathbb{O} \end{bmatrix} R$. The simple minded choice $W = \mathbb{O}$ does not work, as the reader may verify in the example we have been running above. However, we sort of know what to do. Our development of the pseudoinverse through full rank factorization says look at $R = AG^+$. Then $F = R^{-1} \begin{bmatrix} G \\ \cdots \\ \mathbb{O} \end{bmatrix} G^+ = R^{-1} \begin{bmatrix} GG^+ \\ \cdots \\ \mathbb{O} \end{bmatrix} = \begin{bmatrix} R_1 & \vdots & R_2 \end{bmatrix} \begin{bmatrix} I_r \\ \cdots \\ \mathbb{O} \end{bmatrix} = R_1$, which is just the first r columns of an invertible matrix R^{-1}. Hence, F has full column rank and so $F^+ = (F^*F)^{-1}F^*$. Get A^+ from G^+F^+. In our example,

$A^+G^+ = \begin{bmatrix} 1 & 2 \\ 3 & 6 \end{bmatrix} \begin{bmatrix} 1/5 \\ 2/5 \end{bmatrix} = \begin{bmatrix} 1 \\ 3 \end{bmatrix}$, so $F^*F = \begin{bmatrix} 1 & 3 \end{bmatrix} \begin{bmatrix} 1 \\ 3 \end{bmatrix} = [10]$, so $F^+ = \frac{1}{10}\begin{bmatrix} 1 & 3 \end{bmatrix}$. Now $G^+F^+ = \begin{bmatrix} 1/5 \\ 2/5 \end{bmatrix} \frac{1}{10} \begin{bmatrix} 1 & 3 \end{bmatrix} = \begin{bmatrix} 1/50 & 3/50 \\ 2/50 & 6/50 \end{bmatrix}$.

Let us summarize with a table.

TABLE 5.1

1. Given A in $\mathbb{C}_r^{m\times n}$, form $RA = \begin{bmatrix} G \\ \cdots \\ \mathbb{O} \end{bmatrix}$. Then $A^{g_1} =$

$\left[G^+ + (I - G^+G)X \vdots V\right] R \in A\{1\}$, where X and V are arbitrary and $G^+ = G^*(GG^*)^{-1}$.

2. $A^{g_{14}} = \left[G^+ \vdots V\right] R \in A\{1,4\}$, where X was chosen to be $\mathbb{O}$ and V is still arbitrary.

3. $A^{g_{124}} = \left[G^+ \vdots G^+W\right] R \in A\{1,2,4\}$, where V is chosen as G^+W, where W is arbitrary.

4. $A^{g_{1234}} = G^+F^+$, where $F = AG^+$ and $F^+ = (F^*F)^{-1}F^*$.

Now let's play the same game with A^* instead of A. Let's now reduce A^* and be a little clever with notation

$S^*A^* = \begin{bmatrix} F^* \\ \cdots \\ \mathbb{O} \end{bmatrix}$, so $A^* = (S^*)^{-1}\begin{bmatrix} F^* \\ \cdots \\ \mathbb{O} \end{bmatrix}$ and $A = \left[F \vdots \mathbb{O}\right] S^{-1}$. Now F has full column rank, so $F^+ = (F^*F)^{-1}F^*$ and clearly $F^+F = I$. Now we take

$A^{g_1} = S\begin{bmatrix} F^+ + X(I - FF^+) \\ \cdots\cdots\cdots\cdots\cdots \\ V \end{bmatrix}$, where X and V are arbitrary of appropriate size. Then $AA^{g_1}A = [F \vdots \mathbb{O}]S^{-1}S\begin{bmatrix} F^+ + X(I - FF^+) \\ \cdots\cdots\cdots\cdots\cdots \\ V \end{bmatrix}[F \vdots \mathbb{O}]S^{-1}$

$= [FF^+ + FX(I - FF^+)]\,[F : \mathbb{O}]\,S^{-1} =$

$= [FF^+F + \mathbb{O} \vdots \mathbb{O}]S^{-1} = [F \vdots \mathbb{O}]S^{-1} = A$ so $A^{g_1} \in A\{1\}$. Continuing with our example,

$A = \begin{bmatrix} 1 & 2 \\ 3 & 6 \end{bmatrix}$ so $A^* = \begin{bmatrix} 1 & 3 \\ 2 & 6 \end{bmatrix}$ and $\begin{bmatrix} 1 & 0 \\ -2 & 1 \end{bmatrix}\begin{bmatrix} 1 & 3 \\ 2 & 6 \end{bmatrix} = \begin{bmatrix} 1 & 3 \\ 0 & 0 \end{bmatrix}$.

Taking $*$, $\begin{bmatrix} 1 & 2 \\ 3 & 6 \end{bmatrix}\begin{bmatrix} 1 & -2 \\ 0 & 1 \end{bmatrix} = \begin{bmatrix} 1 & \vdots & 0 \\ & & \\ 3 & \vdots & 0 \end{bmatrix}$, so

$A = \begin{bmatrix} 1 & 0 \\ 3 & 0 \end{bmatrix}\begin{bmatrix} 1 & 2 \\ 0 & 1 \end{bmatrix}$. Now $F = \begin{bmatrix} 1 \\ 3 \end{bmatrix}$, $F^* = \begin{bmatrix} 1 & 3 \end{bmatrix}$,

$F^*F = \begin{bmatrix} 1 & 0 \end{bmatrix}$, $F^* = \begin{bmatrix} 1/10 & 3/10 \end{bmatrix}$, $FF^* = \begin{bmatrix} 1/10 & 3/10 \\ 3/10 & 9/10 \end{bmatrix}$,

and $I - FF^+ = \begin{bmatrix} 9/10 & -3/10 \\ -3/10 & 1/10 \end{bmatrix}$. Thus $A^{g_1} =$

$$\begin{bmatrix} 1 & -2 \\ 0 & 1 \end{bmatrix} \begin{bmatrix} \begin{bmatrix} \frac{1}{10} & \frac{3}{10} \end{bmatrix} + \begin{bmatrix} r & u \end{bmatrix} \begin{bmatrix} 9/10 & -3/10 \\ -3/10 & 1/10 \end{bmatrix} \\ \cdots\cdots\cdots\cdots\cdots\cdots \\ x \qquad\qquad y \end{bmatrix}$$

$$= \begin{bmatrix} 1/10 + \frac{9r}{10} - \frac{3u}{10} - 2x & 3/10 - 3r/10 + \frac{u}{10} - 24 \\ x & y \end{bmatrix}.$$

Choosing $r = u = 1$ and $x = 1/10$, $y = -1/10$ we get $A^{g_1} = \begin{bmatrix} 5/10 & 3/10 \\ 1/10 & -1/10 \end{bmatrix}$ is a $\{1\}$-inverse of A that satisfies none of the other MP-equations. Making the special choice $X = \mathbb{O}$, we find $AA^{g_{13}} = [F \vdots \mathbb{O}]S^{-1}S \begin{bmatrix} F^+ \\ V \end{bmatrix} = FF^+$, which is a projection. So, in our example, $A^{g_{13}} = \begin{bmatrix} \frac{1}{10} - 2x & \frac{3}{10} - 2y \\ x & y \end{bmatrix}$, a $\{1, 3\}$-inverse of A. With x and y as above, $A^{g_{13}} = \begin{bmatrix} -1/10 & 5/10 \\ 1/10 & -1/10 \end{bmatrix}$ is in $A\{1, 3\}$ but not $A\{2, 4\}$. Reasoning as before, we see the special choice $V = WF^+$ yields $A^{g_{123}} = S \begin{bmatrix} F^+ \\ \cdots \\ WF^+ \end{bmatrix}$ as a $\{1, 2, 3\}$-inverse of A. In our example, $A^{g_{123}} = \begin{bmatrix} \frac{1}{10} - \frac{2a}{10} & \frac{3}{10} - \frac{6a}{10} \\ 9/10 & 3a/10 \end{bmatrix}$ so, with $a = 1$, we have $\begin{bmatrix} -1/10 & -3/10 \\ 1/10 & 3/10 \end{bmatrix}$ as a specific $\{1, 2, 3\}$-inverse of A. To get A^+, we look at $G = F^+A = F^+ \begin{bmatrix} F \vdots \mathbb{O} \end{bmatrix} S^{-1} = \begin{bmatrix} F^+F \vdots \mathbb{O} \end{bmatrix} \begin{bmatrix} S_1 \\ \cdots \\ S_2 \end{bmatrix} = S_1$, which has full row rank so $G^+ = G^*(GG^*)^{-1}$ The reader may verify our example reproduces the same A^+ as above. We summarize with a table.

TABLE 5.2

1. Given A in $\mathbb{C}_r^{m\times n}$, form $A^* = (S^*)^{-1} \begin{bmatrix} F^* \\ \cdots \\ \mathbb{O} \end{bmatrix}$ to get $A = [F \vdots \mathbb{O}]S^{-1}$.

   Then
   $$A^{g_1} = S \begin{bmatrix} F^+ + X(I - FF^+) \\ \cdots\cdots\cdots\cdots \\ V \end{bmatrix} \in A\{1\},$$ where X and V are arbitrary of appropriate size and $F^+ = (F^*F)^{-1}F^*$.

2. $A^{g_{13}} = S \begin{bmatrix} F^+ \\ \cdots \\ V \end{bmatrix} \in A\{1, 3\}$, where X was chosen to be $\mathbb{O}$ and V is still arbitrary.

3. $A^{g_{123}} = S \begin{bmatrix} F^+ \\ \cdots \\ WF^+ \end{bmatrix} \in A\{1,2,3\}$, where V is chosen as WF^+, where W is arbitrary.

4. $A^{g_{1234}} = A^+ = G^+F^+$, where $G = F^+A$ and $G^+ = G^*(GG^*)^{-1}$.

We indicate next how to get generalized inverses of a specified rank. We use the notation $A\{i, j, k, l\}_s$ for the set of all $\{i, j, k, l\}$ inverses of rank s. We begin with $\{2\}$-inverses.

THEOREM 5.9 (G.W. Stewart, R.E. Funderlic)
Let $A \in \mathbb{C}_r^{m\times n}$ and $O < s \leq r$. Then $A\{2\}_s = \{X \mid X = YZ$, where $Y \in \mathbb{C}^{n\times s}, Z \in \mathbb{C}^{s\times m}, ZAY = I_s)$.

PROOF Let X be in the right-hand set. We must show X is a $\{2\}$-inverse of A. But $XAX = YZAYZ = YIZ = YZ = X$. Conversely, let $X \in A\{2\}_s$. Write $X = FG$ in full rank factorization. Then $F \in \mathbb{C}_s^{n\times s}$ and $G \in \mathbb{C}_s^{s\times m}$. Then $X = XAX$ so $FG = FGAFG$. But then, $F^+FGG^+ = F^+FGAFGG^+$ so $I_s = GAF$. $\square$

COROLLARY 5.7
Let $A \in \mathbb{C}_r^{m\times n}$. Then $A\{1, 2\} = \{FG \mid F \in \mathbb{C}^{n\times r}, G \in \mathbb{C}^{r\times m}, GAF = I_r\}$.

PROOF $A\{1, 2\} = A\{2\}_r$. $\square$

COROLLARY 5.8
If $GAF = I_s$, then $G \in (AF)\{1, 2, 4\}$.

THEOREM 5.10
Let $A \in \mathbb{C}_r^{m\times n}$ and $o < s < r$. Then $A\{2, 3\}_s = \{Y(AY)^+ \mid AY \in \mathbb{C}_s^{m\times s}\}$.

PROOF Let $X = Y(AY)^+$, where $AY \in \mathbb{C}_s^{m\times s}$. Then $AX = AY(AY)^+$, so $(AX) = (AX)^*$ Also, $XAX = Y(AY)^+AY(AY)^+ = Y(AY)^+ = X$. Moreover, $s = rank(AY) = rank(AX) = rank(X)$. Conversely, suppose $X \in A\{2, 3\}_s$. Then AX is a projection of rank s. Thus $(AX)^+ = AX$ and so $X(AX)^+ = XAX = X$, and X plays the role of Y. $\square$

THEOREM 5.11
Let $A \in \mathbb{C}_r^{m\times n}$ and $0 < s \leq r$. Then $A\{2,4\}_s = \{(YA)^+Y \mid YA \in \mathbb{C}_s^{s\times n}\}$.

PROOF The proof goes along the lines of the previous theorem and is left as an exercise. □

Researchers have found uses for generalized inverses of type $\{1,2,3\}$ and $\{1,2,4\}$ (see Goldman and Zelen [$G\&Z$, 1964]).

Exercise Set 21

1. Argue that $A\{1,2,3,4\} \subseteq A\{1,2,3\} \subseteq A\{1,\ 2\} \subseteq A\{1\}$, with equality holding throughout if and only if A is invertible.

2. Suppose G is a $\{1,2,3\}$-inverse of A. Argue $rank(G) = rank(A^+) = rank(A)$.

3. Argue that the following statements are all equivalent:
   (i) $A^*B = \mathbb{O}$.
   (ii) $GB = \mathbb{O}$, where $G \in A\{1,2,3\}$.
   (iii) $HA = \mathbb{O}$, where $H \in B\{1,2,3\}$.

4. Argue that a matrix G is in $A\{1,2,3\}$ if and only if $G = HA^*$, where H is in $A^*A\{1\}$.

5. Argue that a matrix G is in $A\{1,2,4\}$ if and only if $G = A^*H$, where H is in $AA^*\{1\}$.

6. Construct various generalized inverses of $A = \begin{bmatrix} 1 & 2 & 1 & i \\ -1 & 0 & 1 & 2i \end{bmatrix}$.

7. Let $B = A^*(AA^*)^{g_{12}}$. Argue that B is a $\{1,2,4\}$-inverse of A.

8. Let $C = (A^*A)^{g_{12}}A^*$. Argue that C is a $\{1,2,3\}$-inverse of A.

9. Let $B \in A\{1,2,4\}$ and $C \in A\{1,2,3\}$. Argue that $BAC = A^+$. Is it good enough to assume $B \in A\{1,4\}$ and $C \in A\{1,3\}$?

10. Suppose $A \in \mathbb{C}_r^{m\times n}$ and $SAT = \begin{bmatrix} B & \mathbb{O} \\ \mathbb{O} & \mathbb{O} \end{bmatrix}$, where $B \in \mathbb{C}^{r\times r}$ is invertible. Then let $G = TNS$, where $N = \begin{bmatrix} Z & X \\ Y & W \end{bmatrix}$.

Then argue

(i) $G \in A\{1\}$ iff $Z = B^{-1}$.

(ii) $G \in A\{1, 2\}$ iff $Z = B^{-1}$ and $W = YBX$.

(iii) $G \in A\{1, 2, 3\}$ iff $Z = B^{-1}$, $X = -B^{-1}S_1S_2$, and $W = -YS_1S_2$, where $S = \begin{bmatrix} S_1 \\ \cdots \\ S_2 \end{bmatrix}$.

(iv) $G \in A\{1, 2, 4\}$ iff $Z = B^{-1}$,Y$= -T_2 + T_1B^{-1}$, and $W = -T_2 + T_1X$, where $T = [T_1 \vdots T_2]$.

(v) $G = A^+$ iff $Z = B^{-1}$, $X = -B^{-1}S_1S_2^+$, $Y = -T_2^+T_1B^{-1}$, and $W = T_2 + T_1B^{-1}S_1S_2^+$.

(vi) Let $G \in A^*A\{1\}$. Argue that $GA^* \in A\{1, 2, 3\}$. Let $H \in AA^*\{1\}$. Argue that $A^*H \in A\{1, 2, 4\}$.

11. (Urguhart) Let $G \in A\{1, 4\}$ and $H \in A\{1, 3\}$. Prove that $GAH = A^+$.

5.4 {2}-Inverses

In this section, we discuss the 2-inverses of a matrix. The problem of finding 2-inverses is a bit more challenging than that of describing 1-inverses because it is a "quadratic" problem in the unknowns. To understand what this means, let's look at the 2-by-2 case. Given $A = \begin{bmatrix} a & b \\ c & d \end{bmatrix}$, find $X = \begin{bmatrix} x & y \\ u & v \end{bmatrix}$ such that $XAX = X$. That is, solve $\begin{bmatrix} x & y \\ u & v \end{bmatrix}\begin{bmatrix} a & b \\ c & d \end{bmatrix}\begin{bmatrix} x & y \\ u & v \end{bmatrix} = \begin{bmatrix} x & y \\ u & v \end{bmatrix}$ for x, y, u, and v. The reader may check that this amounts to solving the following equations for x, y, u and v

$$\begin{aligned} x &= x^2a + ycx + xbu + ydu \\ y &= xay + y^2c + xbv + ydv \\ u &= uax + vcx + u^2b + vdu \\ v &= uay + vcy + ubv + v^2d, \end{aligned}$$

which are quadratic in the unknowns. The reader may also verify (just interchange letters) that solving for 1-inverses is linear in the unknowns.

We will approach 2-inverses using the rank normal form. First, we need a theorem.

THEOREM 5.12

Let $A \in \mathbb{C}^{m\times n}$ and suppose S and T are invertible matrices of appropriate size. Also suppose X is a 2-inverse of A. Then $T^{-1}XS^{-1}$ is a 2-inverse of SAT. Moreover, every 2-inverse of A is of this form.

PROOF Let S and T be invertible and let X be a 2-inverse of A. We claim $T^{-1}XS^{-1}$ is a 2-inverse of SAT. To prove this, we compute $T^{-1}XS^{-1}(SAT)T^{-1}XS^{-1} = T^{-1}XAXS^{-1} = T^{-1}XS^{-1}$, since $XAX = X$ by assumption. Conversely, let K be any 2-inverse of SAT. Then $K(SAT)K = K$. Note that $TKSS^{-1}SATT^{-1}TKS^{-1} = TKSATKS = TKS$, so if $L = TKS$, then $LAL = (TKS)(S^{-1}SATT^{-1})(TKS) = TKSATKS = TKS = L$. In other words, L is a 2-inverse of A and $K = T^{-1}LS^{-1}$, as claimed. $\square$

The next theorem shows the structure that every 2-inverse takes relative to the rank normal form of its matrix.

THEOREM 5.13 [Bailey, 2002]

Let $A \in \mathbb{C}_r^{m\times n}$. Suppose $RNF(A) = SAT = \begin{bmatrix} I_r & \mathbb{O} \\ \mathbb{O} & \mathbb{O} \end{bmatrix}$ for appropriate invertible matrices S and T. A matrix X is a 2-inverse of A if and only if $X = TRS$, where $R = \begin{bmatrix} M & Y \\ W & WY \end{bmatrix}$ where M is idempotent of size r-by-r and M is a left inverse of Y and right inverse of W, where all these matrices are of appropriate size.

PROOF Suppose $SAT = \begin{bmatrix} I_r & \mathbb{O} \\ \mathbb{O} & \mathbb{O} \end{bmatrix}$, where S and T are invertible, and let $X = TRS = T\begin{bmatrix} M & Y \\ W & WY \end{bmatrix}S$, where $M = M^2$, $MY = Y$, $WM = W$, and M, W, Y are all of appropriate size. Then $XAX = T\begin{bmatrix} M & Y \\ W & WY \end{bmatrix}SAT\begin{bmatrix} M & Y \\ W & WY \end{bmatrix}S = T\begin{bmatrix} M & Y \\ W & WY \end{bmatrix}SS^{-1}\begin{bmatrix} I_r & \mathbb{O} \\ \mathbb{O} & \mathbb{O} \end{bmatrix}T^{-1}T\begin{bmatrix} M & Y \\ W & WY \end{bmatrix}S = T\begin{bmatrix} M & Y \\ W & WY \end{bmatrix}\begin{bmatrix} I_r & \mathbb{O} \\ \mathbb{O} & \mathbb{O} \end{bmatrix}\begin{bmatrix} M & Y \\ W & WY \end{bmatrix}S = T\begin{bmatrix} M & \mathbb{O} \\ W & \mathbb{O} \end{bmatrix}\begin{bmatrix} M & Y \\ W & WY \end{bmatrix}S = T\begin{bmatrix} M^2 & MY \\ WM & WY \end{bmatrix}S = T\begin{bmatrix} M & Y \\ W & WY \end{bmatrix}S = X$. Conversely, suppose X is a 2-inverse of A. Then $R = T^{-1}XS^{-1}$ is a 2-inverse of $SAT = \begin{bmatrix} I_r & \mathbb{O} \\ \mathbb{O} & \mathbb{O} \end{bmatrix}$ by the previous theorem. Partition $R = \begin{bmatrix} M & Y \\ W & Z \end{bmatrix}$, where M is r-by-r. Then $\begin{bmatrix} M & Y \\ W & Z \end{bmatrix}\begin{bmatrix} I_r & \mathbb{O} \\ \mathbb{O} & \mathbb{O} \end{bmatrix}\begin{bmatrix} M & Y \\ W & Z \end{bmatrix} = \begin{bmatrix} M & \mathbb{O} \\ W & \mathbb{O} \end{bmatrix}\begin{bmatrix} M & Y \\ W & Z \end{bmatrix} = \begin{bmatrix} M^2 & MY \\ WM & WY \end{bmatrix} = \begin{bmatrix} M & Y \\ W & Z \end{bmatrix}$.

The conclusions follow by comparing blocks (i.e., $M = M^2$, $MY = Y$, $WM = W$, and $Z = WY$). □

Let's look at an example. Let $A = \begin{bmatrix} 1 & 2 & 3 & 4 \\ 2 & 4 & 6 & 7 \\ 1 & 2 & 3 & 6 \end{bmatrix}$. By performing elementary row and column operations on A, we find $RNF(A) = \begin{bmatrix} 1 & 0 & 0 & 0 \\ 0 & 1 & 0 & 0 \\ 0 & 0 & 0 & 0 \end{bmatrix} = \begin{bmatrix} -7 & 4 & 0 \\ 2 & -1 & 0 \\ -5 & 2 & 1 \end{bmatrix} A \begin{bmatrix} 1 & 0 & -2 & -3 \\ 0 & 0 & 1 & 0 \\ 0 & 0 & 0 & 1 \\ 0 & 1 & 0 & 0 \end{bmatrix}$. According to the previous theorem, we need a 2-by-2 idempotent matrix M. So choose $M = \begin{bmatrix} \frac{1}{2} & \frac{1}{2} \\ \frac{1}{2} & \frac{1}{2} \end{bmatrix}$.
Next we need Y such that $MY = Y$. That is, $\begin{bmatrix} \frac{1}{2} & \frac{1}{2} \\ \frac{1}{2} & \frac{1}{2} \end{bmatrix} \begin{bmatrix} y_1 \\ y_2 \end{bmatrix} = \begin{bmatrix} y_1 \\ y_2 \end{bmatrix}$.
The reader may verify that this equation implies $y_1 = y_2$, so we might just as well take them to equal 1. We also need $WM = W$; that is, we need $[w_1\ w_2] \begin{bmatrix} \frac{1}{2} & \frac{1}{2} \\ \frac{1}{2} & \frac{1}{2} \end{bmatrix} = [w_1\ w_2]$. Once again the reader may verify this entails that $w_1 = w_2$; again why not use 1? Now we can put R together; $R = \begin{bmatrix} \frac{1}{2} & \frac{1}{2} & 1 \\ \frac{1}{2} & \frac{1}{2} & 1 \\ 1 & 1 & 2 \\ 1 & 1 & 2 \end{bmatrix}$. Finally we compute TRS, which will be a 2-inverse of A.
$TRS = \begin{bmatrix} \frac{135}{2} & -\frac{63}{2} & -9 \\ -15 & 7 & 2 \\ -15 & 7 & 2 \\ -\frac{15}{2} & \frac{7}{2} & 1 \end{bmatrix}$. The final verification left to the reader is that this matrix is indeed a 2-inverse of A as promised.

Next, we look at the connection of the 2-inverse to other generalized inverses of a matrix. Let's fix an m-by-n matrix A of rank r over $\mathbb{C}$. Note that if we have a 2-inverse X for A, then $rank(X) = rank(XAX) \leq rank(A) = r$. Now choose arbitrary matrices E in $\mathbb{C}^{n\times k}$ and H in $\mathbb{C}^{k\times m}$ and form the k-by-k matrix HAE. Suppose we can find a 2-inverse of HAE, say $(HAE)^{g_2}$. For example, $(HAE)^+$ would certainly be one such. Then, if we form $X = E(HAE)^{g_2}H$, we find $XAX = (E(HAE)^{g_2}H)A(E(HAE)^{g_2}H) = E((HAE)^{g_2})(HAE)((HAE)^{g_2})H = E((HAE)^{g_2})H = X$. In other words, X is a 2-inverse of A. But do all 2-inverses look like this? Once again we appeal to full rank factorizations of A. Suppose $A = FG$ is a full rank factorization of A. Then our X above looks like $X = E((HFGE)^{g_2})H$. Note HF is k-by-r and GE is r-by-k.

THEOREM 5.14
Suppose $k = r$ =rank(A). Then HAE is invertible iff HF and GE are invertible.

PROOF Note that HF, GE, and HAE are all r-by-r, and $HAE = (HF)(GE)$. Thus $det(HAE) = det(HF)det(GE)$. The theorem follows from this formula and the familiar fact that a matrix is invertible iff it has a nonzero determinant. ▯

THEOREM 5.15
Let $k = r =$ rank(A) and choose H and E so that HAE is invertible. Then $X = E(HAE)^{-1}H$ is a $\{1, 2\}$-inverse of A.

PROOF We already know X is a 2-inverse of A. We compute, $AXA = AE(HAE)^{-1}HA = AE(HFGE)^{-1}HA = AE(GE)^{-1}(HF)^{-1}HA = FGE(GE)^{-1}(HF)^{-1}HFG = FG = A$, making X a 1-inverse as well. ▯

THEOREM 5.16
With the hypotheses of the previous theorem, add that we choose $H = F^*$. Then X is a $\{1, 2, 3\}$-inverse of A.

PROOF We need only that $(AX)^* = AX$. But $AX = AE(HAE)^{-1}H = AE(F^*FGE)^{-1}F^* = (FGE)(GE)^{-1}(F^*F)^{-1}F^* = FF^+$, which we know to be self-adjoint. ▯

THEOREM 5.17
In the previous theorem, choose $E = G^*$ instead of $H = F^*$. Then X is a $\{1, 2, 4\}$-inverse of A.

PROOF We only need that $(XA)^* = XA$. But $XA = E(GE)^{-1}(HF)^{-1}HFG = E(GE)^{-1}(HF)^{-1}HFG = E(GE)^{-1}G = G^*(GG^*)^{-1}G = G^+G$, which we know to be self-adjoint. ▯

THEOREM 5.18
In the previous theorem, choose both $H = F^*$ and $E = G^*$. Then $X = A^+$.

PROOF In this case, $X = G^*(GG^*)^{-1}(F^*F)^{-1} = G^+F^+ = A^+$. ▯

Can we find a way to control the rank of a 2-inverse of a given matrix? The next theorem gives some insight into this question.

THEOREM 5.19

Let $A \in \mathbb{C}_r^{m\times n}$. Suppose $0 < s \leq r$. Then

1. *The rank r 2-inverses of A are exactly the $\{1,\ 2\}$-inverses of A.*
2. *If $0 < s < r$, the rank s 2-inverses of A form the set $\{YZ \mid Y \in \mathbb{C}^{n\times s}, Z \in \mathbb{C}^{s\times m}$, and $ZAY = I_s\}$.*

PROOF The proof of Theorem 5.19 (1) is left as an exercise. For (2), let X be a rank s 2-inverse of A. Let $X = YZ$ be a full rank factorization of X so that then $Y \in \mathbb{C}_s^{n\times s}$ and $Z \in \mathbb{C}_s^{s\times m}$. Now $YZ = X = XAX = YZAYZ$. But $Y^+Y = I_s = ZZ^+$, so $I_s = ZAY$. Conversely, let $X = YZ$, where $ZAY = I_s$. Then $XAX = YZAYZ = YI_sZ = X$. □

We have avoided a serious issue until now. Above we wrote 2-inverses as $X = E(HAE)^{-1}H$. But how did we know we could ever find any matrices E and H so that HAE is invertible? We have the following theorem.

THEOREM 5.20

Let $A \in \mathbb{C}^{m\times n}$. Then X is a 2-inverse of A if and only if there exists E and H where E has full column rank, H has full row rank, $Col(X) = Col(E)$, $Col(X^*) = Col(H^*)$, HAE is invertible, and $X = E(HAE)^{-1}H$.

Before we leave this section, we can actually determine all 1-inverses of a given rank. Suppose $G \in A\{1\}$, where $A \in \mathbb{C}_r^{m\times n}$. We know $r \leq rank(G) \leq \min\{m, n\}$. We claim all the rank s 1-inverses of A are in the set $\{YZ \mid Y \in \mathbb{C}_s^{n\times s},\ Z \in \mathbb{C}_s^{s\times m}$, where $ZAY = \begin{bmatrix} I_r & \mathbb{O} \\ \mathbb{O} & \mathbb{O} \end{bmatrix}\}$. Indeed, suppose G belongs to this set. Then $G = YZ$ and $rank(G) = s$. Partition Y with a block of its first r columns and Z with its first r rows. Say $Y = [Y_1 \mid Y_2]$ and $Z = \begin{bmatrix} Z_1 \\ \cdots \\ Z_2 \end{bmatrix}$. It follows that $Z_1AY_1 = I_r$ and $Z_1AY_2 = \mathbb{O}$. Let $G_1 = Y_1Z_1$. Then $G_1AG_1 = Y_1Z_1AY_1Z_1 = Y_1IZ_1 = Y_1Z_1 = G$ so $G \in A\{2\}$. But $rank(G_1) = r = rank(A)$ so by (5.8), G_1 is also a 1-inverse of A. Thus, $AG_1AGA = AY_1Z_1AYZA = AY_1[I_r \mid \mathbb{O}] \begin{bmatrix} Z_1 \\ \cdots \\ Z_2 \end{bmatrix} A = AY_1Z_1A = AG_1A = A$. Conversely, let G be a 1-inverse of A of rank s. Let $G = FH$ be a full rank factorization of G so $F \in \mathbb{C}_s^{n\times s}$ and $H \in \mathbb{C}_s^{s\times m}$. Then $HAFHAF = HAGAF = HAF$ so HAF is an idempotent of rank r. Thus there exists an invertible matrix S with $SHAFS^{-1} = \begin{bmatrix} I_r & \mathbb{O} \\ \mathbb{O} & \mathbb{O} \end{bmatrix}$. Let $Y = FS^{-1}$, and

$Z = SH$. Then $Y \in \mathbb{C}_s^{n\times s}$, $Z \in \mathbb{C}_s^{s\times m}$ and $ZAY = SHAFS^{-1} = \begin{bmatrix} I_r & \mathbb{O} \\ \mathbb{O} & \mathbb{O} \end{bmatrix} =$ $YZ = FH = G$.

Exercise Set 22

1. Find a 2-inverse for $D = \begin{bmatrix} 1 & 0 & 0 \\ 0 & -1 & 0 \\ 0 & 0 & 0 \end{bmatrix}$, following the example worked above.

2. Find a 2-inverse for $N = \begin{bmatrix} 0 & 1 & 0 \\ 0 & 0 & 1 \\ 0 & 0 & 0 \end{bmatrix}$, following the example worked above.

3. Verify the equations for the 2-by-2 case listed at the beginning of this section.

4. Let $A = FG$ be a full rank factorization of $A \in \mathbb{C}_r^{m\times n}$. Argue that $G^{g_1}F^{g_1} \in A\{1\}$, $G^{g_2}F^{g_1} \in A\{2\}$, $G^{g_4}F^{g_1} \in A\{4\}$, $G^{g_1}F^{g_2} \in A\{2\}$, and $G^{g_1}F^{g_1} \in A\{1\}$. Finally, argue that $A^+ = G^+F^{g_{13}} = G^{g_{14}}F^+$.

5. Suppose $A = FG$ is a full rank factorization of A. Argue that $F(GF)^{-1}G$ and $F(GF)^+G$ are 2-inverses of A.

Further Reading

[Bailey, 2003] Chelsey Elaine Bailey, *The* $\{2\}$*-Inverses of a Matrix*, Masters Thesis, Baylor University, May, (2003).

[Greville, 1974] T. N. E. Greville, Solutions of the Matrix Equation $XAX = X$ and Relations Between Oblique and Orthogonal Projectors, SIAM J. Appl. Math., Vol. 26, No. 4, June, (1974), 828–831.

[Schott, 1997] James R. Schott, *Matrix Analysis for Statistics*, John Wiley & Sons, New York, 1997.

5.5 The Drazin Inverse

We have looked at various kinds of generalized inverses, most dealing with the problem of solving systems of linear equations. However, other inverses have also been found to be useful. The one we consider next was introduced by **M. P. Drazin** [Drazin, 1958] in a more abstract setting. This inverse is intimately connected with the index of a matrix. It is defined only for square matrices and like the MP-inverse, is unique.

DEFINITION 5.2 *(Drazin inverse)*
Let $A \in \mathbb{C}^{n\times n}$, with $index(A) = q$. Then $X \in \mathbb{C}^{n\times n}$ is called a ***Drazin inverse*** *(D-inverse for short) of A iff X satisfies the following equations:*

$$(D1)\ XAX = X$$

$$(D2)\ AX = XA$$

$$(D3)\ A^{q+1}X = A^q.$$

We see that a D-inverse of A is, in particular, a 2-inverse of A that commutes with A. Well, the zero matrix does that! So it must be (D3) that gives the D-inverse its punch. Let's settle the issues of existence and uniqueness right away.

THEOREM 5.21
Let $A \in \mathbb{C}^{n\times n}$, with $index(A) = q$. Then there exists one and only one matrix $X \in \mathbb{C}^{n\times n}$ that satisfies (D1), (D2), and (D3). We shall denote this unique matrix by A^D and call it the ***Drazin inverse*** *of A.*

PROOF For uniqueness, suppose X and Y both satisfy (D1) through (D3). Then $A^{q+1}X = A^q = A^{q+1}Y$ so $A^qAX = A^qAY$. Thus $A^qXA = A^qYA$. Then $A^qXAX = A^qYAX$ so $A^qX = A^qYAX$. Now $A^{q-1}AX = A^{q-1}AYAX$ so $A^{q-1}XA = A^{q-1}AYAX$ so $A^{q-1}XAX = A^{q-1}AYAX^2$ so $A^{q-1}X = A^{q-1}AYXAX = A^{q-1}AYX = A^{q-1}YAX$. Thus we have peeled off one factor of A on the left to get $A^{q-1}X = A^{q-1}YAX$. Continue this process until all As are peeled away and conclude $X = YAX$. A symmetric argument gives $Y = XAY$. Next, (D3) implies $A^q(AY - I) = \mathbb{O}$, so $XA^q(AY - I) = \mathbb{O}$ whence $A^q(XAY - X) = \mathbb{O}$. Now using (D1), $A^q(XAY - XAX) = \mathbb{O}$ so $A^q(XA)(Y - X) = \mathbb{O}$, hence $XA^qXA(Y - X) = \mathbb{O}$. Using (D2), $A^{q-1}(XAX)A(Y - X) = \mathbb{O}$ so, by (D1), $A^{q-1}(XA)(Y - X) = \mathbb{O}$, so $A^{q-1}(XAY - XAX) = A^{q-1}(XAY - X) = \mathbb{O}$. Again, this shows we can peel off factors of A on the left and conclude $XAY - X = \mathbb{O}$ so $X = XAY = Y$ by the above.

There are various approaches for existence, but since we used the full rank factorization to get the MP-inverse, we do the same for the D-inverse. If $A \in \mathbb{C}^{n\times n}$ has full rank n, then take $A^D = A^{-1}$ and easily check the three equations for the D-inverse. It is a fact that the index q of a matrix is characterized by a sequence of full rank factorizations: $A = F_1G_1$, $G_1F_1 = F_2G_2, \ldots$. Then $index(A) = \begin{cases} q \text{ if } (G_qF_q)^{-1} \text{ exists} \\ q+1 \text{ if } G_qF_q = \mathbb{O} \end{cases}$ for the first time. Define $A^D = \begin{cases} F_1F_2\cdots F_q(G_qF_q)^{-(q+1)}G_qG_{q-1}\cdots G_1, & \text{when } (G_qF_q)^{-1} \text{ exists} \\ \mathbb{O} & \text{if } G_qF_q = \mathbb{O} \end{cases}$.

Note, if $G_qF_q = \mathbb{O}$, then A is nilpotent of index $q+1$. It is straightforward to verify that A^D so defined actually satisfies (D1) through (D3). □

A number of facts are easily deduced about the D-inverse.

THEOREM 5.22

Let $A \in \mathbb{C}^{n\times n}$, with $index(A) = q$. Then

1. $\mathcal{C}ol(A^D) = \mathcal{C}ol(A^q)$.
2. $\mathcal{N}ull(A^D) = \mathcal{N}ull(A^q)$.
3. $(AA^D)^2 = AA^D$ *is the projector of* $\mathbb{C}^n$ *onto* $\mathcal{C}ol(A^q)$ *along* $\mathcal{N}ull(A^q)$.
4. $(I - AA^D)^2 = (I - AA^D)$ *is the projector onto* $\mathcal{N}ull(A^q)$ *along* $\mathcal{C}ol(A^q)$.
5. $A^{q+p}(A^D)^p = A^q$.
6. $A^{p+1}A^D = A^p$ *iff* $p \geq q$, *where p and q are positive integers.*

PROOF The proofs are easy and left as exercises. □

We could have used the core-nilpotent factorization to get the Drazin inverse.

THEOREM 5.23

Let $A \in \mathbb{C}^{n\times n}$, with $index(A) = q > 0$. If $A = S\begin{bmatrix} C & \mathbb{O} \\ \mathbb{O} & N \end{bmatrix}S^{-1}$ is a core-nilpotent factorization of A, then $A^D = S\begin{bmatrix} C^{-1} & \mathbb{O} \\ \mathbb{O} & \mathbb{O} \end{bmatrix}S^{-1}$.

PROOF We leave it as an exercise to verify (D1) through (D3). □

Campbell and Meyer [$C\&M$, 1979] talk about the *core* of a matrix A using the D-inverse. It goes like this.

DEFINITION 5.3 *Let $A \in \mathbb{C}^{n\times n}$. Then AA^DA is defined to be the* **core** *of the matrix A and is denoted C_A; we also define $N_A = A - C_A$.*

THEOREM 5.24
Let $A \in \mathbb{C}^{n\times n}$. Then $N_A = A - C_A$ is a nilpotent matrix of index $q = index(A)$.

PROOF If the $index(A) = 0$, then $A = C_A$, so suppose $index(A) \geq 1$. Then $(N_A)^q = (A - AA^DA)^q = [A(I - AA^D)]^q = A^q(I - AA^D)^q = A^q - A^q = \mathbb{O}$. If $p < q$, $A^p = A^{p+1}A^D \neq \mathbb{O}$, so $index(N) = q$. □

DEFINITION 5.4 *(core-nilpotent decomposition)*
Let $A \in \mathbb{C}^{n\times n}$. The matrix $N_A = A - C_A = (I - AA^D)A$ is called the ***nilpotent part of A****, and $A = C_A + N_A$ is called the* ***core-nilpotent decomposition*** *of A.*

THEOREM 5.25
Let $A \in \mathbb{C}^{n\times n}$ and let $A = S\begin{bmatrix} C & \mathbb{O} \\ \mathbb{O} & N \end{bmatrix}S^{-1}$ be a core-nilpotent factorization of A. Then $C_A = S\begin{bmatrix} C & \mathbb{O} \\ \mathbb{O} & \mathbb{O} \end{bmatrix}S^{-1}$ and $N_A = S\begin{bmatrix} \mathbb{O} & \mathbb{O} \\ \mathbb{O} & N \end{bmatrix}S^{-1}$.

PROOF The proof is left as an exercise. □

There is some uniqueness here.

THEOREM 5.26
Let $A \in \mathbb{C}^{n\times n}$. Then A has a unique decomposition $A = X + Y$, where $XY = YX = \mathbb{O}$. The $index(X) \leq 1$ and Y is nilpotent of index $q = index(A)$. Moreover, the unique decomposition is $A = C_A + N_A$.

PROOF If $index(A) = 0$, then $Y = \mathbb{O}$ and A is invertible. If $index(X) = 1$, write $X = S\begin{bmatrix} C & \mathbb{O} \\ \mathbb{O} & \mathbb{O} \end{bmatrix}S^{-1}$. Then $Y = S\begin{bmatrix} \mathbb{O} & \mathbb{O} \\ \mathbb{O} & Y_2 \end{bmatrix}S^{-1}$, since $XY = YX = \mathbb{O}$ and C is invertible. Thus Y_2 is nilpotent with $index(Y_2) = index(A)$. Now $A = X + Y = S\begin{bmatrix} C & \mathbb{O} \\ \mathbb{O} & Y_2 \end{bmatrix}S^{-1}$, so $X = C_A$ and $Y = N_A$. □

COROLLARY 5.9
Let $A \in \mathbb{C}^{n\times n}$. Then $C_A^p = C_{A^p}$, $N_A^p = N_{A^p}$, and $A^p = C_{A^p} + N_{A^p}$. If $p \geq index(A)$, then $A^p = C_A^p$.

Next we list additional properties of the D-inverse, leaving the proofs as exercises.

THEOREM 5.27

Let $A \in \mathbb{C}^{n\times n}$. Then

1. $index(A) = index(C_A) = \begin{cases} 1 \text{ if } index(A) \geq 1 \\ 0 \text{ if } index(A) = 0 \end{cases}$
2. $N_A C_A = C_A N_A = \mathbb{O}$
3. $N_A A^D = A^D N_A = \mathbb{O}$
4. $C_A A A^D = A A^D C_A = C_A$
5. $(A^D)^D = C_A$
6. $A = C_A$ *iff* $index(A) \leq 1$
7. $((A^D)^D)^D = A^D$
8. $A^D = C_A^D$
9. $(A^D)^* = (A^*)^D$
10. $A^D = A^+$ *iff* $AA^+ = A^+A$.

There is a way to do hand computations of the D-inverse. Begin with $A \in \mathbb{C}^{n\times n}$. Chances are slim that you will know the index q of A, so compute A^n. If $A^n = \mathbb{O}$, then $A^D = \mathbb{O}$, so suppose $A^n \neq \mathbb{O}$. Find the Hermite echelon form of A, $HEF(A)$. The basic columns $\mathbf{v}_1, \mathbf{v}_2, \ldots, \mathbf{v}_r$ form a basis for $\mathcal{C}ol(A^q)$. Form $I - HEF(A)$ and call the nonzero columns $\mathbf{v}_{r+1}, \ldots, \mathbf{v}_n$. They form a basis for $\mathcal{N}ull(A^q)$. Form $S = [\mathbf{v}_1 \mid \cdots \mid \mathbf{v}_r \mid \mathbf{v}_{r+1} \mid \cdots \mid \mathbf{v}_n]$. Then $S^{-1}AS = \begin{bmatrix} C & \mathbb{O} \\ \mathbb{O} & N \end{bmatrix}$ is a core-nilpotent factorization of A. Compute $C^{-1}$and form $A^D = S \begin{bmatrix} C^{-1} & \mathbb{O} \\ \mathbb{O} & \mathbb{O} \end{bmatrix} S^{-1}$.

For example, suppose $A = \begin{bmatrix} 1 & 5 & 4 & 3 \\ -2 & 0 & 7 & 6 \\ 5 & 15 & 5 & 3 \\ 6 & 10 & -6 & -6 \end{bmatrix}$. The index of A is not apparent, so we compute that $A^4 = \begin{bmatrix} 5974 & 19570 & 8446 & 5562 \\ 14214 & 31930 & -1854 & -4326 \\ 3708 & 26780 & 27192 & 21012 \\ -16480 & -24720 & 20600 & 19776 \end{bmatrix}$

and note $HEF(A) = \begin{bmatrix} 1 & 0 & -\frac{7}{2} & -3 \\ 0 & 1 & \frac{3}{2} & \frac{6}{5} \\ 0 & 0 & 0 & 0 \\ 0 & 0 & 0 & 0 \end{bmatrix}$. Next, we form $I - HEF(A)$ and find $P = \begin{bmatrix} 5974 & 19570 & 7 & 15 \\ 14214 & 31930 & -3 & -6 \\ 3708 & 26780 & 2 & 0 \\ -16480 & -24720 & 0 & 5 \end{bmatrix}$, where we have eliminated some fractions in the last two columns. Now $P^{-1}AP = \begin{bmatrix} -\frac{69}{2} & -\frac{155}{2} & 0 & 0 \\ \frac{127}{10} & \frac{69}{2} & 0 & 0 \\ 0 & 0 & 0 & 0 \\ 0 & 0 & 0 & 0 \end{bmatrix}$.

We see $C = \begin{bmatrix} -\frac{69}{2} & -\frac{155}{2} \\ \frac{127}{10} & \frac{69}{2} \end{bmatrix}$, so $A^D = P \begin{bmatrix} -\frac{69}{412} & -\frac{155}{412} & 0 & 0 \\ \frac{127}{2060} & \frac{69}{412} & 0 & 0 \\ 0 & 0 & 0 & 0 \\ 0 & 0 & 0 & 0 \end{bmatrix} P^{-1} =$

$$\begin{bmatrix} \frac{1}{206} & \frac{5}{206} & \frac{2}{103} & \frac{3}{206} \\ -\frac{1}{103} & 0 & \frac{7}{206} & \frac{3}{103} \\ \frac{5}{206} & \frac{15}{206} & \frac{5}{206} & \frac{3}{206} \\ \frac{3}{103} & \frac{5}{103} & -\frac{3}{103} & -\frac{3}{103} \end{bmatrix}.$$

We have shown earlier that A^{-1} can always be expressed as a polynomial in A. The MP-inverse of a square matrix may not be expressible as a polynomial in A. The D-inverse, on the other hand, is always expressible as a polynomial in the matrix.

THEOREM 5.28

Let $A \in \mathbb{C}^{n\times n}$. Then there is a polynomial $p(x)$ in $\mathbb{C}[x]$ such that $p(A) = A^D$.

PROOF Write $A = S \begin{bmatrix} C & \mathbb{O} \\ \mathbb{O} & N \end{bmatrix} S^{-1}$ in a core-nilpotent factorization. Now, C is invertible, so there exists a polynomial $q(x)$ with $q(C) = C^{-1}$. Let $p(x) = x^q[q(x)]^{q+1}$, where $q = index(A)$. Then $p(A) = A^q[q(A)]^{q+1}$
$= S \begin{bmatrix} C^q & \mathbb{O} \\ \mathbb{O} & \mathbb{O} \end{bmatrix} \begin{bmatrix} q(C) & \mathbb{O} \\ \mathbb{O} & q(N) \end{bmatrix}^{q+1} S^{-1} = S \begin{bmatrix} C^q[q(C)]^{q+1} & \mathbb{O} \\ \mathbb{O} & \mathbb{O} \end{bmatrix} S^{-1} =$
$S \begin{bmatrix} C^{-1} & \mathbb{O} \\ \mathbb{O} & \mathbb{O} \end{bmatrix} S^{-1} = A^D$. □

COROLLARY 5.10

Let A, B be in $\mathbb{C}^{n\times n}$ and $AB = BA$. Then

1. $(AB)^D = B^D A^D = A^D B^D$
2. $A^D B = B A^D$

3. $AB^D = B^D A$

4. *index*$(AB) \leq max\{index(A), index(B)\}$.

As long as we are talking about polynomials, let's take another look at the minimum polynomial. Recall that for each A in $\mathbb{C}^{n\times n}$ there is a monic polynomial of least degree $\mu_A(x) \in \mathbb{C}[x]$ such that $\mu_A(A) = \mathbb{O}$. Say $\mu_A(x) = x^d + \alpha_{d-1}x^{d-1} + \cdots + \alpha_1 x + \alpha_o$. We have seen that A is invertible iff the constant term $\alpha_o \neq 0$ and, in this case, $A^{-1} = -\frac{1}{\alpha_o}\left[A^{d-1} + \alpha_{d-1}A^{d-2} + \cdots + \alpha_2 A + \alpha_1 I\right]$. Now suppose that i is the least positive integer with $0 = \alpha_o = \alpha_1 = \cdots = \alpha_{i-1}$ but $\alpha_i \neq 0$. Then i is the index of A.

THEOREM 5.29

Let $A \in \mathbb{C}^{n\times n}$ and $\mu_A(x) = x^d + \alpha_{d-1}x^{d-1} + \cdots + \alpha_i x^i$, with $\alpha_i \neq 0$. Then $i = index(A)$.

PROOF Write $A = S\begin{bmatrix} C & \mathbb{O} \\ \mathbb{O} & N \end{bmatrix}S^{-1}$ in a core-nilpotent factorization where $q = index(A) = index(N)$. Since $\mu_A(A) = \mathbb{O}$, we see $\mathbb{O} = \mu_A(N) = N^d + \cdots + \alpha_i N^i = (N^{d-i} + \alpha_{d-i}N^{d-i-1} + \cdots + \alpha_i I)N^i$. Since $(N^{d-i} + \alpha_{d-i}N^{d-i-1} + \cdots + \alpha_i I)$ is invertible, we get $N^i = \mathbb{O}$. Thus $i \geq q$. Suppose $q < i$. Then $A^D A^i = A^{i-1}$. Write $\mu_A(x) = x^i q(x)$ so $\mathbb{O} = \mu_A(A) = A^{i-1}q(A)$. Multiply by AD. Then $\mathbb{O} = A^{i-1}q(A)$. Thus $r(x) = x^{i-1}q(x)$ is a polynomial that annihilates A and $deg(r(x)) < deg(\mu_A(x))$, a contradiction. Therefore $i = q$. ∎

There is a nice connection between the D-inverse and Krylov subspaces. Given a square matrix A and a column vector $\mathbf{b}$, we defined in a previous homework exercise the Krylov subspace, $\mathbf{K}_s(A, b) = span\{\mathbf{b}, A\mathbf{b}, \ldots, A^{s-1}\mathbf{b}\}$. We have the following results that related a system $A\mathbf{x} = \mathbf{b}$ to solutions in a Krylov subspace where A is n-by-n. The proofs are left as exercises.

THEOREM 5.30

The following statements are equivalent:

1. *$A^D\mathbf{b}$ is a solution of $A\mathbf{x} = \mathbf{b}$.*

2. *$\mathbf{b} \in Col(A^q)$, where q is the index of A.*

3. *$A\mathbf{x} = \mathbf{b}$ has a solution in the Krylov subspace $\mathbf{K}_n(A, b)$.*

THEOREM 5.31
Suppose m is the degree of the minimal polynomial of A and q is the index of A. If $\mathbf{b} \in Col(A^q)$, then the linear system $A\mathbf{x} = \mathbf{b}$ has a unique solution $\mathbf{x} = A^D\mathbf{b} \in \mathbf{K}_{m-q}(A, b)$. If $\mathbf{b} \notin Col(A^q)$, then $A\mathbf{x} = \mathbf{b}$ does not have a solution in $\mathbf{K}_n(A, b)$.

Exercise Set 23

1. Find the D-inverse of $\begin{bmatrix} 1 & 0 & 1 \\ 0 & 0 & 1 \\ 0 & 0 & 0 \end{bmatrix}$ and $\begin{bmatrix} 1 & 2 & 1 \\ 0 & 5 & 4 \\ 2 & -1 & -2 \end{bmatrix}$.

2. Fill in the proof of Theorem 5.22.

3. Fill in the proof of Theorem 5.23.

4. Fill in the proof of Theorem 5.25.

5. Fill in the proof of Theorem 5.27.

Further Reading

[B-I&G, 2003] Adi Ben-Israel and Thomas N. E. Greville, *Generalized Inverses: Theory and Applications*, Springer-Verlag, New York, (2003).

[C&G, 1980] R. E. Cline and Thomas N. E. Greville, A Drazin Inverse for Rectangular Matrices, Linear Algebra and Appl., Vol. 29, (1980), 53–62.

[C&M, 1991] S. L. Campbell and C. D. Meyer, Jr., *Generalized Inverses of Linear Transformations*, Dover Publications, New York, (1991).

[C&M&R, 1976] S. L. Campbell, C. D. Meyer, Jr., and N. J. Rose, Applications of the Drazin Inverse to Linear Systems of Differential Equations with Singular Constant Coefficients, SIAM J. Appl. Math., Vol. 31, No. 3, (1976), 411–425.

[Drazin, 1958] M. P. Drazin, Pseudo-Inverses in Associative Rings and Semi-Groups, The American Mathematical Monthly, Vol. 65, (1958), 506–514.

[H&M&S, 2004] Olga Holtz, Volker Mehrmann, and Hans Schneider, Potter, Wielandt, and Drazin on the Matrix Equation $AB = \omega BA$: New Answers to Old Questions, The American Mathematical Monthly, Vol. 111, No. 8, October, (2004), 655–667.

[I&M, 1998] Ilse C. F. Ipsen and Carl D. Meyer, The Idea Behind Krylov Methods, The American Mathematical Monthly, Vol. 105, No. 10, December, (1998), 889–899.

[M&P, 1974] C. D. Meyer, Jr. and R. J. Plemmons, Limits and the Index of a Square Matrix, SIAM J. Appl. Math., Vol. 26, (1974), 469–478.

[M&P, 1977] C. D. Meyer, Jr. and R. J. Plemmons, Convergent Powers of a Matrix with Applications to Iterative Methods for Singular Linear Systems, SIAM J. Numer. Anal., Vol. 36, (1977).

[M&R, 1977] C. D. Meyer, Jr. and Nicholas J. Rose, The Index and the Drazin Inverse of Block Triangular Matrices, SIAM J. Applied Math., Vol. 33, (1977).

[Zhang, 2001] Liping Zhang, A Characterization of the Drazin Inverse, Linear Algebra and its Applications, Vol. 335, (2001), 183–188.

5.6 The Group Inverse

There is yet another kind of inverse that has been found to be useful.

DEFINITION 5.5 *(group inverse)*
*We say a matrix X is a **group inverse** of A iff it satisfies*

$$(MP1)\; AXA = A$$
$$(MP2)\; XAX = X$$
$$(3)\quad AX = XA.$$

Thus, a group inverse of A is a $\{1, 2\}$-inverse of A that commutes with A. First, we establish the uniqueness.

THEOREM 5.32 (uniqueness of the group inverse)
If a matrix A has a group inverse, it is unique.

PROOF Suppose X and Y satisfy the three equations given above. Then $X = XAX = AXX = AYAXX = YAX = YYA = YAY = Y$. ▯

It turns out that not every matrix has a group inverse. We use our favorite, the full rank factorization, to see when we do. When a group inverse does exist, we denote it be $A^{\#}$.

THEOREM 5.33 [Cline, 1964]
Suppose the square matrix $A = FG$ is in full rank factorization. Then A has a group inverse if and only if GF is nonsingular. In this case, $A^{\#} = F(GF)^{-2}G$.

PROOF Suppose $r = rank(A)$ and $A = FG$ is a full rank factorization of A. Then GF is in $\mathbb{C}^{r\times r}$ and $A^2 = FGFG = F(GF)G$, so $rank(A^2) = rank(GF)$, since F has full column rank and G has full row rank. Thus, $rank(A^2) = rank(A)$ if and only if GF is nonsingular. Let $X = F(GF)^{-2}G$. We compute $AXA = FGF(GF)^{-2}GFG = FG = A$, $XAX = F(GF)^{-2}GFGF(GF)^{-2}G = F(GF)^{-2}G = X$ and $XA = FGF(GF)^{-2}G = F(GF)^{-2}G = F(GF)^{-1}(GF)^{-1}GFG = F(GF)^{-2}GFG = XA$. ▯

COROLLARY 5.11
A square matrix A has a group inverse if and only if $index(A) = 1$ and if and only if $rank(A) = rank(A^2)$.

Exercise Set 24

1. If A is nonsingular, argue that $A^{\#} = A^{-1}$.
2. Prove that $A^{\#\#} = A$, if $A^{\#}$ exists.
3. Prove that $A^{*\#} = A^{\#*}$, if $A^{\#}$ exists.
4. Prove that $A^{n\#} = A^{\#n}$ for any positive integer n.

5. Show that $A^{\#} = AGA$ for any G in $A^3\{1\}$.

6. Create a group inverse of rank 2 for a 4-by-4 matrix and a D-inverse of rank 2 for the same matrix. What are the differences?

Further Reading

[C&M, 1991] S. L. Campbell and C. D. Meyer, Jr., *Generalized Inverses of Linear Transformations*, Dover Publications, New York, (1991).

[Hartwig, 1987] Robert E. Hartwig, The Group Inverse of a Block Triangular Matrix, in *Current Trends in Matrix Theory,* Elsevier Science Publishing Co., Inc., F. Uhlig and R. Grone, Editors, (1987).

[Meyer, 1975] C. D. Meyer, Jr., The Role of the Group Generalized Inverse in the Theory of Finite Markov Chains, SIAM Review, Vol. 17, (1975), 443–464.

Chapter 6

Norms

length, norm, unit ball, unit sphere, norm equivalence theorem, distance function, convergent sequence of vectors, Hölder's inequality, Cauchy-Schwarz inequality, Minkowski inequality

6.1 The Normed Linear Space $\mathbb{C}^n$

We have seen how to view $\mathbb{C}^n$ as a space of vectors where you can study the algebra of addition and scalar multiplication. However, there is more you can do with vectors than just add them and multiply them by scalars. Vectors also have properties of a geometric nature. For example, vectors have *length* (some people say *magnitude*). This is the concept we study in this chapter. We are motivated by our experience in $\mathbb{R}^n$, specifically $\mathbb{R}^2$.

By the famous theorem about right triangles (remember which one?), the length square of $\mathbf{x} = (x_1, x_2)$ is $\|\mathbf{x}\|^2 = x_1^2 + x_2^2$. It is natural to define the length of $\mathbf{x}$ as $\|\mathbf{x}\| = \sqrt{x_1^2 + x_2^2}$. However, when dealing with complex vectors, we have a problem. Consider $\mathbf{x} = (1, i)$ in $\mathbb{C}^2$. Then, following the previous formula, $\|\mathbf{x}\|^2 = 1^2 + i^2 = 1 - 1 = 0$. Unless you are doing relativity in physics, this is troublesome. It says a nonzero vector in $\mathbb{C}^2$ can have zero length! That does not seem right, so we need to remedy this problem. We can use the idea of the magnitude of a complex number to make things work out. Recall that if the complex number $z = a + bi$, then $|z|^2 = a^2 + b^2 = z\bar{z} = \bar{z}z$. Let's define $\|\mathbf{x}\|^2 = |1|^2 + |i|^2 = 1 + 1 = 2$; then $\|\mathbf{x}\| = \sqrt{2}$, which seems much more reasonable. Therefore, we define $\|\mathbf{x}\|^2 = |x_1|^2 + |x_2|^2$ on $\mathbb{C}^2$.

With this in mind, we define what we mean by the "length" or "norm" of a vector and give a number of examples. Common sense tells us that lengths of nonzero vectors should be positive real numbers. The other properties we adopt also make good common sense. We can also use our knowledge of the "absolute value" concept as a guide.

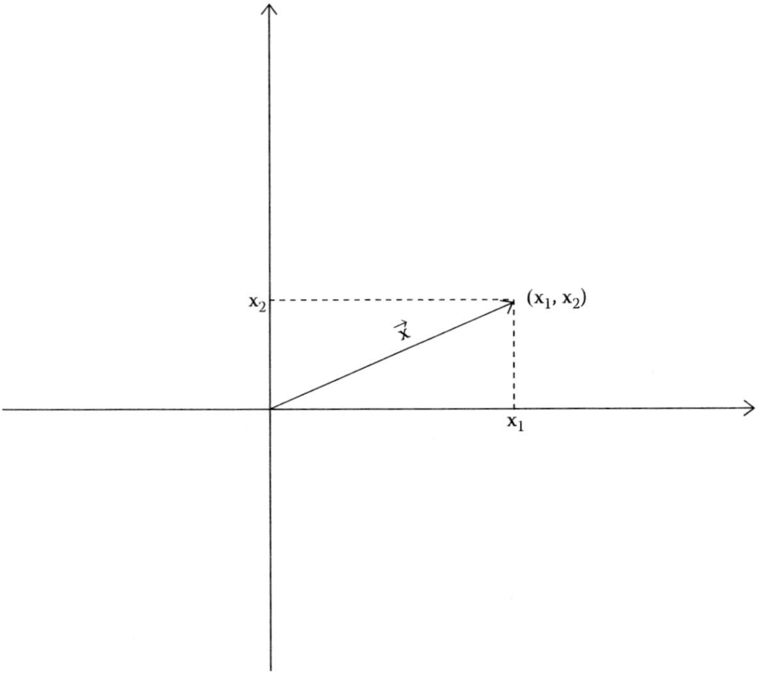

Figure 6.1: Norm of a vector.

DEFINITION 6.1 *(norm)*

A real valued function $\|\,\|: \mathbb{C}^n \to \mathbb{R}$ *is called a* norm *on* $\mathbb{C}^n$ *iff*

1. $\|\mathbf{x}\| \geq 0$ *for all* $\mathbf{x}$ *in* $\mathbb{C}^n$ *and* $\|\mathbf{x}\| = 0$ *if and only if* $\mathbf{x} = \vec{0}$

2. $\|\alpha\mathbf{x}\| = |\alpha|\, \|\mathbf{x}\|$ *for all* $\mathbf{x}$ *in* $\mathbb{C}^n$ *and all* α *in* $\mathbb{C}$

3. $\|\mathbf{x}+\mathbf{y}\| \leq \|\mathbf{x}\| + \|\mathbf{y}\|$ *for all* $\mathbf{x}, \mathbf{y}$ *in* $\mathbb{C}^n$ *(triangle inequality).*

It turns out that there are many examples of norms on $\mathbb{C}^n$. Here are just a few. Each gives a way of talking about the "size" of a vector. Is it "big" or is it "small"? Our intuition has been formed by Euclidean notions of length, but there are many other ways to measure length that may seem strange at first. Some norms may actually be more convenient to use than others in a given situation.

Example 6.1

Let $\mathbf{x} = (x_1, x_2, \ldots, x_n)$ be a vector in $\mathbb{C}^n$.

1. Define $\|\mathbf{x}\|_1 = \sum_{i=1}^{n}|x_i|$. This is a norm called the ℓ_1 *norm, sum norm*, or even the *taxicab norm* (can you see why?).

2. Define $\|\mathbf{x}\|_2 = \left(\sum_{i=1}^{n}|x_i|^2\right)^{\frac{1}{2}}$. This is a norm called the ℓ_2 *norm* or *Euclidean norm.*

3. More generally, define $\|\mathbf{x}\|_p = \left(\sum_{i=1}^{n}|x_i|^p\right)^{1/p}$, where $1 \leq p < \infty$. This is a norm called the ℓ_p *norm* or just the *p-norm* for short and includes the previous two examples as special cases.

4. Define $\|\mathbf{x}\|_\infty = \max_{1\leq i\leq n}\{|x_i|\}$. This norm is called the l_∞ *norm* or the *max norm.*

5. Suppose B is a matrix where $B = S^2$ with $S = S^* \neq \mathbb{O}$. Then define $\|\mathbf{x}\|_B = (\mathbf{x}^* B\mathbf{x})^{\frac{1}{2}}$. This turns out to be a norm also.

There are two subsets that are interesting to look at for any norm, their *unit ball* and *unit sphere.*

DEFINITION 6.2 *(unit ball, unit sphere)*

Let $\|\,\|$ be a norm on $\mathbb{C}^n$. Then, the set $B_1 = \{\mathbf{x} \in \mathbb{C}^n \mid \|\mathbf{x}\| \leq 1\}$ is called the unit ball *for the norm, and $S_1 = \{\mathbf{x} \in \mathbb{C} \mid \|\mathbf{x}\| = 1\}$ is called the* unit sphere *for the norm.*

It may be helpful to visualize some of these sets for various p-norms in $\mathbb{R}^2$ (see Figure 6.2).

THEOREM 6.1 (basic facts about all norms)

Let $\|\,\|$ be a norm on $\mathbb{C}^n$ with $\mathbf{x}$ and $\mathbf{y}$ in $\mathbb{C}^n$. Then

1. $\|\mathbf{x}\| = \|-\mathbf{x}\|$ *for all* $\mathbf{x}$ *in* $\mathbb{C}^n$.

2. $|\|\mathbf{x}\| - \|\mathbf{y}\|| \leq \|\mathbf{x} - \mathbf{y}\| \leq \|\mathbf{x}\| + \|\mathbf{y}\|$.

3. *For any invertible matrix S in $\mathbb{C}^{n\times n}$, the function $\|\mathbf{x}\|_S = \|S\mathbf{x}\|$ defines a norm $\|\,\|_S$ on $\mathbb{C}^n$.*

4. *The unit ball is always a closed, bounded, convex set that contains the origin of $\mathbb{C}^n$.*

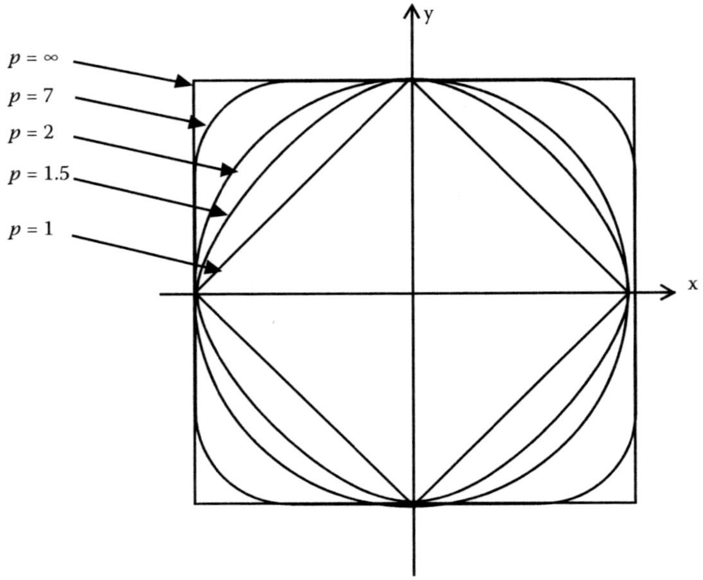

Figure 6.2: Unit spheres in various norms.

5. *Every norm on* $\mathbb{C}^n$ *is uniformly continuous. In other words, given any* $\varepsilon > 0$ *there exists* $\delta > 0$ *such that* $|x_i - y_i| < \delta$ *for* $1 \leq i \leq n$ *implies* $|\|\mathbf{x}\| - \|\mathbf{y}\|| < \varepsilon$*, where* $\mathbf{x} = (x_1, x_2, \ldots, x_n)$ *and* $\mathbf{y} = (y_1, y_2, \ldots, y_n)$*.*

PROOF The proofs are left as exercises. ▯

Recall that a subset K of $\mathbb{C}^n$ is called *convex* if $\mathbf{x}$ and $\mathbf{y}$ in K implies that $t\mathbf{x} + (1-t)\mathbf{y}$ also belongs to K for any real t with $0 < t < 1$. In other words, if $\mathbf{x}$ and $\mathbf{y}$ are in K, the *line segment* from $\mathbf{x}$ to $\mathbf{y}$ lies entirely in K. We remark that given any closed, bounded, convex set K of $\mathbb{C}^n$ that contains the origin, there is a norm on $\mathbb{C}^n$ for which K is the unit ball. This result is proved in Householder [1964]. This and Theorem 6.1(3) suggest that there is an enormous supply of norms to choose from on $\mathbb{C}^n$. However, in a sense, it does not matter which one you choose. The next theorem pursues this idea.

THEOREM 6.2 (the norm equivalence theorem)
Let $\|\,\|$ *and* $\|\,\|'$ *be any two norms on* $\mathbb{C}^n$*. Then, there are real constants* $C_2 \geq C_1 > 0$ *such that* $C_1\,\|\mathbf{x}\| \leq \|\mathbf{x}\|' \leq C_2\,\|\mathbf{x}\|$ *for all* $\mathbf{x}$ *in* $\mathbb{C}^n$*.*

PROOF Define the function $f(\mathbf{x}) = \|\mathbf{x}\|'$ from $\mathbb{C}^n$ to $\mathbb{R}$. This is a continuous function on $\mathbb{C}^n$. Let $S = \{\mathbf{y} \in \mathbb{C}^n \mid \|\mathbf{y}\| = 1\}$. This is a closed and bounded subset of $\mathbb{C}^n$. By a famous theorem of Weierstrass, a continuous function on a closed and bounded subset of $\mathbb{C}^n$ assumes its minimum and its maximum value on that set. Thus, there exists a $\mathbf{y}_{\min}$ in S and a $\mathbf{y}_{\max}$ in S such that $f(\mathbf{y}_{\min}) \leq f(\mathbf{y}) \leq f(\mathbf{y}_{\max})$ for all $\mathbf{y}$ in S. Note $\mathbf{y}_{\min} \neq \vec{0}$ since $\|\mathbf{y}_{\min}\| = 1$ so $f(\mathbf{y}_{\min}) > 0$. Now take any $\mathbf{x}$ in $\mathbb{C}^n$ other than $\vec{0}$, and form $\frac{\mathbf{x}}{\|\mathbf{x}\|}$. Note that this vector is in S, so $\|\mathbf{x}\|' = \|\mathbf{x}\| \left\| \frac{\mathbf{x}}{\|\mathbf{x}\|} \right\|' \geq \|\mathbf{x}\| \, f(\mathbf{y}_{\min})$. In fact, the inequality $\|\mathbf{x}\|' \geq \|\mathbf{x}\| \, f(\mathbf{y}_{\min})$ holds even if $\mathbf{x} = \vec{0}$ since all it says then is $0 \geq 0$. Similarly, $\|\mathbf{x}\|' = \|\mathbf{x}\| \left\| \frac{\mathbf{x}}{\|\mathbf{x}\|} \right\|' \leq \|\mathbf{x}\| \, f(\mathbf{y}_{\max})$. Putting this together, we have $\|\mathbf{x}\| \, f(\mathbf{y}_{\min}) \leq \|\mathbf{x}\|' \leq \|\mathbf{x}\| \, f(\mathbf{y}_{\max})$. Clearly, $f(\mathbf{y}_{\min}) \leq f(\mathbf{y}_{\max})$. Take $C_1 = f(\mathbf{y}_{\min})$ and $C_2 = f(\mathbf{y}_{\max})$. ▯

Since we do not really need this result, you may ignore its proof altogether because it uses ideas from advanced calculus. However, it is important to know that whenever you have a norm — that is, a notion of length — you automatically have a notion of "distance."

DEFINITION 6.3 *(distance function)*
Let $\| \, \|$ be a norm on $\mathbb{C}^n$. For $\mathbf{x}$ and $\mathbf{y}$ in $\mathbb{C}^n$, define the distance from $\mathbf{x}$ to $\mathbf{y}$ by $d(\mathbf{x}, \mathbf{y}) = \|\mathbf{y} - \mathbf{x}\|$.

THEOREM 6.3 (basic facts about any distance function)
Let d be the distance function derived from the norm $\| \, \|$. Then

1. *$d(\mathbf{x}, \mathbf{y}) \geq 0$ for all $\mathbf{x}, \mathbf{y}$ in $\mathbb{C}^n$ and $d(\mathbf{x}, \mathbf{y}) = 0$ iff $\mathbf{x} = \mathbf{y}$.*
2. *$d(\mathbf{x}, \mathbf{y}) = d(\mathbf{y}, \mathbf{x})$ for all $\mathbf{x}, \mathbf{y}$ in $\mathbb{C}^n$ (symmetry).*
3. *$d(\mathbf{x}, \mathbf{z}) \leq d(\mathbf{x}, \mathbf{y}) + d(\mathbf{y}, \mathbf{z})$ for all $\mathbf{x}, \mathbf{y}, \mathbf{z}$ in $\mathbb{C}^n$ (triangle inequality for distance).*
4. *$d(\mathbf{x}, \mathbf{y}) = d(\mathbf{x} + \mathbf{z}, \mathbf{y} + \mathbf{z})$ for all $\mathbf{x}, \mathbf{y}, \mathbf{z}$ in $\mathbb{C}^n$ (translation invariance).*
5. *$d(\alpha\mathbf{x}, \alpha\mathbf{y}) = |\alpha| \, d(\mathbf{x}, \mathbf{y})$ for all $\mathbf{x}, \mathbf{y}$ in $\mathbb{C}^n$ and all $\alpha \in \mathbb{C}$.*

PROOF The proofs here all refer back to basic properties of a norm and so we leave them as exercises. ▯

A huge door has just been opened to us. Once we have a notion of distance, all the concepts of calculus are available. That is because you now have a way of saying that vectors are "close" to each other. However, this is too big a story to tell here. Besides, our main focus is on the algebraic properties of matrices. Even so, just to get the flavor of where we could go with this, we look at the concept of a *convergent sequence of vectors.*

DEFINITION 6.4 *(convergent sequence of vectors)*

Let $(\mathbf{x}_n) = (\mathbf{x}_1, \mathbf{x}_2, \dots)$ *be a sequence of vectors in* $\mathbb{C}^n$. *This sequence is said to* converge *to the vector* $\mathbf{x}$ *in* $\mathbb{C}^n$ *with respect to the norm* $\| \|$ *iff for each* $\varepsilon > 0$ *there is a natural number* $N > 0$ *such that if* $n > N$, *then* $\|\mathbf{x}_n - \mathbf{x}\| < \varepsilon$.

One final remark along these lines. In view of the norm equivalence theorem, the convergence of a sequence of vectors with respect to one norm implies the convergence with respect to any norm. So, for convergence ideas, it does not matter which norm is used. Once we have convergence, it is not difficult to formulate the concept of "continuity." How would you do it?

The astute reader will have noticed we did not offer any proofs that the norms of Example 6.1 really are norms. You probably expected that to show up in the exercises. Well, we next give a little help establishing that the ℓ_p norms really are norms. When you see what is required, you should be thankful. To get this proved, we need a fundamental result credited to **Ludwig Otto Hölder** (22 December, 1859–29 August, 1937).

THEOREM 6.4 (Hölder's inequality)

Let $a_1, a_2, \dots, a_n$ *and* $b_1, b_2, \dots, b_n$ *be any complex numbers and let* p *and* q *be real numbers with* $1 < p, q < \infty$. *Suppose* $\frac{1}{p} + \frac{1}{q} = 1$. *Then*

$$\sum_{i=1}^{n} |a_i b_i| \leq \left(\sum_{i=1}^{n} |a_i|^p \right)^{1/p} \left(\sum_{i=1}^{n} |b_i|^q \right)^{1/q}.$$

PROOF If either all the a_is are zero or all the b_is are zero, the inequality above holds trivially. So let's assume that not all the a_is are zero and not all the b_is are zero. Recall from calculus that the function $f(x) = \ln(x)$ is always concave down on the positive real numbers (look at the second derivative f''). Thus, if α and β are positive real numbers and λ is a real number with $0 < \lambda < 1$, then $\lambda \ln(\alpha) + (1 - \lambda)\ln(\beta) \leq \ln(\lambda\alpha + (1 - \lambda)\beta)$ (see Figure 6.3).

To see this, note that the equation of the chord from $(\alpha, \ell n(\alpha))$ to $(\beta, \ell n(\beta))$ is $y - \ell n(\beta) = \dfrac{\ell n(\beta) - \ell n(\alpha)}{\beta - \alpha}(x - \beta)$, so if $x = \lambda\alpha + (1 - \lambda)\beta$, $y = \ell n(\beta) +$

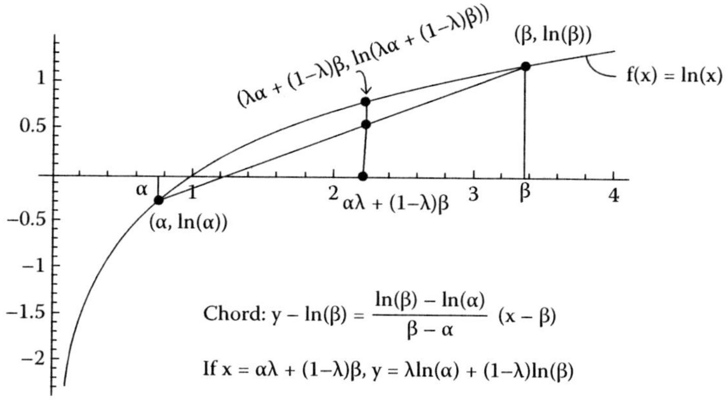

Figure 6.3: Proof of Hölder's Theorem.

$\frac{\ell n(\beta) - \ell n(\alpha)}{\beta - \alpha}(\lambda\alpha + (1-\lambda)\beta - \beta) = \ell n(\beta) + \frac{\ell n(\beta) - \ell n(\alpha)}{\beta - \alpha}(\lambda(\alpha - \beta)) = \ell n(\beta) + \lambda(\ell n(\beta) - \ell n(\alpha))$. From the picture it is clear that $\lambda \ln(\alpha) + (1-\lambda)\ln(\beta) \leq \ln(\lambda\alpha + (1-\lambda)\beta)$.

Now, by laws of logarithm, $\ln(\alpha^{\lambda}\beta^{1-\lambda}) \leq \ln(\lambda\alpha + (1-\lambda)\beta)$. Then, since $\ln(x)$ is strictly increasing on positive reals (just look at the first derivative f'), we get $\alpha^{\lambda}\beta^{1-\lambda} \leq \lambda\alpha + (1-\lambda)\beta$. Now choose

$$\alpha_i = \frac{|a_i|^p}{\sum_{j=1}^{n} |a_j|^p}, \ \beta_i = \frac{|b_i|^q}{\sum_{j=1}^{n} |b_j|^q} \quad \text{and} \quad \lambda = \frac{1}{p}.$$

$$\text{Let } A = \sum_{j=1}^{n} |a_j|^p \quad \text{and} \quad B = \sum_{j=1}^{n} |b_j|^q .$$

Note that to even form α_i and β_i, we need the fact that not all the a_j and not all the b_j are zero.

Then using $\alpha_i^{\lambda}\beta_i^{1-\lambda} \leq \lambda\alpha_i + (1-\lambda)\beta_i$ and $\lambda = \frac{1}{p}$, we get

$$\left(\frac{|a_i|^p}{A}\right)^{1/p}\left(\frac{|b_i|^q}{B}\right)^{1-1/p} \leq \frac{1}{p}\left(\frac{|a_i|^p}{A}\right) + \left(1 - \frac{1}{p}\right)\left(\frac{|b_i|^q}{B}\right)$$

or

$$\frac{|a_i|}{(A)^{1/p}}\frac{|b_i|}{(B)^{1/q}} \leq \frac{1}{p}\left(\frac{|a_i|^p}{A}\right) + \frac{1}{q}\left(\frac{|b_i|^q}{B}\right).$$

This is true for each i, $1 \leq i \leq n$, so we may sum on i and get

$$\sum_{i=1}^{n} \frac{|a_i|\,|b_i|}{(A)^{1/p}(B)^{1/q}} = \frac{1}{(A)^{1/p}(B)^{1/q}} \sum_{i=1}^{n} |a_i b_i|$$

$$\leq \frac{1}{p} \sum_{i=1}^{n} \frac{|a_i|^p}{(A)} + \frac{1}{q} \sum_{i=1}^{n} \frac{|b_i|^q}{(B)} = \frac{1}{p}\frac{1}{(A)} \sum_{i=1}^{n} |a_i|^p + \frac{1}{q}\frac{1}{(B)} \sum_{i=1}^{n} |b_i|^q$$

$$= \frac{1}{p}\frac{1}{(A)}A + \frac{1}{q}\frac{1}{(B)}B = \frac{1}{p} + \frac{1}{q} = 1.$$

Therefore, $\frac{1}{(A)^{1/p}(B)^{1/q}} \sum_{i=1}^{n} |a_i b_i| \leq 1.$

By clearing denominators, we get $\sum_{i=1}^{n} |a_i b_i| \leq (A)^{1/p}(B)^{1/q}$

$$= \left(\sum_{j=1}^{n} |a_j|^p \right)^{1/p} \left(\sum_{j=1}^{n} |b_j|^q \right)^{1/q}$$

and we are done. □

Wow! Do you get the feeling we have really proved something here? Actually, though the symbol pushing was intense, all we really used was the fact that the natural logarithm function is concave down.

As a special case, we get a famous inequality that apparently dates back to 1821 when **Augustin Louis Cauchy** (21 August 1789–23 May 1857) first proved it. This inequality has been generalized by a number of mathematicians, notably **Hermann Amandus Schwarz** (25 January 1843–30 November 1921).

COROLLARY 6.1 (Cauchy–Schwarz inequality in $\mathbb{C}^n$)
Take $p = q = 2$ above. Then

$$\sum_{i=1}^{n} |a_i b_i| \leq \left(\sum_{i=1}^{n} |a_i|^2 \right)^{1/2} \left(\sum_{i=1}^{n} |b_i|^2 \right)^{1/2}.$$

COROLLARY 6.2
The function $\| \|_p : \mathbb{C}^n \to \mathbb{R}$ defined by $\|\mathbf{x}\|_p = \left(\sum_{i=1}^{n} |x_i|^p \right)^{1/p}$ is a norm on $\mathbb{C}^n$ where $\mathbf{x} = (x_1, x_2, \ldots, x_n)$ and $1 < p < \infty$.

PROOF We must establish that this function satisfies the three defining properties of a norm. The first two are easy and left to the reader. The challenge is the triangle inequality. We shall prove $\|\mathbf{x}+\mathbf{y}\|_p \leq \|\mathbf{x}\|_p + \|\mathbf{y}\|_p$ for all $\mathbf{x}$, $\mathbf{y}$ in $\mathbb{C}^n$. Choose any $\mathbf{x}$, $\mathbf{y}$ in $\mathbb{C}^n$. If $\mathbf{x}+\mathbf{y} = \overrightarrow{0}$, the inequality is trivially true, so suppose $\mathbf{x}+\mathbf{y} \neq \overrightarrow{0}$. Then $\|\mathbf{x}+\mathbf{y}\|_p \neq 0$. Thus

$$\begin{aligned}
\|\mathbf{x}+\mathbf{y}\|_p^p &= \sum_{i=1}^{n} |x_i + y_i|^p \\
&= \sum_{i=1}^{n} |x_i + y_i|\,|x_i + y_i|^{p-1} \leq \sum_{i=1}^{n} (|x_i| + |y_i|)\,|x_i + y_i|^{p-1} \\
&= \sum_{i=1}^{n} |x_i|\,|x_i + y_i|^{p-1} + \sum_{i=1}^{n} |y_i|\,|x_i + y_i|^{p-1} \\
&\leq \left(\sum_{i=1}^{n} |x_i|^p\right)^{1/p} \left(\sum_{i=1}^{n} |x_i + y_i|^{(p-1)q}\right)^{1/q} + \left(\sum_{i=1}^{n} |y_i|^p\right)^{1/p} \\
&\times \left(\sum_{i=1}^{n} |x_i + y_i|^{(p-1)q}\right)^{1/q} = \left(\|\mathbf{x}\|_p + \|\mathbf{y}\|_p\right) \left(\sum_{i=1}^{n} |x_i + y_i|^{(p-1)q}\right)^{1/q} \\
&= \left(\|\mathbf{x}\|_p + \|\mathbf{y}\|_p\right) \left(\sum_{i=1}^{n} |x_i + y_i|^{p}\right)^{p/pq} \\
&= \left(\|\mathbf{x}\|_p + \|\mathbf{y}\|_p\right) \left(\|\mathbf{x}+\mathbf{y}\|_p\right)^{p/q}.
\end{aligned}$$

Dividing both sides by $\|\mathbf{x}+\mathbf{y}\|_p^{p/q}$, we get $\|\mathbf{x}+\mathbf{y}\|_p = \|\mathbf{x}+\mathbf{y}\|_p^{p-p/q} \leq \|\mathbf{x}\|_p + \|\mathbf{y}\|_p$ which is what we want.

Notice we needed $\|\mathbf{x}+\mathbf{y}\|_p \neq 0$ to divide above. The reader should be able to explain all the steps above. It may be useful to note that $pq - q = (p-1)q = p$ follows from $\frac{1}{p} + \frac{1}{q} = 1$. ▯

This last inequality is sometimes called the *Minkowski inequality* after **Hermann Minkowski** (22 June 1864–12 January 1909).

Exercise Set 25

1. Give direct arguments that ℓ_1 and ℓ_∞ are norms on $\mathbb{C}^n$.

2. For $\mathbf{x}$ in $\mathbb{C}^n$, define $\|\mathbf{x}\| = \begin{cases} 1 & if\ \mathbf{x} \neq \overrightarrow{0} \\ 0 & if\ \mathbf{x} = \overrightarrow{0} \end{cases}$. Is this a norm on $\mathbb{C}^n$?

3. Let d be a function satisfying (1), (2), and (3) of Theorem 6.3. Does $\|\mathbf{x}\| = d(\mathbf{x}, \vec{0})$ define a norm on $\mathbb{C}^n$? Suppose d satisfies (1), (3), and (5). Do you get a norm then?

4. A vector $\mathbf{x}$ is called a *unit vector* for norm $\|\,\|$ if $\|\mathbf{x}\| = 1$. Suppose $\|\mathbf{x}\| \neq 1$ and $\mathbf{x} \neq \vec{0}$. Can you (easily) find a unit vector in the same direction as $\mathbf{x}$?

5. If you have had some analysis, how would you define a *Cauchy sequence* of vectors in $\mathbb{C}^n$? Can you prove that every Cauchy sequence in $\mathbb{C}^n$ converges? How would you define a convergent series of vectors in $\mathbb{C}^n$?

6. Define $\|\,\| : \mathbb{C}^n \to \mathbb{R}$ by $\|\mathbf{x}\| = |x_1| + |x_2| + \cdots + \frac{1}{n}|x_n|$. Show that $\|\,\|$ is a norm on $\mathbb{C}^n$ and sketch the unit ball in $\mathbb{R}^2$.

7. Let $\mathbf{x}$ be any vector in $\mathbb{C}^n$. Show that $\|\mathbf{x}\|_\infty = \lim_{p\to\infty} \|\mathbf{x}\|_p$. (*Hint:* $n^{1/p} \to 1$ as $p \to \infty$ and $\|\mathbf{x}\|_p \leq n^{1/p}\|\mathbf{x}\|_\infty$.)

8. Compute the p-norm for the vectors $(1, 1, 4, 3)$ and $(i, i, 4 - 2i, 5)$ for $p = 1, 2$, and ∞.

9. Argue that a sequence $(\mathbf{x}_n)$ converges to $\mathbf{x}$ in the $\infty-$norm iff each component in $\mathbf{x}_n$ converges to the corresponding entry in $\mathbf{x}$. What is the limit vector of the sequence $(\frac{1}{n}, 4 + \frac{1}{n^2}, \frac{\sin(n)}{n})$ in the ∞-norm?

10. Prove $\|\mathbf{x}\|_\infty \leq \|\mathbf{x}\|_2 \leq \sqrt{n}\,\|\mathbf{x}\|_\infty$ for all $\mathbf{x}$ in $\mathbb{C}^n$. Why does this say that if a sequence converges in the ∞-norm, then the sequence must also converge to the same result in the 2-norm? Also argue that $\|\mathbf{x}\|_1 \leq \sqrt{n}\,\|\mathbf{x}\|_2$ and $\|\mathbf{x}\|_1 \leq n\,\|\mathbf{x}\|_\infty$ for all $\mathbf{x}$ in $\mathbb{C}^n$.

11. Discuss how the 2-norm can be realized as a matrix multiplication. (*Hint:* If $\mathbf{x}$ is in $\mathbb{C}^{n\times 1}$, $\|\mathbf{x}\|_2 = \sqrt{\mathbf{x}^*\mathbf{x}}$.)

12. Argue that $\mathbf{x}_n \to \mathbf{x}$ iff the sequence of scalars $\|\mathbf{x}_n - \mathbf{x}\| \to 0$ for any norm.

13. Argue that $\left(\sum_{i=1}^{n} x_i\right)^2 \leq n\sum_{i=1}^{n} x_i^2$, where the x_i are all real.

14. Determine when equality holds in Hölder's inequality.

15. Can $\|\mathbf{x} - \mathbf{y}\|_2 = \|\mathbf{x} + \mathbf{y}\|_2$ for two vectors $\mathbf{x}$ and $\mathbf{y}$ in $\mathbb{C}^n$?

16. Argue that $|\|\mathbf{x}\| - \|\mathbf{y}\|| \leq \min(\|\mathbf{x} - \mathbf{y}\|, \|\mathbf{x} + \mathbf{y}\|)$ for all $\mathbf{x}$ and $\mathbf{y}$ in $\mathbb{C}^n$.

17. How does the triangle inequality generalize to n vectors?

18. Argue $\|\mathbf{x}\|_\infty \leq \|\mathbf{x}\|_2 \leq \|\mathbf{x}\|_1$ for all $\mathbf{x}$ in $\mathbb{C}^n$.

19. Prove all the parts of Theorem 6.1.

20. Prove all the parts of Theorem 6.3.

21. Prove if $p < q$, then $n^{-\frac{1}{p}} \|\mathbf{x}\|_p \leq \|\mathbf{x}\|_q \leq \|\mathbf{x}\|_p$ for any $\mathbf{x} \in \mathbb{C}^n$.

22. Let $(0, 0)$ and $(1, 2)$ be points in the (x, y)-plane and let L_i be the line that is equidistant from these two points in the ℓ_1, ℓ_2, and ℓ_∞ norms. Draw these lines.

23. Compute the 1, 2, and ∞ distances between $(-10, 11, 12i)$ and $(4, 1 + i, -35i)$.

24. Prove that if a sequence converges in one norm, it also converges with respect to any norm equivalent to that norm.

25. Can you find a nonzero vector $\mathbf{x}$ in $\mathbb{C}^2$ such that $\|\mathbf{x}\|_\infty = \|\mathbf{x}\|_2 = \|\mathbf{x}\|_1$?

26. Suppose $\|\mathbf{x}\|_a$, $\|\mathbf{x}\|_b$ are norms on $\mathbb{C}^n$. Argue that $\max\{\|\mathbf{x}\|_a, \|\mathbf{x}\|_b\}$ is also an norm on $\mathbb{C}^n$. Suppose λ_1 and λ_2 are positive real numbers. Argue that $\lambda_1 \|\mathbf{x}\|_a + \lambda_2 \|\mathbf{x}\|_b$ is also a norm.

27. Could the p-norm $\|\mathbf{x}\|_p$ be a norm if $0 < p < 1$?

28. Consider the standard basis vectors $\mathbf{e}_i$ in $\mathbb{C}^n$. Argue that for any norm $\| \|$ on $\mathbb{C}^n$, $\|\mathbf{x}\| \leq \sum_{i=1}^{n} |x_i| \|\mathbf{e}_i\|$, where $\mathbf{x} = (x_1, x_2, \ldots, x_n)$.

29. Suppose U is a unitary matrix($U^* = U^{-1}$). Find all norms from the examples given above that have the property $\|U\mathbf{x}\| = \|\mathbf{x}\|$ for all $\mathbf{x} = (x_1, x_2, \ldots, x_n)$ in $\mathbb{C}^n$.

30. A matrix V is called an *isometry* for a norm $\| \|$ iff $\|V\mathbf{x}\| = \|\mathbf{x}\|$ for all $\mathbf{x} = (x_1, x_2, \ldots, x_n)$ in $\mathbb{C}^n$. Argue that the isometries for a norm on $\mathbb{C}^n$ form a subgroup of $GL(n, \mathbb{C})$. Determine the isometry group of the Euclidean norm, the sum norm, and the max norm.

31. Statisticians sometimes like to put "weights" on their data. Suppose $w_1, w_2, \ldots, w_n$ is a set of positive real numbers (i.e., weights). Does $\|\mathbf{x}\| = (\sum_{i=1}^{n} w_i |x_i|^p)^{1/p}$ define a norm on $\mathbb{C}^n$ where $1 < p < \infty$? How about $\|\mathbf{x}\| = \max_i\{w_1 |x_1|, w_2 |x_2|, \ldots, w_n |x_n|\}$?

32. Verify that Example (6.1)(5) actually defines a norm.

33. Argue that the intersection of two convex sets is either empty or again a convex set.

34. Prove that for the 2-norm and any three vectors $\mathbf{z}_1, \mathbf{z}_2, \mathbf{z}_3$ $\|\mathbf{z}_1 - \mathbf{z}_2\|^2 + \|\mathbf{z}_2 - \mathbf{z}_3\|^2 + \|\mathbf{z}_3 - \mathbf{z}_1\|^2 + \|\mathbf{z}_1 + \mathbf{z}_2 + \mathbf{z}_3\|^2 = 3\{\|\mathbf{z}_1\|^2 + \|\mathbf{z}_2\|^2 + \|\mathbf{z}_3\|^2\}$.

35. There is another common proof of the key inequality in Hölder's inequality. We used a logarithm, but you could define a function $f(t) = \lambda(t-1) - t^\lambda + 1$, $0 < \lambda < 1$. Argue that $f(1) = 0$ and $f^1(t) \geq 0$. Conclude that $t^\lambda \leq \lambda t + (1-\lambda)$. Assume a and b are nonnegative real numbers with $a \geq b$. Put $t = \frac{a}{b}$, $\lambda = \frac{1}{p}$. If $a < b$, put $t = \frac{b}{a}$ and $\lambda = \frac{1}{q}$ where $\frac{1}{p} + \frac{1}{q} = 1$. In each case, argue that $a^{\frac{1}{p}} b^{\frac{1}{q}} \leq \frac{a}{p} + \frac{b}{q}$.

36. As a fun problem, draw the lines in the plane $\mathbb{R}^2$ that are equidistant from $(0,0)$ and $(1,2)$ if distance is defined in the $1, 2$ and ∞ norms. Try $(0,0)$ and $(2,1)$ and $(0,0)$ and $(2,2)$.

37. Prove that $2\|\mathbf{x} - \mathbf{y}\| \geq (\|\mathbf{x}\| + \|\mathbf{y}\|) \left\| \frac{\mathbf{x}}{\|\mathbf{x}\|} + \frac{\mathbf{y}}{\|\mathbf{y}\|} \right\|$ for nonzero vectors $\mathbf{x}$ and $\mathbf{y}$.

Further Reading

[Hille, 1972] Einar Hille, *Methods in Classical and Functional Analysis*, Addison-Wesley Publishing Co., Reading, MA, (1972).

[House, 1964] Alston S. Householder, *The Theory of Matrices in Numerical Analysis*, Dover Publications, Inc., New York, (1964).

[Schattsch, 1984] D. J. Schattschneider, The Taxicab Group, The American Mathematical Monthly, Vol. 91, No. 7, 423–428, (1984).

matrix norm, multiplicative norm, Banach norm, induced norm, condition number

---

6.2 Matrix Norms

We recall that the set of all m-by-n matrices $\mathbb{C}^{m\times n}$ with the usual addition and scalar multiplication is a complex vector space, the "vectors" being matrices. It is natural to inquire about norms for matrices. In fact, a matrix can be "strung out" to make a "long" vector in $\mathbb{C}^{mn}$ in various ways, say by putting one row after the other. You could also stack the columns to make one very "tall" column vector out of the matrix. This will give us some clues on how to define norms for matrices since we have so many norms on $\mathbb{C}^p$. In particular, we will see that all the vector norms of the previous section will lead to matrix norms. There are, however, other norms that have also been found useful.

DEFINITION 6.5 *(matrix norm)*

A real valued function $\|\,\|:\mathbb{C}^{m\times n}\to\mathbb{R}$ *is called a* matrix norm *iff it satisfies*

1. $\|A\|\geq 0$ *for all* $A\in\mathbb{C}^{m\times n}$ *and* $\|A\|=0$ *iff* $A=\mathbb{O}$
2. $\|\alpha A\|=|\alpha|\,\|A\|$ *for all* $\alpha\in\mathbb{C}$ *and* $A\in\mathbb{C}^{m\times n}$
3. $\|A+B\|\leq\|A\|+\|B\|$ *for all* $A,B\in\mathbb{C}^{n\times n}$. *(Triangle inequality)*

   We say we have a strong *matrix norm or a* Banach norm *iff we have in addition* $n=m$ *and*

4. $\|AB\|\leq\|A\|\,\|B\|$ *for all* A,B,C *in* $\mathbb{C}^{n\times n}$.

   The terms multiplicative and submultiplicative are also used to describe (4).

As with vector norms, there are many examples of matrix norms.

Example 6.2

1. $\|A\|_1=\sum_{i=1}^{m}\sum_{j=1}^{n}|a_{ij}|$. This is a norm called the *sum norm.*
2. $\|A\|_{col}=\max_{1\leq j\leq n}\sum_{i=1}^{m}|a_{ij}|$. This is a norm called the *maximum column sum norm.*
3. $\|A\|_F=\left(\sum_{i=1}^{m}\sum_{j=1}^{n}|a_{ij}|^2\right)^{1/2}=tr(A^*A)^{1/2}$. This is a norm called the *Frobenius norm.*
4. $\|A\|_p=\left(\sum_{i=1}^{m}\sum_{j=1}^{n}|a_{ij}|^p\right)^{1/p}$. This is a norm called the *Minkowski p-norm* or *Hölder norm.* It generalizes (1) and (3).

5. $\|A\|_\infty = \max_{i,j}\{\left|a_{ij}\right|\}$. This is a norm called the *max norm.*

6. $\|A\|_{row} = \max_{1\le i\le m}\left(\sum_{j=1}^{n}\left|a_{ij}\right|\right)$. This is a norm called the *maximum row sum norm.*

The reader is now invited to make up more norms motivated by the examples above.

We remark that unit balls and unit spheres are defined in the same way as for vector norms: $B_1 = \{A \in \mathbb{C}^{m\times n} \mid \|A\| \le 1\}$ and $S_1 = \{A \in \mathbb{C}^{m\times n} \mid \|A\| = 1\}$. Also, most of the facts we derived for vector norms hold true for matrix norms. We collect a few below.

THEOREM 6.5 (basic facts about matrix norms)

Let $\|\,\|$ be a norm on $\mathbb{C}^{m\times n}$ and $A, B \in \mathbb{C}^{m\times n}$.

Then

1. $|\ \|A\| - \|B\|\ | \le \|A - B\| \le \|A\| + \|B\|$ *for all* $A, B \in \mathbb{C}^{m\times n}$.

2. *The unit ball in $\mathbb{C}^{m\times n}$ is a closed, bounded, convex subset of $\mathbb{C}^{m\times n}$ containing the zero matrix.*

3. *Every matrix norm is uniformly continuous.*

4. *If $\|\,\|$ and $\|\,\|'$ are matrix norms, there exist constants $C_2 \ge C_1 > 0$ such that $C_1\|A\| \le \|A\|' \le C_2\|A\|$ for all A in $\mathbb{C}^{m\times n}$ (norm equivalence theorem).*

5. *Let S be invertible in $\mathbb{C}^{m\times m}$. Then $\|A\|_S = \left\|SAS^{-1}\right\|$ defines another matrix norm on $\mathbb{C}^{m\times m}$.*

PROOF The proofs are left as exercises. ▯

Now we have a way to discuss the convergence of matrices and the distance between matrices. We have a way to address the issue if two matrices are "close" or not. In other words, we could develop analysis in this context. However, we will not pursue these ideas here but rather close our current discussion looking at connections between matrix norms and vector norms. It turns out vector norms induce norms on matrices, as the following theorem shows. These induced norms are also called *operator norms*.

THEOREM 6.6

Let $\| \|_a$ and $\| \|_b$ be vector norms on $\mathbb{C}^m$ and $\mathbb{C}^n$, respectively. Then the function $\| \|_{a,b}$ on $\mathbb{C}^{m\times n}$ defined by $\|A\|_{a,b} = \max_{\mathbf{x}\neq\vec{0}} \frac{\|A\mathbf{x}\|_a}{\|\mathbf{x}\|_b} = \max_{\|\mathbf{x}\|_b=1} \|A\mathbf{x}\|_a$ is a matrix norm on $\mathbb{C}^{m\times n}$. Moreover, $\|A\mathbf{x}\|_a \leq \|A\|_{a,b}\|\mathbf{x}\|_b$ for all $\mathbf{x}$ in $\mathbb{C}^n$ and all A in $\mathbb{C}^{m\times n}$. Finally, if $n = m$, this norm is multiplicative.

PROOF We sketch the proof. The fact that $\|A\|_{a,b}$ exists depends on a fact from advanced calculus, which we shall omit. For each nonzero vector $\mathbf{x}$, $\frac{\|A\mathbf{x}\|_a}{\|\mathbf{x}\|_b} \geq 0$ so the max of all these numbers must also be greater than or equal to zero. Moreover, if $\|A\|_{a,b} = 0$, then $A\mathbf{x} = \vec{0}$ for every vector $\mathbf{x}$, which we know implies $A = \mathbb{O}$. Of course, if $A = \mathbb{O}$, it is obvious that $\|A\|_{a,b} = 0$. For the second property of norm, $\|\alpha A\|_{a,b} = \max_{\mathbf{x}\neq\vec{0}} \frac{\|\alpha A\mathbf{x}\|_a}{\|\mathbf{x}\|_b} = |\alpha| \max_{\mathbf{x}\neq\vec{0}} \frac{\|A\mathbf{x}\|_a}{\|\mathbf{x}\|_b} = |\alpha|\, \|A\|_{a,b}$. Finally, for $\mathbf{x}$ nonzero, $\|A + B\|_{a,b} = \max_{\mathbf{x}\neq\vec{0}} \frac{\|(A + B)\mathbf{x}\|_a}{\|\mathbf{x}\|_b} = \max_{\mathbf{x}\neq\vec{0}} \frac{\|A\mathbf{x}+B\mathbf{x}\|_a}{\|\mathbf{x}\|_b} \leq \max_{\mathbf{x}\neq\vec{0}} \frac{\|A\mathbf{x}\|_a + \|B\mathbf{x}\|_a}{\|\mathbf{x}\|_b} \leq \max_{\mathbf{x}\neq\vec{0}} \frac{\|A\mathbf{x}\|_a}{\|\mathbf{x}\|_b} + \max_{\mathbf{x}\neq\vec{0}} \frac{\|B\mathbf{x}\|_a}{\|\mathbf{x}\|_b} = \|A\|_{a,b} + \|B\|_{a,b}$. Moreover, for any nonzero vector $\mathbf{x}$, $\frac{\|A\mathbf{x}\|_a}{\|\mathbf{x}\|_b} \leq \max_{\mathbf{x}\neq\vec{0}} \frac{\|A\mathbf{x}\|_a}{\|\mathbf{x}\|_b} = \|A\|_{a,b}$ so $\|A\mathbf{x}\|_a \leq \|A\|_{a,b}\|\mathbf{x}\|_b$. Finally, $\|AB\mathbf{x}\|_a \leq \|A\|_{a,b}\, \|B\mathbf{x}\|_a \leq \|A\|_{a,b}\, \|B\|_{a,b}\, \|\mathbf{x}\|_b$. Thus, $\|AB\|_{a,b} = \frac{\|AB\mathbf{x}\|_a}{\|\mathbf{x}\|_b} \leq \|A\|_{a,b}\, \|B\|_{a,b}$. □

DEFINITION 6.6 *(induced norm)*

The matrix norm $\| \|_{a,b}$ in the above theorem is called the norm induced *by the vector norms $\| \|_a$ and $\| \|_b$. If $n = m$ and $\| \|_a = \| \|_b$, we simply say the matrix norm $\| \|$ is induced by the vector norm $\| \|_a$.*

We remark that the geometric interpretation for an induced matrix norm $\| \|$ is that $\|A\|$ is the maximum length of a unit vector (i.e., a vector in S_1) after it was transformed by A. This is clear from the formula $\|A\| = \max_{\|\mathbf{x}\|_b=1} \|A\mathbf{x}\|_a$. Here the terms "length" and "unit" must be understood in terms of the underlying vector norms. Next, we look at some examples of induced norms.

Example 6.3

We will only consider the case where $m = n$ and where the two vector norms are the same. We present the examples in Table 6.2.

TABLE 6.2.1:

Vector Norm	Induced Matrix Norm
$\ell_1 : \|\mathbf{x}\|_1 = \sum_{i=1}^{n} \|x_i\|$	$\|A\|_{1,1} = \max_{1\le j\le n} \sum_{i=1}^{n} \|a_{ij}\| = \|A\|_{col}$
$\ell_2 : \|\mathbf{x}\|_2 = \left(\sum_{i=1}^{n} \|x_i\|^2\right)^{1/2}$	$\|A\|_{2,2} = \max_{\|x\|_2=1} \|A\mathbf{x}\|_2$
$\ell_\infty : \|\mathbf{x}\|_\infty = \max_{1\le i\le n} \|x_i\|$	$\|A\|_{\infty,\infty} = \max_{1\le i\le n} \sum_{j=1}^{n} \|a_{ij}\| = \|A\|_{row}$

You might have thought that the ℓ_2 vector norm would induce the Frobenius norm, but that is not so. Generally, $\|A\|_F \neq \|A\|_{2,2}$. Next we investigate what is so nice about induced norms.

THEOREM 6.7 (basic properties of induced norms)

1. *Let $\|\,\|_{a,b}$ be the matrix norm on $\mathbb{C}^{m\times n}$ induced by the vector norms $\|\,\|_a$ on $\mathbb{C}^m$ and $\|\,\|_b$ on $\mathbb{C}^n$. Let N be any other matrix norm on $\mathbb{C}^{m\times n}$ such that $\|\mathbf{x}\|_a \le N(A)\,\|\mathbf{x}\|_b$ for all $\mathbf{x}$ in $\mathbb{C}^n$, all A in $\mathbb{C}^{m\times n}$. Then $\|A\|_{a,b} \le N(A)$ for all $A \in \mathbb{C}^{m\times n}$.*

2. *Let $\|\,\|_a$, $\|\,\|_b$, $\|\,\|_c$ be vector norms on $\mathbb{C}^n$, $\mathbb{C}^m$, and $\mathbb{C}^p$ respectively and let $\|\,\|_{a,b}$ be the matrix norm on $\mathbb{C}^{m\times n}$ induced by $\|\,\|_a$ and $\|\,\|_b$, $\|\,\|_{b,c}$ be the matrix norm on $\mathbb{C}^{p\times m}$ induced by $\|\,\|_b$ and $\|\,\|_c$. Finally let $\|\,\|_{a,c}$ be the matrix norm on $\mathbb{C}^{p\times n}$ induced by $\|\,\|_a$ and $\|\,\|_c$. Then $\|AB\|_{a,c} \le \|A\|_{a,b}\,\|B\|_{b,c}$ for all $A \in \mathbb{C}^{m\times n}$, $B \in \mathbb{C}^{p\times m}$. The particular case where $m = n = p$ and all vector norms are the same is particularly nice. In this case, the result reads, after dropping all the subscripts.*

$$\|AB\| \le \|A\|\,\|B\|\,.$$

3. *Let $\|\,\|$ be a Banach norm on $\mathbb{C}^{n\times n}$. Then*
   (i) $\|AB\| \le \|A\|\,\|B\|$ *for all A, B in $\mathbb{C}^{n\times n}$*
   (ii) $\|I\| = 1$
   (iii) $\|A^n\| \le \|A\|^n$ *for $A \in \mathbb{C}^{n\times n}$*
   (iv) $\left\|A^{-1}\right\| \ge \frac{1}{\|A\|}$ *for all invertible A in $\mathbb{C}^{n\times n}$.*

PROOF The proofs are left as exercises. ∎

Of course, we have only scratched the surface of the theory of norms. They play an important role in matrix theory. One application is in talking about the convergence of series of matrices. A very useful matrix associated to a square matrix A is e^A. We may wish to use our experience with Taylor series in calculus to write $e^A = \sum_{n=0}^{\infty} \frac{1}{n!} A^n$. For this to make sense, we must be able to deal with the question of convergence, which entails a norm. In applied numerical linear algebra, norms are important in analyzing convergence issues and how well behaved a matrix or system of linear equations might be. There is a number called the *condition number* of a matrix, which is defined as $c(A) = \begin{cases} \infty \text{ if } A \text{ is singular} \\ \left\| A^{-1} \right\| \|A\| \text{ if } A \text{ is nonsingular} \end{cases}$ and measures in some sense "how close" a nonsingular matrix is to being singular. The folks in numerical linear algebra are interested in condition numbers. Suppose we are interested in solving the system of linear equations $A\mathbf{x} = \mathbf{b}$ where A is square and invertible. Theoretically, this system has the unique solution $\mathbf{x} = A^{-1}\mathbf{b}$. Suppose some computer reports the solution as $\widehat{\mathbf{x}}$. Then the *error vector* is $\mathbf{e} = \mathbf{x} - \widehat{\mathbf{x}}$. If we choose a vector norm, we can measure the *absolute error* $\|\mathbf{e}\| = \left\|\mathbf{x} - \widehat{\mathbf{x}}\right\|$ and the *relative error* $\frac{\|\mathbf{e}\|}{\|\mathbf{x}\|}$. But this supposes we know $\mathbf{x}$, making it pointless to talk about the error anyway. Realistically, we do not know $\mathbf{x}$ and would like to know how good an approximation $\widehat{\mathbf{x}}$ is. This leads us to consider something we can compute, namely the *residual vector* $\mathbf{r} = \mathbf{b} - A\widehat{\mathbf{x}}$. We will consider this vector again later and ask how to minimize its length. Anyway, we can now consider the *relative residual* $\frac{\left\|\mathbf{b} - A\widehat{\mathbf{x}}\right\|}{\|\mathbf{b}\|}$, which we can compute with some accuracy. Now choose a matrix norm that is compatible with the vector norm we are already using. Then $\mathbf{r} = \mathbf{b} - A\widehat{\mathbf{x}} = A\mathbf{x} - A\widehat{\mathbf{x}} = A\mathbf{e}$ so $\|\mathbf{r}\| = \|A\mathbf{e}\| \leq \|A\| \|\mathbf{e}\|$ and $\mathbf{e} = A^{-1}\mathbf{r}$ and so $\|\mathbf{e}\| \leq \left\|A^{-1}\right\| \|\mathbf{r}\|$. Putting these together we get

$$\frac{\|\mathbf{r}\|}{\|A\|} \leq \|\mathbf{e}\| \leq \left\|A^{-1}\right\| \|\mathbf{r}\|.$$

Similarly,

$$\frac{\|\mathbf{b}\|}{\|A\|} \leq \|\mathbf{x}\| \leq \left\|A^{-1}\right\| \|\mathbf{b}\|.$$

From these it follows that

$$\frac{1}{\|A\| \left\|A^{-1}\right\|} \frac{\|\mathbf{r}\|}{\|\mathbf{b}\|} \leq \frac{\|\mathbf{e}\|}{\|\mathbf{x}\|} \leq \|A\| \left\|A^{-1}\right\| \frac{\|\mathbf{r}\|}{\|\mathbf{b}\|}$$

or

$$\frac{1}{c(A)} \frac{\|\mathbf{r}\|}{\|\mathbf{b}\|} \leq \frac{\|\mathbf{e}\|}{\|\mathbf{x}\|} \leq c(A) \frac{\|\mathbf{r}\|}{\|\mathbf{b}\|}.$$

Now we have the relative error trapped between numbers we can compute (if we make a good choice of norms). Notice that if the condition number of a matrix is very close to 1, there is not much difference between the relative error and the relative residual.

To learn more about norms and their uses, we point the reader to the references on page 252.

Exercise Set 26

1. Let $\| \|$ be a matrix norm on $\mathbb{C}^{n\times n}$ such that $\|AB\| \leq \|A\| \|B\|$. Show that there exists a vector norm N on $\mathbb{C}^n$ such that $N(A\mathbf{x}) \leq \|A\| N(\mathbf{x})$ for all $A \in \mathbb{C}^{m\times n}$ and all $\mathbf{x}$ in $\mathbb{C}^n$.

2. Show $\|A\|_2^2 \leq \|A\|_{1,1} \|A\|_{\infty,\infty}$ for all $A \in \mathbb{C}^{m\times n}$.

3. Show that the matrix norm $\| \|$ on $\mathbb{C}^{n\times n}$ induced by the vector norm $\| \|_p$, $1 < p < \infty$ satisfies

$$\|A\| \leq \left[\sum_{i=1}^{n} (\sum_{j=1}^{n} \mid a_{ij} \mid^{q})^{\frac{p}{q}} \right]^{1/p} \quad \text{for all } A \in \mathbb{C}^{n\times n}.$$

4. Let $\| \|_p$ and $\| \|_q$ be matrix norms on $\mathbb{C}^{n\times n}$ of Hölder type with $\frac{1}{p} + \frac{1}{q} = 1$. Show that for $p \geq 2$, we have $\|AB\|_p \leq \min(\|A\|_p \|B\|_q, \|A\|_q \|B\|_p)$.

5. Prove $\|A + B\|_F^2 + \|A - B\|_F^2 = 2\|A\|_F^2 + 2\|B\|_F^2$.

6. Find the 1-norm and ∞-norm of $\begin{bmatrix} 8 & 4 & 2 \\ -1 & 3 & 7 \\ 1 & 9 & 1 \end{bmatrix}$.

7. Does $\|A\| = \max_{1\leq i,j\leq n} \{|a_{ij}|\}$ define a Banach norm for n-by-n matrices?

8. Argue directly that $\|A\overrightarrow{x}\|_2 \leq \|A\|_F \|\overrightarrow{x}\|_2$ and $\|AB\|_F \leq \|A\|_F \|B\|_F$.

9. Prove that when A is nonsingular, $\min_{\|\mathbf{x}\|=1} \|A\mathbf{x}\| = \dfrac{1}{\|A^{-1}\|}$ measures how much A can shrink vectors on the unit sphere.

10. Argue that $\|A\|_{2,2} = \|A^*\|_{2,2}$ and $\|A\|_F = \|A^*\|_F$. Also, $\|A^*A\|_{2,2} = \|A\|_{2,2}^2$.

11. Prove that if $SS^* = I$ and $T^*T = I$, then $\|S^*AT\|_{2,2} = \|A\|_{2,2}$.

12. Argue that $\|A\|_F = \left\|A^T\right\|_F$.

13. Prove that if $A = A^T$, then $\|A\|_\infty = \|A\|_1$.

14. Argue that $\|A\mathbf{x}\|_\infty \leq \sqrt{n}\,\|A\|_2\,\|\mathbf{x}\|_\infty$ for all $\mathbf{x}$ in $\mathbb{C}^n$.

15. Prove that $\|A\mathbf{x}\|_2 \leq \sqrt{n}\,\|A\|_\infty\,\|\mathbf{x}\|_2$ for all $\mathbf{x}$ in $\mathbb{C}^n$.

16. Argue that $\frac{1}{\sqrt{n}}\,\|A\|_2 \leq \|A\|_\infty \leq \sqrt{n}\,\|A\|_2$.

17. Suppose A is m-by-n and Q is m-by-m with $Q^{-1} = Q^T$. Argue that $\|QA\|_F = \|A\|_F$.

18. Argue that there cannot exist a norm $\|\,\|$ on $\mathbb{C}^n$ such that $\|A\mathbf{x}\| \leq \|A\|_F\,\|\mathbf{x}\|$ for all $\mathbf{x}$ in $\mathbb{C}^n$.

19. Suppose U and V are unitary matrices, $U^* = U^{-1}$ and $V^* = V^{-1}$. For which norms that we have discussed does $\|UAV\| = \|A\|$. For all A in $\mathbb{C}^{m\times n}$?

20. Suppose $\|\,\|$ is a Banach norm. Suppose $\mathbb{O} \neq E = E^2$.
    (a) Prove that $\|E\| \geq 1$.
    (b) Prove that if A is singular, $\|I - A\| \geq 1$.
    (c) Prove that if A is nonsingular, $\left\|A^{-1}\right\| \geq \dfrac{\|I\|}{\|A\|}$.
    (d) Prove that if $\|I - A\| < 1$, then A is nonsingular.
    (e) Prove that if A is nonsingular, $A + B$ nonsingular the $\left\|A^{-1} - (A+B)^{-1}\right\| \leq \left\|A^{-1}\right\|\,\left\|(A+B)^{-1}\right\|\,\|B\|$.

21. Argue that $\|\,\|_\infty$ is not a Banach norm but $\|\,\|'$ where $\|A\|' = m\,\|A\|_\infty$ is on $\mathbb{C}^{m\times m}$.

22. Suppose Ω is a positive definite m-by-m matrix . Argue that the Mahalanobis norm $\|A\|_\Omega = \sqrt{tr(A^*\Omega A)}$ is in fact a norm on $\mathbb{C}^{m\times m}$.

23. Prove that for an m-by-n matrix A, $\|A\|_2 \leq \|A\|_1 \leq \sqrt{mn}\,\|A\|_2$.

24. Prove that for an m-by-n matrix A, $\|A\|_\infty \leq \|A\|_1 \leq mn\,\|A\|_\infty$.

25. Prove that for an m-by-n matrix A, $\|A\|_\infty \leq \|A\|_2 \leq \sqrt{mn}\,\|A\|_\infty$.

26. Prove Hölder's inequality for m-by-n matrices. Suppose $1 \leq p, q < \infty$ and $\dfrac{1}{p} + \dfrac{1}{q} = 1$. Then $\|AB\|_1 \leq \|A\|_p\,\|B\|_q$.

Further Reading

[B&L, 1988] G. R. Belitskii and I. Yu. Lyubich, Matrix Norms and Their Applications, Oper. Theory Adv. Appl., Vol. 36, Birkhäuser, (1988).

[H&J, 1985] Roger A. Horn and Charles R. Johnson, *Matrix Analysis*, Cambridge University Press, (1985).

[Lü, 1996] Helmut Lütkepohl, *Handbook of Matrices*, John Wiley & Sons, New York, (1996).

[Noble, 1969] Ben Noble, *Applied Linear Algebra*, Prentice Hall, Englewood Cliffs, NJ, (1969).

6.2.1 MATLAB Moment

6.2.1.1 Norms

MATLAB has built in commands to compute certain vector norms. If $\mathbf{v} = (v_1, v_2, \ldots, v_n)$ is a vector in $\mathbb{C}^n$, recall that the p-norm is

$$\|\mathbf{v}\|_p = \left(\sum_{i=1}^{n} |v_i|^p \right)^{1/p}, \textit{ where } 1 \leq p < \infty$$

and

$$\|\mathbf{v}\|_\infty = \max_{1 \leq j \leq n} \{|v_j|\}.$$

In MATLAB, the function is

norm(**v**, p).

Note that $p = -inf$ is acceptable and norm(**v**,-inf) returns $\min_{1 \leq j \leq n} \{|v_j|\}$.

For example, let's generate a random vector in $\mathbb{C}^5$.

```
v=10*rand(5,1)+i*10*rand(5,1)

v =

        6.1543 + 4.0571i
        7.9194 + 9.3547i
        9.2181 + 9.1690i
        7.3821 + 4.1027i
        1.7627 + 8.9365i
```

We can make a list of various norms by

```
>> [norm(v,1), norm(v,2), norm(v,3), norm(v,-inf),norm(v,inf)]
ans =
  50.1839   22.9761   17.9648   7.3713   13.0017
```

For fun, let's try the vector $\mathbf{w} = (i, i, i, i, i)$.

```
>> w = [i i i i i]
w =
   Column 1 through Column 4
     0 + 1.0000i   0 + 1.0000i   0 + 1.0000i   0 + 1.0000i
Column 5
0 + 1.0000i
[norm(w,1), norm(w,2), norm(w,3), norm(w,-inf),norm(w,inf)]
ans =
     5.0000   2.2361   1.7100   1.0000   1.0000
```

MATLAB can also compute certain matrix norms. Recall that the p-norm of a matrix A is

$$\|A\|_p = \max_{v \neq \vec{0}} \frac{\|Av\|_p}{\|\mathbf{v}\|_p}, \quad where\ 1 \leq p < \infty.$$

Also recall that

$$\|A\|_\infty = \max_{1 \leq i \leq n} \sum_{j=1}^{n} |a_{ij}|,$$

which is the maximum row sum norm. Of course, we cannot forget the Frobenius norm

$$\|A\|_F = \left(\sum_{i=1}^{m} \sum_{j=1}^{n} |a_{ij}|^2 \right)^{1/2} = tr(A^* A)^{1/2}.$$

Let's look at an example with a randomly generated 4-by-5 complex matrix. The only matrix norms available in MATLAB are the p-norms where p is 1, 2, inf, or "fro."

```
>> A=10*rand(4,5)+i*10*rand(4,5)

A =

Column 1 through Column 4

   9.5013 + 0.5789i  8.9130 + 103889i   8.2141 + 2.7219i  9.2181 + 4.4510i
   2.3114 + 3.5287i  7.6210 + 2.20277i  4.4470 + 1.9881i  7.3821 + 9.3181i
   6.0684 + 8.1317i  4.5647 + 1.9872i   6.1543 + 0.1527i  1.7627 + 4.6599i
   4.8598 + 0.0986i  0.1850 + 6.0379i   7.9194 + 7.4679i  4.0571 + 4.1865i

Column 5

   9.3547 + 8.4622i
   9.1690 + 5.2515i
   4.1027 + 2.0265i
   8.9365 + 6.7214i

>> [norm(A,1), norm(A,2), norm(A,-inf),norm(A,inf),norm(A,'fro')]

ans =

   38.9386 35.0261 50.0435 50.0435 37.6361
```

The 2-norm of a very large matrix may be difficult to compute, so it can be estimated with the command

normest(A, tol),

where tol is a given error tolerance. The default tolerance is 10^{-6}.

Recall that the *condition number* of a nonsingular matrix A relative to some p-norm is defined as

$$c_p(A) = \|A\|_p \left\|A^{-1}\right\|_p .$$

This number measures the sensitivity to small changes in A as they affect solutions to $A\mathbf{x} = \mathbf{b}$. A matrix is called "well conditioned" if $c_p(A)$ is "small" and "ill conditioned" if $c_p(A)$ is "large." The MATLAB command is

cond(A, p),

where p can be 1, 2, inf, or 'fro.' Rectangular matrices are only accepted for $p = 2$, which is the default. For example, for the matrix A above

>> cond(A, 2)

ans =

8.4687

Again, for large matrices, computing the condition number can be difficult so MATLAB offers two ways to estimate the 1-norm of a square matrix. The commands are rcond and condest. To illustrate, we shall call on a well-known ill conditioned matrix called the Hilbert matrix. This is the square matrix whose (i, j) entry is $\frac{1}{i+j+1}$. It is built into MATLAB with the call hilb(n). We will use the rat format to view the entries as fractions and not decimals. We will also compute the determinant to show just how small it is making this matrix very "close" to being singular.

>> format rat

>> H = hilb(5)

H =

$$\begin{matrix} 1 & \frac{1}{2} & \frac{1}{3} & \frac{1}{4} & \frac{1}{5} \\ \frac{1}{2} & \frac{1}{3} & \frac{1}{4} & \frac{1}{5} & \frac{1}{6} \\ \frac{1}{3} & \frac{1}{4} & \frac{1}{5} & \frac{1}{6} & \frac{1}{7} \\ \frac{1}{4} & \frac{1}{5} & \frac{1}{6} & \frac{1}{7} & \frac{1}{8} \\ \frac{1}{5} & \frac{1}{6} & \frac{1}{7} & \frac{1}{8} & \frac{1}{9} \end{matrix}$$

Chapter 7

Inner Products

Hermitian inner product, parallelogram law, polarization identity, Appolonius identity, Pythagorean theorem

7.1 The Inner Product Space $\mathbb{C}^n$

In this section, we extend the idea of dot product from $\mathbb{R}^n$ to $\mathbb{C}^n$. Let's go back to $\mathbb{R}^2$, where we recall how the notion of dot product was motivated. The idea was to try to get a hold of the angle between two nonzero vectors. We have the notion of norm, $\|\mathbf{x}\| = \sqrt{x_1^2 + x_2^2}$, and thus of distance, $d(\mathbf{x}, \mathbf{y}) = \|\mathbf{y} - \mathbf{x}\|$. Let's picture the situation:

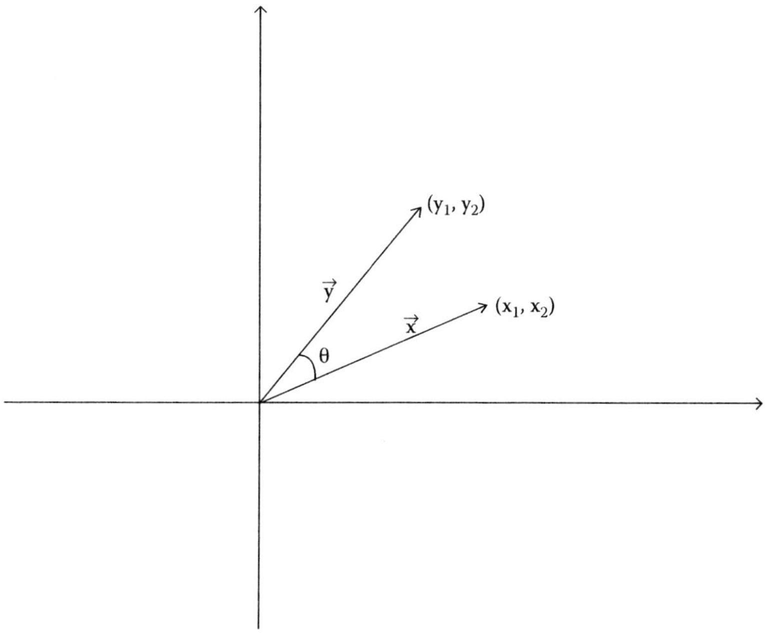

Figure 7.1: The angle between two vectors.

By the famous law of cosines, we have $\|\mathbf{y}-\mathbf{x}\|^2 = \|\mathbf{x}\|^2 + \|\mathbf{y}\|^2 - 2\,\|\mathbf{x}\|\,\|\mathbf{y}\|\cos\theta$. That is, $(y_1 - x_1)^2 + (y_2 - x_2)^2 = x_2^2 + x_2^2 + y_1^2 + y_2^2 - 2\,\|\mathbf{x}\|\,\|\mathbf{y}\|\cos\theta$. Using the good old "foil" method from high school, we get $y_1^2 - 2x_1y_1 + x_1^2 + y_2^2 - 2x_2y_2 + x_2^2 = x_1^2 + x_2^2 + y_1^2 + y_2^2 - 2\,\|\mathbf{x}\|\,\|\mathbf{y}\|\cos\theta$. Now cancel and get $x_1y_1 + x_2y_2 = \|\mathbf{x}\|\,\|\mathbf{y}\|\cos\theta$ or

$$\cos\theta = \frac{x_1y_1 + x_2y_2}{\|\mathbf{x}\|\,\|\mathbf{y}\|}.$$

Now, the length of the vectors $\mathbf{x}$ and $\mathbf{y}$ does not affect θ, the angle between them, so the numerator must be the key quantity determining the angle. This leads us to define the dot product of $\mathbf{x}$ and $\mathbf{y}$ as $\mathbf{x}\cdot\mathbf{y} = x_1y_1 + x_2y_2$. This easily extends to $\mathbb{R}^n$. Suppose we copy this definition over to $\mathbb{C}^n$. In $\mathbb{R}^2$, we have $(x_1, x_2)\cdot(x_1, x_2) = x_1^2 + x_2^2$; if this dot product is zero, $x_1 = x_2 = 0$, so (x_1, x_2) is the zero vector. However, in $\mathbb{C}^2$, $(1, i)\cdot(1, i) = 1^2 + i^2 = 0$; but we dotted a nonzero vector with itself! Do you see the problem? We have seen it before. A French mathematician by the name of **Charles Hermite** (24 December 1822–14 January 1901) came up with an idea on how to solve this problem. Use complex conjugates! In other words, he would define$(1, i)\cdot(1, i) = 1\cdot\overline{1} + i\cdot\overline{i} = 1\cdot 1 + i(-i) = 1 - i^2 = 2$. That feels a lot better. Now you can understand why the next definition is the way it is.

DEFINITION 7.1 *(Hermitian inner product)*
Let $\mathbf{x}$ *and* $\mathbf{y}$ *be in* $\mathbb{C}^n$, *where* $\mathbf{x} = (x_1, x_2, \ldots, x_n)$ *and* $\mathbf{y} = (y_1, y_2, \ldots y_n)$. *Then the Hermitian inner product of* $\mathbf{x}$ *and* $\mathbf{y}$ *is* $\langle \mathbf{x} \mid \mathbf{y}\rangle = x_1\overline{y}_1 + x_2\overline{y}_2 + \cdots + x_n\overline{y}_n = \sum_{j=1}^{n} x_j\overline{y}_j$.

We are using the notation of our physicist friends that **P. A. M. Dirac** (8 August 1902–22 October 1984) pioneered. There is a notable difference however. We put the complex conjugation on the ys, whereas they put it on the xs. That is not an essential difference. We get the same theories but they are a kind of "mirror image" to each other. In matrix notation, we have

$$\langle \mathbf{x} \mid \mathbf{y}\rangle = [\overline{y}_1\overline{y}_2\ldots\overline{y}_n]\begin{bmatrix} x_1 \\ x_2 \\ \vdots \\ x_n \end{bmatrix} = \mathbf{y}^*\mathbf{x}.$$

Note that we are viewing the vectors as columns instead of n-tuples. The next theorem collects some of the basic computational facts about inner products. The proofs are routine and thus left to the reader.

THEOREM 7.1 (basic facts about inner product)
Let $\mathbf{x}, \mathbf{y}, \mathbf{z}$ *be in* $\mathbb{C}^n$, $\alpha, \beta \in \mathbb{C}$. *Then*

1. $\langle \mathbf{x}+\mathbf{y} \mid \mathbf{z} \rangle = \langle \mathbf{x} \mid \mathbf{z} \rangle + \langle \mathbf{y} \mid \mathbf{z} \rangle$
2. $\langle \alpha\mathbf{x} \mid \mathbf{z} \rangle = \alpha \langle \mathbf{x} \mid \mathbf{z} \rangle$
3. $\langle \mathbf{x} \mid \mathbf{y} \rangle = \overline{\langle \mathbf{y} \mid \mathbf{x} \rangle}$
4. $\langle \mathbf{x} \mid \mathbf{x} \rangle \geq 0$ *and* $\langle \mathbf{x} \mid \mathbf{x} \rangle = 0$ *iff* $\mathbf{x} = \vec{0}$
5. $\langle \mathbf{x} \mid \beta\mathbf{y} \rangle = \bar{\beta} \langle \mathbf{x} \mid \mathbf{y} \rangle$.

Notice that for the physicists, scalars come out of the first slot decorated with a complex conjugate, whereas they come out of the second slot unscathed. Next, using the properties established in Theorem 7.1, we can derive additional computational rules. See if you can prove them without mentioning the components of the vectors involved.

THEOREM 7.2
Let $\mathbf{x}, \mathbf{y}, \mathbf{z}$ *be in* $\mathbb{C}^n$, $\alpha, \beta \in \mathbb{C}$. *Then*

1. $\langle \mathbf{x} \mid \mathbf{y}+\mathbf{z} \rangle = \langle \mathbf{x} \mid \mathbf{y} \rangle + \langle \mathbf{x} \mid \mathbf{z} \rangle$
2. $\left\langle \vec{0} \mid \mathbf{y} \right\rangle = \left\langle \mathbf{x} \mid \vec{0} \right\rangle = 0$ *for any* $\mathbf{x}, \mathbf{y}$ *in* $\mathbb{C}^n$
3. $\langle \mathbf{x} \mid \mathbf{y} \rangle = 0$ *for all* $\mathbf{y}$ *in* $\mathbb{C}^n$ *implies* $\mathbf{x} = \vec{0}$
4. $\langle \mathbf{x} \mid \mathbf{z} \rangle = \langle \mathbf{y} \mid \mathbf{z} \rangle$ *for all* $\mathbf{z}$ *in* $\mathbb{C}^n$ *implies* $\mathbf{x} = \mathbf{y}$
5. $\left\langle \sum_{j=1}^{m} \alpha_j \mathbf{x}_j \mid \mathbf{y} \right\rangle = \sum_{j=1}^{m} \alpha_j \langle \mathbf{x}_j \mid \mathbf{y} \rangle$
6. $\left\langle \mathbf{x} \mid \sum_{k=1}^{p} \beta_k \mathbf{y}_k \right\rangle = \sum_{k=1}^{p} \bar{\beta}_k \langle \mathbf{x} \mid \mathbf{y}_k \rangle$
7. $\left\langle \sum_{j=1}^{m} \alpha_j \mathbf{x}_j \mid \sum_{k=1}^{p} \beta_k \mathbf{y}_k \right\rangle = \sum_{j=1}^{m} \sum_{k=1}^{p} \alpha_j \bar{\beta}_k \langle \mathbf{x}_j \mid \mathbf{y}_k \rangle$
8. $\langle \mathbf{x}-\mathbf{y} \mid \mathbf{z} \rangle = \langle \mathbf{x} \mid \mathbf{z} \rangle - \langle \mathbf{y} \mid \mathbf{z} \rangle$
9. $\langle \mathbf{x} \mid \mathbf{y}-\mathbf{z} \rangle = \langle \mathbf{x} \mid \mathbf{y} \rangle - \langle \mathbf{x} \mid \mathbf{z} \rangle$.

Next, we note that this inner product is intimately connected to the l_2 norm on $\mathbb{C}^n$. Namely, for $\mathbf{x} = (x_1, x_2, \ldots x_n)$, $\langle \mathbf{x} \mid \mathbf{x} \rangle = x_1\overline{x}_1 + x_2\overline{x}_2 + \cdots + x_n\overline{x}_n = |x_1|^2 + |x_2|^2 + \cdots + |x_n|^2 = \|\mathbf{x}\|_2^2$. Thus $\|\mathbf{x}\|_2 = \sqrt{\langle \mathbf{x} \mid \mathbf{x} \rangle}$. Note this makes perfect sense, since $\langle \mathbf{x} \mid \mathbf{x} \rangle$ is a positive real number.

We end with four facts that have a geometric flavor. We can characterize the perpendicularity of two vectors through the inner product. Namely, we define $\mathbf{x}$ *orthogonal* to $\mathbf{y}$, in symbols $\mathbf{x} \perp \mathbf{y}$, iff $< \mathbf{x} \mid \mathbf{y} >= 0$.

THEOREM 7.3

Let $\mathbf{x}, \mathbf{y}, \mathbf{z}$ *be in* $\mathbb{C}^n$. *Then*

1. $\|\mathbf{x} + \mathbf{y}\|^2 + \|\mathbf{x} - \mathbf{y}\|^2 = 2\|\mathbf{x}\|^2 + 2\|\mathbf{y}\|^2$ *(parallelogram law)*
2. $\langle \mathbf{x} \mid \mathbf{y} \rangle = 1/4\{\|\mathbf{x} + \mathbf{y}\|^2 - \|\mathbf{x} - \mathbf{y}\|^2 + i\,\|\mathbf{x} + i\mathbf{y}\|^2 - i\,\|\mathbf{x} - i\mathbf{y}\|^2\}$ *(polarization identity)*
3. $\|\mathbf{z} - \mathbf{x}\|^2 + \|\mathbf{z} - \mathbf{y}\|^2 = \frac{1}{2}\|\mathbf{x} - \mathbf{y}\|^2 + 2\left\|\mathbf{z} - \frac{1}{2}(\mathbf{x} + \mathbf{y})\right\|^2$ *(Appolonius identity)*
4. *If* $\mathbf{x} \perp \mathbf{y}$, *then* $\|\mathbf{x} + \mathbf{y}\|^2 = \|\mathbf{x}\|^2 + \|\mathbf{y}\|^2$ *(Pythagorean theorem).*

PROOF The norm here is of course the l_2 norm. The proofs are computational and left to the reader. ▯

Exercise Set 27

1. Explain why $\langle \mathbf{x} \mid \mathbf{y} \rangle = \mathbf{y}^*\mathbf{x}$.
2. Establish the claims of Theorem 7.1, Theorem 7.2, and Theorem 7.3.
3. Argue that $|< \mathbf{x} \mid \mathbf{y} >| \leq < \mathbf{x} \mid \mathbf{x} >< \mathbf{y} \mid \mathbf{y} >$. This is the Cauchy-Schwarz inequality. Actually, you can do a little better. Prove that $|< \mathbf{x} \mid \mathbf{y} >| \leq \sum_{i=1}^{n} |x_i y_i| \leq < \mathbf{x} \mid \mathbf{x} >< \mathbf{y} \mid \mathbf{y} >$. Make up an example where both inequalities are strict.
4. Let $\mathbf{x} = (1, 1, 1, 1)$. Find as many independent vectors that you can such that $< \mathbf{x} \mid \mathbf{y} >= 0$.
5. Argue that $< A\mathbf{x} \mid \mathbf{y} >=< \mathbf{x} \mid A^*\mathbf{y} >$ for all $\mathbf{x}, \mathbf{y}$ in $\mathbb{C}^n$.

6. Prove that $< \mathbf{x} \mid \alpha\mathbf{y} + \beta\mathbf{z} >= \bar{\alpha} < \mathbf{x} \mid \mathbf{y} > +\bar{\beta} < \mathbf{x} \mid \mathbf{z} >$ for all $\mathbf{x}, \mathbf{y}, \mathbf{z}$ in $\mathbb{C}^n$ and all $\alpha, \beta \in \mathbb{C}$.

7. Argue that $\left\langle \mathbf{x} \mid \vec{0} \right\rangle = 0 = \left\langle \vec{0} \mid \mathbf{y} \right\rangle$ for all $\mathbf{x}, \mathbf{y}$ in $\mathbb{C}^n$.

8. Prove that if $\langle \mathbf{x} \mid \mathbf{y} \rangle = 0$, then $\|\mathbf{x} \pm \mathbf{y}\|^2 = \|\mathbf{x}\|^2 + \|\mathbf{y}\|^2$.

9. Argue that if $\langle A\mathbf{x} \mid \mathbf{y} \rangle = \langle \mathbf{x} \mid B\mathbf{y} \rangle$ for all $\mathbf{x}, \mathbf{y}$ in $\mathbb{C}^n$, then $B = A^*$.

10. Prove that $A\mathbf{x} = \mathbf{b}$ is solvable iff $\mathbf{b}$ is orthogonal to every vector in $\mathcal{N}ull(A^*)$.

11. Let $\mathbf{z} = (z_1, z_2) \in \mathbb{C}^2$ and $\mathbf{w} = (w_1, w_2) \in \mathbb{C}^2$. Which of the following defines an inner product on $\mathbb{C}^2$?
    (a) $\langle \mathbf{z} \mid \mathbf{w} \rangle = z_1\overline{w_2}$
    (b) $\langle \mathbf{z} \mid \mathbf{w} \rangle = z_1\overline{w_2} - z_2\overline{w_1}$
    (c) $\langle \mathbf{z} \mid \mathbf{w} \rangle = z_1 w_1 + z_2 w_2$
    (d) $\langle \mathbf{z} \mid \mathbf{w} \rangle = 2z_1\overline{w_1} + i(z_2\overline{w_1} - z_1\overline{w_2}) + 2z_2\overline{w_2}$.

12. Suppose $\|\mathbf{x} + \mathbf{y}\| = 9$, $\|\mathbf{x} - \mathbf{y}\| = 7$, and $\|\mathbf{x}\| = 8$. Can you determine $\|\mathbf{y}\|$?

13. Suppose $\langle \mathbf{z} \mid \mathbf{w} \rangle = \mathbf{w}^* A\mathbf{z}$ defines an inner product on $\mathbb{C}^n$. What can you say about the matrix A? (*Hint:* What can you say about A^* regarding the diagonal elements of A?)

14. Suppose f is a linear map from $\mathbb{C}^n$ to $\mathbb{C}$. Argue that that there is a unique vector $\mathbf{y}$ such that $f(\mathbf{x}) = \langle \mathbf{x} \mid \mathbf{y} \rangle$.

15. Prove $4\langle A\mathbf{x} \mid \mathbf{y} \rangle = \langle A(\mathbf{x}+\mathbf{y}) \mid \mathbf{x}+\mathbf{y} \rangle - \langle A(\mathbf{x}-\mathbf{y}) \mid \mathbf{x}-\mathbf{y} \rangle + i\,\langle A(\mathbf{x}+i\mathbf{y}) \mid \mathbf{x}+i\mathbf{y} \rangle - \langle A(\mathbf{x}-i\mathbf{y}) \mid \mathbf{x}-i\mathbf{y} \rangle$.

16. Prove that for all vectors $\mathbf{x}$ and $\mathbf{y}$, $\|\mathbf{x}+\mathbf{y}\|^2 - i\,\|i\mathbf{x}+\mathbf{y}\|^2 = \|\mathbf{x}\|^2 + \|\mathbf{y}\|^2 - i(\|\mathbf{x}\|^2 + \|\mathbf{y}\|^2) + 2\langle \mathbf{x} \mid \mathbf{y} \rangle$.

Further Reading

[Rassias, 1997] T. M. Rassias, *Inner Product Spaces and Applications*, Chapman & Hall, Boca Raton, FL, (1997).

[Steele, 2004] J. Michael Steele, *The Cauchy-Schwarz Master Class: An Introduction to the Art of Mathematical Inequalities*, Cambridge University Press, Cambridge, and the Mathematical Association of America, Providence, RI, (2004).

orthogonal set of vectors, M-perp, unit vector, normalized, orthonormal set, Fourier expansion, Fourier coefficients, Bessel's inequality, Gram-Schmidt process

---

7.2 Orthogonal Sets of Vectors in $\mathbb{C}^n$

To ask that a set of vectors be an independent set is to ask much. However, in applied mathematics, we often need a stronger condition, namely an orthogonal set of vectors. Remember, we motivated the idea of dot product in an attempt to get a hold of the idea of the angle between two vectors. What we are saying is that the most important angle is a *right angle*. Thus, if you believe $\cos\theta = \dfrac{\mathbf{x}\cdot\mathbf{y}}{\|\mathbf{x}\|\,\|\mathbf{y}\|}$ in $\mathbb{R}^n$ then, for $\theta = 90^\circ$, we must have $\mathbf{x}\cdot\mathbf{y} = 0$. This leads us to the next definition.

DEFINITION 7.2 *(orthogonal vectors and $\perp$)*
Let $\mathbf{x}, \mathbf{y} \in \mathbb{C}^n$. We say $\mathbf{x}$ and $\mathbf{y}$ are ***orthogonal*** *and write $\mathbf{x} \perp \mathbf{y}$ iff $\langle \mathbf{x} \mid \mathbf{y}\rangle = 0$. That is $\mathbf{y}^*\mathbf{x} = 0$. For any subset $M \subseteq \mathbb{C}^n$, define $M^\perp = \{\mathbf{x} \in \mathbb{C}^n \mid \mathbf{x} \perp \mathbf{m}$ for all $\mathbf{m}$ in $M\}$. $M^\perp$ is read "**M-perp**." A set of vectors $\{\mathbf{x}_j\}$ in $\mathbb{C}^n$ is called an* ***orthogonal set*** *iff $< \mathbf{x}_j \mid \mathbf{x}_k >= 0$ if $j \neq k$.*

As usual, there are some easy consequences of the definitions and, as usual, we leave the proofs to the reader.

THEOREM 7.4
Let $\mathbf{x}, \mathbf{y} \in \mathbb{C}^n$, $\alpha \in \mathbb{C}$, $M, N \subseteq \mathbb{C}^n$. Then

1. $\mathbf{x} \perp \mathbf{y}$ *iff* $\mathbf{y} \perp \mathbf{x}$
2. $\mathbf{x} \perp \vec{0}$ *for all* $\mathbf{x}$ *in* $\mathbb{C}^n$
3. $\mathbf{x} \perp \mathbf{y}$ *iff* $\mathbf{x} \perp \alpha\mathbf{y}$ *for all* α *in* $\mathbb{C}$
4. $M^\perp$ *is always a subspace of* $\mathbb{C}^n$

5. $M^{\perp} = (\text{span}(M))^{\perp}$

6. $M \subseteq M^{\perp\perp}$

7. $M^{\perp\perp\perp} = M^{\perp}$

8. $(\mathbb{C}^n)^{\perp} = (\overrightarrow{0})$

9. $(\overrightarrow{0})^{\perp} = \mathbb{C}^n$

10. $M \cap M^{\perp} \subseteq (\overrightarrow{0})$

11. *$M \subseteq N$ implies $N^{\perp} \subseteq M^{\perp}$.*

Sometimes, it is convenient to have an orthogonal set of vectors to be of unit length. This part is usually easy to achieve. We introduce additional language next.

DEFINITION 7.3 *(unit vector, orthonormal set)*
*Let $\mathbf{x} \in \mathbb{C}^n$. We call $\mathbf{x}$ a **unit vector** iff $\langle \mathbf{x} \mid \mathbf{x} \rangle = 1$. This is the same as saying $\|\mathbf{x}\| = 1$. Note that any nonzero vector $\mathbf{x}$ in $\mathbb{C}^n$ can be **normalized** to a unit vector $\mathbf{u} = \dfrac{\mathbf{x}}{\|\mathbf{x}\|}$. A set of vectors $\{\mathbf{x}_j\}$ in $\mathbb{C}^n$ is called an **orthonormal set** iff $\langle \mathbf{x}_j \mid \mathbf{x}_k \rangle = \begin{cases} 0 \text{ if } j \neq k \\ 1 \text{ if } j = k \end{cases}$. In other words, an orthonormal set is just a set of pairwise orthogonal unit vectors. Note that any orthogonal set can be made into an orthonormal set just by normalizing each vector. As we said earlier, orthogonality is a strong demand on a set of vectors.*

THEOREM 7.5
Let D be an orthogonal set of nonzero vectors in $\mathbb{C}^n$. Then D is an independent set.

PROOF This is left as a nice exercise. □

The previous theorem puts a significant limit to the number of mutually orthogonal vectors in a set; you cannot have more than n in $\mathbb{C}^n$. For example, $(i, 2i, 2i)$, $(2i, i, -2i)$, and $(2i, -2i, i)$ form an orthogonal set in $\mathbb{C}^3$ and being, necessarily, independent, form a basis of $\mathbb{C}^3$. The easiest orthonormal basis for $\mathbb{C}^n$ is the standard basis $\overrightarrow{\mathbf{e}}_1 = (1, 0, \ldots, 0)$, $\overrightarrow{\mathbf{e}}_2 = (0, 1, 0, \ldots, 0)$, $\ldots$, $\overrightarrow{\mathbf{e}}_n = (0, 0, \ldots, 0, 1)$. One reason orthonormal sets are so nice is that they associate a particularly nice set of scalars to any vector in their span.

If you have had some analysis, you may have heard of *Fourier expansions* and *Fourier coefficients*. If you have not, do not worry about it.

THEOREM 7.6 (Fourier expansion)
Let $\{\mathbf{u}_1, \mathbf{u}_2, \ldots, \mathbf{u}_m\}$ *be an orthonormal set of vectors. Suppose* $\mathbf{x} = \alpha_1\mathbf{u}_1 + \alpha_2\mathbf{u}_2 + \ldots + \alpha_m\mathbf{u}_m$. *Then* $\alpha_j = \langle \mathbf{x} \mid \mathbf{u}_j \rangle$ *for all* $j = 1, 2, \ldots, m$.

PROOF Again, this is a good exercise. □

DEFINITION 7.4 *(Fourier coefficients)*
Let $\{\mathbf{e}_j\}$ *be an orthonormal set of vectors in* $\mathbb{C}^n$ *and* $\mathbf{x}$ *a vector in* $\mathbb{C}^n$. *The set of scalars* $\{\langle \mathbf{x} \mid \mathbf{e}_j \rangle\}$ *is called the set of* ***Fourier coefficients*** *of* $\mathbf{x}$ *with respect to this orthonormal set.*

In view of Theorem 7.3, if you have an orthonormal basis of $\mathbb{C}^n$ and if you know the Fourier coefficients of a vector, you can reconstruct the vector.

THEOREM 7.7 (Bessel's inequality)
Let $\{\mathbf{e}_1, \mathbf{e}_2, \ldots, \mathbf{e}_m\}$ *be an orthonormal set in* $\mathbb{C}^n$. *If* $\mathbf{x}$ *is any vector in* $\mathbb{C}^n$, *then*

1. $\left\| \mathbf{x} - \sum_{k=1}^{m} \langle \mathbf{x} \mid \mathbf{e}_k \rangle \mathbf{e}_k \right\|^2 = \|\mathbf{x}\|^2 - \sum_{k=1}^{m} |\langle \mathbf{x} \mid \mathbf{e}_k \rangle|^2$

2. $\left(\mathbf{x} - \sum_{k=1}^{m} \langle \mathbf{x} \mid \mathbf{e}_k \rangle \mathbf{e}_k \right) \perp \mathbf{e}_j$ *for each* $j = 1, \ldots, m$

3. $\sum_{k=1}^{m} |\langle \mathbf{x} \mid \mathbf{e}_k \rangle|^2 \leq \|\mathbf{x}\|^2$.

PROOF We will sketch the proof of (1) and leave the other two statements as exercises. We will use the computational rules of inner product quite intensively:

$$\left\| \mathbf{x} - \sum_{k=1}^{m} \langle \mathbf{x} \mid \mathbf{e}_k \rangle \mathbf{e}_k \right\|^2 = \left\langle \left(\mathbf{x} - \sum_{k=1}^{m} \langle \mathbf{x} \mid \mathbf{e}_k \rangle \mathbf{e}_k \right) \;\middle|\; \left(\mathbf{x} - \sum_{k=1}^{m} \langle \mathbf{x} \mid \mathbf{e}_k \rangle \mathbf{e}_k \right) \right\rangle$$

$$= \langle \mathbf{x} \mid \mathbf{x} \rangle - \left\langle \mathbf{x} \;\middle|\; \sum_{k=1}^{m} \langle \mathbf{x} \mid \mathbf{e}_k \rangle \mathbf{e}_k \right\rangle - \left\langle \sum_{k=1}^{m} \langle \mathbf{x} \mid \mathbf{e}_k \rangle \mathbf{e}_k \;\middle|\; \mathbf{x} \right\rangle$$

$$+ \left\langle \sum_{k=1}^{m} \langle \mathbf{x} \mid \mathbf{e}_k \rangle \mathbf{e}_k \;\middle|\; \sum_{k=1}^{m} \langle \mathbf{x} \mid \mathbf{e}_k \rangle \mathbf{e}_k \right\rangle$$

$$= \langle \mathbf{x} \mid \mathbf{x} \rangle - \sum_{k=1}^{m} \overline{\langle \mathbf{x} \mid \mathbf{e}_k \rangle} \langle \mathbf{x} \mid \mathbf{e}_k \rangle - \sum_{k=1}^{m} \langle \mathbf{x} \mid \mathbf{e}_k \rangle \langle \mathbf{e}_k \mid \mathbf{x} \rangle$$

$$+\sum_{k=1}^{m}\sum_{j=1}^{m} \langle \mathbf{x} \mid \mathbf{e}_k\rangle \left\langle \mathbf{e}_k \mid \mathbf{e}_j\right\rangle \overline{\left\langle \mathbf{x} \mid \mathbf{e}_j\right\rangle}$$

$$= \|\mathbf{x}\|^2 - 2\sum_{k=1}^{m} \mid \langle \mathbf{x} \mid \mathbf{e}_k\rangle \mid^2 + \sum_{k=1}^{m} \mid \langle \mathbf{x} \mid \mathbf{e}_k\rangle \mid^2$$

$$= \|\mathbf{x}\|^2 - \sum_{k=1}^{m} \mid \langle \mathbf{x} \mid \mathbf{e}_k\rangle \mid^2 .$$

□

The reader should be sure to understand each of the steps above.

Again there is some nice geometry here. Bessel's inequality may be interpreted as saying that the sum of the squared magnitudes of the Fourier coefficients (i.e., of the components of the vector in various perpendicular directions) never exceeds the square of the length of the vector itself.

COROLLARY 7.1

Suppose $\{\mathbf{u}_1, \mathbf{u}_2, \ldots, \mathbf{u}_n\}$ is an orthonormal basis of $\mathbb{C}^n$. Then for any $\mathbf{x}$ in $\mathbb{C}^n$, we have

1. $\|\mathbf{x}\|^2 = \sum_{k=1}^{n} |\langle \mathbf{x} \mid \mathbf{u}_k\rangle|^2$ *(Parseval's identity)*

2. $\mathbf{x} = \sum_{k=1}^{n} \langle \mathbf{x} \mid \mathbf{u}_k\rangle \mathbf{u}_k$ *(Fourier expansion)*

3. $\langle \mathbf{x} \mid \mathbf{y}\rangle = \sum_{k=1}^{n} \langle \mathbf{x} \mid \mathbf{u}_k\rangle \langle \mathbf{u}_k \mid \mathbf{y}\rangle$ *for all* $\mathbf{x}$, $\mathbf{y}$ *in* $\mathbb{C}^n$.

Our approach all along has been constructive. Now we address the issue of how to generate orthogonal sets of vectors using the pseudoinverse. Begin with an arbitrary nonzero vector in $\mathbb{C}^n$, call it $\mathbf{x}_1$. Form $\mathbf{x}_1^*$. Then we seek to solve the matrix equation $\mathbf{x}_1^*\mathbf{x}_2 = 0$. But we know all solutions of this equation are of the form $\mathbf{x}_2 = \left[I - \mathbf{x}_1\mathbf{x}_1^+\right]\mathbf{v}_1$, where $\mathbf{v}_1$ is arbitrary. We do not want $\mathbf{x}_2$ to be $\vec{0}$, so we must choose $\mathbf{v}_1$ specifically to be outside *span*($\mathbf{x}_1$). Let's suppose we have done so. Then we have $\mathbf{x}_1, \mathbf{x}_2$, orthogonal vectors, both nonzero. We wish a third nonzero vector orthogonal to $\mathbf{x}_1$ and $\mathbf{x}_2$. The form of $\mathbf{x}_2$ leads us to make a guess for $\mathbf{x}_3$, namely $\mathbf{x}_3 = \left[I - \mathbf{x}_1\mathbf{x}_1^+ - \mathbf{x}_2\mathbf{x}_2^+\right]\mathbf{v}_2$. Then $\mathbf{x}_3^* = \mathbf{v}_2^*\left[I - \mathbf{x}_1\mathbf{x}_1^+ - \mathbf{x}_2\mathbf{x}_2^+\right]$. Evidently, $\mathbf{x}_3^*\mathbf{x}_1$ and $\mathbf{x}_3^*\mathbf{x}_2$ are zero, so $\mathbf{x}_1$ and $\mathbf{x}_2$ are orthogonal to $\mathbf{x}_3$. Again we need to take our arbitrary vector $\mathbf{v}_2$ outside *span* $\{\mathbf{x}_1, \mathbf{x}_2\}$ to insure we have not taken $\mathbf{x}_3 = \vec{0}$. The pattern is now clear to produce additional vectors that are pairwise orthogonal. This process must terminate since orthogonal vectors are independent and you cannot have more than n independent vectors in $\mathbb{C}^n$.

To get an orthonormal set of vectors, simply normalize the vectors obtained by the process described above. Let's illustrate with an example.

Example 7.1

In $\mathbb{C}^3$, take $\mathbf{x}_1 = \begin{bmatrix} i \\ i \\ i \end{bmatrix}$. Then $\mathbf{x}_1^+ = \frac{\mathbf{x}_1^*}{\mathbf{x}_1^*\mathbf{x}_1} = \frac{\mathbf{x}_1^*}{\|\mathbf{x}_1\|^2} = \frac{\begin{bmatrix} -i & -i & -i \end{bmatrix}}{3} =$ $\begin{bmatrix} -i/3 & -i/3 & -i/3 \end{bmatrix}$. Then $I - \mathbf{x}_1\mathbf{x}_1^+ = I_3 - \begin{bmatrix} i \\ i \\ i \end{bmatrix} \begin{bmatrix} -i/3 & -i/3 & -i/3 \end{bmatrix}$

$$= I_3 - \begin{bmatrix} 1/3 & 1/3 & 1/3 \\ 1/3 & 1/3 & 1/3 \\ 1/3 & 1/3 & 1/3 \end{bmatrix} = \begin{bmatrix} 2/3 & -1/3 & -1/3 \\ -1/3 & 2/3 & -1/3 \\ -1/3 & -1/3 & 2/3 \end{bmatrix}.$$

Evidently, $\mathbf{v}_1 = \begin{bmatrix} 1 \\ 0 \\ 0 \end{bmatrix}$ is not in span $\left(\begin{bmatrix} i \\ i \\ i \end{bmatrix}\right)$, so

$$\mathbf{x}_2 = \begin{bmatrix} 1 & 0 & 0 \end{bmatrix} \begin{bmatrix} 2/3 & -1/3 & -1/3 \\ -1/3 & 2/3 & -1/3 \\ -1/3 & -1/3 & 2/3 \end{bmatrix} = \begin{bmatrix} 2/3 \\ -1/3 \\ -1/3 \end{bmatrix}.$$

You may verify $\mathbf{x}_1 \perp \mathbf{x}_2$ if you are skeptical. Next, $\mathbf{x}_2\mathbf{x}_2^+ =$
$\begin{bmatrix} 2/3 \\ -1/3 \\ -1/3 \end{bmatrix} \frac{\begin{bmatrix} 2/3 & -1/3 & -1/3 \end{bmatrix}}{2/3} = \begin{bmatrix} 2/3 \\ -1/3 \\ -1/3 \end{bmatrix} \begin{bmatrix} 1 & -1/2 & -1/2 \end{bmatrix} =$
$\begin{bmatrix} 2/3 & -1/3 & -1/3 \\ -1/3 & 1/6 & 1/6 \\ -1/3 & 1/6 & 1/6 \end{bmatrix}$, so $I_3 - \mathbf{x}_1\mathbf{x}_1^+ - \mathbf{x}_2\mathbf{x}_2^+ = \begin{bmatrix} 0 & 0 & 0 \\ 0 & 1/2 & -1/2 \\ 0 & -1/2 & 1/2 \end{bmatrix}$.

Now $\mathbf{v}_2 = \begin{bmatrix} 0 \\ 0 \\ 1 \end{bmatrix}$ is not in span $\left\{\begin{bmatrix} i \\ i \\ i \end{bmatrix}, \begin{bmatrix} 2/3 \\ -1/3 \\ -1/3 \end{bmatrix}\right\}$ (why not?), so

$\mathbf{x}_3 = \begin{bmatrix} 0 \\ -1/2 \\ 1/2 \end{bmatrix}$. (Note, we could have just as well chosen $\mathbf{v}_2 = \begin{bmatrix} 0 \\ 1 \\ 0 \end{bmatrix}$.)

Therefore $\begin{bmatrix} i \\ i \\ i \end{bmatrix}, \begin{bmatrix} +2/3 \\ -1/3 \\ -1/3 \end{bmatrix}, \begin{bmatrix} 0 \\ -1/2 \\ 1/2 \end{bmatrix}$ is an orthogonal set of vectors in

$\mathbb{C}^3$ and necessarily a basis for $\mathbb{C}^3$. Note, the process must stop at this point since

$$\mathbf{x}_1\mathbf{x}_1^+ + \mathbf{x}_2\mathbf{x}_2^+ + \mathbf{x}_3\mathbf{x}_3^+ = \begin{bmatrix} 1/3 & 1/3 & 1/3 \\ 1/3 & 1/3 & 1/3 \\ 1/3 & 1/3 & 1/3 \end{bmatrix} + \begin{bmatrix} 2/3 & -1/3 & -1/3 \\ -1/3 & 1/6 & 1/6 \\ -1/3 & 1/6 & 1/6 \end{bmatrix} + \begin{bmatrix} 0 & 0 & 0 \\ 0 & 1/2 & -1/2 \\ 0 & -1/2 & 1/2 \end{bmatrix}$$
$$= \begin{bmatrix} 1 & 0 & 0 \\ 0 & 1 & 0 \\ 0 & 0 & 1 \end{bmatrix} = I_3.$$

We note a matrix connection here. Form the matrix Q using our orthogonal vectors as columns. Then $Q^*Q = \begin{bmatrix} -i & -i & -i \\ 2/3 & -1/3 & -1/3 \\ 0 & -1/2 & 1/2 \end{bmatrix}$ $\begin{bmatrix} i & 2/3 & 0 \\ i & -1/3 & -1/2 \\ i & -1/3 & 1/2 \end{bmatrix} = \begin{bmatrix} 3 & 0 & 0 \\ 0 & 2/3 & 0 \\ 0 & 0 & 1/2 \end{bmatrix}$ is a diagonal matrix with positive entries on the diagonal. What happens when you compute QQ^*?

Finally, let's create an orthonormal basis by normalizing the vectors obtained above: $\begin{bmatrix} i/\sqrt{3} \\ i/\sqrt{3} \\ i/\sqrt{3} \end{bmatrix}$, $\begin{bmatrix} \sqrt{2/3} \\ -\frac{\sqrt{3}}{3\sqrt{2}} \\ -\frac{\sqrt{3}}{3\sqrt{2}} \end{bmatrix}$, $\begin{bmatrix} 0 \\ -1/\sqrt{2} \\ 1/\sqrt{2} \end{bmatrix}$. Now form $U =$ $\begin{bmatrix} i/\sqrt{3} & 2/\sqrt{6} & 0 \\ i/\sqrt{3} & -\frac{\sqrt{6}}{6} & -1/\sqrt{2} \\ i/\sqrt{3} & -\frac{\sqrt{6}}{6} & +1/\sqrt{2} \end{bmatrix}$ and compute U^*U and UU^*.

Next, we tackle a closely related problem. Suppose we are given $\mathbf{y}_1, \mathbf{y}_2, \mathbf{y}_3, \ldots$, independent vectors in $\mathbb{C}^n$. We wish to generate an orthonormal sequence $\mathbf{q}_1, \mathbf{q}_2, \mathbf{q}_3 \ldots$ so that $span\{\mathbf{q}_1, \mathbf{q}_2, \ldots, \mathbf{q}_j\} = span\{\mathbf{y}_1, \mathbf{y}_2, \ldots, \mathbf{y}_j\}$ for all j. Begin by setting $\mathbf{q}_1 = \dfrac{\mathbf{y}_1}{\sqrt{\mathbf{y}_1^*\mathbf{y}_1}}$. Being independent vectors, none of the $\mathbf{y}$s are zero. Clearly $span(\mathbf{q}_1) = span(\mathbf{y}_1)$ and $\mathbf{q}_1^*\mathbf{q}_1 = \dfrac{\mathbf{y}_1^*}{\sqrt{\mathbf{y}_1^*\mathbf{y}_1}}\dfrac{\mathbf{y}_1}{\sqrt{\mathbf{y}_1^*\mathbf{y}_1}} = 1$, so $\mathbf{q}_1$ is a unit vector.

Following the reasoning above, set $\mathbf{q}_2^* = \mathbf{v}_1^*\left[I - \mathbf{q}_1\mathbf{q}_1^+\right] = \mathbf{v}_1^*\left[I - \mathbf{q}_1\mathbf{q}_1^*\right]$. A good choice of $\mathbf{v}_1^*$ is $\dfrac{\mathbf{y}_2^*}{(\mathbf{y}_2^*\left[I - \mathbf{q}_1\mathbf{q}_1^+\right]\mathbf{y}_2)^{1/2}}$. For one thing, we know $\mathbf{y}_2 \notin span(\mathbf{y}_1)$, so any multiple of $\mathbf{y}_2$ will also not be in there. Clearly $\mathbf{q}_2^*\mathbf{q}_1 = 0$, so we get our orthogonality. Moreover, $\mathbf{q}_2^*\mathbf{q}_2 = \dfrac{\mathbf{y}_2^*\left[I - \mathbf{q}_1\mathbf{q}_1^*\right]}{(\mathbf{y}_2^*\left[I - \mathbf{q}_1\mathbf{q}_1^+\right]\mathbf{y}_2)^{\frac{1}{2}}}$ $\dfrac{\left[I - \mathbf{q}_1\mathbf{q}_1^*\right]\mathbf{y}_2}{(\mathbf{y}_2^*\left[I - \mathbf{q}_1\mathbf{q}_1^+\right]\mathbf{y}_2)^{\frac{1}{2}}} = \dfrac{\mathbf{y}_2^*\left[I - \mathbf{q}_1\mathbf{q}_1^*\right]\mathbf{y}_2}{\mathbf{y}_2^*\left[I - \mathbf{q}_1\mathbf{q}_1^*\right]\mathbf{y}_2} = 1$. Now the pattern should be clear;

choose $\mathbf{q}_3^* = \dfrac{\mathbf{y}_3^*\left[I - \mathbf{q}_1\mathbf{q}_1^+ - \mathbf{q}_2\mathbf{q}_2^+\right]}{\left(\mathbf{y}_3^*\left[I - \mathbf{q}_1\mathbf{q}_1^+ - \mathbf{q}_2\mathbf{q}_2^+\right]\mathbf{y}_3\right)^{\frac{1}{2}}}$. Note $\mathbf{y}_3 \notin span\{\mathbf{y}_1, \mathbf{y}_2\}$. Generally

then $$\mathbf{q}_k^* = \frac{\mathbf{y}_k^*\left[I - \sum_{j=1}^{k-1}\mathbf{q}_j\mathbf{q}_j^+\right]}{\left(\mathbf{y}_k^*\left[I - \sum_{j=1}^{k-1}\mathbf{q}_j\mathbf{q}_j^+\right]\mathbf{y}_k\right)^{\frac{1}{2}}}.$$

This is known as the *Gram-Schmidt orthogonalization process,* named after **Erhardt Schmidt** (13 January 1876–6 December 1959), a German mathematician who described the process in 1907, and **Jőrgen Peterson Gram** (27 June 1850–29 April 1916), a Danish actuary. It produces orthogonal vectors starting with a list of independent vectors, without disturbing the spans along the way.

Exercise Set 28

1. Let $\mathbf{x} = \begin{bmatrix} 1 \\ -1 \\ 1 \\ -1 \end{bmatrix}$. Find a basis for $\{\mathbf{x}\}^\perp$.

2. Let $\mathbf{x}_1, \mathbf{x}_2, ..., \mathbf{x}_r$ be any vectors in $\mathbb{C}^n$. Form the r-by-r matrix G by $G = \left[\langle \mathbf{x}_j \mid \mathbf{x}_k \rangle\right]$ for $j, k = 1, 2, ..., r$. This is called the *Gram matrix* after J. P. Gram, mentioned above. Argue that G is Hermitian. The determinant of G is called the *Grammian*. Prove that $\mathbf{x}_1, \mathbf{x}_2, ..., \mathbf{x}_r$ form an independent set if the Grammian is positive and a dependent set if the Grammian is zero.

3. For complex matrices A and B, argue that $AB^* = \mathbb{O}$ iff $Col(A) \perp Col(B)$.

4. Argue that $\mathbb{C}^n = M \oplus M^\perp$ for any subspace M of $\mathbb{C}^n$. In particular, prove that $\dim(M^\perp) = n - \dim(M)$. Also, argue that $M = M^{\perp\perp}$ for any subspace M of $\mathbb{C}^n$.

5. Prove that every orthonormal subset of vectors in $\mathbb{C}^n$ can be extended to an orthonormal basis of $\mathbb{C}^n$.

6. Apply the Gram-Schmidt processes to $\{(i, 0, -i, 2i), (2i, 2i, 0, 2i), (i, -2i, 0, 0)\}$ and extend the orthonormal set you get to a basis of $\mathbb{C}^4$.

7. Prove the claims of Theorem 7.3.

8. Suppose M and N are subspaces of $\mathbb{C}^n$. Prove that $(M \cap N)^{\perp} = M^{\perp} + N^{\perp}$ and $(M + N)^{\perp} = M^{\perp} \cap N^{\perp}$.

7.2.1 MATLAB Moment

7.2.1.1 The Gram-Schmidt Process

The following M-file produces the usual Gram-Schmidt process on a set of linearly independent vectors. These vectors should be input as the columns of a matrix.

```
function GS = grmsch(A)
[m n] = size(A);
Q(:,1)=A(:,1)/norm(A(:,1));
for i = 2:n
Q(:,i) = A(:,i)-Q(:,1:i-1)*Q(:,1:i-1)'*A(:,i);
Q(:,i) = Q(:,i)/norm(Q(:,i));
end
Q
```

You can see that this is the usual algorithm taught in elementary linear algebra. There is a better algorithm for numerical reasons called the *modified Gram-Schmidt process*. Look this up and make an M-file for it. In the meantime, try out the code above on some nice matrices, such as

$$A = \begin{bmatrix} 1 & 1 & 1 \\ 1 & 1 & 0 \\ 1 & 0 & 0 \\ 1 & 0 & 0 \end{bmatrix}.$$

QR factorization, unitary matrix, Kung's algorithm, orthogonal full rank factorization

---

7.3 QR Factorization

The orthogonalization procedures described in the previous section have some interesting consequences for matrix theory. First, if $A \in \mathbb{C}^{m \times n}$, then $A^* \in \mathbb{C}^{n \times m}$, where $A^* = \overline{A}^T$. If we look at A^*A, which is n-by-n, and partition

A by columns, we see

$$A^*A = \begin{bmatrix} \cdots & \overline{\mathbf{a}}_1^T & \cdots \\ \cdots & \overline{\mathbf{a}}_2^T & \cdots \\ \cdots & \overline{\mathbf{a}}_n^T & \cdots \end{bmatrix} \begin{bmatrix} \vdots & \vdots & \vdots & \vdots \\ \mathbf{a}_1 & \mathbf{a}_2 & \dots & \mathbf{a}_n \\ \vdots & \vdots & \vdots & \vdots \end{bmatrix} = \left[\overline{\mathbf{a}_i}^T \mathbf{a}_j\right] = \left[\langle \mathbf{a}_j \mid \mathbf{a}_i \rangle\right].$$

In other words, the matrix entries of A^*A are just the inner products of the columns of A. Similarly, the matrix entries of AA^* are just the inner products of the rows of A. It should now be easy for the reader to prove the following theorem.

THEOREM 7.8
Let $A \in \mathbb{C}^{m\times n}$. Then

1. *The columns of A are orthogonal iff A^*A is a diagonal matrix.*
2. *A^* is a left inverse of A ($i.e.$, $A^*A = I$) iff the columns of A are orthonormal.*
3. *The rows of A are orthogonal iff AA^* is a diagonal matrix.*
4. *A^* is a right inverse of A ($i.e.$, $AA^* = I$) iff the rows of A are orthonormal.*

From this, we get the following.

COROLLARY 7.2
Let $U \in \mathbb{C}^{n\times n}$. Then the following are equivalent.

1. *The columns of U are orthonormal.*
2. $U^*U = I_n$.
3. $U^* = U^{-1}$.
4. $UU^* = I_n$.
5. *The rows of U are orthonormal.*

This leads us to a definition.

DEFINITION 7.5 *(unitary matrix)*
*The square matrix $U \in \mathbb{C}^{n\times n}$ is called **unitary** iff U satisfies any one (and hence all) of the conditions in Corollary 7.2 above.*

Unitary matrices have many useful features that we will pursue later. But first we want to have another look at the Gram-Schmidt procedure. It leads to a useful factorization of matrices. Suppose, for the purpose of illustration, we have three independent vectors $\mathbf{a}_1, \mathbf{a}_2, \mathbf{a}_3$ that can be considered the columns of a matrix A (i.e., $A = [\mathbf{a}_1 \mid \mathbf{a}_2 \mid \mathbf{a}_3]$). Say, to be even more concrete, $A = \begin{bmatrix} 1 & 1 & 1 \\ 1 & 1 & 0 \\ 1 & 0 & 0 \\ 1 & 0 & 0 \end{bmatrix}$.

We begin the process by taking $\mathbf{q}_1 = \dfrac{\mathbf{a}_1}{\sqrt{\mathbf{a}_1^*\mathbf{a}_1}} = \dfrac{\mathbf{a}_1}{\|\mathbf{a}_1\|}$. Then $\mathbf{q}_1$ is a unit vector, $\mathbf{q}_1^*\mathbf{q}_1 = \dfrac{\mathbf{a}_1^*}{\sqrt{\mathbf{a}_1^*\mathbf{a}_1}}\dfrac{\mathbf{a}_1}{\sqrt{\mathbf{a}_1^*\mathbf{a}_1}} = 1$. But the new wrinkle here is we can solve back for $\mathbf{a}_1$. Namely, $\mathbf{a}_1 = \sqrt{\mathbf{a}_1^*\mathbf{a}_1}\mathbf{q}_1 = \|\mathbf{a}_1\|\,\mathbf{q}_1$. But note, $\mathbf{a}_1^*\mathbf{q}_1 = \dfrac{\mathbf{a}_1^*\mathbf{a}_1}{\sqrt{\mathbf{a}_1^*\mathbf{a}_1}} = \sqrt{\mathbf{a}_1^*\mathbf{a}_1}$ $= \|\mathbf{a}_1\|$, so $\mathbf{a}_1 = (\mathbf{a}_1^*\mathbf{q}_1)\mathbf{q}_1$. In our numerical example, $\mathbf{q}_1 = \begin{bmatrix} \frac{1}{2} \\ \frac{1}{2} \\ \frac{1}{2} \\ \frac{1}{2} \end{bmatrix}$ and

$\mathbf{a}_1 = 2\mathbf{q}_1$ since $\mathbf{a}_1^*\mathbf{q}_1 = \begin{bmatrix} 1 & 1 & 1 & 1 \end{bmatrix}\begin{bmatrix} \frac{1}{2} \\ \frac{1}{2} \\ \frac{1}{2} \\ \frac{1}{2} \end{bmatrix} = 2 = \sqrt{\mathbf{a}_1^*\mathbf{a}_1} = \|\mathbf{a}_1\|$. Note $\mathbf{a}_1^*\mathbf{q}_1$ is real and positive. By the way, we can now write $A = [2\mathbf{q}_1 \mid \mathbf{a}_2 \mid \mathbf{a}_3]$. Next, we solve back for $\mathbf{a}_2$. Let $\mathbf{c}_2 = \mathbf{a}_2 - \overline{(\mathbf{a}_2^*\mathbf{q}_1)}\mathbf{q}_1 = \mathbf{a}_2 - \langle \mathbf{a}_2 \mid \mathbf{q}_1 \rangle \mathbf{q}_1$ and $\mathbf{q}_2 = \dfrac{\mathbf{c}_2}{\sqrt{\mathbf{c}_2^*\mathbf{c}_2}} = \dfrac{\mathbf{c}_2}{\|\mathbf{c}_2\|}$. Since $\mathbf{c}_2 = \mathbf{a}_2 - \overline{(\mathbf{a}_2^*\mathbf{q}_1)}\mathbf{q}_1 = \mathbf{a}_2 - \overline{(\mathbf{a}_2^*\mathbf{q}_1)}\mathbf{q}_1$ we see $\mathbf{c}_2^*\mathbf{q}_1 = \mathbf{a}_2^*\mathbf{q}_1 - (\mathbf{a}_2^*\mathbf{q}_1)\mathbf{q}_1^*\mathbf{q}_1$. But $\mathbf{q}_1^*\mathbf{q}_1 = 1$ so $\mathbf{c}_2^*\mathbf{q}_1 = 0 = \langle \mathbf{q}_1 \mid \mathbf{c}_2 \rangle$ making $\mathbf{c}_2 \perp \mathbf{q}_1$. Therefore, $\mathbf{q}_2 \perp \mathbf{q}_1$, since $\mathbf{q}_2$ is just a scalar multiple of $\mathbf{c}_2$. In our numerical example, $\mathbf{c}_2 = \begin{bmatrix} 1 \\ 1 \\ 0 \\ 0 \end{bmatrix} - 1\begin{bmatrix} \frac{1}{2} \\ \frac{1}{2} \\ \frac{1}{2} \\ \frac{1}{2} \end{bmatrix} = \begin{bmatrix} \frac{1}{2} \\ \frac{1}{2} \\ -\frac{1}{2} \\ -\frac{1}{2} \end{bmatrix}$, so $\mathbf{q}_2 =$

$\begin{bmatrix} \frac{1}{2} \\ \frac{1}{2} \\ -\frac{1}{2} \\ -\frac{1}{2} \end{bmatrix}$. Now we do the new thing and solve for $\mathbf{a}_2$. We see $\mathbf{a}_2 = \mathbf{c}_2 + \overline{(\mathbf{a}_2^*\mathbf{q}_1)}\mathbf{q}_1 = \sqrt{\mathbf{c}_2^*\mathbf{c}_2}\,\mathbf{q}_2 + \overline{(\mathbf{a}_2^*\mathbf{q}_1)}\mathbf{q}_1 = \|\mathbf{c}_2\|\,\mathbf{q}_2 + \langle \mathbf{a}_2 \mid \mathbf{q}_1 \rangle \mathbf{q}_1$ and so $\mathbf{a}_2^*\mathbf{q}_2 = \sqrt{\mathbf{c}_2^*\mathbf{c}_2}\mathbf{q}_2^*\mathbf{q}_2 + (\mathbf{a}_2^*\mathbf{q}_1)\mathbf{q}_1^*\mathbf{q}_2 = \sqrt{\mathbf{c}_2^*\mathbf{c}_2} = \|\mathbf{c}_2\|$. Again, $\mathbf{a}_2^*\mathbf{q}_2$ is real and positive, and $\mathbf{a}_2 = \overline{(\mathbf{a}_2^*\mathbf{q}_1)}\mathbf{q}_1 + (\mathbf{a}_2^*\mathbf{q}_2)\mathbf{q}_2 = \langle \mathbf{a}_2 \mid \mathbf{q}_1 \rangle \mathbf{q}_1 + \|\mathbf{c}_2\|\,\mathbf{q}_2$. In our numerical example, $\mathbf{a}_2 = 1.\mathbf{q}_1 + 1.\mathbf{q}_2$ so $A = [2\mathbf{q}_1 \mid \mathbf{q}_1 + \mathbf{q}_2 \mid \mathbf{a}_3]$. Finally we take $\mathbf{c}_3 = \mathbf{a}_3 - \overline{(\mathbf{a}_3^*\mathbf{q}_1)}\mathbf{q}_1 - \overline{(\mathbf{a}_3^*\mathbf{q}_2)}\mathbf{q}_2 = \mathbf{a}_3 - \langle \mathbf{a}_3 \mid \mathbf{q}_1 \rangle \mathbf{q}_1 - \langle \mathbf{a}_3 \mid \mathbf{q}_2 \rangle \mathbf{q}_2$. We compute $\mathbf{c}_3^*\mathbf{q}_1 =$

$(\mathbf{a}_3^* - (\mathbf{a}_3^*\mathbf{q}_1)\mathbf{q}_1^* - (\mathbf{a}_3^*\mathbf{q}_2^*)\mathbf{q}_2)\mathbf{q}_1 = \mathbf{a}_3^*\mathbf{q}_1 - \mathbf{a}_3^*\mathbf{q}_1 - \overrightarrow{0} = \overrightarrow{0}$ and $\mathbf{c}_3^*\mathbf{q}_2 = \mathbf{a}_3^*\mathbf{q}_2 - \overrightarrow{0} - \mathbf{a}_3^*\mathbf{q}_2 = \overrightarrow{0}$ and conclude $\mathbf{c}_3 \perp \mathbf{q}_1, \mathbf{q}_2$. Take $\mathbf{q}_3 = \dfrac{\mathbf{c}_3}{\sqrt{\mathbf{c}_3^*\mathbf{c}_3}} = \dfrac{\mathbf{c}_3}{\|\mathbf{c}_3\|}$ and get $\mathbf{q}_3 \perp \mathbf{q}_1, \mathbf{q}_2$. In our example, $\mathbf{c}_3 = \begin{bmatrix} 1 \\ 0 \\ 0 \\ 0 \end{bmatrix} - 1/2 \begin{bmatrix} \frac{1}{2} \\ \frac{1}{2} \\ \frac{1}{2} \\ \frac{1}{2} \end{bmatrix} - 1/2 \begin{bmatrix} \frac{1}{2} \\ \frac{1}{2} \\ \frac{1}{2} \\ \frac{1}{2} \end{bmatrix} = \begin{bmatrix} \frac{1}{2} \\ -\frac{1}{2} \\ 0 \\ 0 \end{bmatrix}$

so $\mathbf{q}_3 = \begin{bmatrix} \frac{\sqrt{2}}{2} \\ -\frac{\sqrt{2}}{2} \\ 0 \\ 0 \end{bmatrix}$. Solving for $\mathbf{a}_3$, we see $\mathbf{a}_3 = \mathbf{c}_3 + \overline{(\mathbf{a}_3^*\mathbf{q}_1)}\mathbf{q}_1 + \overline{(\mathbf{a}_3^*\mathbf{q}_2)}\mathbf{q}_2 = \sqrt{\mathbf{c}_3^*\mathbf{c}_3}\mathbf{q}_3 + \overline{(\mathbf{a}_3^*\mathbf{q}_1)}\mathbf{q}_1 + \overline{(\mathbf{a}_3^*\mathbf{q}_2)}\mathbf{q}_2 = \|\mathbf{c}_3\|\,\mathbf{q}_3 + \langle \mathbf{a}_3 \mid \mathbf{q}_1 \rangle\, \mathbf{q}_1 + \langle \mathbf{a}_3 \mid \mathbf{q}_2 \rangle\, \mathbf{q}_2$ and $\mathbf{a}_3^*\mathbf{q}_3 = \sqrt{\mathbf{c}_3^*\mathbf{c}_3} = \|\mathbf{c}_3\|$. So finally, $\mathbf{a}_3 = \overline{(\mathbf{a}_3^*\mathbf{q}_1)}\mathbf{q}_1 + \overline{(\mathbf{a}_3^*\mathbf{q}_2)}\mathbf{q}_2 + (\mathbf{a}_3^*\mathbf{q}_3)\mathbf{q}_3 = \langle \mathbf{a}_3 \mid \mathbf{q}_1 \rangle\, \mathbf{q}_1 + \langle \mathbf{a}_3 \mid \mathbf{q}_2 \rangle\, \mathbf{q}_2 + \|\mathbf{c}_3\|\,\mathbf{q}_3$. In our example

$$\mathbf{a}_3 = \begin{bmatrix} 1 \\ 0 \\ 0 \\ 0 \end{bmatrix} = (\tfrac{1}{2}) \begin{bmatrix} \frac{1}{2} \\ \frac{1}{2} \\ \frac{1}{2} \\ \frac{1}{2} \end{bmatrix} + (\tfrac{1}{2}) \begin{bmatrix} \frac{1}{2} \\ \frac{1}{2} \\ -\frac{1}{2} \\ -\frac{1}{2} \end{bmatrix} + (\tfrac{\sqrt{2}}{2}) \begin{bmatrix} \frac{\sqrt{2}}{2} \\ -\frac{\sqrt{2}}{2} \\ 0 \\ 0 \end{bmatrix} \text{ so}$$

$$A = \left[2\mathbf{q}_1 \mid \mathbf{q}_1 + \mathbf{q}_2 \mid \tfrac{1}{2}\mathbf{q}_1 + \tfrac{1}{2}\mathbf{q}_2 + \tfrac{\sqrt{2}}{2}\mathbf{q}_3\right] = [\mathbf{q}_1 \mid \mathbf{q}_2 \mid \mathbf{q}_3] \begin{bmatrix} 2 & 1 & \frac{1}{2} \\ 0 & 1 & \frac{1}{2} \\ 0 & 0 & \frac{\sqrt{2}}{2} \end{bmatrix}$$

$$= \begin{bmatrix} \frac{1}{2} & \frac{1}{2} & \frac{\sqrt{2}}{2} \\ \frac{1}{2} & \frac{1}{2} & -\frac{\sqrt{2}}{2} \\ \frac{1}{2} & -\frac{1}{2} & 0 \\ \frac{1}{2} & -\frac{1}{2} & 0 \end{bmatrix} \begin{bmatrix} 2 & 1 & \frac{1}{2} \\ 0 & 1 & \frac{1}{2} \\ 0 & 0 & \frac{\sqrt{2}}{2} \end{bmatrix} = QR, \text{ where } Q^*Q = I, \text{ since the}$$

columns of Q are orthonormal and R is upper triangular with positive real entries on the main diagonal. This is the famous *QR factorization*. Notice that the columns of A were independent for this to work. We could generalize the procedure a little bit by agreeing to allow dependencies. If we run into one, agree to replace that column with a column of zeros in Q. For example,

$$A = \begin{bmatrix} 1 & 3 & 1 & 4 & 1 \\ 1 & 3 & 1 & 4 & 0 \\ 1 & 3 & 0 & 0 & 0 \\ 1 & 3 & 0 & 0 & 0 \end{bmatrix} = \begin{bmatrix} \frac{1}{2} & 0 & \frac{1}{2} & 0 & \frac{\sqrt{2}}{2} \\ \frac{1}{2} & 0 & \frac{1}{2} & 0 & \frac{\sqrt{2}}{2} \\ \frac{1}{2} & 0 & -\frac{1}{2} & 0 & 0 \\ \frac{1}{2} & 0 & -\frac{1}{2} & 0 & 0 \end{bmatrix} \begin{bmatrix} 2 & 6 & 1 & 4 & \frac{1}{2} \\ 0 & 0 & 0 & 0 & 0 \\ 0 & 0 & 1 & 4 & \frac{1}{2} \\ 0 & 0 & 0 & 0 & 0 \\ 0 & 0 & 0 & 0 & \frac{\sqrt{2}}{2} \end{bmatrix}.$$

We lose a bit since now Q^*Q is diagonal with zeros and ones but R is still upper triangular square with nonnegative real entries on the main diagonal. We summarize our findings in a theorem.

THEOREM 7.9 (QR factorization)

Let $A \in \mathbb{C}^{m\times n}$, with $n \leq m$. Then there is a matrix $Q \in \mathbb{C}^{m\times n}$ with orthonormal columns and an upper triangular matrix R in $\mathbb{C}^{n\times n}$ such that $A = QR$. Moreover, if $n = m$, Q is square and unitary. Even more, if A is square and nonsingular, R may be selected so as to have positive real numbers on the diagonal. In this case, the factorization is unique.

PROOF Suppose A has full column rank n. Then the columns form an independent set in $\mathbb{C}^m$. Apply the Gram-Schmidt procedure as previously illustrated and write $A = QR$, with $Q^*Q = I$ and R upper triangular with positive real entries on the main diagonal. If the columns of A are dependent, we proceed with the generalization indicated immediately above. Let $A = [\mathbf{a}_1 \mid \mathbf{a}_2 \mid \ldots \mid \mathbf{a}_n]$. If $\mathbf{a}_1 = \overrightarrow{0}$, take $\mathbf{q}_1 = \overrightarrow{0}$; otherwise take $\mathbf{q}_1 = \dfrac{\mathbf{a}_1}{\sqrt{\mathbf{a}_1^*\mathbf{a}_1}} = \dfrac{\mathbf{a}_1}{\|\mathbf{a}_1\|}$.

Next compute $\mathbf{c}_2 = \mathbf{a}_2 - (\overline{\mathbf{a}_{2^*}\mathbf{q}_1})\mathbf{q}_1 = \mathbf{a}_2 - \langle \mathbf{a}_2 \mid \mathbf{q}_1\rangle\, \mathbf{q}_1$. If $\mathbf{c}_2 = \overrightarrow{0}$, which happens iff $\mathbf{a}_2$ depends on $\mathbf{a}_1$, set $\mathbf{q}_2 = \overrightarrow{0}$. If $\mathbf{c}_2 \neq \overrightarrow{0}$ take $\mathbf{q}_2 = \dfrac{\mathbf{c}_2}{\sqrt{\mathbf{c}_2^*\mathbf{c}_2}} = \dfrac{\mathbf{c}_2}{\|\mathbf{c}_2\|}$. Generally, for $k = 2, 3, \ldots, n$, compute $\mathbf{c}_k = \mathbf{a}_k - \sum_{j=1}^{k-1}(\overline{\mathbf{a}_k^*\mathbf{q}_j})\mathbf{q}_j = \mathbf{a}_k - \sum_{j=1}^{k-1}\langle \mathbf{a}_k \mid \mathbf{q}_j\rangle\, \mathbf{q}_j$. If $\mathbf{c}_k = \overrightarrow{0}$ (iff $\mathbf{a}_k$ depends on the previous $\mathbf{a}'s$), set $\mathbf{q}_k = \overrightarrow{0}$. Else, take $\mathbf{q}_k = \dfrac{\mathbf{c}_k}{\sqrt{\mathbf{c}_{k^*}\mathbf{c}_k}} = \dfrac{\mathbf{c}_k}{\|\mathbf{c}_k\|}$. This constructs a list of vectors $\mathbf{q}_1, \mathbf{q}_2, \ldots, \mathbf{q}_n$ that is an orthogonal set consisting of unit vectors and the zero vector wherever a dependency was detected. Now, by construction, each $\mathbf{q}_j$ is a linear combination of $\mathbf{a}_1, \ldots, \mathbf{a}_j$, but also, each $\mathbf{a}_j$ is a linear combination of $\mathbf{q}_1, \ldots, \mathbf{q}_j$. Thus, there exist scalars α_{kj} for $j = 1, 2, \ldots, n$ such that $\mathbf{a}_j = \sum_{k=1}^{j}\alpha_{kj}\mathbf{q}_k$.

Indeed $\alpha_{kj} = (\overline{\mathbf{a}_k^*\mathbf{q}_j}) = \langle \mathbf{a}_k \mid \mathbf{q}_j\rangle$. To fill out the matrix R we take $\alpha_{kj} = 0$ if $k > j$. We also take $\alpha_{ij} = 0$ for all $j = 1, 2, \ldots, n$ for each i where $\mathbf{q}_i = \overrightarrow{0}$. In this way, we have $A = QR$ where the columns of Q are orthogonal to each other and R is upper triangular. But we promised more! Namely that Q is supposed to have orthonormal columns. All right, suppose some zero columns occurred. Take the columns that are not zero and extend to a basis of $\mathbb{C}^m$. Say we get $\mathbf{q}'_1, \mathbf{q}'_2, \ldots, \mathbf{q}'_p$ as additional orthonormal vectors. Replace each zero column in turn by $\mathbf{q}'_1, \mathbf{q}'_2, \ldots, \mathbf{q}'_p$. Now we have a new $\widetilde{Q}$ matrix with all orthonormal columns. Moreover, $\widetilde{Q}R$ is still A since each new $\mathbf{q}'_j$ matches a zero row of R. This gives the promised factorization.

If A is nonsingular, $m = n = rank(A)$, so Q had no zero columns. Moreover, $Q^*A = R$ and Q necessarily unitary says R cannot have any zero entries on its main diagonal since R is necessarily invertible. Since the columns of Q form an orthonormal basis for $\mathbb{C}^m$, the upper triangular part of R is uniquely determined and the process puts lengths of nonzero vectors on the main diagonal of R. The uniqueness proof is left as an exercise. This will complete the proof. □

One final version of QR says that if we have a matrix in $\mathbb{C}^{m\times n}$ of rank r, we can always permute the columns of A to find a basis of the column space. Then we can apply QR to those r columns.

COROLLARY 7.3
Suppose $A \in \mathbb{C}_r^{m\times n}$. Then there exists a permutation matrix P such that $AP = [QR \mid M]$, where $Q^*Q = I_r$, R is r-by-r upper triangular with positive elements on its main diagonal.

We illustrate with our matrix from above.

$$AP = \left[\begin{array}{ccc|cc} 1 & 1 & 1 & 3 & 4 \\ 1 & 1 & 0 & 3 & 4 \\ 1 & 0 & 0 & 3 & 0 \\ 1 & 0 & 0 & 3 & 0 \end{array}\right]$$

$$= \left[\begin{array}{ccc|cc} \frac{1}{2} & \frac{1}{2} & \frac{\sqrt{2}}{2} & 0 & 0 \\ \frac{1}{2} & \frac{1}{2} & -\frac{\sqrt{2}}{2} & 0 & 0 \\ \frac{1}{2} & -\frac{1}{2} & 0 & 0 & 0 \\ \frac{1}{2} & -\frac{1}{2} & 0 & 0 & 0 \end{array}\right] \left[\begin{array}{ccc|cc} 2 & 1 & \frac{1}{2} & 6 & 4 \\ 0 & 1 & \frac{1}{2} & 0 & 4 \\ 0 & 0 & \frac{\sqrt{2}}{2} & 0 & 0 \\ 0 & 0 & 0 & 0 & 0 \\ 0 & 0 & 0 & 0 & 0 \end{array}\right].$$

So, what good is this QR factorization? Well, our friends in applied mathematics really like it. So do the folks in statistics. For example, suppose you have a system of linear equations $A\mathbf{x} = \mathbf{b}$. You form the so-called "normal equations" $A^*A\mathbf{x} = A^*\mathbf{b}$. Suppose you are lucky enough to get $A = QR$. Then $A^*A\mathbf{x} = R^*Q^*Q\mathbf{x} = A^*\mathbf{b}$ so $R^*R\mathbf{x} = R^*Q^*\mathbf{b}$. But R^* is invertible, so we are reduced to solving $R\mathbf{x} = Q^*\mathbf{b}$. But this is an upper triangular system that can be solved by back substitution.

There is an interesting algorithm due to **S. H. Kung** [2002] that gives a way of finding the QR factorization by simply using elementary row and column operations. We demonstrate this next.

7.3.1 Kung's Algorithm

Suppose $A \in \mathbb{C}^{m\times n}$. Then $A^* \in \mathbb{C}^{n\times m}$ and $A^*A \in \mathbb{C}^{n\times n}$. We see A^*A is square and Hermitian and its diagonal elements are real nonnegative numbers.

Suppose A has n linearly independent columns. Then A^*A is invertible. Thus we can use $\frac{n(n-1)}{2}$ pairs of transvections $(T_{ij}(c)^*, T_{ij}(c))$to "clean out" the off diagonal elements and obtain $(AE)^*(AE) = E^*A^*AE = diag[d_1, d_2, ..., d_n] = D$, a diagonal matrix with strictly positive elements. This says the columns of AE are orthogonal. Let $C = diag[\sqrt{d_1}, ..., \sqrt{d_n}]$. Then $Q = AEC^{-1}$ has orthonormal columns. Also, E is upper triangular, so E^{-1} is also as is $R = CE^{-1}$. Finally note, $A = QR$.

Let's work through an example. Suppose $A = \begin{bmatrix} i & 0 & 3 \\ 2 & 0 & 0 \\ 0 & 2-i & 0 \\ 4i & 0 & 0 \end{bmatrix}$. Then

$$A^*A = \begin{bmatrix} 21 & 0 & -3i \\ 0 & 5 & 0 \\ 3i & 0 & 9 \end{bmatrix} \text{ and } T_{31}(\tfrac{-3i}{21})A^*AT_{13}(\tfrac{3i}{21}) = \begin{bmatrix} 21 & 0 & 0 \\ 0 & 5 & 0 \\ 0 & 0 & \frac{60}{7} \end{bmatrix}.$$

Thus, $C = \begin{bmatrix} \sqrt{21} & 0 & 0 \\ 0 & \sqrt{5} & 0 \\ 0 & 0 & \sqrt{\frac{60}{7}} \end{bmatrix}$ and $E = \begin{bmatrix} 1 & 0 & \frac{3i}{21} \\ 0 & 1 & 0 \\ 0 & 0 & 1 \end{bmatrix}$. Now $Q =$

$$AEC^{-1} = \begin{bmatrix} \frac{i}{\sqrt{21}} & 0 & 2\sqrt{\frac{5}{21}} \\ \frac{2}{\sqrt{21}} & 0 & \frac{i}{\sqrt{105}} \\ 0 & \frac{2-i}{\sqrt{5}} & 0 \\ \frac{4i}{\sqrt{21}} & 0 & -\frac{2}{\sqrt{105}} \end{bmatrix} \text{ and } R = CE^{-1} = \begin{bmatrix} \sqrt{21} & 0 & -i\sqrt{\frac{3}{7}} \\ 0 & \sqrt{5} & 0 \\ 0 & 0 & 2 \end{bmatrix}.$$

The reader may verify that $A = QR$.

We finish this section with an important application to full rank factorizations.

THEOREM 7.10

Every matrix $A \in \mathbb{C}^{m\times n}$ has a full rank factorization $A = FG$, where $F^* = F^+$.

PROOF Indeed take any full rank factorization of A, say $A = FG$. Then the columns of F form a basis for $Col(A)$. Write $F = QR$. Then $A = (QR)G = Q(RG)$ is also a full rank factorization of A. Take $F_1 = Q$, $G_1 = RG$. Then $F_1^*F_1 = Q^*Q = I$, but then $F_1^+ = (F_1^*F_1)^{-1}F_1^* = F_1^*$. □

DEFINITION 7.6 *(orthogonal full rank factorizations)*

If $A = FG$ is a full rank factorization and $F^* = F^+$, call the factorization an orthogonal full rank factorization.

Exercise Set 29

1. Find an orthogonal full rank factorization for $\begin{bmatrix} 1 & 1 & 0 \\ 1 & 0 & 1 \\ 1 & 1 & 0 \\ 1 & 0 & 1 \end{bmatrix}$.

Further Reading

[Kung, 2002] Sidney H. Kung, Obtaining the QR Decomposition by Pairs of Row and Column Operations, The College Mathematics Journal, Vol. 33, No. 4, September, (2002), 320–321.

7.3.2 MATLAB Moment

7.3.2.1 The QR Factorization

If A is an m-by-n complex matrix, then A can be written $A = QR$, where Q is unitary and R is upper triangular the same size as A. Sometimes people use a permutation matrix P to permute columns of A so that the magnitudes of the diagonal of R appear in decreasing order. Then $AP = QR$.

MATLAB offers four versions of the QR-factorization, full size or economy size, with or without column permutations. The command for the full size is

$$[Q, R] = qr(A).$$

If a portion of R is all zeros, part of Q is not necessary so the economy QR is obtained as

$$[Q, R] = qr(A, 0).$$

To get the diagonal of R lined up in decreasing magnitudes, we use

$$[Q, R, P] = qr(A).$$

Let's look at some examples.

```
>> A=pascal(4)
A =
    1   1   1    1
    1   2   3    4
    1   3   6   10
    1   4  10   20
>> [Q,R]=qr(A)
Q =
   -0.5000    0.6708    0.5000    0.2236
   -0.5000    0.2236   -0.5000   -0.6708
   -0.5000   -0.2236   -0.5000    0.6708
   -0.5000   -0.6708    0.5000   -0.2236
R =
   -2.000    -5.000    -10.000   -17.5000
     0       -2.2361   -6.7082   -14.0872
     0         0        1.000      3.5000
     0         0          0       -0.2236
>> format rat
>>[Q,R]=qr(A)
Q =
   -1/2     646/963      1/2    646/2889
   -1/2     646/2889    -1/2   -646/963
   -1/2    -646/2889    -1/2    646/963
   -1/2    -646/963      1/2   -646/2889
R =
   -2         -5            -10          -35/2
    0     -2889/1292    -2207/329     -4522/321
    0          0             1            7/2
    0          0             0        -646/2889
To continue the example,
>>[Q,R,P]=qr(A)
Q =
   -202/4593   -1414/2967    1125/1339   -583/2286
   -263/1495   -1190/1781   -577/3291    1787/2548
   -263/598    -217/502     -501/1085   -621/974
   -263/299     699/1871     975/4354    320/1673
R =
   -4693/202   -1837/351    -1273/827   -1837/153
       0       -2192/1357   -846/703    -1656/1237
       0           0         357/836    -281/1295
       0           0            0       -301/4721
```

P =

0	0	1	0
0	1	0	0
0	0	0	1
1	0	0	0

Note that as a freebee, we get an orthonormal basis for the column space of A, namely, the columns of Q.

7.4 A Fundamental Theorem of Linear Algebra

We have seen already how to associate subspaces to a matrix. The rank-plus-nullity theorem gives us an important formula relating the dimension of the null space and the dimension of the column space of a matrix. In this section, we develop further the connections between the fundamental subspaces of a matrix. We continue the picture of a matrix transforming vectors to vectors. More specifically, let A be an m-by-n matrix of rank r. Then A transforms vectors from $\mathbb{C}^n$ to vectors in $\mathbb{C}^m$ by the act of multiplication. A similar view applies to $\overline{A}$, A^*, A^T, and A^+. We then have the following visual representation of what is going on.

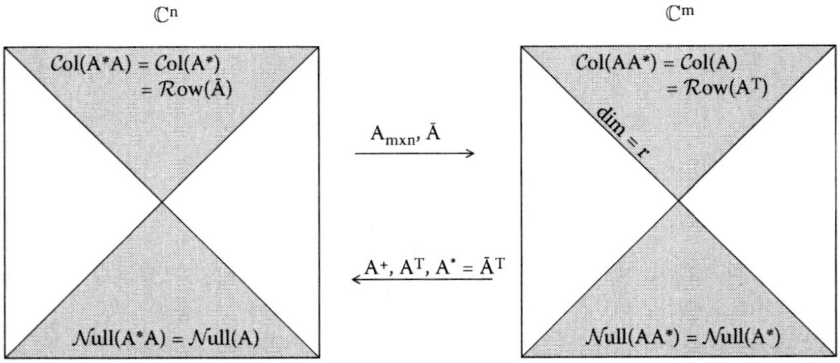

Figure 7.2: The fundamental subspaces of a matrix.

The question now is, how do the other subspaces fit into this picture? To answer this question we will use the inner product and orthogonality ideas. Recall $\langle \mathbf{x} \mid \mathbf{y} \rangle = \mathbf{y}^*\mathbf{x} = \sum_{i=1}^{n} x_i \overline{y_i}$ and $M^\perp$ denote the set of all vectors that are orthogonal to each and every vector of M. Recall that $M = M^{\perp\perp}$ and $M \oplus M^{\perp} = \mathbb{C}^n$ for all subspaces M of $\mathbb{C}^n$. Also, if $N \subseteq M$, then $M^{\perp} \subseteq N^{\perp}$.

We will make repeated use of a formula that is remarkably simple to prove.

THEOREM 7.11

Let A be in $\mathbb{C}^{m\times n}$, $\mathbf{x} \in \mathbb{C}^n$, $\mathbf{y} \in \mathbb{C}^m$. Then

$$\langle A\mathbf{x} \mid \mathbf{y}\rangle = \left\langle \mathbf{x} \mid A^*\mathbf{y}\right\rangle .$$

PROOF We compute

$$\begin{aligned}\langle \mathbf{x} \mid A^*\mathbf{y}\rangle &= (A^*\mathbf{y})^*\mathbf{x} = (\mathbf{y}^*A^{**})\mathbf{x} \\ &= (\mathbf{y}^*A)\mathbf{x} = \mathbf{y}^*(A\mathbf{x}) = \langle A\mathbf{x} \mid \mathbf{y}\rangle .\end{aligned}$$

□

Next, we develop another simple fact — one with important consequences. First, we need some notation. We know what it means for a matrix to multiply a single vector. We can extend this idea to a matrix, multiplying a whole collection of vectors.

DEFINITION 7.7 *Let $A \in \mathbb{C}^{m\times n}$ and let M be a subset of $\mathbb{C}^n$*
Let $A(M) = \{A\mathbf{x} \mid \mathbf{x} \in M\}$. Naturally, $A(M) \subseteq \mathbb{C}^m$.

This is not such a wild idea since $A(\mathbb{C}^n) = Col(A)$ and $A(\mathcal{N}ull(A)) = \left\{\overrightarrow{0}\right\}$. Now for that simple but useful fact.

THEOREM 7.12

Let $A \in \mathbb{C}^{m\times n}$, $M \subseteq \mathbb{C}^n$, and $N \subseteq \mathbb{C}^m$. Then

$$A(M) \subseteq N \text{ iff } A^*(N^{\perp}) \subseteq M^{\perp}.$$

PROOF We prove the implication from left to right. Suppose $A(M) \subseteq N$. We need to show $A^*(N^{\perp}) \subseteq M^{\perp}$. Take a vector $\mathbf{y}$ in $A^*(N^{\perp})$. We must argue that $\mathbf{y} \in M^{\perp}$. That is, we must show $\mathbf{y} \perp \mathbf{m}$ for all vectors $\mathbf{m}$ in M. Fix a vector $\mathbf{m} \in M$. We will compute $\langle \mathbf{y} \mid \mathbf{m}\rangle$ and hope to get 0. But $\mathbf{y} = A^*\mathbf{x}$ for some $\mathbf{x} \in N^{\perp}$ so $\langle \mathbf{y} \mid \mathbf{m}\rangle = \langle A^*\mathbf{x} \mid \mathbf{m}\rangle = \langle \mathbf{x} \mid A\mathbf{m}\rangle$. But $\mathbf{x} \in N^{\perp}$, and $A\mathbf{m} \in N$, so this inner product is zero, as we hoped. To prove the converse, just apply the result just proved to A^*. Conclude that $A^{**}(M^{\perp\perp}) \subseteq N^{\perp\perp}$ (i.e., $A(M) \subseteq N$). This completes the proof. □

Now we get to the really interesting results.

THEOREM 7.13
Let $A \in \mathbb{C}^{m \times n}$. Then

1. $\mathcal{N}ull(A) = \mathcal{C}ol(A^*)^{\perp}$
2. $\mathcal{N}ull(A)^{\perp} = \mathcal{C}ol(A^*)$
3. $\mathcal{N}ull(A^*) = \mathcal{C}ol(A)^{\perp}$
4. $\mathcal{N}ull(A^*)^{\perp} = \mathcal{C}ol(A)$.

PROOF It is pretty clear that if we can prove any one of the four statements above by replacing A by A^* and taking "perps," we get all the others. We will focus on (1). Clearly $A^*(\mathcal{N}ull(A^*)) \subseteq (\overrightarrow{0})$. So, by 7.12, $A((\overrightarrow{0})^{\perp}) \subseteq \mathcal{N}ull(A^*)^{\perp}$. But $(\overrightarrow{0})^{\perp} = \mathbb{C}^n$; so this says $A(\mathbb{C}^n) \subseteq \mathcal{N}ull(A^*)^{\perp}$, However, $A(\mathbb{C}^n) = \mathcal{C}ol(A)$, as we have already noted, so we conclude $\mathcal{C}ol(A) \subseteq \mathcal{N}ull(A^*)^{\perp}$. We would like the other inclusion as well. It is a triviality that $A(\mathbb{C}^n) \subseteq A(\mathbb{C}^n)$. But look at this the right way: $A(\mathbb{C}^n) \subseteq A(\mathbb{C}^n) = \mathcal{C}ol(A)$. Then again, appealing to 7.12, $A^*(\mathcal{C}ol(A)^{\perp}) \subseteq \mathbb{C}^{n\perp} = (\overrightarrow{0})$. Thus, $\mathcal{C}ol(A)^{\perp} \subseteq \mathcal{N}ull(A^*)$, which gets us $\mathcal{N}ull(A^*)^{\perp} \subseteq \mathcal{C}ol(A)$. ▯

COROLLARY 7.4
Let $A \in \mathbb{C}^{m \times n}$. Then

1. $\mathcal{N}ull(\overline{A}) = \mathcal{C}ol(A^T)^{\perp} = \mathcal{R}ow(A)^{\perp}$
2. $\mathcal{C}ol(\overline{A}) = \mathcal{N}ull(A^T)^{\perp}$
3. $\mathcal{N}ull(\overline{A})^{\perp} = \mathcal{R}ow(A)$
4. $Null(A^T) = Col(\overline{A})^{\perp}$.

We now summarize with a theorem and a picture.

THEOREM 7.14 (fundamental theorem of linear algebra)
Let $A \in \mathbb{C}_r^{m \times n}$. Then

1. $\dim(\mathcal{C}ol(A)) = r$.
2. $\dim(\mathcal{N}ull(A)) = n - r$.
3. $\dim(\mathcal{C}ol(A^*)) = r$.
4. $\dim(\mathcal{N}ull(A^*)) = m - r$.

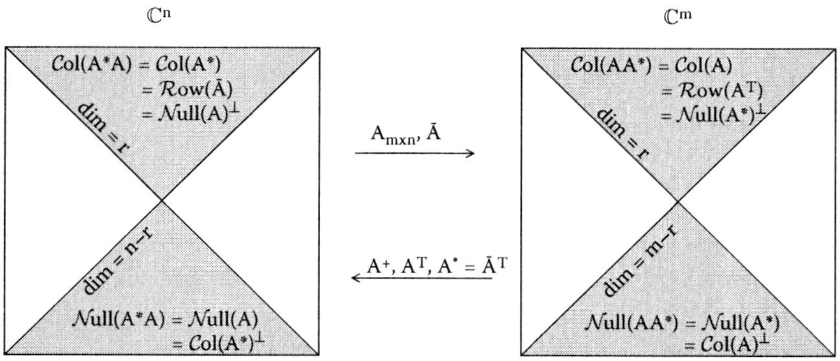

Figure 7.3: Fundamental theorem of linear algebra.

5. *$\mathcal{N}ull(A)$ is the orthogonal complement of $\mathcal{C}ol(A^*)$.*
6. *$\mathcal{N}ull(A^*)$ is the orthogonal complement of $\mathcal{C}ol(A)$.*
7. *$\mathcal{N}ull(\overline{A})$ is the orthogonal complement of $\mathcal{R}ow(A)$.*

There is a connection to the problem of solving systems of linear equations.

COROLLARY 7.5
Consider the system of linear equation $A\mathbf{x} = \mathbf{b}$. Then the following are equivalent:

1. *$A\mathbf{x} = \mathbf{b}$ has a solution.*
2. *$\mathbf{b} \in \mathcal{C}ol(A)$ (i.e., $\mathbf{b}$ is a linear combination of the columns of A).*
3. *$A^*\mathbf{y} = \overrightarrow{0}$ implies $\mathbf{b}^*\mathbf{y} = 0$.*
4. *$\mathbf{b}$ is orthogonal to every vector that is orthogonal to all the columns of A.*

Exercise Set 30

1. If $A \in \mathbb{C}^{m\times n}$, $B \in \mathbb{C}^{n\times p}$, and $C \in \mathbb{C}^{n\times p}$, then $A^*AB = A^*AC$ if and only if $AB = AC$.
2. Prove that $\mathcal{N}ull(A^*A) = \mathcal{N}ull(A)$.

3. Prove that $Col(A^*A) = Col(A^*)$.

4. Fill in the details of the proof of Corollary 7.4.

5. Fill in the details of the proof of Theorem 7.14.

Further Reading

[Strang, 1988] Gilbert Strang, *Linear Algebra and its Applications,* 3rd Edition, Harcourt Brace Jovanovich Publishers, San Diego, (1988).

[Strang, 1993] Gilbert Strang, The Fundamental Theorem of Linear Algebra, The American Mathematical Monthly, Vol. 100, No. 9, (1993), 848–855.

[Strang, 2003] Gilbert Strang, *Introduction to Linear Algebra,* 3rd Edition, Wellesley-Cambridge Press, Wellesley, MA, (2003).

7.5 Minimum Norm Solutions

We have seen that a consistent system of linear equations $A\mathbf{x} = \mathbf{b}$ can have many solutions; indeed, there can be infinitely many solutions and they form an affine subspace.

Now we are in a position to ask, among all of these solutions, is there a shortest one? That is, is there a solution of minimum norm? The first question is, which norm? For this section, we choose our familiar Euclidean norm, $\|\mathbf{x}\|_2 = tr(\mathbf{x}^*\mathbf{x})^{\frac{1}{2}}$.

DEFINITION 7.8 *(minimum norm)*
We say $\mathbf{x}_0$ *is a minimum norm solution of* $A\mathbf{x} = \mathbf{b}$ *iff* $\mathbf{x}_0$ *is a solution and* $\|\mathbf{x}_0\| \leq \|\mathbf{x}\|$ *for all solutions* $\mathbf{x}$ *of* $A\mathbf{x} = \mathbf{b}$.

Recall that 1-inverses are the "equation solvers." We have established the consistency condition: $A\mathbf{x} = \mathbf{b}$ is consistent iff $AG\mathbf{b} = \mathbf{b}$ for some $G \in A\{1\}$. In this case, all solutions can be described by $\mathbf{x} = G\mathbf{b} + (I - GA)\mathbf{z}$, where $\mathbf{z}$ is

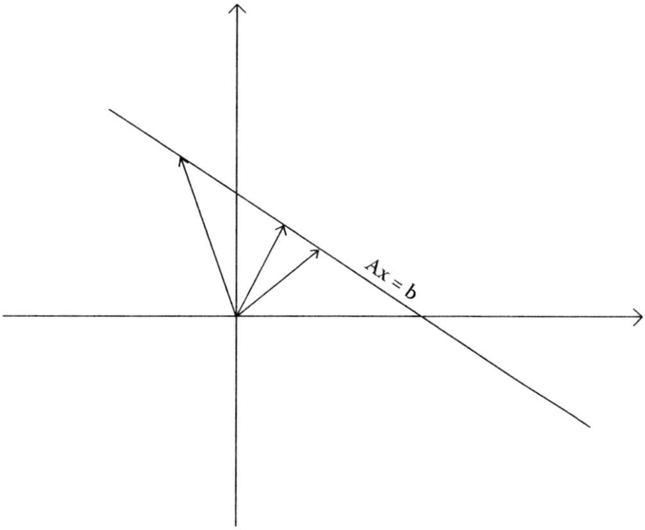

Figure 7.4: Vector solution of various length (norm).

arbitrary in $\mathbb{C}^m$. Of course, we could use A^+ for G. The first thing we establish is that there is, in fact, a minimum norm solution to any consistent system of linear equations and it is unique.

THEOREM 7.15

Suppose $A\mathbf{x} = \mathbf{b}$ is a consistent system of linear equations (i.e., $\mathbf{b} \in Col(A)$). Then there exists a unique solution of $A\mathbf{x} = \mathbf{b}$ of minimum norm. In fact, it lies in $Col(A^*)$.

PROOF For existence, choose $\mathbf{x}_0 = A^+\mathbf{b}$. Take any solution $\mathbf{x}$ of $A\mathbf{x} = \mathbf{b}$. Then $\mathbf{x} = A^+\mathbf{b} + (I - A^+A)\mathbf{z}$ for some $\mathbf{z}$. Thus, $\|\mathbf{x}\|^2 = \langle \mathbf{x} \mid \mathbf{x} \rangle = \langle A^+\mathbf{b} + (I - A^+A)\mathbf{z} \mid A^+\mathbf{b} + (I - A^+A)\mathbf{z} \rangle = \langle A^+\mathbf{b} \mid A^+\mathbf{b} \rangle + \langle A^+\mathbf{b} \mid (I - A^+A)\mathbf{z} \rangle + \langle (I - A^+A)\mathbf{z} \mid A^+\mathbf{b} \rangle + \langle (I - A^+A)\mathbf{z} \mid (I - A^+A)\mathbf{z} \rangle \; \|A^+\mathbf{b}\|^2 + \langle (I - A^+A)A^+\mathbf{b} \mid \mathbf{z} \rangle + \langle \mathbf{z} \mid (I - A^+A)A^+\mathbf{b} \rangle + \|(I - A^+A)\mathbf{z}\|^2 = \|A^+\mathbf{b}\|^2 + 0 + 0 + \|(I - A^+A)\mathbf{z}\|^2 \geq \|A^+\mathbf{b}\|^2$, with equality holding iff $\|(I - A^+A)\mathbf{z}\| = 0$ iff $(I - A^+A)\mathbf{z} = \overrightarrow{0}$ iff $\mathbf{x} = A^+\mathbf{b}$. Thus, $A^+\mathbf{b}$ is the unique minimum norm solution. Since $A^+ = A^*(AA^*)^+$, we have $A^+\mathbf{b} \in Col(A^*)$. □

Once again, we see the prominence of the Moore-Penrose inverse. It turns out that the minimum norm issue is actually intimately connected with $\{1,4\}$-inverses. Recall, $G \in A\{1, 4\}$ iff $AGA = A$ and $(GA)^* = GA$.

THEOREM 7.16
Let $G \in A\{1,4\}$. *Then*

1. $GA = A^+A$

2. $(I - GA)^* = (I - GA) = (I - GA)^2$

3. $A(I - GA)^* = \mathbb{O}$

4. $(I - GA)A^* = \mathbb{O}$.

PROOF Compute that $GA = GAA^+A = (GA)^*(A^+A)^* = A^*G^*A^*A^{+*} = (AGA)^*A^{+*} = A^*A^{+*} = (A^+A)^* = A^+A$. The other claims are now easy and left to the reader. □

So, {1,4}-inverses G have the property that no matter which one you choose, GA is always the same, namely A^+A. In fact, more can be said.

COROLLARY 7.6
$G \in A\{1,4\}$ *iff* $GA = A^+A$. *In particular, if* $G \in A\{1,4\}$, *then* $G\mathbf{b} = A^+\mathbf{b}$ *for any* $\mathbf{b} \in Col(A)$.

Thus {1,4}-inverses are characterized by giving the minimum norm solutions.

THEOREM 7.17
Suppose $A\mathbf{x} = \mathbf{b}$ *is consistent and* $G \in A\{1,4\}$. *Then* $G\mathbf{b}$ *is the unique solution of minimum norm. Conversely, suppose* $H \in \mathbb{C}^{n\times m}$ *and, whenever* $A\mathbf{x} = \mathbf{b}$ *is consistent,* $AH\mathbf{b} = \mathbf{b}$ *and* $\|H\mathbf{b}\| < \|\mathbf{z}\|$ *for all solutions* $\mathbf{z}$ *other than* $H\mathbf{b}$; *then* $H \in A\{1,4\}$.

PROOF The details are left to the reader. □

Exercise Set 31

1. Suppose $G \in A\{1,4\}$. Argue that $A\{1,4\} = \{H \mid HA = GA\}$.

2. Argue that $A\{1,4\} = \{H \mid HA = A^+A\}$.

3. Suppose $G \in A\{1,4\}$. Argue that $A\{1,4\} = \{G + W(I - AG) \mid W$ arbitrary$\}$.

4. Suppose $G \in A\{1,3\}$ and $H \in A\{1,4\}$. Argue that $HAG = A^+$.

5. Let $\mathbf{u}$ and $\mathbf{v}$ be in $\mathcal{C}ol(A^*)$ with $A\mathbf{u} = A\mathbf{v}$. Prove that $\mathbf{u} = \mathbf{v}$.

7.6 Least Squares

Finally, the time has come to face up to a reality raised in the very first chapter. A system of linear equations $A\mathbf{x} = \mathbf{b}$ may not have any solutions at all. The realities of life sometime require us to come up with a "solution" even in this case. Again, we face a minimization problem. Once again, we use the Euclidean norm in this section. Suppose $A\mathbf{x} = \mathbf{b}$ is inconsistent. It seems reasonable to seek out a vector in the column space of A that is closest to $\mathbf{b}$. In other words, if $\mathbf{b} \notin \mathcal{C}ol(A)$, the vector $r(\mathbf{x}) = A\mathbf{x} - \mathbf{b}$, which we call the *residual vector*, is never zero. We shall try to minimize the length of this vector in the Euclidean norm. Statisticians do this all the time under the name "least squares."

DEFINITION 7.9 *(least squares solutions)*
A vector $\mathbf{x}_0$ is called a least squares solution of the system of linear equations $A\mathbf{x} = \mathbf{b}$ iff $\|A\mathbf{x}_0 - \mathbf{b}\| \leq \|A\mathbf{x} - \mathbf{b}\|$ for all vectors $\mathbf{x}$.

Remarkably, the connection here is with $\{1,3\}$-inverses.

THEOREM 7.18
Let $A \in \mathbb{C}^{m \times n}$ and $G \in A\{1,3\}$. Then

1. $AG = AA^+$

2. $(I - AG)^* = I - AG = (I - AG)^2$

3. $(I - AG)^*A = \mathbb{O}$

4. $A^*(I - AG) = \mathbb{O}$.

PROOF For (1), we compute $AG = AA^+AG = AA^{+*}(AG)^* = A^{+*} \times A^*G^*A^* = A^{+*}(AGA)^* = A^{+*}A^* = (AA^+)^* = AA^+$. The other claims are now clear. □

COROLLARY 7.7
$G \in A\{1,3\}$ *iff* $AG = AA^+$.

THEOREM 7.19
Suppose $G \in A\{1,3\}$. Then $\mathbf{x}_0 = G\mathbf{b}$ is a least squares solution of the linear system $A\mathbf{x} = \mathbf{b}$.

PROOF Suppose $G \in A\{1,3\}$. We use the old add-and-subtract trick. $\|A\mathbf{x} - \mathbf{b}\|^2 = \|A\mathbf{x} - AG\mathbf{b} - \mathbf{b} + AG\mathbf{b}\|^2 = \left\|A(\mathbf{x} - A^+\mathbf{b})) + (AA^+\mathbf{b} - \mathbf{b})\right\|^2 = \left\langle A(\mathbf{x} - A^+\mathbf{b})) + (AA^+\mathbf{b} - \mathbf{b}) \mid A(\mathbf{x} - A^+\mathbf{b})) + (AA^+\mathbf{b} - \mathbf{b})\right\rangle = \langle A(\mathbf{x} - A^+\mathbf{b})) \mid A(\mathbf{x} - A^+\mathbf{b}))\rangle + \left\langle A(\mathbf{x} - A^+\mathbf{b}) \mid (AA^+\mathbf{b} - \mathbf{b})\right\rangle + \langle(AA^+\mathbf{b} - \mathbf{b}) \mid A(\mathbf{x} - A^+\mathbf{b}))\rangle + \left\langle(AA^+\mathbf{b} - \mathbf{b}) \mid (AA^+\mathbf{b} - \mathbf{b})\right\rangle = \left\|A(\mathbf{x} - A^+\mathbf{b}))\right\|^2 + \|(AA^+\mathbf{b} - \mathbf{b})\|^2 \geq \left\|(AA^+\mathbf{b} - \mathbf{b})\right\|^2 = \|(AG\mathbf{b} - \mathbf{b})\|^2$, and equality holds iff $\left\|A(\mathbf{x} - A^+\mathbf{b}))\right\|^2 = 0$ iff $\mathbf{x} - A^+\mathbf{b} \in \mathcal{N}ull(A)$. ▯

THEOREM 7.20
Let G be any element of $A\{1,3\}$. Then $\mathbf{x}_1$ is a least squares solution of the linear system $A\mathbf{x} = \mathbf{b}$ iff $\|A\mathbf{x}_1 - \mathbf{b}\| = \|\mathbf{b} - AG\mathbf{b}\|$.

PROOF Suppose $\|A\mathbf{x}_1 - \mathbf{b}\| = \|\mathbf{b} - A\mathbf{x}_0\|$, where $\mathbf{x}_0 = G\mathbf{b}$. By our theorem above, $\|\mathbf{b} - A\mathbf{x}_0\| \leq \|\mathbf{b} - A\mathbf{x}\|$ for all $\mathbf{x}$, so $\|A\mathbf{x}_1 - \mathbf{b}\| = \|\mathbf{b} - A\mathbf{x}_0\| \leq \|\mathbf{b} - A\mathbf{x}\|$ for all $\mathbf{x}$, making $\mathbf{x}_1$ a least squares solution. Conversely, suppose $\mathbf{x}_1$ is a least squares solution of $A\mathbf{x} = \mathbf{b}$. Then, by definition, $\|A\mathbf{x}_1 - \mathbf{b}\| \leq \|A\mathbf{x} - \mathbf{b}\|$ for all choices of $\mathbf{x}$. Choose $\mathbf{x} = G\mathbf{b}$. Then $\|A\mathbf{x}_1 - \mathbf{b}\| \leq \|AG\mathbf{b} - \mathbf{b}\|$. But $G\mathbf{b}$ is a least squares solution, so $\|AG\mathbf{b} - \mathbf{b}\| \leq \|A\mathbf{x} - \mathbf{b}\|$ for all $\mathbf{x}$, so if we take $\mathbf{x}_1$ for $\mathbf{x}$, $\|AG\mathbf{b} - \mathbf{b}\| \leq \|A\mathbf{x}_1 - \mathbf{b}\|$. Hence, equality must hold. ▯

THEOREM 7.21
Let G be any element of $A\{1,3\}$. Then, $\mathbf{x}_0$ is a least squares solution of $A\mathbf{x} = \mathbf{b}$ iff $A\mathbf{x}_0 = AG\mathbf{b} = AA^+\mathbf{b}$.

PROOF Note $AG(AG\mathbf{b}) = (AGA)(AG\mathbf{b}) = AG\mathbf{b}$, so the system on the right is consistent. Moreover, $\|A\mathbf{x}_0 - \mathbf{b}\| = \|AG\mathbf{b} - \mathbf{b}\|$, so $\mathbf{x}_0$ is a least squares solution of the left-hand system. Conversely, suppose $\mathbf{x}_0$ is a least squares solution. Then $\|A\mathbf{x}_0 - \mathbf{b}\| = \|\mathbf{b} - AG\mathbf{b}\|$. But $\|A\mathbf{x}_0 - \mathbf{b}\|^2 = \|A\mathbf{x}_0 - AG\mathbf{b} + AG\mathbf{b} - \mathbf{b}\|^2 = \|A\mathbf{x}_0 - AG\mathbf{b}\|^2 + \|\mathbf{b} - AG\mathbf{b}\|^2$, which says $\|A\mathbf{x}_0 - AG\mathbf{b}\| = 0$. Thus, $A\mathbf{x}_0 = AG\mathbf{b}$. ▯

As we have noted, a linear system may have many least squares solutions. However, we can describe them all.

THEOREM 7.22
Let G be any element of $A\{1,3\}$. Then all least squares solutions of $A\mathbf{x} = \mathbf{b}$ are of the form $G\mathbf{b} + (I - GA)\mathbf{z}$ for $\mathbf{z}$ arbitrary.

PROOF Let $\mathbf{y} = G\mathbf{b} + (I - GA)\mathbf{z}$ and compute that $A(I - GA)\mathbf{y} = \overrightarrow{0}$ so $A\mathbf{y} = AG\mathbf{b}$. Hence, $\mathbf{y}$ is a least squares solution. Conversely, suppose $\mathbf{x}$ is a least squares solution. Then $A\mathbf{x} = AG\mathbf{b}$ so $\overrightarrow{0} = -GA(\mathbf{x} - G\mathbf{b})$ whence $\mathbf{x} = \mathbf{x} - GA(\mathbf{x} - G\mathbf{b}) = G\mathbf{b} + \mathbf{x} - G\mathbf{b} - GA(\mathbf{x} - G\mathbf{b}) = G\mathbf{b} + (I - GA)(\mathbf{x} - G\mathbf{b})$. Take $\mathbf{z} = \mathbf{x} - G\mathbf{b}$ to get the desired form. ▯

It is nice that we can characterize when a least squares solution is unique. This often happens in statistical examples.

THEOREM 7.23
Suppose $A \in \mathbb{C}^{m \times n}$. The system of linear equations $A\mathbf{x} = \mathbf{b}$ has a unique least squares solution iff $rank(A) = n$.

PROOF Evidently, the least squares solution is unique iff $I - GA = \mathbb{O}$ iff $GA = I$. This says A has a left inverse, which we know to be true iff $rank(A) = n$ (i.e., A has full column rank). ▯

Finally, we can put two ideas together. There can be many least squares solutions to an inconsistent system of linear equations. We may ask, among all of these, is there one of minimum norm? The answer is very nice indeed.

THEOREM 7.24 (Penrose)
Among the least squares solutions of $A\mathbf{x} = \mathbf{b}$, $A^+\mathbf{b}$ is the one of minimum norm. Moreover, if G has the the property that $G\mathbf{b}$ is the minimum norm least squares solution for all $\mathbf{b}$, then $G = A^+$.

PROOF The proof is left to the reader. ▯

Isn't the Moore-Penrose inverse great? When $A\mathbf{x} = \mathbf{b}$ has a solution, $A^+\mathbf{b}$ is the solution of minimum norm. When $A\mathbf{x} = \mathbf{b}$ does not have a solution, $A^+\mathbf{b}$ gives the least squares solution of minimum norm. You cannot lose computing $A^+\mathbf{b}$!

We end this section with an example. Consider the linear system

$$\begin{cases} x_1 + x_2 = 1 \\ x_1 - x_2 = 0 \, . \\ \quad\;\; x_2 = \frac{3}{4} \end{cases}$$

In matrix notation, this system is expressed as

$$A\mathbf{x} = \begin{bmatrix} 1 & 1 \\ 1 & -1 \\ 0 & 1 \end{bmatrix} \begin{bmatrix} x_1 \\ x_2 \end{bmatrix} = \begin{bmatrix} 1 \\ 0 \\ \frac{3}{4} \end{bmatrix} = \mathbf{b}.$$

Evidently, A has rank 2, so a full rank factorization trivially is $A = FG = \begin{bmatrix} 1 & 1 \\ 1 & -1 \\ 0 & 1 \end{bmatrix} \begin{bmatrix} 1 & 0 \\ 0 & 1 \end{bmatrix}$ so $A^+ = (F^*F)^{-1}F^* = \begin{bmatrix} 2 & 0 \\ 0 & 3 \end{bmatrix}^{-1} \begin{bmatrix} 1 & 1 & 0 \\ 1 & -1 & 0 \end{bmatrix} = \begin{bmatrix} \frac{1}{2} & \frac{1}{2} & 0 \\ \frac{1}{3} & -\frac{1}{3} & \frac{1}{3} \end{bmatrix}$. We compute $AA^+\mathbf{b} = \begin{bmatrix} 1 & 1 \\ 1 & -1 \\ 0 & 1 \end{bmatrix} \begin{bmatrix} \frac{1}{2} & \frac{1}{2} & 0 \\ \frac{1}{3} & -\frac{1}{3} & \frac{1}{3} \end{bmatrix} \begin{bmatrix} 1 \\ 0 \\ \frac{3}{4} \end{bmatrix} = \begin{bmatrix} \frac{4}{6} & \frac{1}{6} & \frac{2}{6} \\ \frac{1}{6} & \frac{4}{6} & \frac{-2}{6} \\ \frac{2}{6} & \frac{-2}{6} & \frac{2}{6} \end{bmatrix} \begin{bmatrix} 1 \\ 0 \\ \frac{3}{4} \end{bmatrix} = \begin{bmatrix} \frac{11}{12} \\ \frac{-1}{6} \\ \frac{7}{12} \end{bmatrix} \neq \mathbf{b} = \begin{bmatrix} 1 \\ 0 \\ \frac{3}{4} \end{bmatrix}$. Thus, the system is inconsistent. The best approximate solution is $\mathbf{x}_0 = A^+\mathbf{b} = \begin{bmatrix} \frac{1}{2} \\ \frac{7}{12} \end{bmatrix}$. Note that one can compute a measure of error: $\|A\mathbf{x}_0 - \mathbf{b}\| = \left\|AA^+\mathbf{b} - \mathbf{b}\right\| = \left\| \begin{bmatrix} \frac{11}{12} \\ \frac{-1}{6} \\ \frac{7}{12} \end{bmatrix} - \begin{bmatrix} 1 \\ 0 \\ \frac{3}{4} \end{bmatrix} \right\| = \left[\left(\frac{1}{12}\right)^2 + \left(\frac{2}{12}\right)^2 + \left(\frac{2}{12}\right)^2\right]^{\frac{1}{2}} = \left[\frac{9}{144}\right]^{\frac{1}{2}} = \frac{3}{12} = \frac{1}{4}$.

Note that the error vector is given by $E = [I - A^+A]\mathbf{b}$.

We have chosen this example so we can draw a picture to try to gain some intuition as to what is going on here.

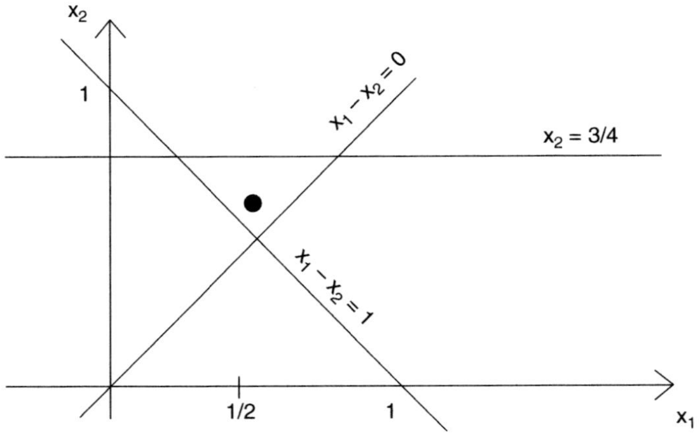

Figure 7.5: A least squares solution.

The best approximate solution is $(\frac{1}{2}, \frac{7}{12})$.

Exercise Set 32

1. Consider the system $\begin{bmatrix} 1 & 0 & 2 \\ 0 & 1 & -2 \\ 2 & 2 & 0 \\ 1 & 2 & -2 \end{bmatrix} \begin{bmatrix} x_1 \\ x_2 \\ x_3 \end{bmatrix} = \begin{bmatrix} -1 \\ -1 \\ 2 \\ 1 \end{bmatrix}$. Verify that this system is inconsistent. Find all least squares solutions. (*Hint:* There will be infinitely many.)

2. We have previously seen that $G = A^{+} + (I - A^{+}A)W + V(I - AA^{+})$ is a 1-inverse of A for any choice of V and W of appropriate size. If you did not complete exercise 9 on page 205, do it now. Argue that choosing $V = \mathbb{O}$ makes G a 3-inverse as well. If instead we choose $W = \mathbb{O}$, argue that we get a 4-inverse as well.

3. Argue that if $G \in AA^{*}\{1, 2\}$, then $A^{*}G$ is a $\{1, 3\}$-inverse of A.

4. Argue that if $H \in A^{*}A\{1, 2\}$, then HA^{*} is a $\{1, 4\}$-inverse of A.

5. Let $G \in A\{1, 3, 4\}$. Argue that $A\{1, 3, 4\} = \{G + (I - GA)W(I - AG) \mid W$ arbitrary of appropriate size$\}$.

6. Let $H \in A\{1, 2, 4\}$. Argue that $A\{1, 2, 4\} = \{H + HV(I - AH) \mid V$ arbitrary of appropriate size$\}$.

7. Let $K \in A\{1, 2, 3\}$. Argue that $A\{1, 2, 3\} = \{K + (I - KA)WK \mid W$ arbitrary of appropriate size$\}$.

8. Argue that $\mathbf{x}_0$ is the minimum norm least squares solution of $A\mathbf{x} = \mathbf{b}$ iff a) $\|A\mathbf{x}_0 - \mathbf{b}\| \leq \|A\mathbf{x} - \mathbf{b}\|$ and b)$\|\mathbf{x}_0\| < \|\mathbf{x}\|$ for any $\mathbf{x} \neq \mathbf{x}_0$.

9. Suppose $A = FG$ is a full rank factorization of A. Then $A\mathbf{x} = \mathbf{b}$ iff $FG\mathbf{x} = \mathbf{b}$ iff a) $F\mathbf{y} = \mathbf{b}$ and b) $G\mathbf{x} = \mathbf{y}$. Argue that $\mathbf{y} = F^{+}\mathbf{b} = (F^{*}F)^{-1}F^{*}\mathbf{b}$ and $\|F\mathbf{y} - \mathbf{b}\|$ is minimal. Also argue that $G\mathbf{x} = F^{+}\mathbf{b}$ is always consistent and has minimum norm.

10. Argue that $\mathbf{x}_0$ is a least squares solution of $A\mathbf{x} = \mathbf{b}$ iff $\mathbf{x}_0$ is a solution to the always consistent (prove this) system $A^{*}A\mathbf{x} = A^{*}\mathbf{b}$. These latter are often called the *normal equations*. Prove this latter is equivalent to $A\mathbf{x} - \mathbf{b} \in \mathcal{N}ull(A^{*})$.

11. Suppose $A = FG$ is a full rank factorization. Then the normal equations are equivalent to $F^{*}A\mathbf{x} = F^{*}\mathbf{b}$.

Further Reading

[Albert, 1972] A. Albert, *Regression and the Moore-Penrose Pseudoinverse,* Academic Press, New York, NY, (1972).

[Björck, 1967] A. Björck, Solving Linear Least Squares Problems by Gram-Schmidt Orthogonalization, Nordisk Tidskr. Informations-Behandling, Vol. 7, (1967), 1–21.

Chapter 8

Projections

idempotent, self-adjoint, projection, the approximation problem

8.1 Orthogonal Projections

We begin with some geometric motivation. Suppose we have two nonzero vectors **x** and **y** and we hold a flashlight directly over the tip of **y**. We want to determine the shadow **y** casts on **x**. The first thing we note is

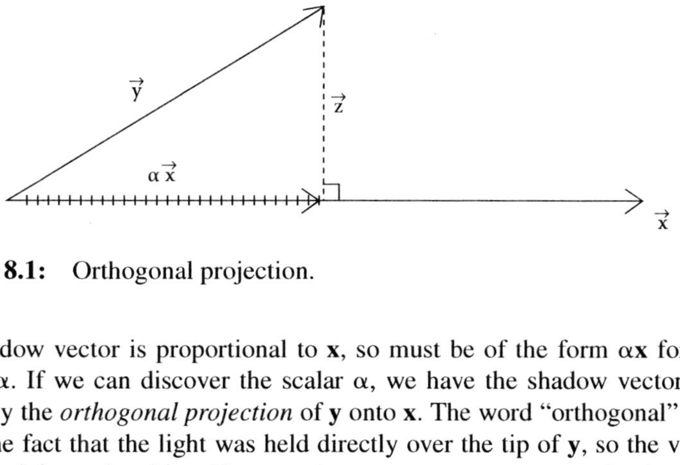

Figure 8.1: Orthogonal projection.

the shadow vector is proportional to **x**, so must be of the form $\alpha\mathbf{x}$ for some scalar α. If we can discover the scalar α, we have the shadow vector, more formally the *orthogonal projection* of **y** onto **x**. The word "orthogonal" comes from the fact that the light was held directly over the tip of **y**, so the vector **z** forms a right angle with **x**. Note, $\alpha\mathbf{x} + \mathbf{z} = \mathbf{y}$, so $\mathbf{z} = \mathbf{y} - \alpha\mathbf{x}$ and $\mathbf{z} \perp \mathbf{x}$. Thus, $0 = <\mathbf{z}|\mathbf{x}> = <\mathbf{y} - \alpha\mathbf{x}|\mathbf{x}> = <\mathbf{y}|\mathbf{x}> - \alpha <\mathbf{x}|\mathbf{x}>$ so $\alpha = \dfrac{<\mathbf{y}|\mathbf{x}>}{<\mathbf{x}|\mathbf{x}>}$. This is great! It gives a formula to compute the shadow vector.

DEFINITION 8.1 *Let* $\mathbf{x}, \mathbf{y}$ *be vectors in* $\mathbb{C}^n$ *with* $\mathbf{x} \neq \vec{0}$. *We define the* orthogonal projection *of* **y** *onto* **x** *by* $P_{\mathbf{x}}(\mathbf{y}) = \dfrac{<\mathbf{y}|\mathbf{x}>}{<\mathbf{x}|\mathbf{x}>}\mathbf{x}$.

First, we note that the formula can be written as $P_{\mathbf{x}}(\mathbf{y}) = \frac{\mathbf{x}^*\mathbf{y}}{\mathbf{x}^*\mathbf{x}}\mathbf{x} = \mathbf{x}\frac{\mathbf{x}^*\mathbf{y}}{\mathbf{x}^*\mathbf{x}} = (\mathbf{x}\mathbf{x}^+)\mathbf{y}$. Here comes the Moore-Penrose (MP) inverse again! This suggests that $P_{\mathbf{x}}$ can be viewed as the matrix $\mathbf{x}\mathbf{x}^+$, and orthogonal projection of $\mathbf{y}$ onto $\mathbf{x}$ can be achieved by the appropriate matrix multiplication. The next thing we note is that $P_{\mathbf{x}}$ is unchanged if we multiply $\mathbf{x}$ by a nonzero scalar. Thus $P_{\mathbf{x}}$ depends on the "line" (i.e., one dimensional subspace) $span(\mathbf{x})$ and not just $\mathbf{x}$.

LEMMA 8.1
For any nonzero scalar β, $P_{\beta\mathbf{x}}(\mathbf{y}) = P_{\mathbf{x}}(\mathbf{y})$.

PROOF
$P_{\beta\mathbf{x}}(\mathbf{y}) = \frac{<\mathbf{y}|\beta\mathbf{x}>}{<\beta\mathbf{x}|\beta\mathbf{x}>}\beta\mathbf{x} = \frac{<\mathbf{y}|\mathbf{x}>\bar{\beta}}{\beta<\mathbf{x}|\mathbf{x}>\bar{\beta}}\beta\mathbf{x} = \frac{<\mathbf{y}|\mathbf{x}>}{<\mathbf{x}|\mathbf{x}>}\mathbf{x} = P_{\mathbf{x}}(\mathbf{y}).$ □

So, from now on, we write $P_{sp(\mathbf{x})}$ instead of $P_{\mathbf{x}}$, indicating the connection of $P_{sp(\mathbf{x})}$ with the one dimensional subspace $span(\mathbf{x})$. Next, to solidify this connection with $sp(\mathbf{x})$, we show the vectors in $sp(\mathbf{x})$ are exactly those vectors left fixed by $P_{sp(\mathbf{x})}$.

LEMMA 8.2
$\{\mathbf{y}|P_{sp(\mathbf{x})}(\mathbf{y}) = \mathbf{y}\} = sp(\mathbf{x})$.

PROOF If $\mathbf{y} = P_{sp(\mathbf{x})}(\mathbf{y})$, then $\mathbf{y} = \frac{<\mathbf{y}|\mathbf{x}>}{<\mathbf{x}|\mathbf{x}>}\mathbf{x}$, a multiple of $\mathbf{x}$ so $\mathbf{y} \in sp(\mathbf{x})$. Conversely, if $\mathbf{y} \in sp(\mathbf{x})$, then $\mathbf{y} = \beta\mathbf{x}$ for some scalar β, so $P_{sp(\mathbf{x})}(\mathbf{y}) = P_{sp(\mathbf{x})}(\beta\mathbf{x}) = \frac{<\beta\mathbf{x}|\mathbf{x}>}{<\mathbf{x}|\mathbf{x}>}\mathbf{x} = \beta\frac{<\mathbf{x}|\mathbf{x}>}{<\mathbf{x}|\mathbf{x}>}\mathbf{x} = \beta\mathbf{x} = \mathbf{y}$. □

Next, we establish the geometrically obvious fact that if we project twice, we do not get anything new the second time.

LEMMA 8.3
For all vectors $\mathbf{y}$, $P_{sp(\mathbf{x})}(P_{sp(\mathbf{x})}(\mathbf{y})) = P_{sp(\mathbf{x})}(\mathbf{y})$.

PROOF $P_{sp(\mathbf{x})}(P_{sp(\mathbf{x})}(\mathbf{y})) = P_{sp(\mathbf{x})}\left(\frac{<\mathbf{y}|\mathbf{x}>}{<\mathbf{x}|\mathbf{x}>}\mathbf{x}\right) = \frac{\left\langle\frac{<\mathbf{y}|\mathbf{x}>}{<\mathbf{x}|\mathbf{x}>}\mathbf{x}|\mathbf{x}\right\rangle}{<\mathbf{x}|\mathbf{x}>}\mathbf{x} = \frac{\frac{<\mathbf{y}|\mathbf{x}>}{<\mathbf{x}|\mathbf{x}>}<\mathbf{x}|\mathbf{x}>}{<\mathbf{x}|\mathbf{x}>}\mathbf{x} = \frac{<\mathbf{y}|\mathbf{x}>}{<\mathbf{x}|\mathbf{x}>}\mathbf{x} = P_{sp(\mathbf{x})}(\mathbf{y})$ □

Taking the matrix point of view, $P_{sp(\mathbf{x})} = \mathbf{x}\mathbf{x}^+$, we have $(P_{sp(\mathbf{x})})^2 = \mathbf{x}\mathbf{x}^+\mathbf{x}\mathbf{x}^+ = \mathbf{x}\mathbf{x}^+ = P_{sp(\mathbf{x})}$. This says the matrix $P_{sp(\mathbf{x})}$ is idempotent. Let's make it official.

DEFINITION 8.2 *A matrix P in $\mathbb{C}^{n\times n}$ is called* ***idempotent*** *iff $P^2 = P$.*

Next, we note an important relationship with the inner product.

LEMMA 8.4
For all $\mathbf{y}, \mathbf{z}$ in $\mathbb{C}^n$, $< P_{sp(\mathbf{x})}(\mathbf{z})|\mathbf{y} > = < \mathbf{z}|P_{sp(\mathbf{x})}(\mathbf{y}) >$.

PROOF We compute both sides:

$$< P_{sp(\mathbf{x})}(\mathbf{z})|\mathbf{y} >= \left\langle \frac{< \mathbf{z}|\mathbf{x} >}{< \mathbf{x}|\mathbf{x} >}\mathbf{x}|\mathbf{y} \right\rangle = \frac{< \mathbf{z}|\mathbf{x} >}{< \mathbf{x}|\mathbf{x} >} < \mathbf{x}|\mathbf{y} >, \text{ but also,}$$

$$< \mathbf{z}|P_{sp(\mathbf{x})}(\mathbf{y}) > = < \mathbf{z}|\frac{< \mathbf{y}|\mathbf{x} >}{< \mathbf{x}|\mathbf{x} >}\mathbf{x} >=< \mathbf{z}|\mathbf{x} > \frac{\overline{< \mathbf{y}|\mathbf{x} >}}{< \mathbf{x}|\mathbf{x} >} = \frac{< \mathbf{z}|\mathbf{x} >}{< \mathbf{x}|\mathbf{x} >}$$

$< \mathbf{x}|\mathbf{y} >$. ▯

Okay let's make another definition.

DEFINITION 8.3 *(self-adjoint)*
A matrix P is ***self adjoint*** *iff $< P\mathbf{x}|\mathbf{y} >=< \mathbf{x}|P\mathbf{y} >$ for all $\mathbf{x}, \mathbf{y}$.*

In view of our fundamental formula $< A\mathbf{x}|\mathbf{y} >=< \mathbf{x}|A^*\mathbf{y} >$ for all $\mathbf{x}, \mathbf{y}$, we see self-adjoint is the same as Hermitian symmetric (i.e., $P = P^*$). This property of $P_{sp(\mathbf{x})}$ is obvious because of the MP-equation: $(P_{sp(\mathbf{x})})^* = (\mathbf{x}\mathbf{x}^+)^* = \mathbf{x}\mathbf{x}^+ = P_{sp(\mathbf{x})}$.

Next, we establish the "linearity" properties of $P_{sp(x)}$.

LEMMA 8.5

1. $P_{sp(\mathbf{x})}(\lambda\mathbf{y}) = \lambda P_{sp(\mathbf{x})}(\mathbf{y})$ *for λ any scalar and $\mathbf{y}$ any vector.*
2. $P_{sp(\mathbf{x})}(\mathbf{y}_1 + \mathbf{y}_2) = P_{sp(\mathbf{x})}(\mathbf{y}_1) + P_{sp(\mathbf{x})}(\mathbf{y}_2)$ *for any vectors $\mathbf{y}_1$ and $\mathbf{y}_2$.*

PROOF
By now, it should be reasonable to leave these computations to the reader. ▯

Next, we note a connection to the orthogonal complement of $sp(\mathbf{x})$, $sp(\mathbf{x})^\perp$.

LEMMA 8.6

1. $\mathbf{y} - P_{sp(\mathbf{x})}(\mathbf{y}) \perp P_{sp(\mathbf{x})}(\mathbf{y})$ *for all vectors* $\mathbf{y}$.

2. $\mathbf{y} - P_{sp(\mathbf{x})}(\mathbf{y}) \in sp(\mathbf{x})^{\perp}$ *for all vectors* $\mathbf{y}$.

PROOF Again we leave the details to the reader. □

Finally, we end with a remarkable geometric property of orthogonal projections. They solve a minimization problem without using any calculus. Given a vector $\mathbf{y}$ not in $sp(\mathbf{x})$, can we find a vector in $sp(\mathbf{x})$ that has the least distance to $\mathbf{x}$?

That is, we want to minimize the distance $d(\mathbf{y}, \mathbf{z}) = \|\mathbf{y} - \mathbf{z}\|$ as we let $\mathbf{z}$ vary through $sp(\mathbf{x})$. But $\|\mathbf{y} - \mathbf{z}\|^2 = \left\|\left(\mathbf{y} - P_{sp(\mathbf{x})}(\mathbf{y})\right) + (P_{sp(\mathbf{x})}(\mathbf{y}) - \mathbf{z})\right\|^2 = \left\|\mathbf{y} - P_{sp(\mathbf{x})}(\mathbf{y})\right\|^2 + \left\|P_{sp(\mathbf{x})}(\mathbf{y}) - \mathbf{z}\right\|^2$ by the Pythagorean theorem. Note $(P_{sp(\mathbf{x})}(\mathbf{y}) - \mathbf{y}) \perp (\mathbf{y} - P_{sp(\mathbf{x})}(\mathbf{z}))$ since $\mathbf{y} - P_{sp(\mathbf{x})}(\mathbf{z}) \in sp(\mathbf{x})^{\perp}$. Now conclude $\left\|P_{sp(\mathbf{x})}(\mathbf{y}) - \mathbf{y}\right\| \leq \|\mathbf{y} - \mathbf{z}\|$. So, among all vectors $\mathbf{z}$ in $sp(\mathbf{x})$, $P_{sp(\mathbf{x})}(\mathbf{y})$ is closest to $\mathbf{y}$. We have proved the following.

LEMMA 8.7

$d(P_{sp(\mathbf{x})}(\mathbf{y}), \mathbf{y}) \leq d(\mathbf{z}, \mathbf{y})$ *for all* $\mathbf{z} \in sp(\mathbf{x})$.

We can generalize this idea.

DEFINITION 8.4 *(the approximation problem)*

Let M be a subspace of $\mathbb{C}^n$ and let $\mathbf{x}$ be a vector in $\mathbb{C}^n$. Then by the ***approximation problem for*** $\mathbf{x}$ ***and*** M *we mean the problem of finding a vector $\mathbf{m}_0$ in M such that $\|\mathbf{x} - \mathbf{m}_0\| \leq \|\mathbf{x} - \mathbf{m}\|$ for all $\mathbf{m} \in M$.*

We note that if $\mathbf{x}$ is in M, then clearly $\mathbf{x}$ itself solves the approximation problem for $\mathbf{x}$ and M. The next theorem gives some significant insight into the approximation problem.

THEOREM 8.1

1. *A vector $\mathbf{m}_0 \in M$ solves the approximation problem for $\mathbf{x}$ and M if and only if $(\mathbf{x} - \mathbf{m}_0) \in M^{\perp}$.*

   Moreover,

2. *If $\mathbf{m}_1$ and $\mathbf{m}_2$ both solve the approximation problem for $\mathbf{x}$ and M, then $\mathbf{m}_1 = \mathbf{m}_2$. That is, if the approximation problem has a solution, the solution is unique.*

PROOF

1. First, suppose that $\mathbf{m}_0 \in M$ solves the approximation problem for $\mathbf{x}$ and M. We show $\mathbf{z} = \mathbf{x} - \mathbf{m}_0$ is orthogonal to every $\mathbf{m}$ in M. Without loss of generality, we may assume $\|\mathbf{m}\| = 1$. For any $\lambda \in \mathbb{C}$, we have

$$\|\mathbf{z} - \lambda\mathbf{m}\|^2 =< \mathbf{z} - \lambda\mathbf{m}|z - \lambda\mathbf{m} >=< \mathbf{z}|\mathbf{z} > - < \mathbf{z}|\mathbf{m} > \bar{\lambda} - \lambda < \mathbf{m}|\mathbf{z} > +\lambda < \mathbf{m}|\mathbf{m} > \bar{\lambda} = \|\mathbf{z}\|^2 - | < \mathbf{z}|\mathbf{m} > |^2 + | < \mathbf{z}|\mathbf{m} > |^2 - < \mathbf{z}|\mathbf{m} > \|\mathbf{z}\|^2 - |< \mathbf{z}|\mathbf{m} >|^2 + (< \mathbf{z}|\mathbf{m} > -\lambda)(\overline{< \mathbf{z} \mid \mathbf{m} > -\lambda}) = \|\mathbf{z}\|^2 - | < \mathbf{z}|\mathbf{m} > |^2 + | < \mathbf{z}|\mathbf{m} > -\lambda|^2.$$

We may now choose λ to be what we wish. We wish $\lambda =< \mathbf{z}|\mathbf{m} >$. Then, our computation above reduces to $\|\mathbf{z} - \lambda\mathbf{m}\|^2 = \|\mathbf{z}\|^2 - | < \mathbf{z}|\mathbf{m} > |^2$. Now $\mathbf{z} - \lambda\mathbf{m} = (\mathbf{x} - \mathbf{m}_0) - \lambda\mathbf{m} = \mathbf{x} - (\mathbf{m}_0 + \lambda\mathbf{m})$. Since $\mathbf{m}_0$ and $\mathbf{m}$ lie in M, so does $\mathbf{m}_0 + \lambda\mathbf{m}$, and so, by assumption, $\|\mathbf{x} - \mathbf{m}_0\| \leq \|\mathbf{x} - (\mathbf{m}_0 + \lambda\mathbf{m})\|$. We can translate back to $\mathbf{z}$ and get $\|\mathbf{z}\| \leq \|\mathbf{z} - \lambda\mathbf{m}\|$, so $\|\mathbf{z}\|^2 \leq \|\mathbf{x} - \lambda\mathbf{m}\|^2$. Therefore, we can conclude $\|\mathbf{z}\|^2 \leq \|\mathbf{z}\|^2 - | < \mathbf{z}|\mathbf{m} > |^2$. By cancellation, we get $0 \leq -| < \mathbf{z}|\mathbf{m} > |^2$. But the only way this can happen is that $| < \mathbf{z}|\mathbf{m} > | = 0$, hence $< \mathbf{z}|\mathbf{m} >= 0$. Thus, $\mathbf{z} \perp \mathbf{m}$ (i.e., $(\mathbf{x} - \mathbf{m}_0) \perp \mathbf{m}$), as we claim.

   Conversely, suppose $\mathbf{m}_0$ is a vector in M such that $\mathbf{x} - \mathbf{m}_0 \in M^{\perp}$. We claim $\mathbf{m}_0$ solves the approximation problem for $\mathbf{x}$ and M. By the Pythagorean theorem, $\|(\mathbf{x} - \mathbf{m}_0) + \mathbf{m}\|^2 = \|\mathbf{x} - \mathbf{m}_0\|^2 + \|\mathbf{m}\|^2 \geq \|\mathbf{x} - \mathbf{m}_0\|^2$ for any $\mathbf{m} \in M$. This says $\|\mathbf{x} - \mathbf{m}_0\|^2 \leq \|\mathbf{x} - \mathbf{m}_0 + \mathbf{m}\|^2$. Let $\mathbf{n} \in M$ and take $\mathbf{m} = \mathbf{m}_0 - \mathbf{n}$ which, since M is a subspace, still belongs to M. Then $\|\mathbf{x} - \mathbf{m}_0\| \leq \|(\mathbf{x} - \mathbf{m}_0) + (\mathbf{m}_0 - \mathbf{n})\| = \|\mathbf{x} - \mathbf{n}\|$. Since $\mathbf{n}$ was arbitrary from M, we see $\mathbf{m}_0$ solves the approximation problem for $\mathbf{x}$ and M.

2. Now let's argue the uniqueness by taking two solutions of the approximation problem and showing they must have been equal all along. Suppose $\mathbf{m}_1$ and $\mathbf{m}_2$ both solve the approximation problem for $\mathbf{x}$ and M. Then both $\mathbf{x} - \mathbf{m}_1 \in M^{\perp}$ and $\mathbf{x} - \mathbf{m}_2 \in M^{\perp}$. Let $\mathbf{m}_3 = \mathbf{m}_1 - \mathbf{m}_2$. We hope of course that $\mathbf{m}_3 = \vec{0}$. For this, we measure the length of $\mathbf{m}_3$. Then $\|\mathbf{m}_3\|^2 = < \mathbf{m}_3|\mathbf{m}_3 >=< \mathbf{m}_3|(\mathbf{x}-\mathbf{m}_2)-\mathbf{x}+\mathbf{m}_1 >=< \mathbf{m}_3|(\mathbf{x}-\mathbf{m}_2)-(\mathbf{x}-\mathbf{m}_1) >= < \mathbf{m}_3|(\mathbf{x}-\mathbf{m}_2) > - < \mathbf{m}_3|(\mathbf{x}-\mathbf{m}_1) >= 0-0 = 0$. Great! Now conclude $\|\mathbf{m}_3\| = 0$ so $\mathbf{m}_3 = \vec{0}$; hence, $\mathbf{m}_1 = \mathbf{m}_2$ and we are done. ∎

In particular, this theorem says $P_{sp(\mathbf{x})}(\mathbf{y})$ is the unique vector in $sp(\mathbf{x})$ of minimum distance from $\mathbf{y}$. We remark that the approximation problem is of great interest to statisticians. It may not be recognized in the form we stated it above, but be patient. We will get there. Also, you may have forgotten the problem set out in Chapter 1 of finding "best" approximate solutions to systems of linear equations that have no solutions. We have not and we have been working toward this problem since we introduced the notion of inner product. Finally, note that the approximation problem is always solvable in $\mathbb{C}^n$. That is, let $\mathbf{x} \in \mathbb{C}^n$ and M be any subspace. Now, we showed awhile back that $\mathbb{C}^n = M \oplus M^{\perp}$, so $\mathbf{x}$ is uniquely expressible $\mathbf{x} = \mathbf{m} + \mathbf{n}$, where $\mathbf{m} \in M$, $\mathbf{n} \in M^{\perp}$. But then, $\mathbf{x} - \mathbf{m} = \mathbf{n} \in M^{\perp}$, so by our theorem, $\mathbf{m}$ solves the approximation problem for $\mathbf{x}$ and M.

We have seen how to project onto a line. The question naturally arises, can we project onto a plane (i.e., a two-dimensional subspace). So now take a two-dimensional subspace M of $\mathbb{C}^n$. Take any orthonormal basis $\{\mathbf{u}_1, \mathbf{u}_2\}$ for M. Guided by the matrix formula above, we guess $P_M = [\mathbf{u}_1|\mathbf{u}_2][\mathbf{u}_1|\mathbf{u}_2]^+$. From the MP-equations, we have $P_M^2 = P_M = P_M^*$. For any matrix A, recall $Fix(A) = \{\mathbf{x}|A\mathbf{x} = \mathbf{x}\}$. Clearly, $\overrightarrow{0}$ is in $Fix(A)$ for any matrix A. We claim $M = Fix(P_M) = Col(P_M)$. Let $\mathbf{y} \in Fix(P_M)$. Then $\mathbf{y} = P_M\mathbf{y} \in Col(P_M)$. Conversely, if $\mathbf{y} \in Col(P_M)$, then $\mathbf{y} = P_M(\mathbf{x})$ for some $\mathbf{x}$. But then $P_M(\mathbf{y}) = P_M P_M(\mathbf{x}) = P_M(\mathbf{x}) = \mathbf{y}$. This establishes $Fix(P_M) = Col(P_M)$. To get that these are M, we need a clever trick. For any matrix A, $A^+ = A^*(AA^*)^+$. Then $P_M = [\mathbf{u}_1|\mathbf{u}_2][\mathbf{u}_1|\mathbf{u}_2]^+ = [\mathbf{u}_1|\mathbf{u}_2]([\mathbf{u}_1|\mathbf{u}_2]^*([\mathbf{u}_1|\mathbf{u}_2][\mathbf{u}_1|\mathbf{u}_2]^*)^+ = [\mathbf{u}_1|\mathbf{u}_2]$ $\left(\left[\frac{\mathbf{u}_1^*}{\mathbf{u}_2^*}\right]\left([\mathbf{u}_1|\mathbf{u}_2]\left[\frac{\mathbf{u}_1^*}{\mathbf{u}_2^*}\right]\right)^+\right) = [\mathbf{u}_1|\mathbf{u}_2]\left(\left[\frac{\mathbf{u}_1^*}{\mathbf{u}_2^*}\right](\mathbf{u}_1\mathbf{u}_1^* + \mathbf{u}_2\mathbf{u}_2^*)^+\right)$. However, $(\mathbf{u}_1\mathbf{u}_1^* + \mathbf{u}_2\mathbf{u}_2^*)^2 = (\mathbf{u}_1\mathbf{u}_1^* + \mathbf{u}_2\mathbf{u}_2^*)(\mathbf{u}_1\mathbf{u}_1^* + \mathbf{u}_2\mathbf{u}_2^*) = (\mathbf{u}_1\mathbf{u}_1^*\mathbf{u}_1\mathbf{u}_1^* + \mathbf{u}_1\mathbf{u}_1^*\mathbf{u}_2\mathbf{u}_2^* + \mathbf{u}_2\mathbf{u}_2^*\mathbf{u}_1\mathbf{u}_1^* + \mathbf{u}_2\mathbf{u}_2^*\mathbf{u}_2\mathbf{u}_2^* = \mathbf{u}_1\mathbf{u}_1^* + \mathbf{u}_2\mathbf{u}_2^*$. Also, $(\mathbf{u}_1\mathbf{u}_1^* + \mathbf{u}_2\mathbf{u}_2^*)^* = \mathbf{u}_1\mathbf{u}_1^* + \mathbf{u}_2\mathbf{u}_2^*$. Therefore, $(\mathbf{u}_1\mathbf{u}_1^* + \mathbf{u}_2\mathbf{u}_2^*)^+ = \mathbf{u}_1\mathbf{u}_1^* + \mathbf{u}_2\mathbf{u}_2^*$. Thus, $P_M = \left([\mathbf{u}_1|\mathbf{u}_2]\left[\frac{\mathbf{u}_1^*}{\mathbf{u}_2^*}\right]\right)(\mathbf{u}_1\mathbf{u}_1^* + \mathbf{u}_2\mathbf{u}_2^*) = (\mathbf{u}_1\mathbf{u}_1^* + \mathbf{u}_2\mathbf{u}_2^*)^2 = \mathbf{u}_1\mathbf{u}_1^* + \mathbf{u}_2\mathbf{u}_2^*$. This rather nice formula implies $P_M(\mathbf{u}_1) = \mathbf{u}_1$ and $P_M(\mathbf{u}_2) = \mathbf{u}_2$, so $P_M\mathbf{x} = \mathbf{x}$ for all $\mathbf{x}$ in M since P_M fixes a basis. This evidently says $M \subseteq Fix(P_M)$. But, if $\mathbf{x} \in Fix(P_M)$, then $\mathbf{x} = P_M(\mathbf{x}) = (\mathbf{u}_1\mathbf{u}_1^* + \mathbf{u}_2\mathbf{u}_2^*)\mathbf{x} = \mathbf{u}_1\mathbf{u}_1^*\mathbf{x} + \mathbf{u}_2\mathbf{u}_2^*\mathbf{x} = (\mathbf{u}_1^*\mathbf{x})\mathbf{u}_1 + (\mathbf{u}_2^*\mathbf{x})\mathbf{u}_2 \in sp\{\mathbf{u}_1, \mathbf{u}_2\} = M$. This completes the argument that $M = Fix(P_M) = Col(P_M)$. Finally, we show that $P_M(\mathbf{x})$ solves the approximation problem for $\mathbf{x}$ and M. It suffices to show $\mathbf{x} - P_M(\mathbf{x}) \in M^{\perp}$. So take $\mathbf{m} \in M$ and compute the inner product $< \mathbf{x} - P_M(\mathbf{x})|\mathbf{m} >=< \mathbf{x}|\mathbf{m} > - < P_M(\mathbf{x})|\mathbf{m} >=< \mathbf{x}|\mathbf{m} > - < \mathbf{x}|P_M(\mathbf{m}) >= < \mathbf{x}|\mathbf{m} > - < \mathbf{x}|\mathbf{m} >= 0$. Note that we used $P_M^* = P_M$ and $M = Fix(P_M)$ in this calculation.

Now it should be clear to the reader how to proceed to projecting on a three-dimensional space. We state a general theorem.

THEOREM 8.2

Let M be an m-dimensional subspace of $\mathbb{C}^n$. Then there exists a matrix P_M such that

1. $P_M^2 = P_M = P_M^*$
2. $M = Fix(P_M) = Col(P_M)$
3. *For any $\mathbf{x} \in \mathbb{C}^n$, $P_M(\mathbf{x})$ solves the approximation problem for $\mathbf{x}$ and M.*

Indeed, select an orthonormal basis $\{\mathbf{u}_1, \mathbf{u}_2, \ldots, \mathbf{u}_m\}$ for M and form the matrix $P_M = [\mathbf{u}_1|\mathbf{u}_2|\cdots|\mathbf{u}_m][\mathbf{u}_1|\mathbf{u}_2|\cdots|\mathbf{u}_m]^+ = \mathbf{u}_1\mathbf{u}_1^* + \mathbf{u}_2\mathbf{u}_2^* + \cdots + \mathbf{u}_m\mathbf{u}_m^*$. Then for any $\mathbf{x} \in \mathbb{C}^n$, $P_M(\mathbf{x}) = <\mathbf{x}|\mathbf{u}_1>\mathbf{u}_1 + <\mathbf{x}|\mathbf{u}_2>\mathbf{u}_2 + \ldots + <\mathbf{x}|\mathbf{u}_m>\mathbf{u}_m$.

PROOF The details are an extension of our discussion above and are left to the reader. ▯

Actually a bit more can be said here. If $Q^2 = Q = Q^*$ and $Fix(Q) = Col(Q) = M$, then $Q = P_M$. In particular, this says it does not matter which orthonormal basis of M you choose to construct P_M. For simplicity say $P_M = [\mathbf{u}_1|\mathbf{u}_2][\mathbf{u}_1|\mathbf{u}_2]^+$. Then $QP_M = Q[\mathbf{u}_1|\mathbf{u}_2][\mathbf{u}_1|\mathbf{u}_2]^+ = [Q\mathbf{u}_1|Q\mathbf{u}_2][\mathbf{u}_1|\mathbf{u}_2]^+ = [\mathbf{u}_1|\mathbf{u}_2][\mathbf{u}_1|\mathbf{u}_2]^+ = P_M$. Thus, $QP_M = P_M$ since Q leaves vectors in M fixed. Also $P_M Q = [\mathbf{u}_1|\mathbf{u}_2][\mathbf{u}_1|\mathbf{u}_2]^+ Q = [\mathbf{u}_1\mathbf{u}_1^* + \mathbf{u}_2\mathbf{u}_2^*]Q = [\mathbf{u}_1\mathbf{u}_1^* + \mathbf{u}_2\mathbf{u}_2^*][\mathbf{q}_1|\mathbf{q}_2] = [\mathbf{u}_1\mathbf{u}_1^*\mathbf{q}_1 + \mathbf{u}_2\mathbf{u}_2^*\mathbf{q}_1|\mathbf{u}_1\mathbf{u}_1^*\mathbf{q}_2 + \mathbf{u}_2\mathbf{u}_2^*\mathbf{q}_2] = [<\mathbf{q}_1|\mathbf{u}_1>\mathbf{u}_1 + <\mathbf{q}_1|\mathbf{u}_2>\mathbf{u}_2| <\mathbf{q}_2|\mathbf{u}_1>\mathbf{u}_1 + <\mathbf{q}_2|\mathbf{u}_2>\mathbf{u}_2] = [\mathbf{q}_1|\mathbf{q}_2] = Q$. Thus $P_M Q = Q$. Therefore, $Q = Q^* = (P_M Q)^* = Q^* P_M^* = QP_M = P_M$. Now it makes sense to call P_M *the orthogonal projection onto the subspace M*.

Exercise Set 33

1. Let $\mathbf{v} = \begin{bmatrix} 1 \\ 2 \\ -2 \end{bmatrix}$. Compute the projection onto $sp(\mathbf{v})$ and $sp(\mathbf{v})^\perp$.

2. Compute the projection onto the span of $\{(1, 1, 1), (-1, 0, 1)\}$.

3. Prove that if $UU^* = I$, then U^*U is a projection.

4. Let $P_{M,N}$ denote the projector of $\mathbb{C}^n$ onto M along N. Argue that $(P_{M,N})^* = P_{N^\perp, M^\perp}$.

5. Argue that P is a projection iff $P = PP^*$.

6. Suppose $\mathbb{C}^n = M \oplus N$. Argue that $N = M^\perp$ iff $P_{M,N} = (P_{M,N})^*$.

7. Argue that for all $\mathbf{x}$ in $\mathbb{C}^n$, $\|P_M\mathbf{x}\| \leq \|\mathbf{x}\|$, with equality holding iff $\mathbf{x} \in M$.

8. (Greville) Argue that $G \in A\{2\}$ iff $G = (EAF)^+$ for some projections E and F.

9. (Penrose) Argue that E is idempotent iff $E = (FG)^+$ for some projections F and G.

10. (Greville) Argue that the projector $P_{M,N}$ is expressible with projections through the use of the MP-inverse, namely, $P_{M,N} = (P_{N^\perp} P_M)^+ = ((I - P_N)P_M)^+$.

11. A typical computation of the projection P_M onto the subspace M is to obtain an orthonormal basis for M. Here is a neat way to avoid that: Take any basis at all of M, say $\{\mathbf{b}_1, \mathbf{b}_2, \ldots, \mathbf{b}_m\}$, and form the matrix $F = [\mathbf{b}_1 \mid \mathbf{b}_2 \mid \cdots \mid \mathbf{b}_m]$. Argue that $P_M = FF^+ = F(F^*F)^{-1}F^*$.

12. Suppose E is a projection. Argue that $G \in E\{2, 3, 4\}$ iff G is a projection and $\mathcal{C}ol(G) \subseteq \mathcal{C}ol(E)$.

13. Let M and N be subspaces of $\mathbb{C}^n$. Prove that the following statements are all equivalent:
    (i) $P_M - P_N$ is invertible
    (ii) $\mathbb{C}^n = M \oplus N$
    (iii) there is a projector Q with $\mathcal{C}ol(Q) = M$ and $\mathcal{N}ull(Q) = N$.

    In fact, when one of the previous holds, prove that $(P_M - P_N)^{-1} = Q + Q^* - I$.

14. Suppose $P = P^2$. Argue that $P = P^*$ iff $\mathcal{N}ull(P) = (Fix(P))^\perp$.

15. Argue that $Q = \begin{bmatrix} Cos^2(\theta) & Sin(x)Cos(\theta) \\ Sin(x)Cos(\theta) & Sin^2(\theta) \end{bmatrix}$ is a projection of the plane $\mathbb{R}^2$. What is the geometry behind this projection?

Further Reading

[Banerjee, 2004] Sudipto Banerjee, Revisiting Spherical Trigonometry with Orthogonal Projections, The College Mathematics Journal, Vol. 35, No. 5, November, (2004), 375–381.

[Gross, 1999] Jürgen Gross, On Oblique Projection, Rank Additivity and the Moore-Penrose Inverse of the Sum of Two Matrices, Linear and Multilinear Algebra, Vol. 46, (1999), 265–275.

8.2 The Geometry of Subspaces and the Algebra of Projections

In the previous section, we showed that, starting with a subspace M of $\mathbb{C}^n$, we can construct a matrix P_M such that $P_M^2 = P_M = P_M^*$ and $M = Col(P_M) = Fix(P_M)$. Now we want to go farther and establish a one-to-one correspondence between subspaces of $\mathbb{C}^n$ and idempotent self-adjoint matrices. Then we can ask how the relationships between subspaces is reflected in the algebra of these special matrices. We begin with a definition.

DEFINITION 8.5 *(orthogonal projection matrix)*
A matrix P is called an ***orthogonal projection matrix*** *(or just* ***projection,*** *for short) if it is self-adjoint and idempotent (i.e., $P^2 = P = P^*$). Let $\mathbb{P}(\mathbb{C}^{n\times n})$ denote the collection of all n-by-n projections.*

We use the notation $Lat(\mathbb{C}^n)$ to denote the collection of all subspaces of $\mathbb{C}^n$.

THEOREM 8.3
There is a one-to-one correspondence between $\mathbb{P}(\mathbb{C}^{n\times n})$ and $Lat(\mathbb{C}^n)$ given as follows: to P in $\mathbb{P}(\mathbb{C}^{n\times n})$, assign $\varphi(P) = Fix(P) = Col(P)$ and to M in $Lat(\mathbb{C}^n)$, assign $\psi(M) = P_M$.

PROOF It suffices to prove $\psi\varphi(P) = P$ and $\varphi\psi(M) = M$. We note that ψ is well defined by the discussion at the end of the previous section. First, $\psi\varphi(P) = \psi(Col(P)) = P_{Col(P)} = P_M$, where $M = Col(P)$. But then, $P = P_M$ since P_M is the only projection with $M = Fix(P_M) = Col(P_M)$. Next, $\varphi\psi(M) = \varphi(P_M) = Col(P_M) = M$. □

Now this correspondence between subspaces and projections opens up a whole world of questions. For example, how can you tell if one subspace is contained in another one? How does $P_{M^\perp}$ relate to P_M? How do P_{M+N} and $P_{M\cap N}$ relate to P_M and P_N?

THEOREM 8.4
For any subspace M of $\mathbb{C}^n$, $P_{M^\perp} = I - P_M$.

PROOF $Col(I - P_M) = Fix(I - P_M) = \{\mathbf{x} | (I - P_M)\mathbf{x} = \mathbf{x}\} = \{\mathbf{x} | P_M\mathbf{x} = \vec{0}\} = \mathcal{N}ull(P_M) = Col(P_M^*)^\perp = Col(P_M)^\perp = M^\perp$. Note $I - P_M$ is a projection since $(I - P_M)^2 = I - P_m - P_m + P_m^2 = I - P_m$ and $(I - P_m)^* = I^* - P_m^* = I - P_m$. Thus, $P_{M^\perp} = P_{Col(I-P_M)} = I - P_M$. ▯

As a shorthand, write $P^\perp = I - P$ when $P \in \mathbb{P}(\mathbb{C}^{n\times n})$.

THEOREM 8.5

$M \subseteq N$ iff $P_M = P_N P_M$ iff $P_M = P_M P_N$.

PROOF Suppose $M \subseteq N$. Then $P_M\mathbf{x} \in M \subseteq N = Fix(P_N)$, so $P_N(P_M\mathbf{x}) = P_M\mathbf{x}$. Since this is true for all $\mathbf{x}$ in $\mathbb{C}^n$, we get $P_N P_M = P_M$. Conversely, if $P_N P_M = P_M$, then $Col(P_M) \subseteq Col(P_N)$. That is, $M \subseteq N$. The second 'iff' follows by taking $*$. ▯

DEFINITION 8.6 *(partial order)*

For P, Q in $\mathbb{P}(\mathbb{C}^{n\times n})$ define $P \leq Q$ iff $P = PQ$.

This gives us a way of saying when one projection is "smaller" than another one.

THEOREM 8.6

Let P, Q, and R be projections.

1. For all P, $P \leq P$.
2. If $P \leq Q$ and $Q \leq P$, then $P = Q$.
3. If $P \leq Q$ and $Q \leq R$, then $P \leq R$.
4. $P \leq Q$ iff $Col(P) \subseteq Col(Q)$.
5. $\mathbb{O}$ and I are projections and $\mathbb{O} \leq P \leq I$ for all P in $\mathbb{P}(\mathbb{C}^{n\times n})$.
6. $P \leq Q$ iff $P = QP$.
7. If $P \in \mathbb{P}(\mathbb{C}^{n\times n})$, then $I - P \in \mathbb{P}(\mathbb{C}^{n\times n})$.
8. $PQ \in \mathbb{P}(\mathbb{C}^{n\times n})$ iff $PQ = QP$.
9. $P + Q \in \mathbb{P}(\mathbb{C}^{n\times n})$ iff $PQ = QP = \mathbb{O}$.
10. $P \leq Q$ iff $Q^\perp \leq P^\perp$.

PROOF The proofs are routine and left to the reader. □

THEOREM 8.7

Let M and N be subspaces of $\mathbb{C}^n$.

1. *P_{M+N} is the projection uniquely determined by (i) $P_M \leq P_{M+N}$ and $P_N \leq P_{M+N}$ and (ii) if $P_M \leq Q$ and $P_N \leq Q$, then $P_{M+N} \leq Q$, where Q is a projection.*

2. *$P_{M\cap N}$ is the projection uniquely determined by (i) $P_{M\cap N} \leq P_M$ and $P_{M\cap N} \leq P_N$ and if $Q \leq P_M$ and $Q \leq P_N$, then $Q \leq P_{M\cap N}$.*

PROOF

1. First $M, N \subseteq M+N$, so P_M, $P_N \leq P_{M+N}$. Now suppose Q is a projection with $P_M \leq Q$ and $P_N \leq Q$. Then, $M \subseteq Col(Q)$ and $N \subseteq Col(Q)$, so $M + N \subseteq Col(Q)$, making $P_{M+N} \leq P_{Col(Q)} = Q$. For the uniqueness, suppose H is a projection satisfying the same two properties P_{M+N} does. That is, P_M, $P_N \leq H$ and if $P_M \leq Q$ and $P_N \leq Q$, then $H \leq Q$. Then, with H playing the role of Q, we get $P_{M+N} \leq H$. But with P_{M+N} playing the role of Q, we get $H \leq P_{M+N}$. Therefore, $H = P_{M+N}$.

2. The proof is analogous to the one above. □

Our next goal is to derive formulas for P_{M+N} and $P_{M\cap N}$ for any subspaces M and N of $\mathbb{C}^n$. First, we must prepare the way. The MP-equations are intimately connected with projections. We will explore this more fully in the next section but, for now, recall $(MP1)$ $AA^+A = A$ and $(MP3)$ $(AA^+)^* = AA^+$. The first equation says $AA^+AA^+ = AA^+$. In other words $MP1$ and $MP3$ imply $P = AA^+$ is a projection. But what does P project onto? We need $Fix(P) = Fix(AA^+) = \{\mathbf{x} | AA^+\mathbf{x} = \mathbf{x}\}$. But if $\mathbf{y} \in Col(A)$, then $\mathbf{y} = A\mathbf{x}$ for some $\mathbf{x}$, so $AA^+\mathbf{y} = AA^+A\mathbf{x} = A\mathbf{x} = \mathbf{y}$. Thus we see $Col(A) \subseteq Fix(AA^+)$. On the other hand, if $\mathbf{y} \in Fix(AA^+)$, then $A(A^+\mathbf{y}) = \mathbf{y}$ so $\mathbf{y} \in Col(A)$. We conclude $Fix(AA^+) = Col(A)$, so that AA^+ is the projection onto the column space of A. Let's record our findings.

LEMMA 8.8

For any matrix $A \in \mathbb{C}^{m\times n}$, $AA^+ \in \mathbb{C}^{m\times m}$ is the projection onto the column space of A.

Next, we need the following Lemma.

LEMMA 8.9
For any matrix A, $\mathcal{C}ol(A) = \mathcal{C}ol(AA^*)$

PROOF If $\mathbf{x} \in \mathcal{C}ol(AA^*)$ then $\mathbf{x} = AA^*\mathbf{y}$ for some $\mathbf{y}$, so $\mathbf{x} = A(A^*\mathbf{y})$ is evidently in $\mathcal{C}ol(A)$. Conversely, if $\mathbf{x} \in \mathcal{C}ol(A)$, then $\mathbf{x} = A\mathbf{y}$ for some $\mathbf{y}$. Now $\mathbf{x} = AA^+\mathbf{x} = A(A^*(AA^*)^+)\mathbf{x}$. In other words, $AA^*\mathbf{z} = \mathbf{x}$, where $\mathbf{z} = (AA^*)^+\mathbf{x}$. This puts $\mathbf{x}$ in $\mathcal{C}ol(AA^*)$. □

Another fact we will need is in the following Lemma.

LEMMA 8.10
Let $A \in \mathbb{C}^{m\times n}$, $B \in \mathbb{C}^{m\times k}$, *and* $M = [A \vdots B]$, *the augmented matrix in* $\mathbb{C}^{m\times(n+k)}$. *Then* $\mathcal{C}ol(M) = \mathcal{C}ol(A) + \mathcal{C}ol(B)$.

PROOF The proof is left as an exercise. □

Now we come to a nice result that is proved in a bit more generality than we need.

THEOREM 8.8
Let $A \in \mathbb{C}^{m\times n}$ *and* $B \in \mathbb{C}^{m\times k}$. *Then*

1. $\mathcal{C}ol(AA^* + BB^*) = \mathcal{C}ol(A) + \mathcal{C}ol(B)$
2. $\mathcal{N}ull(AA^* + BB^*) = \mathcal{N}ull(AA^*) \cap \mathcal{N}ull(BB^*)$.

PROOF

1. Let $M = [A \vdots B]$ be the m-by-$(n+k)$ augmented matrix. Then $\mathcal{C}ol(M) = \mathcal{C}ol(A) + \mathcal{C}ol(B)$ by (2.10). But also, $\mathcal{C}ol(M) = \mathcal{C}ol(MM^*)$ by (8.9). However, $MM^* = [A \vdots B]\begin{bmatrix} A^* \\ \cdots \\ B^* \end{bmatrix} = AA^* + BB^*$. Hence, $\mathcal{C}ol(AA^* + BB^*) = \mathcal{C}ol(MM^*) = \mathcal{C}ol(M) = \mathcal{C}ol(A) + \mathcal{C}ol(B)$.

2. Let $\mathbf{x} \in \mathcal{N}ull(AA^*) \cap \mathcal{N}ull(BB^*)$. Then $AA^*\mathbf{x} = \overrightarrow{0}$ and $BB^*\mathbf{x} = \overrightarrow{0}$, so $(AA^* + BB^*)\mathbf{x} = \overrightarrow{0}$, putting $\mathbf{x}$ in $\mathcal{N}ull(AA^* + BB^*)$. Now let $\mathbf{x} \in \mathcal{N}ull(AA^* + BB^*)$. Then $[AA^* + BB^*](\mathbf{x}) = \overrightarrow{0}$ so $AA^*\mathbf{x} + BB^*\mathbf{x} = \overrightarrow{0}$.

Thus $\mathbf{x}^* AA^*\mathbf{x} + \mathbf{x}^* BB^*\mathbf{x} = \overrightarrow{0}$. That is, $\|A^*\mathbf{x}\|^2 + \|B^*\mathbf{x}\|^2 = 0$ so $\|A^*\mathbf{x}\| = 0 = \|B^*\mathbf{x}\|$. We conclude $\mathbf{x} \in \mathcal{N}ull(A^*) = \mathcal{N}ull(AA^*)$ and $\mathbf{x} \in \mathcal{N}ull(B^*) = \mathcal{N}ull(BB^*)$. Therefore, $\mathbf{x} \in \mathcal{N}ull(AA^*) \cap \mathcal{N}ull(BB^*)$.

□

Now we apply these results to projections. Note that if P and Q are projections, then $P + Q = PP^* + QQ^*$. Also note that while $P + Q$ is self-adjoint, it need not be a projection.

COROLLARY 8.1

Let P and Q be in $\mathbb{P}(\mathbb{C}^{n\times n})$. Then

1. $Col(P + Q) = Col(P) + Col(Q)$
2. $\mathcal{N}ull(P + Q) = \mathcal{N}ull(P) \cap \mathcal{N}ull(Q)$.

Let $M = Col(A)$ for some matrix A. We have noted above that AA^+ is the projection onto M. Thus, if M and N are any subspaces, $[P_M + P_N][P_M + P_N]^+$ is the projection on $Col(P_M + P_N) = Col(P_M) + Col(P_N) = M + N$ using (8.8)(1). This is part of the next theorem. Recall our shorthand notation $P^\perp = I - P$ for a projection P.

THEOREM 8.9

Let M and N be subspaces of $\mathbb{C}^n$. Let $P = P_M$ and $Q = P_N$. Then all the following expressions are equal to the orthogonal projection onto $M + N$.

1. P_{M+N}
2. $[P + Q][P + Q]^+$
3. $[P + Q]^+[P + Q]$
4. $Q + [(PQ^\perp)^+(PQ^\perp)]$
5. $P + [(QP^\perp)^+(QP^\perp)]$
6. $P + P^\perp[P^\perp Q]^+$
7. $Q + Q^\perp[Q^\perp P]^+$.

PROOF There is clearly lots of symmetry in these formulas, since $M + N = N + M$. That (2) equals (3) follows from the fact that $P + Q$ is self-adjoint. Take the $*$ of (2) and you get (3), and vice-versa. That (1) and (2) are equal follows from the discussion just ahead of the theorem. We will give an order theoretic argument that (1) equals (4). The equality of (1) and (5) will then follow by symmetry. Let $H = Q + (PQ^{\perp})^{+}(PQ^{\perp})$. First note $(PQ^{\perp})^{+}(PQ^{\perp})Q = \mathbb{O}$, so that H is in fact a projection. Also, $HQ = Q + \mathbb{O} = Q$ so $Q \leq H$. Next, $PH = P(Q + (PQ^{\perp})^{+}(PQ^{\perp})] = PQ + P(PQ^{\perp})^{+}(PQ^{\perp}) = PQ + P[I - (I - (PQ^{\perp})^{+}(PQ^{\perp})] = PQ + P - P(I - (PQ^{\perp})^{+}(PQ^{\perp})) = P + [PQ - P(I - (PQ^{\perp})^{+}(PQ^{\perp}))]$. But $\mathbb{O} = (PQ^{\perp})[I - (PQ^{\perp})^{+}(PQ^{\perp})] = P(I - Q)[I - (PQ^{\perp})^{+}(PQ^{\perp})] = [P - PQ][I - (PQ^{\perp})^{+}(PQ^{\perp})] = P(I - (PQ^{\perp})^{+}(PQ^{\perp})) - PQ$, since $Q \leq I - (PQ^{\perp})^{+}(PQ^{\perp})$. Thus, $PQ = P(I - (PQ^{\perp})^{+}(PQ^{\perp}))$ and, consequently, $PH = PQ + P - PQ = P$. Thus, $P \leq H$. Now let K be any projection with $K \geq P$ and $K \geq Q$. Then $KH = K[Q + (PQ^{\perp})^{+}(PQ^{\perp})] = KQ + K(PQ^{\perp})^{+}(PQ^{\perp})) = Q + K(PQ^{\perp})^{+}(PQ^{\perp})$. But $Q \leq K$ so $QK^{\perp} = 0$ whence $PQ^{\perp}K^{\perp} = PK^{\perp} = \mathbb{O}$, so $K^{\perp} = [I - (PQ^{\perp})^{+}(PQ^{\perp})]K^{\perp}$. This says $K^{\perp} \leq I - (PQ^{\perp})^{+}(PQ^{\perp})$ or, equivalently, $(PQ^{\perp})^{+}(PQ^{\perp}) \leq K$. Thus, $K(PQ^{\perp})^{+}(PQ^{\perp}) = (PQ^{\perp})^{+}(PQ^{\perp})$ and so $KH = H$ putting $H \leq K$. Therefore, $H = P_{M+N}$ by (8.7)(1).

Now we have (1) through (5) all equal. So let's look at (6). Let $U = P^{\perp}Q$. Then $UQ = Q$ so $U^{+}UQ = U^{+}U$. This says $U^{+}UQP^{\perp} = U^{+}UP^{\perp}$. By taking the * of both sides, we get $P^{\perp}QU^{+}U = P^{\perp}U^{+}U$ so $U = UU^{+}U = P^{\perp}U^{+}U$. Thus, $UU^{+} = P^{\perp}U^{+}UU^{+} = P^{\perp}U^{+}$. Therefore, $UU^{+} = P^{\perp}U^{+}$; that is, $(P^{\perp}Q)(P^{\perp}Q) = P^{\perp}(P^{\perp}Q)^{+}$. But $(P^{\perp}Q)(P^{\perp}Q)^{+} = [(P^{\perp}Q)(P^{\perp}Q)^{+}]^{*} = (P^{\perp}Q)^{+*}(P^{\perp}Q)^{*} = (QP^{\perp})(QP^{\perp})$ so $(QP^{\perp})^{+}(QP^{\perp}) = P^{\perp}(P^{\perp}Q)^{+}$. Now (6) follows from (5). Of course (7) follows by symmetry. □

Thus our first goal is accomplished; namely, in finite dimensions, the projection onto the linear sum of two subspaces is computable in many ways in terms of the individual projections.

We now turn to our second goal of representing the projection on the intersection of two subspaces in terms of the individual projections.

THEOREM 8.10

Let M and N be subspaces of $\mathbb{C}^n$. Let $P = P_M$ and $Q = P_N$. Then all the following expressions are equal to the orthogonal projection onto $M \cap N$.

1. $P_{M \cap N}$
2. $2Q(Q + P)^{+}P$
3. $2P(Q + P)^{+}Q$
4. $2[P - P(Q + P)^{+}P]$
5. $2[Q - Q(Q + P)^{+}Q]$

6. $P - (P - QP)^+(P - QP) = P - (Q^\perp P)^+(Q^\perp P)$

7. $Q - (Q - PQ)^+(Q - PQ) = Q - (P^\perp Q)^+(P^\perp Q)$

8. $P - P(PQ^\perp)^+$

9. $Q - Q(QP^\perp)^+$.

PROOF As before. we have much symmetry since $M \cap N = N \cap M$. We begin by showing (2) equals (3). To do this, we first show $Q(Q + P)^+P - P(Q + P)^+Q = \mathbb{O}$. But $Q(Q + P)^+P = P(Q + P)^+Q = Q(Q + P)^+P + Q(Q + P)^+Q - Q(Q + P)^+Q + P(Q + P)^+Q = Q(Q + P)^+(Q + P) - (Q + P)(Q + P)^+Q$. By (2.12), $\mathcal{C}ol(Q) \subseteq \mathcal{C}ol(Q) + \mathcal{C}ol(P) = \mathcal{C}ol(Q + P)$, so $Q(Q + P)^+(Q + P) = Q = (Q + P)(Q + P)^+Q$. Thus, $Q(Q + P)^+P - P(Q + P)^+Q = Q - Q = \mathbb{O}$, and so $Q(Q + P)^+P = P(Q + P)^+Q$, and it follows that (2) equals (3).

Next, we argue that (2), and hence (3), also equals (1). We use the uniqueness characterization of $P_{M\cap N}$. Let $H = Q(Q+P)^+P + P(Q+P)^+Q = 2Q(Q+P)^+P = 2P(Q+P)^+Q$. Now $HP = [2Q(Q+P)^+P]P = 2Q(Q+P)^+P^2 = 2Q(Q + P)^+P = H$ and, similarly, $HQ = H$. Thus, $\mathcal{C}ol(H) \subseteq M \cap N$. But also $H = HP_{M\cap N} = [Q(Q + P)^+P + P(Q + P)^+Q]P_{M\cap N} = Q(Q + P)^+PP_{M\cap N} + P(Q+P)^+QP_{M\cap N} = Q(Q+P)^+P_{M\cap N} + P(Q+P)^+P_{M\cap N} = [Q(Q + P)^+ + P(Q + P)^+]P_{M\cap N} = (Q + P)(Q + P)^+P_{M\cap N} = P_{M\cap N}$. This last equality follows because $M \cap N \subseteq \mathcal{C}ol(Q + P)$ and so we conclude $H = P_{M\cap N}$.

Next, we show (4) and (5) equal (1) by showing $P(Q + P)^+Q = P - P(Q + P)^+P$. The argument that $Q(Q + P)^+P = Q - (Q + P)^+Q$ is similar and will be left as an exercise. Now $P(Q + P)^+Q - (P - P(Q + P)^+P) = P(Q + P)^+Q - P + P(Q + P)^+P = P(Q + P)^+Q + P(Q + P)^+P - P = P(Q + P)^+(Q + P) - P = P - P = \mathbb{O}$.

To see that (6) and (7) equal (1), note first that $M \cap N \subseteq M$ and $M \cap N \subseteq N$, so $PP_{M\cap N} = P_{M\cap N}$ and $QP_{M\cap N} = P_{M\cap N}$ and $QPP_{M\cap N} = P_{M\cap N}$. Now let $S = P - (P - QP)^+(P - QP)$. We claim S is a projection. Clearly $S^* = S$ and $SP = S$. It follows $PS = S$. In particular, $\mathcal{C}ol(S) \subseteq \mathcal{C}ol(P) = M$. Next, $S^2 = P^2 - P(P - QP)^+(P - QP) - (P - QP)^+(P - QP)P + ((P - QP)^+(P - QP))^2 = P[P - (P - QP)^+(P - QP)] = PS = S$. Now $P - QP = (P - QP)(P - QP)^+(P - QP) = P(P - QP)^+(P - QP) - QP(P - QP)^+(P - QP)$ so $S = PS = P[P - (P - QP)^+(P - QP)] = P - P(P - QP)^+(P - QP) = QP - QP(P - QP)^+(P - QP) = Q[P - P(P - QP)^+(P - QP)] = QPS = QS$. Thus, $QS = S$ and $SQ = S$, so $\mathcal{C}ol(S) \subseteq \mathcal{C}ol(Q) = N$. Therefore, $\mathcal{C}ol(S) \subseteq M \cap N$ and so $P_{M\cap N}S = S$. But $SP_{M\cap N} = [P-(P-QP)^+(P-QP)]P_{M\cap N} = PP_{M\cap N} - (P-QP)^+(P-QP)P_{M\cap N} = P_{M\cap N} - 0 = P_{M\cap N}$. Thus, $S = P_{M\cap N}$. The argument for (8) and

(9) are similar to the ones given for (6). They will be left as an exercise. This completes our theorem. □

Next, we look at some special cases whose proofs are more or less immediate.

COROLLARY 8.2
Let $A \in \mathbb{C}^{m\times n}$ and $B \in \mathbb{C}^{m\times k}$. Then

1. $[A \vdots B][A \vdots B]^+ = P_{Col(A)+Col(B)}$
2. $2AA^+[AA^+ + BB^+]^+BB^+ = P_{Col(A)\cap Col(B)}$.

COROLLARY 8.3
Let M and N be subspaces of $\mathbb{C}^n$ with $P = P_M$ and $Q = P_N$. Then if $PQ = QP$,

1. $P_{M+N} = P + Q - PQ$
2. $P_{M\cap N} = PQ = QP$.

In particular, if $PQ = \mathbb{O}$ (i.e., if $P \perp Q$),

1. $P_{M+N} = P + Q$
2. $P_{M\cap N} = \mathbb{O}$.

We end this section with an example. Suppose $M = span\{\mathbf{a}_1, \mathbf{a}_2, \ldots, \mathbf{a}_r\}$ and $N = span\{\mathbf{b}_1, \mathbf{b}_2, \ldots, \mathbf{b}_s\}$ are subspaces of $\mathbb{C}^n$. We form the matrices $A = [\mathbf{a}_1 \mid \mathbf{a}_2 \mid \ldots \mid \mathbf{a}_r]$ and $[\mathbf{b}_1 \mid \mathbf{b}_2 \mid \ldots \mid \mathbf{b}_s]$, which are n-by-r and n-by-s, respectively. Of course, $AA^+ = P_{Col(A)} = P_M$ and $BB^+ = P_{Col(B)} = P_N$. Now form the augmented matrix $M = [A \vdots B]$, which is n-by-$(r+s)$ and, with a left multiplication by a suitable invertible matrix R, we produce the row reduced echelon form of M; say $RREF(M) = RM = R[A \vdots B] = [RA \vdots RB] = \begin{bmatrix} E_{11} & E_{12} \\ \mathbb{O} & E_{22} \\ \mathbb{O} & \mathbb{O} \end{bmatrix}$. Now $B = R^{-1}\begin{bmatrix} E_{12} \\ E_{22} \\ \mathbb{O} \end{bmatrix} = R^{-1}\begin{bmatrix} E_{12} \\ \mathbb{O} \\ \mathbb{O} \end{bmatrix} + R^{-1}\begin{bmatrix} \mathbb{O} \\ E_{22} \\ \mathbb{O} \end{bmatrix}$ and $\begin{bmatrix} E_{11} \\ \mathbb{O} \\ \mathbb{O} \end{bmatrix}^+ \begin{bmatrix} E_{12} \\ \mathbb{O} \\ \mathbb{O} \end{bmatrix} = \begin{bmatrix} E_{11} \\ \mathbb{O} \\ \mathbb{O} \end{bmatrix} [E_{11}^+ \ \mathbb{O} \ \mathbb{O}] \begin{bmatrix} E_{12} \\ \mathbb{O} \\ \mathbb{O} \end{bmatrix} = \begin{bmatrix} E_{12} \\ \mathbb{O} \\ \mathbb{O} \end{bmatrix}$. Thus, $R^{-1}\begin{bmatrix} E_{12} \\ \mathbb{O} \\ \mathbb{O} \end{bmatrix}$ has columns in $M \cap N$. If we let $R^{-1}\begin{bmatrix} E_{12} \\ \mathbb{O} \\ \mathbb{O} \end{bmatrix}$, then WW^+ is the projection on $M \cap N$. For example, suppose

$M = span\{(1,1,1,1,0), (1,2,3,4,0), (3,5,7,9,0), (0,1,2,2,0)\}$ and $N = span\{(2,3,4,7,0), (1,0,1,0,0), (3,3,5,7,0), (0,1,0,3,0)\}$. Then $RM =$

$$\begin{bmatrix} 0 & 2 & 0 & -1 & 0 \\ \frac{1}{2} & -1 & -\frac{1}{2} & 1 & 0 \\ -1 & 1 & 1 & -1 & 0 \\ \frac{1}{2} & -1 & \frac{1}{2} & 0 & 0 \\ 0 & 0 & 0 & 0 & 1 \end{bmatrix} \left[\begin{array}{cccc|cccc} 1 & 1 & 3 & 0 & 2 & 1 & 3 & 0 \\ 1 & 2 & 5 & 1 & 3 & 0 & 3 & 1 \\ 1 & 3 & 7 & 2 & 4 & 1 & 5 & 0 \\ 1 & 4 & 9 & 2 & 7 & 0 & 7 & 3 \\ 0 & 0 & 0 & 0 & 0 & 0 & 0 & 0 \end{array}\right] =$$

$$\left[\begin{array}{cccc|cccc} 1 & 0 & 1 & 0 & -1 & 0 & -1 & -1 \\ 0 & 1 & 2 & 0 & 3 & 0 & 3 & 2 \\ 0 & 0 & 0 & 1 & -2 & 0 & -2 & -2 \\ 0 & 0 & 0 & 0 & 0 & 1 & 1 & -1 \\ 0 & 0 & 0 & 0 & 0 & 0 & 0 & 0 \end{array}\right] = RREF(M).$$

Now $W = R^{-1} = \begin{bmatrix} E_{12} \\ \mathbb{O} \\ \mathbb{O} \end{bmatrix} \begin{bmatrix} 1 & 1 & 0 & 1 & 0 \\ 1 & 2 & 1 & 0 & 0 \\ 1 & 3 & 2 & 1 & 0 \\ 1 & 4 & 2 & 0 & 0 \\ 0 & 0 & 0 & 0 & 1 \end{bmatrix} \begin{bmatrix} -1 & 0 & -1 & -1 \\ 3 & 0 & 3 & 2 \\ -2 & 0 & -2 & -2 \\ 0 & 0 & 0 & 0 \\ 0 & 0 & 0 & 0 \end{bmatrix} =$

$\begin{bmatrix} 2 & 0 & 2 & 1 \\ 3 & 0 & 3 & 1 \\ 4 & 0 & 4 & 1 \\ 7 & 0 & 7 & 3 \\ 0 & 0 & 0 & 0 \end{bmatrix}$ and $W^+ = \begin{bmatrix} -1/12 & 1/12 & 1/4 & -1/12 & 0 \\ 0 & 0 & 0 & 0 & 0 \\ -1/12 & 1/12 & 1/4 & -1/12 & 0 \\ 1/2 & -1/3 & -7/6 & 2/3 & 0 \end{bmatrix}$. The

projection onto $M \cap N$ is $WW^+ = \begin{bmatrix} 1/6 & 0 & -1/6 & 1/3 & 0 \\ 0 & 1/6 & 1/3 & 1/6 & 0 \\ -1/6 & 1/3 & 5/6 & 0 & 0 \\ 1/3 & 1/6 & 0 & 5/6 & 0 \\ 0 & 0 & 0 & 0 & 0 \end{bmatrix}$. The trace of this projection is 2, which gives the dimension of $M \cap N$. A basis for $M \cap N$ is $\{(2,3,4,7,0), (1,1,1,3,0)\}$.

Exercise Set 34

1. Prove that $tr((P_M + P_N)(P_M + P_N)^+) = tr(P_M) + tr(P_M) - tr(P_{M \cap N})$.

2. (G. Trenkler) Argue that A is a projection iff $3tr(A^*A) + tr(AAA^*A^*) = 2Re(tr(AA + AA^*A))$.

3. Suppose P_M and P_N are projections. Argue that $P_M + P_N$ is a projection iff $M \perp N$.

4. Suppose P and Q are projections. Argue that PQ is a projection iff $PQ = QP$. Indeed, argue that the following statements are equivalent for projections P and Q:

   (a) PQ is idempotent
   (b) $tr(PQPQ) = tr(PQ)$
   (c) PQ is self-adjoint
   (d) $PQ = QP$.

5. Suppose P_M and P_N are projections. Argue that $P_M P_N = P_N P_M$ iff $M = (M \cap N) \oplus (M \cap N^{\perp})$.

6. Suppose P_M and P_N are projections. Argue that the following statements are all equivalent:

   (a) $P_M - P_N$ is a projection
   (b) $P_N \leq P_M$
   (c) $\|P_N\mathbf{x}\| \leq \|P_M\mathbf{x}\|$ for all $\mathbf{x}$
   (d) $N \subseteq M$
   (e) $P_M P_N = P_N$
   (f) $P_N P_M = P_N$.

7. Suppose P is an idempotent. Argue that P is a projection iff $\|P\mathbf{x}\| \leq \|\mathbf{x}\|$ for all $\mathbf{x}$.

Further Reading

[P&O&H, 1999] R. Piziak, P. L. Odell, and R. Hahn, Constructing Projections on Sums and Intersections, Computers and Mathematics with Applications, Vol. 37, (1999), 67–74.

[S&Y, 1998] Henry Stark and Yongyi Yang, *Vector Space Projections,* John Wiley & Sons, New York, (1998).

8.3 The Fundamental Projections of a Matrix

We have seen how to take a matrix A and associate with it four subspaces we called "fundamental": $\mathcal{C}ol(A)$, $\mathcal{N}ull(A)$, $\mathcal{C}ol(A^*)$ and $\mathcal{N}ull(A^*)$. All these subspaces generate projections. Remarkably, all these subspaces are related to the MP-inverse of A.

THEOREM 8.11
Let $A \in \mathbb{C}^{m\times n}$. Then

1. $AA^+ = P_{\mathcal{C}ol(A)} = P_{\mathcal{N}ull(A^*)^\perp}$
2. $A^+A = P_{\mathcal{C}ol(A^*)} = P_{\mathcal{N}ull(A)^\perp}$
3. $I - AA^+ = P_{\mathcal{C}ol(A)^\perp} = P_{\mathcal{N}ull(A^*)}$
4. $I - A^+A = P_{\mathcal{C}ol(A^*)^\perp} = P_{\mathcal{N}ull(A)}$.

PROOF We have already argued (1) in the previous section. A similar argument shows A^+A is a projection. Then $I - AA^+$ and $I - A^+A$ are projections. The only real question is what does A^+A project onto? Well $A^+A = A^+A^{++}$ so, by (1), A^+A projects onto $\mathcal{C}ol(A^+)$. So the only thing left to show is that $\mathcal{C}ol(A^+) = \mathcal{C}ol(A^*)$. This follows from two identities. First, suppose $\mathbf{x} \in \mathcal{C}ol(A^*)$. Then, $\mathbf{x} = A^*\mathbf{z}$ for some $\mathbf{z}$. But $A^* = A^+AA^*$, so $\mathbf{x} = A^*\mathbf{z} = A^+AA^*\mathbf{z} = A^+(AA^*\mathbf{z})$, putting $\mathbf{x} \in \mathcal{C}ol(A^+)$. Next, suppose $\mathbf{x} \in \mathcal{C}ol(A^+)$. Then $\mathbf{x} = A^+\mathbf{z}$ for some z. But $A^+ = A^*A^{+*}A^+$, so $\mathbf{x} = A^+\mathbf{z} = A^*(A^{+*}A^+\mathbf{z})$, putting $\mathbf{x}$ in $\mathcal{C}ol(A^*)$. $\square$

COROLLARY 8.4
For any matrix $A \in \mathbb{C}^{m\times n}$, $\mathcal{C}ol(A^*) = \mathcal{C}ol(A^+)$ and $\mathcal{N}ull(A) = \mathcal{C}ol(A^+)^\perp$.

Next, we relate the fundamental projections to a full rank factorization of A.

THEOREM 8.12
Let $A = FG$ be a full rank factorization of $A \in \mathbb{C}_r^{m\times n}$. Then

1. $AA^+ = FF^+$
2. $A^+A = G^+G$

3. $I_m - AA^+ = I_m - FF^+$

4. $I_n - A^+A = I_n - G^+G$.

PROOF The proof is easy and left to the reader. □

Let's look at an example.

Example 8.1

Let $A = \begin{bmatrix} 3 & 6 & 13 \\ 2 & 4 & 9 \\ 1 & 2 & 3 \end{bmatrix} = FG = \begin{bmatrix} 3 & 13 \\ 2 & 9 \\ 1 & 3 \end{bmatrix} \begin{bmatrix} 1 & 2 & 0 \\ 0 & 0 & 1 \end{bmatrix}$. *As before, we compute* $G^+ = \begin{bmatrix} 1/5 & 0 \\ 2/5 & 0 \\ 0 & 1 \end{bmatrix}$ *and* $F^+ = \begin{bmatrix} -3/26 & -11/13 & 79/26 \\ 1/13 & 3/13 & -9/13 \end{bmatrix}$. *Then* $G^+G = \begin{bmatrix} 1/5 & 2/5 & 0 \\ 2/5 & 4/5 & 0 \\ 0 & 0 & 1 \end{bmatrix}$ *is a rank 2 projection onto* $\mathcal{C}ol(A^*)$; $I - G^+G = \begin{bmatrix} 4/5 & -2/5 & 0 \\ -2/5 & 1/5 & 0 \\ 0 & 0 & 0 \end{bmatrix}$ *is a rank 1 projection onto* $\mathcal{N}ull(A)$. *Next* $FF^+ = \begin{bmatrix} 17/26 & 6/13 & 3/26 \\ 6/13 & 5/13 & -2/13 \\ 3/26 & -2/13 & 25/26 \end{bmatrix}$ *is a rank 2 projection onto* $\mathcal{C}ol(A)$, *and* $I - FF^+ = \begin{bmatrix} 9/26 & -6/13 & -3/26 \\ -6/13 & 8/13 & 2/13 \\ -3/26 & 2/13 & 1/26 \end{bmatrix}$ *is a rank 1 projection onto* $\mathcal{N}ull(A^*)$.

Next, we develop an assignment of a projection to a matrix that will prove useful later. The most crucial property is (3), which we use heavily later. This property uniquely characterizes A'.

DEFINITION 8.7 *(the prime mapping)*

For any matrix $A \in \mathbb{C}^{m\times n}$, *we assign the projection* $A' = I - A^+A \in \mathbb{P}(\mathbb{C}^{n\times n})$. *That is,* A' *is the projection onto the null space of* A.

We next collect some formulas involving A'.

THEOREM 8.13

Let $A \in \mathbb{C}^{m\times n}$. Then

1. $AA' = \mathbb{O}$; $(A^*A)' = A'$; $(A^*)'A = \mathbb{O}$.
2. If $P \in \mathbb{P}(\mathbb{C}^{n\times n})$, then $P' = P^{\perp} = I - P$.
3. $AB = \mathbb{O}$ iff $B = A'B$. In fact, A' is the unique projection with this property.
4. If $B = B^*$ and $AB = BA$, then $AB' = B'A$ and $A'B = BA'$ and $A'B = BA'$.
5. If $P,Q \in \mathbb{P}(\mathbb{C}^{n\times n})$, then $P \wedge Q = (Q'P)'P$ and $P \vee Q = (P' \wedge Q')'$.
6. If $P \leq Q$, then $Q = P \vee (Q \wedge P')$.

PROOF

1. $AA' = A(I - A^+A) = A - AA^+A = \mathbb{O}$.
2. Suppose P is a projection. Then $P^+ = P$ so $P' = I - P^+P = I - P^2 = I - P = P^{\perp}$.
3. Suppose $AB = \mathbb{O}$. Then $A'B = (I - A^+A)B = B - A^+AB = B - \mathbb{O} = B$. Conversely, if $B = A'B$, then $AB = AA'B = \mathbb{O}B = \mathbb{O}$. Now let P be any projection with the property $AB = \mathbb{O}$ iff $B = PB$. Then $AA' = \mathbb{O}$, so $A' = PA'$, making $A' \leq P$. But also, $P = PP$ so $AP = \mathbb{O}$ whence $A^+AP = \mathbb{O}$, so $A'P = (I - A^+A)P = P - A^+AP = P$, so $P \leq A'$. Thus, $P = A'$.
4. Suppose $B = B^*$ and $AB = BA$. Now $BB' = \mathbb{O}$, so $ABB' = \mathbb{O}$, so $BAB' = \mathbb{O}$, so $AB' = B'(AB')$. Similarly, $A^*B = BA^*$, so $A^*B' = B'A^*B'$. Taking $*$, we get $B'A = B'AB'$. Thus, $AB' = B'AB' = B'A$. Also, $AA' = \mathbb{O}$, so $BAA' = \mathbb{O}$, so $ABA' = \mathbb{O}$, so $BA' = A'BA'$. Taking $*$ we get $A'B = A'BA'$. Therefore, $A'B = BA'$.
5. $(Q'P)'P = (Q^{\perp}P)'P = (I - (Q^{\perp}P)^+(Q^{\perp}P))P = P - (Q^{\perp}P)^+(Q^{\perp}P) = P \wedge Q$ by (8.10)(6). Now $(P' \wedge Q')' = ((Q''P')'P')' = ((QP^{\perp})'P^{\perp})' = ((I - (QP^{\perp})^+(QP^{\perp}))P^{\perp})' = (P^{\perp} - (QP^{\perp})^+(QP^{\perp}))' = I - (P^{\perp} - (QP^{\perp})^+(QP^{\perp})) = P + (QP^{\perp})^+(QP^{\perp}) = P \vee Q$ by (8.9)(5). Note we have used that $P^{\perp} - (QP^{\perp})^+(QP^{\perp})$ is a projection since $P^{\perp} - (QP^{\perp})^+(QP^{\perp}) \leq P^{\perp}$.

6. Suppose $P \leq Q$. Then $P \vee (Q \wedge P') = P \vee (Q'P')'P' = P \vee [(I - Q)(I - P)]'P^{\perp} = P \vee [I - Q - P + PQ]'(I - P) = P \vee (Q(I - P)) = P \vee (Q - QP) = P \vee (Q - P)$. But $P(Q - P) = \mathbb{O}$ so $P \vee (Q - P) = P + (Q - P) = Q$. ∎

Well if one prime is so good, what happens if you prime twice? It must be twice as good don't you think? Again A'' is a projection we assign to A.

THEOREM 8.14

1. $A'' = A^{+}A = P_{Col(A^{*})} = P_{\mathcal{N}ull(A)^{\perp}}$.
2. *If* $P \in \mathbb{P}(\mathbb{C}^{n \times n})$, *then* $P'' = P$.
3. $A = AA'' = (A^{*})''A$.
4. *If* $AB = A$, *then* $A'' \leq B''$.
5. *If* P *is a projection,* $AP = A$ *iff* $A'' \leq P$. *That is,* A'' *is the smallest projection* P *such that* $AP = A$.
6. $(AB)'' \leq B''$.
7. $(AB)'' = (A''B)''$.
8. $(A^{*}A)'' = A''$; $(AA^{*})'' = A^{*''}$.
9. $((AB)'B^{*})'' \leq A'$.
10. $AB^{*} = \mathbb{O}$ *iff* $A'' \perp B''$ *iff* $A''B'' = \mathbb{O}$.
11. *If* $AB = AC$, *then* $A''B = A''C$.

Exercise Set 35

1. Fill in the proofs of Theorem 8.12.
2. Let $\mathbf{a} = \begin{bmatrix} 1 \\ i \end{bmatrix}$. Compute $\mathbf{a}^{+}$ and $P = \mathbf{a}\mathbf{a}^{+}$. Verify that P is an orthogonal projection. Onto what does P project?
3. Fill in the proofs of Theorem 8.14.

Further Reading

[Greville, 1974] T. N. E. Greville, Solutions of the Matrix Equation XAX = X and Relations Between Oblique and Orthogonal Projectors, SIAM J. Appl. Math., Vol. 26, No. 4, June, (1974), 828–831.

[B-I & D, 1966] A. Ben-Israel and D. Dohen, On Iterative Computation of Generalized Inverses and Associated Projections, J. SIAM Numer. Anal., III, (1966), 410–419.

8.3.1 MATLAB Moment

8.3.1.1 The Fundamental Projections

It hardly seems necessary to define M-files to compute the fundamental projections. After we have input a matrix A, we can easily compute the four projections onto the fundamental subspaces:

A*pinv(A)	the projection onto the column space of A
pinv(A)*A	the projection onto the column space of A*
eye(m)-A*pinv(A)	the projection onto the null space of A*
eye(n)-pinv(A)*A	the projection onto the null space of A.

Since "prime mapping" plays a crucial role later, we could create a file as follows:

```
1  function P=prime(A)
2  [m n] = size(A)
3  P = eye(n)-pinv(A)*A.
```

Experiment with a few matrices. Compute their fundamental projections.

8.4 Full Rank Factorizations of Projections

We have seen that every matrix A in $\mathbb{C}_r^{m\times n}$ with $r > 0$ has infinitely many full rank factorizations $A = FG$, where $F \in \mathbb{C}_r^{m\times r}$ and $G \in \mathbb{C}_r^{r\times n}$. The columns of F form a basis for column space of A. Applying the Gram-Schmidt process, we can make these columns of F orthonormal. Then $F^*F = I_r$. But then, it is easy to check that F^* satisfies the four MP-equations. Thus, $F^* = F^+$, which leads us to what we called an *orthogonal full rank factorization,* as we defined in (7.6) of Chapter 7. Indeed, if U is unitary, $A = (FU)(U^*G)$ is again an orthogonal full rank factorization if $A = FG$ is. We summarize with a theorem.

THEOREM 8.15
Every matrix A in $\mathbb{C}_r^{m\times n}$ with $r > 0$ has infinitely many orthogonal full rank factorizations.

Next, we consider the special case of a projection $P \in \mathbb{C}_r^{n\times n}$. We have already noted that $P^+ = P$. But now take P in a full rank factorization that is orthogonal. $P = FG$, where $F^+ = F^*$. Then $P = P^* = (FG)^* = G^*F^* = G^*F^+$. But $P = P^+ = (FG)^+ = G^+F^+ = G^+F^*$ and $GP = GG^+F^* = F^*$. But then $P = PP = FGP = FF^*$. This really is not a surprise since $FF^* = FF^+$ is the projection onto $\mathcal{C}ol(P)$. But that is P! Again we summarize.

THEOREM 8.16
Every projection P in $\mathbb{C}_r^{n\times n}$ has a full rank factorization $P = FG$ where $G = F^* = F^+$.

For example, consider the rank 1 projection $\begin{bmatrix} 9/26 & -6/13 & -3/26 \\ -6/13 & 8/13 & 2/13 \\ -3/26 & 2/13 & 1/26 \end{bmatrix}$ $= \begin{bmatrix} 9/26 \\ -6/13 \\ -3/26 \end{bmatrix} \begin{bmatrix} 1 & -4/3 & -1/3 \end{bmatrix}$. This is a full rank factorization but it is not orthogonal. But Gram-Schmidt is easy to apply here. Just normalize the column vector and then $\begin{bmatrix} 9/26 & -6/13 & -3/26 \\ -6/13 & 8/13 & 2/13 \\ -3/26 & 2/13 & 1/26 \end{bmatrix} = \begin{bmatrix} 3/\sqrt{26} \\ -4/\sqrt{26} \\ -1/26 \end{bmatrix}$ $\begin{bmatrix} 3/\sqrt{26} & -4/\sqrt{26} & -1/\sqrt{26} \end{bmatrix}$. We can use this factorization of a projection to create invertible matrices. This will be useful later. Say F is m-by-r of rank r, that is, F has full column rank. Then $F^+ = (F^*F)^{-1}F^*$, as we have seen. This says $F^+F = I_r$. Now write $I - FF^+ = F_1F_1^+$ in orthogonal full rank factorization. Form the m-by-m matrix $S = \begin{bmatrix} F^+ \\ \cdots \\ F_1^+ \end{bmatrix}$. We claim $S^{-1} = [F \vdots F_1]$. We compute $SS^{-1} = \begin{bmatrix} F^+ \\ \cdots \\ F_1^+ \end{bmatrix} [F \vdots F_1] = \begin{bmatrix} F^+F & F^+F_1 \\ F_1^+F & F_1^+F \end{bmatrix} = \begin{bmatrix} I_r & F^+F_1 \\ F_1^+F & I_{m-r} \end{bmatrix}$

$= \begin{bmatrix} I_r & \mathbb{O} \\ \mathbb{O} & I_{m-r} \end{bmatrix}$ since $F_1^+ = F_1^+(F_1F_1^+) = F_1^+(I - FF^+)$ so $F_1^+F = F_1^+(I - FF^+)F = \mathbb{O}$ and $F_1 = F_1F_1^+F_1 = (I - FF^+)F_1$ so $F^+F_1 = \mathbb{O}$ as well.

To illustrate, let $F = \begin{bmatrix} 3 & 13 \\ 2 & 9 \\ 1 & 3 \end{bmatrix}$, which has rank 2. We saw above $I - FF^+ = \begin{bmatrix} \frac{3}{\sqrt{26}} \\ \frac{-4}{\sqrt{26}} \\ \frac{-1}{\sqrt{26}} \end{bmatrix} \begin{bmatrix} \frac{3}{\sqrt{26}} & \frac{-4}{\sqrt{26}} & \frac{-1}{\sqrt{26}} \end{bmatrix} = F_1F_1^+$ so $S = \begin{bmatrix} F^+ \\ \cdots \\ F_1^+ \end{bmatrix}$

$= \begin{bmatrix} \frac{-3}{26} & \frac{-11}{13} & \frac{79}{26} \\ \frac{1}{13} & \frac{3}{13} & \frac{-9}{13} \\ \cdots & \cdots & \cdots \\ \frac{3}{\sqrt{26}} & \frac{-4}{\sqrt{26}} & \frac{-1}{\sqrt{26}} \end{bmatrix}$ is invertible with inverse $S^{-1} = \begin{bmatrix} 3 & 13 & \vdots & \frac{3}{\sqrt{26}} \\ 2 & 9 & \vdots & \frac{-4}{\sqrt{26}} \\ 1 & 3 & \vdots & \frac{-1}{\sqrt{26}} \end{bmatrix}$.

Similarly, if G has full row rank, we write $I = G^+G = F_2F_2^+$ in orthogonal full rank factorization and form $T = \begin{bmatrix} G^+ & \vdots & F_2 \end{bmatrix}$. Then $T^{-1} = \begin{bmatrix} G \\ \cdots \\ F_2^+ \end{bmatrix}$.

Exercise Set 36

1. Find an orthogonal full rank projection of $\begin{bmatrix} \frac{1}{2} & \frac{1}{2} \\ \frac{1}{2} & \frac{1}{2} \end{bmatrix}$.

8.5 Affine Projections

We have seen how to project (orthogonally) onto a linear subspace. In this section, we shall see how to project onto a subspace that has been moved away from the origin. These are called *affine subspaces* of $\mathbb{C}^n$.

DEFINITION 8.8 *(affine subspace)*

By an ***affine subspace*** *of* $\mathbb{C}^n$*, we mean any set of vectors of the form* $M(\mathbf{a}, U) = \{\mathbf{a} + \mathbf{u} | \mathbf{u} \in U\}$*, where* U *is a subspace of* $\mathbb{C}^n$ *and* $\mathbf{a}$ *is a fixed vector from* $\mathbb{C}^n$*. The notation* $M(\mathbf{a}, U) = \mathbf{a} + U$ *is very convenient.*

We draw a little picture to try to give some meaning to this idea.

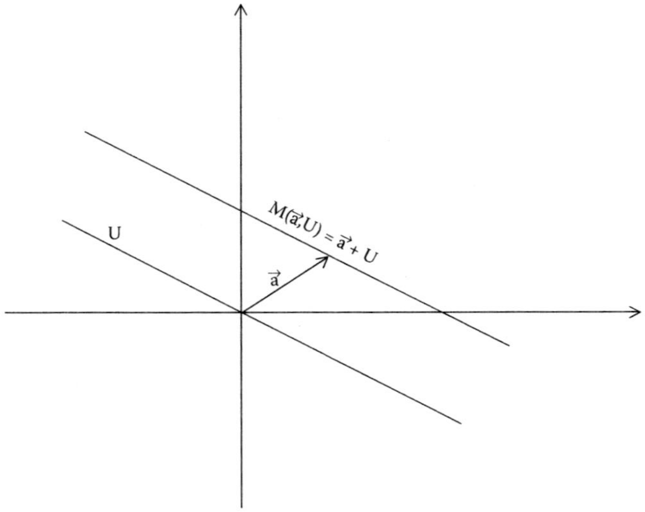

Figure 8.2: Affine subspace.

The following facts are readily established and are left to the reader.

1. $\mathbf{a} \in M(\mathbf{a}, U)$.
2. $M(\vec{0}, U) = U$.
3. $M(\mathbf{a}, U) = U$ iff $\mathbf{a} \in U$.
4. $M(\mathbf{a}, U) \subseteq M(\mathbf{b}, W)$ iff $U \subseteq W$ and $\mathbf{a} - \mathbf{b} \in W$.
5. $M(\mathbf{a}, U) = M(\mathbf{a}, W)$ iff $U = W$.
6. $M(\mathbf{a}, U) = M(\mathbf{b}, U)$ iff $\mathbf{a} - \mathbf{b} \in U$.
7. $M(\mathbf{a}, U)$ is a convex subset of $\mathbb{C}^n$.
8. $M(\mathbf{a}, U) \cap M(\mathbf{b}, W) \neq \emptyset$ iff $\mathbf{a} - \mathbf{b} \in U + W$.
9. If $\mathbf{z} \in M(\mathbf{a}, U) \cap M(\mathbf{b}, W)$ then $M(\mathbf{a}, U) \cap M(\mathbf{b}, W) = M(\mathbf{z}, U \cap W)$.

In view of (5), we see that the linear subspace associated with an affine subspace is uniquely determined by the affine subspace. Indeed, given the affine subspace, M, $U = \{\mathbf{y} - \mathbf{a} | y \in M\}$ is the uniquely determined linear subspace called the *direction space* of M. We call the affine subspaces $M(\mathbf{a}, U)$ and $M(\mathbf{b}, W)$ *parallel* iff $U \subseteq W$. If $M(\mathbf{a}, U)$ and $M(\mathbf{b}, W)$ have a point $\mathbf{c}$ in common, then $M(\mathbf{a}, U) = M(\mathbf{c}, U) \subseteq M(\mathbf{c}, W) = M(\mathbf{b}, W)$. Thus, parallel affine subspaces are either totally disjoint or one of them is contained in the other. Note that through any point $\mathbf{x}$ in $\mathbb{C}^n$, there is one and only one affine subspace with given direction U parallel to W, namely $\mathbf{x} + U$. Does all this sound familiar from geometry? Finally, we note by (8) that if $U \oplus W = \mathbb{C}^n$, then $M(\mathbf{a}, U) \cap M(\mathbf{b}, W)$ is a singleton set.

Next, we recall the correspondence we have made between linear subspaces and orthogonal projections; $U \leftrightarrow P_U = P_U^2 = P_U^*$, where $U = Col(P_U) = Fix(P_U)$. We should like to have the same correspondence for affine subspaces. The idea is to translate, project, and then translate back.

DEFINITION 8.9 *(affine projection)*
For $\mathbf{x}$ in $\mathbb{C}^n$, define $\Pi_{M(\mathbf{a},U)}\mathbf{x} = \mathbf{a} + P_U(\mathbf{x} - \mathbf{a}) = \mathbf{a} + P_U(\mathbf{x}) - P_U(\mathbf{a}) = P_U(\mathbf{x}) + P_{U^\perp}(\mathbf{a})$ as the projection of $\mathbf{x}$ onto the affine subspace $M(\mathbf{a}, U)$.

The following are easy to compute.

1. $\Pi^2_{M(\mathbf{a},U)}(\mathbf{x}) = \Pi_{M(\mathbf{a},U)}(\mathbf{x})$ for all $\mathbf{x}$ in $\mathbb{C}^n$.
2. $M(\mathbf{a}, U) = \{\mathbf{y} | \mathbf{y} = \Pi_{M(\mathbf{a},U)}(\mathbf{x})\}$ for all $\mathbf{x}$ in $\mathbb{C}^n$.
3. $M(\mathbf{a}, U) = \{\mathbf{x} | \Pi_{M(\mathbf{a},U)}(\mathbf{x}) = \mathbf{x}\}$.
4. $\Pi_{M(\mathbf{a},U^\perp)}(\mathbf{x}) = \mathbf{x} - \Pi_{M(\mathbf{a},U)}(\mathbf{x})$.
5. $\Pi_{M(\mathbf{a},U)}(\overrightarrow{0}) = \mathbf{a} - P_U(\mathbf{a}) = P_{U^\perp}(\mathbf{a})$.
6. If $\mathbf{a} \in U$, $\Pi_{M(\mathbf{a},U)}(\mathbf{x}) = P_U(\mathbf{x})$.
7. $\left\|\Pi_{M(\mathbf{a},U)}(\mathbf{y}) - \Pi_{M(\mathbf{a},U)}(\mathbf{z})\right\| = \|P_U(\mathbf{z}) - P_U(\mathbf{y})\|$.
8. $\left\|\Pi_{M(\mathbf{a},U)}(\mathbf{x})\right\|^2 = \|P_U(\mathbf{x})\|^2 + \|P_{U^\perp}(\mathbf{a})\|^2$.
9. $\mathbf{x} = \Pi_{M(\mathbf{a},U)}(\mathbf{x}) = (\mathbf{x} - \mathbf{a}) - P_U(\mathbf{x} - \mathbf{a}) = P_{U^\perp}(\mathbf{x} - \mathbf{a})$.

As a concrete illustration, we take the case of projecting on a line. Let $\mathbf{b} \neq \overrightarrow{0}$ be in $\mathbb{C}^n$ and let $U = sp(\mathbf{b})$ the one dimensional subspace spanned by $\mathbf{b}$. We shall compute the affine projection onto $M(\mathbf{a}, sp(\mathbf{b}))$. From the definition, $\Pi_{M(\mathbf{a},sp(\mathbf{b}))}(\mathbf{x}) = \mathbf{a} + \dfrac{<\mathbf{x} - \mathbf{a}|\mathbf{b}>}{<\mathbf{b}|\mathbf{b}>}\mathbf{b} = \mathbf{a} + \dfrac{\mathbf{b}^*(\mathbf{x} - \mathbf{a})\mathbf{b}}{\mathbf{b}^*\mathbf{b}} = \mathbf{a} + \mathbf{b}\mathbf{b}^+(\mathbf{x} - \mathbf{a}) = (I - \mathbf{b}\mathbf{b}^+)(\mathbf{a}) + \mathbf{b}\mathbf{b}^+(\mathbf{x})$ and $\Pi_{M(\mathbf{a},sp(\mathbf{b})^\perp}(\mathbf{x}) = (\mathbf{x} - \mathbf{a}) + \dfrac{<\mathbf{a} - \mathbf{x}|\mathbf{b}>}{<\mathbf{b}|\mathbf{b}>}\mathbf{b} = (\mathbf{x} - \mathbf{a}) + \mathbf{b}\mathbf{b}^+(\mathbf{a} - \mathbf{x}) = (I - \mathbf{b}\mathbf{b}^+)\mathbf{x} - (I - \mathbf{b}\mathbf{b}^+)\mathbf{a} = (I - \mathbf{b}\mathbf{b}^+)(\mathbf{x} - \mathbf{a})$. We see that, geometrically, there are several ways to resolve the vector $\Pi_{M(\mathbf{a},U)}(\mathbf{x})$ as the algebraic formulas have indicated.

Now, if A is a matrix in $\mathbb{C}^{m\times n}$, then we write $M(\mathbf{a}, A)$ for the affine subspace $M(\mathbf{a}, Col(A))$. Consider a line L of slope m in $\mathbb{R}^2$. Then $L = \{(x, mx + y_0)|x \in \mathbb{R}\} \subseteq \mathbb{R}^2$. We associate the matrix $\begin{pmatrix} 1 & 0 \\ m & 0 \end{pmatrix}$ of rank 1 with this line. Then $L = \left\{\begin{pmatrix} 1 & 0 \\ m & 0 \end{pmatrix}\begin{bmatrix} x \\ y \end{bmatrix} + \begin{bmatrix} 0 \\ y_0 \end{bmatrix} |x, y \in \mathbb{R}\right\}$. We can then use the pseudoinverse to compute the orthogonal projection onto the direction space of this line, which is just the line parallel passing through the origin. The affine projection can then be computed as $\Pi_L \begin{bmatrix} x \\ y \end{bmatrix} = \begin{bmatrix} 0 \\ y_0 \end{bmatrix} + \begin{bmatrix} 1 & 0 \\ m & 0 \end{bmatrix}\begin{bmatrix} 1 & 0 \\ m & 0 \end{bmatrix}^+ \left(\begin{bmatrix} x \\ y \end{bmatrix} - \begin{bmatrix} 0 \\ y_0 \end{bmatrix}\right) = \begin{bmatrix} \frac{1}{1+m^2} & \frac{m}{1+m^2} \\ \frac{m}{1+m^2} & \frac{m^2}{1+m^2} \end{bmatrix}\begin{bmatrix} x \\ y - y_0 \end{bmatrix} + \begin{bmatrix} 0 \\ y_0 \end{bmatrix} = \begin{bmatrix} \frac{x+m(y-y_0)}{1+m^2} \\ \frac{mx+m^2(y-y_0)}{1+m^2} + y_0 \end{bmatrix}$. So, for example, say $y = -3x + 4$. Then $\Pi_L \begin{bmatrix} x \\ y \end{bmatrix} = \begin{bmatrix} \frac{x-3(y-4)}{10} \\ \frac{-3x+9y-36}{10} + 4 \end{bmatrix} = \begin{bmatrix} \frac{x-3y+12}{10} \\ \frac{-3x+9y+4}{10} \end{bmatrix}$ so $\Pi_L \begin{bmatrix} 10 \\ 0 \end{bmatrix} = \begin{bmatrix} \frac{10+30+12}{10} \\ \frac{-30+90+4}{10} \end{bmatrix} = \begin{bmatrix} .8 \\ 6.4 \end{bmatrix} \in L$.

Next, consider a plane in $\mathbb{R}^3$ given by $ax + by + cz = d$ with $c \neq 0$. We associate the rank 2 matrix $\begin{bmatrix} 1 & 0 & 0 \\ 0 & 1 & 0 \\ \frac{-a}{c} & \frac{-b}{c} & 0 \end{bmatrix}$ with this plane. Then the plane is described by the set of vectors

$$\left\{\begin{bmatrix} 1 & 0 & 0 \\ 0 & 1 & 0 \\ \frac{-a}{c} & \frac{-b}{c} & 0 \end{bmatrix}\begin{bmatrix} x \\ y \\ z \end{bmatrix} + \begin{bmatrix} 0 \\ 0 \\ \frac{d}{c} \end{bmatrix} |x, y, z \in \mathbb{R}\right\}.$$

We compute the affine projection $\Pi \begin{bmatrix} x \\ y \\ z \end{bmatrix} =$

$$\begin{bmatrix} 0 \\ 0 \\ \frac{d}{c} \end{bmatrix} + \begin{bmatrix} 1 & 0 & 0 \\ 0 & 1 & 0 \\ \frac{-a}{c} & \frac{-b}{c} & 0 \end{bmatrix}\begin{bmatrix} 1 & 0 & 0 \\ 0 & 1 & 0 \\ \frac{-a}{c} & \frac{-b}{c} & 0 \end{bmatrix}^+ \left(\begin{bmatrix} x \\ y \\ z \end{bmatrix} - \begin{bmatrix} 0 \\ 0 \\ \frac{d}{c} \end{bmatrix}\right).$$

The pseudoinverse above is easily computed and the product $\begin{bmatrix} 1 & 0 & 0 \\ 0 & 1 & 0 \\ \alpha & \beta & 0 \end{bmatrix}$

$$\begin{bmatrix} 1 & 0 & 0 \\ 0 & 1 & 0 \\ \alpha & \beta & 0 \end{bmatrix}^+ = \begin{bmatrix} \frac{1+\beta^2}{1+\alpha^2+\beta^2} & \frac{-\beta\alpha}{1+\alpha^2+\beta^2} & \frac{\alpha}{1+\alpha^2+\beta^2} \\ \frac{-\alpha\beta}{1+\alpha^2+\beta^2} & \frac{1+\alpha^2}{1+\alpha^2+\beta^2} & \frac{\beta}{1+\alpha^2+\beta^2} \\ \frac{\alpha}{1+\alpha^2+\beta^2} & \frac{\beta}{1+\alpha^2+\beta^2} & \frac{\alpha^2+\beta^2}{1+\alpha^2+\beta^2} \end{bmatrix}.$$

For example, to the plane $3x+2y+6z = 6$, we associate $\begin{bmatrix} 1 & 0 & 0 \\ 0 & 1 & 0 \\ \frac{-1}{2} & \frac{-1}{3} & 0 \end{bmatrix}$, so the plane consists of the vectors in $\left\{ \begin{bmatrix} 1 & 0 & 0 \\ 0 & 1 & 0 \\ \frac{-1}{2} & \frac{-1}{3} & 0 \end{bmatrix} \begin{bmatrix} x \\ y \\ z \end{bmatrix} + \begin{bmatrix} 0 \\ 0 \\ 1 \end{bmatrix} \mid x, y, z \in \mathbb{R} \right\}$. The affine projection

$$\Pi \begin{bmatrix} x \\ y \\ z \end{bmatrix} = \begin{bmatrix} 1 & 0 & 0 \\ 0 & 1 & 0 \\ \frac{-1}{2} & \frac{-1}{3} & 0 \end{bmatrix} \begin{bmatrix} 1 & 0 & 0 \\ 0 & 1 & 0 \\ \frac{-1}{2} & \frac{-1}{3} & 0 \end{bmatrix}^+ \left(\begin{bmatrix} x \\ y \\ z \end{bmatrix} - \begin{bmatrix} 0 \\ 0 \\ 1 \end{bmatrix} \right) + \begin{bmatrix} 0 \\ 0 \\ 1 \end{bmatrix} = \begin{bmatrix} \frac{40}{49} & \frac{-6}{49} & \frac{-18}{49} \\ \frac{-6}{49} & \frac{45}{49} & \frac{-12}{49} \\ \frac{-18}{49} & \frac{-12}{49} & \frac{13}{49} \end{bmatrix} \begin{bmatrix} x \\ y \\ z-1 \end{bmatrix} + \begin{bmatrix} 0 \\ 0 \\ 1 \end{bmatrix} =$$

$$\begin{bmatrix} \frac{40}{49}x & \frac{-6}{49}y & \frac{-18}{49}(z-1) \\ \frac{-6}{49}x & \frac{45}{49}y & \frac{-12}{49}(z-1) \\ \frac{-18}{49}x & \frac{-12}{49}y & \frac{13}{49}(z-1)+1 \end{bmatrix}.$$ So, to project $\begin{bmatrix} 1 \\ 1 \\ 3 \end{bmatrix}$ onto this plane, we simply compute $\Pi \begin{bmatrix} 1 \\ 1 \\ 3 \end{bmatrix} = \begin{bmatrix} \frac{-2}{49} \\ \frac{15}{49} \\ \frac{45}{49} \end{bmatrix}$. As one last illustration, suppose we wish to project $\begin{bmatrix} 1 \\ 1 \\ 5 \end{bmatrix}$ onto the line in $\mathbb{R}^3$ given by $L(t) = (1, 3, 0)+t(1, -1, 2)$.

We associate the matrix $\begin{bmatrix} 1 & 0 & 0 \\ -1 & 0 & 0 \\ 2 & 0 & 0 \end{bmatrix}$ of rank 1 to this line so the line becomes the set of vectors $\left\{ \begin{bmatrix} 1 & 0 & 0 \\ -1 & 0 & 0 \\ 2 & 0 & 0 \end{bmatrix} \begin{bmatrix} x \\ y \\ z \end{bmatrix} + \begin{bmatrix} 1 \\ 3 \\ 0 \end{bmatrix} \mid x, y, z \in \mathbb{R} \right\}$.

The affine projection onto this line is $\Pi \begin{bmatrix} x \\ y \\ z \end{bmatrix} = \begin{bmatrix} \frac{1}{6} & \frac{-1}{6} & \frac{2}{6} \\ \frac{-1}{6} & \frac{1}{6} & \frac{-2}{6} \\ \frac{2}{6} & \frac{-2}{6} & \frac{4}{6} \end{bmatrix}$

$$\left(\begin{bmatrix} x \\ y \\ z \end{bmatrix} - \begin{bmatrix} 1 \\ 3 \\ 0 \end{bmatrix}\right) + \begin{bmatrix} 1 \\ 3 \\ 0 \end{bmatrix} = \begin{bmatrix} \frac{1}{6}(x-1) & \frac{-1}{6}(y-3) & \frac{2}{6}z+1 \\ \frac{-1}{6}(x-1) & \frac{1}{6}(y-3) & \frac{-2}{6}z+3 \\ \frac{2}{6}(x-1) & \frac{-2}{6}(y-3) & \frac{4}{6}z \end{bmatrix}.$$

So, for example, $\Pi \begin{bmatrix} 1 \\ 1 \\ 5 \end{bmatrix} = \begin{bmatrix} 3 \\ 1 \\ 4 \end{bmatrix}$.

Since orthogonal projections solve a minimization problem so do affine projections. If we wish to solve the approximation problem for $\mathbf{x}$ and $M(\mathbf{a}, U)$ — that is, we wish to minimize $\|\mathbf{x} - \mathbf{y}\|$ for $\mathbf{y} \in M(\mathbf{a}, U)$ — we see that this is the same as minimizing $\|(\mathbf{x} - \mathbf{a}) - (\mathbf{y} - \mathbf{a})\|$, as $\mathbf{y} - \mathbf{a}$ runs through the linear subspace U. But this problem is solved by $P_U(\mathbf{x}-\mathbf{a})$. Thus $\left\|\mathbf{x} - \Pi_{M(\mathbf{a},U)}(\mathbf{x})\right\| = \|(\mathbf{x} - \mathbf{a}) - P_U(\mathbf{x} - \mathbf{a})\|$ solves the minimization problem for $M(a, U)$. To illustrate, suppose we want to minimize the distance between $\begin{bmatrix} 1 \\ 1 \\ 3 \end{bmatrix}$ and the plane $3x + 2y + 6z = 6$. What vector should we use? $\Pi \begin{bmatrix} 1 \\ 1 \\ 3 \end{bmatrix}$ of course!

Then $\left\|\begin{bmatrix} 1 \\ 1 \\ 3 \end{bmatrix} - \Pi \begin{bmatrix} 1 \\ 1 \\ 3 \end{bmatrix}\right\| = \left\|\begin{bmatrix} 1 \\ 1 \\ 3 \end{bmatrix} - \begin{bmatrix} \frac{-2}{49} \\ \frac{19}{49} \\ \frac{45}{49} \end{bmatrix}\right\| = \frac{119}{49} = \frac{17}{7}$. Similarly, the minimum distance from $\begin{bmatrix} 1 \\ 1 \\ 5 \end{bmatrix}$ to the line $L(t) = (1, 3, 0) + t(1, -1, 2)$ is

$$\left\|\begin{bmatrix} 1 \\ 1 \\ 5 \end{bmatrix} - \Pi \begin{bmatrix} 1 \\ 1 \\ 5 \end{bmatrix}\right\| = \left\|\begin{bmatrix} 1 \\ 1 \\ 5 \end{bmatrix} - \begin{bmatrix} 3 \\ 1 \\ 4 \end{bmatrix}\right\| = \left\|\begin{bmatrix} -2 \\ 0 \\ 1 \end{bmatrix}\right\| = \sqrt{5}.$$

Our last task is to project onto the intersection of two affine subspaces when this intersection is not empty. You may recall it was not that easy to project orthogonally onto linear subspaces. It takes some work for affine subspaces as well. We offer a more computational approach this time.

Suppose matrices A and B determine affine subspaces that intersect (i.e., $M(\mathbf{a}, A) \cap M(\mathbf{b}, B) \neq \emptyset$). By (8), we have $\mathbf{b} - \mathbf{a} \in \mathcal{C}ol(A) + \mathcal{C}ol(B) = \mathcal{C}ol([A \vdots B])$. Thus, $\mathbf{b} - \mathbf{a} = A\mathbf{x}_1 + B\mathbf{x}_2 = [A \vdots B] \begin{bmatrix} \mathbf{x}_1 \\ \cdots \\ \mathbf{x}_2 \end{bmatrix}$. One useful fact about pseudoinverses is $Y^+ = (Y^*Y)^+Y^*$, so we compute that $[A \vdots B]^+ = \begin{bmatrix} A^* \\ B^* \end{bmatrix} \left([A \vdots B] \begin{bmatrix} A^* \\ B^* \end{bmatrix}\right)^+ = \begin{bmatrix} A^*(AA^* + BB^*)^+ \\ B^*(AA^* + BB^*)^+ \end{bmatrix}$. Let $D = AA^* + BB^*$. Then the projection onto $\mathcal{C}ol(A) + \mathcal{C}ol(B)$ is $[A \vdots B][A \vdots B]^+ =$

$[A\vdots B]\begin{bmatrix} A^*D^+ \\ B^*D^+ \end{bmatrix} = AA^*D^+ + BB^*D^+ = DD^+$. So $M(\mathbf{a}, A) \cap M(\mathbf{b}, B) \neq \emptyset$ iff $\mathbf{b} - \mathbf{a} = DD^+(\mathbf{b} - \mathbf{a})$. Next, $\mathbf{b} - \mathbf{a} = DD^+(\mathbf{b} - \mathbf{a})$ iff $\mathbf{b} - \mathbf{a} = AA^*D^+(\mathbf{b} - \mathbf{a}) + BB^*D^+(\mathbf{b} - \mathbf{a})$ iff $AA^*D^+(\mathbf{b} - \mathbf{a}) + \mathbf{a} = -BB^*D^+(\mathbf{b} - \mathbf{a}) + \mathbf{b}$. Also $\mathcal{C}ol(A)$, $\mathcal{C}ol(B) \subseteq \mathcal{C}ol(A) + \mathcal{C}ol(B)$ so $DD^+A = A$ and $DD^+B = B$. But $DD^+A = A$ implies $(AA^* + BB^*)D^+A = A$, so $BB^*D^+A = A - AA^*D^+A$ whence $BB^*D^+AA^* = AA^* - AA^*D^+AA^*$. But this matrix is self-adjoint, so $BB^*D^+AA^* = (BB^*D^+AA^*)^* = AA^*D^+BB^*$.

THEOREM 8.17

$M(\mathbf{a}, A) \cap M(\mathbf{b}, B) \neq \emptyset$ iff $\mathbf{b} - \mathbf{a} = DD^*(\mathbf{b} - \mathbf{a})$, where $D = AA^* + BB^*$. In that case, $M(\mathbf{a}, A) \cap M(\mathbf{b}, B) = M(\mathbf{c}, C)$, where $\mathbf{c} = AA^*D^+(\mathbf{b} - \mathbf{a}) + \mathbf{a}$ and $C = [BB^*D^+A\vdots AA^*D^+B]$.

PROOF Let $\mathbf{y} \in M(\mathbf{a}, A) \cap M(\mathbf{b}, B)$. Then $\mathbf{y} = A\mathbf{x}_1 + \mathbf{a}$ and $\mathbf{y} = B\mathbf{x}_2 + \mathbf{b}$ for suitable $\mathbf{x}_1$ and $\mathbf{x}_2$. Then $A\mathbf{x}_1 + \mathbf{a} = B\mathbf{x}_2 + \mathbf{b}$ so $\mathbf{b} - \mathbf{a} = A\mathbf{x}_1 - B\mathbf{x}_2$. In matrix notation we write this as $[A\vdots - B]\begin{bmatrix} \mathbf{x}_1 \\ \cdots \\ \mathbf{x}_2 \end{bmatrix} = \mathbf{b} - \mathbf{a}$. The solution set to this equation is $\begin{bmatrix} \mathbf{x}_1 \\ \mathbf{x}_2 \end{bmatrix} = [A\vdots - B]^+(\mathbf{b} - \mathbf{a}) + [I - [A\vdots - B][A\vdots - B]^+]\mathbf{z} =$ $\begin{bmatrix} A^*D^+(\mathbf{b} - \mathbf{a}) \\ -B^+D^+(\mathbf{b} - \mathbf{a}) \end{bmatrix} + \begin{bmatrix} I - A^*D^+A & A^*D^+B \\ B^*D^+A & I - B^*D^+B \end{bmatrix}\begin{bmatrix} \mathbf{z}_1 \\ \mathbf{z}_2 \end{bmatrix}$, where $\mathbf{z} = \begin{bmatrix} \mathbf{x}_1 \\ \mathbf{x}_2 \end{bmatrix}$. Then $\mathbf{y} = A\mathbf{x}_1 + \mathbf{a} = \vdots B - BB^*D^+B]\mathbf{z} = B\ [AA^*D(\mathbf{b} - \mathbf{a}) + \mathbf{a}] +$

$[A - DD^*D^+A\vdots A^*D^+B]\mathbf{z}_1 = \mathbf{c} + [BB^*D^+A\vdots AA^*D^+B]\mathbf{z}_1$. This puts $\mathbf{y}$ in $M(\mathbf{c}, C)$.

Conversely, suppose $\mathbf{y} \in M(\mathbf{c}, C)$. Then $\mathbf{y} = \mathbf{c} + [BB^*D^+A\vdots AA^*D^+B]\mathbf{z}$ for some $\mathbf{z} \in \mathcal{C}ol(C)$. But then $\mathbf{y} = (AA^*D^+(\mathbf{b} - \mathbf{a}) + \mathbf{a}) + [A - AA^*D^+A\vdots AA^*D^+B]\mathbf{z} = A[A^*D^+(\mathbf{b} - \mathbf{a})] + A[I - A^*D^+A\vdots A^*D^+B]\mathbf{z} + \mathbf{a} = A\mathbf{w}_1 + \mathbf{a} \in M(\mathbf{a}, A)$. Similarly, $\mathbf{y} = (-BB^*D^+(\mathbf{b} - \mathbf{a}) + \mathbf{b}) + [BB^*D^+A\vdots B - BB^*D^+B]\mathbf{z} = B[-B^*D^+(\mathbf{b} - \mathbf{a}) + B[B^*D^+A\vdots I - B^*D^+B]\mathbf{z} + \mathbf{b} = B\mathbf{w}_2 + \mathbf{b} \in M(\mathbf{b}, B)$. This puts $\mathbf{y} \in M(\mathbf{a}, A) \cap M(\mathbf{b}, B)$ and completes the proof. □

Before we write down the projection onto the intersection of two affine subspaces, we obtain a simplification.

THEOREM 8.18

$[BB^*D^+A\dot{:}AA^*D^+B][BB^*D^+A\dot{:}AA^*D^+B]^+ = [BB^*D^+AA^*][BB^*D^+AA^*]^+$.

PROOF Compute $CC^+ = [BB^*D^+A\dot{:}AA^*D^+B][BB^*D^+A\dot{:}AA^*D^+B]^+ = CC^*(CC^*)^+ = CC^*[BB^*D^+A\dot{:}AA^*D^+B][BB^*D^+A\dot{:}AA^*D^+B]^+ = CC^*[BB^*D^+AA^*D^+BB^* + AA^*D^+BB^*D^+AA^*]^+ = CC^*[(BB^* + AA^*)D^+BB^*D^+AA^*]^+ = CC^*[DD^+BB^*D^+AA^*]^+ = CC^*[BB^*D^+AA^*]^+ = [BB^*D^+AA^*][BB^*D^+AA^*]^+$, which is what we want. ∎

We can now exhibit a concrete formula for the affine projection onto the intersection of affine subspaces: $\Pi_{M(\mathbf{a},A)\cap M(\mathbf{b},B)}(\mathbf{x}) = [BB^*D^+AA^*][BB^*D^+AA^*]^+(\mathbf{x} - \mathbf{c}) + \mathbf{c}$, where $\mathbf{c} = AA^*D^+(\mathbf{b} - \mathbf{a}) + \mathbf{a}$. We note with interest that $BB^*D^+AA^* = BB^*[AA^* + BB^*]^+AA^*$ is just the parallel sum of BB^* and AA^*. In particular the orthogonal projection onto $\mathcal{C}ol(A)\cap\mathcal{C}ol(B)$ can be computed as $P_{\mathcal{C}ol(A)\cap\mathcal{C}ol(B)} = [BB^*D^+AA^*][BB^*D^+AA^*]^+$. This adds yet another formula to our list.

We illustrate with an example. Consider the plane in $\mathbb{R}^3$, $3x - 6y - 2z = 15$, to which we associate the matrix $A = \begin{bmatrix} 1 & 0 & 0 \\ 0 & 1 & 0 \\ \frac{3}{2} & -3 & 0 \end{bmatrix}$ and vector $\mathbf{a} = \begin{bmatrix} 0 \\ 0 \\ \frac{-15}{2} \end{bmatrix}$. Also consider the plane $2x + y - 2z = 5$, to which we associate the matrix $B = \begin{bmatrix} 1 & 0 & 0 \\ 0 & 1 & 0 \\ 1 & \frac{1}{2} & 0 \end{bmatrix}$ and vector $\mathbf{b} = \begin{bmatrix} 0 \\ 0 \\ \frac{-5}{2} \end{bmatrix}$. We shall compute the projection onto the intersection of these planes. First, $AA^* = \begin{bmatrix} 1 & 0 & \frac{3}{2} \\ 0 & 1 & -3 \\ \frac{3}{2} & -3 & \frac{45}{2} \end{bmatrix}$, $BB^* = \begin{bmatrix} 1 & 0 & 1 \\ 0 & 1 & \frac{1}{2} \\ 1 & \frac{1}{2} & \frac{5}{4} \end{bmatrix}$, and $D^+ = (AA^* + BB^*)^+ = \begin{bmatrix} \frac{3}{4} & \frac{-1}{4} & \frac{-1}{5} \\ \frac{-1}{4} & \frac{3}{4} & \frac{1}{5} \\ \frac{-1}{5} & \frac{1}{5} & 25 \end{bmatrix}$. Now $DD^+ = I$ ensuring that these planes do intersect without appealing to dimension considerations. Next, $\mathbf{c} = \begin{bmatrix} \frac{1}{5} \\ \frac{-7}{5} \\ 3 \end{bmatrix}$ and $C = \begin{bmatrix} \frac{49}{100} & \frac{7}{100} & \frac{21}{40} \\ \frac{7}{100} & \frac{1}{100} & \frac{3}{40} \\ \frac{21}{40} & \frac{3}{40} & \frac{9}{16} \end{bmatrix}$ and so $\Pi\begin{bmatrix} x \\ y \\ z \end{bmatrix} = \begin{bmatrix} \frac{196}{425}(x-\frac{1}{5})+\frac{28}{425}(y+\frac{7}{5})+\frac{42}{85}(z+3)+\frac{1}{5} \\ \frac{28}{425}(x-\frac{1}{5})+\frac{4}{425}(y+\frac{7}{5})+\frac{6}{85}(z+3)+\frac{7}{5} \\ \frac{42}{85}(x-\frac{1}{5})+\frac{6}{85}(y+\frac{7}{5})+\frac{9}{17}(z+3)-3 \end{bmatrix} = \begin{bmatrix} \frac{1}{5} \\ \frac{-7}{5} \\ -3 \end{bmatrix} +$ $\begin{bmatrix} 14t \\ 2t \\ 15t \end{bmatrix}$, where $t = \frac{14}{215}(x-\frac{1}{5})+\frac{2}{415}(y+\frac{7}{5})+\frac{15}{425}(z+3)$.

The formula we have demonstrated can be extended to any number of affine subspaces that intersect nontrivially.

Exercise Set 37

1. Consider $A = \begin{bmatrix} 1 & 0 & 1 & 1 & 0 & 1 \\ 0 & 1 & 1 & 1 & 1 & 0 \\ 1 & 1 & 0 & 1 & 0 & 1 \\ 1 & 1 & 0 & 0 & 1 & 1 \\ 1 & 1 & 0 & 0 & 1 & 1 \\ 0 & 0 & 1 & 0 & 1 & 1 \end{bmatrix}$ and $B = \begin{bmatrix} 1 & 0 & 1 & 0 & 1 & 1 \\ 0 & 1 & 0 & 1 & 0 & 1 \\ 0 & 1 & 1 & 0 & 1 & 0 \\ 1 & 1 & 0 & 1 & 0 & 1 \\ 1 & 1 & 0 & 1 & 0 & 0 \\ 0 & 0 & 1 & 0 & 0 & 1 \end{bmatrix}$.

   Let $\mathbf{a} = \begin{bmatrix} 1 \\ 0 \\ 0 \\ 0 \\ 0 \\ 1 \end{bmatrix}$ and $\mathbf{b} = \begin{bmatrix} 0 \\ 1 \\ 0 \\ 0 \\ 0 \\ 1 \end{bmatrix}$. Compute AA^*, BB^*, and $D^+ = (AA^* + BB^*)$, and $\mathbf{c}$. Finally, compute $\prod(x, y, z)$, the projection onto the intersection of $M(\mathbf{a}, A)$ and $M(\mathbf{b}, B)$.

Further Reading

[P&O, 2004] R. Piziak and P. L. Odell, Affine Projections, Computers and Mathematics with Applications, 48, (2004), 177–190.

[Rock, 1970] R. Tyrrell Rockafellar, *Convex Analysis*, Princeton University Press, Princeton, NJ, (1970).

8.6 Quotient Spaces (optional)

In this section, we take a more sophisticated look at affine subspaces. This section may be omitted without any problems. Let V be a vector space and M a subspace of V. Let V/M (read "$V mod M$") denote the collection of all affine subspaces of V generated by M

$$V/M = \{\mathbf{v} + M \mid \mathbf{v} \in V\}.$$

Our goal is to make this set V/M into a vector space in its own right using the same scalars that V has. In other words, affine subspaces can be viewed as vectors. First, we review some basics.

THEOREM 8.19
Let M be a subspace of V. Then T.A.E.:

1. $\mathbf{u} + M = \mathbf{v} + M$
2. $\mathbf{v} \in \mathbf{u} + M$
3. $\mathbf{u} \in \mathbf{v} + M$
4. $\mathbf{u} - \mathbf{v} \in M$.

PROOF The proof is left as an exercise. □

Let's talk about the vector operations first. There are some subtle issues here. Suppose $\mathbf{u} + M$ and $\mathbf{v} + M$ are affine subspaces. (By the way, some people call these sets *cosets*.) How would you add them? The most natural approach would be

$$(\mathbf{u} + M) \boxplus (\mathbf{v} + M) := (\mathbf{u} + \mathbf{v}) + M.$$

The problem is that the vector $\mathbf{u}$ does not uniquely determine the affine space $\mathbf{u} + M$. Indeed, $\mathbf{u} + M$ could equal $\mathbf{u}_1 + M$ and $\mathbf{v} + M$ could equal $\mathbf{v}_1 + M$. How do we know $(\mathbf{u} + \mathbf{v}) + M = (\mathbf{u}_1 + \mathbf{v}_1) + M$? The fancy language is, how do we know $\boxplus$ is "well defined"? Suppose $\mathbf{u} + M = \mathbf{u}_1 + M$ and $\mathbf{v} + M = \mathbf{v}_1 + M$. Then $\mathbf{u} - \mathbf{u}_1 \in M$ and $\mathbf{v} - \mathbf{v}_1 \in M$. But M is a subspace, so the sum $(\mathbf{u}-\mathbf{u}_1)+(\mathbf{v}-\mathbf{v}_1) \in M$. A little algebra then says $(\mathbf{u}+\mathbf{v})-(\mathbf{u}_1+\mathbf{v}_1) \in M$ whence $(\mathbf{u}+\mathbf{v})+M = (\mathbf{u}_1+\mathbf{v}_1)+M$. This says $(\mathbf{u}+M)\boxplus(\mathbf{v}+M) = (\mathbf{u}_1+M)\boxplus(\mathbf{v}_1+M)$. In other words, it does not matter what vectors we use to determine the affine space; the sum is the same.

The next issue is scalar multiplication. The same issue of "well definedness" must be confronted. Again, the most natural definition seems to be

$$\alpha(\mathbf{u} + M) = \alpha\mathbf{u} + M \text{ for } \alpha \text{ any scalar.}$$

Suppose $\mathbf{u} + M = \mathbf{v} + M$. Then $\mathbf{u} - \mathbf{v} \in M$, so $\alpha(\mathbf{u} - \mathbf{v}) \in M$ since M is a subspace. But then $\alpha\mathbf{u} - \alpha\mathbf{v} \in M$, so $\alpha\mathbf{u} + M = \alpha\mathbf{v} + M$, and again this definition of scalar multiplication does not depend on the choice of vector to represent the affine subspace. Note, the zero vector in V/M is $\overrightarrow{0} + M = M$ itself! We can visualize V/M as being created by a great collapsing occurring in V where whole sets of vectors are condensed and become single vectors in a new space.

$$V = \begin{bmatrix} M & \mathbf{u}_1 + M & \mathbf{u}_2 + M \\ \mathbf{u}_3 + M & \cdots & \cdots \end{bmatrix} \rightarrow V/M = \begin{bmatrix} \overrightarrow{0} & \cdot & \cdot \\ \cdot & \cdots & \cdots \end{bmatrix}.$$

THEOREM 8.20

Let V be a vector space and M a subspace. Then, with the operations defined above, V/M becomes a vector space in its own right.

PROOF The hard part, proving the well definedness of the operations, was done above. The vector space axioms can now be checked by the reader. □

There is a *natural mapping* η from V to V/M for any subspace M of V defined by $\eta(\mathbf{u}) = \mathbf{u} + M$.

THEOREM 8.21

With the notation above, $\eta : V \rightarrow V/M$ is a linear transformation that is onto. Moreover, $ker(\eta) = M$.

PROOF We supply a quick proof with details left to the reader. First, compute $\eta(\alpha\mathbf{u}+\beta\mathbf{v}) = (\alpha\mathbf{u}+\beta\mathbf{v})+M = (\alpha\mathbf{u}+M)\boxplus(\beta\mathbf{v}+M) = \alpha(\mathbf{u}+M)\boxplus\beta(\mathbf{v}+M) = \alpha\eta(\mathbf{u}) + \beta\eta(\mathbf{v})$ and $\mathbf{v} \in ker(\eta)$ iff $\eta(\mathbf{v}) = \overrightarrow{0}$ iff $\mathbf{v} + M = M$ iff $\mathbf{v} \in M$ so $ker(\eta) = M$. □

In finite dimensions, we can get a nice relationship between the dimension of V and that of V/M.

THEOREM 8.22

Suppose V is a finite dimensional vector space and M is a subspace of V. If $\{\boldsymbol{m}_1, \boldsymbol{m}_2, \ldots, \boldsymbol{m}_k\}$ is a basis for M and $\{\boldsymbol{v}_1 + M, \ldots, \boldsymbol{v}_n + M\}$ is a basis for V/M, the $\mathcal{B} = \{\boldsymbol{m}_1, \boldsymbol{m}_2, \ldots, \boldsymbol{m}_k, \boldsymbol{v}_1, \ldots, \boldsymbol{v}_n\}$ is a basis for V.

PROOF Suppose first that $\{\mathbf{v}_1+M, \ldots, \mathbf{v}_n+M\}$ is a linearly independent set of vectors. Then, if $\sum \alpha_i \mathbf{v}_i = \overrightarrow{0}$, we have $\overrightarrow{0}_{V/M} = \eta(\overrightarrow{0}) = \eta(\sum \alpha_i \mathbf{v}_i) = \sum \eta(\alpha_i \mathbf{v}_i) = \sum \alpha_i(\mathbf{v}_i + M)$. Thus, by independence, all the $\alpha_i' s$ are zero. This says $n \leq dim(V)$ so that V/M must be finite dimensional. Moreover, consider a linear combination of the **m**s and **v**s, say $\sum \alpha_i \mathbf{v}_i + \sum \beta_j \mathbf{m}_j$. Then $\eta(\sum \alpha_i \mathbf{v}_i + \sum \beta_j \mathbf{m}_j) = \sum \alpha_i(\mathbf{v}_i + M)$, which implies all the αs are zero. But then, we are left with $\sum \beta_j \mathbf{m}_j = \overrightarrow{0}$. Again, by the assumption of independence, all the βs are zero as well. Hence we have the independence we need. For spanning, let $\mathbf{v}$ be in V. Then $\eta(\mathbf{v})$ is in V/M, so there must exist $\alpha_i' s$ such that $\mathbf{v}+M = \sum \alpha_i(\mathbf{v}_i + M) = (\sum \alpha_i \mathbf{v}_i) + M$, and so $\mathbf{v} - (\sum \alpha_i \mathbf{v}_i) \in M$. But since we have a basis for M, there must exist $\beta_j' s$ with $\mathbf{v} - (\sum \alpha_i \mathbf{v}_i) = \sum \beta_j \mathbf{m}_j$. Therefore, $\mathbf{v} = \sum \alpha_i \mathbf{v}_i + \sum \beta_j \mathbf{m}_j$, which is in the span of $\mathcal{B}$. This completes the argument. ▯

COROLLARY 8.5
In finite dimensions, dim(V)=dim(M) + dim(V/M) for any subspace M of V.

PROOF The proof is left as an exercise. ▯

We end with a theorem about linear transformations and their relation to quotient spaces.

THEOREM 8.23
Suppose $T : V \to V$ *is a linear transformation and M is an invariant subspace for* T. *Then* $\hat{T} : V/M \to V/M$, *defined by* $\hat{T}(\mathbf{v} + M) = T(\mathbf{v}) + M$, *is a well-defined linear map whose minimal polynomial divides that of* T.

PROOF The proof is left as an exercise. ▯

Exercise Set 38

1. Prove Theorem 8.19.

2. Finish the proof of Theorem 8.20.

3. Let M be a subspace of V. For vectors $\mathbf{v}$, $\mathbf{u}$ in V, define $\mathbf{u} \sim \mathbf{v}$ iff $\mathbf{u} - \mathbf{v} \in M$. Argue that $\sim$ is an equivalence relation on V and the equivalence classes are exactly the affine subspaces generated by M.

4. Prove Corollary 8.5.

5. Fill in the details of Theorem 8.21 and Theorem 8.23.

6. Suppose $T : V \to W$ is a linear map and K is a subspace of V contained in $ker(T)$. Prove that there exists a unique linear map $\widehat{T}$ so that $\widehat{T}:V/K \to W$, with the property that $\widehat{T} \circ \eta = T$. Moreover, $ker(\widehat{T}) = \{\mathbf{v} + K \mid \mathbf{v} \in \ker(T)\}$ and $im(\widehat{T}) = im(T)$. (*Hint*: Begin by proving that $\widehat{T}$ is well defined.) Then compute $im(\widehat{T})$ and $ker(\widehat{T})$.

7. (first isomorphism theorem) Let $T : V \to W$ be a linear map. Argue that $\widehat{T}:V/ker(\widehat{T}) \to W$ defined by $\widehat{T}(\mathbf{v} + ker(T)) = T(\mathbf{v})$ is not only linear but is actually one-to-one. Conclude that $V/ker(T)$ is isomorphic to $im(T)$.

8. Suppose $V = M \oplus N$. Argue that N is isomorphic to V/M.

9. Suppose $T : V \to W$ is a linear map. Prove that $dim(ker(T)) + dim(im(T)) = dim(V)$, where V is finite dimensional.

10. (second isomorphism theorem) Let V be a vector space with subspaces M and N. Prove that $(M + N)/N$ is isomorphic to $M/(M \cap N)$. (*Hint*: Define $T : M+N \to M/(M\cap N)$ by $T(\mathbf{m}+\mathbf{n}) = \mathbf{m}+(M\cap N)$. Prove T is well defined and determine the kernel. Then use the first isomorphism theorem.)

11. (third isomorphism theorem) Let V be a vector space with subspaces M and N, and suppose $M \subseteq N$. Prove that $(V/M)/(N/M)$ is isomorphic to V/N. (*Hint:* Define $T : V/M \to V/N$ by $T(\mathbf{v} + M) = \mathbf{v} + N$. Prove T is well defined and determine the kernel. Then use the first isomorphism theorem.)

12. Suppose you have a basis $\mathbf{b}_1, \ldots, \mathbf{b}_k$ of a subspace M of V. How could you use it to construct a basis of V/M?

Chapter 9

Spectral Theory

eigenvalue, eigenvector, eigenspace, spectrum, geometric multiplicity, eigenprojection, algebraic multiplicity

9.1 Eigenstuff

In this chapter, we deal only with square matrices. Suppose we have a matrix A in $\mathbb{C}^{n\times n}$ and a subspace $M \subseteq \mathbb{C}^{n\times n}$. Recall, M is an *invariant subspace* for A if $A(M) \subseteq M$. That is, when A multiplies a vector $\mathbf{v}$ in M, the result $A\mathbf{v}$ remains in M. It is easy to see, for example, that the null space of A is an invariant subspace for A. For the moment, we restrict to one-dimensional subspaces. Suppose $\mathbf{v}$ is a nonzero vector in $\mathbb{C}^n$ and $sp(\mathbf{v})$ is the one-dimensional subspace spanned by $\mathbf{v}$. Then, to have an invariant subspace of A, we would need to have $A(sp(\mathbf{v})) \subseteq sp(\mathbf{v})$. But $\mathbf{v}$ is in $sp(\mathbf{v})$, so $A\mathbf{v}$ must be in $sp(\mathbf{v})$ also. Therefore, $A\mathbf{v}$ would have to be a scalar multiple of $\mathbf{v}$. Let's say $A\mathbf{v} = \lambda\mathbf{v}$. Conversely, if we can find a nonzero vector $\mathbf{v}$ and a scalar λ with $A\mathbf{v} = \lambda\mathbf{v}$, then $sp(\mathbf{v})$ is an invariant subspace for A. So the search for one-dimensional invariant subspaces for a matrix A boils down to solving $A\mathbf{v} = \lambda\mathbf{v}$ for a nonzero vector $\mathbf{v}$. This leads to some language.

DEFINITION 9.1 *(eigenvalue, eigenvector, eigenspace)*
A complex number λ is called an ***eigenvalue*** *of A in $\mathbb{C}^{n\times n}$ if and only if there is a nonzero vector $\mathbf{v}$ in $\mathbb{C}^n$ with $A\mathbf{v} = \lambda\mathbf{v}$. The vector $\mathbf{v}$ is called an* ***eigenvector****. The set of all possible eigenvalues of A is called the* ***spectrum*** *of A and is denoted $\lambda(A)$. For λ, an eigenvalue of A, we associate the subspace $Eig(A, \lambda) = M_\lambda = \mathcal{N}ull(\lambda I - A)$ and call M_λ the* ***eigenspace*** *associated with λ. The dimension of M_λ is called the* ***geometric multiplicity*** *of λ. Finally, M_λ, being a subspace, has a projection P_{M_λ} associated with it called the λth* ***eigenprojection****. In fact, $P_{M_\lambda} = (\lambda I - A)'$.*

Eigenvalues were used in 1772 by **Pierre Simon Laplace** (23 March 1749–5 March 1827) to discuss variations in planetary motions. Let's look at some examples to help understand what is going on.

Example 9.1

1. *Consider the identity matrix I_n. Then $I_n\mathbf{v} = \mathbf{v}$ for all $\mathbf{v}$, so 1 is the only eigenvalue of I_n. Therefore, $\lambda(I_n) = \{1\}$. Also, $Eig(I, 1) = M_1 = \mathcal{N}ull(1I_n - I_n) = \mathcal{N}ull(\mathbb{O}) = \mathbb{C}^n$, so the geometric multiplicity of 1 is n. Similarly, the zero matrix $\mathbb{O}$ has $\mathbb{O}\mathbf{v} = \overrightarrow{0} = 0\mathbf{v}$ for all $\mathbf{v}$ so $\lambda(\mathbb{O}) = \{0\}$ and $M_0 = \mathcal{N}ull(0I - I) = \mathcal{N}ull(I) = (\overrightarrow{0})$. The geometric multiplicity of 0 is 0.*

2. *For a slightly more interesting example, consider* $A = \begin{bmatrix} 1 & 0 & 0 \\ 0 & 1 & 0 \\ 0 & 0 & 0 \end{bmatrix}$.

   Then 1 is an eigenvalue of A since $\begin{bmatrix} 1 & 0 & 0 \\ 0 & 1 & 0 \\ 0 & 0 & 0 \end{bmatrix}\begin{bmatrix} 1 \\ 0 \\ 0 \end{bmatrix} = \begin{bmatrix} 1 \\ 0 \\ 0 \end{bmatrix} = 1\begin{bmatrix} 1 \\ 0 \\ 0 \end{bmatrix}$. *We see an eigenvector for 1 is* $\begin{bmatrix} 1 \\ 0 \\ 0 \end{bmatrix}$. *But also,* $\begin{bmatrix} 1 & 0 & 0 \\ 0 & 1 & 0 \\ 0 & 0 & 0 \end{bmatrix}\begin{bmatrix} 0 \\ 1 \\ 0 \end{bmatrix} = \begin{bmatrix} 0 \\ 1 \\ 0 \end{bmatrix} = 1\begin{bmatrix} 0 \\ 1 \\ 0 \end{bmatrix}$, *so* $\begin{bmatrix} 0 \\ 1 \\ 0 \end{bmatrix}$ *is an eigenvector for 1, independent of the other eigenvector* $\begin{bmatrix} 1 \\ 0 \\ 0 \end{bmatrix}$. *Also* $\begin{bmatrix} 1 & 0 & 0 \\ 0 & 1 & 0 \\ 0 & 0 & 0 \end{bmatrix}\begin{bmatrix} 0 \\ 0 \\ 1 \end{bmatrix} = \begin{bmatrix} 0 \\ 0 \\ 0 \end{bmatrix} = 0\begin{bmatrix} 0 \\ 0 \\ 1 \end{bmatrix}$ *so 0 is an eigenvalue for A as well. Indeed,* $M_0 = \mathcal{N}ull(-A) = \left\{\mathbf{x} \mid \begin{bmatrix} -1 & 0 & 0 \\ 0 & -1 & 0 \\ 0 & 0 & 0 \end{bmatrix}\begin{bmatrix} x_1 \\ x_2 \\ x_3 \end{bmatrix} = \begin{bmatrix} 0 \\ 0 \\ 0 \end{bmatrix}\right\} = \left\{\mathbf{x} \mid \begin{bmatrix} -x_1 \\ -x_2 \\ x_3 \end{bmatrix} = \begin{bmatrix} 0 \\ 0 \\ 0 \end{bmatrix}\right\} = \left\{\begin{bmatrix} 0 \\ 0 \\ \alpha \end{bmatrix} \mid \alpha \in \mathbb{C}\right\}$ *and* $dim(M_0) = 1$. *Meanwhile,* $M_1 = \mathcal{N}ull(I - A) = \mathcal{N}ull\left(\begin{bmatrix} 0 & 0 & 0 \\ 0 & 0 & 0 \\ 0 & 0 & 1 \end{bmatrix}\right) = \left\{\mathbf{x} \mid \begin{bmatrix} 0 & 0 & 0 \\ 0 & 0 & 0 \\ 0 & 0 & 1 \end{bmatrix}\begin{bmatrix} x_1 \\ x_2 \\ x_3 \end{bmatrix} = \begin{bmatrix} 0 \\ 0 \\ 0 \end{bmatrix}\right\} = \left\{\begin{bmatrix} \beta \\ \gamma \\ 0 \end{bmatrix} \mid \beta, \gamma \in \mathbb{C}\right\}$ *so* $dim(M_1) = 2$.

3. *Let* $A = \begin{bmatrix} 1 & 0 & 0 \\ 0 & 2 & 0 \\ 0 & 0 & 3 \end{bmatrix}$. *Then* $A\mathbf{v} = \lambda\mathbf{v}$ *iff* $\begin{bmatrix} 1 & 0 & 0 \\ 0 & 2 & 0 \\ 0 & 0 & 3 \end{bmatrix} \begin{bmatrix} x_1 \\ x_2 \\ x_3 \end{bmatrix} = \lambda \begin{bmatrix} x_1 \\ x_2 \\ x_3 \end{bmatrix}$ *iff* $\begin{bmatrix} x_1 \\ 2x_2 \\ 3x_3 \end{bmatrix} = \begin{bmatrix} \lambda x_1 \\ \lambda x_2 \\ \lambda x_3 \end{bmatrix}$. *Thus* $\begin{bmatrix} 1 \\ 0 \\ 0 \end{bmatrix}$ *has eigenvalue* $\lambda = 1$, $\begin{bmatrix} 0 \\ 1 \\ 0 \end{bmatrix}$ *has* $\lambda = 2$, *and* $\begin{bmatrix} 0 \\ 0 \\ 1 \end{bmatrix}$ *has* $\lambda = 3$. *Thus* $\lambda(A) = \{1, 2, 3\}$ *and each eigenvalue has geometric multiplicity equal to 1. Can you generalize this example to any diagonal matrix?*

4. *Let* $A = \begin{bmatrix} 1 & 2 & 3 \\ 0 & 4 & 5 \\ 0 & 0 & 6 \end{bmatrix}$. *Then* $A\mathbf{v} = \lambda\mathbf{v}$ *iff* $\begin{bmatrix} 1 & 2 & 3 \\ 0 & 4 & 5 \\ 0 & 0 & 6 \end{bmatrix} \begin{bmatrix} x_1 \\ x_2 \\ x_3 \end{bmatrix} = \lambda \begin{bmatrix} x_1 \\ x_2 \\ x_3 \end{bmatrix}$ *iff* $\begin{bmatrix} x_1 + 2x_2 + 3x_3 \\ 4x_2 + 5x_3 \\ 6x_3 \end{bmatrix} = \begin{bmatrix} \lambda x_1 \\ \lambda x_2 \\ \lambda x_3 \end{bmatrix}$ *iff* $\begin{matrix} (1-\lambda)x_1 + 2x_2 + 3x_3 = 0 \\ (4-\lambda)x_2 + 5x_3 = 0 \\ (6-\lambda)x_3 = 0 \end{matrix}$. *Solving this system we find that* $\begin{bmatrix} \frac{8}{5} \\ \frac{5}{2} \\ 1 \end{bmatrix}$ *is an eigenvector for* $\lambda = 6$, $\begin{bmatrix} 2 \\ 3 \\ 0 \end{bmatrix}$ *has eigenvalue* $\lambda = 4$, *and* $\begin{bmatrix} 1 \\ 0 \\ 0 \end{bmatrix}$ *has eigenvalue* $\lambda = 1$. *Thus,* $\lambda(A) = \{1, 4, 6\}$. *Do you see how to generalize this example?*

5. *(Jordan blocks) For* $\lambda \in \mathbb{C}$ *and m any integer, we can form the m-by-m matrix* $J_m(\lambda)$ *formed, by taking the m-by-m diagonal matrix* $diag(\lambda, \lambda, \ldots, \lambda)$ *and adding to it the matrix with all zeros except ones on the super diagonal. So, for example,* $J_1(\lambda) = [\lambda]$, $J_2(\lambda) = \begin{bmatrix} \lambda & 1 \\ 0 & \lambda \end{bmatrix}$, $J_3(\lambda) = \begin{bmatrix} \lambda & 1 & 0 \\ 0 & \lambda & 1 \\ 0 & 0 & \lambda \end{bmatrix}$, $J_4(\lambda) = \begin{bmatrix} \lambda & 1 & 0 & 0 \\ 0 & \lambda & 1 & 0 \\ 0 & 0 & \lambda & 1 \\ 0 & 0 & 0 & \lambda \end{bmatrix}$, *and so on. It is relatively easy to see that* λ *is the only eigenvalue of* $J_m(\lambda)$.

THEOREM 9.1

Let $A \in \mathbb{C}^{n \times n}$. *Then these are all equivalent statements:*

1. $\lambda \in \lambda(A)$.

2. $\mathcal{N}ull(\lambda I - A)$ *is not the trivial subspace.*

3. *$\lambda I - A$ is not an invertible matrix.*

4. *The system of equations $(\lambda I - A)\mathbf{v} = \overrightarrow{0}$ has a nonzero solution.*

5. *The projection $(\lambda I - A)'$ is not $\mathbb{O}$.*

6. $det(\lambda I - A) = 0$.

PROOF The proof is left as an exercise. $\square$

We note that it really does not matter whether we use $\lambda I - A$ or $A - \lambda I$ above. It is strictly a matter of convenience which we use.

For all we know, the spectrum of a matrix A could be empty. At this point, we need a very powerful fact about complex numbers to prove that this never happens for complex matrices. We use the polynomial connection.

THEOREM 9.2
Every matrix A in $\mathbb{C}^{n\times n}$ has at least one eigenvalue.

PROOF Take any nonzero vector $\mathbf{v}$ in $\mathbb{C}^n$ and consider the vectors $\mathbf{v}$, $A\mathbf{v}$, $A^2\mathbf{v}, \ldots, A^n\mathbf{v}$. This is a set of n + 1 vectors in a vector space of dimension n, hence they must be dependent. This means there exist scalars $\alpha_0, \alpha_1, \ldots, \alpha_n$, not all zero, so that $\alpha_0\mathbf{v} + \alpha_1 A\mathbf{v} + \cdots + \alpha_n A^n\mathbf{v} = \overrightarrow{0}$. Choose the largest subscript j such that $\alpha_j \neq 0$. Call it m. Consider the polynomial $p(z) = \alpha_0 + \alpha_1 z + \cdots + \alpha_m z^m$. Note $\alpha_m \neq 0$ but $\alpha_{m+1} = \alpha_{m+2} = \cdots = \alpha_n = 0$ by our choice of m. This polynomial factors completely in $\mathbb{C}[z]$; say $p(z) = \gamma(r_1 - z)(r_2 - z)\cdots(r_m - z)$ where $\gamma \neq 0$. Then $\overrightarrow{0} = (\alpha_0 I + \alpha_1 A + \cdots + \alpha_m A^m)(\mathbf{v}) = [\gamma(r_1 I - A)(r_2 I - A)\cdots(r_m I - A)]\,\mathbf{v}$. If $\mathcal{N}ull(r_m I - A) \neq (\overrightarrow{0})$, then r_m is an eigenvalue and we are done. If not, $(r_m I - A)\mathbf{v} \neq \overrightarrow{0}$. If $r_{m-1}I - A$ has a nontrivial nullspace, then r_{m-1} is an eigenvalue and we are done. If not, $(r_{m-1}I - A)(r_m I - A)\mathbf{v} \neq \overrightarrow{0}$. Since we ultimately get the zero vector, some nullspace must be nontrivial. Say $(r_j I - A)$ has nontrivial nullspace, then r_j is an eigenvalue and we are done. $\square$

Now we know the spectrum of a complex matrix is never empty. At the other extreme, we could wonder if it is infinite. That is also not the case, as the following will show.

THEOREM 9.3
The eigenvectors corresponding to distinct eigenvalues of $A \in \mathbb{C}^{n\times n}$ are linearly independent.

PROOF Suppose $\lambda_1, \lambda_2, \ldots, \lambda_m$ are a set of distinct eigenvalues of A with corresponding eigenvectors $\mathbf{v}_1, \mathbf{v}_2, \ldots, \mathbf{v}_m$. Set $\alpha_1\mathbf{v}_1 + \alpha_2\mathbf{v}_2 + \cdots + \alpha_m\mathbf{v}_m = \overrightarrow{0}$. Our hope is that all the α's are zero. Let $A_1 = (\lambda_2 I - A)\cdots(\lambda_m I - A)$. Then $A_1\overrightarrow{0} = \overrightarrow{0} = A_1(\alpha_1\mathbf{v}_1 + \cdots + \alpha_m\mathbf{v}_m) = \alpha_1 A_1\mathbf{v}_1 + \cdots + \alpha_m A_1\mathbf{v}_m = \alpha_1(\lambda_2 - \lambda_1)(\lambda_3 - \lambda_1)\cdots(\lambda_m - \lambda_1)\mathbf{v}_1 + \overrightarrow{0} + \cdots + \overrightarrow{0}$, so α_1 must equal zero. In a similar fashion we show all the α's are zero. □

COROLLARY 9.1

Every matrix in $\mathbb{C}^{n\times n}$ has at most n distinct eigenvalues. In particular, the spectrum of an n-by-n complex matrix is a finite set of complex numbers.

PROOF You cannot have more than n distinct eigenvalues, else you would have $n + 1$ independent vectors in $\mathbb{C}^n$. This is an impossibility. □

COROLLARY 9.2

If $A \in \mathbb{C}^{n\times n}$ has n distinct eigenvalues, then the corresponding eigenvectors form a basis of $\mathbb{C}^n$.

We end this section with a remark about another polynomial that is associated with a matrix. We have noted that λ is an eigenvalue of A iff the system of equations $(\lambda I - A)\mathbf{v} = \overrightarrow{0}$ has a nontrivial solution. This is so iff $det(\lambda I - A) = 0$. But if we write out $det(\lambda I - A)$, we see a polynomial in λ. This polynomial is just the *characteristic polynomial.* The roots of this polynomial are the eigenvalues of A. For small textbook examples, this polynomial is a neat way to get the eigenvalues. For matrices up to five-by-five, we can in theory always get the eigenvalues from the characteristic polynomial. But who would try this with an eight-by-eight matrix? No one in their right mind! Now the characteristic polynomial does have interesting theoretical properties. For example, the number of times an eigenvalue appears as a root of the characteristic polynomial is called its *algebraic multiplicity,* and this can be different from the *geometric multiplicity* we defined above. The coefficients of the characteristic polynomial are quite interesting as well.

Exercise Set 39

1. Prove that $\lambda \in \lambda(A)$ iff $\bar{\lambda} \in \lambda(A^*)$.
2. Argue that A is invertible iff $0 \notin \lambda(A)$.
3. Prove that if A is an upper (lower) triangular matrix, the eigenvalues of A are exactly the diagonal elements of A.

4. Argue that if S is invertible, then $\lambda(S^{-1}AS) = \lambda(A)$.

5. Prove that if $A = A^*$, then $\lambda(A) \subseteq \mathbb{R}$.

6. Argue that if $U^* = U^{-1}$, then $\lambda \in \lambda(U)$ implies $|\lambda| = 1$.

7. Prove that if $\lambda \in \lambda(A)$, then $\lambda^n \in \lambda(A^n)$.

8. Argue that if A is invertible and $\lambda \in \lambda(A)$, then $\lambda^{-1} \in \lambda(A^{-1})$.

9. Suppose λ and μ are two distinct eigenvalues of A. Let $\mathbf{v}$ be an eigenvector for μ. Argue that $\mathbf{v}$ belongs to $Col(\lambda I - A)$.

10. As above, if $\mathbf{v}$ is an eigenvector for μ, then $\mathbf{v}$ is an eigenvector for $\lambda I - A$, with eigenvalue $\lambda - \mu$.

11. Suppose A is an n-by-n matrix and $\{\mathbf{v}_1, \mathbf{v}_2, \ldots, \mathbf{v}_n\}$ is a basis of $\mathbb{C}^n$ consisting of eigenvectors of A. Argue that the eigenvectors belonging to λ, one of A's eigenvalues, form a basis of $\mathcal{N}ull(\lambda I - A)$, while those not belonging to λ form a basis of $Col(\lambda I - A)$.

12. Prove Theorem 9.1.

13. Prove that if A is nilpotent, then 0 is its only eigenvalue. What does that say about the minimum polynomial of A?

14. Solve example (9.1.4) in detail.

15. Find explicit formulas for the eigenvalues of $\begin{bmatrix} a & b \\ c & d \end{bmatrix}$.

16. If A is a 2-by-2 matrix and you know its trace and determinant, do you know its eigenvalues?

17. Prove that if $U \in \mathbb{C}^{n\times n}$ is unitary, the eigenvectors of U belonging to distinct eigenvalues are orthogonal.

18. Argue that if the characteristic polynomial $\chi_A(x) = det(xI - A)$ has real coefficients, then the complex eigenvalues of A occur in complex conjugate pairs. Do the eigenvectors also occur this way if A has real entries?

19. Suppose A and B have a common eigenvector $\mathbf{v}$. Prove that $\mathbf{v}$ is also an eigenvector of any matrix of the form $\alpha A + \beta B$, $\alpha, \beta \in \mathbb{C}$.

20. Suppose $\mathbf{v}$ is an eigenvector belonging to $\lambda \neq 0$. Argue that $\mathbf{v} \in Col(A)$. Conclude that $Eig(A, \lambda) \subseteq Col(A)$.

21. If $\lambda \neq 0$ is an eigenvalue of AB, argue that λ is also an eigenvalue of BA.

22. Suppose P is an idempotent. Argue that $\lambda(P) \subseteq \{0, 1\}$.

23. Suppose p is a polynomial and $\lambda \in \lambda(A)$. Argue that $p(\lambda) \in \lambda(p(A))$.

24. How would you define all the eigenstuff for a linear transformation T: $\mathbb{C}^n \to \mathbb{C}^n$?

25. Suppose $A = \begin{bmatrix} A_{11} & A_{12} \\ \mathbb{O} & A_{22} \end{bmatrix}$. Argue that $\chi_A = \chi_{A_{11}} \cdot \chi_{A_{22}}$. Conclude that $\lambda(A) = \lambda(A_{11}) \cup \lambda(A_{22})$.

26. Find an explicit matrix with eigenvalues 2, 4, 6 and eigenvectors (1, 0, 0), (1, 1, 0), (1, 1, 1).

27. Argue that the product of all the eigenvalues of $A \in \mathbb{C}^{n\times n}$ is the determinant of A and the sum is the trace of A.

28. Suppose $A \in \mathbb{C}^{n\times n}$ has characteristic polynomial $\chi_A(x) = \sum_{i=0}^{n} c_i x_i$. If you are brave, argue that c_r is the sum of all principal minors of order n-r of A times $(-1)^{n-r}$. At least find the trace and determinant of A among the cs.

29. Do A and A^T have the same eigenvalues? How about the same eigenvectors? How about A and A^*?

30. Suppose $B = S^{-1}AS$. Argue that $\chi_A = \chi_B$. Do A and B necessarily share the same eigenvectors?

31. Suppose $\mathbf{v}$ is an eigenvector of A. Argue that $S^{-1}\mathbf{v}$ is an eigenvector of $S^{-1}A$.

32. Here is another matrix/polynomial connection. Suppose $p(x) = x^n + a_{n-1}x^{n-1} + \cdots + a_1 x + a_0$. We associate a matrix to this polynomial called the *companion matrix* as follows: $C(p(x)) =$

$$\begin{bmatrix} 0 & 1 & 0 & 0 & \cdots & \cdots & \vdots & \vdots \\ 0 & 0 & 1 & 0 & \cdots & \cdots & \vdots & \vdots \\ \vdots & \vdots & \vdots & \vdots & \cdots & \cdots & \ddots & 1 \\ -a_0 & -a_1 & -a_2 & \cdots & \cdots & \cdots & -a_{n-2} & -a_{n-1} \end{bmatrix}$$

. What is the characteristic polynomial of $C(p(x))$? What are the eigenvalues?

33. Suppose A is a matrix whose rows sum up to m. Argue that m is an eigenvalue of A. Can you determine the corresponding eigenvector?

34. Suppose $C = C(x^n - 1)$. Suppose $A = a_1 I + a_2 C + \cdots + a_n C^{n-1}$. Determine the eigenvalues of A.

35. Argue that AB and BA have the same characteristic polynomial and hence the same eigenvalues. More generally, if A is m-by-n and B is n-by-m, then AB and BA have the same nonzero eigenvalues counting multiplicity. (*Hint:* $\begin{bmatrix} I & A \\ \mathbb{O} & I \end{bmatrix}^{-1} \begin{bmatrix} AB & \mathbb{O} \\ B & \mathbb{O} \end{bmatrix} \begin{bmatrix} I & A \\ \mathbb{O} & I \end{bmatrix} = \begin{bmatrix} \mathbb{O} & \mathbb{O} \\ B & BA \end{bmatrix}$ so $\begin{bmatrix} AB & \mathbb{O} \\ B & \mathbb{O} \end{bmatrix}$ and $\begin{bmatrix} \mathbb{O} & \mathbb{O} \\ B & BA \end{bmatrix}$ are similar matrices.)

36. Argue that if $AB = BA$, then A and B have at least one common eigenvector. Must they have a common eigenvalue?

37. Suppose M is a nontrivial subspace of $\mathbb{C}^n$, which is invariant for A. Argue that M contains at least one eigenvector of A.

38. Argue that $Eig(A, \lambda)$ is always an invariant subspace for A.

39. (J. Gross and G. Trenkler) Suppose P and Q are orthogonal projections of the same size. Prove that PQ is an orthogonal projection iff all nonzero eigenvalues of $P + Q$ are greater or equal to one.

40. Argue that $\lambda \in \lambda(A)$ implies $\alpha\lambda \in \lambda(\alpha A)$.

41. Argue that if $\lambda \in \lambda(A)$ with eigenvector $\mathbf{v}$, then $\bar{\lambda} \in \lambda(\overline{A})$ with eigenvector $\overline{\mathbf{v}}$.

42. Prove that if $\lambda \in \lambda(A)$, then $\lambda + \tau \in \lambda(A + \tau I)$.

43. Prove that $\lambda \in \lambda(A)$ iff $\lambda \in \lambda(S^{-1}AS)$, where S is invertible.

44. Consider the matrix $P_N = \begin{bmatrix} \frac{1}{N} & \frac{N-1}{N} & 0 & 0 & 0 & \cdots & 0 \\ 0 & \frac{2}{N} & \frac{N-2}{N} & 0 & 0 & \cdots & 0 \\ \vdots & \vdots & \vdots & \vdots & \vdots & \vdots & 0 \\ 0 & 0 & \cdots & \frac{k}{N} & \frac{N-k}{N} & \cdots & 0 \\ \vdots & \vdots & & & & & 0 \\ 0 & 0 & \cdots & 0 & 0 & \frac{N-1}{N} & \frac{1}{N} \\ 0 & 0 & \cdots & & & 0 & 1 \end{bmatrix}$.

Find all the eigenvalues of P_N and corresponding eigenvectors.

Further Reading

[A-S&A, 2005] Rhaghib Abu-Saris and Wajdi Ahmad, Avoiding Eigenvalues in Computing Matrix Powers, The American Mathematical Monthly, Vol. 112, No. 5., May, (2005), 450–454.

[Axler, 1996] Sheldon Axler, *Linear Algebra Done Right*, Springer, New York, (1996).

[B&R, 1986(2)] T. S. Blyth and E. F. Robertson, *Matrices and Vector Spaces*, Vol. 2, Chapman & Hall, New York, (1986).

[Holland, 1997] Samuel S. Holland, Jr., The Eigenvalues of the Sum of Two Projections, in *Inner Product Spaces and Applications*, T. M. Rassias, Editor, Longman, (1997), 54–64.

[J&K, 1998] Charles R. Johnson and Brenda K. Kroschel, Clock Hands Pictures for 2×2 Real Matrices, The College Mathematics Journal, Vol. 29, No. 12, March, (1998), 148–150.

[Olš, 2003] Gregor Olšavský, The Number of 2 by 2 Matrices over $\mathbb{Z}/p\mathbb{Z}$ with Eigenvalues in the Same Field, Mathematics Magazine, Vol. 76, No. 4, October, (2003), 314–317.

[Scho, 1995] Steven Schonefeld, Eigenpictures: Picturing the Eigenvector Problem, The College Mathematics Journal, Vol. 26, No. 4, September, (1995), 316–319.

[Tr&Tr,2003] Dietrich Trenkler and Götz Trenkler, On the Square Root of $aa^T + bb^T$, The College Mathematics Journal, Vol. 34, No. 1, January, (2003), 39–41.

[Zizler, 1997] Peter Zizler, Eigenpictures and Singular Values of a Matrix, The College Mathematics Journal, Vol. 28, No. 1, January, (1997), 59–62.

9.1.1 MATLAB Moment

9.1.1.1 Eigenvalues and Eigenvectors in MATLAB

The eigenvalues of a square matrix A are computed from the function

$$eig(A).$$

More generally, the command

$$[V, D] = eig(A)$$

returns a diagonal matrix D and a matrix V whose columns are the corresponding eigenvectors such that $AV = VD$. For example,

```
>> B = [1 + i 2 + 2i 3 + i; 2 + 2i 4 + 4i 9i; 3 + 3i 6 + 6i 8i]
B =
   1.0000 + 1.0000i   2.0000 + 2.0000i   3.0000 + 1.0000i
   2.0000 + 2.0000i   4.0000 + 4.0000i        0 + 9.0000i
   3.0000 + 3.0000i   6.0000 + 6.0000i        0 + 8.0000i
>> eig(B)
ans =
   6.3210 + 14.1551i
   0.0000 - 0.0000i
   -1.3210 - 1.1551i
>> [V, D] = eig(B)
V =
   0.2257 - 0.1585i        0.8944               0.6839
   0.6344 + 0.0693i   -0.4472 - 0.0000i   -.5210 + 0.3564i
        0.7188         0.0000 + 0.0000i   -0.0643 - 0.3602i
D =
   6.3210 + 14.1551i          0                   0
          0           0.0000 + 0.0000i            0
          0                   0           -1.3210 - 1.1551i
```

A really cool thing to do with eigenvalues is to plot them using the "plot" command. Do a "help plot" to get an idea of what this command can do. Or, just try

$$plot(eig(A), 'o'), grid\ on$$

and experiment with a variety of matrices.

9.2 The Spectral Theorem

In this section, we derive a very nice theorem about the structure of certain square matrices that are completely determined by their eigenvalues. Our approach is motivated by the theory of rings of operators pioneered by **Irving Kaplansky** (22 March 1917–25 June 2006). First, we present some definitions.

DEFINITION 9.2 *(nilpotent, normal)*

A matrix $A \in \mathbb{C}^{n\times n}$ is called ***nilpotent*** *iff there is some power of A that produces the zero matrix. That is, $A^n = \mathbb{O}$ for some $n \in \mathbb{N}$. The least such n is called the* index of nilpotency.

A matrix A is called ***normal*** *iff $AA^* = A^*A$. That is, A is normal iff it commutes with its conjugate transpose.*

Example 9.2

1. If $A = A^*$ (i.e., A is self-adjoint or Hermitian), then A is normal.
2. If $U^* = U^{-1}$ (i.e., U is unitary), then U is normal.
3. If $A^* = -A$ (i.e., A is skew Hermitian), then A is normal.
4. $N = \begin{bmatrix} 0 & 1 & 0 \\ 0 & 0 & 1 \\ 0 & 0 & 0 \end{bmatrix}$ is nilpotent. What is its index?

Recall that we have proved that $AA^* = \mathbb{O}$ iff $A = \mathbb{O}$ for $A \in \mathbb{C}^{n\times n}$. The first fact we need is that there are no nonzero normal nilpotent matrices.

THEOREM 9.4

If A is normal and nilpotent, then $A = \mathbb{O}$.

PROOF First, we show the theorem is true for A self-adjoint. The argument is by induction. Suppose $A = A^*$. Suppose $A^i = \mathbb{O}$ implies $A = \mathbb{O}$ as long as $i < n$. We want the same implication to hold for n. We look at two cases: n is even, say $n = 2m$. Then $\mathbb{O} = A^n = A^{2m} = A^m A^m = A^m A^{*m} = A^m(A^m)^*$. But then, $A^m = \mathbb{O}$, so applying induction ($m < n$), we conclude $A = \mathbb{O}$. On the other hand, if n is odd other than 1, say $n = 2m + 1$, then $\mathbb{O} = A^n = A^{n+1} = A^{2m+2} = A^{m+1}A^{m+1} = A^{m+1}(A^*)^{m+1} = (A^{m+1})(A^{m+1})^*$ so again $A^{m+1} = \mathbb{O}$. Applying induction, $A = \mathbb{O}$ since $m + 1 < n$.

Now, if A is normal and $A^n = \mathbb{O}$, then $(AA^*)^n = A^n(A^*)^n = \mathbb{O}$ so, since AA^* is self-adjoint, $AA^* = \mathbb{O}$. But then $A = \mathbb{O}$ and our proof is complete. ∎

Next, recall that for any matrix A, we defined $A' = I - A^+A$; that is, A' is the projection onto the null space of A. This projection was characterized as the unique projection P that satisfies $AX = \mathbb{O}$ iff $X = PX$. Finally, recall the minimal polynomial $\mu_A(x)$, which is the monic polynomial of least degree such that $\mu_A(A) = \mathbb{O}$.

THEOREM 9.5

Let A be normal. Then A' is a polynomial in A and hence $\mathbb{O} = AA' = A'A$.

PROOF We first dispose of some trivial cases. If $A = \mathbb{O}$, $\mathbb{O}' = I$, so A' is trivially a polynomial in A. Let's assume $A \neq \mathbb{O}$. Suppose $\mu_A(x) = \alpha_0 + \alpha_1 x + \alpha_2 x^2 + \cdots + x^m$. If $\alpha_0 \neq 0$, then by a result from Chapter 1, we have A^{-1} exists so $A' = \mathbb{O}$, again a trivial case. Now let's assume $\alpha_0 = 0$. Thus $\mu_A(x) = \alpha_1 x + \alpha_2 x^2 + \cdots + x^m$. Of course, α_1 could be zero or α_2 could be zero along with α_1. But this could not go on because if $\alpha_2 = \alpha_3 = \alpha_{m-1}$ were all zero, we would have $A^m = \mathbb{O}$ hence $A^m = \mathbb{O}$, whence $A = \mathbb{O}$ by (9.4) above. So some α_j in this list cannot be zero. Take the least j with $\alpha_j \neq 0$. Call this j, i. Then $\alpha_i \neq 0$ but $\alpha_{i-1} = \alpha_{i-2} = \cdots = \alpha_1 = 0$. Note $i \leq m - 1$ so $i < m$. Thus, $\mu_A(x) = \alpha_i x^i + \alpha_{i+1} x^{i+1} + \cdots + x^m$. Form $E = I + \sum_{j=1}^{m-i} \frac{\alpha_{i+j}}{\alpha_i} A^j = I + \frac{\alpha_{i+1}}{\alpha_i} A + \frac{\alpha_{i+2}}{\alpha_i} A^2 + \cdots + \frac{1}{\alpha_i} A^{m-i}$. Now E is a polynomial in A, so clearly $AE = EA$. Also E is normal, $EE^* = E^*E$. Next, suppose $AX = \mathbb{O}$ for some matrix X. Then $EX = X$ since X kills off any term with an A in it. Conversely, suppose $X = EX$ for some matrix X. Now $\mathbb{O} = \mu_A(A) = A^i E$, so $(AE)^i = A^i E^i = (A^i E) E^{i-1} = \mathbb{O}$, so by (9.4) above $AE = \mathbb{O}$. Note AE is a polynomial in the normal matrix A, so is itself normal. But then $AX = AEX = \mathbb{O}$. Hence we have the right property: $AX = \mathbb{O}$ iff $X = EX$. The only question left is if E is a projection.

But $AE = \mathbb{O}$ so $EE = E$, making E idempotent. Moreover, $(AE^*)(AE^*)^* = AE^*EA^* = E^*AEA^* = \mathbb{O}$, using that E^* is a polynomial in A^* and A commutes with A^*. Thus $AE^* = \mathbb{O}$, so $E^* = EE^*$. But now we have it since $E^* = EE^* = (EE^*)^* = (E^*)^* = E$. Thus E is a projection and by uniqueness must equal A'. ∎

Let us establish some notation for our big theorem. Let $A \in \mathbb{C}^{n \times n}$. Let $P_\lambda = (\lambda I - A)'$ denote the λth *eigenprojection* of A, where λ is any complex number. Recall λ is in the spectrum of A, $\lambda(A)$, iff $P_\lambda \neq \mathbb{O}$.

THEOREM 9.6

Suppose A is normal. Then

1. $AP_\lambda = P_\lambda A = \lambda P_\lambda$.
2. $(\mu I - A)P_\lambda = (\mu - \lambda)P_\lambda$.

3. $\{P_\lambda | \lambda \in \lambda(A)\}$ is a finite set of pairwise orthogonal projections.

4. If $\lambda(A) = \{0\}$, then $A = \mathbb{O}$.

PROOF

1. Since $(\lambda I - A)P_\lambda = (\lambda I - A)(\lambda I - A)' = \mathbb{O}$, we have $AP_\lambda = \lambda P_\lambda$. Also, $\lambda I - A$ is normal so by (9.5), $P_\lambda(\lambda I - A) = (\lambda I - A)P_\lambda$ so $P_\lambda \lambda - P_\lambda A = \lambda P_\lambda - AP_\lambda$. Canceling λP_λ, we get $P_\lambda A = AP_\lambda$.

2. $(\mu I - A)P_\lambda = \mu P_\lambda - AP_\lambda = \mu P_\lambda - AP_\lambda = (\mu - \lambda)P_\lambda$.

3. We know by Theorem 9.3 that $\lambda(A)$ is a finite set. Moreover, if $\lambda_1 \neq \lambda_2$ in $\lambda(A)$, then $P_{\lambda_2} = \dfrac{(A - \lambda_1 I)}{\lambda_2 - \lambda_1} P_{\lambda_2}$. Thus, $P_{\lambda_1} P_{\lambda_2} = P_{\lambda_1}(A - \lambda_1 I)\dfrac{P_{\lambda_2}}{\lambda_1 - \lambda_2} = (P_{\lambda_1} A - \lambda_1 P_{\lambda_1})\dfrac{P_{\lambda_2}}{\lambda_2 - \lambda_1} = \mathbb{O}$, making $P_{\lambda_1} \perp P_{\lambda_2}$ whenever λ_1 and λ_2 are distinct eigenvalues.

4. Suppose 0 is the only eigenvalue of A. Look at the minimum polynomial of A, $\mu_A(x) = \alpha_0 + \alpha_1 x + \cdots + x^m$. Since we are working over complex numbers, we can factor μ_A completely as linear factors $\mu_A(x) = (x - \beta_1)(x - \beta_2)\cdots(x - \beta_m)$. If all the βs are zero, $\mu_A(A) = \mathbb{O} = A^m$, so $A^m = \mathbb{O}$. By Theorem 9.4, $A = \mathbb{O}$ and we are done. Suppose not all the βs are zero. Relabeling if necessary, suppose the first k, βs are not zero. Then $\mu_A(x) = (x - \beta_1)(x - \beta_2)\cdots(x - \beta_k)x^{m-k}$. Then $\mathbb{O} = \mu_A(A) = (A - \beta_1 I)\cdots(A - \beta_k I)A^{m-k}$. Now $\beta_1 \neq 0$ so $\beta_1 \notin \lambda(A)$ so $P_{\beta_1} = \mathbb{O}$. But $(A - \beta_1 I)[(A - \beta_2 I)\cdots(A - \beta_k I)A^{m-k}] = \mathbb{O}$ implies $[(A - \beta_2 I)\cdots(A - \beta_k I)A^{m-k}] = (A - \beta_1 I)'[(A - \beta_2 I)\cdots A^{m-k}] = \mathbb{O}$. The same is true for β_2 so $(A - \beta_3 I)\cdots(A - \beta_k I)A^{m-k} = \mathbb{O}$. Proceeding inductively, we peal off each $A - \beta I$ term until we have only $A^{m-k} = \mathbb{O}$. But then using (9.4) again we have $A = \mathbb{O}$. ▯

We need one more piece of evidence before we go after the spectral theorem.

THEOREM 9.7

Suppose A is normal and F is a projection that commutes with A (i.e., $AF = FA$). Then

1. AF is normal.

2. $(\lambda I - AF)' = (\lambda I - A)'F = F(\lambda I - A)'$ if $\lambda \neq 0$.

PROOF

1. First $(AF)(AF)^* = AFF^*A^* = AF^2A^* {=} FAA^*F = F^*A^*AF = (AF)^*(AF)$.

2. Suppose $\lambda \neq 0$ and X is any matrix. Then $(\lambda I - FA)X = \mathbb{O}$ iff $\lambda X = FAX$ iff $X = F(\lambda^{-1}AX)$ iff $X = FX$ and $X = \lambda^{-1}AX$ iff $X = FX$ and $\lambda X = AX$ iff $X = FX$ and $(\lambda I - A)X = \mathbb{O}$ iff $X = FX$ and $X = (\lambda I - A)'X$ iff $X = F(\lambda I - A)'X$. By uniqueness, $F(\lambda I - A)' = (\lambda I - FA)'$. Also, $(\lambda I - A)'$ is a polynomial in $\lambda I - A$ and hence F commutes with $(\lambda I - A)$. ▯

Now we are ready for the theorem. Let's establish the notation again. Let $A \in \mathbb{C}^{n\times n}$ and suppose the distinct eigenvalues of A are $\lambda_1, \lambda_2, \ldots, \lambda_m$. Suppose the eigenspaces are $M_{\lambda_1}, M_{\lambda_2}, \ldots, M_{\lambda_m}$, and the eigenprojections are $P_{\lambda_1}, P_{\lambda_2}, \ldots, P_{\lambda_m}$.

THEOREM 9.8 (the spectral theorem)
With the notations as above, the following statements are equivalent:

1. A is normal.

2. $A = \lambda_1 P_{\lambda_1} + \lambda_2 P_{\lambda_2} + \cdots + \lambda_m P_{\lambda_m}$, $P_{\lambda_1} + \cdots + P_{\lambda_m} = I$, and $P_{\lambda_i} \perp P_{\lambda_j}$ for all $\lambda_i \neq \lambda_j$.

3. The eigenspaces M_{λ_i} are pairwise orthogonal and $M_{\lambda_1} \oplus M_{\lambda_2} \oplus \cdots \oplus M_{\lambda_m} = \mathbb{C}^n$.

PROOF Suppose A is normal. Let $F = I - \sum_{i=1}^{m} P_{\lambda_i}$. Then $F^* = (I - \sum_{i=1}^{m} P_{\lambda_i})^* = I^* - \sum_{i=1}^{m} P_{\lambda_i}^* = I - \sum_{i=1}^{m} P_{\lambda_i} = F$, so F is self-adjoint. Next, $F^2 = F(I - \sum_{i=1}^{m} P_{\lambda_i}) = F - \sum_{i=1}^{m} FP_{\lambda_i}$. But $FP_{\lambda_j} = P_{\lambda_j} - \sum_{i=1}^{m} P_{\lambda_i}P_{\lambda_j} = P_{\lambda_j} - P_{\lambda j}^2 = \mathbb{O}$ for each j, so $F^2 = F$, using that the eigenprojections are orthogonal. Thus, F is a projection. Moreover, $AF = FA$ by (9.7) (1) so $(\lambda I - AF)' = (\lambda I - A)'F = F(\lambda I - A)'$ by (9.6) for all $\lambda \neq 0$. Thus, for all $\lambda \neq 0$, $(\lambda I - A)'(I - F) = \sum(\lambda I - A)'(P_{\lambda_i}) = ((\lambda I - A)')^2 = (A - \lambda I)'$. Then $(\lambda I - FA)' = (\lambda I - A)'F = \mathbb{O}$ for all $\lambda \neq 0$. This says that the only eigenvalue possible for FA is zero so $FA = AF = \mathbb{O}$ by (9.6)(4). Hence $F = A'F = P_0F = \mathbb{O}$ for, if $0 \in \lambda(A)$, then $P_0F = FP_0 = \mathbb{O}$ and if $0 \notin \lambda(A)$, then $P_0 = \mathbb{O}$. Thus, in any case, $I = \sum_{i=1}^{m} P_{\lambda_i}$.

Next, $A = AI = A(\sum_{i=1}^{m} P_{\lambda_i}) = \sum_{i=1}^{m} AP_{\lambda_i} = \sum_{i=1}^{m} \lambda_i P_{\lambda_i}$ by (9.7)(1). Thus, we have the hard part proved (i.e., (1) implies (2)). Next, we show (2) implies (1). Suppose (2) is true. Then $AA^* = (\sum_{i=1}^{m} \lambda_i P_{\lambda_i})(\sum_{i=1}^{m} \bar{\lambda}_i P_{\lambda_i}^*) = \sum_{i=1}^{m} |\lambda_i|^2 P_{\lambda_i}$ and $A^*A = (\sum_{i=1}^{m} \bar{\lambda}_i P_{\lambda_i}^*)(\sum_{i=1}^{m} \lambda_i P_{\lambda_i}) = \sum_{i=1}^{m} |\lambda_i|^2 P_{\lambda_i}$. Thus $AA^* = A^*A$. □

Again suppose (2) is true. The eigenspaces are orthogonal pairwise because their projections are. Moreover if $\mathbf{v} \in \mathbb{C}^n$, then $\mathbf{v} = I\mathbf{x} = (P_{\lambda_i} + \cdots + P_{\lambda_m})(\mathbf{v}) = P_{\lambda_i}\mathbf{v} + P_{\lambda_2}\mathbf{v} + \cdots + P_{\lambda_m}\mathbf{v} \in M_{\lambda_1} \oplus^{\perp} M_{\lambda_2} \oplus^{\perp} \cdots \oplus^{\perp} M_{\lambda_m}$, so $\mathbb{C}^n = M_{\lambda_1} \oplus^{\perp} M_{\lambda_2} \oplus^{\perp} \cdots \oplus^{\perp} M_{\lambda_m}$. Conversely, if we assume (3), then since $M_{\lambda_i} \perp M_{\lambda_j}$ we have $P_{\lambda_i} \perp P_{\lambda_j}$ when $i \neq j$. Moreover if $\mathbf{v} \in \mathbb{C}^n$, then $\mathbf{v} = \mathbf{v}_1 + \mathbf{v}_2 + \cdots + \mathbf{v}_m$, where $\mathbf{v}_i \in M_{\lambda_i}$. Then $A\mathbf{v} = A(\mathbf{v}_1 + \mathbf{v}_2 + \cdots + \mathbf{v}_m) = A\mathbf{v}_1 + A\mathbf{v}_2 + \cdots + A\mathbf{v}_m = \lambda_1\mathbf{v}_1 + \lambda_2\mathbf{v}_2 + \cdots + \lambda_m\mathbf{v}_m$. But $P_{\lambda_i}\mathbf{v}_i = \mathbf{v}_i$ since $Fix(P_{\lambda_i}) = M_{\lambda_i}$. $P_{\lambda_j}\mathbf{v}_i = \overrightarrow{0}$ for $i \neq j$. Thus $P_{\lambda_i}\mathbf{v}_i = \mathbf{v}_i$ so $A\mathbf{v} = \lambda_1 P_{\lambda_i}\mathbf{v} + \lambda_2 P_{\lambda_2}\mathbf{v} + \cdots + \lambda_m P_{\lambda_m}\mathbf{v} = (\lambda_1 P_{\lambda_i} + \cdots + \lambda_m P_{\lambda_m})\mathbf{v}$. Similarly $I\mathbf{v} = \mathbf{v} = \mathbf{v}_1 + \mathbf{v}_2 + \cdots + \mathbf{v}_m = P_{\lambda_i}\mathbf{v} + P_{\lambda_2}\mathbf{v} + \cdots + P_{\lambda_m}\mathbf{v} = (P_{\lambda_i} + P_{\lambda_2} + \cdots + P_{\lambda_m})\mathbf{v}$ and so (2) follows. This completes the proof of the spectral theorem.

There is uniqueness in this spectral representation of a normal matrix, as the next theorem shows.

THEOREM 9.9

Suppose Λ is a nonempty finite set of complex numbers and $\{Q_\lambda | \lambda \in \Lambda\}$ is an orthogonal set of projections such that $\sum_{\lambda\in\Lambda} Q_\lambda = I$. If $A = \sum_{\lambda\in\Lambda} \lambda Q_\lambda$, then A is normal, $Q_\lambda = (\lambda I - A)'$ for all $\lambda \in \Lambda$, and $\lambda(A) \subseteq \Lambda$. Moreover, $\Lambda = \lambda(A)$ iff $Q_\lambda \neq \mathbb{O}$ for each nonzero $\lambda \in \Lambda$.

PROOF We compute $A^*Q_{\lambda_0} = (\sum_{\lambda\in\Lambda} \lambda Q_\lambda)^* Q_{\lambda_0} = (\sum_{\lambda\in\Lambda} \bar{\lambda} Q_\lambda) Q_{\lambda_0} = \sum_{\lambda\in\Lambda} \bar{\lambda} Q_\lambda Q_{\lambda_0} = \bar{\lambda}_0 Q_{\lambda_0}$ and $AQ_{\lambda_0} = (\sum_{\lambda\in\Lambda} \lambda Q_\lambda) Q_{\lambda_0} = \sum_{\lambda\in\Lambda} \lambda Q_\lambda Q_{\lambda_0} = \lambda_0 Q_{\lambda_0}$. Thus, $A^*A = A^*(\Sigma\lambda Q_\lambda) = \Sigma\lambda A^* Q_\lambda = \Sigma\lambda\bar{\lambda} Q_\lambda = \Sigma\bar{\lambda} A Q_\lambda = A(\Sigma\bar{\lambda} Q_\lambda) = AA^*$. Thus, A is normal. We adopt the convention that $Q_\lambda = \mathbb{O}$ if $\lambda \in \Lambda$. Since for any $\mu \in \mathbb{C}$, $\mu I - A = \mu I - \Sigma\lambda Q_\lambda = \Sigma(\mu - \lambda)Q_\lambda$, $(\mu I - A)Q_\mu = \sum_{\lambda\in\Lambda}(\mu - \lambda)Q_\lambda Q_\mu = \mathbb{O}$. Then, for all $\mu \in \mathbb{C}$ with $\lambda \neq \mu$, $P_\lambda \leq (P_\mu)' \leq Q_\mu'$. Then, for any $\mu \in \mathbb{C}$, $Q_\mu' = I - Q_\mu = \sum_{\lambda\in\Lambda'\{\mu\}} Q_\lambda \leq P_\mu'$. Thus, $P_\mu \leq Q_\mu$ for each $\mu \in \mathbb{C}$ so that $P_\mu = (\mu I - A)' = Q_\mu$ for all $\mu \in \mathbb{C}$. Now, if $\mu \notin \Lambda$, then $P_\mu = Q_\mu = \mathbb{O}$, which says $\mu \notin \lambda(A)$. Thus, $\lambda(A) \subseteq \Lambda$. If $\mu \in \Lambda$ such that $Q_\mu \neq 0$, then $P_\mu = Q_\mu \neq 0$, which means $\mu \in \lambda(A)$. Thus, $\Lambda = \lambda(A)$ if $Q_\mu \neq 0$ for every $\mu \in \Lambda$, which completes our proof. □

One of the nice consequences of the spectral theorem is that certain functions of a matrix are easily computed.

THEOREM 9.10

Let A be normal with spectral representation $A = \sum_{\lambda_i \in \lambda(A)} \lambda_i P_{\lambda_i}$. *Then*

1. $A^2 = \sum_{\lambda_i \in \lambda(A)} \lambda_i^2 P_{\lambda_i}$.
2. $A^n = \sum_{\lambda_i \in \lambda(A)} \lambda_i^n P_{\lambda_i}$.
3. If $q(x)$ is any polynomial in $\mathbb{C}[x]$, then $q(A) = \sum_{\lambda_i \in \lambda(A)} q(\lambda_i) P_{\lambda_i}$.

Moreover, if $\lambda \in \lambda(A)$, there exists a polynomial $P_\lambda(x)$ in $\mathbb{C}[x]$ such that $P_\lambda = P_\lambda(A)$.

PROOF

1. Compute $A^2 = (\Sigma \lambda P_\lambda)(\Sigma \mu P_\mu) = \sum_\lambda \sum_\mu \lambda\mu P_\lambda P_\mu = \Sigma \lambda^2 P_\lambda^2 = \Sigma \lambda^2 P_\lambda$.
2. Use induction.
3. Compute $\mu A^n + \upsilon A^m = \sum_\lambda \mu\lambda^n P_\lambda + \sum_\lambda \mu\lambda^m P_\lambda = \sum_\lambda (\mu\lambda^n + \upsilon\lambda^m) P_\lambda$ so clearly, $q(A) = \sum_\lambda q(\lambda) P_\lambda$. ▯

To get the polynomial $P_\lambda(x)$, we recall the Lagrange interpolation polynomial: $P_{\lambda_j}(x) = \frac{(x-\lambda_1)(x-\lambda_2)\ldots(x-\lambda_{j-1})(x-\lambda_{j+1})\ldots(x-\lambda_m)}{(\lambda_j-\lambda_1)(\lambda_j-\lambda_2)\ldots(\lambda_j-\lambda_{j-1})(\lambda_j-\lambda_{j+1})\ldots(\lambda_j-\lambda_m)}$. Then $P_{\lambda_j}(\lambda_i) = \delta_{ij}$, the Kronecker δ. Then $P_{\lambda_j}(A) = \sum_{i=1}^m P_{\lambda_j}(\lambda_i) P_{\lambda_i} = P_{\lambda_j} = (\lambda_j I - A)'$.

Recall that if λ is a complex number, $\lambda^+ = \begin{cases} \lambda^{-1} & \text{if } \lambda \neq 0 \\ 0 & \text{if } \lambda = 0 \end{cases}$. For normal matrices, we can capture the pseudoinverse from its eigenvalues and eigenprojections.

THEOREM 9.11

Suppose A is normal with spectral representation $A = \sum_{\lambda \in \lambda(A)} \lambda P_\lambda$; *then* $A^+ = \sum_{\lambda \in \lambda(A)} \lambda^+ P_\lambda$. *In particular,* A^+ *is normal also.*

PROOF The proof is left as an exercise. ▯

Exercise Set 40

1. Give an example of a matrix that is not self-adjoint, not unitary, and not skew adjoint and yet is normal.

2. The spectral theorem is sometimes formulated as follows: A is normal iff $A = \sum \lambda_i u_i u_i^*$, where the u_i are an orthonormal basis formed from the eigenspaces of A. Prove this form.

3. Suppose A is normal. Argue that A^k is normal for all $k \in \mathbb{N}$; moreover, if $p(x) \in \mathbb{C}[x]$, argue that $p(A)$ is normal.

4. Argue that A is normal iff the real part of A commutes with the imaginary part of A. (Recall the Cartesian decomposition of a matrix.)

5. Suppose $A \in \mathbb{C}^{n \times n}$. Define $R_\lambda = (\lambda I - A)^{-1}$ for $\lambda \notin \lambda(A)$. Prove that $R_\lambda - R_\mu = (\mu - \lambda) R_\lambda R_\mu$.

6. Argue that an idempotent P is normal iff $P = P^*$.

7. Argue that A is normal iff A^* is normal and the eigenspaces associated with the mutually conjugate eigenvalues of A and A^* are the same.

8. Suppose A is normal and λ is an eigenvalue of A. Write $\lambda = a + bi$, where a and b are real. Argue that a is an eigenvalue of the real part of A and b is an eigenvalue for the imaginary part of A.

9. Argue that a normal matrix is skew-Hermitian iff all it eigenvalues are pure imaginary complex numbers.

10. Write a matrix A in its Cartesian decomposition, $A = B + Ci$, where B and C have only real entries. Argue that if A is normal, so is $\begin{bmatrix} B & -C \\ C & B \end{bmatrix}$. Can you discover any other connections between A and this 2-by-2 block matrix?

11. Prove that A is normal iff every eigenvector of A is also an eigenvector of A^*.

12. Argue that A is normal iff A^* is expressible as a polynomial in A.

13. Prove that A is normal iff A commutes with A^*A.

14. Argue that A is normal iff A commutes with $A + A^*$ iff A commutes with $A - A^*$ iff A commutes with $AA^* - A^*A$.

15. Prove the following are all equivalent statements.
    (a) A is normal
    (b) A commutes with $A + A^*$
    (c) A commutes with $A - A^*$
    (d) A commutes with $AA^* - A^*A$

16. Prove that a matrix B in $\mathbb{C}^{m\times n}$ has full row rank iff the mapping $A \mapsto AB^* + B^*A$ takes $\mathbb{C}^{m\times n}$ onto the set of m-by-m Hermitian matrices.

17. Let $A \in \mathbb{C}^{n\times n}$ be normal and suppose the distinct eigenvalues of A are $\lambda_1, \lambda_2, \ldots, \lambda_m$. Suppose $\chi_A(x) = \prod_{k=1}^{m}(x - \lambda_k)^{d_k}$. Argue that $\mu_A(x) = \prod_{k=1}^{m}(x - \lambda_k)$ and $dim(Eig(A, \lambda_k)) = d_k$.

18. Give an example of 2-by-2 normal matrices A and B such that neither $A + B$ nor AB is normal.

Group Project

A long-standing problem on how the eigenvalues of Hermitian matrices A and B relate to the eigenvalues of $A + B$ has been solved. Read about this and write a paper. See [Bhatia, 2001].

Further Reading

[Bhatia, 2001] Rajendra Bhatia, Linear Algebra to Quantum Cohomology: The Story of Alfred Horn's Inequalities, The American Mathematical Monthly, Vol. 108, No. 4, (2001), 289–318.

[Brown, 1988] William C. Brown, *A Second Course in Linear Algebra*, John Wiley & Sons, New York, (1988).

[Fisk, 2005] Steve Fisk, A Very Short Proof of Cauchy's Interlace Theorem, The American Mathematical Monthly, Vol. 112, No. 2, February, (2005), 118.

[Hwang, 2004] Suk-Geun Hwang, Cauchy's Interlace Theorem for Eigenvalues of Hermitian Matrices, The American Mathematical Monthly, Vol. 111, No. 2, February, (2004), 157–159.

[J&S, 2004] Charles R. Johnson and Brian D. Sutton, Hermitian Matrices, Eigenvalue Multiplicities, and Eigenvector Components, SIAM J. Matrix Anal. Appl., Vol. 26, No. 2, (2004), 390–399.

[Zhang 1999] Fuzhen Zhang, *Matrix Theory: Basic Results and Techniques*, Springer, New York, (1999).

9.3 The Square Root and Polar Decomposition Theorems

The spectral theorem has many important consequences. In this section, we develop two of them. We also further our analogies with complex numbers and square matrices.

THEOREM 9.12
Let $A \in \mathbb{C}^{n\times n}$. If $AA^* = \lambda P$ where P is a projection and $A \neq \mathbb{O}$, then λ is a positive real number.

PROOF Suppose $AA^* = \lambda P$, where P is a projection and $A \neq \mathbb{O}$. Then AA^* is not $\mathbb{O}$, so there must exist $\mathbf{v} \in \mathbb{C}^n$ with $AA^*\mathbf{v} \neq \overrightarrow{0}$. Then $P\mathbf{v} \neq \overrightarrow{0}$ either. But $\|A^*\mathbf{v}\|^2 =< A^*\mathbf{v}|A^*\mathbf{v} >= < AA^*\mathbf{v}|\mathbf{v} >=< \lambda P\mathbf{v}|\mathbf{v} >= \lambda < P\mathbf{v}|\mathbf{v} >= \lambda < P\mathbf{v}|P\mathbf{v} >= \lambda \|P\mathbf{v}\|^2$. Now $A^*\mathbf{v} \neq 0$ so $\lambda = \dfrac{\|A^*\mathbf{v}\|^2}{\|P\mathbf{v}\|^2}$ is positive and real. □

COROLLARY 9.3
The eigenvalues of AA^* are nonnegative real numbers.

PROOF Suppose $A \neq \mathbb{O}$. Since AA^* is normal, it has a spectral representation $AA^* = \sum_{\lambda_i \in \sigma(AA^*)} \lambda_i(\lambda_i I - AA^*)'$. Let $E = (\mu I - AA^*)'$ for any μ in $\lambda(AA^*)$. Then $(EA)(EA)^* = EAA^*E = \sum_{\lambda_i \in \sigma(AA^*)} \lambda_i E(\lambda_i I - AA^*)'E = \mu E$. Now if $EA = \mathbb{O}$ then $\mu E = \mathbb{O}$, so $\mu = 0$ since $E \neq \mathbb{O}$ because $\mu \in \lambda(AA^*)$. Finally, if $EA \neq \mathbb{O}$, then $\mu > 0$ by Theorem 9.12 above. □

THEOREM 9.13 (the square root theorem)
For each $A \in \mathbb{C}^{n\times n}$, there exists $B = B^*$ such that $B^2 = AA^*$.

PROOF By the spectral theorem, $AA^* = \sum_{\lambda_i \in \sigma(AA^*)} \lambda_i(\lambda_i I - AA^*)'$. By Corollary 9.3, we have $\lambda \geq 0$ for all $\lambda \in \lambda(AA^*)$. Put $B = \sum_{\lambda_i \in \sigma(AA^*)} \sqrt{\lambda_i}(\lambda_i I - AA^*)'$. Then clearly $B = B^*$ and $B^2 = AA^*$. ▯

One of the useful ways of representing a complex number $z = a + bi$ is in its polar form $z = re^{i\theta}$. Here $r \geq 0$ is real, $e^{i\theta}$ has magnitude 1, and $(e^{i\theta})^* = e^{-i\theta} = (e^{i\theta})^{-1}$. The analog for matrices is the next theorem.

THEOREM 9.14 (polar decomposition theorem)
For any $A \in \mathbb{C}^{n\times n}$, there exists $B = B^*$ in $\mathbb{C}^{n\times n}$ and U with $U^* = U^+$ such that $A = BU$. Moreover, $U'' = A''$, $U^{*\prime\prime} = A^{*\prime\prime}$, and $B^2 = AA^*$.

PROOF By the square root theorem, there exists $B = B^*$ in $\mathbb{C}^{n\times n}$ such that $B^2 = AA^*$. Let $U = B^+A$. Then $BU = BB^+A = (BB^+)^*A = B^{+*}B^*A = (B^{*+}B^*)A = B^{*\prime\prime}A = B''A = (B^*B)''A = (B^2)''A = (AA^*)''A = A^{*\prime\prime}A = A$. Also, since $B^{*\prime\prime} = (BB^*)'' = (AA^*)'' = A^{*\prime\prime}$, $U'' = (B^+A)'' = (B^{+\prime\prime}A)'' = (B^{*\prime\prime}A)'' = (A^{*\prime\prime}A)'' = A''$ and $U^{*\prime\prime} = (A^*B^{+*})'' = (A^*B^{*+})'' = (A^*B^+)'' = (A^{*\prime\prime}B^+)'' = (B''B^+)'' = [(B^{+*})''B^+]'' = B^{+\prime\prime} = B^{*\prime\prime} = A^{*\prime\prime}$. Finally, we claim $U^* = U^+$. But $UU^* = (B^+A)(A^*B^+) = B^+(AA^*)B^+ = B^+B^2B^+ = (B^+B)(BB^+) = B''B^{*\prime\prime} = B^{*\prime\prime}B^{*\prime\prime} = B^{*\prime\prime} = A^{*\prime\prime} = U^{*\prime\prime}$. Also, $U^*U = (A^*B^+)(B^+A) = A^*(B^+)^2A = A^*[(B^*B)^+B^*B^+]A = A^*[(B^2)^+BB^+]A = A^*[(B^2)^+B^{*\prime\prime}]A = A^*[(B^2)^+B'']A = A^*(B^2)^+B'']A = A^*[(B^+)^2(B''A)] = A^*(B^+)^2A = A^*(AA^*)^+A = A^*A^{*+} = A'' = U''$. Finally, $UU^*U = UU'' = U$ and $U^*UU^* = U''U^* = (U^{+*})''U^* = U^*$. Thus, U^* satisfies the four Moore-Penrose equations, and so $U^* = U^+$. ▯

We end this section with an interesting result about Hermitian matrices.

THEOREM 9.15
Suppose $A = A^*$, $B = B^*$, and $AB = BA$. Then there exists $C = C^*$ and polynomials p and q with real coefficients such that $A = p(C)$ and $B = q(C)$.

PROOF Use the spectral theorem to write $A = \lambda_1 E_1 + \cdots + \lambda_r E_r$ and $B = \mu_1 F_1 + \cdots + \mu_s F_s$. Now $AB = BA$ implies $BE_j = E_j B$ for all j, so $E_j F_k = F_k E_j$ for all j, k. Let $C = \sum_{j,k} c_{jk} E_j F_k$. Note $C = C^*$. Choose polynomials such that $p(c_{jk}) = \lambda_j$ and $q(c_{jk}) = \mu_k$ for all j, k. Then $C^n = \sum c_{jk}^n E_j F_k$ and

$p(C) = \sum \lambda_j E_j F_k = \sum \left(\sum \lambda_j E_j\right) F_k = \sum A F_k = A \sum F_k = A$ and $q(C) = \sum \mu_k E_j F_k = \sum \left(\sum Ej(\mu_k F_k)\right) = B$. □

Exercise Set 41

1. Write A in its polar form $A = BU$. Argue that A is normal iff B and U commute.

2. Write A in its polar form $A = BU$. Argue that λ is an eigenvalue of $\sqrt{B}$ iff λ^2 is an eigenvalue of B.

3. Write A in its polar form $A = BU$. Suppose λ is an eigenvalue of A written in its polar form $\lambda = re^{i\theta}$. Argue that r is an eigenvalue of B and $e^{i\theta}$ is an eigenvalue of U.

4. Prove that A is normal iff in a polar decomposition $A = BU$ and $AU = UA$ iff $AB = BA$.

Further Reading

[Higham, 1986a] N. J. Higham, Computing the Polar Decomposition–With Applications, SIAM J. Sci. Statist. Comput., Vol. 7, (1986), 1160–1174.

[Higham, 1986b] N. J. Higham, Newton's Method for the Matrix Square Root, Math. Comp., Vol. 46, (1986), 537–549.

[Higham, 1987] N. J. Higham, Computing Real Square Roots of a Real Matrix, Linear Algebra and Applications, 88/89, (1987), 405–430.

[Higham, 1994] N. J. Higham, The Matrix Sign Decomposition and Its Relation to the Polar Decomposition, Linear Algebra and Applications, 212/213, (1994), 3–20.

Chapter 10

Matrix Diagonalization

10.1 Diagonalization with Respect to Equivalence

Diagonal matrices are about the nicest matrices around. It is easy to add, subtract, and multiply them; find their inverses (if they happen to be invertible); and find their pseudoinverses. The question naturally arises, is it possible to express an arbitrary matrix in terms of diagonal ones? Let's illustrate just how easy this is to do. Consider $A = \begin{bmatrix} 1 & 1 & 2 \\ 1 & 2 & 2 \\ 1 & 1 & 2 \\ 1 & 2 & 2 \end{bmatrix}$. Note A is 4-by-3 with rank 2. First, write A in a full rank factorization; $A = \begin{bmatrix} 1 & 1 \\ 1 & 2 \\ 1 & 1 \\ 1 & 2 \end{bmatrix} \begin{bmatrix} 1 & 0 & 2 \\ 0 & 1 & 0 \end{bmatrix} = FG$. Now the columns of F and the rows of G are independent, so there are not be any zero columns in F or zero rows in G. We could use any nonzero scalars really, but just to be definite, take the lengths of the columns of F, normalize the columns, and form the diagonal matrix of column lengths. So $A = \begin{bmatrix} \frac{1}{2} & \frac{1}{\sqrt{10}} \\ \frac{1}{2} & \frac{2}{\sqrt{10}} \\ \frac{1}{2} & \frac{1}{\sqrt{10}} \\ \frac{1}{2} & \frac{2}{\sqrt{10}} \end{bmatrix} \begin{bmatrix} 2 & 0 \\ 0 & \sqrt{10} \end{bmatrix} \begin{bmatrix} 1 & 0 & 2 \\ 0 & 1 & 0 \end{bmatrix}$. Now do the same with the rows of G. Then $A = \begin{bmatrix} \frac{1}{2} & \frac{1}{\sqrt{10}} \\ \frac{1}{2} & \frac{1}{\sqrt{10}} \\ \frac{1}{2} & \frac{1}{\sqrt{10}} \\ \frac{1}{2} & \frac{1}{\sqrt{10}} \end{bmatrix} \begin{bmatrix} 2 & 0 \\ 0 & \sqrt{10} \end{bmatrix} \begin{bmatrix} \sqrt{5} & 0 \\ 0 & 1 \end{bmatrix} \begin{bmatrix} \frac{1}{\sqrt{5}} & 0 & \frac{2}{\sqrt{5}} \\ 0 & 1 & 0 \end{bmatrix} =$

$$\begin{bmatrix} \frac{1}{2} & \frac{1}{\sqrt{10}} \\ \frac{1}{2} & \frac{1}{\sqrt{10}} \\ \frac{1}{2} & \frac{1}{\sqrt{10}} \\ \frac{1}{2} & \frac{1}{\sqrt{10}} \end{bmatrix} \begin{bmatrix} 2\sqrt{5} & 0 \\ 0 & \sqrt{10} \end{bmatrix} \begin{bmatrix} \frac{1}{\sqrt{5}} & 0 & \frac{2}{\sqrt{5}} \\ 0 & 1 & 0 \end{bmatrix}$$

. It is clear that you can do this with any m-by-n matrix of rank r: $A = F_1 D G_1$, where F_1 has independent columns, D is r-by-r diagonal with nonzero entries down the main diagonal, and G_1 has independent rows. All this is fine, but so what? The numbers along the diagonal of D seem to have little to do with the original matrix A. Only the size of D seems to be relevant because it reveals the rank of A. All right, suppose someone insists on writing $A = F_1 D G_1$, where F_1 and G_1 are invertible. Well, this imposes some size restrictions; namely, F_1 and G_1 must be square. If A is in $\mathbb{C}_r^{m\times n}$, then F_1 would have to be m-by-m of rank m, G_1 would have to be n-by-n of rank n, and D would have to be m-by-n of rank r, meaning exactly r elements on the diagonal of D are nonzero. Now theory tells us we can always extend an independent set to a basis. Let's do this for the columns of F_1 and the rows of G_1. This will get us invertible matrices. Let's do this for our example above:

$$\begin{bmatrix} \frac{1}{2} & \frac{1}{\sqrt{10}} & \vdots & a_1 & b_1 \\ \frac{1}{2} & \frac{1}{\sqrt{10}} & \vdots & a_2 & b_2 \\ \frac{1}{2} & \frac{1}{\sqrt{10}} & \vdots & a_3 & b_3 \\ \frac{1}{2} & \frac{1}{\sqrt{10}} & \vdots & a_4 & b_4 \end{bmatrix} \begin{bmatrix} 2\sqrt{5} & 0 & 0 \\ 0 & \sqrt{10} & 0 \\ 0 & 0 & 0 \\ 0 & 0 & 0 \end{bmatrix} \begin{bmatrix} \frac{1}{\sqrt{5}} & 0 & \frac{2}{\sqrt{5}} \\ 0 & 1 & 0 \\ \cdots & \cdots & \cdots \\ c_1 & c_2 & c_3 \end{bmatrix} = \begin{bmatrix} 1 & 1 & 2 \\ 1 & 2 & 2 \\ 1 & 1 & 2 \\ 1 & 2 & 2 \end{bmatrix}$$

$= A$. Notice it does not matter how we completed those bases. Note the rank of A is revealed by the number of nonzero elements on the diagonal of D. In essence, we have found invertible matrices S and T such that $SAT = D$, where D is a diagonal matrix. The entries on the diagonal of D as yet do not seem to be meaningfully related to A. Anyway, this leads us to a definition.

DEFINITION 10.1 *(diagonalizable with respect to equivalence)*

Let $A \in \mathbb{C}_r^{m\times n}$. We say A is ***diagonalizable with respect to equivalence*** *iff there exist invertible matrices S in $\mathbb{C}^{m\times m}$ and T in $\mathbb{C}^{n\times n}$ such that $SAT = \begin{bmatrix} D_r & \mathbb{O} \\ \mathbb{O} & \mathbb{O} \end{bmatrix}$, where $D_r \in \mathbb{C}_r^{r\times r}$ is diagonal of rank r. Note that if $r = m$, we write $SAT = [D_r \vdots \mathbb{O}]$, and if $r = n$, we set $SAT = \begin{bmatrix} D_r \\ \cdots \\ \mathbb{O} \end{bmatrix}$.*

THEOREM 10.1

Let $A \in \mathbb{C}_r^{m\times n}$. Then A is diagonalizable with respect to equivalence if and only if there exists nonsingular matrices $S \in \mathbb{C}^{m\times m}$ and $T \in \mathbb{C}^{n\times n}$ such that $SAT = \begin{bmatrix} I_r & \mathbb{O} \\ \mathbb{O} & \mathbb{O} \end{bmatrix}$.

PROOF Suppose first such an S and T exist. Choose an arbitrary diagonal matrix D_r in $\mathbb{C}_r^{r\times r}$. Then $D = \begin{bmatrix} D_r & \mathbb{O} \\ \mathbb{O} & I_{m-r} \end{bmatrix}$ is nonsingular and so is DS. Let $S_1 = DS$ and $T_1 = T$. Then $S_1AT_1 = DSAT = D\begin{bmatrix} I_r & \mathbb{O} \\ \mathbb{O} & \mathbb{O} \end{bmatrix} = \begin{bmatrix} D_r & \mathbb{O} \\ \mathbb{O} & \mathbb{O} \end{bmatrix}$. Conversely, suppose A is diagonalizable with respect to equivalence. Then there exist S_1, T_1 invertible with $S_1AT_1 = \begin{bmatrix} D_r & \mathbb{O} \\ \mathbb{O} & \mathbb{O} \end{bmatrix} = \begin{bmatrix} D_r & \mathbb{O} \\ \mathbb{O} & I_{m-r} \end{bmatrix}\begin{bmatrix} I_r & \mathbb{O} \\ \mathbb{O} & \mathbb{O} \end{bmatrix} = D\begin{bmatrix} I_r & \mathbb{O} \\ \mathbb{O} & \mathbb{O} \end{bmatrix}$. Choose $S = D^{-1}S_1$ and $T = T_1$. Then $SAT = D^{-1}S_1AT_1 = D^{-1}D\begin{bmatrix} I_r & \mathbb{O} \\ \mathbb{O} & \mathbb{O} \end{bmatrix} = \begin{bmatrix} I_r & \mathbb{O} \\ \mathbb{O} & \mathbb{O} \end{bmatrix}$. This completes the proof. $\square$

This theorem says there is no loss in generality in using I_r in the definition of diagonalizability with respect to equivalence. In fact, there is nothing new here. We have been here before. This is just matrix equivalence and we proved in Chapter 4 (4.4) that every matrix $A \in \mathbb{C}_r^{m\times n}$ is equivalent to a matrix of the form $\begin{bmatrix} I_r & \mathbb{O} \\ \mathbb{O} & \mathbb{O} \end{bmatrix}$, namely rank normal form, and hence, in our new terminology, that *every* matrix $A \in \mathbb{C}_r^{m\times n}$ is diagonalizable with respect to equivalence. What we did not show was how to actually produce invertible matrices S and T that bring A to this canonical form where only rank matters from a full rank factorization of A.

Let $A \in \mathbb{C}_r^{m\times n}$ and let $A = FG$ be a full rank factorization of A. Then

$$\begin{bmatrix} F^+_{r\times m} \\ \cdots \\ \underset{(m-r)\times m}{W_1}\ [I_m - FF^+]_{m\times m} \end{bmatrix} A[\underset{n\times r}{G^+} \vdots \underset{n\times n}{(I_n - G^+G)}\ \underset{n\times(n-r)}{W_2}\ ] =$$

$$\begin{bmatrix} F^+_{r\times m} \\ \cdots \\ W_1[I_m - FF^+] \end{bmatrix} FG[G^+ \vdots (I_n - G^+G)W_2] =$$

$$\begin{bmatrix} F^+AG^+ & F^+A(I - G^+G)W_2 \\ W_1(I_m - FF^+)AG^+ & W_1(I - FF^+)A(I - G^+G)W_2 \end{bmatrix} = \begin{bmatrix} I_r & \mathbb{O} \\ \mathbb{O} & \mathbb{O} \end{bmatrix}.$$

Note that the arbitrary matrices $W_1 \in \mathbb{C}^{(m-r)\times m}$ and W_2 in $\mathbb{C}^{n\times(n-r)}$ were needed to achieve the appropriate sizes of matrices in the product. At the moment, they are quite arbitrary. Even more generally, take an arbitrary matrix $M \in \mathbb{C}^{r\times r}$. Then $\begin{bmatrix} F^+ \\ \cdots \\ W_1(I - FF^+) \end{bmatrix} A[G^+M \vdots (I - G^+G)W_2] = \begin{bmatrix} M & \mathbb{O} \\ \mathbb{O} & \mathbb{O} \end{bmatrix}$. We see that every full rank factorization of A leads to a "diagonal reduction"of A.

Of course, there is no reason to believe that the matrices flanking A above are nonsingular. Indeed, if we choose $W_1 = \mathbb{O}$ and $W_2 = \mathbb{O}$, they definitely will not be invertible. We must choose our W's much more carefully. We again appeal to full rank factorizations. Begin with any full rank factorization of A, say $A = FG$. Then take positive full rank factorizations of the projections $I_m - FF^+ = F_1G_1$ and $I_n - G^+G = F_2G_2$. That is, $F_1^* = G_1$ and $F_2^* = G_2$. Then we have the following:

1. $F_1 = G_1^+$
2. $G_1 = F_1^+$
3. $F_2 = G_2^+$
4. $G_2 = F_2^+$
5. $G_1F = G_1G_1^+G_1F = G_1(I - FF^+)F = \mathbb{O}$
6. $F^+G_1^+ = F^+F_1 = F^+F_1F_1^+F_1 = F^+F_1G_1F_1 = F^+(I - FF^+) F_1 = \mathbb{O}$
7. $G_2G^+ = G_2G_2^+G_2G^+ = G_2F_2G_2G^+ = G_2(I - G^+G)G^+ = \mathbb{O}$
8. $GG_2^+ = GF_2 = GF_2F_2^+F_2 = GF_2G_2F_2 = G(I - G^+G)F_2 = \mathbb{O}$.

Now we shall make $\begin{bmatrix} F^+ \\ \cdots \\ W_1(I - FF^+) \end{bmatrix}$ invertible by a judicious choice of W_1. Indeed, we compute that $\begin{bmatrix} F^+ \\ \cdots \\ W_1(I - FF^+) \end{bmatrix} [F \vdots (I - FF^+)W_1^*] =$

$$\begin{bmatrix} F^+F & F^+(I - FF^+)W_1^* \\ W_1(I - FF^+)F & W_1(I - FF^+)W_1^* \end{bmatrix} = \begin{bmatrix} I_r & \mathbb{O} \\ \mathbb{O} & W_1(I - FF^+)W_1^* \end{bmatrix}.$$

We badly need $W_1(I - FF^+)W_1^* = I_{m-r}$. All right, choose $W_1 = F_1^+ = G_1$. Then $W_1(I - FF^+)W_1^* = W_1F_1G_1W_1^* = F_1^+F_1G_1F_1^{+*} = IG_1F_1^{+*} = G_1G_1^* = I$. Thus, if we take $S = \begin{bmatrix} F^+ \\ \cdots \\ F_1^+(I - FF^+) \end{bmatrix}$, then S is invertible and

$S^{-1} = [F \vdots (I - FF^+)F_1^{+*}]$. But $F_1^+(I - FF^+) = F_1^+F_1G_1 = G_1 = F_1^+$, so we can simplify further. If $S = \begin{bmatrix} F^+ \\ \cdots \\ F_1^+ \end{bmatrix}$, then $S^{-1} = [F \vdots F_1]$. Similarly, if $T = [G^+ \vdots (I - G^+G)W_2]$, we choose $W_2 = F_2$ and find that $T^{-1} = \begin{bmatrix} G \\ \cdots \\ F_2^+ \end{bmatrix}$. These invertible matrices S and T are not unique. Take $S = \begin{bmatrix} F^+ + W_3(I - FF^+) \\ \cdots \\ F_1^+ \end{bmatrix}$. Then $S^{-1} = [F \vdots (I - FW_3)F_1]$, as the reader may verify. Here W_3 is completely arbitrary other than it has to be the right size. Similarly for $T = [G^+ + (I - G^+G)W_4 \vdots F_2^+]$, we have $T^{-1} = \begin{bmatrix} G \\ \cdots \\ F_2(I - W_4G) \end{bmatrix}$, where again W_4 is arbitrary of appropriate size. Let us summarize our findings in a theorem.

THEOREM 10.2

Every matrix A in $\mathbb{C}_r^{m\times n}$ is equivalent to $\begin{bmatrix} I_r & \mathbb{O} \\ \mathbb{O} & \mathbb{O} \end{bmatrix}$. If $A = FG$ is any full rank factorization of A and $I - FF^+ = F_1F_1^*$ and $I - G^+G = F_2F_2^*$ are positive full rank factorizations, then $S = \begin{bmatrix} F^+ + W_3(I - FF^+) \\ \cdots \\ F_1^+ \end{bmatrix}$ and $T = [G^+ + (I - G^+G)W_4 \vdots F_2^+]$ are invertible with $S^{-1} = [F \vdots (I - FW_3)F_1]$ and $T^{-1} = \begin{bmatrix} G \\ \cdots \\ F_2(I - W_4G) \end{bmatrix}$ and $SAT = \begin{bmatrix} I_r & \mathbb{O} \\ \mathbb{O} & \mathbb{O} \end{bmatrix}$.

Let's look at an example.

Example 10.1

Let $A = \begin{bmatrix} 1 & 2 & 2 \\ 7 & 6 & 10 \\ 4 & 4 & 6 \\ 1 & 0 & 1 \end{bmatrix} = FG = \begin{bmatrix} 1 & 2 \\ 7 & 6 \\ 4 & 4 \\ 1 & 0 \end{bmatrix} \begin{bmatrix} 1 & 0 & 1 \\ 0 & 1 & \frac{1}{2} \end{bmatrix}$ *in* $\mathbb{C}_2^{4\times 3}$. *First,*

$I - FF^+ = \begin{bmatrix} \frac{17}{38} & \frac{1}{38} & \frac{-10}{38} & \frac{16}{38} \\ \frac{1}{38} & \frac{9}{38} & \frac{-14}{38} & \frac{-8}{38} \\ \frac{-10}{38} & \frac{-14}{38} & \frac{26}{38} & \frac{5}{38} \\ \frac{16}{38} & \frac{-8}{38} & \frac{4}{38} & \frac{24}{38} \end{bmatrix}$. *Next, we compute a positive full rank factorization of this projection as described previously:* $I - FF^+ = \begin{bmatrix} \frac{-1}{\sqrt{6}} & \frac{4}{\sqrt{57}} \\ \frac{-1}{\sqrt{6}} & \frac{-2}{\sqrt{57}} \\ \frac{2}{\sqrt{6}} & \frac{1}{\sqrt{57}} \\ 0 & \frac{6}{\sqrt{57}} \end{bmatrix} \begin{bmatrix} \frac{-1}{\sqrt{6}} & \frac{-1}{\sqrt{6}} & \frac{2}{\sqrt{6}} & 0 \\ \frac{4}{\sqrt{57}} & \frac{-2}{\sqrt{57}} & \frac{1}{\sqrt{57}} & \frac{6}{\sqrt{57}} \end{bmatrix} = F_1 F_1^*$. *Then we find* $F_1^+ = \begin{bmatrix} \frac{-32}{76} & \frac{16}{76} & \frac{-8}{76} & \frac{28}{76} \\ \frac{37}{76} & \frac{-9}{76} & \frac{14}{76} & \frac{-30}{76} \end{bmatrix}$. *Now we can put S together;*

$$S = \begin{bmatrix} \frac{2}{76} & \frac{16}{76} & \frac{-8}{76} & \frac{28}{76} \\ \frac{37}{76} & \frac{-9}{76} & \frac{14}{76} & \frac{-30}{76} \\ \cdots & \cdots & \cdots & \cdots \\ \frac{-1}{\sqrt{6}} & \frac{-1}{\sqrt{6}} & \frac{2}{\sqrt{6}} & 0 \\ \frac{4}{\sqrt{57}} & \frac{-2}{\sqrt{57}} & \frac{1}{\sqrt{57}} & \frac{6}{\sqrt{57}} \end{bmatrix} \text{ and } S^{-1} = \begin{bmatrix} 1 & 2 & \vdots & \frac{-1}{\sqrt{6}} & \frac{4}{\sqrt{57}} \\ 7 & 6 & \vdots & \frac{-1}{\sqrt{6}} & \frac{-2}{\sqrt{57}} \\ 4 & 4 & \vdots & \frac{2}{\sqrt{6}} & \frac{1}{\sqrt{57}} \\ 1 & 0 & \vdots & 0 & \frac{6}{\sqrt{57}} \end{bmatrix}.$$

T is computed in a similar manner. We find $T = \begin{bmatrix} \frac{5}{9} & \frac{-2}{9} & \frac{-6}{9} \\ \frac{-2}{9} & \frac{8}{9} & \frac{-3}{9} \\ \frac{4}{9} & \frac{2}{9} & \frac{6}{9} \end{bmatrix}$ *and*

$$SAT = \begin{bmatrix} 1 & 0 & 0 \\ 0 & 1 & 0 \\ 0 & 0 & 0 \\ 0 & 0 & 0 \end{bmatrix} \text{ as the reader can verify.}$$

Actually, a little more is true. The matrix S above can always be chosen to be unitary, not just invertible. To gain this, we begin with an orthogonal full rank factorization of A, $A = FG$, where $F^* = F^+$. That is, the columns of F form an orthonormal basis of the column space of A. Then $SS^* = \begin{bmatrix} F^+ \\ \cdots \\ W_1(I - FF^+) \end{bmatrix} [(F^+)^* \vdots (I - FF^+)W_1^*] = \begin{bmatrix} F^+F^{**} & \mathbb{O} \\ \mathbb{O} & W_1(I - FF^+)W_1^* \end{bmatrix} = \begin{bmatrix} I_r & \mathbb{O} \\ \mathbb{O} & W_1(I - FF^+)W_1^* \end{bmatrix}$. As before, select $W_1 = F_1^+$ and get $SS^* = I$.

Exercise Set 42

1. Verify the formulas labeled (1) through (8) on page 354.

Further Reading

[Enegren, 1995] Disa Enegren, On Simultaneous Diagonalization of Matrices, Masters Thesis, Baylor University, May, (1995).

[Wielenga, 1992] Douglas G. Wielenga, Taxonomy of Necessary and Sufficient Conditions for Simultaneously Diagonalizing a Pair of Rectangular Matrices, Masters Thesis, Baylor University, August, (1992).

similar matrices, principal idempotents, minimal polynomial relative to a vector, primary decomposition theorem

10.2 Diagonalization with Respect to Similarity

In this section, we demand even more of a matrix. In the previous section, every matrix was diagonalizable in the weak sense of equivalence. That will not be the case in this section. Here we deal only with square matrices.

DEFINITION 10.2

*Two matrices $A, B \in \mathbb{C}^{n\times n}$ are **similar** (in symbols $A \sim B$) iff there exists an invertible matrix $S \in \mathbb{C}^{n\times n}$ with $S^{-1}AS = B$. A matrix A is **diagonalizable with respect to similarity** iff A is similar to a diagonal matrix.*

Let's try to get a feeling for what is going on here. The first thing we note is that this notion of diagonalizability works only for square matrices. Let's just look at a simple 3-by-3 case. Say $S^{-1}AS = D$ or, what is the same, $AS = SD$, where $D = \begin{bmatrix} \lambda_1 & 0 & 0 \\ 0 & \lambda_2 & 0 \\ 0 & 0 & \lambda_3 \end{bmatrix}$ and $S = [\mathbf{s}_1|\mathbf{s}_2|\mathbf{s}_3]$. Then $A[\mathbf{s}_1|\mathbf{s}_2|\mathbf{s}_3] =$

$[\mathbf{s}_1|\mathbf{s}_2|\mathbf{s}_3]\begin{bmatrix}\lambda_1 & 0 & 0\\ 0 & \lambda_2 & 0\\ 0 & 0 & \lambda_3\end{bmatrix}$ or $[A\mathbf{s}_1|A\mathbf{s}_2|A\mathbf{s}_3] = [\lambda_1\mathbf{s}_1|\lambda_2\mathbf{s}_2|\lambda_3\mathbf{s}_3]$. This says $A\mathbf{s}_1 = \lambda_1\mathbf{s}_1$, $A\mathbf{s}_2 = \lambda_2\mathbf{s}_2$ and $A\mathbf{s}_3 = \lambda_3\mathbf{s}_3$. Thus, the diagonal elements of D must be the eigenvalues of A and the columns of S must be eigenvectors corresponding to those eigenvalues. Moreover, S has full rank, being invertible, so these eigenvectors are independent. But then they form a basis of $\mathbb{C}^3$. The next theorem should not now surprise you.

THEOREM 10.3
Let $A \in \mathbb{C}^{n\times n}$. Then A is diagonalizable with respect to similarity iff A has n linearly independent eigenvectors (i.e., $\mathbb{C}^n$ has a basis consisting entirely of eigenvectors of A).

PROOF The proof is left as an exercise. Just generalize the example above. ▯

Thus, if you can come up with n linearly independent eigenvectors of $A \in \mathbb{C}^{n\times n}$, you can easily construct S to bring A into diagonal form.

THEOREM 10.4
Let $A \in \mathbb{C}^{n\times n}$ and let $\lambda_1, \lambda_2, \ldots \lambda_s$ be distinct eigenvalues of A. Suppose $\mathbf{v}_1, \mathbf{v}_2, \ldots, \mathbf{v}_s$ are eigenvectors corresponding to each of the λs. Then $\{\mathbf{v}_1, \mathbf{v}_2, \ldots, \mathbf{v}_s\}$ is a linearly independent set.

PROOF Suppose to the contrary that the set is dependent. None of these eigenvectors is zero. So let t be the largest index so that $\mathbf{v}_1, \mathbf{v}_2, \ldots \mathbf{v}_t$ is an independent set. It could be $t = 1$, but it cannot be that $t = s$. So $1 \leq t < s$. Now $\mathbf{v}_1, \ldots \mathbf{v}_t, \mathbf{v}_{t+1}$ is a dependent set by definition of how we chose t so there exists scalars $\alpha_1, \alpha_2, \ldots, \alpha_t, \alpha_{t+1}$, not all zero with $\alpha_1\mathbf{v}_1 + \alpha_2\mathbf{v}_2 + \ldots + \alpha_t\mathbf{v}_t + \alpha_{t+1}\mathbf{v}_{t+1} = \vec{0}$. Multiply by A and get $\alpha_1\lambda_1\mathbf{v}_1 + \alpha_2\lambda_2\mathbf{v}_2 + \ldots + \alpha_{t+1}\lambda_{t+1}\mathbf{v}_{t+1} = \vec{0}$. Multiply the original dependency by λ_{t+1} and get $\alpha_1\lambda_{t+1}\mathbf{v}_1 + \ldots + \alpha_{t+1}\lambda_{t+1}\mathbf{v}_{t+1} = \vec{0}$. Subtracting the two equations yields $\alpha_1(\lambda_1 - \lambda_{t+1})\mathbf{v}_1 + \ldots + \alpha_z(\lambda_t - \lambda_{t+1})\mathbf{v}_1 = \vec{0}$. But $\mathbf{v}_1, \ldots, \mathbf{v}_t$ is an independent set, so all the coefficients must be zero. But the λs are distinct so the only way out is for all the αs to be zero. But this contradicts the fact that not all the αs are zero. This completes our indirect proof. ▯

COROLLARY 10.1
If $A \in \mathbb{C}^{n\times n}$ has n distinct eigenvalues then A is diagonalizable with respect to similarity.

Matrices that are diagonalizable with respect to a similarity have a particularly nice form. The following is a more general version of the spectral theorem.

THEOREM 10.5

Suppose $A \in \mathbb{C}^{n\times n}$ and $\lambda_1, \lambda_2, \ldots, \lambda_k$ are its distinct eigenvalues. Then A is diagonalizable with respect to a similarity iff there exist unique idempotents $E_1, E_2, \ldots, E_k$, such that

1. $E_i E_j = \mathbb{O}$ *whenever* $i \neq j$
2. $\sum_{i=1}^{k} E_i = I_n$
3. $A = \sum_{i=1}^{k} \lambda_i E_i$.

Moreover, the E_i are expressible as polynomials in A.

PROOF Suppose first such idempotents E_i exist satisfying (1), (2), and (3). Then we know from earlier work that these idempotents effect a direct sum decomposition of $\mathbb{C}^n$, namely, $\mathbb{C}^n = Col(E_1) \oplus Col(E_2) \oplus \cdots \oplus Col(E_k)$. Let $r_i = dim(Col(E_i))$. Choose a basis of each summand and union these to get a basis of the whole space. Then create an invertible matrix S by arranging the basis vectors as columns. Then $AS = (\sum_{i=1}^{k}\lambda_i E_i)S = \sum_{i=1}^{k}\lambda_i E_i S =$

$$S\begin{bmatrix} \lambda_1 I_{r_1} & & & \\ & \lambda_2 I_{r_2} & & \\ & & \ddots & \\ & & & \lambda_k I_{r_k} \end{bmatrix}$$

. Thus A is diagonalizable with respect to a similarity. Conversely, suppose $AS = S\begin{bmatrix} \lambda_1 I_{r_1} & & & \\ & \lambda_2 I_{r_2} & & \\ & & \ddots & \\ & & & \lambda_k I_{r_k} \end{bmatrix}$.

Define $E_i = [\mathbb{O} \cdots \mathbb{O}\mathcal{B}_i\mathbb{O} \cdots \mathbb{O}]S^{-1}$, where $\mathcal{B}_i$ is a collection of column vectors from a basis of $Col(E_i)$. Then the E_i satisfy (1), (2), and (3), as we ask the reader to verify. For uniqueness, suppose idempotents F_i satisfy (1), (2), and (3). Then $E_i A = AE_i = \lambda_i E_i$ and $F_j A = AF_j = \lambda_j F_j$. Thus $E_i(AF_j) = \lambda_j E_i F_j$ and $(E_i A)F_j = \lambda_i E_i F_j$, which imply $E_i F_j = \mathbb{O}$ when $i \neq j$. Now $E_i = E_i \sum F_j = E_i Fi = (\sum E_j)F_i = F_i$. Finally, consider the polynomial $p_i(x) = \prod_{\substack{j=1\\ j\neq i}}^{k} (x - \lambda_j)$. Note $p_i(\lambda_i) \neq 0$. Let $F_i = \dfrac{p_i(A)}{p_i(\lambda_i)}$. Compute that $F_i E_j = E_i$ if $i = j$ and $\mathbb{O}$ if $i \neq j$. Then $F_i = F_i(\sum E_j) = E_i$. ☐

The E_is above are called the **principal idempotents** of A. For example, they allow us to define functions of a matrix that is diagonalizable with respect to a similarity; namely, given f and A, define $f(A) = \sum f(\lambda_i)E_i$.

The minimum polynomial of a matrix has something to say about diagonalizability. Let's recall some details. Let $A \in \mathbb{C}^{n\times n}$. Then $I, A, A^2, \ldots, A^{n^2}$ is a list of $n^2 + 1$ matrices (vectors) in the n^2 dimensional vector space $\mathbb{C}^{n\times n}$. These then must be linearly dependent. Thus, there is a uniquely determined integer $s \leq n^2$ such that $I, A, A^2, \ldots, A^{s-1}$ are linearly independent but $I, A, A^2, \ldots, A^{s-1}, A^s$ are linearly dependent. Therefore, A^s is a linear combination of the "vectors" that precede A^s in this list, say $A^s = \beta_0 I + \beta_1 A + \ldots + \beta_{s-1}A^{s-1}$. The minimum polynomial of A is $\mu_A(x) = x^s - \beta_{s-1}x^{s-1} - \ldots - \beta_0$ in $\mathbb{C}[x]$.

THEOREM 10.6

Let $A \in \mathbb{C}^{n\times n}$. Then

1. $\mu_A(x) \neq 0$ in $\mathbb{C}[x]$ yet $\mu_A(A) = \mathbb{O}$ in $\mathbb{C}^{n\times n}$.
2. μ_A is the unique monic polynomial (i.e., leading coefficient 1) of least degree that annihilates A.
3. If $p(x) \in \mathbb{C}[x]$ and $p(A) = \mathbb{O}$, then $\mu_A(x)$ divides $p(x)$.

PROOF

1. The proof follows from the remarks preceding the theorem.
2. The proof is pretty clear since if $p(x)$ were another monic polynomial with $deg(p(x)) = s$ and $p(A) = \mathbb{O}$, then $\mu_A(x) - p(x)$ is a polynomial of degree less than s, which still annihilates A, a contradiction.
3. Suppose $p(x) \in \mathbb{C}[x]$ and $p(A) = \mathbb{O}$. Apply the division algorithm in $\mathbb{C}[x]$ and write $p(x) = \mu_A(x)q(x) + r(x)$ where $r(x) = 0$ or $deg r(x) < s = deg \mu_A(x)$. Then $r(A) = p(A) - \mu_A(A)q(A) = \mathbb{O}$. This contradicts the definition of μ_A unless $r(x) = 0$. Thus, $p(x) = \mu_A(x)q(x)$, as was to be proved. ☐

Let's look at some small examples.

Example 10.2

1. *Let* $A = \begin{bmatrix} a & b \\ c & d \end{bmatrix}$. *In theory, we should compute* $\{I, A, A^2, A^3, A^4\}$ *and look for a dependency. But* $A^2 = \begin{bmatrix} a^2+bc & ab+bd \\ ca+cd & cb+d^2 \end{bmatrix} = \begin{bmatrix} a^2+bc & b(a+d) \\ c(a+d) & cb+d^2 \end{bmatrix}$, *so* $A^2-(a+d)A = \begin{bmatrix} bc-ad & 0 \\ 0 & bc-ad \end{bmatrix}$ *so* $A^2 - tr(A)A + det(A)I = \mathbb{O}$. *Thus, the minimum polynomial of A,* $\mu_A(x)$ *divides* $p(x) = x^2 - tr(A)x + det(A)$. *Hence either* $\mu_A(x) = x - \lambda$ *or* $\mu_A(x) = x^2 - tr(A)x + det(A)$. *In the first instance, A must be a scalar multiple of the identity matrix. So, for 2-by-2 matrices, the situation is pretty well nailed down:* $A = \begin{bmatrix} 5 & 0 \\ 0 & 5 \end{bmatrix}$ *has minimal polynomial* $\mu_A(x) = x - 5$, *while* $A = \begin{bmatrix} 5 & 1 \\ 0 & 5 \end{bmatrix}$ *has minimal polynomial* $\mu_A(x) = x^2 - 10x + 25 = (x-5)^2$. *Meanwhile,* $A = \begin{bmatrix} 1 & 2 \\ 3 & 4 \end{bmatrix}$ *has minimum polynomial* $\mu_A(x) = x^2 - 5x - 2$.

2. *The 3-by-3 case is a bit more intense. Let* $A = \begin{bmatrix} a & b & c \\ d & e & f \\ g & h & j \end{bmatrix}$. *Then*
$A^2 = \begin{bmatrix} a^2+bd+cg & \cdots \end{bmatrix}$
$A^3 = \begin{bmatrix} a(a^2+bd+cg)+ \\ d(a^2+bd+cg)+ & \cdots \\ g(a^2+bd+cg)+ \end{bmatrix}$ *so* $A^3 - tr(A)A^2 + (ea + ja + je - db - cq - fh)A - \det(A)I = \mathbb{O}$. *Thus, the minimum polynomial* μ_A *is of degree 3, 2, or 1 and divides (or equals)*

$$x^3 - tr(A)x^2 - (ea + ja + je - db - cg - fh)x - det(A) = 0.$$

Well, that is about as far as brute force can take us. Now you see why we have theory. Let's look at another approach to computing the minimum polynomial of a matrix.

Let $A \in \mathbb{C}^{n\times n}$. For any $\mathbf{v} \in \mathbb{C}^n$, we can construct the list of vectors $\mathbf{v}, A\mathbf{v}, A^2\mathbf{v}, \ldots$. As before, there is a least nonnegative integer d with $\mathbf{v}, A\mathbf{v}, \ldots, A^d\mathbf{v}$ linearly dependent. Evidently, $d \leq n$ and $d = 0$ iff $v = \overrightarrow{0}$ and $d = 1$ iff $\mathbf{v}$ is an eigenvector of A. Say $A^d\mathbf{v} = \beta_0\mathbf{v} + \beta_1 A\mathbf{v} + \cdots + \beta_{d-1}A^{d-1}\mathbf{v}$. Then define the **minimal polynomial of A relative to v** as $\mu_{A,\mathbf{v}}(x) = x^d - \beta_{d-1}x^{d-1} - \cdots - \beta_1 x - \beta_0$. Clearly, $\mathbf{v} \in \mathcal{N}ull(\mu_{A,\mathbf{v}}(A))$ so $\mu_{A,\mathbf{v}}$ is the unique monic polynomial p of least degree so that $\mathbf{v} \in \mathcal{N}ull(p(A))$. Now, if $\mu_{A,\mathbf{v}_1}(x)$ and $\mu_{A,\mathbf{v}_2}(x)$ both divide a polynomial $p(x)$, then $\mathbf{v}_1$ and $\mathbf{v}_2$ belong to $\mathcal{N}ull(p(A))$. So if you are lucky enough to have a basis $\mathbf{v}_1, \mathbf{v}_2, \cdots, \mathbf{v}_n$ and a polynomial $p(x)$ divided by each $\mu_{A,\mathbf{v}_i}(x)$, then the basis $\{\mathbf{v}_1, \ldots, \mathbf{v}_n\} \subseteq \mathcal{N}ull(p(A))$ so $\mathcal{N}ull(p(A)) = \mathbb{C}^n$ whence $p(A) = \mathbb{O}$.

THEOREM 10.7

Let $A \in \mathbb{C}^{n\times n}$ and let $\{\mathbf{b}_1, \mathbf{b}_2, \ldots, \mathbf{b}_n\}$ be a basis of $\mathbb{C}^n$. Then $\mu_A(x) = LCM(\mu_{A,\mathbf{b}_1}(x), \ldots, \mu_{A,\mathbf{b}_n}(x))$. That is, the minimum polynomial of A is the least common multiple of the minimum polynomials of A relative to a basis.

PROOF Let $p(x) = LCM(\mu_{A,\mathbf{b}_1}(x), \mu_{A,\mathbf{b}_2}(x), \ldots, \mu_{A,\mathbf{b}_n}(x))$. Then $p(A) = \mathbb{O}$ as we noted above. Thus $\mu_A(x)$ divides $p(A)$. Conversely, for each j, $1 \leq j \leq n$, apply the division algorithm and get $\mu_A(x) = q_j(x)\mu_{A,\mathbf{b}_j}(x) + r_j(x)$, where $r_j(x) = 0$ or $degr_j(x) < deg(\mu_{A,\mathbf{b}_j}(x))$. Then $\overrightarrow{0} = \mu_A(A)\mathbf{b}_j = q_j(A)(\mu_{A,\mathbf{b}_j}(A)\mathbf{b}_j) + r_j(A)\mathbf{b}_j = r_j(A)\mathbf{b}_j$.

By minimality of the degree of $\mu_{A,\mathbf{b}_j}$, we must have $r_j(x) = 0$. This says every $\mu_{A,\mathbf{b}_j}$ divides $\mu_A(x)$. Thus the LCM of the $\mu_{A,\mathbf{b}_j}(x)$ divides $\mu_A(x)$ but that is $p(x)$. Therefore, $p(x) = \mu_A(x)$ and our proof is done. ∎

For example, let $A = \begin{bmatrix} -1 & -1 & 2 \\ -1 & 0 & 1 \\ 0 & -1 & 1 \end{bmatrix}$. The standard basis is probably the easiest one to think of, so $A\mathbf{e}_1 = \begin{bmatrix} -1 & -1 & 2 \\ -1 & 0 & 1 \\ 0 & -1 & 1 \end{bmatrix}\begin{bmatrix} 1 \\ 0 \\ 0 \end{bmatrix} = \begin{bmatrix} -1 \\ -1 \\ 0 \end{bmatrix}$, $A^2\mathbf{e}_1 = \begin{bmatrix} -1 & -1 & 2 \\ -1 & 0 & 1 \\ 0 & -1 & 1 \end{bmatrix}\begin{bmatrix} -1 \\ -1 \\ 0 \end{bmatrix} = \begin{bmatrix} 2 \\ 1 \\ 1 \end{bmatrix}$, $A^3\mathbf{e}_1 = \begin{bmatrix} -1 & -1 & 2 \\ -1 & 0 & 1 \\ 0 & -1 & 1 \end{bmatrix}\begin{bmatrix} 2 \\ 1 \\ 1 \end{bmatrix} = \begin{bmatrix} -1 \\ -1 \\ 0 \end{bmatrix} = A\mathbf{e}_1$, so $(A^3 - A)\mathbf{e}_1 = \overrightarrow{0}$. Thus $\mu_{A,\mathbf{e}_1}(x) = x^3 - x = x(x-1)(x+1)$. $A\mathbf{e}_2 = \begin{bmatrix} -1 & -1 & 2 \\ -1 & 0 & 1 \\ 0 & -1 & 1 \end{bmatrix}\begin{bmatrix} 0 \\ 1 \\ 0 \end{bmatrix} = \begin{bmatrix} -1 \\ 0 \\ -1 \end{bmatrix}$, $A^2\mathbf{e}_2 = \begin{bmatrix} -1 & -1 & 2 \\ -1 & 0 & 1 \\ 0 & -1 & 1 \end{bmatrix}\begin{bmatrix} -1 \\ 0 \\ -1 \end{bmatrix} = \begin{bmatrix} -1 \\ 0 \\ -1 \end{bmatrix} = A\mathbf{e}_2$, so $\mu_{A,e_2}(x) = x^2 - x = x(x-1)$. Finally, $A\mathbf{e}_3 = \begin{bmatrix} -1 & -1 & 2 \\ -1 & 0 & 1 \\ 0 & -1 & 1 \end{bmatrix}\begin{bmatrix} 0 \\ 0 \\ 1 \end{bmatrix} = \begin{bmatrix} 2 \\ 1 \\ 1 \end{bmatrix}$, $A^2\mathbf{e}_3 =$

$\begin{bmatrix} -1 & -1 & 2 \\ -1 & 0 & 1 \\ 0 & -1 & 1 \end{bmatrix} \begin{bmatrix} 2 \\ 1 \\ 1 \end{bmatrix} = \begin{bmatrix} -1 \\ -1 \\ 0 \end{bmatrix}$, $A^3\mathbf{e}_3 = \begin{bmatrix} -1 & -1 & 2 \\ -1 & 0 & 1 \\ 0 & -1 & 1 \end{bmatrix} \begin{bmatrix} -1 \\ -1 \\ 0 \end{bmatrix} = \begin{bmatrix} 2 \\ 1 \\ 1 \end{bmatrix} = A\mathbf{e}_3$, so $\mu_{A,\mathbf{e}_3}(x) = x^3 - x = x(x-1)(x+1)$ and $\mu_A(x) = LCM(x^3 - x, x^2 - x) = x(x-1)(x+1) = x^3 - x$. Did you notice we never actually computed any powers of A in this example?

The next theorem makes an important connection with eigenvalues of a matrix and its minimum polynomial.

THEOREM 10.8

Let $A \in \mathbb{C}^{n\times n}$. Then the eigenvalues of A are exactly the roots of the minimum polynomial $\mu_A(x)$.

PROOF Suppose λ is a root of $\mu_A(x)$. Then $\mu_A(x) = (x-\lambda)q(x)$ where $deg q < deg \mu_A$. But $\mathbb{O} = (A - \lambda I)q(A)$ and $q(A) \neq \mathbb{O}$. Thus there exists a vector $\mathbf{v}$, necessarily nonzero, such that $q(A)\mathbf{v} \neq \vec{0}$. Let $\mathbf{w} = q(A)\mathbf{v}$. Then $\vec{0} = (A - \lambda I)\mathbf{w}$ so λ is an eigenvalue of A.

Conversely, suppose λ is an eigenvalue of A with eigenvector $\mathbf{v}$. Then $A^2\mathbf{v} = A\lambda\mathbf{v} = \lambda A\mathbf{v} = \lambda^2\mathbf{v}$. Generally, $A^j\mathbf{v} = \lambda^j\mathbf{v}$. So if $\mu_A(x) = x^s - \alpha_{s-1}x^{s-1} - \cdots - \alpha_1 x + \alpha_0$ then $\vec{0} = \mu_A(A)\mathbf{v} = (A^s - \alpha_{s-1}A^{s-1} - \cdots - \alpha_1 A - \alpha_0 I)\mathbf{v} = \lambda^s\mathbf{v} - \alpha_{s-1}\lambda^{s-1}\mathbf{v} - \cdots - \alpha\lambda\mathbf{v} - \alpha_0\mathbf{v} = (\lambda^s - \alpha_{s-1}\lambda^{s-1} - \cdots - \alpha_0)\mathbf{v} = \mu_A(\lambda)\mathbf{v}$. But $\mathbf{v} \neq \vec{0}$ so $\mu_A(\lambda) = 0$. □

THEOREM 10.9

Let $A \in \mathbb{C}^{n\times n}$ with distinct eigenvalues $\lambda_1, \lambda_2, \ldots, \lambda_s$. Then the following are equivalent:

1. *A is diagonalizable with respect to similarity.*
2. *There is a basis of $\mathbb{C}^n$ consisting of eigenvectors of A.*
3. *$\mu_A(x) = (x - \lambda_1)(x - \lambda_2)\cdots(x - \lambda_s)$.*
4. *$GCD(\mu_A, \mu'_A) = 1$.*
5. *μ_A and μ'_A do not have a common root.*

PROOF We already have (1) is equivalent to (2) by (10.3). (4) and (5) are equivalent by general polynomial theory. Suppose (2). Then there is a basis $\mathbf{b}_1, \mathbf{b}_2, \ldots, \mathbf{b}_n$ of eigenvectors of A. Then $\mu_A(x) = LCM(\mu_{A,\mathbf{b}_1}(x), \ldots, \mu_{A,\mathbf{b}_n}(x))$. But $\mu_{A,\mathbf{b}_i}(x) = x - \beta_0$ since the $\mathbf{b}_i$ are eigenvectors and β_0 is some eigenvalue. Thus (3) follows. Conversely, if we have (3), consider $\mathbf{b}_1, \mathbf{b}_2, \cdots, \mathbf{b}_s$

eigenvectors that belong to $\lambda_1, \lambda_2, \cdots, \lambda_s$, respectively. Then $\{\mathbf{b}_1, \mathbf{b}_2, \cdots, \mathbf{b}_s\}$ is an independent set. If it spans $\mathbb{C}^n$, we have a basis of eigenvectors and (2) follows. If not, extend this set to a basis, $\{\mathbf{b}_1, \cdots, \mathbf{b}_s, \mathbf{w}_{s+1}, \cdots, \mathbf{w}_n\}$. If, by luck, all the **w**s are eigenvectors of A, (2) follows again. So suppose some **w** is not an eigenvector. Then $\mu_{A,\mathbf{w}_j}(x)$ has degree at least 2, a contradiction. □

Now we come to a general result about all square matrices over $\mathbb{C}$.

THEOREM 10.10 (primary decomposition theorem)

Let $A \in \mathbb{C}^{n\times n}$. Suppose $\lambda_1, \lambda_2, \ldots, \lambda_s$ are the distinct eigenvalues of A. Let $\mu_A(x) = (x-\lambda_1)^{e_1}(x-\lambda_2)^{e_2}\cdots(x-\lambda_s)^{e_s}$. Then

$$\mathbb{C}^n = \mathcal{N}ull((A-\lambda_1 I)^{e_1}) \oplus \mathcal{N}ull((A-\lambda_2 I)^{e_2}) \oplus \cdots \oplus \mathcal{N}ull((A-\lambda_s I)^{e_s})$$

Moreover, each $\mathcal{N}ull((A-\lambda_i I)^{e_i})$ is A-invariant. Also if $\chi_A(x) = (x-\lambda_1)^{d_1}(x-\lambda_2)^{d_2}\cdots(x-\lambda_s)^{d_s}$, then $dim(\mathcal{N}ull((A-\lambda_i I)^{e_i})) = d_i$. Moreover, an invertible matrix S exists with $S^{-1}AS = \begin{bmatrix} A_1 & & & \\ & A_2 & & \\ & & \ddots & \\ & & & A_s \end{bmatrix}$ where each A_i is d_i-by-d_i.

PROOF Let $q_i(x) = (x-\lambda_1)^{e_1}\cdots(x-\lambda_{i-1})^{e_{i-1}}(x-\lambda_{i+1})^{e_{i+1}}\cdots(x-\lambda_s)^{e_s}$. In other words, delete the ith factor in the minimum polynomial. Then $deg(q_i(x)) < deg(\mu_A(x))$ and the polynomials $q_1(x), q_2(x), \ldots, q_s(x)$ are relatively prime (i.e., they have no common prime factors). Thus, there exist polynomials $p_1(x), p_2(x), \ldots, p_s(x)$ with $1 = q_1(x)p_1(x) + q_2(x)p_2(x) + \ldots + q_s(x)p_s(x)$. But then $I = q_1(A)p_1(A) + q_2(A)p_2(A) + \ldots + q_s(A)p_s(A)$. Let $h_i(x) = q_i(x)p_i(x)$ and let $E_i = h_i(A)$ $i = 1, \ldots, s$. Already we see $I = E_1 + E_2 + \cdots + E_s$. Moreover, for $i \neq j$, $E_iE_j = q_i(A)p_i(A)q_j(A)p_j(A) = \mu_A(A)\, times$ (something) $= \mathbb{O}$. Next, $E_i = E_i(I) = E_i(E_1 + \cdots + E_s) = E_iE_i$, so each E_i is idempotent. Therefore, already we have $\mathbb{C}^n = Col(E_1) + Col(E_2) + \cdots + Col(E_s)$. Could any of these column spaces be zero? Say $Col(E_i) = (\overrightarrow{0})$. Then $\mathbb{C}^n = Col(E_1) + \cdots + Col(E_{i-1}) + Col(E_{i+1}) + \cdots + Col(E_s)$. Now $q_i(A)E_j = q_i(A)q_j(A)p_j(A) = \mathbb{O}$ for $j \neq i$ so, since every vector in $\mathbb{C}^n$ is a sum of vectors from the column spaces $Col(E_1), \ldots, Col(E_s)$ without $Col(E_i)$, it follows $q_i(A) = \mathbb{O}$. But this contradicts the fact that $\mu_A(x)$ *is* the minimum polynomial of A so $Col(E_i) \neq (\overrightarrow{0})$ for $i = 1, \cdots, s$. Thus, we have a direct sum $\mathbb{C}^n = Col(E_1) \oplus \cdots \oplus Col(E_s)$. To finish we need to identify these column spaces. First, $(A-\lambda_i I)^{e_i}E_i = (A-\lambda_i I)^{e_i}q_i(A)p_i(A) = \mu_A(A)p_i(A) = \mathbb{O}$ so $Col(E_i) \subseteq \mathcal{N}ull(A-\lambda_i I)^{e_i}$. Next, take $\mathbf{v} \in \mathcal{N}ull(A-\lambda_i I)^{e_i}$ and write

$\mathbf{v} = E_1\mathbf{v}_1 + E_2\mathbf{v}_2 + \cdots + E_s\mathbf{v}_s$ where $E_j\mathbf{v}_j \in \mathcal{C}ol(E_j)$, $j = 1, \ldots, s$. Then $\vec{0} = (A - \lambda_i I)^{e_i}\mathbf{v} = (A - \lambda_i I)^{e_i} E_1\mathbf{v}_1 + \ldots + (A - \lambda_i I)^{e_i} E_s\mathbf{v}_s = E_1(A - \lambda_i I)^{e_i}\mathbf{v}_1 + \ldots + E_s(A - \lambda_i I)^{e_i}\mathbf{v}_s$.

Note the Es are polynomials in A, so we get to commute them with $(A - \lambda_i I)^{e_i}$, which is also a polynomial in A. Since the sum is direct we have each piece must be zero — in other words, $(A - \lambda_i I)^{e_i} E_j\mathbf{v}_j = \vec{0}$. Thus for $j \neq i$, $(A - \lambda_i I)^{e_i} E_j\mathbf{v}_j = (A - \lambda_j I)^{e_j} E_j\mathbf{v}_j = \vec{0}$. Now $GCD((x - \lambda_i)^{e_i}, (x - \lambda_j)^{e_j}) = 1$ so there exist polynomials $a(x)$ and $b(x)$ with $1 = a(x)(x - \lambda_i)^{e_i} + b(x)(x - \lambda_j)^{e_j}$. Therefore, $E_j\mathbf{v}_j = IE_j\mathbf{v}_j = (a(A)(A - \lambda_i I)^{e_i} + b(A)(A - \lambda_j)^{e_j})E_j\mathbf{v}_j = a(A)(A - \lambda_i I)^{e_i} E_j\mathbf{v}_j + b(A)(A - \lambda_j)^{e_j} E_j\mathbf{v}_j = \vec{0} + \vec{0} = \vec{0}$. Thus $\mathbf{v} = E_1\mathbf{v}_1 + \ldots + E_s\mathbf{v}_s = E_i\mathbf{v}_i$, putting $\mathbf{v}$ in the column space of E_i. Now, by taking a basis of each $\mathcal{N}ull((A - \lambda_i I)^{e_i})$, we construct an invertible matrix S so that

$$S^{-1}AS = \begin{bmatrix} A_1 & & & \\ & A_2 & & \\ & & \ddots & \\ & & & A_s \end{bmatrix}.$$

Let $\mu_{A_i}(x)$ be the minimum polynomial of A_i. We know that $\mu_A(x) = \mu_{S^{-1}AS}(x) = LCM(\mu_{A_1}(x), \ldots, \mu_{A_s}(x))$ and $\chi_A(x) = \chi_{S^{-1}AS}(x) = \chi_{A_1}(x)\chi_{A_2}(x) \cdots \chi_{A_s}(x)$. Now $(A - \lambda_i I)^{e_i} = \mathbb{O}$ on $\mathcal{N}ull\,[(A - \lambda_i I)^{e_i}]$ so $\mu_{A_i}(x)$ divides $(x - \lambda_i)^{e_i}$ and so the $\mu_{A_i}(x)'s$ are relatively prime. This means that $\mu_A(x) = \mu_{A_1}(x) \cdots \mu_{A_s}(x) = (x - \lambda_1)^{e_1}(x - \lambda_2)^{e_2} \cdots (x - \lambda_s)^{e_s}$. Thus, for each i, $\mu_{A_i}(x) = (x - \lambda_i)^{e_i}$. Now $\chi_{A_i}(x)$ must equal $(x - \lambda_i)^{r_i}$, where $r_i \geq e_i$. But $(x - \lambda_1)^{r_1}(x - \lambda_2)^{r_2} \cdots (x - \lambda_s)^{r_s} = \chi_A(x) = (x - \lambda_1)^{d_1}(x - \lambda_2)^{d_2} \cdots (x - \lambda_s)^{d_s}$. By the uniqueness of the prime decomposition of polynomials, $r_i = d_i$ for all i. That wraps up the theorem. □

This theorem has some rather nice consequences. With notation as above, let $D = \lambda_1 E_1 + \lambda_2 E_2 + \ldots + \lambda_s E_s$. We claim D is diagonalizable with respect to similarity. Now, since $I = E_1 + E_2 + \ldots + E_s$, any $\mathbf{v} \in \mathbb{C}^n$ can be written as $\mathbf{v} = \mathbf{v}_1 + \mathbf{v}_2 + \ldots + \mathbf{v}_s$, $\mathbf{v}_i \in \mathcal{C}ol(E_i) = Fix(E_i)$. But $D\mathbf{v}_i = DE_i\mathbf{v}_i = (\lambda_1 E_1 + \cdots + \lambda_s E_s)E_i\mathbf{v}_i = \lambda_i E_i^2\mathbf{v}_i = \lambda_i\mathbf{v}_i$. Thus every nonzero vector in the column space of E_i is an eigenvector and every vector is a sum of these. So the eigenvectors span and hence a basis of eigenvectors of D can be selected. Moreover, $A = AE_1 + AE_2 + \cdots + AE_s$, so if we take $N = A - D = (A - \lambda_1 I)E_1 + \cdots + (A - \lambda_s I)E_s$, we see $N^2 = (A - \lambda_1 I)^2 E_1 + \cdots + (A - \lambda_s I)^2 E_s$, $N^3 = (A - \lambda_1 I)^3 E_1 + \cdots + (A - \lambda_s I)^3 E_s$, and so on. Eventually, since $(A - \lambda_1 I)^{e_i} E_i = \mathbb{O}$, we get $N^r = \mathbb{O}$, where $r \geq$ all e_is. That is, N is nilpotent. Note that both D and N are polynomials in A, so they commute (i.e., $DN = ND$). Thus we have the next theorem that holds for any matrix over $\mathbb{C}$.

COROLLARY 10.2 (Jordan decomposition)

Let $A \in \mathbb{C}^{n \times n}$. Then $A = D + N$, where D is diagonalizable and N is nilpotent. Then there exists S invertible, with $S^{-1}AS = \widetilde{D} + \widetilde{N}$, where $\widetilde{D}$ is a diagonal matrix and $\widetilde{N}$ is nilpotent.

PROOF The details are left to the reader. ▯

COROLLARY 10.3

Let $A \in \mathbb{C}^{n \times n}$. The D and the N of (10.2) commute and are unique. Indeed, each of them is a polynomial in A.

PROOF Now $D = \lambda_1 E_1 + \lambda_2 E_2 + \cdots + \lambda_s E_s$ and all the E_is are polynomials in A and so $N = A - D$ is a polynomial in A. Suppose $A = D_1 + N_1$, where D_1 is diagonalizable with respect to similarity, and N_1 is nilpotent and $D_1 N_1 = N_1 D_1$. We shall argue $D = D_1$, $N = N_1$. Now $D + N = A = D_1 + N_1$ so $D - D_1 = N_1 - N$. Now D and D_1 commute and so are simultaneously diagonalizable; hence $D - D_1$ is diagonalizable. Also N and N_1 commute and so $N_1 - N$ is nilpotent. To see this, look at $(N_1 - N)^s = \sum_{j=0}^{s} \binom{s}{j} N_1^{s-j} N^j$ by the binomial expansion, which you remember works for commuting matrices. By taking an s large enough — say, $s = 2n$ or more — we annihilate the right-hand side. Now this means $D - D_1$ is a diagonalizable nilpotent matrix. Then $S^{-1}(D - D_1)S$ is a diagonal nilpotent matrix. But this must be the zero matrix. ▯

We could diagonalize all matrices in $\mathbb{C}^{n \times n}$ if it were not for that nasty nilpotent matrix that gets in the way. We shall have more to say about this representation later when we discuss the Jordan canonical form. It is of interest to know if two matrices can be brought to diagonal form by the same invertible matrix. The next theorem gives necessary and sufficient conditions.

THEOREM 10.11

Let $A, B \in \mathbb{C}^{n \times n}$ be diagonalizable with respect to similarity. Then there exists S invertible that diagonalizes both A and B if and only if $AB = BA$.

PROOF Suppose S exists with $S^{-1}AS = D_1$ and $S^{-1}BS = D_2$. Then $A = SD_1S^{-1}$ and $B = SD_2S^{-1}$. Then $AB = SD_1S^{-1}SD_2S^{-1} = SD_1D_2S^{-1} = SD_2D_1S^{-1} = SD_2S^{-1}SD_1S^{-1} = BA$ using the fact that diagonal matrices commute. Conversely, suppose $AB = BA$. Since A is diagonalizable, its minimal polynomial is of the form $\mu_A(x) = (x - \lambda_1)(x - \lambda_2) \cdots (x - \lambda_s)$, where the

λ, s are the distinct eigenvalues of A. By the primary decomposition theorem, $\mathbb{C}^n = \mathcal{N}ull(A - \lambda_1 I) \oplus \cdots \oplus \mathcal{N}ull(A - \lambda_s I)$. Now for any $\mathbf{v}_i$ in $\mathcal{N}ull(A - \lambda_i I)$, $AB\mathbf{v}_i = BA\mathbf{v}_i = B(\lambda_i \mathbf{v}_i) = \lambda_i B\mathbf{v}_i$. This says each $\mathcal{N}ull(A - \lambda_i I)$ is B-invariant. We can therefore find a basis of $\mathcal{N}ull(A - \lambda_i I)$, say $\mathcal{B}_i$, consisting of eigenvectors of B. Then $\bigcup_{i=1}^{s} \mathcal{B}_i$ is a basis of $\mathbb{C}^n$ consisting of eigenvectors of both A and of B. Forming this basis into a matrix S creates an invertible matrix that diagonalizes both A and B. □

It follows from the primary decomposition theorem that any square complex matrix is similar to a block diagonal matrix. Actually, we can do even better. For this, we first refine our notion of a nilpotent matrix, relativising this idea to a subspace.

DEFINITION 10.3 *(Nilpotent on V) Let V be a subspace of $\mathbb{C}^n$ and let $A \in \mathbb{C}^{n \times n}$. We say that A is nilpotent on V iff there exists a positive integer k such that $A^k v = \vec{0}$ for all $\mathbf{v}$ in V. The least such k is called* ***the index of A on V***, *$Ind_V(A)$.*

Of course, $A^k v = \vec{0}$ for all $\mathbf{v}$ in V is equivalent to saying $V \subseteq \mathcal{N}ull(A^k)$. Thus, $Ind_V(A) = q$ iff $V \subseteq \mathcal{N}ull(A^q)$ but $V \nsubseteq \mathcal{N}ull(A^{q-1})$. For example, consider

$$\mathbb{C}^4 = V \oplus W = span\{\mathbf{e}_1, \mathbf{e}_2\} \oplus span\{\mathbf{e}_3, \mathbf{e}_4\}$$

and let $A = \begin{bmatrix} 0 & 1 & 0 & 0 \\ 0 & 0 & 0 & 0 \\ 0 & 0 & 1 & 0 \\ 0 & 0 & 0 & 1 \end{bmatrix}$. Note that

$$\begin{aligned} A\mathbf{e}_1 &= \vec{0} \\ A\mathbf{e}_2 &= \mathbf{e}_1 \\ A\mathbf{e}_3 &= \mathbf{e}_3 \\ A\mathbf{e}_4 &= \mathbf{e}_4 \end{aligned}$$

and $A^2 = \begin{bmatrix} 0 & 0 & 0 & 0 \\ 0 & 0 & 0 & 0 \\ 0 & 0 & 1 & 0 \\ 0 & 0 & 0 & 1 \end{bmatrix}$, so A is nilpotent of index 2 on $V = span\{\mathbf{e}_1, \mathbf{e}_2\}$.

Now we come to the crucial theorem to take the next step.

THEOREM 10.12
Suppose $\mathbb{C}^n = V \oplus W$, where V and W are A-invariant for $A \in \mathbb{C}^{n \times n}$ and suppose $dim(V) = r$. If A is nilpotent of index q on V, then there exists a basis

of V, $\{\mathbf{v}_1,\mathbf{v}_2,\ldots,\mathbf{v}_r\}$ *such that*

$$\begin{aligned} A\mathbf{v}_1 &= \overrightarrow{0} \\ A\mathbf{v}_2 &\in span\{\mathbf{v}_1\} \\ A\mathbf{v}_3 &\in span\{\mathbf{v}_1,\mathbf{v}_2\} \\ &\vdots \\ A\mathbf{v}_{r-1} &\in span\{\mathbf{v}_1,\mathbf{v}_2,\ldots,\mathbf{v}_{r-2}\} \\ A\mathbf{v}_r &\in span\{\mathbf{v}_1,\mathbf{v}_2,\ldots,\mathbf{v}_{r-1}\}. \end{aligned}$$

PROOF The case when A is the zero matrix is trivial and left to the reader. By definition of q, $V \subsetneqq \mathcal{N}ull(A^{q-1})$, so there exists $\mathbf{v} \neq \overrightarrow{0}$ such that $A^{q-1}\mathbf{v} \neq \overrightarrow{0}$. Let $\mathbf{v}_1 = A^{q-1}\mathbf{v} \neq \overrightarrow{0}$ but $A\mathbf{v}_1 = \overrightarrow{0}$. Now if $span\{\mathbf{v}_1\} = V$, we are done. If not, $span\{\mathbf{v}_1\} \subsetneqq V$. Look at $\mathcal{C}ol(A) \cap V$. If $\mathcal{C}ol(A) \cap V \subseteq span\{\mathbf{v}_1\}$, choose any $\mathbf{v}_2$ in $V \backslash span\{\mathbf{v}_1\}$. Then $\{\mathbf{v}_1,\mathbf{v}_2\}$ is independent and $A\mathbf{v}_1 = \overrightarrow{0}$, $A\mathbf{v}_2 \in \mathcal{C}ol(A) \cap V \subseteq span\{\mathbf{v}_1\}$. If $\mathcal{C}ol(A) \cap V \subsetneqq span\{\mathbf{v}_1\}$, then we first notice that $\mathcal{C}ol(A^q) \cap V = (\overrightarrow{0})$. The proof goes like this. Suppose $\mathbf{x} \in \mathcal{C}ol(A^q) \cap V$. Then $\mathbf{x} \in V$ and $\mathbf{x} = A^q\mathbf{y}$ for some $\mathbf{y}$. But $\mathbf{y} = \mathbf{v} + \mathbf{w}$ for some $\mathbf{v} \in V$, $\mathbf{w} \in W$. Thus $\mathbf{x} = A^q(\mathbf{v}) + A^q(\mathbf{w}) = A^q(\mathbf{w}) \in V \cap W = (\overrightarrow{0})$. Therefore $\mathbf{x} = \overrightarrow{0}$. Now we can consider a chain

$$\mathcal{C}ol(A) \cap V \supseteq \mathcal{C}ol(A^2) \cap V \supseteq \cdots \supseteq \mathcal{C}ol(A^q) \cap V = (\overrightarrow{0}).$$

Therefore, there must be a first positive integer j such that $\mathcal{C}ol(A^j) \cap V \subsetneqq span(\mathbf{v}_1)$ but $\mathcal{C}ol(A^{j+1}) \cap V \subseteq span(\mathbf{v}_1)$. Choose $\mathbf{v}_2 \in \left(\mathcal{C}ol(A^j) \cap V\right) \backslash span(\mathbf{v}_1)$. Then in this case, we have achieved $A\mathbf{v}_1 = \overrightarrow{0}$ and $A\mathbf{v}_2 \in span\{\mathbf{v}_1\}$. Continue in this manner until a basis of V of the desired type is obtained. □

This theorem has some rather nice consequences.

COROLLARY 10.4

Suppose N is an n-by-n nilpotent matrix. There is an invertible matrix S such that $S^{-1}NS$ is strictly upper triangular. That is,

$$S^{-1}NS = \begin{bmatrix} 0 & a_{12} & a_{13} & \cdots & a_{1n} \\ 0 & 0 & a_{23} & \cdots & a_{2n} \\ \vdots & \vdots & \vdots & & \vdots \\ 0 & 0 & 0 & \cdots & a_{n-1n} \\ 0 & 0 & 0 & \cdots & 0 \end{bmatrix}.$$

We can now improve on the primary decompostion theorem. Look at the matrix $A_i - \lambda_i I_{d_i}$. Clearly this matrix is nilpotent on $\mathcal{N}ull(A - \lambda_i I)^{e_i}$, so we can get a basis of $\mathcal{N}ull(A - \lambda_i I)^{e_i}$ and make an invertible matrix S_i with S_i^{-1}

$(A_i - \lambda_i I_{d_i})S_i = \begin{bmatrix} 0 & a_{12} & a_{13} & \cdots & a_{1d_i} \\ 0 & 0 & a_{23} & \cdots & a_{2d_i} \\ \vdots & \vdots & \vdots & & \vdots \\ 0 & 0 & 0 & \cdots & a_{d_i-1d_i} \\ 0 & 0 & 0 & \cdots & 0 \end{bmatrix}$. But then $S_i^{-1}(A_i - \lambda_i I_{d_i})S_i =$

$S_i^{-1}A_iS_i - \lambda_i I_{d_i} = \begin{bmatrix} 0 & a_{12} & a_{13} & \cdots & a_{1d_i} \\ 0 & 0 & a_{23} & \cdots & a_{2d_i} \\ \vdots & \vdots & \vdots & & \vdots \\ 0 & 0 & 0 & \cdots & a_{d_i-1d_i} \\ 0 & 0 & 0 & \cdots & 0 \end{bmatrix}$ so $S_i^{-1}A_iS_i =$

$\begin{bmatrix} \lambda_i & a_{12} & a_{13} & \cdots & a_{1d_i} \\ 0 & \lambda_i & a_{23} & \cdots & a_{2d_i} \\ \vdots & \vdots & \vdots & & \vdots \\ 0 & 0 & 0 & \cdots & a_{d_i-1d_i} \\ 0 & 0 & 0 & \cdots & \lambda_i \end{bmatrix}$. By piecing together a basis for each primary component in this way, we get the following theorem.

THEOREM 10.13 (triangularization theorem)
Let $A \in \mathbb{C}^{n\times n}$. Suppose $\lambda_1, \lambda_2, \ldots, \lambda_s$ are the distinct eigenvalues of A. Let $\mu_A(x) = (x-\lambda_1)^{e_1}(x-\lambda_2)^{e_2}\cdots(x-\lambda_s)^{e_s}$. Also let $\chi_A(x) = (x-\lambda_1)^{d_1}(x-\lambda_2)^{d_2}\cdots(x-\lambda_s)^{d_s}$. Then there exists an invertible matrix S such that $S^{-1}AS$ is block diagonal with upper triangular blocks. Moreover, the ith block has the eigenvalue λ_i on its diagonal. In particular, this says that A is similar to an upper triangular matrix whose diagonal elements consist of the eigenvalues of A repeated according to their algebraic multiplicity.

It turns out, we can make an even stronger conclusion, and that is the subject of the next section.

Exercise Set 43

1. Argue that similarity is an equivalence relation on n-by-n matrices.

2. Prove if A is similar to $\mathbb{O}$, then $A = \mathbb{O}$.

3. Prove that if A is similar to P and P is idempotent, then A is idempotent. Does the same result hold if "idempotent" is replaced by "nilpotent"?

4. Suppose A has principal idempotents E_i. Argue that a matrix B commutes with A iff it commutes with every one of the E_i.

5. Suppose A has principal idempotents E_i. Argue that $\mathcal{C}ol(E_i) = Eig(A, \lambda_i)$ and $\mathcal{N}ull(E_i)$ is the direct sum of the eigenspaces of A not associated with λ_i.

6. Suppose $A \sim B$ and A is invertible. Argue that B must also be invertible and $A^{-1} \sim B^{-1}$.

7. Suppose $A \sim B$. Argue that $A^T \sim B^T$.

8. Suppose $A \sim B$. Argue that $A^k \sim B^k$ for all positive integers k. Also argue then that $p(A) \sim p(B)$ for any polynomial $p(x)$.

9. Suppose $A \sim B$. Argue that $det(A) = det(B)$.

10. Suppose $A \sim B$ and $\lambda \in \mathbb{C}$. Argue that $(A - \lambda I) \sim (B - \lambda I)$ and so $det(A - \lambda I) = det(B - \lambda I)$.

11. Find matrices A and B with $tr(A) = tr(B)$, $det(A) = det(B)$, and $rank(A) = rank(B)$, but A is not similar to B.

12. Suppose $A \sim B$. Suppose A is nilpotent of index q. Argue that B is nilpotent with index q.

13. Suppose $A \sim B$ and A is idempotent. Argue that B is idempotent.

14. Suppose A is diagonalizable with respect to similarity and $B \sim A$. Argue that B is also diagonalizable with respect to a similarity.

15. Give an example of two 2-by-2 matrices that have identical eigenvalues but are not similar.

16. Argue that the intersection of any family of A-invariant subspaces of a matrix A is again A-invariant. Why is there a smallest A-invariant subspace containing any given set of vectors?

17. Let $A \in \mathbb{C}^{n \times n}$ and $\mathbf{v} \neq \overrightarrow{0}$ in $\mathbb{C}^n$. Let $\mu_{A,\mathbf{v}}(x) = x^d - \beta_{d-1}x^{d-1} - \cdots - \beta_1 x - \beta_0$.

    (a) Prove that $\{\mathbf{v}, A\mathbf{v}, \ldots, A^{d-1}\mathbf{v}\}$ is a linearly independent set.

    (b) Let $\mathcal{K}_d(A, \mathbf{v}) = span\{\mathbf{v}, A\mathbf{v}, \ldots, A^{d-1}\mathbf{v}\}$. Argue that $\mathcal{K}_d(A, \mathbf{v})$ is the smallest A-invariant subspace of $\mathbb{C}^n$ that contains $\mathbf{v}$. Moreover, $dim(\mathcal{K}_d(A, \mathbf{v})) = \deg(\mu_{A,\mathbf{v}}(x))$.

    (c) Prove that $\mathcal{K}_d(A, \mathbf{v}) = \{p(A)\mathbf{v} \mid p \in \mathbb{C}[x]\}$.

    (d) Let $A = \begin{bmatrix} 0 & -1 & 1 \\ 1 & 0 & 1 \\ 0 & 0 & 2 \end{bmatrix}$. Compute $\mathcal{K}_d(A, \mathbf{v})$, where $\mathbf{v} = \mathbf{e}_1$.

(e) Extend the independent set $\{\mathbf{v}, A\mathbf{v}, \ldots, A^{d-1}\mathbf{v}\}$ to a basis of $\mathbb{C}^n$, say $\{\mathbf{v}, A\mathbf{v}, \ldots, A^{d-1}\mathbf{v}, \mathbf{w}_1, \ldots, \mathbf{w}_{d-n}\}$. Form the invertible matrix $S = [\mathbf{v} \mid A\mathbf{v} \mid \cdots \mid A^{d-1}\mathbf{v} \mid \mathbf{w}_1 \mid \cdots \mid \mathbf{w}_{d-n}]$. Argue that $S^{-1}AS =$

$$\begin{bmatrix} C & ? \\ \mathbb{O} & ? \end{bmatrix}, \text{ where } C = \begin{bmatrix} 0 & 0 & 0 & \cdots & \beta_0 \\ 1 & 0 & 0 & \cdots & \beta_1 \\ 0 & 1 & 0 & \cdots & \beta_2 \\ \vdots & \vdots & \vdots & & \vdots \\ 0 & 0 & 0 & \cdots & \beta_{d-1} \end{bmatrix}.$$

(f) Suppose $\mu_{A,\mathbf{v}}(x) = (x - \lambda)^d$. Argue that $\mathbf{v}, (A - \lambda I)\mathbf{v}, \ldots, (A - \lambda I)^{d-1}\mathbf{v}$ is a basis for $\mathcal{K}_d(A, \mathbf{v})$.

18. Suppose $A \in \mathbb{C}^{n\times n}$. Prove that the following statements are all equivalent:
    (a) There is an invertible matrix S such that $S^{-1}AS$ is upper triangular.
    (b) There is a basis $\{\mathbf{v}_1, \ldots, \mathbf{v}_n\}$ of $\mathbb{C}^n$ such that $A\mathbf{v}_k \in span\{\mathbf{v}_1, \ldots, \mathbf{v}_k\}$ for all $k = 1, 2, \ldots, n$.
    (c) There is a basis $\{\mathbf{v}_1, \ldots, \mathbf{v}_n\}$ of $\mathbb{C}^n$ such that $span\{\mathbf{v}_1, \ldots, \mathbf{v}_k\}$ is A-invariant for all $k = 1, 2, \ldots, n$.

Further Reading

[Abate, 1997] Marco Abate, When is a Linear Operator Diagonalizable?, The American Mathematical Monthly, Vol. 104, No. 9, November, (1997), 824–830.

[B&R, 2002] T. S. Blyth and E. F. Robertson, *Further Linear Algebra*, Springer, New York, (2002).

unitarily equivalent, Schur's lemma, Schur decomposition

10.3 Diagonalization with Respect to a Unitary

In this section, we demand even more. We want to diagonalize a matrix using a unitary matrix, not just an invertible one. For one thing, that relieves us of having to invert a matrix.

DEFINITION 10.4 *Two matrices $A, B \in \mathbb{C}^{n\times n}$ are called* ***unitarily similar*** *(or* ***unitarily equivalent****) iff there exists a unitary matrix U, ($U^* = U^{-1}$) such that $U^{-1}AU = B$. A matrix A is* ***diagonalizable with respect to a unitary*** *iff A is unitarily similar to a diagonal matrix.*

We begin with a rather fundamental result about complex matrices that goes back to the Russian mathematician **Issai Schur** (10 January 1875–10 January 1941).

THEOREM 10.14 (Schur's lemma, Math. Annalen, 66, (1909), 408–510)
Let $A \in \mathbb{C}^{n\times n}$. Then there is a unitary matrix U such that U^*AU is upper triangular. That is, A is unitarily equivalent to an upper triangular matrix T. Moreover, the eigenvalues of A (possibly repeated) comprise the diagonal elements of T.

PROOF The proof is by induction on the size of the matrix n. The result is clearly true for $n = 1$. (Do you notice that all books seem to say that when doing induction?) Assume the theorem is true for matrices of size k-by-k. Our job is to show the theorem is true for matrices of size ($k+1$)-by-($k+1$). So suppose A is a matrix of this size. Since we are working over the complex numbers, we know A has an eigenvalue, say λ_1 and an eigenvector of unit length $\mathbf{w}_1$ belonging to it. Extend $\mathbf{w}_1$ to an orthonormal basis of $\mathbb{C}^{k+1}$ (using Gram-Schmidt, for example). Say $\{\mathbf{w}_1, \mathbf{w}_2, \ldots, \mathbf{w}_{k+1}\}$ is an orthonormal basis of $\mathbb{C}^{k+1}$. Form the matrix W with columns $\mathbf{w}_1, \mathbf{w}_2, \ldots, \mathbf{w}_{k+1}$. Then W is unitary and $W^*AW =$

$$W^*[A\mathbf{w}_1, \mid A\mathbf{w}_2, \mid \ldots \mid A\mathbf{w}_{k+1}] = \begin{bmatrix} - & \mathbf{w}_1^* & - \\ - & \mathbf{w}_2^* & - \\ - & \mathbf{w}_{k+1}^* & \end{bmatrix} [\lambda_1\mathbf{w}_1 \mid A\mathbf{w}_2 \mid \ldots \mid$$

$$A\mathbf{w}_{k+1}] = \begin{bmatrix} \lambda_1 & * & * & \ldots & * \\ 0 & & & & \\ 0 & & C & & \\ \vdots & & & & \\ 0 & & & & \end{bmatrix}$$

, where C is k-by-k. Now the induction hypothesis provides a k-by-k unitary matrix V_1 such that $V_1^*CV_1 = T_1$ is upper triangular. Let $V = \begin{bmatrix} 1 & 0 & \ldots & 0 \\ 0 & & & \\ \vdots & & V_1 & \\ 0 & & & \end{bmatrix}$ and note that V is unitary. Let $U = WV$. Then $V^*W^*AWV = \begin{bmatrix} \lambda_1 & * & * & \dots & * \\ 0 & & & & \\ 0 & & V_1^*CV_1 & & \\ 0 & & & & \end{bmatrix} =$

$\begin{bmatrix} \lambda_1 & * & * & \dots & * \\ 0 & & & & \\ 0 & & T_1 & & \\ \vdots & & & & \\ 0 & & & & \end{bmatrix}$, which is an upper triangular matrix as we hoped. $\square$

DEFINITION 10.5 *(Schur decomposition)*
The theorem above shows that every $A \in \mathbb{C}^{n\times n}$ can be written as $A = UTU^*$ where T is upper triangular. This is called a ***Schur decomposition*** *of A.*

THEOREM 10.15
Let $A \in \mathbb{C}^{n\times n}$. Then the following are equivalent:

1. *A is diagonalizable with respect to a unitary.*
2. *There is an orthonormal basis of $\mathbb{C}^n$ consisting of eigenvectors of A.*
3. *A is normal.*

PROOF Suppose (1). Then there exists a unitary matrix U with $U^{-1}AU = U^*AU = D$, a diagonal matrix. But then $AA^* = UDU^*UD^*U^* = UDD^*U^* = UD^*DU^* = (UD^*U^*)(UDU^*) = A^*A$ so A has to be normal. Thus (1) implies (3).

Next suppose A is normal. Write A in a Schur decomposition $A = UTU^*$, where T is upper triangular. Then $A^* = UT^*U^*$ and so $TT^* = U^*AUU^*A^*U = U^*AA^*U = U^*A^*AU = U^*A^*UU^*AU = T^*T$. This says T is normal. But if T is normal and upper triangular it must be diagonal. To see this compute

$$\begin{bmatrix} t_{11} & & \dots & t_{1n} \\ 0 & t_{22} & \dots & t_{2n} \\ \vdots & \vdots & & \vdots \\ 0 & 0 & & \dots t_{nn} \end{bmatrix} \begin{bmatrix} \bar{t}_{11} & 0 & \dots & 0 \\ \bar{t}_{12} & & & 0 \\ \vdots & & & 0 \\ \bar{t}_{1n} & & & \bar{t}_{nn} \end{bmatrix} = \begin{bmatrix} \bar{t}_{11} & 0 & \dots & 0 \\ \bar{t}_{12} & \bar{t}_{22} & \dots & \vdots \\ \vdots & & & \vdots \\ \bar{t}_{1n} & & & \bar{t}_{nn} \end{bmatrix}$$

$$\begin{bmatrix} t_{11} & & t_{1n} \\ 0 & & \vdots \\ \vdots & & \vdots \\ 0 & 0 & t_{nn} \end{bmatrix}$$. Now compare the diagonal entries:

$$\begin{aligned} |t_{11}|^2 + |t_{12}|^2 + |t_{13}|^2 + \ldots + |t_{1n}|^2 &= (1,1)\ entry = |t_{11}|^2 \\ |t_{22}|^2 + |t_{23}|^2 + \ldots |t_{2n}|^2 &= (2,2)\ entry = |t_{12}|^2 + |_{22}|^2 + |t_{22}|^2 \\ &\vdots \\ |t_{nn}|^2 &= (n,n)\ entry = |t_{1n}|^2 + t_{2n}^2 + \ldots + |t_{nn}|^2 . \end{aligned}$$

We see that $\left|t_{ij}\right|^2 = 0$ whenever $i \neq j$. Thus (3) implies (1). If (1) holds there exists a unitary U such that $U^{-1}AU = D$, a diagonal matrix. Then $AU = UD$ and the usual argument gives the columns of U as eigenvectors of A. Thus the columns of U yield an orthonormal basis of $\mathbb{C}^n$ consisting entirely of eigenvectors of A.

Conversely, if such a basis exists, form a matrix U with these vectors as columns. Then U is unitary and diagonalizes A. ∎

COROLLARY 10.5
Let $A \in \mathbb{C}^{n\times n}$.

1. *If A is Hermitian, A is diagonalizable with respect to a unitary.*
2. *If A is skew Hermitian ($A^* = -A$), then A is diagonalizable with respect to a unitary.*
3. *If A is unitary, then A is diagonalizable with respect to a unitary.*

It is sometimes of interest to know when a family of matrices over $\mathbb{C}$ can be simultaneously triangularized by a single invertible matrix. We state the following theorem without proof.

THEOREM 10.16
Suppose $\mathcal{F}$ is a family of n-by-n matrices over $\mathbb{C}$ that commute pairwise. Then there exists an invertible matrix S in $\mathbb{C}^{n\times n}$ such that $S^{-1}AS$ is upper triangular for every A in $\mathcal{F}$.

Exercise Set 44

1. Is $U = \begin{bmatrix} \frac{1+i}{2} & \frac{\sqrt{2}}{2} \\ \frac{1-i}{2} & \frac{\sqrt{2}i}{2} \end{bmatrix}$ a unitary matrix?

2. Argue that U is a unitary matrix iff it transforms an orthonormal basis to an orthonormal basis.

3. Find a unitary matrix U that diagonalizes $A = \begin{bmatrix} 1 & 1 & 1 & 1 \\ 1 & 1 & 1 & 1 \\ 1 & 1 & 1 & 1 \\ 1 & 1 & 1 & 1 \end{bmatrix}$.

4. Argue that unitary similarity is an equivalence relation.

5. Prove that the unitary matrices form a subgroup of $GL(n,\mathbb{C})$, the group of invertible n-by-n matrices.

6. Argue that the eigenvalues of a unitary matrix must be complex numbers of magnitude 1 — that is, numbers of the form $e^{i\theta}$.

7. Prove that a normal matrix is unitary iff all its eigenvalues are of magnitude 1.

8. Argue that the diagonal entries of an Hermitian matrix must be real.

9. Let $A = A^*$ and suppose the eigenvalues of A are lined up as $\lambda_1 \geq \lambda_2 \geq \cdots \geq \lambda_n$, with corresponding orthonormal eigenvectors $\mathbf{u}_1, \mathbf{u}_2, \ldots, \mathbf{u}_n$. For $\mathbf{v} \neq 0$. Define $\rho(\mathbf{v}) = \dfrac{\langle A\mathbf{v} \mid \mathbf{v}\rangle}{\langle \mathbf{v} \mid \mathbf{v}\rangle}$. Argue that $\lambda_n \leq \rho(\mathbf{v}) \leq \lambda_1$. Even more, argue that $\underset{\mathbf{v}\neq 0}{max}\,\rho(\mathbf{v}) = \lambda_1$ and $\underset{\mathbf{v}\neq 0}{min}\,\rho(\mathbf{v}) = \lambda_n$.

10. Prove the claims of Corollary 10.5.

11. Argue that the eigenvalues of a skew-Hermitian matrix must be pure imaginary.

12. Suppose A is 2-by-2 and $U^*AU = T = \begin{bmatrix} t_1 & t_3 \\ 0 & t_2 \end{bmatrix}$ be a Schur decomposition. Must t_1 and t_2 be eigenvalues of A?

13. Use Schur triangularization to prove that the determinant of a matrix is the product of its eigenvalues and the trace is the sum.

14. Use Schur triangularization to prove the Cayley-Hamilton theorem for complex matrices.

Further Reading

[B&H 1983] Ā. Björck and S. Hammarling, A Schur Method for the Square Root of a Matrix, Linear Algebra and Applications, 52/53, (1983), 127–140.

[Schur, 1909] I. Schur, Über die characteristischen Wurzeln einer linearen Substitution mit einer Anwendung auf die Theorie der Integralgleichungen, Math. Ann., 66, (1909), 488–510.

10.3.1 MATLAB Moment

10.3.1.1 Schur Triangularization

The Schur decomposition of a matrix A is produced in MATLAB as follows:

$$[Q, T] = schur(A)$$

Here Q is unitary and T is upper triangular with $A = QTQ^*$. For example,

```
>> B = [1 + i2 + 2i3 + i; 2 + 2i4 + 4i9i; 3 + 3i6 + 6i8i]
B =
      1.0000 + 1.0000i   2.0000 + 2.0000i   3.0000 + 1.0000i
      2.0000 + 2.0000i   4.0000 + 4.0000i        0 + 9.0000i
      3.0000 + 3.0000i   6.0000 + 6.0000i        0 + 8.0000i
>> [U,T] = schur(B)
U =
   0.2447 - .01273i    0.7874 + 0.4435i   -0.0787 + 0.3179i
   0.6197 + 0.1525i   -0.2712 - 0.3003i   -0.1573 + 0.6358i
   0.7125 + 0.0950i    0.1188 - 0.0740i    0.1723 - 0.6588i
T =
   6.3210 + 14.1551i   2.7038 + 1.3437i    1.6875 + 5.4961i
           0           0.0000 - 0.0000i   -3.0701 - 0.0604i
           0                   0          -1.3210 - 1.1551i
>> format rat
>> B = [1 + i2 + 2i3 + i; 2 + 2i4 + 4i9i; 3 + 3i6 + 6i8i]
```

B =

$1+1i$	$2+2i$	$3+1i$
$2+2i$	$4+4i$	$0+9i$
$3+3i$	$6+6i$	$0+8i$

U =

$185/756 - 1147/9010i$	$663/842 + 149/336i$
$1328/2143 + 329/2157i$	$-739/2725 - 2071/6896i$
$280/393 + 130/1369i$	$113/951 - 677/9147i$

$-227/2886 + 659/2073i$
$-227/1443 + 1318/2073i$
$339/1967 - 421/639i$

T =

$6833/1081 + 4473/316i$	$2090/773 + 1079/803i$
0	$* - *$
0	0

$20866/12365 + 709/129i$
$-5477/1784 - 115/1903i$
$-1428/1081 - 1311/1135i$

10.4 The Singular Value Decomposition

In the previous sections, we increased our demands on diagonalizing a matrix. In this section we will relax our demands and in some sense get a better result. The theorem we are after applies to *all* matrices, square or not. So suppose $A \in \mathbb{C}_r^{m\times n}$. We have already seen how to use a full rank factorization of A to get $SAT = \begin{bmatrix} I_r & \mathbb{O} \\ \mathbb{O} & \mathbb{O} \end{bmatrix}$, where S is unitary and T is invertible. The natural question arises as to whether we can get T unitary as well.

So let's try! Let's begin with an orthogonal full rank factorization $A = FG$ with $F^+ = F^*$. We will also need the factorizations $I - FF^+ = F_1F_1^+ = F_1F_1^*$ and $I - G^+G = F_2F_2^+ = F_2F_2^*$. Then $\begin{bmatrix} F^+ \\ \cdots \\ F_1^+ \end{bmatrix} A \left[G^+D \,\vdots\, (I - G^+G)W_2\right] = \begin{bmatrix} D & \mathbb{O} \\ \mathbb{O} & \mathbb{O} \end{bmatrix}$. The matrix $U^* = \begin{bmatrix} F^+ \\ \cdots \\ F_1^+ \end{bmatrix}$ is unitary. At the moment, W_2 is arbitrary. Next, consider $V = \left[G^+D \,\vdots\, (I - G^+G)W_2\right]$. We would like to make V unitary and we can fiddle with D and W_2 to make it happen.

But $V^*V = \begin{bmatrix} (G^+D)^* \\ [(I - G^+G)W_2]^* \end{bmatrix} \begin{bmatrix} G^+D \vdots (I - G^+G)W_2 \end{bmatrix} =$

$$\begin{bmatrix} D^*G^{+*}G^+D & D^*G^{+*}(I - G^+G)W_2 \\ W_2^*(I - G^+G)G^+D & W_2^*(I - G^+G)W_2 \end{bmatrix} =$$

$$\begin{bmatrix} D^*G^{+*}G^+D & \mathbb{O} \\ & \\ \mathbb{O} & W_2^*(I - G^+G)W_2 \end{bmatrix}.$$

We have already seen that by choosing $W_2 = F_2$ we can achieve I_{n-r} in the lower right. So the problem is, how can you get I_r in the upper left by choice of D? By the way, in case you missed it, $G^{+*} = (G^+GG^+)^* = G^{+*}(G^+G)^* = G^{+*}G^+G$ so that $D^*G^{+*}(I - G^+G)W_2 = \mathbb{O}$. Anyway our problem is to get $D^*G^{+*}G^+D = I_r$. But $G = F^+A = F^*A$ so $D^*G^{+*}G^+D = D^*(GG^*)^{-1}D = D^*(F^*AA^*F)^{-1}D$ since $G^+ = G^*(GG^*)^{-1}$ and so $G^{+*} = (GG^*)^{-1}G$. So to achieve the identity, we need $F^*AA^*F = DD^*$, where D is invertible. Equivalently, we need $AA^*F = FDD^*$. Let's take stock so far.

THEOREM 10.17

Let $A \in \mathbb{C}_r^{m\times n}$, $A = FG$ be an orthogonal full rank factorization of A. If there exists $D \in \mathbb{C}_r^{r\times r}$ with $GG^* = DD^*$, then there exist unitary matrices S and T with $SAT = \begin{bmatrix} D & \mathbb{O} \\ \mathbb{O} & \mathbb{O} \end{bmatrix}$.

One way to exhibit such a matrix D for a given A is to choose for the columns of F an orthonormal basis consisting of eigenvectors of AA^* corresponding to nonzero eigenvalues. We know the eigenvalues of AA^* are nonnegative. Then $AA^*F = FE$ where E is the r-by-r diagonal matrix of real positive eigenvalues of AA^*. Let D be the diagonal matrix of the positive square roots of these eigenvalues Then $D^*[F^*AA^*F]D = I$.

Thus we have the following theorem.

THEOREM 10.18 (singular value decomposition)

Let $A \in \mathbb{C}_r^{m\times n}$, $A = FG$ where the columns of F are an orthonormal basis of $Col(A) = Col(AA^*)$ consisting of eigenvectors of AA^* corresponding to nonzero eigenvalues. Suppose $I - FF^+ = F_1F_1^* = F_1F_1^+$ and $I - G^+G = F_2F_2^* = F_2F_2^+$. Then there exist unitary matrices U and V with $U^*AV = \begin{bmatrix} D_r & \mathbb{O} \\ \mathbb{O} & \mathbb{O} \end{bmatrix}$, where D_r is an r-by-r diagonal matrix whose diagonal entries are the positive square roots of the nonzero eigenvalues of AA^*. The matrices U and V can be constructed explicitly from $U^* = \begin{bmatrix} F^* \\ \cdots \\ F_1^+ \end{bmatrix}$ and $V = \begin{bmatrix} G^+D \vdots F_2 \end{bmatrix}$.

It does not really matter if we use the eigenvalues of AA^* or A^*A, as the next theorem shows.

THEOREM 10.19

Let $A \in \mathbb{C}_r^{m\times n}$. The eigenvalues of A^*A and AA^* differ only by the geometric multiplicity of the zero eigenvalue, which is $n-r$ for A^*A and $m-r$ for AA^*, where $r = r(A^*A) = r(AA^*)$.

PROOF Since A^*A is self-adjoint, there is an orthonormal basis of eigenvectors $\mathbf{v}_1, \mathbf{v}_2, \ldots, \mathbf{v}_n$ corresponding to eigenvalues $\lambda_1, \lambda_2, \ldots, \lambda_n$ (not necessarily distinct). Then $\langle A^*A\mathbf{v}_i \mid \mathbf{v}_j\rangle = \lambda_i \langle \mathbf{v}_i \mid \mathbf{v}_j\rangle = \lambda_i\delta_{ij}$ for $1 \leq i, j \leq n$. So $\langle A\mathbf{v}_i \mid A\mathbf{v}_j\rangle = \langle A^*A\mathbf{v}_i \mid \mathbf{v}_j\rangle = \lambda_i\delta_{ij}$. Thus $\langle A\mathbf{v}_i \mid A\mathbf{v}_i\rangle = \lambda_i$, *for* $i = 1, 2, \ldots, n$. Thus $A\mathbf{v}_i = \overrightarrow{0}$ iff $\lambda_i = 0$. Also $AA^*(A\mathbf{v}_i) = A(A^*A\mathbf{v}_i) = \lambda_i(A\mathbf{v}_i)$ so for $\lambda_i \neq 0$, $A\mathbf{v}_i$ is an eigenvector of AA^*. Thus, if λ is an eigenvalue of A^*A, then λ is an eigenvalue of AA^*. A symmetric argument gives the other implication and we are done. $\square$

Let's look at an example next.

Example 10.3

Consider $A = \begin{bmatrix} 1 & 0 & 1 & 1 \\ 0 & 1 & -1 & 0 \\ 1 & 1 & 0 & 1 \end{bmatrix}$. Then $AA^* = \begin{bmatrix} 3 & -1 & 2 \\ -1 & 2 & 1 \\ 2 & 1 & 3 \end{bmatrix}$ with eigenvalues 5, 3, and 0 and corresponding eigenvectors $\begin{bmatrix} 1 \\ 0 \\ 1 \end{bmatrix}$, $\begin{bmatrix} -1 \\ 2 \\ 1 \end{bmatrix}$ and $\begin{bmatrix} -1 \\ -1 \\ 1 \end{bmatrix}$. With the help of Gram-Schmidt, we get $F = \begin{bmatrix} \frac{1}{\sqrt{2}} & \frac{-1}{\sqrt{6}} \\ 0 & \frac{2}{\sqrt{6}} \\ \frac{1}{\sqrt{2}} & \frac{1}{\sqrt{6}} \end{bmatrix}$. Then $F^*AA^*F = \begin{bmatrix} 5 & 0 \\ 0 & 3 \end{bmatrix}$ so $D = \begin{bmatrix} \sqrt{5} & 0 \\ 0 & \sqrt{3} \end{bmatrix}$ and $D^*\left[F^*AA^*F\right]^{-1} D =$
$\begin{bmatrix} \sqrt{5} & 0 \\ 0 & \sqrt{3} \end{bmatrix}\begin{bmatrix} 1/5 & 0 \\ 0 & 1/5 \end{bmatrix}\begin{bmatrix} \sqrt{5} & 0 \\ 0 & \sqrt{3} \end{bmatrix} = \begin{bmatrix} 1 & 0 \\ 0 & 1 \end{bmatrix}$.

The singular value decomposition really is quite remarkable. Not only is it interesting mathematically, but it has many important applications as well. It deserves to be more widely known. So let's take a closer look at it. Our theorem says that any matrix A can be written as $A = U\Sigma V^*$, where U and V are unitary and Σ is a diagonal matrix with nonnegative diagonal entries. Such a decomposition of A is called a *singular value decomposition* (SVD) of A. Our proof of the existence rested heavily on the diagonalizability of

AA^* and the fact that the eigenvalues of AA^* are nonnegative. In a sense, we did not have much choice in the matter since, if $A = U\Sigma V^*$, then $AA^* = U\Sigma V^* V\Sigma^* U^* = U\Sigma^2 U^*$ so $AA^*U = U\Sigma^2$, which implies that the columns of U are eigenvectors of AA^* and the eigenvalues of AA^* are the squares of the diagonal entries of Σ. Note here U is m-by-m while $A^*A = V\Sigma^*U^*U\Sigma V^* = V\Sigma^2 V^*$ says the columns of V, which is n-by-n, form an orthonormal basis for $\mathbb{C}^n$. Now since permutation matrices are unitary, there is no loss of generality is assuming the diagonal entries of Σ can be arranged to be in decreasing order and since A and Σ have the same rank we have $\sigma_1 \geq \sigma_2 \ldots \geq \sigma_r > 0 = \sigma_{r+1} = \sigma_{r+2} = \sigma_{r+3} = \cdots = \sigma_n$. Some more language, the columns of U are called *left singular vectors* and the columns of V are called *right singular vectors* of A.

So what is the geometry behind the SVD? We look at $A \in \mathbb{C}^{m\times n}$ as taking vectors from $\mathbb{C}^n$ and changing them into vectors in $\mathbb{C}^m$. We have $A =$

$$[\mathbf{u}_1 \mid \mathbf{u}_2 \mid \ldots] \begin{bmatrix} \sigma_1 & & & & & & \\ & \sigma_2 & & & & & \\ & & \sigma_r & & & & \\ & & & 0 & & & \\ & & & & 0 & & \\ & & & & & \vdots & \\ & & & & & & 0 \end{bmatrix} \begin{bmatrix} \mathbf{v}_1^* \\ \mathbf{v}_2^* \\ \vdots \end{bmatrix} \text{ or } A[\mathbf{v}_1 \mid \mathbf{v}_2 \mid \ldots$$

$\mid \mathbf{v}_n] = [\sigma_1\mathbf{u}_1 \mid \sigma_2\mathbf{u}_2 \mid \ldots \mid \sigma_r\mathbf{u}_r \ldots]$ so $A\mathbf{v}_1 = \sigma_1\mathbf{u}_1, \ldots, A\mathbf{v}_r = \sigma_r\mathbf{u}_r$, $A\mathbf{v}_{r+1} = \vec{0}, \ldots, A\mathbf{v}_m = \vec{0}$. So, when we express a vector $\mathbf{v} \in \mathbb{C}^n$ in terms of the orthonormal basis $\{\mathbf{v}_1, \mathbf{v}_2, \ldots, \mathbf{v}_n\}$, we see A contracts some components and dilates others depending on the magnitudes of the singular values. Then the change in dimension causes A to discard components or append zeros. Note the vectors $\{\mathbf{v}_{r+1}, \ldots, \mathbf{v}_n\}$ provide an orthonormal bases for the null space of A. So what a great pair of bases the SVD provides!

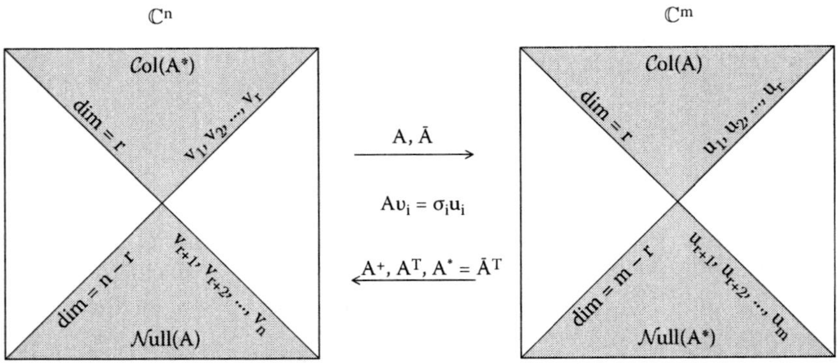

Figure 10.1: SVD and the fundamental subspaces.

Let's try to gain some insight as to the choice of these bases. If all you want is a diagonal representation of A, select an orthonormal basis of $\mathcal{N}ull(A)$, $\{\mathbf{v}_{r+1}, \dots, \mathbf{v}_n\}$ and extend it to an orthonormal basis of $\mathcal{N}ull(A)^{\perp} = \mathcal{C}ol(A^*)$ so that $\{\mathbf{v}_1, \mathbf{v}_2, \dots, \mathbf{v}_r, \mathbf{v}_{r+1}, \dots, \mathbf{v}_n\}$ becomes an orthonormal basis of $\mathbb{C}^n$. Define $\{\mathbf{u}_1, \dots, \mathbf{u}_r\} = \left\{\frac{A\mathbf{v}_1}{\|A\mathbf{v}_1\|}, \dots, \frac{A\mathbf{v}_r}{\|A\mathbf{v}_r\|}\right\}$. Then extend with a basis of $\mathcal{N}ull(A^*) = \mathcal{C}ol(A)^{\perp}$. Then $AV = A[\mathbf{v}_1 \mid \dots \mid \mathbf{v}_r \mid \dots] = [A\mathbf{v}_1 \mid \dots \mid A\mathbf{v}_r \mid \dots] = [\mathbf{u}_2 \mid \dots \mid \mathbf{u}_r \mid \dots] \begin{bmatrix} \|A\mathbf{v}_1\| & & \\ & \|A\mathbf{v}_2\| & \\ & & \ddots \end{bmatrix}$. But the problem is the **u**s do not have to be orthonormal! However, this is where the eigenvectors of A^*A come in to save the day. Take the eigenvalues of A^*A written $\lambda_1 \geq \lambda_2 \geq \dots \lambda_r > \lambda_{r+1} = \dots = \lambda_n = 0$. Take a corresponding orthonormal basis of eigenvectors $\mathbf{v}_1, \mathbf{v}_2, \dots, \mathbf{v}_n$ of A^*A. Define $\mathbf{u}_1 = \frac{A\mathbf{v}_1}{\sqrt{\lambda_1}}, \dots, \mathbf{u}_r = \frac{A\mathbf{v}_r}{\sqrt{\lambda_r}}$. The key point is that the **u**s are orthonormal. To see this, compute $\langle \mathbf{u}_i \mid \mathbf{u}_j \rangle = \frac{\lambda_i}{\sqrt{\lambda_i\lambda_j}} < \mathbf{v}_i \mid \mathbf{v}_j >= \delta_{ij}$. Now just extend the **u**s to an orthonormal basis of $\mathbb{C}^m$. Then $AV = U\Sigma$ where the $\sqrt{\lambda_i}$ and zeros appear on the diagonal of Σ zeros appear everywhere else.

We note in passing that if $A = A^*$, then $A^*A = A^2 = AA^*$, so if λ is an eigenvalue of A, then λ^2 is an eigenvalue of $A^*A = AA^*$. Thus the left and right singular vectors for A are just the eigenvectors of A and the singular values of A are the magnitudes of its eigenvalues. Thus, for positive semidefinite matrices A, the SVD coincides with the spectral diagonalization. Thus for any matrix A, AA^* has SVD the same as the spectral (unitary) diagonalization. There is a connection between the SVD and various norms of a matrix.

THEOREM 10.20

Let $A \in \mathbb{C}^{m\times n}$ with singular values $\sigma \geq \sigma_2 \geq \dots \geq \sigma_n > 0$. Then $\|A\|_2 = \sigma_1$ where $\|A\|_{op} = \max_{\|\mathbf{v}\|=1} \|A\mathbf{v}\|_2$ and $\|A\|_2 = (\sigma_1^2 + \sigma_2^2 + \cdots + \sigma_n^2)^{\frac{1}{2}}$. If A is invertible $\sigma_n > 0$ and $\left\|A^{-1}\right\|_{op} = 1/\sigma_n$. Moreover, $\sigma_k =$

$$\min_{dim(M)=n-k+1} \left(\max_{\mathbf{v}\in M\setminus\{\vec{0}\}} \frac{\|A\mathbf{v}\|_2}{\|\mathbf{v}\|_2} \right).$$

There is yet more information packed into the SVD. Write $A = U\Sigma V^* =$

$$[U_1 \mid U_2]\begin{bmatrix} \sigma_1 & & & 0 \\ & \ddots & & \\ & & \sigma_r & 0 \\ 0 & & & 0 \end{bmatrix}\left[\frac{V_1^*}{V_2^*}\right].$$ Then

(i) $U_1 U_1^* = AA^+ =$ the projection onto $\mathcal{C}ol(A)$.
(ii) $U_2 U_2^* = I - AA^+ =$ the projection onto $\mathcal{C}ol(A)^\perp = \mathcal{N}ull(A^*)$.
(iii) $V_1 V_1^* = A^+ A =$ the projection onto $\mathcal{N}ull(A)^\perp = \mathcal{C}ol(A^*)$.
(iv) $V_2 V_2^* = I - A^+ A =$ the projection onto $\mathcal{N}ull(A)$.

Indeed, we can write $A = U_1 \begin{bmatrix} \sigma_1 & & 0 \\ & \ddots & \\ & & \ddots & \\ 0 & & \sigma_r \end{bmatrix} V_1^*$. Setting $U_1 = F$, $G = \begin{bmatrix} \sigma_1 & & 0 \\ & \ddots & \\ & & \ddots & \\ 0 & & \sigma_r \end{bmatrix} V_1^*$, we get a full rank factorization of A. In fact, $A^+ = V_1 \begin{bmatrix} 1/\sigma_1 & & \\ & \ddots & \\ & & 1/\sigma_r \end{bmatrix} U_1^*$. Even more, write $A = [\mathbf{u}_1 \mid \ldots \mid \mathbf{u}_r] \begin{bmatrix} \sigma_1 & & \\ & \ddots & \\ & & \sigma_r \end{bmatrix} \begin{bmatrix} \mathbf{v}_{1^*} \\ \vdots \\ \mathbf{v}_{r^*} \end{bmatrix} = \sigma_1 \mathbf{u}_1 \mathbf{v}_1^* + \ldots + \sigma_r \mathbf{u}_r \mathbf{v}_r^*$ which is the outer product form of matrix multiplication, yielding A as a sum of rank 1 matrices.

The applications of the SVD seem to be endless, especially when the matrices involved are large. We certainly cannot mention them all here. Let's just mention a few.

One example is in computing the eigenvalues of A^*A. This is important to statisticians in analyzing *covariance matrices* and doing something called *principal component analysis*. Trying to compute those eigenvalues directly can be a numerical catastrophe. If A has singular values in the range .001 to 100 of magnitude, then A^*A has eigenvalues that range from .000001 to 10000. So computing directly with A can have a significant advantage. With large data matrices, measurement error can hide dependencies. An effective measure of rank can be made by counting the number of singular values larger than the size of the measurement error.

The SVD can help solve the approximation problem. Consider $A\mathbf{x} - \mathbf{b} = U\Sigma V^*\mathbf{x} - \mathbf{b} = U(\Sigma U^*\mathbf{x}) - U(U^*\mathbf{b}) = U(\Sigma\mathbf{y} - \mathbf{c})$, where $\mathbf{y} = V^*\mathbf{x}$ and $\mathbf{c} = U^*\mathbf{b}$.

Also $\|A\mathbf{x}-\mathbf{b}\|_2 = \|U(\Sigma\mathbf{y}-\mathbf{c})\| = \|\Sigma\mathbf{y}-\mathbf{c}\|$ since U is unitary. This second minimization is easy. Indeed $\mathbf{y} = \Sigma^+\mathbf{c} + \begin{bmatrix} c_1/\sigma_1 \\ \vdots \\ c_r/\sigma_r \\ 0 \\ 0 \\ 0 \end{bmatrix}$ so $\widehat{\mathbf{x}} = V\Sigma^+U^*\mathbf{b} = A^+\mathbf{b}$ gives the minimum length solution as we already knew. There is a much longer story to tell, but we have exhausted our allotted space for the SVD.

Exercise Set 45

1. Argue that the rank of a matrix is equal to the number of nonzero singular values.

2. Prove that matrices A and B are unitarily equivalent iff they have the same singular values.

3. Argue that U is unitary iff all its singular values are equal to one.

4. For matrices A and B, argue that A and B are unitarily equivalent iff A^*A and B^*B are similar.

5. Suppose $A = U\Sigma V^*$ is a singular value decomposition of A. Argue that $\|A\|_F = \|\Sigma V^*\|_F = \|\Sigma^*\|_F = (|\sigma_1|^2 + |\sigma_2|^2 + \cdots + |\sigma_n|^2)^{\frac{1}{2}}$.

6. Compare the SVDs of A with the SVDs of A^*.

7. Suppose $A = U\Sigma V^*$ is an SVD of A. Argue that $\|A\|_2 = |\sigma_1|$.

8. Suppose $A = QR$ is a QR factorization of A. Argue that A and R have the same singular values.

9. Argue that the magnitude of the determinant of a square matrix A is the product of its singular values.

10. Argue that if a matrix is normal, its singular values are just the magnitudes of its eigenvalues.

11. Argue that the singular values of a square matrix are invariant under unitary transformations. That is, the singular values of A, AU, and UA are the same if U is unitary.

12. Suppose $A = U\Sigma V^*$ is a singular value decomposition of A. Prove that $\mathbf{x} = V\Sigma^+ U^* \mathbf{b}$ minimizes $\|\mathbf{b} - A\mathbf{x}\|_2$.

13. Suppose $A = U\Sigma V^*$ is a singular value decomposition of A. Argue that $\sigma_n \|\mathbf{v}\|_2 \leq \|A\mathbf{v}\|_2 \leq \sigma_1 \|\mathbf{v}\|_2$.

14. Suppose A is an m-by-n matrix of rank r and $0 < k < r$. Find a matrix of rank k that is closest to A in the Frobenius norm. (*Hint*: Suppose $A = U\Sigma V^*$ is a singular value decomposition of A. Take $U\Sigma_k V^*$, where $\Sigma_k = \begin{bmatrix} \sigma_1 & & & 0 \\ & \ddots & & \\ & & \sigma_k & 0 \\ 0 & & & 0 \end{bmatrix}$.)

15. Suppose A is an m-by-n matrix. Argue that the 2-norm of A is its largest singular value.

16. Suppose A is a matrix. Argue that the condition number of A relative to the 2-norm is the ratio of its largest singular value to its smallest singular value.

17. Argue that $BB^* = CC^*$ iff there exists U with $UU^* = I$ such that $C = BU$.

18. Suppose $A = U \begin{bmatrix} D_r & \mathbb{O} \\ \mathbb{O} & \mathbb{O} \end{bmatrix} V^*$ is the singular value decomposition of A. Argue that $B = V[\begin{bmatrix} D_r^{-1} & E \\ F & G \end{bmatrix}]U^*$ is a 1-inverse of A and all 1-inverses look like this where E, F, and G are arbitrary.

Further Reading

[B&D, 1993] Z. Bai and J. Demmel, Computing the Generalized Singular Value Decomposition, SIAM J. Sce. Comput., Vol. 14, (1993), 1464–1486.

[B&K&S, 1989] S. J. Blank, Nishan Krikorian, and David Spring, A Geometrically Inspired Proof of the Singular Value Decomposition, The American Mathematical Monthly, Vol. 96, March, (1989), 238–239.

[C&J&L&R, 2005] John Clifford, David James, Michael Lachance, and Joan Remski, A Constructive Approach to Singular Value Decomposition and Symmetric Schur Factorization, The American Mathematical Monthly, Vol. 112, April, (2005), 358–363.

[M&R, 1998] Colm Mulcahy and John Rossi, A Fresh Approach to the Singular Value Decomposition, The College Mathematics Journal, Vol. 29, No. 3, (1998), 199–206.

[Good, 1969] I. J. Good, Some Applications of the Singular Decomposition of a Matrix, Technometrics, Vol. 11, No. 4, Nov., (1969), 823–831.

[Long, 1983] Cliff Long, Visualization of Matrix Singular Value Decomposition, Mathematics Magazine, Vol. 56, No. 3, May, (1983), 161–167.

[M&M, 1983] Cleve Moler and Donald Morrison, Singular Value Analysis of Cryptograms, The American Mathematical Monthly, February, (1983), 78–86.

[Strang, 1980] G. Strang, *Linear Algebra and Its Applications*, Academic Press, New York, (1980).

[T&B, 1997] L. N. Trefethen and D. Bau III, *Numerical Linear Algebra*, Society of Industrial and Applied Mathematicians, Philadelphia, (1997).

[G&L, 1983] G. H. Golub and C. Van Loan, *Matrix Computations*, Johns Hopkins University Press, Baltimore, (1983).

[H&O, 1996] Roger A. Horn and Ingram Olkin, When Does A*A=B*B and Why Does One Want to Know?, The American Mathematical Monthly, Vol. 103, No 6., June-July, (1996), 470–481.

[Kalman, 1996] Dan Kalman, A Singularly Valuable Decomposition: The SVD of a Matrix, The College Mathematics Journal, Vol. 27, No. 1, January, (1996), 2–23.

[Koranyi, 2001] S. Koranyi, Around the Finite-Dimensional Spectral Theorem, The American Mathematical Monthly, Vol. 108, (2001) 120–125.

10.4.1 MATLAB Moment

10.4.1.1 The Singular Value Decomposition

Of course, MATLAB has a built-in command to produce the singular value decomposition of a matrix. Indeed, many computations in MATLAB are based on the SVD of a matrix. The command is

$$[U, S, V] = svd(A).$$

This returns a unitary matrix U, a unitary matrix V, and a diagonal matrix S with the singular values of A down its diagonal such that

$$A = USV^*$$

up to roundoff. There is an economy size decomposition. The command is

$$[U, S, V] = svd(A, 0).$$

If all you want are the singular values of A, just type

$$svd(A).$$

Let's look at some examples.

```
>> A = vander(1:5)
A =
      1     1     1     1     1
     16     8     4     2     1
     81    27     9     3     1
    256    64    16     4     1
    625   125    25     5     1
>> [U,S,V] = svd(A)
U =
   -0.0018   -0.0739    0.6212   -0.7467   -0.2258
   -0.0251   -0.3193    0.6273    0.3705    0.6055
   -0.1224   -0.6258    0.1129    0.3594   -0.6720
   -0.3796   -0.6165   -0.4320   -0.4046    0.3542
   -0.9167    0.3477    0.1455    0.1109   -0.0732
S =
   695.8418         0         0         0         0
          0   18.2499         0         0         0
          0         0    2.0018         0         0
          0         0         0    0.4261         0
          0         0         0         0    0.0266


V =
   -0.9778    0.1996    0.0553    0.0294   -0.0091
   -0.2046   -0.8500   -0.3898   -0.2675    0.1101
   -0.0434   -0.4468    0.4348    0.6292   -0.4622
   -0.0094   -0.1818    0.6063    0.0197    0.7739
   -0.0021   -0.0706    0.5369   -0.7289   -0.4187
```

```
>> format rat
>> [U,S,V] = svd(A)
U =
      -18/10123     -476/6437     579/932     -457/612   -240/1063
     -298/11865     -692/2167    685/1092    379/1023    637/1052
      -535/4372    -2249/3594   297/2630   1044/2905    -465/692
      -976/2571    -3141/5095  -499/1155  -547/1352    559/1578
   -39678/43285      491/1412    657/4517    207/1867   -100/1367
S =
     13221/19          0            0           0          0
        0         33306/1825        0           0          0
        0             0         1097/548        0          0
        0             0            0        268/629        0
        0             0            0           0       23/865
V =
    -2029/2075     395/1979    157/2838   367/12477  -218/24085
     -593/2898   -2148/2527    -313/803   -811/3032      82/745
    -439/10117    -915/2048   1091/2509     246/391    -403/872
      -79/8430    -313/1722     596/983   644/32701   2091/2702
     -87/41879    -286/4053    792/1475    -277/380   -675/1612
>> svd(A)
ans =
    13221/19
    33306/1825
    1097/548
    268/629
    23/865
```

Now try the complex matrix B = [1+i 2+2i 3+i;2+2i 4+4i 9i;3+3i 6+6i 8i].

The SVD is very important for numerical linear algebra. Unitary matrices preserve lengths, so they tend not to magnify errors. The MATLAB functions of rank, null, and orth are based on the SVD.

Chapter 11

Jordan Canonical Form

11.1 Jordan Form and Generalized Eigenvectors

We know that there are matrices that cannot be diagonalized. However, there is a way to almost diagonalize all complex matrices. We now develop this somewhat long story.

11.1.1 Jordan Blocks

There are matrices that can be considered the basic building blocks of all square complex matrices. They are like the prime numbers in $\mathbb{N}$. Any $n \in \mathbb{N}$ is uniquely a product of primes except for the order of the prime factors that can be permuted. An analogous fact holds for complex matrices. The basic building blocks are the *Jordan blocks*.

DEFINITION 11.1 *(λ-Jordan block)*

For $\lambda \in \mathbb{C}$, $k \in \mathbb{N}$, define $J_k(\lambda) = \begin{bmatrix} \lambda & 1 & 0 & 0 & \cdots & 0 & 0 \\ 0 & \lambda & 1 & 0 & \cdots & 0 & 0 \\ 0 & 0 & \lambda & 1 & \cdots & 0 & 0 \\ \vdots & \vdots & \vdots & \vdots & \vdots & \vdots & 1 \\ 0 & 0 & 0 & 0 & \cdots & 0 & \lambda \end{bmatrix} \in \mathbb{C}^{k \times k}$.

This is a ***λ-Jordan block of order k****.*

For example, $J_1(\lambda) = [\lambda]$, $J_2(\lambda) = \begin{bmatrix} \lambda & 1 \\ 0 & \lambda \end{bmatrix}$, $J_3(\lambda) = \begin{bmatrix} \lambda & 1 & 0 \\ 0 & \lambda & 1 \\ 0 & 0 & \lambda \end{bmatrix}$, and so on.

More concretely, $J_2(5) = \begin{bmatrix} 5 & 1 \\ 0 & 5 \end{bmatrix}$, $J_3(4) = \begin{bmatrix} 4 & 1 & 0 \\ 0 & 4 & 1 \\ 0 & 0 & 4 \end{bmatrix}$.

Now we collect some basic facts about Jordan blocks. First we illustrate some important results.

Note that $\begin{bmatrix} \lambda & 1 & 0 \\ 0 & \lambda & 1 \\ 0 & 0 & \lambda \end{bmatrix}\begin{bmatrix} 1 \\ 0 \\ 0 \end{bmatrix} = \begin{bmatrix} \lambda \\ 0 \\ 0 \end{bmatrix} = \lambda\begin{bmatrix} 1 \\ 0 \\ 0 \end{bmatrix}$, so λ is an eigenvalue of $J_3(\lambda)$ with eigenvector the standard basis vector $\mathbf{e}_1$. Moreover, $\begin{bmatrix} x_1 \\ x_2 \\ x_3 \end{bmatrix} \in Eig(J_3(\lambda); \lambda)$ iff $\begin{bmatrix} x_1 \\ x_2 \\ x_3 \end{bmatrix} \in \mathcal{N}ull(J_3(\lambda) - \lambda I)$ iff $\begin{bmatrix} 0 & 1 & 0 \\ 0 & 0 & 1 \\ 0 & 0 & 0 \end{bmatrix}\begin{bmatrix} x_1 \\ x_2 \\ x_3 \end{bmatrix} = \begin{bmatrix} 0 \\ 0 \\ 0 \end{bmatrix}$ iff $x_2 = 0$ and $x_3 = 0$. Thus $Eig(J_3(\lambda); \lambda) = span(\mathbf{e}_1) = \mathbb{C}\mathbf{e}_1$, so the geometric multiplicity of λ is 1. Moreover, the characteristic polynomial of $J_3(\lambda)$, $\chi_{J_3(\lambda)}(x) = det(xI - J_3(\lambda)) = det\left(\begin{bmatrix} x & 0 & 0 \\ 0 & x & 0 \\ 0 & 0 & x \end{bmatrix} - \begin{bmatrix} \lambda & 1 & 0 \\ 0 & \lambda & 1 \\ 0 & 0 & \lambda \end{bmatrix}\right) = det\begin{pmatrix} x-\lambda & -1 & 0 \\ 0 & x-\lambda & -1 \\ 0 & 0 & x-\lambda \end{pmatrix} = (x-\lambda)^3$. Also, $J_3(\lambda) - \lambda I = \begin{bmatrix} 0 & 1 & 0 \\ 0 & 0 & 1 \\ 0 & 0 & 0 \end{bmatrix}$, which is nilpotent of index 3, so the minimal polynomial $\mu_{J_3(\lambda)}(x) = (x - \lambda)^3$ as well. In particular, the spectrum of $J_3(\lambda)$ is $\{\lambda\}$. We extend and summarize our findings with a general theorem.

THEOREM 11.1

Consider a general Jordan block $J_k(\lambda)$:

1. *$J_k(\lambda)$ is an upper triangular matrix.*
2. *$det(J_k(\lambda)) = \lambda^k$ and $Tr(J_k(\lambda)) = k\lambda$.*
3. *$J_k(\lambda)$ is invertible iff $\lambda \neq 0$.*
4. *The spectrum $\lambda(J_k(\lambda)) = \{\lambda\}$; that is, λ is the only eigenvalue of $J_k(\lambda)$ and the standard basis vector $\mathbf{e}_1$ spans the eigenspace $Eig(J_k(\lambda); \lambda)$.*
5. *$dim(Eig(J_k(\lambda); \lambda)) = 1$; that is, the geometric multiplicity of λ is 1.*
6. *The characteristic polynomial of $J_k(\lambda)$ is $\chi_{J_k(\lambda)}(x) = (x - \lambda)^k$.*
7. *The minimal polynomial of $J_k(\lambda)$ is $\mu_{J_k(\lambda)}(x) = (x - \lambda)^k$.*

8. $J_k(\lambda) = J_k(0) + \lambda I_k$ *and* $J_k(0)$ *is nilpotent of index k.*

9. $rank(\, J_k(0)) = k - 1.$

10. $J_k(\lambda)^n = \begin{bmatrix} \lambda^n & \binom{n}{1}\lambda^{n-1} & \binom{n}{2}\lambda^{n-2} & \cdots & \cdots \\ 0 & \lambda^n & \binom{n}{1}\lambda^{n-1} & \cdots & \cdots \\ 0 & 0 & \lambda^n & \cdots & \cdots \\ \vdots & \vdots & \vdots & \vdots & \vdots \\ 0 & 0 & 0 & \cdots & \lambda^n \end{bmatrix}$ *for* $n \in \mathbb{N}$.

11. *If* $\lambda \neq 0,\ J_k(\lambda)^{-1} =$
$$\begin{bmatrix} \frac{1}{\lambda} & -\frac{1}{\lambda^2} & \frac{1}{\lambda^3} & -\frac{1}{\lambda^4} & \cdots & (-1)^{i+j}\frac{1}{\lambda^j} & \cdots & (-1)^{1+k}\frac{1}{\lambda^k} \\ 0 & \frac{1}{\lambda} & -\frac{1}{\lambda^2} & \frac{1}{\lambda^3} & \cdots & \cdots & \cdots & \cdots \\ 0 & 0 & \frac{1}{\lambda} & -\frac{1}{\lambda^2} & \cdots & \cdots & \cdots & \cdots \\ \vdots & \vdots & \vdots & \vdots & \vdots & \vdots & \vdots & \vdots \\ 0 & 0 & 0 & 0 & \cdots & \cdots & \cdots & \frac{1}{\lambda} \end{bmatrix} =$$
$\frac{1}{\lambda}J_k(0) - \frac{1}{\lambda^2}J_k(0)^2 + \frac{1}{\lambda^3}J_k(0)^3 - \cdots .$

12. *Suppose* $p(x) \in \mathbb{C}[x]$ *; then* $p(J_k(\lambda)) = \begin{bmatrix} p(\lambda) & \frac{p'(\lambda)}{1!} & \frac{p''(\lambda)}{2!} & \cdots & \frac{p^{(r-1)}(\lambda)}{(r-1)!} \\ 0 & p(\lambda) & \frac{p'(\lambda)}{1!} & \cdots & \vdots \\ 0 & 0 & p(\lambda) & \ddots & \vdots \\ \vdots & \vdots & \vdots & \ddots & \frac{p'(\lambda)}{1!} \\ 0 & 0 & 0 & \cdots & p(\lambda) \end{bmatrix}.$

13. $J_k(0)J_k(0)^T = \begin{bmatrix} I_{k-1} & \mathbf{0} \\ \mathbf{0} & 0 \end{bmatrix}.$

14. $J_k(0)^T J_k(0) = \begin{bmatrix} 0 & \mathbf{0} \\ \mathbf{0} & I_{k-1} \end{bmatrix}.$

15. $J_k(0)^+ = J_k(0)^T.$

16. $J_k(0)\mathbf{e}_{i+1} = \mathbf{e}_i$ *for* $i = 1, 2, \ldots, k\text{-}1.$

17. *If* $\lambda \neq 0$, $rank(J_k(\lambda))^m = rank(J_k(\lambda))^{m+1} = k$ *for* $m = 1,2,3, \ldots .$

18. $rank(J_k(0)^m) - rank(J_k(0)^{m+1}) = 0$ *if* $m \geq k$.

19. $rank(J_k(0)^m) - rank(J_k(0)^{m+1}) = 1$ *for* $m = 1, 2, 3, \ldots, k\text{-}1.$

PROOF The proofs are left as exercises. □

Exercise Set 46

1. Consider $J = J_6(4) = \begin{bmatrix} 4 & 1 & 0 & 0 & 0 & 0 \\ 0 & 4 & 1 & 0 & 0 & 0 \\ 0 & 0 & 4 & 1 & 0 & 0 \\ 0 & 0 & 0 & 4 & 1 & 0 \\ 0 & 0 & 0 & 0 & 4 & 1 \\ 0 & 0 & 0 & 0 & 0 & 4 \end{bmatrix}$. What are the determinant and trace of J? Is J invertible? If so, what is J^{-1}. What is the spectrum of J and what is the eigenspace? Compute the minimal and characteristic polynomials. What is J^7? Suppose $p(x) = x^2 + 3x - 8$. What is $p(J)$?

2. Prove all the parts of Theorem 11.1.

3. Show that $J_k(\lambda) = \lambda I_n + \sum e_i e_{i+1}^T$ for $n \geq 2$.

4. Argue that $J_k(\lambda)$ is not diagonalizable for $n > 1$.

11.1.1.1 MATLAB Moment

11.1.1.1.1 Jordan Blocks It is easy to write an M-file to create a matrix that is a Jordan block. The code is

```
function J = jordan(lambda, n)
if n == 0, J = [], else
J = nilp(n) + lambda * eye(n); end
```

Experiment making some Jordan blocks. Compute their ranks and characteristic polynomials with MATLAB.

11.1.2 Jordan Segments

Now we need to complicate our structure by allowing matrices that are block diagonal with Jordan blocks of varying sizes down the main diagonal. For the time being, we hold λ fixed.

DEFINITION 11.2 *(λ-Jordan segment)*

*A **λ-Jordan segment of length k** is a block diagonal matrix consisting of k λ-Jordan blocks of various sizes. In symbols, we write, $J(\lambda; p_1, p_2, \ldots, p_k) = BlockDiagonal[J_{p_1}(\lambda), J_{p_2}(\lambda), \ldots, J_{p_k}(\lambda)]$. Note $J(\lambda; p_1, p_2, \ldots, p_k) \in \mathbb{C}^{t \times t}$, where $t = \sum_{j=1}^{k} p_j$. Let's agree to write the p_j's in descending order. The sequence $Segre(\lambda) = (p_1, p_2, \ldots, p_k)$ is called the **Segre sequence** of*

$J(\lambda; p_1, p_2, \ldots, p_k)$. Clearly, given λ and the Segre sequence of λ, the λ-Jordan Segment is completely determined.

Let's look at some examples for clarification. $J(5;3,2) = \begin{bmatrix} 5 & 1 & 0 & 0 & 0 \\ 0 & 5 & 1 & 0 & 0 \\ 0 & 0 & 5 & 0 & 0 \\ 0 & 0 & 0 & 5 & 1 \\ 0 & 0 & 0 & 0 & 5 \end{bmatrix} \in \mathbb{C}^{5\times 5}$, $J(7;2,1,1,1) = \begin{bmatrix} 7 & 1 & 0 & 0 & 0 \\ 0 & 7 & 0 & 0 & 0 \\ 0 & 0 & 7 & 0 & 0 \\ 0 & 0 & 0 & 7 & 0 \\ 0 & 0 & 0 & 0 & 7 \end{bmatrix}$, $J(0;3,2,2) = \begin{bmatrix} 0 & 1 & 0 & 0 & 0 & 0 & 0 \\ 0 & 0 & 1 & 0 & 0 & 0 & 0 \\ 0 & 0 & 0 & 0 & 0 & 0 & 0 \\ 0 & 0 & 0 & 0 & 1 & 0 & 0 \\ 0 & 0 & 0 & 0 & 0 & 0 & 0 \\ 0 & 0 & 0 & 0 & 0 & 0 & 1 \\ 0 & 0 & 0 & 0 & 0 & 0 & 0 \end{bmatrix}$.

Note $rank(J(0;3,2,2)) = 4$ and $nullity = 3$.

To understand the basics about Jordan segments, we need some generalities about block diagonal matrices.

THEOREM 11.2

Let $A = \begin{bmatrix} B & \mathbb{O} \\ \mathbb{O} & C \end{bmatrix}$, where B and C are square matrices.

1. *$A^2 = \begin{bmatrix} B^2 & \mathbb{O} \\ \mathbb{O} & C^2 \end{bmatrix}$; more generally, $A^k = \begin{bmatrix} B^k & \mathbb{O} \\ \mathbb{O} & C^k \end{bmatrix}$ for k, any positive integer.*

2. *If $p(x) \in \mathbb{C}[x]$ and $p(A) = \mathbb{O}$, then $p(B) = \mathbb{O}$ and $p(C) = \mathbb{O}$. Conversely, if $p(B) = \mathbb{O}$ and $p(C) = \mathbb{O}$, then $p(A) = \mathbb{O}$.*

3. *The minimal polynomial $\mu_A(x) = LCM(\mu_B(x), \mu_C(x))$; moreover, if $(x-\lambda)$ occurs k_B times in $\mu_B(x)$ and k_C times in $\mu_C(x)$, then $(x-\lambda)$ occurs $max\{k_B, k_C\}$ times in $\mu_A(x)$.*

4. *The characteristic polynomial $\chi_A(x) = \chi_B(x)\chi_C(x)$; so if $(x-\lambda)$ occurs k_B times in $\chi_B(x)$ and k_C times in $\chi_C(x)$, then $(x-\lambda)$ occurs $k_B + k_C$ times in $\chi_A(x)$.*

5. *$\lambda(A) = \lambda(B) \cup \lambda(C)$; that is, λ is an eigenvalue of A iff λ is an eigenvalue of either B or C.*

6. *The geometric multiplicity of λ in $\lambda(A)$ equals the geometric multiplicity of λ in B plus the geometric multiplicity of λ in C (of course, one of these could be zero).*

PROOF Once again, the details are left to the reader. □

These properties extend to matrices with a finite number of square blocks down the diagonal.

COROLLARY 11.1

Suppose $A = BlockDiagonal[A_1, A_2, \ldots, A_k]$. *Then*

1. $\mu_A(x) = LCM[\mu_{A_1}(x), \mu_{A_2}(x), \ldots, \mu_{A_k}(x)]$.
2. $\chi_A(x) = \prod_{j=1}^{k} \chi_{A_j}(x)$.
3. $\lambda(A) = \bigcup_{j=1}^{k} \lambda(A_j)$.

Now, we can apply these ideas to λ-Jordan segments.

COROLLARY 11.2

Consider a λ*-Jordan segment of length k,* $J = J(\lambda; p_1, p_2, \ldots, p_k)$.

1. λ *is the only eigenvalue of* $J(\lambda; p_1, p_2, \ldots, p_k)$.
2. *The geometric multiplicity of* λ *is k with eigenvectors* $\mathbf{e}_1, \mathbf{e}_{p_1+1}, \ldots, \mathbf{e}_{p_1+p_2+\cdots p_{k-1}+1}$.
3. $\mu_J(x) = (x - \lambda)^{\max\{p_1, p_2, \ldots, p_k\}}$.
4. $\chi_J(x) = (x - \lambda)^t$, *where* $t = \sum_{j=1}^{k} p_j$.
5. $rank(J(0; p_1, p_2, p_3, \ldots, p_k)) = \left(\sum_{j=1}^{k} p_j\right) - k$.
6. $index(J(0; p_1, p_2, p_3, \ldots, p_k)) = max\{p_1, p_2, p_3, \ldots, p_k\}$.

PROOF Once again, the proofs are left as exercises. □

To illustrate, consider $J(5; 3, 2, 2) = \begin{bmatrix} 5 & 1 & 0 & & & & \\ 0 & 5 & 1 & & & & \\ 0 & 0 & 5 & & & & \\ & & & 5 & 1 & & \\ & & & 0 & 5 & & \\ & & & & & 5 & 1 \\ & & & & & 0 & 5 \end{bmatrix}$. Clearly, the only eigenvalue is 5; its geometric multiplicity is 3 with eigenvectors $\mathbf{e}_1, \mathbf{e}_4, \mathbf{e}_6$; the minimal polynomial is $\mu(x) = (x - 5)^3$, and the characteristic polynomial is $\chi(x) = (x - 5)^7$. Consider $J(0; 3, 2, 2) = \begin{bmatrix} 0 & 1 & 0 & & & & \\ 0 & 0 & 1 & & & & \\ 0 & 0 & 0 & & & & \\ & & & 0 & 1 & & \\ & & & 0 & 0 & & \\ & & & & & 0 & 1 \\ & & & & & 0 & 0 \end{bmatrix}$.

Zero is the only eigenvalue with geometric multiplicity 3, eigenvectors $\mathbf{e}_1$,$\mathbf{e}_4$,$\mathbf{e}_6$, minimal polynomial $\mu(x) = x^3$, and characteristic polynomial $\chi(x) = x^7$. The rank is 4 and index is 3.

Exercise Set 47

1. Consider $J = J(4; 3, 3, 2, 1, 1)$. Write out J explicitly. Compute the minimal and characteristic polynomials of J. What is the geometric multiplicity of 4 and what is a basis for the eigenspace? What is the trace and determinant of J? Is J invertible? If so, find J^{-1}.

2. Prove all the parts of Theorem 11.2.

3. Prove all the parts of Corollary 11.1 and Corollary 11.2.

11.1.2.1 MATLAB Moment

11.1.2.1.1 Jordan Segments We have to be a little bit more clever to write a code to generate Jordan segments, but the following should work:

```
functionJ = jordseg(lambda, p)
k = length(p)
J = jordan(lambda, p(k))
fori = k - 1 : -1 : 1
J = blkdiag(jordan(lambda, p(i)), J)
end
```

We need to be sure to enter p as a vector.For example,

```
>>jordseg(3, [2, 2, 4])

ans =
 3  1  0  0  0  0  0  0
 0  3  0  0  0  0  0  0
 0  0  3  1  0  0  0  0
 0  0  0  3  0  0  0  0
 0  0  0  0  3  1  0  0
 0  0  0  0  0  3  1  0
 0  0  0  0  0  0  3  1
 0  0  0  0  0  0  0  3
```

11.1.3 Jordan Matrices

Now we take the next and final step in manufacturing complex matrices up to similarity.

DEFINITION 11.3 *(Jordan matrix)*

*A **Jordan matrix** is a block diagonal matrix whose blocks are Jordan segments. The notation will be* $J = J((\lambda_1; p_{11}, p_{12}, \ldots, p_{1k(1)}), (\lambda_2; p_{21}, p_{22}, \ldots, p_{2k(2)}), \ldots, (\lambda_s; p_{s1}, p_{s2}, \ldots, p_{sk(s)})) = BlockDiagonal[J(\lambda_1; p_{11}, p_{12}, \ldots, p_{1k(1)}), \ldots, J(\lambda_s; p_{s1}, p_{s2}, \ldots, p_{sk(s)})]$. *The data* $(\lambda_i, Segre(\lambda_i))$ *for* $i = 1,2, \ldots, s$ *is called the **Segre characteristic** of* J *and clearly completely determines the structure of* J.

For example,
$J = J((3; 2, 1, 1), (-2; 3), (4; 3, 2), (i; 1, 1)) =$

$$\begin{bmatrix}
3 & 1 & & & & & & & & & & & & \\
0 & 3 & & & & & & & & & & & & \\
 & & 3 & & & & & & & & & & & \\
 & & & 3 & & & & & & & & & & \\
 & & & & -2 & 1 & 0 & & & & & & & \\
 & & & & 0 & -2 & 1 & & & & & & & \\
 & & & & 0 & 0 & -2 & & & & & & & \\
 & & & & & & & 4 & 1 & 0 & & & & \\
 & & & & & & & 0 & 4 & 1 & & & & \\
 & & & & & & & 0 & 0 & 4 & & & & \\
 & & & & & & & & & & 4 & 1 & & \\
 & & & & & & & & & & 0 & 4 & & \\
 & & & & & & & & & & & & i & \\
 & & & & & & & & & & & & & i
\end{bmatrix}.$$

The Segre characteristic of J is $\{(3; 2, 1, 1), (-2; 3), (4; 3, 2), (i; 1, 1)\}$. Now, using the general theory of block diagonal matrices developed above, we can deduce the following.

THEOREM 11.3

Suppose $J = J((\lambda_1; p_{11}, p_{12}, \ldots, p_{1k(1)}), (\lambda_2; p_{21}, p_{22}, \ldots, p_{2k(2)}), \ldots, (\lambda_s; p_{s1}, p_{s2}, \ldots, p_{sk(s)}))$.

Then

1. *the eigenvalues of* J *are* $\lambda_1, \lambda_2, \ldots, \lambda_s$
2. *the geometric multiplicity of* λ_i *is* $k(i)$

3. *the minimum polynomial of J is* $\mu(x) = \prod_{i=1}^{s}(x - \lambda_i)^{\max(p_{i1}, p_{i2}, \dots, p_{ik(i)})}$

4. *the characteristic polynomial of J is* $\chi(x) = \prod_{i=1}^{s}(x - \lambda_i)^{t_i}$, *where* $t_i = \sum_{j=1}^{k(i)} p_{ij}$.

PROOF The proofs are left as exercises. □

Now the serious question is whether any given matrix A is similar to a Jordan matrix and is this matrix unique in some sense. The remarkable fact is that this is so. Proving that is a nontrivial task. We work on that next.

Exercise Set 48

1. Consider $J = J((4;5), (\pi; 2, 2, 1), (2+2i; 3))$. What are the eigenvalues of J and what are their geometric multiplicities? Compute the minimal and characteristic polynomials of J.

2. Prove all the parts of Theorem 11.3.

11.1.3.1 MATLAB Moment

11.1.3.1.1 Jordan Matrices Programming Jordan matrices is a bit of a challenge. We are grateful to our colleague Frank Mathis for the following code:

```
functionJmat = jordform(varargin)
n = nargin
lp = varargin(n)
lp = lp{:}
k = length(lp)
Jmat = jordseg(lp(1), lp(2 : k))
fori = n − 1 : −1 : 1
lp = narargin(i)
lp = lp{:}
k = length(lp)
Jmat = blkdiag(jordseg(lp(1), lp(2 : k)), Jmat
end
```

Try

$$>>jordform([8,[4,2,2]],[9,[3,1]]).$$

Note the format of the entry.

11.1.4 Jordan's Theorem

There are many proofs of Jordan's theorem in the literature. We will try to present an argument that is as elementary and constructive as possible. Induction is always a key part of the argument. Let's take an arbitrary A in $\mathbb{C}^{n\times n}$. We shall argue that we can get A into the form of a Jordan matrix by using similarities. That is, we shall find an invertible matrix S so that $S^{-1}AS = J$ is a Jordan matrix. This matrix will be unique up to the order in which the blocks are arranged on the diagonal of J. This will be our canonical form for the equivalence relation of similarity. We can then learn the properties of A by looking at its Jordan canonical form (JCF).

The proof we present is broken down into three steps. The first big step is to get the matrix into triangular form with eigenvalues in a prescribed order down the main diagonal.This is established by Schur's theorem, which we have already proved. As a reminder, Schur's theorem states: Let $A \in \mathbb{C}^{n\times n}$ with eigenvalues $\lambda_1, \lambda_2, \ldots, \lambda_n$ (not necessarily distinct) be written in any prescribed order. Then there exists a unitary matrix $U \in \mathbb{C}^{n\times n}$ with $U^{-1}AU = T$, where T is upper triangular and $t_{ii} = \lambda_i$ for i = 1, 2, ... , n. Hence, we may assume $U^{-1}AU$ is upper triangular with equal diagonal entries occurring consecutively down the diagonal. Suppose now, $\lambda_1, \lambda_2, \ldots, \lambda_m$ is a list of the distinct eigenvalues of A.

For the second stage, we show that transvections can be used to "zero out" portions of the upper triangular matrix T when a change in eigenvalue on the diagonal occurs. This can be done without disturbing the diagonal elements or the upper triangular nature of T.

Suppose $r < s$ and consider $T_{rs}(\alpha)^{-1}TT_{rs}(\alpha) = T_{rs}(-\alpha)TT_{rs}(\alpha)$, where α is any complex scalar. Notice that this similarity transformation changes entries of T only in the rth row to the right of the sth column and in the sth column above the rth row and replaces t_{rs} by $t_{rs} + \alpha(t_{rr} - t_{ss})$.

By way of illustration, consider

$$\begin{bmatrix} 1 & -\alpha & 0 \\ 0 & 1 & 0 \\ 0 & 0 & 1 \end{bmatrix} \begin{bmatrix} t_{11} & t_{12} & t_{13} \\ 0 & t_{22} & t_{23} \\ 0 & 0 & t_{33} \end{bmatrix} \begin{bmatrix} 1 & \alpha & 0 \\ 0 & 1 & 0 \\ 0 & 0 & 1 \end{bmatrix} = \begin{bmatrix} 1 & -\alpha & 0 \\ 0 & 1 & 0 \\ 0 & 0 & 1 \end{bmatrix} \begin{bmatrix} t_{11} & t_{11}\alpha + t_{12} & t_{13} \\ 0 & t_{22} & t_{23} \\ 0 & 0 & t_{33} \end{bmatrix}$$

$$= \begin{bmatrix} t_{11} & t_{11}\alpha + t_{12} - t_{22} & t_{13} - \alpha t_{23} \\ 0 & t_{22} & t_{23} \\ 0 & 0 & t_{33} \end{bmatrix} = \begin{bmatrix} t_{11} & t_{12} + \alpha(t_{11} - t_{22}) & t_{13} - \alpha t_{23} \\ 0 & t_{22} & t_{23} \\ 0 & 0 & t_{33} \end{bmatrix}.$$

By choosing $\alpha_{rs} = \dfrac{-t_{rs}}{t_{rr} - t_{ss}}$, we can zero out the (r, s) entry as long as $t_{rr} \neq t_{ss}$. Thus, working with the sequence of transvections corresponding to positions

$(n-1,n), \underbrace{(n-2,n-1),(n-2,n)}, \underbrace{(n-3,n-2),(n-3,n-2),(n-3,n)},$ $\underbrace{(n-4,n-3),(n-4,n-2),\ldots}$ etc., we form the invertible matrix $Q = UT_{n-1,n}(\alpha_{n-1,n})T_{n-2,n-1}(\alpha_{n-2,n-1})\cdots$ and note that $Q^{-1}AQ = \begin{bmatrix} T_1 & & & \\ & T_2 & & \\ & & \ddots & \\ & & & T_m \end{bmatrix}$, a block diagonal matrix where $T_i = \lambda_i I + M_i$, where M_i is strictly upper triangular (i.e., upper triangular with zeros on the main diagonal). Before going on to the final stage, let us illustrate the argument above to be sure it is clear.

Consider $A = \begin{bmatrix} 2 & 2 & 3 & 4 \\ 0 & 2 & 5 & 6 \\ 0 & 0 & 3 & 9 \\ 0 & 0 & 0 & 3 \end{bmatrix}$. Being already upper triangular, we do not need Schur. We see $a_{11} = a_{22}$ and $a_{33} = a_{44}$ so we begin by targeting a_{23}: $T_{23}(-5)AT_{23}(5) = \begin{bmatrix} 2 & 2 & 13 & 4 \\ 0 & 2 & 0 & -39 \\ 0 & 0 & 3 & 9 \\ 0 & 0 & 0 & 3 \end{bmatrix}$. The reader should now continue on with positions $(3,4)$, $(1,3)$, and $(1,4)$ to obtain $Q^{-1}AQ = \begin{bmatrix} 2 & 2 & 0 & 0 \\ 0 & 2 & 0 & 0 \\ 0 & 0 & 3 & 9 \\ 0 & 0 & 0 & 3 \end{bmatrix}$. Now we go on to the final stage to produce JCF.

In view of the previous stage, it suffices to work with matrices of the form $\lambda I + M$ where M is strictly upper triangular. Say $T = \begin{bmatrix} \lambda & * & * & \cdots & * \\ 0 & \lambda & * & \cdots & * \\ 0 & 0 & \lambda & \cdots & * \\ \vdots & \vdots & \vdots & \ddots & \vdots \\ 0 & 0 & 0 & \cdots & \lambda \end{bmatrix}$. We induct on n, the size of the matrix. For $n = 1$, $T = [\lambda] = J_1[\lambda]$, which is already in JCF. That seemed a little too easy, so assume $n = 2$. Then $T = \begin{bmatrix} \lambda & b \\ 0 & \lambda \end{bmatrix}$. If $b = 0$, again we have JCF, so assume $b \neq 0$. Then, using a dilation, we can make b to be 1; $D_1(b^{-1})TD_1(b) = \begin{bmatrix} b^{-1} & 0 \\ 0 & 1 \end{bmatrix}\begin{bmatrix} \lambda & b \\ 0 & \lambda \end{bmatrix}\begin{bmatrix} b & 0 \\ 0 & 1 \end{bmatrix} = \begin{bmatrix} b^{-1} & 0 \\ 0 & 1 \end{bmatrix}\begin{bmatrix} \lambda b & b \\ 0 & \lambda \end{bmatrix} = \begin{bmatrix} b^{-1}\lambda b & 1 \\ 0 & \lambda \end{bmatrix} = \begin{bmatrix} \lambda & 1 \\ 0 & \lambda \end{bmatrix}$, which is JCF. Now for the induction, suppose $n > 2$. Consider the leading principal submatrix of T, which is size $(n-1)$-by-$(n-1)$. By the induction hypothesis, this matrix can be brought to JCF by a similarity Q. Then, $\begin{bmatrix} Q^{-1} & \mathbb{O} \\ \mathbb{O} & 1 \end{bmatrix} T \begin{bmatrix} Q & \mathbb{O} \\ \mathbb{O} & 1 \end{bmatrix} =$

$$T_1 = \begin{bmatrix} & & & & * \\ & F & & & * \\ & & & & \vdots \\ & & & & * \\ 0 & 0 & \cdots & 0 & \lambda \end{bmatrix}$$ and F is a block diagonal matrix of λ-*Jordan* blocks. The problem is that there may be nonzero entries in the last column. But any entry in that last column opposite a row of F that contains a one can be zeroed out. We illustrate.

Say $$T_1 = \begin{bmatrix} \lambda & 1 & 0 & 0 & 0 & 0 & a \\ 0 & \lambda & 0 & 0 & 0 & 0 & b \\ 0 & 0 & \lambda & 1 & 0 & 0 & c \\ 0 & 0 & 0 & \lambda & 1 & 0 & d \\ 0 & 0 & 0 & 0 & \lambda & 1 & e \\ 0 & 0 & 0 & 0 & 0 & \lambda & f \\ 0 & 0 & 0 & 0 & 0 & 0 & \lambda \end{bmatrix}.$$ Now, for example, d lies in the (4,7) position across from a one in the fourth row of the leading principal submatrix of T_1 of size 6-by-6. We can use a transvection similarity to zero it out. Indeed,

$$T_{57}(d)T_1T_{57}(-d) = \begin{bmatrix} \lambda & 1 & 0 & 0 & 0 & 0 & a \\ 0 & \lambda & 0 & 0 & 0 & 0 & b \\ 0 & 0 & \lambda & 1 & 0 & 0 & c \\ 0 & 0 & 0 & \lambda & 1 & 0 & d \\ 0 & 0 & 0 & 0 & \lambda & 1 & e+\lambda d \\ 0 & 0 & 0 & 0 & 0 & \lambda & f \\ 0 & 0 & 0 & 0 & 0 & 0 & \lambda \end{bmatrix} T_{57}(-d) =$$

$$\begin{bmatrix} \lambda & 1 & 0 & 0 & 0 & 0 & a \\ 0 & \lambda & 0 & 0 & 0 & 0 & b \\ 0 & 0 & \lambda & 1 & 0 & 0 & c \\ 0 & 0 & 0 & \lambda & 1 & 0 & -d+d=0 \\ 0 & 0 & 0 & 0 & \lambda & 1 & -\lambda d+e+\lambda d=e \\ 0 & 0 & 0 & 0 & 0 & \lambda & f \\ 0 & 0 & 0 & 0 & 0 & 0 & \lambda \end{bmatrix}.$$

In like manner, we eliminate all nonzero elements that live in a row with a superdiagonal one in it. Thus, we have achieved something tantalizingly close to JCF:

$$T_2 = \begin{bmatrix} \lambda & 1 & 0 & 0 & 0 & 0 & 0 \\ 0 & \lambda & 0 & 0 & 0 & 0 & b \\ 0 & 0 & \lambda & 1 & 0 & 0 & 0 \\ 0 & 0 & 0 & \lambda & 1 & 0 & 0 \\ 0 & 0 & 0 & 0 & \lambda & 1 & 0 \\ 0 & 0 & 0 & 0 & 0 & \lambda & f \\ 0 & 0 & 0 & 0 & 0 & 0 & \lambda \end{bmatrix}.$$

Now comes the tricky part. First, if b and f are zero, we are done. We have achieved JCF. Suppose then that $b = 0$ but $f \neq 0$. Then f can be made one using a dilation similarity:

$$D_7(f)T_2D_7(f^{-1}) = \begin{bmatrix} \lambda & 1 & 0 & 0 & 0 & 0 & 0 \\ 0 & \lambda & 0 & 0 & 0 & 0 & 0 \\ 0 & 0 & \lambda & 1 & 0 & 0 & 0 \\ 0 & 0 & 0 & \lambda & 1 & 0 & 0 \\ 0 & 0 & 0 & 0 & \lambda & 1 & 0 \\ 0 & 0 & 0 & 0 & 0 & \lambda & 1 \\ 0 & 0 & 0 & 0 & 0 & 0 & \lambda \end{bmatrix} \text{ and we have JCF.}$$

Now suppose $f = 0$ but $b \neq 0$. Then b can be made one by a dilation similarity as above; $D_7(b)T_2D_7(b^{-1}) = \begin{bmatrix} \lambda & 1 & 0 & 0 & 0 & 0 & 0 \\ 0 & \lambda & 0 & 0 & 0 & 0 & 1 \\ 0 & 0 & \lambda & 1 & 0 & 0 & 0 \\ 0 & 0 & 0 & \lambda & 1 & 0 & 0 \\ 0 & 0 & 0 & 0 & \lambda & 1 & 0 \\ 0 & 0 & 0 & 0 & 0 & \lambda & 0 \\ 0 & 0 & 0 & 0 & 0 & 0 & \lambda \end{bmatrix}$. Now use a permutation similarity to create JCF; swap columns 3 and 7 and rows 3 and 7 to get

$$P(37)D_7(b)T_2D_7(b^{-1})P(37) =$$

$$= P(37) \begin{bmatrix} \lambda & 1 & 0 & 0 & 0 & 0 & 0 \\ 0 & \lambda & 1 & 0 & 0 & 0 & 0 \\ 0 & 0 & 0 & 1 & 0 & 0 & \lambda \\ 0 & 0 & 0 & \lambda & 1 & 0 & 0 \\ 0 & 0 & 0 & 0 & \lambda & 1 & 0 \\ 0 & 0 & 0 & 0 & 0 & \lambda & 0 \\ 0 & 0 & \lambda & 0 & 0 & 0 & 0 \end{bmatrix} = \begin{bmatrix} \lambda & 1 & 0 & 0 & 0 & 0 & 0 \\ 0 & \lambda & 1 & 0 & 0 & 0 & 0 \\ 0 & 0 & \lambda & 1 & 0 & 0 & 0 \\ 0 & 0 & 0 & \lambda & 1 & 0 & 0 \\ 0 & 0 & 0 & 0 & \lambda & 1 & 0 \\ 0 & 0 & 0 & 0 & 0 & \lambda & 0 \\ 0 & 0 & 0 & 0 & 0 & 0 & \lambda \end{bmatrix}.$$

Finally, suppose both b and f are nonzero. We will show that the element opposite the smaller block can be made zero. Consider $D_7(f)T_{15}(-\frac{b}{f})T_{26}(-\frac{b}{f})$ $T_2T_{26}(\frac{b}{f})T_{15}(\frac{b}{f})D_7(f^{-1}) =$

$$T_{15}(-\tfrac{b}{f}) \begin{bmatrix} \lambda & 1 & 0 & 0 & 0 & \frac{b}{f} & 0 \\ 0 & \lambda & 0 & 0 & 0 & 0 & 0 \\ 0 & 0 & \lambda & 1 & 0 & 0 & 0 \\ 0 & 0 & 0 & \lambda & 1 & 0 & 0 \\ 0 & 0 & 0 & 0 & \lambda & 1 & 0 \\ 0 & 0 & 0 & 0 & 0 & \lambda & f \\ 0 & 0 & 0 & 0 & 0 & 0 & \lambda \end{bmatrix} T_{15}(\tfrac{b}{f}) =$$

$$D_7(f) \begin{bmatrix} \lambda & 1 & 0 & 0 & 0 & 0 & 0 \\ 0 & \lambda & 0 & 0 & 0 & 0 & 0 \\ 0 & 0 & \lambda & 1 & 0 & 0 & 0 \\ 0 & 0 & 0 & \lambda & 1 & 0 & 0 \\ 0 & 0 & 0 & 0 & \lambda & 1 & 0 \\ 0 & 0 & 0 & 0 & 0 & \lambda & f \\ 0 & 0 & 0 & 0 & 0 & 0 & \lambda \end{bmatrix} D_7(f^{-1}) =$$

$$\begin{bmatrix} \lambda & 1 & 0 & 0 & 0 & 0 & 0 \\ 0 & \lambda & 0 & 0 & 0 & 0 & 0 \\ 0 & 0 & \lambda & 1 & 0 & 0 & 0 \\ 0 & 0 & 0 & \lambda & 1 & 0 & 0 \\ 0 & 0 & 0 & 0 & \lambda & 1 & 0 \\ 0 & 0 & 0 & 0 & 0 & \lambda & 1 \\ 0 & 0 & 0 & 0 & 0 & 0 & \lambda \end{bmatrix}, \text{ which is JCF.}$$

This completes our argument for getting a matrix into JCF. Remarkably, after the triangularization given by Schur's theorem, all we used were elementary row and column operations. By the way, the Jordan we have been talking about is **Marie Ennemond Camille Jordan** (5 January 1838–22 January 1922), who presented this canonical form in 1870. Apparently, he won a prize for this work.

An alternate approach to stage three above is to recall a deep theorem we worked out in Chapter 3 concerning nilpotent matrices. Note that $N = T - \lambda I$ is nilpotent and Theorem 81 of Chapter 3 says that this nilpotent matrix is similar to BlockDiagonal[Nilp[p_1], Nilp[p_2], ... , Nilp[p_k]] where the total number of blocks in the nullity of N and the number of j-by-j blocks is $rank(N^{j-1}) - 2rank(N^j) + rank(N^{j+1})$. If S is a similarity transformation that affects this transformation then $S^{-1}NS = S^{-1}(T - \lambda I)S = S^{-1}TS - \lambda I = BlockDiagonal[Nilp[p_1], Nilp[p_2], \ldots, Nilp[p_k]]$ so $S^{-1}TS = \lambda I + BlockDiagonal[Nilp[p_1], Nilp[p_2], \ldots, Nilp[p_k]]$ is in JCF. Finally, we remark that the one on the superdiagonal of the JCF are not essential. Using dilation similarities, we can put any scalar on the superdiagonal for each block. What is essential is the number and the sizes of the Jordan blocks.

11.1.4.1 Generalized Eigenvectors

In this section we look at another approach to Jordan's theorem where we show how to construct a special kind of basis. Of course, constructing bases is the same as constructing invertible matrices. We have seen that not all square matrices are diagonalizable. This is because they may not have enough eigenvectors to span the whole space. For example, take $A = \begin{bmatrix} 0 & 0 \\ 1 & 0 \end{bmatrix}$. Being lower triangular makes it easy to see that zero is the only eigenvalue. If we compute the eigenspace, $Eig(A, 0) = \left\{ \begin{bmatrix} x_1 \\ x_2 \end{bmatrix} \mid A \begin{bmatrix} x_1 \\ x_2 \end{bmatrix} = \begin{bmatrix} 0 \\ 0 \end{bmatrix} \right\} = \left\{ \begin{bmatrix} x_1 \\ x_2 \end{bmatrix} \mid \begin{bmatrix} 0 & 0 \\ 1 & 0 \end{bmatrix} \begin{bmatrix} x_1 \\ x_2 \end{bmatrix} = \begin{bmatrix} 0 \\ x_1 \end{bmatrix} = \begin{bmatrix} 0 \\ 0 \end{bmatrix} \right\} = \left\{ \begin{bmatrix} 0 \\ x_2 \end{bmatrix} \mid x_2 \in \mathbb{C} \right\} = sp\left(\begin{bmatrix} 0 \\ 1 \end{bmatrix} \right)$, which is one dimensional. There is no way to generate $\mathbb{C}^2$ with eigenvectors of A. Is there a way to get more "eigenvectors" of A? This leads us to the next definition.

DEFINITION 11.4 *(generalized eigenvectors)*

Let λ be an eigenvalue of $A \in \mathbb{C}^{n\times n}$. A nonzero vector $\mathbf{x}$ is a ***generalized eigenvector of level q belonging to $\boldsymbol{\lambda}$*** *(or just* ***$\boldsymbol{\lambda}$-q-eigenvector*** *for short) if and only if $(A-\lambda I)^q\mathbf{x} = \overrightarrow{0}$ but $(A-\lambda I)^{q-1}\mathbf{x} \neq \overrightarrow{0}$. That is, $\mathbf{x} \in \mathcal{N}ull((A-\lambda I)^q)$ but $\mathbf{x} \notin \mathcal{N}ull((A-\lambda I)^{q-1})$.*

Note that a 1-eigenvector is just an ordinary eigenvector as defined previously. Also, if $\mathbf{x}$ is a λ-q-eigenvector, then $(A-\lambda I)^p\mathbf{x} = \overrightarrow{0}$ for any $p \geq q$. We could have defined a generalized eigenvector of A to be any nonzero vector $\mathbf{x}$ such that $(A-\lambda I)^k\mathbf{x} = \overrightarrow{0}$ for some $k \in \mathbb{N}$. Then the least such k would be the level of $\mathbf{x}$. You might wonder if λ could be any scalar, not necessarily an eigenvalue. But if $(A-\lambda I)^k\mathbf{x} = \overrightarrow{0}$ for $\mathbf{x} \neq \overrightarrow{0}$, then $(A-\lambda I)^k$ is not an invertible matrix whence $(A-\lambda I)$ is not invertible, making λ an eigenvalue. Thus, there are no "generalized eigenvalues."

DEFINITION 11.5 *(Jordan string)*

Let $\mathbf{x}$ be a q-eigenvector of A belonging to the eigenvalue λ. We define the ***Jordan string ending at*** $\mathbf{x}$ *by*

$$\begin{cases}
\mathbf{x}_q = \mathbf{x} & \mathbf{x}_q \\
\mathbf{x}_{q-1} = (A-\lambda I)\mathbf{x} & A\mathbf{x}_q = \lambda\mathbf{x}_q + \mathbf{x}_{q-1} \\
\mathbf{x}_{q-2} = (A-\lambda I)^2\mathbf{x} = (A-\lambda I)\mathbf{x}_{q-1} & A\mathbf{x}_{q-1} = \lambda\mathbf{x}_{q-1} + \mathbf{x}_{q-2}. \\
\vdots & \vdots \\
\mathbf{x}_2 = (A-\lambda I)^{q-2}\mathbf{x} = (A-\lambda I)\mathbf{x}_3 & A\mathbf{x}_3 = \lambda\mathbf{x}_3 + \mathbf{x}_2 \\
\mathbf{x}_1 = (A-\lambda I)^{q-1}\mathbf{x} = (A-\lambda I)\mathbf{x}_2 & A\mathbf{x}_2 = \lambda\mathbf{x}_2 + \mathbf{x}_1 \\
 & A\mathbf{x}_1 = \lambda\mathbf{x}_1
\end{cases}$$

Note what is going on here. If $\mathbf{x}$ is an eigenvector of A belonging to λ, then $(A-\lambda I)\mathbf{x} = \overrightarrow{0}$ (i.e., $(A-\lambda I)$ annihilates $\mathbf{x}$). Now if we do not have enough eigenvectors for A— that is, enough vectors annihilated by $A-\lambda I$ — then it is reasonable to consider vectors annihilated by $(A-\lambda I)^2$, $(A-\lambda I)^3$, and so on. If we have a Jordan string $\mathbf{x}_1, \mathbf{x}_2, \ldots, \mathbf{x}_q$, then $(A-\lambda I)\mathbf{x}_1 = (A-\lambda I)^q\mathbf{x} = \overrightarrow{0}$, $(A-\lambda I)^2\mathbf{x}_2 = \overrightarrow{0}, \ldots, (A-\lambda I)^q\mathbf{x}_q = \overrightarrow{0}$. Note that $\mathbf{x}_1$ is an ordinary eigenvector and $(A-\lambda I)$ is nilpotent on $span\{\mathbf{x}_1, \mathbf{x}_2, \ldots, \mathbf{x}_q\}$, which is A-invariant. The first clue to the usefulness of Jordan strings is given by the next theorem.

THEOREM 11.4

The vectors in any Jordan string $\{\mathbf{x}_1, \mathbf{x}_2, \ldots, \mathbf{x}_q\}$ are linearly independent. Moreover, each $\mathbf{x}_i$ is a generalized eigenvector of level i for $i = 1, 2, \ldots, q$.

PROOF With the notation as above, set $\alpha_1\mathbf{x}_1 + \alpha_2\mathbf{x}_2 + \ldots + \alpha_{q-1}\mathbf{x}_{q-1} + \alpha_q\mathbf{x}_q = \overrightarrow{0}$. Apply $(A - \lambda I)^{q-1}$ to this equation and get $\alpha_1(A - \lambda I)^{q-1}\mathbf{x}_1 + \alpha_2(A - \lambda I)^{q-1}\mathbf{x}_2 + \ldots + \alpha_q(A - \lambda I)^{q-1}\mathbf{x}_q = \overrightarrow{0}$. But all terms vanish except $\alpha_q(A - \lambda I)^{q-1}\mathbf{x}_q$. Thus $\alpha_q(A - \lambda I)^{q-1}\mathbf{x}_q = \alpha_q(A - \lambda I)^{q-1}\mathbf{x}$, so $\alpha_q = 0$ since $(A-\lambda I)^{q-1}\mathbf{x} \neq \overrightarrow{0}$. In a similar manner, each α can be forced to be zero. Clearly, $(A - \lambda I)\mathbf{x}_1 = \overrightarrow{0}$, so $\mathbf{x}_1$ is an eigenvector of A. Next, $(A - \lambda I)\mathbf{x}_2 = \mathbf{x}_1 \neq \overrightarrow{0}$ and yet $(A - \lambda I)^2\mathbf{x}_2 = (A - \lambda I)\mathbf{x}_1 = \overrightarrow{0}$. Thus, $\mathbf{x}_2$ is a generalized eigenvector of level 2. Now apply induction. □

We are off to a good start in generating more independent vectors corresponding to an eigenvalue. Next, we generalize the notion of an eigenspace.

DEFINITION 11.6 *If λ is an eigenvalue of $A \in \mathbb{C}^{n\times n}$, define*

$$G_\lambda(A) = \left\{\mathbf{x} \in \mathbb{C}^n \mid (A - \lambda I)^p\mathbf{x} = \overrightarrow{0} \text{ for some } p \in \mathbb{N}\right\}.$$

This is the ***generalized eigenspace belonging to*** λ.

THEOREM 11.5
For each eigenvalue λ, $G_\lambda(A)$ is an A-invariant subspace of $\mathbb{C}^n$. Indeed, $G_\lambda(A) = \mathcal{N}ull((A - \lambda I)^n)$.

PROOF The proof is left as an exercise. □

Clearly, the eigenspace for λ, $Eig(A, \lambda) = \mathcal{N}ull(A-\lambda I) \subseteq G_\lambda(A)$. Now the idea is something like this. Take a generalized eigenspace $G_\lambda(A)$ and look for a generalized eigenvector $\mathbf{x}$, say with $(A - \lambda I)^{k-1}\mathbf{x} \neq \overrightarrow{0}$ but $(A - \lambda I)^k\mathbf{x} = \overrightarrow{0}$ and such that there is no vector in $G_\lambda(A)$ that is not annihilated at least by $(A - \lambda I)^k$. This gives a generalized eigenvector of level k where k is the length of a longest Jordan string associated with λ. Then $\mathbf{x} = \mathbf{x}_k, \mathbf{x}_{k-1}, \ldots, \mathbf{x}_1$ is a list of independent vectors in $G_\lambda(A)$. If they span $G_\lambda(A)$, we are done. If not we look for another generalized eigenvector $\mathbf{y}$ and create another string independent of the first. We continue this process until (hopefully) we have a basis of $G_\lambda(A)$. We hope to show we can get a basis of $\mathbb{C}^n$ consisting of eigenvectors and generalized eigenvectors of A. There are many details to verify. Before we get into the nitty-gritty, let's look at a little example. Suppose A is 3-by-3 and λ is an eigenvalue of A. Say we have a Jordan string, $\mathbf{x}_3 = \mathbf{x}$, $\mathbf{x}_2 = (A - \lambda I)\mathbf{x} = A\mathbf{x}_3 - \lambda\mathbf{x}_3$ and $\mathbf{x}_1 = (A - \lambda I)^2\mathbf{x} = (A - \lambda I)\mathbf{x}_2 = A\mathbf{x}_2 - \lambda\mathbf{x}_2$ and necessarily $(A - \lambda I)_3\mathbf{x} = \overrightarrow{0}$ and $\mathbf{x}_1$ is an eigenvector for λ. Then solving we see $A\mathbf{x}_3 = \mathbf{x}_2 + \lambda\mathbf{x}_3$, $A\mathbf{x}_2 = \mathbf{x}_1 + \lambda\mathbf{x}_2$ and $A\mathbf{x}_1 = \lambda\mathbf{x}_1$, Then $A[\mathbf{x}_1 \mid \mathbf{x}_2 \mid \mathbf{x}_3] = [A\mathbf{x}_1 \mid A\mathbf{x}_2 \mid A\mathbf{x}_3] = [\lambda\mathbf{x}_1 \mid$

$$\mathbf{x}_1 + \lambda\mathbf{x}_2 \mid \mathbf{x}_2 + \lambda\mathbf{x}_3] = [\mathbf{x}_1 \mid \mathbf{x}_2 \mid \mathbf{x}_3]\begin{bmatrix} \lambda & 1 & 0 \\ 0 & \lambda & 1 \\ 0 & 0 & \lambda \end{bmatrix} = [\mathbf{x}_1 \mid \mathbf{x}_2 \mid \mathbf{x}_3]J_3(\lambda).$$

Theorem 11.4 says $\{\mathbf{x}_1, \mathbf{x}_2, \mathbf{x}_3\}$ is an independent set, so if $S = [\mathbf{x}_1 \mid \mathbf{x}_2 \mid \mathbf{x}_3]$, then $AS = SJ_3(\lambda)$ or $S^{-1}AS = J_3(\lambda)$. Thus, A is similar to $J_3(\lambda)$. We call $\{\mathbf{x}_1, \mathbf{x}_2, \mathbf{x}_3\}$ a *Jordan basis*. The general story is much more complicated than this, but at least you now have a hint of where we are going.

First, a little bad news. Suppose we have a Jordan string $(A - \lambda I)\mathbf{x}_1 = \vec{0}, (A - \lambda I)\mathbf{x}_2 = \mathbf{x}_1, \dots, (A - \lambda I)\mathbf{x}_q = \mathbf{x}_{q-1}$. Then $\mathbf{y}_1 = \mathbf{x}_1, \mathbf{y}_2 = \mathbf{x}_2 + \alpha_1\mathbf{x}_1$, $\mathbf{y}_3 = \mathbf{x}_3 + \alpha_1\mathbf{x}_2 + \alpha_2\mathbf{x}_1, \dots, \mathbf{y}_q = \mathbf{x}_q + \alpha_1\mathbf{x}_{q-1} + \alpha_2\mathbf{x}_{q-2} + \dots + \alpha_{q-1}\mathbf{x}_1$ is also a Jordan string and the subspace spanned by the **y**s is the same as the subspace spanned by the **x**s. Thus, Jordan bases are far from unique. Note that $\mathbf{y}_q$ above contains all the arbitrary constants. So once $\mathbf{y}_q$ is nailed down, the rest of the chain is determined. This is why we want generalized eigenvectors of a high level. If we start building up from an eigenvector, we have to make arbitrary choices, and this is not so easy. Let's look at a couple of concrete examples before we develop more theory.

Consider a Jordan block $A = J_3(5) = \begin{bmatrix} 5 & 1 & 0 \\ 0 & 5 & 1 \\ 0 & 0 & 5 \end{bmatrix}$. Evidently, 5 is the only eigenvalue of A. The eigenspace $Eig(A, 5) = \left\{ \begin{bmatrix} x \\ y \\ z \end{bmatrix} \mid (A - 5I)\begin{bmatrix} x \\ y \\ z \end{bmatrix} = \begin{bmatrix} 0 \\ 0 \\ 0 \end{bmatrix} \right\} = \left\{ \begin{bmatrix} x \\ y \\ z \end{bmatrix} \mid \begin{bmatrix} 0 & 1 & 0 \\ 0 & 0 & 1 \\ 0 & 0 & 0 \end{bmatrix}\begin{bmatrix} x \\ y \\ z \end{bmatrix} = \begin{bmatrix} 0 \\ 0 \\ 0 \end{bmatrix} \right\} = \left\{ \begin{bmatrix} x \\ y \\ z \end{bmatrix} \mid \begin{bmatrix} y \\ z \\ 0 \end{bmatrix} = \begin{bmatrix} 0 \\ 0 \\ 0 \end{bmatrix} \right\} = \left\{ \begin{bmatrix} x \\ 0 \\ 0 \end{bmatrix} \mid \in \mathbb{C} \right\} = sp\left(\begin{bmatrix} 1 \\ 0 \\ 0 \end{bmatrix} \right)$. Thus $Eig(A, 5)$ is one dimensional, so we are short on eigenvectors if we hope to span $\mathbb{C}^3$. Note that $(A - 5I) = \begin{bmatrix} 0 & 1 & 0 \\ 0 & 0 & 1 \\ 0 & 0 & 0 \end{bmatrix}$, $(A - 5I)^2 = \begin{bmatrix} 0 & 0 & 1 \\ 0 & 0 & 0 \\ 0 & 0 & 0 \end{bmatrix}$, and $(A - 5I)^3 = \mathbb{O}$, so we would like a 3-eigenvector $\mathbf{x}_3 = \begin{bmatrix} a \\ b \\ c \end{bmatrix}$. Evidently $(A - 5I)^3\mathbf{x}_3 = \begin{bmatrix} 0 \\ 0 \\ 0 \end{bmatrix}$.

Now $(A - 5I)^2\mathbf{x}_3 = \begin{bmatrix} 0 & 0 & 1 \\ 0 & 0 & 0 \\ 0 & 0 & 0 \end{bmatrix}\begin{bmatrix} a \\ b \\ c \end{bmatrix} = \begin{bmatrix} c \\ 0 \\ 0 \end{bmatrix}$, so we must require $c \neq 0$. Say $c = 1$. Next, $\mathbf{x}_2 = (A - 5I)\mathbf{x}_3 = \begin{bmatrix} 0 & 1 & 0 \\ 0 & 0 & 1 \\ 0 & 0 & 0 \end{bmatrix}\begin{bmatrix} a \\ b \\ 1 \end{bmatrix} = \begin{bmatrix} b \\ 1 \\ 0 \end{bmatrix}$

and $\mathbf{x}_1 = \begin{bmatrix} 0 & 1 & 0 \\ 0 & 0 & 1 \\ 0 & 0 & 0 \end{bmatrix} \begin{bmatrix} b \\ 1 \\ 0 \end{bmatrix} = \begin{bmatrix} 1 \\ 0 \\ 0 \end{bmatrix}$.

So, choosing $a = b = 1$, $\mathbf{x}_3 = \begin{bmatrix} 1 \\ 1 \\ 1 \end{bmatrix}$, $\mathbf{x}_2 = \begin{bmatrix} 1 \\ 1 \\ 0 \end{bmatrix}$, and $\mathbf{x}_1 = \begin{bmatrix} 1 \\ 0 \\ 0 \end{bmatrix}$ is a Jordan string, which is a Jordan basis of $\mathbb{C}^3$. Note $A\mathbf{x}_1 = A \begin{bmatrix} 1 \\ 0 \\ 0 \end{bmatrix} = 5 \begin{bmatrix} 1 \\ 0 \\ 0 \end{bmatrix} = \begin{bmatrix} 5 \\ 0 \\ 0 \end{bmatrix}$, $A\mathbf{x}_2 = A \begin{bmatrix} 1 \\ 1 \\ 0 \end{bmatrix} = \begin{bmatrix} 1 \\ 0 \\ 0 \end{bmatrix} + 5 \begin{bmatrix} 1 \\ 1 \\ 0 \end{bmatrix} = \begin{bmatrix} 6 \\ 5 \\ 0 \end{bmatrix}$ and $A\mathbf{x}_3 = \begin{bmatrix} 1 \\ 1 \\ 0 \end{bmatrix} + 5 \begin{bmatrix} 1 \\ 1 \\ 1 \end{bmatrix} = \begin{bmatrix} 6 \\ 6 \\ 5 \end{bmatrix}$ so $A \begin{bmatrix} 1 & 1 & 1 \\ 0 & 1 & 1 \\ 0 & 0 & 1 \end{bmatrix} = \begin{bmatrix} 5 & 6 & 6 \\ 0 & 5 & 6 \\ 0 & 0 & 5 \end{bmatrix} =$

$\begin{bmatrix} 1 & 1 & 1 \\ 0 & 1 & 1 \\ 0 & 0 & 1 \end{bmatrix} \begin{bmatrix} 5 & 1 & 0 \\ 0 & 5 & 1 \\ 0 & 0 & 5 \end{bmatrix}$. Here $G_5(A) = \mathbb{C}^3$ and a Jordan basis is $\{ \begin{bmatrix} 1 \\ 0 \\ 0 \end{bmatrix}, \begin{bmatrix} 1 \\ 1 \\ 0 \end{bmatrix}, \begin{bmatrix} 1 \\ 1 \\ 1 \end{bmatrix} \}$. Actually, we could have just chosen the standard basis $\{\mathbf{e}_1, \mathbf{e}_2, \mathbf{e}_3\}$, which is easily checked also to be a Jordan basis. Indeed, $A\mathbf{e}_1 = 5\mathbf{e}_1$, $A\mathbf{e}_2 = 5\mathbf{e}_2 + \mathbf{e}_1$, and $A\mathbf{e}_3 = 5\mathbf{e}_3 + \mathbf{e}_2$. Moreover, it is not difficult to see that the standard basis $\{\mathbf{e}_1, \mathbf{e}_2, \ldots, \mathbf{e}_n\}$ is a Jordan basis for the Jordan block $J_n(\lambda)$. Furthermore, the standard basis is a Jordan basis for any Jordan matrix where each block comes from a Jordan string. To see why more than one Jordan block might be needed for a given eigenvalue, we look at another example.

Consider $A = \begin{bmatrix} 0 & 3 & 0 & 3i \\ 2 & 0 & -2i & 0 \\ 0 & -3i & 0 & 3 \\ 2i & 0 & 2 & 0 \end{bmatrix}$. We find the characteristic polynomial of A is $\chi_A(x) = x^4$ and the minimum polynomial of A is $\mu_A(x) = x^2$ since $A^2 = \mathbb{O}$. Thus, 0 is the only eigenvalue of A. The eigenspace of 0 is $Eig(A, 0) =$

$$= \left\{ c \begin{bmatrix} 0 \\ -i \\ 0 \\ 1 \end{bmatrix} + d \begin{bmatrix} i \\ 0 \\ 1 \\ 0 \end{bmatrix} \mid c, d \in \mathbb{C} \right\} = sp\{\mathbf{x}_1 = \begin{bmatrix} 0 \\ -i \\ 0 \\ 1 \end{bmatrix}, \mathbf{x}_3 = \begin{bmatrix} i \\ 0 \\ 1 \\ 0 \end{bmatrix}\}.$$

The geometric multiplicity of $\lambda = 0$ is 2. Since $A^2 = \mathbb{O}$, the highest a level can be for a Jordan string is 2. Thus, the possibilities are two Jordan strings of length two or one of length two and two of length 1. To find a 2-eigenvector we solve the necessary system of equations and find one choice is $\mathbf{x}_2 = \begin{bmatrix} 0 \\ -i \\ 1/2 \\ 1 \end{bmatrix}$.

Then $\mathbf{x}_2 = \begin{bmatrix} 0 \\ 1 \\ 0 \\ 1 \end{bmatrix}$, $\mathbf{x}_1 = \begin{bmatrix} 2+2i \\ 0 \\ 1-i \\ 0 \end{bmatrix}$ is a Jordan string of length 2. Similarly, we find $\mathbf{x}_4 = \begin{bmatrix} i \\ 0 \\ 1 \\ 1/3 \end{bmatrix}$, $\mathbf{x}_3 = \begin{bmatrix} i \\ 0 \\ 1 \\ 0 \end{bmatrix}$. Now $\{\mathbf{x}_1, \mathbf{x}_2, \mathbf{x}_3, \mathbf{x}_4\}$ turns out to be an independent set, hence is a Jordan basis of $\mathbb{C}^4$. This must happen as the next theorem will show. In other words, $G_0(A)$ is four dimensional with a basis consisting of two Jordan strings of length two.

Now $\begin{bmatrix} 0 & 0 & i & i \\ -i & -i & 0 & 0 \\ 0 & 1/2 & 1 & 1 \\ 1 & 1 & 0 & 1/3 \end{bmatrix}^{-1} A \begin{bmatrix} 0 & 0 & i & i \\ -i & -i & 0 & 0 \\ 0 & 1/2 & 1 & 1 \\ 1 & 1 & 0 & 1/3 \end{bmatrix} =$
$\begin{bmatrix} 0 & 1 & 0 & 0 \\ 0 & 0 & 0 & 0 \\ 0 & 0 & 0 & 1 \\ 0 & 0 & 0 & 0 \end{bmatrix}$. Note the right-hand matrix is not a Jordan block matrix but is composed of two blocks in a block diagonal matrix, $\begin{bmatrix} J_2(0) & \mathbb{O} \\ \mathbb{O} & J_2(0) \end{bmatrix}$.

THEOREM 11.6

Let $S_1 = \{\mathbf{x}_1, \mathbf{x}_2, \dots, \mathbf{x}_q\}$ and $S_2 = \{\mathbf{y}_1, \mathbf{y}_2, \dots, \mathbf{y}_p\}$ be two Jordan strings belonging to the same eigenvalue λ. If the eigenvector $\mathbf{x}_1$ and $\mathbf{y}_1$ are independent, then $S_1 \cup S_2$ is a linearly independent set of vectors.

PROOF The proof is left as an exercise. □

We recall that eigenvectors belonging to different eigenvalues are independent. A similar result holds for generalized eigenvectors.

THEOREM 11.7

Let $S_1 = \{\mathbf{x}_1, \mathbf{x}_2, \dots, \mathbf{x}_q\}$, $S_2 = \{\mathbf{y}_1, \mathbf{y}_2, \dots, \mathbf{y}_p\}$ be Jordan strings belonging to distinct eigenvalues λ_1 and λ_2. Then $S_1 \cup S_2$ is a linearly independent, set.

PROOF To show independence, we do the usual thing. Set $\alpha_1\mathbf{x}_1+\ldots+\alpha_q\mathbf{x}_q+\beta_1\mathbf{y}_1+\ldots+\beta_p\mathbf{y}_p = \overrightarrow{0}$. Apply $(A-\lambda_1 I)^q$. Remember $(A-\lambda_1 I)^q\mathbf{x}_i = \overrightarrow{0}$ for $i = 1, \ldots, q$, so we get $((A-\lambda_1 I)^q)(\beta_1\mathbf{y}_1+\beta_2\mathbf{y}_2+\ldots+\beta_p\mathbf{y}_p) = \overrightarrow{0}$. Next, apply $(A-\lambda_2 I)^{p-1}$. Note $(A-\lambda_2 I)^{p-1}(A-\lambda_1 I)^q = (A-\lambda_1 I)^q(A-\lambda_2 I)^{p-1}$ being polynomials in A. Now $(A-\lambda_2 I)^{p-1}\mathbf{y}_j = \overrightarrow{0}$ for $j = 1, \ldots, p-1(A-\lambda_2 I)\mathbf{y}_1 = \overrightarrow{0}$, so $\beta_p(\lambda_2-\lambda_1)^p\mathbf{y}_1 = \overrightarrow{0}$. Thus, $\beta_p = 0$ since $\lambda_1 \neq \lambda_2$ and $\mathbf{y}_1 \neq \overrightarrow{0}$. Continuing in this manner, we kill off all the βs, leaving $\alpha_1\mathbf{x}_1+\ldots+\alpha_q\mathbf{x}_q = \overrightarrow{0}$. But the **x**s are independent by Theorem 11.4, so all the αs are zero as well and we are done. □

The next step is to give a characterization of a generalized eigenspace more useful than that of Theorem 11.5.

THEOREM 11.8

Let $A \in \mathbb{C}^{n\times n}$ with minimum polynomial $\mu_A(x) = (x-\lambda_1)^{e_1}(x-\lambda_2)^{e_2}\cdots(x-\lambda_s)^{e_s}$ and characteristic polynomial $\chi_A(x) = (x-\lambda_1)^{d_1}(x-\lambda_2)^{d_2}\cdots(x-\lambda_s)^{d_s}$. Then $dim(G_{\lambda_i}(A)) = d_i$ so $G_{\lambda_i}(A) = \mathcal{N}ull((A-\lambda_i I)^{e_i})$.

PROOF Let $\mathbf{x} \in \mathcal{N}ull(A-\lambda_i I)^{e_i}$. Then $(A-\lambda_i I)^{e_i}\mathbf{x} = \overrightarrow{0}$, so $\mathbf{x}$ is in $G_{\lambda_i}(A)$. This says $dim(G_{\lambda_i}(A)) \geq d_i$ since $\mathcal{N}ull(A-\lambda_i I)^{e_i} \subseteq G_{\lambda_i}(A)$, and we determined the dimension of $\mathcal{N}ull(A-\lambda_i I)^{e_i}$ in the primary decomposition theorem. Getting equality is surprisingly difficult. We begin with a Schur triangularization of A using a unitary matrix that puts the eigenvalue λ_i on the main diagonal first in our list of eigenvalues down the main diagonal. More precisely, we know there exists U unitary such that $U^{-1}AU = T$ where T is upper triangular and $T = \begin{bmatrix} \lambda_i & * & * & \cdots & * & \\ 0 & \lambda_i & * & \cdots & * & W \\ 0 & 0 & \lambda_i & \cdots & \vdots & \\ \vdots & \vdots & & \ddots & * & \\ 0 & 0 & \cdots & 0 & \lambda_i & \\ & \mathbb{O} & & & & R \end{bmatrix}$. Since the characteristic polynomials of A and T are identical, λ_i must appear exactly d_i times down the main diagonal. Note R is upper triangular with values different from λ_i on its diagonal. Now consider $T-\lambda_i I = \begin{bmatrix} 0 & * & * & \cdots & * & \\ 0 & 0 & * & \cdots & * & W \\ 0 & 0 & 0 & \cdots & \vdots & \\ \vdots & \vdots & & \ddots & * & \\ 0 & 0 & \cdots & 0 & 0 & \\ & \mathbb{O} & & & & \widetilde{R} \end{bmatrix}$, which has exactly d_i zeros down the diagonal, while $\widetilde{R}$ has no zeros on its main diagonal. Now we look at the standard basis vectors. Clearly, $e_1 \in \mathcal{N}ull(T - \lambda_i I)$ and possibly other standard basis vectors as well, but $e_{d_i+1}, \ldots, e_n$ do not. Any of these basis vectors when multiplied by $T - \lambda_i I$ produce a column of $T - \lambda_i I$ that is not the zero column since, at the very least, $\widetilde{R}$ has a nonzero entry on its diagonal. Now $(T - \lambda_i I)^2$ adds a superdiagonal of zeros in the d_i-by-d_i upper left submatrix (remember how zeros migrate in an upper triangular matrix with zeros on the main diagonal when you start raising the matrix to powers?), so $e_1, e_2 \in \mathcal{N}ull(T - \lambda_i I)^2$ for sure and possibly other standard basis vectors as well, but $e_{d_i+1}, \ldots, e_n$ do not. Continuing to raise $T - \lambda_i I$ to powers, we eventually find $e_1, e_2, \ldots, e_{d_i} \in \mathcal{N}ull(T - \lambda_i I)^{d_i}$ but $e_{d_i+1}, \ldots, e_n$ do not. At this point, the d_i-by-d_i upper left submatrix is completely filled with zeros, so for any power k higher than d_i we see $dim(\mathcal{N}ull(T - \lambda_i I)^k) = d_i$. In particular then, $dim(\mathcal{N}ull(T - \lambda_i I)^n) = d_i$. However, $\mathcal{N}ull((T - \lambda_i I)^n) = G_{\lambda_i}(T)$. But $G_{\lambda_i}(T) = G_{\lambda_i}\left(U^{-1}AU\right) = U^{-1}G_{\lambda_i}(A)$. Since U^{-1} is invertible, $dim(G_{\lambda_i}(A)) = d_i$ as well. □

Now, the primary decomposition theorem gives the following.

COROLLARY 11.3

Let $A \in \mathbb{C}^{n\times n}$ with minimum polynomial $\mu_A(x) = (x-\lambda_1)^{e_1}(x-\lambda_2)^{e_2}\ldots(x-\lambda_s)^{e_s}$ and characteristic polynomial $\chi_A(x) = (x-\lambda_1)^{d_1}(x-\lambda_2)^{d_2}\ldots(x-\lambda_s)^{d_s}$. Then $\mathbb{C}^n = G_{\lambda_1}(A) \oplus G_{\lambda_2}(A) \oplus \cdots \oplus G_{\lambda_s}(A)$. In particular, $\mathbb{C}^n$ has a basis consisting entirely of generalized eigenvectors of A.

Unfortunately, our work is not yet done. We must still argue that we can line a basis of generalized eigenvectors up into Jordan strings that will make a basis for each generalized eigenspace. This we do next. Evidently, it suffices to show we can produce a Jordan basis for each generalized eigenspace, for then we can make a basis for the entire space by pasting bases together.

THEOREM 11.9

Consider the generalized eigenspace $G_\lambda(A)$ belonging to the eigenvalue λ of A. Then $G_\lambda(A) = Z_1 \oplus Z_2 \oplus \cdots \oplus Z_g$, where each Z_i has a basis that is a Jordan string.

Now comes the challenge. To get this theorem proved is no small task. We shall do it in bite-sized pieces, revisiting some ideas that have been introduced earlier. We are focusing on this single eigenvalue λ of A. Suppose $\chi_A(x) = (x-\lambda)^d g(x)$ and $\mu_A(x) = (x-\lambda)^e h(x)$, where λ is not a root of either $g(x)$ or $h(x)$. The construction of Jordan strings sitting over an eigenvalue is complicated by the fact that you cannot just choose an arbitrary basis of $Eig(A, \lambda) =$

$\mathcal{N}ull(A-\lambda I) \subseteq G_\lambda(A)$ and build a string over each basis eigenvector. Indeed, we need a basis of a very special form; namely,

$$\{(A-\lambda I)^{m_1-1}\mathbf{v}_1, (A-\lambda I)^{m_2-1}\mathbf{v}_2, \dots, (A-\lambda I)^{m_g-1}\mathbf{v}_g\},$$

where $g = nlty((A-\lambda I)) = \dim(Eig(A,\lambda))$ is the geometric multiplicity of λ and $m_1 \geq m_2 \geq \cdots \geq m_g \geq 1$ with $m_1+m_2+\cdots+m_g = d = \dim(G_\lambda(A))$. Then the Jordan strings

$$\begin{array}{llll}
\mathbf{v}_1 & \mathbf{v}_2 & & \\
(A-\lambda I)\mathbf{v}_1 & (A-\lambda I)\mathbf{v}_2 & \cdots & \mathbf{v}_g \\
(A-\lambda I)^2\mathbf{v}_1 & (A-\lambda I)^2\mathbf{v}_2 & \vdots & (A-\lambda I)\mathbf{v}_g \\
\vdots & \vdots & \vdots & \vdots \\
(A-\lambda I)^{m_1-1}\mathbf{v}_1 & (A-\lambda I)^{m_2-1}\mathbf{v}_2 & & (A-\lambda I)^{m_g-1}\mathbf{v}_g
\end{array}$$

will form a basis of $G_\lambda(A)$ and each column corresponds to a Jordan block for λ. The ith column will be a basis for Z_i. Now that we know where we are headed, let's begin the journey. There is a clue in the basis vectors we seek; namely,

$$(A-\lambda I)^{m_i-1}\mathbf{v}_i \in \mathcal{N}ull(A-\lambda I) \cap \mathcal{C}ol((A-\lambda I)^{m_i-1}).$$

This suggests we look at subspaces

$$N_k = \mathcal{N}ull(A-\lambda I) \cap \mathcal{C}ol((A-\lambda I)^{k-1}).$$

This we will do, but first we look at the matrix $(A-\lambda I)$. We claim the index of this matrix is e, the multiplicity of λ as a root of the minimal polynomial. Clearly,

$$(\overrightarrow{0}) \subseteq \mathcal{N}ull(A-\lambda I) \subseteq \cdots \subseteq \mathcal{N}ull(A-\lambda I)^e \subseteq \mathcal{N}ull(A-\lambda I)^{e+1} \subseteq \cdots.$$

Our first claim is that

$$\mathcal{N}ull(A-\lambda I)^e = \mathcal{N}ull(A-\lambda I)^{e+1}.$$

Suppose $\mathbf{w} \in \mathcal{N}ull(A-\lambda I)^{e+1}$. Then $(A-\lambda I)^{e+1}\mathbf{w} = \overrightarrow{0}$. But $\mu_A(x) = (x-\lambda)^e h(x)$ where $GCD((x-\lambda)^e, h(x)) = 1$. Thus, there exist $a(x), b(x)$ in $\mathbb{C}[x]$ with

$$\begin{aligned}
1 &= a(x)(x-\lambda)^e + b(x)h(x) \text{ and so} \\
(x-\lambda)^e &= a(x)(x-\lambda)^{2e} + b(x)h(x)(x-\lambda)^e \\
&= a(x)(x-\lambda)^{2e} + b(x)\mu_A(x).
\end{aligned}$$

But then

$$(A-\lambda)^e = a(A)(A-\lambda)^{2e} + b(A)\mu_A(A) = a(A)(A-\lambda)^{2e},$$

and thus

$$(A-\lambda)^e \mathbf{w} = a(A)(A-\lambda)^{e-1}(A-\lambda)^{e+1}\mathbf{w} = \overrightarrow{0},$$

putting $\mathbf{w}$ in $\mathcal{N}ull(A-\lambda I)^e$. We have proved

$$\mathcal{N}ull(A-\lambda I)^e = \mathcal{N}ull(A-\lambda I)^{e+1}.$$

An induction argument establishes that equality persists from this point forward in the chain (exercise). But could there be an earlier equality? Note that $\mu_A(x) = (x-\lambda)^e h(x) = (x-\lambda)[(x-\lambda)^{e-1}h(x)]$ and $(x-\lambda)^{e-1}h(x)$ has degree less than that of the minimal polynomial so that $(A-\lambda)^{e-1}h(A)$ cannot be the zero matrix. Hence there exists some nonzero vector $\mathbf{v}$ with $(A-\lambda)^{e-1}h(A)\mathbf{v} \neq \overrightarrow{0}$ whence $h(A)\mathbf{v} \neq \overrightarrow{0}$. Yet $(A-\lambda I)[(A-\lambda)^{e-1}h(A)\mathbf{v}] = \mu_A(A)\mathbf{v} = \overrightarrow{0}$ so

$$\mathcal{N}ull(A-\lambda I)^{e-1} \subsetneqq \mathcal{N}ull(A-\lambda I)^e.$$

Thus the first time equality occurs is at the power e. This says the index of $(A-\lambda I)$ is e and we have a proper chain of subspaces

$$(\overrightarrow{0}) \subsetneqq \mathcal{N}ull(A-\lambda I) \subsetneqq \cdots \subsetneqq \mathcal{N}ull(A-\lambda I)^e = \mathcal{N}ull(A-\lambda I)^{e+1} = \cdots.$$

Though we have worked with null spaces, our real interest is with the column spaces of powers of $(A-\lambda I)$. We see

$$\begin{aligned}&\mathbb{C}^n \supsetneqq Col(A-\lambda I) \supsetneqq Col((A-\lambda I)^2) \supsetneqq \\ &\cdots \supsetneqq Col((A-\lambda I)^{e-1}) \supsetneqq Col((A-\lambda I)^e) = \cdots.\end{aligned}$$

We can now intersect each of these column spaces with $\mathcal{N}ull(A-\lambda I)$ to get

$$\begin{aligned}&\mathcal{N}ull(A-\lambda I) \supseteq \mathcal{N}ull(A-\lambda I) \cap Col(A-\lambda I) \supseteq \\ &\cdots \supseteq \mathcal{N}ull(A-\lambda I) \cap Col((A-\lambda I)^e) = \cdots.\end{aligned}$$

In other words,

$$N_1 \supseteq N_2 \supseteq \cdots \supseteq N_e = N_{e+1} = \cdots,$$

where we can no longer guarantee the inclusions are all proper. There could be some equalities sprinkled about in this chain of subspaces. If we let $n_i = dim(N_i)$ we have

$$n_1 \geq n_2 \geq \cdots \geq n_e.$$

Next, we recall Corollary 3.2, which says

$$n_i = \dim(N_i) = nlty((A - \lambda I)^i) - nlty((A - \lambda I)^{i-1}$$

so that

$$\begin{aligned} n_1 &= nlty(A - \lambda I) \\ n_2 &= nlty(A - \lambda I)^2 - nlty(A - \lambda I) \\ n_3 &= nlty(A - \lambda I)^3 - nlty(A - \lambda I)^2 \\ &\vdots \\ n_e &= nlty(A - \lambda I)^e - nlty(A - \lambda I)^{e-1} \geq 1 \\ n_{e+1} &= nlty(A - \lambda I)^{e+1} - nlty(A - \lambda I)^e = 0. \end{aligned}$$

From this, it is clear what the sum of the n_is is; namely,

$$\begin{aligned} n_1 + n_2 + \cdots + n_e = nlty(A - \lambda I)^e &= \dim(\mathcal{N}ull(A - \lambda I)^e) \\ &= \dim(G_\lambda(A)) = d. \end{aligned}$$

Thus $(n_1, n_2, \ldots, n_e)$ forms a partition of the integer d. This reminds us of the notion of a conjugate partition we talked about earlier. Recall the Ferrer's diagram:

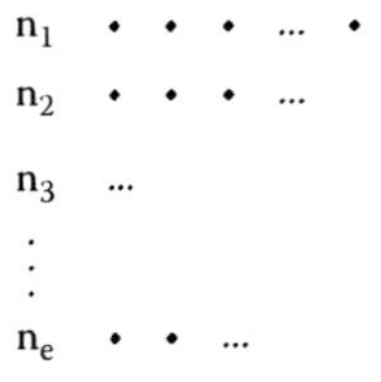

Figure 11.1: Ferrer's diagram.

This diagram has e rows, and the first row has exactly g dots since n_1 is the dimension of $\mathcal{N}ull(A - \lambda I)$. It is time to introduce the conjugate partition $(m_1, m_2, \ldots, m_g)$. This means $m_1 \geq m_2 \geq \cdots \geq m_g \geq 1$ and $m_1 + \cdots + m_g = d$ and $m_1 = e$. It is also clear that $m_1 = m_2 = \cdots = m_{n_e}$.

The next step is to build a basis of $\mathcal{N}ull(A - \lambda I)$ with the help of the conjugate partition. Indeed, we claim there is a basis of $\mathcal{N}ull(A - \lambda I)$ that looks like

$$\{(A - \lambda I)^{m_1 - 1}\mathbf{v}_1, (A - \lambda I)^{m_2 - 1}\mathbf{v}_2, \ldots, (A - \lambda I)^{m_g - 1}\mathbf{v}_g\}.$$

To understand how this is, we start with a basis of N_e and successively extend it to a basis of N_1. Let $\{\mathbf{b}_1, \mathbf{b}_2, \ldots, \mathbf{b}_{n_e}\}$ be a basis of N_e. By definition, the **b**s are eigenvectors of the special form;

$$\mathbf{b}_1 = (A - \lambda I)^{e-1}\mathbf{v}_1, \mathbf{b}_2 = (A - \lambda I)^{e-1}\mathbf{v}_2, \ldots, \mathbf{b}_{n_e} = (A - \lambda I)^{e-1}\mathbf{v}_{n_e}.$$

But remember, $e = m_1 = m_2 = \cdots = m_{n_e}$, so

$$\mathbf{b}_1 = (A - \lambda I)^{m_1 - 1}\mathbf{v}_1, \mathbf{b}_2 = (A - \lambda I)^{m_2 - 1}\mathbf{v}_2, \ldots, \mathbf{b}_{n_e} = (A - \lambda I)^{m_{ne} - 1}\mathbf{v}_{n_e}.$$

Now if $n_e = n_{e-1}$, no additional basis vectors are required at this stage. However, if $n_{e-1} > n_e$, an additional $n_{e-1} - n_e$ basis vectors will be needed. The corresponding "m" values will be $e - 1$ for $n_{e-1} - n_e$ additional basis vectors so that

$$\mathbf{b}_{n_e+1} = (A - \lambda I)^{e-2}\mathbf{v}_{n_e+1} = (A - \lambda I)^{m_{ne+1}-1}\mathbf{v}_{n_e+1}, \quad \text{and so on.}$$

Continuing in this manner, we produce a basis of $\mathcal{N}ull(A-\lambda I)$ of the prescribed form. Next, we focus on the vectors

$$\mathbf{v}_1, \mathbf{v}_2, \ldots, \mathbf{v}_g.$$

The first thing to note is that

$$\mu_{A,\mathbf{v}_i}(x) = (x - \lambda)^{m_i} \quad \text{for } i = 1, 2, \ldots, g.$$

By construction, $(A - \lambda I)^{m_i - 1}\mathbf{v}_i$ belongs to $\mathcal{N}ull(A - \lambda I)$ so $(A - \lambda I)^{m_i}\mathbf{v}_i = \overrightarrow{0}$. This means $\mu_{A,\mathbf{v}_i}(x)$ divides $(x - \lambda)^{m_i}$. However, being a basis vector, $(A - \lambda I)^{m_i - 1}\mathbf{v}_i \neq \overrightarrow{0}$ so we conclude $\mu_{A,\mathbf{v}_i}(x) = (x - \lambda)^{m_i}$.

Next, we characterize the subspaces Z of $G_\lambda(A)$ that correspond to Jordan strings, which in turn correspond to Jordan blocks, giving the Jordan segment belonging to λ in the JCF. Consider

$$Z_{\mathbf{v}} = \{p(A)\mathbf{v} \mid p(x) \in \mathbb{C}[x]\}$$

for any vector $\mathbf{v}$. This is the set of all polynomial expressions in A acting on the vector $\mathbf{v}$. This is a subspace that is A-invariant. Moreover, if $\mathbf{v} \neq \overrightarrow{0}$, $\{\mathbf{v}, A\mathbf{v}, \ldots, A^{k-1}\mathbf{v}\}$ is a basis for $Z_{\mathbf{v}}$, where $k = \deg\mu_{A,v}(x)$. In particular, $\dim(Z_{\mathbf{v}}) = \deg\mu_{A,v}(x)$. This can be established with the help of the division algorithm. Even more is true. If $\mu_{A,v}(x) = (x-\lambda)^k$, then $\{\mathbf{v}, (A-\lambda I)\mathbf{v}, \ldots, (A-\lambda I)^{k-1}\mathbf{v}\}$ is a basis for $Z_{\mathbf{v}}$

We want to consider these subspaces relative to $\mathbf{v}_1, \mathbf{v}_2, \ldots, \mathbf{v}_g$ constructed above. First, we note

$$Z_{\mathbf{v}_i} \subseteq G_\lambda(A) \quad \text{for } i = 1, 2, \ldots, g.$$

From this we conclude

$$Z_{\mathbf{v}_1} + \cdots + Z_{\mathbf{v}_g} \subseteq G_\lambda(A).$$

In view of $\mu_{A,\mathbf{v}_i}(x) = (x - \lambda)^{m_i}$ for $i = 1, 2, \ldots, g$, each $Z_{\mathbf{v}_i}$ has as its basis $\{\mathbf{v}_i, (A - \lambda I)\mathbf{v}_i, \ldots, (A - \lambda I)^{m_i - 1}\mathbf{v}_i\}$. There are two major obstacles to overcome. We must establish that the sum (1) is direct and (2) equals $G_\lambda(A)$. The latter we will do by a dimension argument. The former is done by induction.

Let's do the induction argument first. Assume we have a typical element of the sum set equal to zero:

$$p_1(A)\mathbf{v}_1 + p_2(A)\mathbf{v}_2 + \cdots + p_g(A)\mathbf{v}_g = \overrightarrow{0},$$

where each $p_i(A)\mathbf{v}_i \in Z_{\mathbf{v}_i}$. We would like to conclude $p_i(A)\mathbf{v}_i = \overrightarrow{0}$ for $i = 1, 2, \ldots, g$. This will say the sum is direct. We induct on $m_1 = e$. Suppose $m_1 = 1$. Then, since $m_1 \geq m_2 \geq \cdots \geq m_g \geq 1$, we must be in the case where $m_1 = m_2 = \cdots = m_g = 1$. This says

$$\{(A - \lambda I)^{m_1 - 1}\mathbf{v}_1, (A - \lambda I)^{m_2 - 1}\mathbf{v}_2, \ldots, (A - \lambda I)^{m_g - 1}\mathbf{v}_g\} = \{\mathbf{v}_1, \mathbf{v}_2, \ldots, \mathbf{v}_g\}$$

is an independent set of vectors. By the corollary of the division algorithm called the remainder theorem, we see

$$p_j(x) = (x - \lambda)q_j(x) + r_j(\lambda) \text{ for } j = 1, 2, \ldots, g$$

where $r_j(\lambda)$ is a scalar, possibly zero. Then,

$$\begin{aligned} p_j(A)\mathbf{v}_j &= (A - \lambda I)q_j(A)\mathbf{v}_j + r_j(\lambda)\mathbf{v}_j \\ &= \mu_{A,v_j}(A)q_j(A)\mathbf{v}_j + r_j(\lambda)\mathbf{v}_j \\ &= r_j(\lambda)\mathbf{v}_j. \end{aligned}$$

Thus,

$$p_1(A)\mathbf{v}_1 + p_2(A)\mathbf{v}_2 + \cdots + p_g(A)\mathbf{v}_g = \overrightarrow{0}$$

reduces to

$$r_1(\lambda)\mathbf{v}_1 + r_2(\lambda)\mathbf{v}_2 + \cdots + r_g(\lambda)\mathbf{v}_g = \overrightarrow{0}.$$

which is just a scalar linear combination of the **v**s. By linear independence, all the $r_j(\lambda)$ must be zero. Thus,

$$p_j(x) = (x - \lambda)q_j(x) \text{ for } j = 1, 2, \ldots, g.$$

Therefore,

$$p_j(A) = (A - \lambda I)q_j(A) \text{ for } j = 1, 2, \ldots, g$$

and thus

$$p_j(A)\mathbf{v}_j = (A - \lambda I)q_j(A)\mathbf{v}_j = \overrightarrow{0} \text{ for } j = 1, 2, \ldots, g,$$

as we had hoped.

Now let's assume the independence follows for $m_1 - 1$. We prove independence for m_1. Rather than write out the formal induction, we will look at a concrete example to illustrate the idea. Again, it boils down to showing $\mu_{A,\mathbf{v}_i}(x)$ divides $p_j(x)$. The idea is to push it back to the case $m_1 = 1$. Assume the result holds for $m_1 - 1 = k$. It will be helpful to note $\mu_{A,(A-\lambda I)\mathbf{v}_i}(x) = (x - \lambda)^{m_j - 1}$. For the sake of concreteness, suppose $m_1 = 2, m_2 = 2, m_3 = 1, m_4 = 1$. Then we have $\{(A - \lambda I)\mathbf{v}_1, (A - \lambda I)\mathbf{v}_2, \mathbf{v}_3, \mathbf{v}_4\}$ is an independent set. Suppose

$$p_1(A)\mathbf{v}_1 + p_2(A)\mathbf{v}_2 + p_3(A)\mathbf{v}_3 + p_4(A)\mathbf{v}_4 = \vec{0}.$$

Multiply this equation by $(A - \lambda I)$ and note we can commute this with any polynomial expression in A:

$$\begin{aligned}&p_1(A)(A - \lambda I)\mathbf{v}_1 + p_2(A)(A - \lambda I)\mathbf{v}_2 + p_3(A)(A - \lambda I)\mathbf{v}_3\\ &\quad + p_4(A)(A - \lambda I)\mathbf{v}_4 = \vec{0}.\end{aligned}$$

Since $\mathbf{v}_3$ and $\mathbf{v}_4$ are eigenvectors, this sum reduces to

$$p_1(A)(A - \lambda I)\mathbf{v}_1 + p_2(A)(A - \lambda I)\mathbf{v}_2 = \vec{0}.$$

But $\mu_{A,(A-\lambda I)\mathbf{v}_1}(x) = (x - \lambda)$ and $\mu_{A,(A-\lambda I)\mathbf{v}_2}(x) = (x - \lambda)$, so by the case above, we may conclude that $(x - \lambda)$ divides $p_1(x)$ and $p_2(x)$. Thus, we have $p_1(x) = (x - \lambda)q_1(x)$, $p_2(x) = (x - \lambda)q_2(x)$. Look again at our zero sum:

$$q_1(A)(A - \lambda I)\mathbf{v}_1 + q_2(A)(A - \lambda I)\mathbf{v}_2 + p_3(A)\mathbf{v}_3 + p_4(A)\mathbf{v}_4 = \vec{0}$$

and note

$$\begin{aligned}&\mu_{A,(A-\lambda I)\mathbf{v}_1}(x) = (x - \lambda),\ \mu_{A,(A-\lambda I)\mathbf{v}_2}(x) = (x - \lambda),\ \mu_{A,\mathbf{v}_3}(x)\\ &\quad = (x - \lambda),\ \mu_{A,\mathbf{v}_4}(x) = (x - \lambda).\end{aligned}$$

This is exactly the case we started with, so we can conclude the $(x - \lambda)$ divides $q_1(x)$, $q_2(x)$, $p_3(x)$, and $p_4(x)$. Let's say $q_1(x) = (x - \lambda)h_1(x)$, $q_2(x) = (x - \lambda)h_2(x)$ $p_3(x) = (x - \lambda)h_3(x)$, $p_4(x) = (x - \lambda)h_4(x)$. Then

$$\begin{aligned}p_1(A)\mathbf{v}_1 &= q_1(A)(A - \lambda I)\mathbf{v}_1 = h_1(A)(A - \lambda I)^2\mathbf{v}_1 = \vec{0}\\ p_2(A)\mathbf{v}_2 &= q_2(A)(A - \lambda I)\mathbf{v}_2 = h_2(A)(A - \lambda I)^2\mathbf{v}_2 = \vec{0}\\ p_3(A)\mathbf{v}_3 &= h_3(A)(A - \lambda I)\mathbf{v}_3 = \vec{0}\\ p_4(A) &= h_4(A)(A - \lambda I)\mathbf{v}_4 = \vec{0}.\end{aligned}$$

The general induction argument is left to the reader. Hopefully, the idea is now clear from the example why it works, and it is a notational challenge to write the general argument.

The last step is to get the direct sum to equal the whole generalized eigenspace. This is just a matter of computing the dimension.

$$\begin{aligned}
\dim(Z_{\mathbf{v}_1} \oplus Z_{\mathbf{v}_2} \oplus \cdots \oplus Z_{\mathbf{v}_g}) &= \sum_{i=1}^{g} \dim(Z_{\mathbf{v}_i}) \\
&= \sum_{i=1}^{g} \deg(\mu_{A,\mathbf{v}_i}(x)) \\
&= \sum_{i=1}^{g} m_j \\
&= d \\
&= \dim(G_\lambda(A)).
\end{aligned}$$

Now the general case should be rather clear, though the notation is a bit messy. The idea is to piece together Jordan bases of each generalized eigenspace to get a Jordan basis of the whole space. More specifically, suppose $A \in \mathbb{C}^{n\times n}$ and A has distinct eigenvalues $\lambda_1, \lambda_2, \ldots, \lambda_s$. Suppose the minimal polynomial of A is $\mu_A(x) = (x-\lambda)^{e_1}(x-\lambda)^{e_2}\cdots(x-\lambda)^{e_s}$ and g_i is the geometric multiplicity of λ_i for $i = 1, 2, \ldots, s$. That is, $g_i = dim(\mathcal{N}ull(A - \lambda I))$. The consequence of the discussion above is that there exist positive integers m_{ij} for $i = 1, 2, \ldots, s,\ j = 1, 2, \ldots, g_i$ and vectors $\mathbf{v}_{ij}$ such that $e_i = m_{i1} \geq m_{i2} \geq \cdots \geq m_{ig_i} \geq 1$ and $\mu_{A,\mathbf{v}_{ij}}(x) = (x - \lambda_i)^{m_{ij}}$ such that

$$\mathbb{C}^n = \bigoplus_{i=1}^{s}\bigoplus_{j=1}^{g_i} Z_{\mathbf{v}_{ij}}.$$

Indeed, if we choose the bases

$$\mathcal{B}_{ij} = \{(A - \lambda_i I)^{m_{ij}-1}\mathbf{v}_{ij}, \ldots, (A - \lambda_i I)\mathbf{v}_{ij}, \mathbf{v}_{ij}\}$$

for $Z_{v_{ij}}$ and union these up, we get a Jordan basis

$$\mathcal{B} = \bigcup_{i=1}^{s}\bigcup_{j=1}^{g_i} \mathcal{B}_{ij}$$

for $\mathbb{C}^n$. Moreover, if we list these vectors as columns in a matrix S, we have $A^{-1}AS = J$, where J is a Jordan matrix we call a JCF or *Jordan normal form* of *A(JNF)*. We will consider the uniqueness of this form in a moment but let's consider an example first.

Consider $A = \begin{bmatrix} 0 & 1 & -3i & 0 \\ 0 & 0 & 0 & -3i \\ 3i & 0 & 0 & 1 \\ 0 & 3i & 0 & 0 \end{bmatrix}$. One way or another, we find the characteristic polynomial and factor it:

$$\chi_A(x) = (x-3)^2(x+3)^2.$$

Hence, the eigenvalues of A are -3 and 3. Next, we determine the eigenspaces and the geometric multiplicities in the usual way we do null space calculation:

$$\begin{aligned} Eig(A,3) &= \mathcal{N}ull(A-3I) = \left\{ \begin{bmatrix} a \\ b \\ c \\ d \end{bmatrix} \mid (A-3I) \begin{bmatrix} a \\ b \\ c \\ d \end{bmatrix} = \begin{bmatrix} 0 \\ 0 \\ 0 \\ 0 \end{bmatrix} \right\} \\ &= sp\left\{ \begin{bmatrix} -i \\ 0 \\ 1 \\ 0 \end{bmatrix} \right\} \end{aligned}$$

Thus, the geometric multiplicity of 3 is 1. Similarly,

$$\begin{aligned} Eig(A,-3) &= \mathcal{N}ull(A+3I) = \left\{ \begin{bmatrix} a \\ b \\ c \\ d \end{bmatrix} \mid (A+3I) \begin{bmatrix} a \\ b \\ c \\ d \end{bmatrix} = \begin{bmatrix} 0 \\ 0 \\ 0 \\ 0 \end{bmatrix} \right\} \\ &= sp\left\{ \begin{bmatrix} i \\ 0 \\ 1 \\ 0 \end{bmatrix} \right\}, \end{aligned}$$

so the geometric multiplicity of -3 is also 1. Next, we seek a generalized eigenvector for 3. This means solving

$$A \begin{bmatrix} a \\ b \\ c \\ d \end{bmatrix} = 3 \begin{bmatrix} a \\ b \\ c \\ d \end{bmatrix} + \begin{bmatrix} -i \\ 0 \\ 1 \\ 0 \end{bmatrix}.$$

We find a one parameter family of choices, one being

$$\begin{bmatrix} 0 \\ -i \\ 0 \\ 1 \end{bmatrix}.$$

Thus

$$G_3(A) = sp\left\{\begin{bmatrix} -i \\ 0 \\ 1 \\ 0 \end{bmatrix}, \begin{bmatrix} 0 \\ -i \\ 0 \\ 1 \end{bmatrix}\right\}.$$

Similarly, we find

$$G_{-3}(A) = sp\left\{\begin{bmatrix} i \\ 0 \\ 1 \\ 0 \end{bmatrix}, \begin{bmatrix} 0 \\ i \\ 0 \\ 1 \end{bmatrix}\right\}.$$

Thus

$$\begin{bmatrix} -i & 0 & i & 0 \\ 0 & -i & 0 & i \\ 1 & 0 & 1 & 0 \\ 0 & 1 & 0 & 1 \end{bmatrix}^{-1} A \begin{bmatrix} -i & 0 & i & 0 \\ 0 & -i & 0 & i \\ 1 & 0 & 1 & 0 \\ 0 & 1 & 0 & 1 \end{bmatrix} = \begin{bmatrix} 3 & 1 & 0 & 0 \\ 0 & 3 & 0 & 0 \\ 0 & 0 & -3 & 1 \\ 0 & 0 & 0 & -3 \end{bmatrix}.$$

Finally, we discuss the uniqueness of the JCF. Looking at the example above, we could permute the block for 3 and the block for -3. Hence, there is some wiggle room. Typically, there is no preferred order for the eigenvalues of a matrix. Also, the ones appearing on the superdiagonal could be replaced by any set of nonzero scalars (see Meyers [2000]). Tradition has it to have ones. By the way, some treatments put the ones on the subdiagonal, but again this is only a matter of taste. It just depends on how you order the Jordan bases. Let's make some agreements. Assume all blocks belonging to a given eigenvalue stay together and form a segment, and the blocks are placed in order of decreasing size. The essential uniqueness then is that the number of Jordan segments and the number and sizes of the Jordan blocks is uniquely determined by the matrix. Let's talk through this a bit more. Suppose J_1 and J_2 are Jordan matrices similar to A. Then they are similar to each other and hence have the same characteristic polynomials. This means the eigenvalues are the same and have the same algebraic multiplicity, so the number of times a given eigenvalue appears is the same in J_1 and J_2. Now let's focus on a given eigenvalue λ. The geometric multiplicity $g = dim(\mathcal{N}ull(A - \lambda I))$ determines the number of blocks in the Jordan segment belonging to λ. The largest block in the segment is e, where e is the multiplicity of λ as a root of the minimal polynomial. It seems tantalizingly close to be able to say J_1 and J_2 are the same, except maybe where the segments are placed. However, here is the rub. Suppose the algebraic multiplicity is 4 and the geometric multiplicity is 2. Then λ appears four times and there are two blocks. It could be a 3-by-3 and a 1-by-1 or it could be two 2-by-2s. Luckily, there is a formula for the number of k-by-k blocks in a given segment. This formula only depends on the rank (or nullity) of powers of $(A - \lambda I)$. We have seen this argument before when we characterized nilpotent matrices. It has been a while, so let's sketch it out again. Let's concentrate on the d-by-d λ-segment.

$$\begin{bmatrix} \lambda & & & & \\ & \lambda & & ? & \\ & & \lambda & & \\ & \mathbb{O} & & \ddots & \\ & & & & \lambda \end{bmatrix}.$$

This segment is a block diagonal matrix with the Jordan blocks $J_{m_i}(\lambda)$, where $m_1 \geq m_2 \geq \cdots \geq m_g \geq 1$, say

$$Seg(\lambda) = BlockDiagonal[J_{m_1}(\lambda), J_{m_2}(\lambda), \ldots, J_{m_g}(\lambda)].$$

But then

$$Seg(\lambda) - \lambda I_d = BlockDiagonal[Nilp[m_1], Nilp[m_2], \ldots, Nilp[m_g]] \equiv N.$$

Recall

$$nlty(N^k) = \begin{cases} k & \text{if } 1 \leq k \leq d-1 \\ d & \text{if k} \geq d \end{cases}$$

and so

$$nlty(N^k) - nlty(N^{k-1}) = \begin{cases} 1 & \text{if } 1 \leq k \leq d \\ 0 & \text{if } k > d \end{cases}.$$

Now

$$nlty(Seg(\lambda) - \lambda I_d)^k = \sum_{i=1}^{g} nlty(Nilp[m_i]^k)$$

and so

$$\begin{aligned} &nlty(Seg(\lambda) - \lambda I_d)^k - nlty(Seg(\lambda) - \lambda I_d)^{k-1} \\ &= \sum_{i=1}^{g} nlty(Nilp[m_i]^k) - nlty(Nilp[m_i]^{k-1}) \\ &= \sum_{\substack{i=1 \\ k \leq m_i}}^{g} 1 \end{aligned}$$

.

This difference of nullities thus counts how many blocks have size at least k-by-k since the power k has not killed them off yet. Consequently, the difference

$$[nlty(Seg(\lambda) - \lambda I_d)^k - nlty(Seg(\lambda) - \lambda I_d)^{k-1}] - [nlty(Seg(\lambda) - \lambda I_d)^{k+1} \\ -nlty(Seg(\lambda) - \lambda I_d)^k]$$

counts exactly the number of blocks that are of size k-by-k. This can be restated using ranks:

$$rank(Seg(\lambda) - \lambda I_d)^{k-1}) - 2rank(Seg(\lambda) - \lambda I_d)^k) + rank(Seg(\lambda) - \lambda I_d)^{k+1}).$$

Note that these computations did not depend whether we are in J_1 or J_2, so the number and the sizes of the Jordan blocks in every segment must be the same. Up to ordering the segments, J_1 and J_2 are therefore essentially the same.

Further Reading

[F&I&S, 1979] S. H. Friedberg, A. J. Insel, and L. E. Spence, *Linear Algebra*, Prentice Hall, Englewood Cliffs, NJ, (1979).

[H&K, 1971] K. Hoffman and R. Kunze, *Linear Algebra*, 2nd Edition, Prentice Hall, Englewood Cliffs, NJ, (1979).

[MacDuffee, 1946] C. C. MacDuffee, *The Theory of Matrices*, Chelsea Publishing Company, New York, (1946).

[Perlis, 1958] S. Perlis, *Theory of Matrices*, 2nd Edition, Addison-Wesley, Reading, MA, (1958).

[Väliaho, 1986] H. Väliaho, An Elementary Approach to the Jordan Canonical Form of a Matrix, The American Mathematical Monthly, Vol. 93, (1986), 711–714.

Exercise Set 49

1. How does $\chi_{A^{-1}}$ relate to χ_A? (*Hint:* $\chi_{A^{-1}}(x) = \dfrac{(-x)^n}{det(A)}\chi_A(x)$.)

2. Consider $A = \begin{bmatrix} 1 & 0 \\ 0 & 1 \end{bmatrix}$ and $B = \begin{bmatrix} 1 & 1 \\ 0 & 1 \end{bmatrix}$. Argue that $\chi_A = \chi_B$ but A and B are not similar.

3. Prove all the parts of Theorem 11.5.

4. Prove all the parts of Theorem 11.6.

5. Prove all the parts of Corollary 11.3.

6. Prove all the parts of Theorem 11.9.

7. Fill in the details in all the computational examples in the text.

8. Prove the Cayley-Hamilton theorem using Jordan's theorem.

9. Say everything you can about a matrix A whose JCF is
$$\begin{bmatrix} 5 & 1 & 0 & & & & & & \\ 0 & 5 & 1 & & & & & & \\ & & 5 & & & & & & \\ & & & 5 & 1 & & & & \\ & & & & 5 & & & & \\ & & & & & \pi & 1 & & \\ & & & & & & \pi & & \\ & & & & & & & \sqrt{3} & \\ & & & & & & & & \sqrt{3} \end{bmatrix}.$$

10. Suppose A is a 4-by-4 matrix with eigenvalue 3 of multiplicity 4. List all possible JCFs A might have. How many JCFs can an arbitrary 4-by-4 matrix have? Exhibit them.

11. Prove the claim made about the lack of uniqueness of Jordan strings at the top of page 405.

12. Argue that the standard basis $\{\mathbf{e}_1, \mathbf{e}_2, \ldots, \mathbf{e}_n\}$ is a Jordan basis for the Jordan block $J_n(\lambda)$. Furthermore, the standard basis is a Jordan basis for any Jordan matrix where each block comes from a Jordan string.

13. Argue that a generalized eigenspace for A is A-invariant.

14. If S is invertible, prove that $G_\lambda(S^{-1}AS) = S^{-1}(G_\lambda(A))$.

15. For a vector $\mathbf{v}$, argue that the following two statements are equivalent: (1) there exists a positive integer k with $(A - \lambda I)^k\mathbf{v} = 0$, and (2) there is a sequence $\mathbf{v}_1, \mathbf{v}_2, \ldots, \mathbf{v}_k = \mathbf{v}$ such that $(A - \lambda I)\mathbf{v}_k = \mathbf{v}_{k-1}, (A - \lambda I)\mathbf{v}_{k-1} = \mathbf{v}_{k-2}, \ldots, (A - \lambda I)\mathbf{v}_1 = \overrightarrow{0}$.

16. Suppose $M = sp\{\mathbf{x}_1, A\mathbf{x}_1, \ldots, A^{d-1}\mathbf{x}_1\}$, where $A^d\mathbf{x}_1 = \overrightarrow{0}$ for the first time. Suppose $\mathbf{x}_1$ is a generalized eigenvector of level d for the eigenvalue λ of A. Argue that $\mathbf{w}_1 = (A - \lambda I)^{d-1}\mathbf{x}_1, \mathbf{w}_2 = (A - \lambda I)^{d-2}\mathbf{x}_1, \ldots, \mathbf{w}_d = \mathbf{x}_1$ is a Jordan string that is a basis for M.

17. Do the induction argument indicated in the proof of Theorem 11.9.

18. Argue that $Z_{\mathbf{v}} = \{p(A)\mathbf{v} \mid p(x) \in \mathbb{C}[x]\}$ is a subspace that is A-invariant.

19. Prove that $Z_{\mathbf{v}_i} \subseteq G_{\lambda}(A)$ for $i = 1, 2, \ldots, g$.

11.2 The Smith Normal Form (optional)

There is a more general approach to JCF that extends to more general scalars than complex numbers. For this, we need to be able to work with matrices that have polynomial entries. In symbols, this is $\mathbb{C}[x]^{m\times n}$. The concepts of dealing with scalar matrices extend naturally to matrices with polynomial entries. We define matrix equivalence in the usual way. Two m-by-n matrices $A(x)$ and $B(x)$ in $\mathbb{C}[x]^{m\times n}$ are *equivalent* iff there exist matrices $P(x)$ and $Q(x)$ such that $B(x) = P(x)A(x)Q(x)$, where $P(x)$ and $Q(x)$ are invertible and of appropriate size. Another way to say invertible is $P(x)$ and $Q(x)$ have nonzero scalar determinants. It is easy to see equivalence is indeed an equivalence relation. When dealing with complex matrices, the fundamental result was rank normal form. This said every complex matrix was equivalent to a matrix with ones down the diagonal, as many ones as rank, and zeros elsewhere. When dealing with matrices in $\mathbb{C}[x]^{m\times n}$, there is an analog called *Smith normal form.* This is named for **Henry John Stephen Smith** (2 November 1826–9 February 1883). Actually, Smith was a number theorist and obtained a canonical form for matrices having only integer entries. (Philos. Trans. Roy. Soc. London, 151, (1861), 293-326).It was Frobenius who proved the analogous result for matrices with polynomial entries. (Jour. Reine Angew. Math. (Crelle), 86, (1878), 146-208.) Here we have monic polynomials down the diagonal, as many as rank, each polynomial divides the next and zeros elsewhere.

THEOREM 11.10 (Smith normal form)
Suppose A(x) is in $\mathbb{C}[x]^{m\times n}$ of rank r. Then there is a unique matrix SNF(A(x)) in $\mathbb{C}[x]^{m\times n}$ equivalent to A(x) where SNF(A(x)) is a diagonal, matrix with monic polynomials $s_1(x)$, $s_2(x)$, ... ,$s_r(x)$ on the diagonal, where $s_i(x)$ is divisible by $s_{i-1}(x)$ for $i = 2, \ldots, r$.

PROOF We argue the existence first. Note that the transvections, dilations, and permutation matrices do the same work that they did for scalar matrices, even though we are now allowing polynomial entries in our matrices. Indeed, only the transvections will need polynomial entries in our proof. We note the determinant of all these matrices is a nonzero scalar so, just as before, they are all invertible. The proof goes by induction on m and n. The case $m = n = 1$ is clear, so we consider the case $m = 1, n > 1$. In this case, $A(x) = [a_1(x)\ a_2(x) \cdots a_n(x)]$. If all the $a_i(x)$s are zero, we are done, so let's assume otherwise. Then there must be an $a_i(x)$ with minimal degree. Use an elementary column operation if necessary and put that polynomial in the first position. In other words, we may assume $a_1(x)$ has minimal degree. Now, using elementary matrices, we can replace all other $a_j(x)$s by zero. The key is the division algorithm for polynomials. Take any nonzero $a_j(x)$ in $A(x)$ other than $a_1(x)$. Divide $a_j(x)$ by $a_1(x)$ and get $a_j(x) = q_j(x)a_1(x) + r_j(x)$, where $r_j(x)$ is zero or its degree is strictly less than $deg(a_1(x))$. Multiply the first column by $-q_j(x)$ and add the result to the jth column. That produces $r_j(x)$ in the jth position. Then, if $r_j(x) = 0$, we are happy. If not, swap $r_j(x)$ to the first position. If there still remain nonzero entries, go through the same procedure again (i.e., divide the nonzero entry by $r_j(x)$ and multiply that column by minus the quotient and add to the column of the nonzero entry producing another remainder). Again, if the remainder is zero we are done; if not, go again. Since the degrees of the remainders are strictly decreasing, this process cannot go on forever. It must terminate in a finite number of steps. In fact, this process has no more than $deg(a_1(x))$ steps. This completes the induction since we have all the entries zero except the first. A dilation may be needed to produce a monic polynomial.

The case $m > 1$ and $n = 1$ is similar and is left as an exercise.

Now assume m and n are greater than 1. Suppose the theorem is true for matrices of size (m-1)-by-(n-1). We may assume that the (1, 1) entry of $A(x)$ is nonzero with minimal degree among the nonzero entries of $A(x)$. After all, if $A(x)$ is zero we are done. Then if not, row and column swaps can be used if necessary. Now, using the method described above, we can reduce $A(x)$ using a finite number of elementary matrices to $A_1(x) = \begin{bmatrix} a_{11}^{(1)}(x) & 0 & \cdots & 0 \\ 0 & a_{22}^{(1)}(x) & \cdots & a_{2n}^{(1)}(x) \\ \vdots & \vdots & & \vdots \\ 0 & a_{m2}^{(1)}(x) & \cdots & a_{mn}^{(1)}(x) \end{bmatrix}$.

We would like to get all the entries divisible by $a_{11}^{(1)}(x)$. If for some $i, j > 1$, $a_{ij}^{(1)}(x)$ that is not zero is not divisible by $a_{11}^{(1)}(x)$, then add the ith row to the first row and apply the procedure above again. Then we get a matrix $A_2(x) = \begin{bmatrix} a_{11}^{(2)}(x) & 0 & \cdots & 0 \\ 0 & a_{22}^{(2)}(x) & \cdots & a_{2n}^{(2)}(x) \\ \vdots & \vdots & & \vdots \\ 0 & a_{m2}^{(2)}(x) & \cdots & a_{mn}^{(2)}(x) \end{bmatrix}$ where the degree of $a_{11}^{(2)}(x)$ is strictly less than the degree of $a_{11}^{(1)}(x)$. If there is still an entry $a_{ij}^{(2)}(x)$ not divisible by $a_{11}^{(2)}(x)$, repeat the process again. In a finite number of steps, we must produce a matrix

$$A_3(x) = \begin{bmatrix} a_{11}^{(3)}(x) & 0 & \cdots & 0 \\ 0 & a_{22}^{(3)}(x) & \cdots & a_{2n}^{(3)}(x) \\ \vdots & \vdots & & \vdots \\ 0 & a_{m2}^{(3)}(x) & \cdots & a_{mn}^{(3)}(x) \end{bmatrix}$$ where every entry is divisible by $a_{11}^{(3)}(x)$. We can use a dilation if necessary to make $a_{11}^{(3)}(x)$ monic. Now the induction hypothesis applies to the lower-right corner $\begin{bmatrix} a_{22}^{(3)}(x) & \cdots & a_{2n}^{(3)}(x) \\ \vdots & & \vdots \\ a_{m2}^{(3)}(x) & \cdots & a_{mn}^{(3)}(x) \end{bmatrix}$ and we are essentially done with the existence argument. ▯

Before we prove the uniqueness of the Smith normal form, we recall the idea of a minor. Take any m-by-n matrix A. A *minor of order k* is obtained by choosing k rows and k columns and forming the determinant of the resulting square matrix. Now the minors of a matrix with polynomial entries are polynomials so we can deal with their greatest common divisors (GCDs).

THEOREM 11.11

Let $g_k(x)$ denote the GCD of the order k minors of $A(x)$ and $h_k(x)$ denote the GCD of the order k minors of $B(x)$. Suppose $A(x)$ is equivalent to $B(x)$. Then $g_k(x) = h_k(x)$ for all k.

PROOF Suppose $A(x)$ is equivalent to $B(x)$. Then there exist invertible $P(x)$ and $Q(x)$ such that $B(x) = P(x)A(x)Q(x)$. $P(x)$ and $Q(x)$ are just products of elementary matrices, so we argue by cases.

Suppose $B(x) = E(x)A(x)$, where $E(x)$ is an elementary matrix. We consider the three cases. Let $R(x)$ be an i-by-i minor of $A(x)$ and $S(x)$ the i-by-i minor of $E(x)A(x)$ in the same position. Suppose $E(x) = P_{ij}$, a swap of rows. The effect on $A(x)$ is (1) to leave $R(x)$ unchanged or (2) to interchange two rows of $R(x)$, or (3) to interchange a row of $R(x)$ with a row not in $R(x)$. In case (1) $S(x) = R(x)$; in case (2), $S(x) = -R(x)$; in case(3), $S(x)$ is except possibly for a sign, another i-by-i minor of $A(x)$.

Next, suppose $E(x)$ is a dilation $D_i(x)$ where α is a nonzero scalar. Then either $S(x) = R(x)$ or $S(x) = \alpha R(x)$. Lastly, consider a transvection $E(x) = T_{ij}(f(x))$. The effect on $A(x)$ is (1) to leave $R(x)$ unchanged, (2) to increase one of the rows of $R(x)$ by $f(x)$ times another of row of $R(x)$, or (3) to increase one of the rows of $R(x)$ by $f(x)$ times a row not of $R(x)$. In cases (1) and (2), $S(x) = R(x)$; in case (3), $S(x) = R(x) \pm f(x)C(x)$, where $C(x)$ is an i-by-i minor of $A(x)$.

Thus any i-by-i minor of $E(x)$ is a linear combination of i-by-i minors of $A(x)$. If $g(x)$ is the GCD of all i-by-i minors of $A(x)$ and $h(x)$ is the GCD of all

i-by-i minors of $E(x)A(x)$, then $g(x)$ divides $h(x)$. Now $A(x) = E(x)^{-1}B(x)$ and $E^{-1}(x)$ is a product of elementary matrices, so by a symmetric argument, $h(x)$ divides $g(x)$. Since these are monic polynomials, $g(x) = h(x)$.

Next, suppose $B(x) = E(x)A(x)F(x)$, where $E(x)$ and $F(x)$ are products of elementary matrices. Let $C(x) = E(x)A(x)$ and $D(x) = C(x)F(x)$. Since $D(x)^T = F(x)^T C(x)^T$ and $F(x)^T$ is a product of elementary matrices, the GCD of all i-by-i minors of $D(x)^T$ is the GCD of all i-by-i minors of $C(x)^T$. But the GCD of all i-by-i minors of $D(x)$.The same is true for $C(x)^T$ and $C(x)$ so the GCD of all i-by-i minors of $B(x) = E(x)A(x)F(x)$ is the GCD of all i-by-i minors of $A(x)$. □

We are now in a position to argue the uniqueness of the Smith normal form.

THEOREM 11.12

Suppose A(x) is in $\mathbb{C}[x]^{m\times n}$ of rank r. Let $g_k(x)$ denote the GCD of the order k minors of A(x). Let $g_0(x) = 1$ and $diag[s_1(x), s_2(x), \ldots, s_r(x), 0, \ldots, 0]$ be a Smith normal form of A(x). Then r is the maximal integer with $g_r(x)$ nonzero and $s_i(x) = \dfrac{g_i(x)}{g_{i-1}(x)}$ for i = 1, 2, ..., r.

PROOF We begin by arguing that $A(x)$ is equivalent to $diag[s_1(x), s_2(x), \ldots, s_r(x), 0, \ldots, 0]$. These two matrices have the same GCD of minors of order k by the theorem above. Except for the diagonal matrix, these minors are easy to compute. Namely, $g_k(x) = s_1(x)s_2(x)\cdots s_k(x)$ for $k = 1,2, \ldots, r$. Thus, the $s_i(x)$'s are uniquely determined by $A(x)$. □

The polynomials $s_1(x), s_2(x), \ldots, s_r(x)$ are called the *invariant factors* of $A(x)$.

THEOREM 11.13

A(x) and B(x) in $\mathbb{C}[x]^{m\times n}$ are equivalent iff they have the same invariant factors.

PROOF The proof is left as an exercise. □

Before we go any further, let's look at an example.

Example 11.1

Let $A(x) = \begin{bmatrix} x & x-1 & x+2 \\ x^2+x & x^2 & x^2+2x \\ x^2-2x & x^2-3x+2 & x^2+x-3 \end{bmatrix}$

$$A(x)T_{12}(-1) = \begin{bmatrix} 1 & x-1 & x+2 \\ x & x^2 & x^2 \\ x-2 & x^2-3x+2 & x^2+x-3 \end{bmatrix}$$

$$T_{31}(-x+2)T_{21}(-x)A(x)T_{12}(-1) = \begin{bmatrix} 1 & x-1 & x-12 \\ 0 & x & 0 \\ 0 & 0 & x+1 \end{bmatrix}$$

$$T_{31}(-x+2)T_{21}(-x)A(x)T_{12}(-1)T_{21}(-x+1)T_{31}(-x-2) =$$

$$\begin{bmatrix} 1 & 0 & 0 \\ 0 & x & 0 \\ 0 & 0 & x+1 \end{bmatrix}$$

$$T_{23}(-1)T_{31}(-x+2)T_{21}(-x)A(x)T_{12}(-1)T_{21}(-x+1)T_{31}(-x-2)$$

$$= \begin{bmatrix} 1 & 0 & 0 \\ 0 & x & -x-1 \\ 0 & 0 & x+1 \end{bmatrix}$$

$$T_{23}(-1)T_{31}(-x+2)T_{21}(-x)A(x)T_{12}(-1)T_{21}(-x+1)T_{31}(-x-2)T_{23}(1)$$

$$= \begin{bmatrix} 1 & 0 & 0 \\ 0 & -1 & -x-1 \\ 0 & x+1 & x+1 \end{bmatrix}$$

$T_{32}(x+1)T_{23}(-1)T_{31}(-x+2)T_{21}(-x)A(x)T_{12}(-1)T_{21}(-x+1)T_{31}(-x-2)T_{23}(1) =$

$$\begin{bmatrix} 1 & 0 & 0 \\ 0 & 1 & x+1 \\ 0 & 0 & -x^2-x \end{bmatrix}$$

$D_2(-1)T_{32}(x+1)T_{23}(-1)T_{31}(-x+2)T_{21}(-x)A(x)T_{12}(-1)T_{21}(-x+1)T_{31}(-x-2)T_{23}T_{32}(-x-1)D_3(-1)$

$$= \begin{bmatrix} 1 & 0 & 0 \\ 0 & 1 & 0 \\ 0 & 0 & x(x+1) \end{bmatrix} = SNF(A(x)).$$

There are other polynomials that can be associated with $A(x)$.

DEFINITION 11.7

(elementary divisors)

Let $A(x) \in \mathbb{C}[x]^{n\times n}$. Write each invariant factor in its prime factorization over $\mathbb{C}$, say $s_i(x) = (x-\lambda_{i1})^{e_{i1}}(x-\lambda_{i2})^{e_{i2}}\dots(x-\lambda_{ik_i})^{e_{ik_i}}$, $i = 1, 2, \dots, r$. However, some of the e_{ij} may be zero since $s_i(x)$ divides $s_{i+1}(x)$, $e_{i+1j} \geq e_{ij}$ $i = 1, 2, \dots, r-1$, $j = 1, 2, \dots, k_i$. The nontrivial factors $(x-\lambda_{ij})^{e_{ij}}$ are called the elementary divisors of $A(x)$ over $\mathbb{C}$.

Example 11.2
Suppose

$SNF(A(x)) =$

$$\begin{bmatrix} 1 & 0 & 0 & 0 & 0 \\ 0 & 1 & 0 & 0 & 0 \\ 0 & 0 & (x-1)(x^2+1) & 0 & 0 \\ 0 & 0 & 0 & (x-1)(x^2+1)^2x & 0 \\ 0 & 0 & 0 & 0 & (x-1)^2(x^2+1)^2x^2(x^2-5) \end{bmatrix}.$$

The invariant factors are $s_1(x) = 1$, $s_2(x) = 1$, $s_3(x) = (x-1)(x^2+1)$, $s_4(x) = (x-1)(x^2+1)^2x$, *and* $s_5(x) = (x-1)^2(x^2+1)^2x^2(x^2-5)$. Now the elementary divisors over the complex field $\mathbb{C}$ are $(x-1)^2, x-1, x-1, (x+i)^2, (x+i)^2, x+i, (x-i)^2, (x-i)^2, x-i, x^2, x, x-\sqrt{5}, x+\sqrt{5}$.

Note that the invariant factors determine the rank and the elementary divisors. Conversely, the rank and elementary divisors determine the invariant factors and, hence, the Smith normal form. To see how this goes, suppose $\lambda_1, \lambda_2, \ldots, \lambda_p$ are the distinct complex numbers appearing in the elementary divisors. Let $(x-\lambda_i)^{e_{i1}}, \ldots, (x-\lambda_i)^{e_{ik_k}}$ be the elementary divisors containing λ_i. Agree to order the degrees $e_{i1} \geq \cdots \geq e_{ik_k} > 0$. The number r of invariant factors must be greater or equal to $max(k_1, \ldots, k_p)$. The invariant factors can then be reconstructed by the following formula:

$$s_j(x) = \prod_{i=1}^{p}(x-\lambda_i)^{e_i r+1-j} \text{ for } j = 1, 2, \ldots, r$$

where we agree $(x-\lambda_i)^{e_{ij}} = 1$ when $j > k_i$.

We can learn things about scalar matrices by using the following device. Given a scalar matrix $A \in \mathbb{C}^{n\times n}$, we can associate a matrix with polynomial entries in $\mathbb{C}[x]^{n\times n}$; this is the *characteristic matrix* $xI - A$. So if $A = [a_{ij}] \in \mathbb{C}^{n\times n}$, then

$$xI - A = \begin{bmatrix} x-a_{11} & -a_{12} & \cdots & \\ -a_{21} & x-a_{22} & \cdots & \\ \vdots & \vdots & & \\ -a_{n1} & \vdots & & x-a_{nn} \end{bmatrix}$$ in $\mathbb{C}[x]^{n\times n}$. Of course, the determinant of $xI - A$ is just the characteristic polynomial of A. The main result here is a characterization of the similarity of scalar matrices.

THEOREM 11.14
For A and B in $\mathbb{C}^{n\times n}$, *the following statements are equivalent:*

1. *A and B are similar.*

2. *$xI - A$ and $xI - B$ are equivalent.*

3. *$xI - A$ and $xI - B$ have the same invariant factors.*

PROOF Suppose A and B are similar. Then there exists an invertible matrix S with $B = S^{-1}AS$. Then it is easy to see $S^{-1}(xI - A)S = xI - B$. Conversely, suppose $P(x)$ and $Q(x)$ are invertible with $P(x)(xI - A) = (xI - B)Q(x)$. By dividing (carefully), write $P(x) = (xI - B)P_1(x) + R_1$ and $Q(x) = Q_1(x)(xI - A) + R_2$, where R_1 and R_2 are scalar matrices. Then, by considering degree, we conclude $P_1(x) - Q_1(x) = 0$. Therefore, $R_1 = R_2$ and so $R_1 A = BR_1$. It remains to prove R_1 is invertible. Suppose $S(x)$ is the inverse to $P(x)$. Write $S(x) = (xI - A)Q_2(x) + C$, where C is a scalar matrix. Now $I = (xI - B)Q_3(x) + R_1 C$ since $R_1 A = BR_1$ and $P(x)S(x) = I$. Note $Q_3(x) = P_1(x)(xI - A)Q_2(x) + P_1(x)C + R_1 Q_1(x)$. Now, by considering degrees, conclude $Q_3(x)$ is zero. Thus $R_1 C = I$ and we are done. ▯

We leave it as an exercise that Jordan's theorem follows from the existence and uniqueness of the Smith normal form.

THEOREM 11.15 (Jordan's theorem)
If A is a square complex matrix, then A is similar to a unique Jordan matrix (up to permutation of the blocks).

PROOF The proof is left as an exercise. The uniqueness comes from the uniqueness of the Smith normal form of $xI - A$. ▯

Exercise Set 50

1. With the notation from above, argue that the number of elementary divisors of $A(x)$ is $\sum_{i=1}^{r} k_i$.

2. Suppose $A(x)$ is invertible in $\mathbb{C}[x]^{n\times n}$. Argue that $det(A(x))$ is a nonzero constant and the converse. (Hint: Look at $A(x)B(x) = I$ and take determinants of both sides.)

3. Argue that $A(x)$ is invertible in $\mathbb{C}[x]^{n\times n}$ iff $A(x)$ is a product of elementary matrices in $\mathbb{C}[x]^{n\times n}$.

4. Prove that the characteristic polynomial of $A \in \mathbb{C}^{n\times n}$ is the product of the invariant factors of $xI - A$.

5. Prove that the minimum polynomial of $A \in \mathbb{C}^{n\times n}$ is the invariant factor of $xI - A$ of highest degree.

6. Prove that $A \in \mathbb{C}^{n\times n}$ is similar to a diagonal matrix iff $xI - A$ has linear elementary divisors in $\mathbb{C}[x]$.

7. Prove that if D is a diagonal matrix, the elementary divisors of $xI - D$ are its diagonal elements.

8. Argue that matrix equivalence in $\mathbb{C}[x]^{m\times n}$ is an equivalence relation.

9. Prove Theorem 11.13.

10. Prove Theorem 11.15.

Further Reading

[Brualdi, 1987] Richard A. Brualdi, The Jordan Canonical Form: An Old Proof, The American Mathematical Monthly, Vol. 94, No. 3, 257–267, (1987).

[Filippov, 1971] A. F. Filippov, A Short Proof of the Theorem on Reduction of a Matrix to Jordan Form, Vestnik, Moscow University, No. 2, 18–19, (1971). (Also Moscow University Math. Bull., 26, 70–71, (1971).)

[F&S, 1983] R. Fletcher and D. Sorensen, An Algorithmic Derivation of the Jordan Canonical Form, The American Mathematical Monthly, Vol. 90, No. 1, 12–16, (1983).

[Gantmacher,1959] F. R. Gantmacher, *The Theory of Matrices,* Vol. 1, Chelsea Publishing Company, New York, (1959).

[G&W, 1981] A. Galperin and Z. Waksman, An Elementary Approach to Jordan Theory, The American Mathematical Monthly, Vol. 87, 728–732, (1981).

[G&L&R, 1986] I. Gohberg, P. Lancaster, and L. Rodman, *Invariant Subspaces of Matrices with Applications*, John Wiley & Sons, New York, (1986).

[H&J, 1986] R. Horn and C. R. Johnson, *Introduction to Matrix Analysis*, Cambridge University Press, Cambridge, (1986).

[Jordan, 1870] C. Jordan, *Traité des Substitutions et des Équations Algébriques*, Paris, (1870), 125.

[L&T, 1985] Peter Lancaster and Miron Tismenetsky, *The Theory of Matrices: With Applications*, 2nd Edition, Academic Press, Orlando, (1985).

[Noble, 1969] Ben Noble, *Applied Linear Algebra,* Prentice Hall, Englewood Cliffs, NJ (1969).

[Sobczk, 1997] Garret Sobczyk, The Generalized Spectral Decomposition of a Linear Operator, The College Mathematics Journal, Vol. 28, No. 1, January, (1997), 27–38.

[Strang, 1980] Gilbert Strang, *Linear Algebra and Its Applications*, 2nd Edition Academic Press, New York, (1980).

[T&A 1932] H. W. Turnbull and A. C. Aitken, *An Introduction to the Theory of Canonical Matrices*, Blackie & Son, London, (1932).

Chapter 12

Multilinear Matters

bilinear map, bilinear form, symmetric, skew-symmetric, nondegenerate, quadratic map, alternating

12.1 Bilinear Forms

In this section, we look at a generalization of the idea of an inner product. Let V_1, V_2, and W be vector spaces over $\mathbb{R}$ or $\mathbb{C}$, which we denote by $\mathbb{F}$ when it does not matter which scalars are being used.

DEFINITION 12.1 *(bilinear map)*
*A **bilinear map** φ is a function $\varphi : V_1 \times V_2 \longrightarrow W$ such that*

1. $\varphi(\mathbf{x}+\mathbf{y}, \mathbf{z}) = \varphi(\mathbf{x}, \mathbf{z}) + \varphi(\mathbf{y}, \mathbf{z})$ *for all* $\mathbf{x}, \mathbf{y} \in V_1$ *all* $\mathbf{z} \in V_2$
2. $\varphi(\alpha\mathbf{x}, \mathbf{z}) = \alpha\varphi(\mathbf{x}, \mathbf{z})$ *for all* $\alpha \in \mathbb{F}$ *for all* $\mathbf{x} \in V_1$ *all* $\mathbf{z} \in V_2$
3. $\varphi(\mathbf{x}, \mathbf{y}+\mathbf{z}) = \varphi(\mathbf{x}, \mathbf{y}) + \varphi(\mathbf{x}, \mathbf{z})$ *for all* $\mathbf{x} \in V_1$*,all* $\mathbf{y}, \mathbf{z} \in V_2$
4. $\varphi(\mathbf{x}, \beta\mathbf{y}) = \varphi(\mathbf{x}, \mathbf{y})\beta$ *for all* $\beta \in \mathbb{F}$*, all* $\mathbf{x} \in V_1$*, all* $\mathbf{y} \in V_2$.

*In particular, if $V_1 = V_2$ and $W = \mathbb{F}$, we traditionally call φ a **bilinear form.** We denote $L^2(V_1, V_2; W)$ to be the set of all bilinear maps on V_1 and V_2 with values in W. We write $L^2(V; W)$ for $L^2(V, V; W)$.*

We note that the name "bilinear" makes sense, since a bilinear map is linear in each of its variables. More precisely, if we fix $\mathbf{y}$ in V_2, the map $d_\varphi(\mathbf{y}) : V_1 \rightarrow W$ defined by $d_\varphi(\mathbf{y})(\mathbf{x}) = \varphi(\mathbf{x}, \mathbf{y})$ is linear and the map $s_\varphi(\mathbf{x})$ defined by $s_\varphi(\mathbf{x})(\mathbf{y}) = \varphi(\mathbf{x}, \mathbf{y})$ is linear for each fixed $\mathbf{x}$ in V_1. Here $s_\varphi(\mathbf{x}) : V_2 \longrightarrow W$.

We also note that the zero map $\mathbb{O}(\mathbf{x}, \mathbf{y}) = \overrightarrow{0}$ is a bilinear map and any linear combination of bilinear maps is again bilinear. This says that $L^2(V_1, V_2; W)$ is a vector space over $\mathbb{F}$ in its own right. Now let's look at some examples.

Example 12.1

1. Let V_1 and V_2 be any vector spaces over $\mathbb{F}$ and let $f : V_1 \longrightarrow \mathbb{F}$ and $g : V_2 \longrightarrow \mathbb{F}$ be linear maps (i.e., linear functionals). Then $\varphi(\mathbf{x}, \mathbf{y}) = f(\mathbf{x})g(\mathbf{y})$ is a bilinear form on V_1 and V_2.

2. Fix a matrix A in $\mathbb{F}^{m\times n}$. Define $\varphi_A(X, Y) = X^T AY$, where $X \in \mathbb{F}^{m\times q}$ and $Y \in \mathbb{F}^{n\times p}$. Then $\varphi_A \in L^2(\mathbb{F}^{m\times q}, \mathbb{F}^{n\times p}; \mathbb{F}^{q\times p})$. We can make φ_A into a bilinear form by modifying the definition to $\varphi_A(X, Y) = tr(X^T AY)$.

3. If we take $q = p = 1$ in (2) above, we get

$$\varphi_A\left(\begin{bmatrix} x_1 \\ \vdots \\ x_m \end{bmatrix}, \begin{bmatrix} y_1 \\ \vdots \\ y_n \end{bmatrix}\right)$$
$$= [x_1 \cdots x_m] \begin{bmatrix} a_{11} & a_{12} & \cdots & a_{1m} \\ a_{21} & a_{22} & \cdots & a_{2m} \\ \vdots & \vdots & & \vdots \\ a_{m_1} & a_{m_2} & \cdots & a_{mn} \end{bmatrix} \begin{bmatrix} y_1 \\ \vdots \\ y_n \end{bmatrix}$$
$$= \mathbf{x}^T A\mathbf{y} = \sum_{i=1}^{m}\sum_{j=1}^{n} a_{ij}x_i y_i \in \mathbb{F}.$$

In gory detail,

$$\varphi_A\left(\begin{bmatrix} x_1 \\ \vdots \\ x_m \end{bmatrix}, \begin{bmatrix} y_1 \\ \vdots \\ y_n \end{bmatrix}\right) = \begin{matrix} a_{11}x_1y_1 + a_{12}x_1x_2y_2 + & \cdots & +a_{1n}x_1y_n \\ +a_{21}x_2y_1 + a_{22}x_2y_2 + & \cdots & +a_{2n}x_2y_n \\ +\cdots & & \\ \vdots & & \\ +a_{m1}x_my_1 + a_{m2}x_my_2 + & \cdots & +a_{mn}x_my_n \end{matrix}$$

To be even more concrete,

$$\varphi_A\left(\begin{bmatrix} x_1 \\ x_2 \end{bmatrix}, \begin{bmatrix} y_1 \\ y_2 \\ y_3 \end{bmatrix}\right)$$
$$= [x_1x_2] \begin{bmatrix} a_{11} & a_{12} & a_{13} \\ a_{21} & a_{22} & a_{23} \end{bmatrix} \begin{bmatrix} y_1 \\ y_2 \\ y_3 \end{bmatrix}$$
$$= \begin{bmatrix} x_1a_{11} + x_2a_{21} & x_1a_{12} + x_2a_{22} & x_1a_{13} + x_2a_{23} \end{bmatrix} \begin{bmatrix} y_1 \\ y_2 \\ y_3 \end{bmatrix}$$
$$= (x_1a_{11} + x_2a_{21})\, y_1 + (x_1a_{12} + x_2a_{22})\, y_2 + (x_1a_{13} + x_2a_{23})\, y_3$$
$$= a_{11}x_1y_1 + a_{21}x_2y_1 + a_{12}x_1y_2 + a_{22}x_2y_2 + a_{13}x_1y_3 + a_{23}x_2y_3$$
$$= (a_{11}x_1y_1 + a_{12}x_1y_2 + a_{13}x_1y_3) + (a_{21}x_2y_1 + a_{22}x_2y_2 + a_{23}x_2y_3).$$

We spent extra time on this example because it arises often in practice. As an extreme case, $\varphi_z : \mathbb{F} \times \mathbb{F} \to \mathbb{F}$ by $\varphi_z(x, y) = xzy$ is a bilinear form for each fixed scalar z.

Next we collect some elementary computational facts about bilinear forms.

THEOREM 12.1

Let φ be a bilinear form on V. Then

1. $\varphi(\vec{0}, \mathbf{y}) = 0 = \varphi(\mathbf{x}, \vec{0}) = \varphi(\vec{0}, \vec{0})$ *for all* $\mathbf{x} \in V$
2. $\varphi(\alpha\mathbf{x} + \beta\mathbf{y}, \mathbf{z}) = \alpha\varphi(\mathbf{x}, \mathbf{z}) + \beta\varphi(\mathbf{y}, \mathbf{z})$ *for all* $\alpha, \beta \in \mathbb{F}$, *all* $\mathbf{x}, \mathbf{y}, \mathbf{z} \in V$
3. $\varphi(\sum_{i=1}^{n} \alpha_i \mathbf{x}_i, \mathbf{y}) = \sum_{i=1}^{n} \alpha_i \varphi(\mathbf{x}_i, \mathbf{y})$ *for all* $\alpha_i \in \mathbb{F}$, *all* $\mathbf{x}_i, \mathbf{y} \in V$
4. $\varphi(\mathbf{x}, \alpha\mathbf{y} + \beta\mathbf{z}) = \varphi(\mathbf{x}, \mathbf{y})\alpha + \varphi(\mathbf{x}, \mathbf{z})\beta = \alpha\varphi(\mathbf{x}, \mathbf{y}) + \beta\varphi(\mathbf{x}, \mathbf{z})$ *for all* $\alpha, \beta \in \mathbb{F}$ *all* $\mathbf{x}, \mathbf{y}, \mathbf{z} \in V$
5. $\varphi(\mathbf{x}, \sum_{j=1}^{m} \beta_j \mathbf{y}_j) = \sum_{j=1}^{m} \varphi(\mathbf{x}, \mathbf{y}_j)\beta_j$ *for all* $\beta_j \in \mathbb{F}$, *all* $\mathbf{y}_j, \mathbf{x} \in V$
6. $\varphi(\alpha\mathbf{x} + \beta\mathbf{y}, \gamma\mathbf{w} + \delta\mathbf{z}) = \alpha\varphi(\mathbf{x}, \mathbf{w})\gamma + \alpha\varphi(\mathbf{x}, \mathbf{z})\delta + \beta\varphi(\mathbf{y}, \mathbf{w})\gamma + \beta\varphi(\mathbf{y}, \mathbf{z})\delta$ *for all* $\alpha, \beta, \gamma, \delta$ *in* $\mathbb{F}$ *and all* $\mathbf{x}, \mathbf{y}, \mathbf{z}, \mathbf{w}$ *in* V
7. $\varphi(\sum_{i=1}^{n} \alpha_i \mathbf{x}_i, \sum_{j=1}^{m} \beta_j \mathbf{y}_j) = \sum_{i=1}^{n} \sum_{j=1}^{m} \alpha_i \varphi(\mathbf{x}_i, \mathbf{y}_j)\beta_j$
8. $\varphi(-\mathbf{x}, \mathbf{y}) = -\varphi(\mathbf{x}, \mathbf{y}) = \varphi(\mathbf{x}, -\mathbf{y})$ *for all* $\mathbf{x}, \mathbf{y}$ *in* V
9. $\varphi(-\mathbf{x}, -\mathbf{y}) = \varphi(\mathbf{x}, \mathbf{y})$ *for all* $\mathbf{x}, \mathbf{y}$ *in* V
10. $\varphi(\mathbf{x} + \mathbf{y}, \mathbf{x} + \mathbf{y}) = \varphi(\mathbf{x}, \mathbf{x}) + \varphi(\mathbf{x}, \mathbf{y}) + \varphi(\mathbf{y}, \mathbf{x}) + \varphi(\mathbf{y}, \mathbf{y})$ *for all* $\mathbf{x}, \mathbf{y} \in V$
11. $\varphi(\mathbf{x} - \mathbf{y}, \mathbf{x} - \mathbf{y}) = \varphi(\mathbf{x}, \mathbf{x}) - \varphi(\mathbf{x}, \mathbf{y}) - \varphi(\mathbf{y}, \mathbf{x}) + \varphi(\mathbf{y}, \mathbf{y})$ *for all* $\mathbf{x}, \mathbf{y} \in V$
12. $\varphi(\mathbf{x} + \mathbf{y}, \mathbf{x} + \mathbf{y}) + \varphi(\mathbf{x} - \mathbf{y}, \mathbf{x} - \mathbf{y}) = 2\varphi(\mathbf{x}, \mathbf{x}) + 2\varphi(\mathbf{y}, \mathbf{y})$ *for all* $\mathbf{x}, \mathbf{y} \in V$
13. $\varphi(\mathbf{x} + \mathbf{y}, \mathbf{x} + \mathbf{y}) - \varphi(\mathbf{x} - \mathbf{y}, \mathbf{x} - \mathbf{y}) = 2\varphi(\mathbf{x}, \mathbf{y}) + 2\varphi(\mathbf{y}, \mathbf{x})$ *for all* $\mathbf{x}, \mathbf{y} \in V$
14. $\varphi(\mathbf{x} + \mathbf{y}, \mathbf{x} - \mathbf{y}) = \varphi(\mathbf{x}, \mathbf{x}) - \varphi(\mathbf{x}, \mathbf{y}) + \varphi(\mathbf{y}, \mathbf{x}) - \varphi(\mathbf{y}, \mathbf{y})$
15. $\varphi(\mathbf{x} - \mathbf{y}, \mathbf{x} + \mathbf{y}) = \varphi(\mathbf{x}, \mathbf{x}) - \varphi(\mathbf{y}, \mathbf{x}) + \varphi(\mathbf{x}, \mathbf{y}) - \varphi(\mathbf{y}, \mathbf{y})$
16. $\varphi(\mathbf{x} + \mathbf{y}, \mathbf{x} - \mathbf{y}) + \varphi(\mathbf{x} - \mathbf{y}, \mathbf{x} + \mathbf{y}) = 2\varphi(\mathbf{x}, \mathbf{x}) - 2\varphi(\mathbf{y}, \mathbf{y})$
17. $\varphi(\mathbf{x} - \mathbf{y}, \mathbf{x} + \mathbf{y}) - \varphi(\mathbf{x} + \mathbf{y}, \mathbf{x} - \mathbf{y}) = 2\varphi(\mathbf{x}, \mathbf{y}) - 2\varphi(\mathbf{y}, \mathbf{x})$
18. $\varphi(\mathbf{x}, \mathbf{y}) + \varphi(\mathbf{y}, \mathbf{x}) = \varphi(\mathbf{x} + \mathbf{y}, \mathbf{x} + \mathbf{y}) - \varphi(\mathbf{x}, \mathbf{x}) - \varphi(\mathbf{y}, \mathbf{y})$

PROOF The proofs are routine computations and best left to the reader. □

It turns out arbitrary bilinear forms can be studied in terms of two special kinds of forms. We have the following trivial decomposition of an arbitrary φ:

$$\begin{aligned}\varphi(\mathbf{x}, \mathbf{y}) &= \frac{1}{2}\left[\varphi(\mathbf{x}, \mathbf{y}) + \varphi(\mathbf{y}, \mathbf{x})\right] + \frac{1}{2}\left[\varphi(\mathbf{x}, \mathbf{y}) - \varphi(\mathbf{y}, \mathbf{x})\right] \\ &= \varphi_{sym}(\mathbf{x}, \mathbf{y}) + \varphi_{skew}(\mathbf{x}, \mathbf{y}).\end{aligned}$$

The reader will be asked to verify that $\varphi_{sym}(\mathbf{x}, \mathbf{y}) = \varphi_{sym}(\mathbf{y}, \mathbf{x})$ for all $\mathbf{x}, \mathbf{y}$ and φ_{sym} is bilinear and $\varphi_{skew}(\mathbf{y}, \mathbf{x}) = -\varphi_{skew}(\mathbf{x}, \mathbf{y})$ where again φ_{skew} is bilinear. The first property of φ_{sym} is called *symmetry* and the second is *skew-symmetry*.

DEFINITION 12.2 *(symmetric, skew-symmetric)*
Let φ be a bilinear form on V. Then we say

1. *φ is **symmetric** iff $\varphi(\mathbf{x}, \mathbf{y}) = \varphi(\mathbf{y}, \mathbf{x})$ for all $\mathbf{x}, \mathbf{y} \in V$.*

2. *φ is **skew-symmetric** iff $\varphi(\mathbf{x}, \mathbf{y}) = -\varphi(\mathbf{y}, \mathbf{x})$ for all $\mathbf{x}, \mathbf{y} \in V$.*

The fancy talk says that $L^2(V; \mathbb{F})$ is the direct sum of the subspace consisting of all symmetric bilinear forms and the subspace of all skew-symmetric bilinear forms. So, in a sense, if we know everything about symmetric forms and everything about skew-symmetric forms, we should know everything about all bilinear forms. (Right!)

DEFINITION 12.3 *(nondegenerate)*
*A bilinear form $\varphi : V \times W \to \mathbb{F}$ is called **nondegenerate on the left** iff $\varphi(\mathbf{v}, \mathbf{w}) = 0$ for all $\mathbf{w} \in W$ means $\mathbf{v} = \overrightarrow{0}$. Also, φ is **nondegenerate on the right** iff $\varphi(\mathbf{v}, \mathbf{w}) = 0$ for all $\mathbf{v} \in V$ means $\mathbf{w} = \overrightarrow{0}$. We call φ **nondegenerate** iff φ is nondegenerate on the left and on the right.*

We have seen that $d_\varphi(\mathbf{y}) : V \to \mathbb{F}$ and $s_\varphi(\mathbf{x}) : W \longrightarrow \mathbb{F}$ are linear. That is, for each $y \in W$, $d_\varphi(\mathbf{y}) \in L(V; \mathbb{F}) = V^*$ and for each $\mathbf{x} \in V$, $s_\varphi(\mathbf{x}) \in L(W; \mathbb{F}) = W^*$. We can lift the level of abstraction by considering $d_\varphi : W \to V^*$ and $s_\varphi : V \longrightarrow W^*$ by $\mathbf{y} \longmapsto d_\varphi(\mathbf{y})$ and $\mathbf{x} \longmapsto s_\varphi(\mathbf{x})$. These maps are themselves linear.

A bilinear map φ is a function of two variables. However, there is a natural way to associate a function of one variable by "restricting φ to the diagonal." That is, look at $\varphi(\mathbf{x}, \mathbf{x})$ only.

DEFINITION 12.4 *(quadratic map)*

*Let V and W be vector spaces over $\mathbb{F}$. A **quadratic map** from V to W is a map $Q : V \longrightarrow W$ such that*

1. $Q(\alpha\mathbf{x}) = \alpha^2 Q(\mathbf{x})$ *for all* $\alpha \in \mathbb{F}$, *all* $\mathbf{x} \in V$.

2. $Q(\mathbf{x}+\mathbf{y}) + Q(\mathbf{x}-\mathbf{y}) = 2Q(\mathbf{x}) + 2Q(\mathbf{y})$, *all* $\mathbf{x}, \mathbf{y} \in V$.

If $W = \mathbb{F}$, we, again traditionally, speak of Q as a quadratic form.

THEOREM 12.2

Let Q be a quadratic map. Then

1. $Q(\overrightarrow{0}) = 0$.

2. $Q(-\mathbf{x}) = Q(\mathbf{x})$ *for all* $\mathbf{x} \in V$.

3. $Q(\mathbf{x}-\mathbf{y}) = Q(\mathbf{y}-\mathbf{x})$ *for all* $\mathbf{x}, \mathbf{y} \in V$.

4. $Q(\alpha\mathbf{x}+\beta\mathbf{y}) = Q(\alpha\mathbf{x}) + 2\alpha\beta Q(\mathbf{x}) + Q(\beta\mathbf{x})$.

5. $Q(\alpha\mathbf{x}+\beta\mathbf{y}) + Q(\alpha\mathbf{x}-\beta\mathbf{y}) = 2\alpha^2 Q(\mathbf{x}) + 2\beta^2 Q(\mathbf{y})$.

6. $Q(\frac{1}{2}(\mathbf{x}+\mathbf{y})) + Q(\frac{1}{2}(\mathbf{x}-\mathbf{y})) = \frac{1}{2}Q(\mathbf{x}) + \frac{1}{2}Q(\mathbf{y})$.

7. $Q(\mathbf{x}+\mathbf{y}) - Q(\mathbf{x}-\mathbf{y}) = \frac{1}{2}\left[Q(\mathbf{x}+2\mathbf{y}) + Q(\mathbf{x}-2\mathbf{y})\right]$.

8. $2Q(\mathbf{x}+\mathbf{z}+\mathbf{y}) + 2Q(\mathbf{y}) = Q(\mathbf{x}+\mathbf{z}+2\mathbf{y}) + Q(\mathbf{x}+\mathbf{z})$.

9. $2Q(\mathbf{x}+\mathbf{y}) + 2Q(\mathbf{z}+\mathbf{y}) = Q(\mathbf{x}+\mathbf{z}+2\mathbf{y}) + Q(\mathbf{x}-\mathbf{z})$.

PROOF Again the proofs are computations and referred to the exercises. □

Now beginning with $\varphi : V \times V \longrightarrow W$ bilinear, we can associate a map $\Phi : V \longrightarrow W$ by $\Phi(\mathbf{x}) = \varphi(\mathbf{x}, \mathbf{x})$. Then Φ is a quadratic map called *the quadratic map associated with* φ.

Exercise Set 51

1. Let $A = \begin{bmatrix} 1 & 2 & 3 \\ 4 & 5 & 6 \end{bmatrix}$. Write out explicitly the bilinear form determined by this matrix.

2. Work out all the computational formulas of Theorem 12.1.

3. Verify all the claims made in the previous examples.

4. Let φ be a bilinear form on the $\mathbb{F}$ vector space V. Then φ is called **alternating** iff $\varphi(\mathbf{x}, \mathbf{x}) = 0$ for all $\mathbf{x} \in V$. Argue that φ is alternating if and only if φ is skew-symmetric.

5. Let φ be a bilinear form on V, W and $S : V \longrightarrow V$ and $T : V \longrightarrow V$ be linear maps. Define $\varphi_{S,T}(\mathbf{v}, \mathbf{w}) = \varphi(S\mathbf{v}, T\mathbf{w})$ for $\mathbf{v} \in V$, $\mathbf{w} \in W$. Argue that $\varphi_{S,T}$ is also a bilinear form on V, W.

6. Let φ be a bilinear form on V. Then $\varphi_{sym}(\mathbf{x}, \mathbf{y}) = \frac{1}{2}[\varphi(\mathbf{x}, \mathbf{y}) + \varphi(\mathbf{y}, \mathbf{x})]$ is a symmetric bilinear form on V. Show this. Also show $\varphi_{skew}(\mathbf{x}, \mathbf{y}) = \frac{1}{2}[\varphi(\mathbf{x}, \mathbf{y}) - \varphi(\mathbf{y}, \mathbf{x})]$ is a skew symmetric bilinear form on V. Argue that $\varphi(\mathbf{x}, \mathbf{y}) = \varphi_{sym}(\mathbf{x}, \mathbf{y}) + \varphi_{skew}(\mathbf{x}, \mathbf{y})$ and this way of representing φ as a symmetric plus a skew-symmetric bilinear form is unique.

7. Argue that $L^2(V, W; \mathbb{F})$ is a vector space over $\mathbb{F}$. Also argue that the symmetric forms are a subspace of $L^2(V; \mathbb{F})$, as are the skew-symmetric forms, and that $L^2(V; F)$ is the direct sum of these to subspaces.

8. Let V be the vector space $\mathcal{C}([-\pi, \pi])$ of $\mathbb{F}$ valued functions defined on $[-\pi, \pi]$, which are continuous. Define $\varphi(f, g) = \int_{-\pi}^{\pi} f(x)g(x)dx$. Argue that φ is a symmetric bilinear form on V.

9. You have seen how to define a bilinear map. How would you define a *trilinear* map $\varphi : V_1 \times V_2 \times V_3 \longrightarrow W$. How about a *p-linear* map?

10. Let φ be bilinear on V. Show that $\varphi_{sym}(\mathbf{x}, \mathbf{y}) = \frac{1}{4}\{\varphi(\mathbf{x}+\mathbf{y}, \mathbf{x}+\mathbf{y}) - \varphi(\mathbf{x}-\mathbf{y}, \mathbf{x}-\mathbf{y})\}$.

11. Let φ be a bilinear form on V (i.e., $\varphi \in L^2(V; \mathbb{F})$). Then $s_\varphi(\mathbf{x}) : V \longrightarrow \mathbb{F}$ and $d_\varphi(\mathbf{y}) : V \longrightarrow \mathbb{F}$ are linear functionals. Show $s_\varphi : V \longrightarrow L(V; \mathbb{F}) = V^*$ and $d_\varphi : V \longrightarrow L(V; \mathbb{F}) = V^*$ are linear maps in their own right and φ is nondegenerate on the right (left) iff $d_\varphi(s_\varphi)$ is injective iff $\ker(d_\varphi)(\ker(s_\varphi))$ is trivial. Thus φ is nondegenerate iff both s_φ and d_φ are injective. In other words, φ is degenerate iff at least one of $\ker(s_\varphi)$ or $\ker(d_\varphi)$ is not trivial.

12. Suppose a bilinear form is both symmetric and skew-symmetric. What can you say about it.

13. Verify the computational rules for quadratic maps given in Theorem 12.2.

14. Let $\varphi : V \times V \longrightarrow W$ be bilinear. Verify that the associated quadratic map $\Phi : V \longrightarrow W$ given by $\Phi(\mathbf{x}) = \varphi(\mathbf{x}, \mathbf{x})$ is indeed a quadratic map.

15. A quadratic equation in two variables x and y is an equation of the form $ax^2+2bxy+cy^2+dx+ey+f=0$. Write this equation in matrix form where $\mathbf{x} = \begin{bmatrix} x \\ y \end{bmatrix}$ and $A = \begin{bmatrix} a & b \\ b & c \end{bmatrix}$. Note $\mathbf{x}^T A\mathbf{x}$ is the quadratic form associated to this equation. Generalize this to three variables; generalize it to n variables.

16. Suppose $A \in \mathbb{R}^{n\times n}$ and $A = A^T$. Call A(or Q) **definite** iff $Q(x) = \mathbf{x}^T A\mathbf{x}$ takes only one sign as $\mathbf{x}$ varies over all nonzero vectors $\mathbf{x}$ in $\mathbb{R}^n$, Call A **positive definite** iff $Q(\mathbf{x}) > 0$ for all nonzero $\mathbf{x}$ in $\mathbb{R}^n$, **negative definite** if $Q(\mathbf{x}) < 0$ for all nonzero $\mathbf{x}$ in $\mathbb{R}^n$, and **indefinite** iff Q takes on both positive and negative values. Q is called **positive semidefinite** iff $Q(\mathbf{x}) \geq 0$ and **negative semidefinite** iff $Q(\mathbf{x}) \leq 0$ for all $\mathbf{x}$. These words apply to either A or Q. Argue that A is positive definite iff all its eigenvalues are positive.

17. Prove that if A is positive definite real symmetric, then so is A^{-1}.

18. Argue that if A is positive definite real symmetric, then $det(A) > 0$. Is the converse true?

19. Prove that if A is singular, $A^T A$ is positive semidefinite but not positive definite.

Further Reading

[Lam, 1973] T. Y. Lam, *The Algebraic Theory of Quadratic Forms,* The Benjamin/Cummings Publishing Company, Reading, MA, (1973).

congruence, discriminant

12.2 Matrices Associated to Bilinear Forms

Let φ be a bilinear form on V and $\mathcal{B} = \{\mathbf{b}_1, \mathbf{b}_2, ..., \mathbf{b}_n\}$ be an ordered basis of V. We now describe how to associate a matrix to φ relative to this basis. Let $\mathbf{x}$ and $\mathbf{y}$ be in V. Then $\mathbf{x} = x_1\mathbf{b}_1 + \cdots + x_n\mathbf{b}_n$ and $\mathbf{y} = y_1\mathbf{b}_1 + \cdots + y_n\mathbf{b}_n$, so

$\varphi(\mathbf{x}, \mathbf{y}) = \varphi(\sum_{i=1}^{n} x_i \mathbf{b}_i, \mathbf{y}) = \sum_{i=1}^{n} x_i \varphi(\mathbf{b}_i, \mathbf{y}) = \sum_{i=1}^{n} x_i \varphi(\mathbf{b}_i, \sum_{j=1}^{n} y_j \mathbf{b}_j) = \sum_{i=1}^{n}\sum_{j=1}^{n} x_i y_i \varphi$ $(\mathbf{b}_i, \mathbf{b}_j)$.

Let $a_{ij} = \varphi(\mathbf{b}_i, \mathbf{b}_j)$. These n^2 scalars completely determine the action of φ on any pair of vectors. Thus, we naturally associate the n-by-n matrix $A = [a_{ij}] = [\varphi(\mathbf{b}_i, \mathbf{b}_j)] := Mat(\varphi; \mathcal{B})$. Moreover, $\varphi(\mathbf{x}, \mathbf{y}) = Mat(\mathbf{x}; \mathcal{B})^T Mat(\varphi; \mathcal{B})$ $Mat(\mathbf{y}; \mathcal{B})$.

Conversely, given any n-by-n matrix $A = [a_{ij}]$ and an ordered basis $\mathcal{B}$ of V, then $\varphi(\mathbf{x}, \mathbf{y}) = Mat(\mathbf{x}; \mathcal{B})^T A Mat(\mathbf{y}; \mathcal{B})$ defines a bilinear form on V whose matrix is A relative to $\mathcal{B}$. It is straightforward to verify that, given an ordered basis $\mathcal{B}$ of V, there is a one-to-one correspondence between bilinear forms on V and n-by-n matrices given by $\varphi \longmapsto Mat(\varphi, \mathcal{B})$. Moreover, $Mat(a\varphi_1 + b\varphi_2; \mathcal{B}) = aMat(\varphi_1; \mathcal{B}) + bMat(\varphi_2; \mathcal{B})$.

The crucial question now is what happens if we change the basis. How does the matrix representing φ change and how is this new matrix related to the old one. Let $\mathcal{C}$ be another ordered basis of V. How is $Mat(\varphi; \mathcal{B})$ related to $Mat(\varphi, \mathcal{C})$? Naturally, the key is the change of basis matrix. Let P be the (invertible) change of basis matrix $P = P_{\mathcal{B}\leftarrow\mathcal{C}}$ so that $Mat(x; \mathcal{B}) = PMat(x; \mathcal{C})$. Then for any $\mathbf{x}, \mathbf{y} \in V$, $\varphi(\mathbf{x}, \mathbf{y}) = Mat(\mathbf{x}; \mathcal{B})^T Mat(\varphi; \mathcal{B}) Mat(\mathbf{y}; \mathcal{B}) =$
$(PMat(\mathbf{x}; \mathcal{C}))^T Mat(\varphi; \mathcal{B}) PMat(\mathbf{y}; \mathcal{C}) =$
$Mat(\mathbf{x}; \mathcal{C})^T (P^T Mat(\varphi; \mathcal{B}) P) Mat(\mathbf{y}; \mathcal{C})$. But also $\varphi(\mathbf{x}, \mathbf{y}) =$
$Mat(\mathbf{x}; \mathcal{C})^T (Mat(\varphi; \mathcal{C}) Mat(\mathbf{y}; \mathcal{C})$, so we conclude

$$Mat(\varphi; \mathcal{C}) = P^T Mat(\varphi; \mathcal{B}) P.$$

This leads us to yet another equivalence relation on n-by-n matrices.

DEFINITION 12.5 *(congruence)*

A matrix $A \in \mathbb{F}^{n\times n}$ is ***congruent*** *to a matrix $B \in \mathbb{F}^{n\times n}$ iff there exists an invertible matrix P such that $B = P^T AP$. We write $A \sim B$ to symbolize congruence.*

THEOREM 12.3

$\sim$ is an equivalence relation on $\mathbb{F}^{n\times n}$. That is,

1. *$A \sim A$ for all A.*
2. *If $A \sim B$, then $B \sim A$.*
3. *If $A \sim B$ and $B \sim C$, then $A \sim C$.*

PROOF The proof is left as an exercise. ▯

We note that congruence is a special case of matrix equivalence so, for example, congruent matrices must have the same rank. However, we did not demand that P^T equal P^{-1}, so congruence is not as strong as similarity. Congruent matrices need not have the same eigenvalues or even the same determinant. Any property that is invariant for congruence, such as rank, can be ascribed to the associated bilinear form. Thus, we define the rank of φ to be the rank of any matrix that represents φ. Moreover, we have the following theorem.

THEOREM 12.4
Let φ be a bilinear form on V. Then φ is nondegenerate iff $rank(\varphi) = \dim V$.

PROOF The proof is left as an exercise. ▯

Suppose A and B are congruent. Then $B = P^T AP$ for some nonsingular matrix P. Then $det(B) = det(P^T AP) = det(P)^2 det(A)$. Thus, the determinants of A and B may differ, but in a precise way. Namely, one is a nonzero square scalar times the other. We define the *discriminant* of a bilinear form φ to be $\left\{a^2 det(A) \mid a \neq 0, A \text{ represents } \varphi \text{ in some ordered basis}\right\}$ .This set of scalars is an invariant under congruence. We summarize this section with a theorem.

THEOREM 12.5
Let φ be a bilinear form on V and $\mathcal{B} = \{\mathbf{b}_1, \mathbf{b}_2, \cdots, \mathbf{b}_n\}$, an ordered basis of V. Then $\varphi(\mathbf{x}, \mathbf{y}) = Mat(\mathbf{x}, \mathcal{B})^T Mat(\varphi; \mathcal{B}) Mat(\mathbf{y}; \mathcal{B})$ where $Mat(\varphi, \mathcal{B}) = \left[\varphi(\mathbf{b}_i, \mathbf{b}_j)\right]$. Moreover, if $\mathcal{C}$ is another ordered basis of V, then $Mat(\varphi; \mathcal{C}) = (P_{\mathcal{C}\leftarrow\mathcal{B}})^T Mat(\varphi; \mathcal{B}) P_{\mathcal{C}\leftarrow\mathcal{B}}$. Moreover, if two matrices are congruent, then they represent the same bilinear form on V. Also, φ is symmetric iff $Mat(\varphi; \mathcal{B})$ is a symmetric matrix and φ is skew-symmetric iff $Mat(\varphi; \mathcal{B})$ is a skew-symmetric matrix.

Exercise Set 52

1. Prove Theorem 12.3.

2. Find two congruent matrices A and B that have different determinants.

3. Prove Theorem 12.4.

4. Prove Theorem 12.5.

orthogonal, isotropic, orthosymmetric, radical, orthogonal direct sum

12.3 Orthogonality

One of the most useful geometric concepts that we associate with inner products is to identify when two vectors are *orthogonal* (i.e., *perpendicular*). We can do this with bilinear forms as well. Given a bilinear form φ, we can define the related notion of orthogonality in the natural way. Namely, we say $\mathbf{x}$ *is* φ*−orthogonal to* $\mathbf{y}$ iff $\varphi(\mathbf{x}, \mathbf{y}) = 0$. We can symbolize this by $\mathbf{x} \perp_\varphi \mathbf{y}$. If φ is understood, we will simplify things by just using the word "orthogonal" and the symbol $\perp$. Unlike with inner products, some strange things can happen. It is possible for a nonzero vector to be orthogonal to itself! Such a vector is called *isotropic,* and these vectors actually occur meaningfully in relativity theory in physics. It is also possible that a vector $\mathbf{x}$ is orthogonal to a vector $\mathbf{y}$ but $\mathbf{y}$ is not orthogonal to $\mathbf{x}$. However, we have the following nice result.

THEOREM 12.6
Let φ be a bilinear form on V. Then the orthogonality relation is symmetric (i.e., $\mathbf{x} \perp_\varphi \mathbf{y}$ implies $\mathbf{y} \perp_\varphi \mathbf{x}$) iff φ is either symmetric or skew-symmetric.

PROOF If φ is symmetric or skew-symmetric, then clearly the orthogonality relation is symmetric. So, assume the relation is symmetric. Let $\mathbf{x}, \mathbf{y}, \mathbf{z} \in V$. Let $\mathbf{w} = \varphi(\mathbf{x}, \mathbf{y})\mathbf{z} - \varphi(\mathbf{x}, \mathbf{z})\mathbf{y}$. Then we compute that $\mathbf{x} \perp \mathbf{w}$. By symmetry $\mathbf{w} \perp \mathbf{x}$, which is equivalent to $\varphi(\mathbf{x}, \mathbf{y})\varphi(\mathbf{z}, \mathbf{x}) - \varphi(\mathbf{x}, \mathbf{z})\varphi(\mathbf{y}, \mathbf{x}) = 0$. Set $\mathbf{x} = \mathbf{y}$ and conclude $\varphi(\mathbf{x}, \mathbf{x})\,[\varphi(\mathbf{z}, \mathbf{x}) - \varphi(\mathbf{x}, \mathbf{z})] = 0$ for all $\mathbf{x}, \mathbf{z} \in V$. Swapping $\mathbf{x}$ and $\mathbf{z}$, we can also conclude $\varphi(\mathbf{z}, \mathbf{z})\,[\varphi(\mathbf{z}, \mathbf{x}) - \varphi(\mathbf{x}, \mathbf{z})] = 0$. This seems to be good news since it seems to say $\varphi(\mathbf{z}, \mathbf{z}) = 0$ (i.e., $\mathbf{z}$ is isotropic or $\varphi(\mathbf{z}, \mathbf{x}) = \varphi(\mathbf{x}, \mathbf{z})$). The problem is we might have a mixture of these cases. The rest of the proof says its all one (all isotropic) or all the other (i.e., φ is symmetric).

So suppose φ is not symmetric (if it is, we are finished). Then there must exist vectors $\mathbf{u}$ and $\mathbf{v}$ such that $\varphi(\mathbf{u}, \mathbf{v}) \neq \varphi(\mathbf{v}, \mathbf{u})$. Then, by what we showed above, $\varphi(\mathbf{u}, \mathbf{u}) = \varphi(\mathbf{v}, \mathbf{v}) = 0$. We claim $\varphi(\mathbf{w}, \mathbf{w}) = 0$ for all $\mathbf{w}$ in V. Since $\mathbf{u}$ is isotropic $\varphi(\mathbf{u} + \mathbf{w}, \mathbf{u} + \mathbf{w}) = \varphi(\mathbf{u}, \mathbf{w}) + \varphi(\mathbf{w}, \mathbf{u}) + \varphi(\mathbf{w}, \mathbf{w})$. If $\mathbf{w}$ is not isotropic, then $\varphi(\mathbf{w}, \mathbf{x}) = \varphi(\mathbf{x}, \mathbf{w})$ for all $\mathbf{x} \in V$. In particular, $\varphi(\mathbf{w}, \mathbf{u}) = \varphi(\mathbf{u}, \mathbf{w})$ and $\varphi(\mathbf{w}, \mathbf{v}) = \varphi(\mathbf{v}, \mathbf{w})$. By above, $\varphi(\mathbf{u}, \mathbf{w})\varphi(\mathbf{v}, \mathbf{u}) - \varphi(\mathbf{u}, \mathbf{v})\varphi(\mathbf{w}, \mathbf{u}) = 0$, so $\varphi(\mathbf{u}, \mathbf{w})\,[\varphi(\mathbf{v}, \mathbf{u}) - \varphi(\mathbf{u}, \mathbf{v})] = 0$; but $\varphi(\mathbf{u}, \mathbf{v}) \neq \varphi(\mathbf{v}, \mathbf{u})$, so we must conclude

$\varphi(\mathbf{w}, \mathbf{u}) = \varphi(\mathbf{u}, \mathbf{w}) = 0$. Similarly, $\varphi(\mathbf{w}, \mathbf{v}) = \varphi(\mathbf{v}, \mathbf{w}) = 0$. Thus $\varphi(\mathbf{u}+\mathbf{w}, \mathbf{u}+\mathbf{w}) = \varphi(\mathbf{w}, \mathbf{w})$. But $\varphi(\mathbf{u}+\mathbf{w}, \mathbf{v}) = \varphi(\mathbf{u}, \mathbf{v}) + \varphi(\mathbf{u}, \mathbf{v}) = \varphi(\mathbf{u}, \mathbf{v}) \neq \varphi(\mathbf{v}, \mathbf{u}) = \varphi(\mathbf{v}, \mathbf{u}+\mathbf{w})$. Thus $\mathbf{u}+\mathbf{w}$ is also isotropic. Therefore, $\varphi(\mathbf{w}, \mathbf{w}) = 0$. Therefore, all vectors in V are isotropic and φ is alternating, hence skew-symmetric. □

To cover both of these cases, we simply call φ *orthosymmetric* whenever the associated orthogonality relation is symmetric.

Now let S be any subset of a vector space V with φ an orthosymmetric bilinear form on V. We define $S^{\perp} = \{\mathbf{v} \in V \mid \varphi(\mathbf{v}, \mathbf{s}) = 0 \text{ for all } \mathbf{s} \in S\}$. We define the *radical* of S by $Rad(S) = S \cap S^{\perp}$ so that $Rad(V) = V^{\perp}$. We see that φ is nondegenerate iff $Rad(V) = \{\overrightarrow{0}\}$.

THEOREM 12.7

Let S be any subset of the vector space V.

1. *$S^{\perp}$ is always a subspace of V.*
2. *$S \subseteq (S^{\perp})^{\perp}$.*
3. *If $S_1 \subseteq S_2$, then $S_2^{\perp} \subseteq S_1^{\perp}$.*
4. *If S is a finite dimensional subspace of V, then $S = S^{\perp\perp}$.*

PROOF The proof is relegated to the exercises. □

If V is the direct sum of two subspaces W_1 and W_2 (i.e., $V = W_1 \oplus W_2$), we call the direct sum an *orthogonal direct sum* iff $W_1 \subseteq W_2^{\perp}$ and write $V = W_1 \oplus^{\perp} W_2$. Of course, this idea extends to any finite number of subspaces.

If φ is a bilinear form on V then, by restriction, φ is a bilinear form on any subspace of V. The next theorem says we can restrict our study to nondegenerate orthosymmetric forms.

THEOREM 12.8

Let S be the complement of $Rad(V)$. Then $V = Rad(V) \oplus^{\perp} S$ and φ restricted to S is nondegenerate.

PROOF Since $Rad(V)$ is a subspace of V, it has a complement. Choose one, say M; then $V = Rad(V) \oplus M$. But all vectors are orthogonal to $Rad(V)$, so $V = Rad(V) \oplus^{\perp} M$. Let $\mathbf{v} \in M \cap M^{\perp}$. Then $\mathbf{v} \in M^{\perp}$ so $\mathbf{v} \in Rad(V)$. But $\mathbf{v} \in M$ also, so $\mathbf{v} \in Rad(V) \cap M = (\overrightarrow{0})$. Hence $\mathbf{v} = \overrightarrow{0}$. Thus $M \cap M^{\perp} = \{\overrightarrow{0}\}$. □

Exercise Set 53

1. Prove Theorem 12.7.

2. Consider the bilinear form associated with the matrix $A = \begin{bmatrix} 1 & 0 \\ 0 & -1 \end{bmatrix}$.
Identify all the isotropic vectors.

3. A quadratic function f of n variables looks like $f(x_1, x_2, ..., x_n) = \sum_{i=1}^{n}\sum_{j=1}^{n} q_{ij}x_ix_j + \sum_{i=1}^{n} c_ix_i + d$. What does this formula reduce to if $n = 1$? Argue that f can be written in matrix form $f(\mathbf{v}) = \mathbf{v}^T Q\mathbf{v} + \mathbf{c}^T\mathbf{v} + d$, where Q is a real n-by-n symmetric nonzero matrix. If you have had some calculus, compute the gradient $\nabla f(\mathbf{v})$ in matrix form.

orthogonal basis, orthonormal basis, Sym(n;ℂ), Sylvester's law of inertia, signature, inertia

12.4 Symmetric Bilinear Forms

In this section, we focus on bilinear forms that are symmetric. They have a beautiful characterization.

THEOREM 12.9
Suppose φ is a symmetric bilinear form on the finite dimensional space V. Then there exists an ordered basis $\mathcal{B}$ of V such that $Mat(\varphi; \mathcal{B})$ is diagonal. Such a basis is called an orthogonal basis.

PROOF The proof is by induction We seek a basis $\mathcal{B} = \{\mathbf{b}_1, \mathbf{b}_2, \ldots, \mathbf{b}_n\}$ such that $\varphi(\mathbf{b}_i, \mathbf{b}_j) = 0$ if $i \neq j$. If $\varphi = 0$ or $n = 1$, the theorem holds trivially, so suppose $\varphi \neq 0$ and $n > 1$. We claim $\varphi(\mathbf{x}, \mathbf{x}) \neq 0$ for some $\mathbf{x} \in V$. If $\varphi(\mathbf{x}, \mathbf{x}) = 0$ for all $\mathbf{x} \in V$, then $\varphi(\mathbf{x}, \mathbf{y}) = \frac{1}{4}\{\varphi(\mathbf{x}+\mathbf{y}, \mathbf{x}+\mathbf{y}) - \varphi(\mathbf{x}-\mathbf{y}, \mathbf{x}-\mathbf{y})\} = 0$ for all $\mathbf{x}, \mathbf{y}$, making $\varphi = 0$ against our assumption. Let $W = sp(\mathbf{x})$. We claim $V = W \oplus^{\perp} W^{\perp}$. First, let $\mathbf{z} \in W \cap W^{\perp}$. Then $\mathbf{z} = \alpha\mathbf{x}$ for some α and $\mathbf{z} \in W^{\perp}$ so $\mathbf{z} \perp \mathbf{z}$. Thus, $\varphi(\mathbf{z}, \mathbf{z}) = 0 = \varphi(\alpha\mathbf{x}, \alpha\mathbf{x}) = \alpha^2\varphi(\mathbf{x}, \mathbf{x})$. Since $\varphi(\mathbf{x}, \mathbf{x}) \neq 0$, $\alpha^2 = 0$, so $\alpha = 0$ so $\mathbf{z} = \vec{0}$. We conclude $W \cap W^{\perp} = (\vec{0})$. Now let $\mathbf{v} \in V$ and set $\mathbf{b} = \mathbf{v} - \frac{\varphi(\mathbf{v}, \mathbf{x})}{\varphi(\mathbf{x}, \mathbf{x})}\mathbf{x}$. Then $V = \left(\frac{\varphi(\mathbf{v}, \mathbf{x})}{\varphi(\mathbf{x}, \mathbf{x})}\right)\mathbf{x} + \mathbf{b}$ and

$\varphi(\mathbf{x}, \mathbf{b}) = \varphi(\mathbf{x}, \mathbf{v}) - \dfrac{\varphi(\mathbf{v}, \mathbf{x})}{\varphi(\mathbf{x}, \mathbf{x})}\varphi(\mathbf{x}, \mathbf{x}) = 0$ since φ is symmetric. Thus, $\mathbf{b} \in W^{\perp}$ and so $V = W \oplus W^{\perp}$. Now restrict φ to $W^{\perp}$, which is $n-1$ dimensional. By induction there is a basis $\{\mathbf{b}_1, \mathbf{b}_2, \ldots, \mathbf{b}_{n-1}\}$ with $\varphi(\mathbf{x}_i, \mathbf{b}_j) = 0$ if $i \neq j$. Add $\mathbf{x}$ to this basis. ▯

This theorem has significant consequences for symmetric matrices under congruence.

COROLLARY 12.1

Any symmetric matrix over $\mathbb{F}$ is congruent to a diagonal matrix.

PROOF Let A be an n-by-n symmetric matrix over $\mathbb{F}$. Let φ_A be the symmetric bilinear form determined by A on $\mathbb{F}^n$, (i.e., $\varphi_A(\mathbf{x}, \mathbf{y}) = \mathbf{x}^T A\mathbf{y}$). Then $Mat(\varphi; std) = A$, where std denotes the standard basis of $\mathbb{F}^n$. Now, by the previous theorem, there is a basis $\mathcal{B}$ of $\mathbb{F}^n$ in which $Mat(\varphi_A; \mathcal{B})$ is diagonal. But $Mat(\varphi_A; \mathcal{B})$ and $Mat(\varphi_A; std) = A$ are congruent and so A is congruent to a diagonal matrix. ▯

Unfortunately, two distinct diagonal matrices can be congruent, so the diagonal matrices do not form a set of canonical forms for the equivalence relation of congruence. For example, if $Mat(\varphi, \mathcal{B}) = \begin{bmatrix} d_1 & & 0 \\ & d_2 & \\ & & \vdots \\ 0 & & d_n \end{bmatrix}$, and we select nonzero scalars $\alpha_1, \ldots, \alpha_n$, we can "scale" each basis vector to get a new basis $\mathcal{C} = \{\alpha_1\mathbf{b}_1, \ldots, \alpha_n\mathbf{b}_n)$. Then $Mat(\varphi, \mathcal{C}) = \begin{bmatrix} \alpha_1^2 d_1 & & & 0 \\ & \ddots & & \\ & & \ddots & \\ 0 & & & \alpha_n^2 d_n \end{bmatrix}$, and these two matrices are congruent.

Now we restrict our attention to the complex field $\mathbb{C}$. Here we see congruence reduces to ordinary matrix equivalence.

THEOREM 12.10

Suppose φ is a symmetric bilinear form of rank r over the n-dimensional space V. Then there exists an ordered basis $\mathcal{B} = \{\mathbf{b}_1, \mathbf{b}_2, \ldots, \mathbf{b}_n\}$ of V such that

1. *$Mat(\varphi; \mathcal{B})$ is diagonal*

2. $\varphi(\mathbf{b}_j, \mathbf{b}_j) = \begin{cases} 1 & \textit{for } j = 1, 2, \ldots, r \\ 0 & \textit{for } j > r \end{cases}$.

PROOF Let $\{\mathbf{a}_1, \dots, \mathbf{a}_n\}$ be an orthogonal basis as provided by Theorem 12.9. Then there exist r values of j such that $\varphi(\mathbf{a}_j, \mathbf{a}_j) \neq 0$, else the rank of φ would not be r. Reorder this basis if necessary so these basis vectors become the first r. Then define $\mathbf{b}_j = \begin{cases} \dfrac{1}{\sqrt{\varphi(\mathbf{a}_j, \mathbf{a}_j)}}\mathbf{a}_j \text{ for } j = 1, \dots, r \\ \mathbf{a}_j \text{ for } j > r \end{cases}$. Then we have our basis so that $Mat(\varphi; \mathcal{B}) = \begin{bmatrix} I_r & \mathbb{O} \\ \mathbb{O} & \mathbb{O} \end{bmatrix}$. ▯

COROLLARY 12.2
If φ is a nondegenerate symmetric bilinear form over the n-dimensional space V, then V has an orthonormal basis for φ.

Let $Sym(n; \mathbb{C})$ denote the set of n-by-n symmetric matrices over $\mathbb{C}$. Let $I_{k,m}$ be the matrix with k ones and m zeros on the main diagonal and zeros elsewhere. Then the rank of a matrix is a complete invariant for congruence, and the set of all matrices of the form $I_{k,m}$ for $k + m = n$ is a set of canonical forms for congruence. That is, every matrix in $Sym(n; \mathbb{C})$ is congruent to a unique matrix of the form $I_{k,m}$ for some $k = 0, 1, \dots, n$ and $m = n - k$.

The story over the real field is more interesting. The main result is named for **James Joseph Sylvester** (3 September 1814–15 March 1897).

THEOREM 12.11 (Sylvester's law of inertia)
Suppose φ is a symmetric bilinear form of rank r on the real n-dimensional vector space V. Then there is an ordered basis $\mathcal{B} = \{\mathbf{b}_1, \mathbf{b}_2, \dots, \mathbf{b}_n\}$ such that
$$Mat(\varphi, \mathcal{B}) = \begin{bmatrix} I_k & & \mathbb{O} \\ & -I_m & \\ \mathbb{O} & & \mathbb{O} \end{bmatrix} = I_{k,m,l},$$
where k is the number of ones, $k + m = r$, and k is an invariant for φ under congruence.

PROOF Begin with the basis $\{\mathbf{a}_1, \mathbf{a}_2, \dots, \mathbf{a}_n\}$ given by Theorem 12.9. Reorder the basis if necessary so that $\varphi(\mathbf{a}_j, \mathbf{a}_j) = 0$ for $j > r$ and $\varphi(\mathbf{a}_j, \mathbf{a}_j) \neq 0$ for $1 \leq j \leq r$. Then the basis $\mathcal{B} = \{\mathbf{b}_1, \mathbf{b}_2, \dots, \mathbf{b}_n\}$, where $\mathbf{b}_j = \dfrac{1}{\sqrt{\varphi(a_j, a_j)}}\mathbf{a}_j$, $1 \leq j \leq r$, and $\mathbf{b}_j = \mathbf{a}_j$ for $j > r$ yields a matrix as above. The hard part is to prove that k, m, l do not depend on the basis chosen. Let V^+ be the subspace of V spanned by the basis vectors for which $\varphi(\mathbf{b}_j, \mathbf{b}_j) = 1$ and V^- the subspace spanned by the basis vectors for which $\varphi(\mathbf{b}_j, \mathbf{b}_j) = -1$. Now $k = \dim V^+$. If $\overrightarrow{0} \neq \mathbf{x} \in V^+$, $\varphi(\mathbf{x}, \mathbf{x}) > 0$ on V^+ and if $\overrightarrow{0} \neq \mathbf{x} \in V^-$, then $\varphi(\mathbf{x}, \mathbf{x}) < 0$. Let $V^\perp$ be the subspace spanned by the remaining basis vectors. Note if $\mathbf{x} \in V^{\perp}$, $\varphi(\mathbf{x}, \mathbf{y}) = 0$ for all $\mathbf{y} \in V$. Since $\mathcal{B}$ is a basis, we have $V = V^{+} \oplus V^{-} \oplus V^{\perp}$. Let W be any subspace of V such that $\varphi(\mathbf{w}, \mathbf{w}) > 0$ for all nonzero $\mathbf{w} \in W$. Then $W \cap span\{V^{-}, V^{\perp}\} = (\vec{0})$ for suppose $\mathbf{w} \in W, \mathbf{b} \in V^{-}, \mathbf{c} \in V^{\perp}$ and $\mathbf{w} + \mathbf{b} + \mathbf{c} = \vec{0}$. Then $0 = \varphi(\mathbf{w}, \mathbf{w} + \mathbf{b} + \mathbf{c}) = \varphi(\mathbf{w}, \mathbf{w}) + \varphi(\mathbf{w}, \mathbf{b}) + \varphi(\mathbf{w}, \mathbf{c}) = \varphi(\mathbf{w}, \mathbf{w}) + \varphi(\mathbf{w}, \mathbf{b})$ and $0 = \varphi(\mathbf{b}, \mathbf{w} + \mathbf{b} + \mathbf{c}) = \varphi(\mathbf{b}, \mathbf{b}) + \varphi(\mathbf{b}, \mathbf{w}) + \vec{0} = \varphi(\mathbf{b}, \mathbf{b}) + \varphi(\mathbf{w}, \mathbf{b})$. But then $\varphi(\mathbf{w}, \mathbf{w}) = \varphi(\mathbf{b}, \mathbf{b})$, but $\varphi(\mathbf{w}, \mathbf{w}) \geq 0$ and $\varphi(\mathbf{b}, \mathbf{b}) \leq 0$, so the only way out is $\varphi(\mathbf{w}, \mathbf{w}) = \varphi(\mathbf{b}, \mathbf{b}) = 0$. Therefore $\mathbf{w} = \mathbf{b} = \vec{0}$ so $\mathbf{c} = \vec{0}$ as well. Now $V = V^{+} \oplus V^{-} \oplus V^{\perp}$ and $W, V^{-}, V^{\perp}$ are independent so $\dim W \leq \dim V^{+}$. If W is any subspace of V on which φ takes positive values on nonzero vectors, $\dim W \leq \dim V^{+}$. So if $\mathcal{B}'$ is another ordered basis that gives a matrix I_{k^1, m^1, l^1}, then V_1^{+} has dimension $k^1 \leq \dim V^{+} = k$. But our argument is symmetric in these bases so $k \leq k^1$. Therefore, $k = \dim V^{+} = \dim V_1^{+}$ for any such basis. Since $k + l = r$ and $k^1 + l^1 = r$, it follows $l = l^1$ as well, and since $k + l + m = n$, we must have $m^1 = m$ also. □

Note that $V^{\perp} = Rad(V)$ above and $\dim(V^{\perp}) = \dim V - rank(\varphi)$. The number $\dim V^{+} - \dim V^{-}$ is sometimes called the *signature* of φ.

THEOREM 12.12

Congruence is an equivalence relation on $Sym(n; \mathbb{R})$, the set of n-by-n symmetric matrices over the reals. The set of matrices $I_{k,l,m}$ where $k + l + m = n$ is a set of canonical forms for congruence and the pair of numbers (k, m) or $(k + m, k - m)$ is a complete invariant for congruence.

In other words, two real symmetric matrices are congruent iff they have the same rank and the same signature.

Exercise Set 54

1. Suppose A is real symmetric positive definite. Argue that A must be nonsingular.

2. Suppose A is real symmetric positive definite. Argue that the leading principal submatrices of A are all positive definite.

3. Suppose A is real symmetric positive definite. Argue that A can be reduced to upper triangular form using only transvections and all the pivots will be positive.

4. Suppose A is real symmetric positive definite. Argue that A can be factored as $A = LDL^T$, where L is lower triangular with ones along its diagonal, and D is a diagonal matrix with all diagonal entries positive.

5. Suppose A is real symmetric positive definite. Argue that A can be factored as $A = LL^T$, where L is lower triangular with positive diagonal elements.

6. Suppose A is real and symmetric. Argue that the following statements are all equivalent:
   (a) A is positive definite.
   (b) The leading principal submatrices of A are all positive definite.
   (c) A can be reduced to upper triangular form using only transvections and all the pivots will be positive.
   (d) A can be factored as $A = LL^T$, where L is lower triangular with positive diagonal elements.
   (e) A can be factored as $A = B^T B$ for some nonsingular matrix B.
   (f) All the eigenvalues of A are positive.

7. Suppose A is a real symmetric definite matrix. Do similar results to exercise 6 hold?

8. Suppose A is real symmetric nonsingular. Argue that A^2 is positive definite.

9. Argue that a Hermitian matrix is positive definite iff it is congruent to the identity matrix.

10. Suppose H is a Hermitian matrix of rank r. Argue that H is positive semidefinite iff it is congruent to $\begin{bmatrix} I_r & \mathbb{O} \\ \mathbb{O} & \mathbb{O} \end{bmatrix}$. Conclude $H \geq O$ iff $H = P^*P$ for some matrix P.

11. Argue that two Hermitian matrices A and B are congruent iff they have the same rank and the same number of positive eigenvalues (counting multiplicities).

12. Suppose A and B are Hermitian. Suppose A is positive definite. Argue that there exists an invertible matrix P with $P^*AP = I$ and $P^*BP = D$, where D is a diagonal matrix with the eigenvalues of $A^{-1}B$ down the diagonal.

13. Suppose A is a square n-by-n complex matrix. Define the **inertia** of A to be $(\pi(A), \nu(A), \delta(A))$, where $\pi(A)$ is the number of eigenvalues of A

(counting algebraic multiplicities) in the open right-half plane, $\nu(A)$ is the number of eigenvalues in the open left-half plane, and $\delta(A)$ is the number of eigenvalues on the imaginary axis. Note that $\pi(A)+\nu(A)+\delta(A)=n$. Argue that A is nonsingular iff $\delta(A)=0$. If H is Hermitian, argue that $\pi(H)$ is the number of positive eigenvalues of H, $\nu(H)$ is the number of negative eigenvalues of H, and $\delta(H)$ is the number of times zero occurs as an eigenvalue. Argue that $rank(H)=\pi(H)+\nu(H)$ and $signature(H)=\pi(H)-\nu(H)$.

14. Suppose A and B are n-by-n Hermitian matrices of rank r and suppose $A=MBM^*$ for some M. Argue that A and B have the same inertia.

Further Reading

[Roman, 1992] Steven Roman, *Advanced Linear Algebra*, Springer-Verlag, New York, (1992).

[C&R, 1964] J. S. Chipman and M. M. Rao, Projections, Generalized Inverses and Quadratic Forms, J. Math. Anal. Appl., IX, (1964), 1–11.

12.5 Congruence and Symmetric Matrices

In the previous sections, we have motivated the equivalence relation of congruence on n-by-n matrices over $\mathbb{F}$. Namely, $A, B \in \mathbb{F}^{n\times n}$, $A \sim_C B$ iff there exists an invertible matrix P such that $B = P^T AP$. Evidently congruence is a special case of matrix equivalence, so any conclusions that obtain for equivalent matrices hold for congruent matrices as well. For example, congruent matrices must have the same rank.

Now P being invertible means that P can be expressed as a product of elementary matrices. The same is so for P^T. Indeed if $P^T = E_m E_{m-1}\cdots E_1$, then $P = E_1^T E_2^T \ldots E_m^T$. Therefore, $B = P^T AP = E_m E_{m-1}\cdots E_1 A E_1^T E_2^T \cdots E_m^T$. Thus, B is obtained from A by performing pairs of elementary operations, each pair consisting of an elementary row operation and the *same* elementary column operation. Indeed

$$\begin{bmatrix} 1 & 0 & 0 \\ \lambda & 1 & 0 \\ 0 & 0 & 1 \end{bmatrix} \begin{bmatrix} a & b & c \\ d & e & f \\ g & h & k \end{bmatrix} \begin{bmatrix} 1 & \lambda & 0 \\ 0 & 1 & 0 \\ 0 & 0 & 1 \end{bmatrix} =$$

$$\begin{bmatrix} a & a\lambda + b & c \\ \lambda a + d & \lambda^2 a + 2\lambda b + e & \lambda c + f \\ g & \lambda g + h & k \end{bmatrix}.$$

Thus, we see if A is symmetric, (i.e., $A = A^T$), then $E^T AE$ is still symmetric and, if E^T zeros out the (i, j) entry of A where $i \neq j$, then E zeros out the (j, i) entry simultaneously. We are not surprised now by the next theorem.

THEOREM 12.13

Let A be a symmetric matrix in $\mathbb{F}_r^{n\times n}$. Then A is congruent to a diagonal matrix whose first r diagonal entries are nonzero while the remaining $n - r$ diagonal entries are zero.

The proof indicates an algorithm that will effectively reduce a symmetric matrix to diagonal form. It is just Gauss elimination again: $E_m \cdots E_1 A = P^T A$ produces an upper triangular matrix while $E_1^T E_2^T \cdots E_m^T$ produces a diagonal matrix. These elementary operations have no effect on the rank at each stage.

ALGORITHM 12.1

$$\left[A \vdots I\right] \rightarrow \left[E_1 A E_1^T \vdots I E_1^T\right] \rightarrow \cdots \rightarrow$$

$$\left[E_m \cdots E_1 A E_1^T \cdots E_m^T \vdots E_1^T \cdots E_m^T\right] = \left[D \vdots P\right]$$

For example, if

$$A = \begin{bmatrix} 1 & 2 & 2 \\ 2 & 3 & 5 \\ 2 & 5 & 5 \end{bmatrix}, \textit{ then } \left[\begin{array}{ccc|ccc} 1 & 2 & 2 & 1 & 0 & 0 \\ 2 & 3 & 5 & 0 & 1 & 0 \\ 2 & 5 & 5 & 0 & 0 & 1 \end{array}\right] \rightarrow$$

$$\left[\begin{array}{c|c} T_{21}(-2) A T_{21}(-2)^T & I T_{21}(-2)^T \end{array}\right]$$

$$= \left[\begin{array}{ccc|ccc} 1 & 0 & 2 & 1 & -2 & 0 \\ 0 & -1 & 1 & 0 & 1 & 0 \\ 2 & 1 & 5 & 0 & 0 & 1 \end{array}\right] \rightarrow$$

$$\left[\begin{array}{c:c} T_{31}(-2)A^{(1)}T_{31}(-2)^T & T_{21}(-2)^T T_{31}(-2)^T \end{array}\right]$$

$$= \left[\begin{array}{ccc:ccc} 1 & 0 & 0 & 1 & -2 & -2 \\ 0 & -1 & 1 & 0 & 1 & 0 \\ 0 & 1 & 5 & 0 & 0 & 1 \end{array}\right] \to$$

$$\left[\begin{array}{c:c} T_{23}(-1)A^{(2)}T_{23}(-1)^T & T_{21}(-2)^T T_{31}(-2)^T T_{23}(-1)^T \end{array}\right]$$

$$= \left[\begin{array}{ccc:ccc} 1 & 0 & 0 & 1 & 0 & -2 \\ 0 & -2 & 0 & 0 & 1 & 0 \\ 0 & 0 & 1 & 0 & -1 & 1 \end{array}\right].$$

Then

$$\begin{bmatrix} 1 & 0 & -2 \\ 0 & 1 & 0 \\ 0 & -1 & 1 \end{bmatrix}^{\mathsf{T}} A \begin{bmatrix} 1 & 0 & -2 \\ 0 & 1 & 0 \\ 0 & -1 & 1 \end{bmatrix} = \begin{bmatrix} 1 & 0 & 0 \\ 0 & -2 & 0 \\ 0 & 0 & 1 \end{bmatrix}.$$

A couple of important facts need to be noted here. Since P need not be an orthogonal matrix ($P^{-1} = P^T$), the diagonal entries of $P^T AP$ need not be eigenvalues of A. Indeed, for an elementary matrix E, EAE^T need not be the same as EAE^{-1}. Also, the diagonal matrix to which A has been reduced above is not unique.

Over the complex numbers, congruence reduces to equivalence.

THEOREM 12.14

Let $A \in \mathbb{C}_r^{n\times n}$. Then A is congruent to $\begin{bmatrix} I_r & \mathbb{O} \\ \mathbb{O} & \mathbb{O} \end{bmatrix}$. Therefore, two matrices A, $B \in \mathbb{C}^{n\times n}$ are congruent iff they have the same rank.

Exercise Set 55

1. Find a diagonal matrix congruent to $\begin{bmatrix} 1 & 2 & 3 \\ 2 & 4 & 5 \\ 3 & 5 & 6 \end{bmatrix}$.

2. Argue that if A is congruent to B and A is symmetric, then B must be symmetric.

3. Prove directly that the matrix $\begin{bmatrix} -1 & 0 \\ 0 & 0 \end{bmatrix}$ is not congruent to $\begin{bmatrix} 1 & 0 \\ 0 & 0 \end{bmatrix}$ over $\mathbb{R}$, but they are congruent over $\mathbb{C}$.

12.6 Skew-Symmetric Bilinear Forms

Recall that a bilinear form φ is called *skew-symmetric* iff $\varphi(\mathbf{x}, \mathbf{y}) = -\varphi(\mathbf{y}, \mathbf{x})$ for all $\mathbf{x}, \mathbf{y} \in V$. Any matrix representation of φ gives a skew-symmetric matrix $A^T = -A$ and skew-symmetric matrices must have zero diagonals. Suppose $\mathbf{u}, \mathbf{v}$ are vectors in V with $\varphi(\mathbf{u}, \mathbf{v}) = 1$. Then $\varphi(\mathbf{v}, \mathbf{u}) = -1$. If we restrict φ to $H = span\{\mathbf{u}, \mathbf{v}\}$, its matrix has the form $\begin{bmatrix} 0 & 1 \\ -1 & 0 \end{bmatrix}$. Such a pair of vectors is called a *hyperbolic pair* and H is called a *hyperbolic plane*.

THEOREM 12.15

Let φ be a skew symmetric bilinear form on the $\mathbb{F}$ vector space V of n dimensions. Then there is a basis $\mathcal{B} = \{\mathbf{a}_1, \mathbf{b}_1, \mathbf{a}_2, \mathbf{b}_2, \ldots, \mathbf{a}_k, \mathbf{b}_k, \mathbf{c}_{n-2k}, \ldots, \mathbf{c}_n\}$, where

$$Mat(\varphi, \mathcal{B}) = \begin{bmatrix} \begin{bmatrix} 0 & 1 \\ -1 & 0 \end{bmatrix} & & & & \\ & \begin{bmatrix} 0 & 1 \\ -1 & 0 \end{bmatrix} & & & \\ & & \ddots & & \\ & & & \begin{bmatrix} 0 & 1 \\ -1 & 0 \end{bmatrix} & \\ & & & & \mathbb{O} \end{bmatrix},$$

where $rank(\varphi) = 2k$.

PROOF Suppose φ is nonzero and skew-symmetric on V. Then there exists a pair of vectors $\mathbf{a}, \mathbf{b}$ with $\varphi(\mathbf{a}, \mathbf{b}) \neq 0$ say $\varphi(\mathbf{a}, \mathbf{b}) = \alpha$. Replacing a by $(1/\alpha)\mathbf{a}$, we may assume $\varphi(\mathbf{a}, \mathbf{b}) = 1$. Let $\mathbf{c} = \alpha\mathbf{a} + \beta\mathbf{b}$. Then $\varphi(\mathbf{c}, \mathbf{a}) =$

$\varphi(\alpha\mathbf{a}+\beta\mathbf{b}, \mathbf{a}) = \beta\varphi(\mathbf{b}, \mathbf{a}) = -\beta$ and $\varphi(\mathbf{c}, \mathbf{b}) = \varphi(\alpha\mathbf{a}+\beta\mathbf{b}, \mathbf{b}) = \alpha\varphi(\mathbf{a}, \mathbf{b}) = \alpha$ so $\mathbf{c} = \varphi(\mathbf{c}, \mathbf{b})\mathbf{a} - \varphi(\mathbf{c}, \mathbf{a})\mathbf{b}$. Note $\mathbf{a}$ and $\mathbf{b}$ are independent. Let $W = span\{\mathbf{a}, \mathbf{b}\}$ We claim $V = W \oplus W^{\perp}$. Let $\mathbf{x}$ be any vector in V and $\mathbf{c} = \varphi(\mathbf{x}, \mathbf{b})\mathbf{a} - \varphi(\mathbf{x}, \mathbf{a})\mathbf{b}$ and $\mathbf{d} = \mathbf{x} - \mathbf{c}$. Then $\mathbf{c} \in W$ and $\mathbf{d} \in W^{\perp}$ since $\varphi(\mathbf{d}, \mathbf{a}) = \varphi(\mathbf{x} - \varphi(\mathbf{x}, \mathbf{b})\mathbf{a} + \varphi(\mathbf{x}, \mathbf{a})\mathbf{b}, \mathbf{a}) = \varphi(\mathbf{x}, \mathbf{a}) + \varphi(\mathbf{x}, \mathbf{a})\varphi(\mathbf{b}, \mathbf{a}) = 0$ and, similarly, $\varphi(\mathbf{d}, \mathbf{b}) = 0$. Thus, $V = W + W^{\perp}$. Moreover, $W \cap W^{\perp} = (\vec{0})$, so $V = W \oplus W^{\perp}$. Now φ restricted to $W^{\perp}$ is a skew-symmetric bilinear form . If this restriction is zero, we are done. If not, there exist $\mathbf{a}_2$, $\mathbf{b}_2$ in $W^{\perp}$ with $\varphi(\mathbf{a}_2, \mathbf{b}_2) = 1$. Let $W_2 = span\{\mathbf{a}_2, \mathbf{b}_2\}$. Then $V = W \oplus W_2 \oplus W_0$. This process must eventually cease since we only have finitely many dimensions. So we get $\varphi(\mathbf{a}_j, \mathbf{b}_j) = 1$ for $j = 1, \ldots, k$ and $\varphi(\mathbf{a}_i, \mathbf{a}_j) = \varphi(\mathbf{b}_i, \mathbf{b}_j) = \varphi(\mathbf{a}_i, \mathbf{b}_j) = 0$ if $i \neq j$ and if W_j is the hyperbolic plane spanned by $\{\mathbf{a}_j, \mathbf{b}_j\}$, $V = W_1 \oplus \cdots \oplus W_k \oplus W_0$, where every vector in W_0 is orthogonal to all $\mathbf{a}_j$ and $\mathbf{b}_j$ and φ restricted to W_0 is zero. It is clear the matrix of φ relative to $\{\mathbf{a}_1, \mathbf{b}_1, \mathbf{a}_2, \mathbf{b}_2, \ldots, \mathbf{a}_k, \mathbf{b}_k, \mathbf{c}_1, \ldots, c_p\}$ has the advertised form. □

COROLLARY 12.3

If φ is a nondegenerate skew symmetric bilinear form on V, then $dim(V)$ must be even and V is a direct sum of hyperbolic planes, and $Mat(\varphi; \mathcal{B})$ is

$$\begin{bmatrix} \begin{bmatrix} 0 & 1 \\ -1 & 0 \end{bmatrix} & & & \\ & \begin{bmatrix} 0 & 1 \\ -1 & 0 \end{bmatrix} & & \\ & & \ddots & \\ & & & \begin{bmatrix} 0 & 1 \\ -1 & 0 \end{bmatrix} \end{bmatrix} \text{ for some basis } \mathcal{B}.$$

Exercise Set 56

1. Prove that every matrix congruent to a skew-symmetric matrix is also skew-symmetric.

2. Argue that skew-symmetric matrices over $\mathbb{C}$ are congruent iff they have the same rank.

3. Prove that a skew-symmetric matrix must have zero trace.

4. Suppose A is an invertible symmetric matrix and K is skew-symmetric with $(A + K)(A - K)$ invertible. Prove that $S^T AS = A$, where $S = (A + K)^{-1}(A - K)$.

5. Suppose A is an invertible symmetric matrix and K is such that $(A+K)(A-K)$ is invertible. Suppose $S^T AS = A$, where $S = (A+K)^{-1}(A-K)$ and $I + S$ is invertible. Argue that K is skew-symmetric.

12.7 Tensor Products of Matrices

In this section, we introduce another way to multiply two matrices together to get another matrix.

DEFINITION 12.6 *(tensor or Kronecker product)*
Let $A \in \mathbb{C}^{m\times n}$ and $B \in \mathbb{C}^{p\times q}$. Then $A \otimes B$ is defined to be the mp-by-nq matrix $A \otimes B = \begin{bmatrix} C_{11} & \dots & C_{1n} \\ \vdots & & \vdots \\ C_{m1} & \dots & C_{mn} \end{bmatrix}$, *where C_{ij} is a block matrix of size pq-by-pq defined by $C_{ij} = (ent_{ij}(A))B = a_{ij}B$. In other words,*

$$A \otimes B = \begin{bmatrix} a_{11}B & a_{12}B & \dots & a_{1n}B \\ a_{21}B & a_{22}B & \dots & a_{2n}B \\ \vdots & \vdots & & \vdots \\ a_{m1}B & a_{m2}B & & a_{mn}B \end{bmatrix}.$$

For example,

$$\begin{bmatrix} a_{11} & a_{12} \\ a_{21} & a_{22} \\ a_{31} & a_{32} \end{bmatrix} \otimes \begin{bmatrix} b_{11} & b_{12} & b_{13} \\ b_{21} & b_{22} & b_{23} \end{bmatrix}$$
$$= \begin{bmatrix} a_{11}b_{11} & a_{11}b_{12} & a_{11}b_{13} & a_{12}b_{11} & a_{12}b_{12} & a_{12}b_{13} \\ a_{11}b_{21} & a_{11}b_{22} & a_{11}b_{23} & a_{12}b_{21} & a_{12}b_{22} & a_{12}b_{23} \\ a_{21}b_{11} & a_{21}b_{12} & a_{21}b_{13} & a_{22}b_{11} & a_{22}b_{12} & a_{22}b_{13} \\ a_{21}b_{21} & a_{21}b_{22} & a_{21}b_{23} & a_{22}b_{21} & a_{22}b_{22} & a_{22}b_{23} \\ a_{31}b_{11} & a_{31}b_{12} & a_{31}b_{13} & a_{32}b_{11} & a_{32}b_{12} & a_{32}b_{13} \\ a_{31}b_{21} & a_{31}b_{22} & a_{31}b_{23} & a_{32}b_{21} & a_{32}b_{22} & a_{32}b_{23} \end{bmatrix}.$$

We can view $A \otimes B$ as being formed by replacing each element a_{ij} of A by the p-by-q matrix $a_{ij}B$. So the tensor product of A with B is a partitioned matrix consisting of m rows and n columns of p-by-q blocks. The ij^{th} block is $a_{ij}B$. The element of $A \otimes B$ that appears in the $[p(i-1)+r]^{th}$ row and $[q(j-1)+s]^{th}$ column is the rs^{th} element $a_{ij}b_{rs}$ of $a_{ij}B$.

Note that $A \otimes B$ can always be formed regardless of the size of the matrices, unlike ordinary matrix multiplication. Also note that $A \otimes B$ and $B \otimes A$ have the same size but rarely are equal. In other words, we do not have a commutative multiplication here. Let's now collect some of the basic results about computing with tensor products.

THEOREM 12.16 (basic facts about tensor products)

1. *For any matrices* $A, B, C, (A \otimes B) \otimes C = A \otimes (B \otimes C)$.

2. *For* A, B *m-by-n and* C *p-by-q*,
   $(A + B) \otimes C = A \otimes C + B \otimes C$ *and*
   $C \otimes (A + B) = C \otimes A + C \otimes B$.

3. *If* α *is a scalar (considered as a 1-by-1 matrix),*
   then $\alpha \otimes A = \alpha A = A \otimes \alpha$ *for any matrix* A.

4. $\mathbb{O} \otimes A = \mathbb{O} = A \otimes \mathbb{O}$ *for any matrix* A.

5. *If* α *is a scalar,* $(\alpha A) \otimes B = \alpha(A \otimes B) = A \otimes (\alpha B)$
   for any matrices A *and* B.

6. *If* $D = diag(d_1, d_2, \ldots, d_m)$, *then* $D \otimes A =$
   $diag(d_1 A, d_2 A, \ldots, d_m A)$.

7. $I_m \otimes I_n = I_{mn}$.

8. *For any matrices* $A, B, (A \otimes B)^T = A^T \otimes B^T$.

9. *For any matrices* $A, B, (A \otimes B)^* = A^* \otimes B^*$.

10. *If* A *is m-by-n,* B *p-by-q,* C *n-by-s,* D *q-by-t, then* $(A \otimes B)(C \otimes D) =$ $(AC) \otimes (BD)$.

11. *If* A *and* B *are square, not necessarily the same*
    size, then $tr(A \otimes B) = tr(A)tr(B)$.

12. *If* A *and* B *are invertible, not necessarily the same size, then* $(A \otimes B)^{-1} =$ $A^{-1} \otimes B^{-1}$.

13. *If* A *is m-by-m and* B *n-by-n, then* $\det(A \otimes B) = (\det(A))^n (\det(B))^m$.

14. *If* A *is m-by-n and* B *p-by-q and if* A^{g_1}, B^{g_1} *are one-inverses of* A *and* B, *respectively, then* $A^{g_1} \otimes B^{g_1}$ *is a one-inverse of* $A \otimes B$.

15. *For any matrices A and B,* $rank(A \otimes B) = rank(A)rank(B)$.
Consequently $A \otimes B$ *has full row rank iff A and B have full row rank. A similar statement holds for column rank. In particular,* $A \otimes B$ *is invertible iff both A and B are.*

16. $\overline{A \otimes B} = \overline{A} \otimes \overline{B}$.

PROOF The proofs will be left as exercises. □

All right, so what are tensor products good for? As usual, everything in linear algebra boils down to solving systems of linear equations. Suppose we are trying to solve $AX = B$ where A is m-by-n, B is m-by-p, and X is n-by-p; say $\begin{bmatrix} a_{11} & a_{12} \\ a_{21} & a_{22} \end{bmatrix} \begin{bmatrix} x_1 & x_2 \\ x_3 & x_4 \end{bmatrix} = \begin{bmatrix} b_{11} & b_{12} \\ b_{21} & b_{23} \end{bmatrix}$. If we are clever, we can write this as an ordinary system of linear equations:
$\begin{bmatrix} a_{11} & a_{12} & 0 & 0 \\ a_{21} & a_{22} & 0 & 0 \\ 0 & 0 & a_{11} & a_{12} \\ 0 & 0 & a_{21} & a_{22} \end{bmatrix} \begin{bmatrix} x_1 \\ x_3 \\ x_2 \\ x_4 \end{bmatrix} = \begin{bmatrix} b_{11} \\ b_{21} \\ b_{12} \\ b_{22} \end{bmatrix}$. This is just $(I_2 \otimes A)\mathbf{x} = \mathbf{b}$ where $\mathbf{x}$ and $\mathbf{b}$ are big tall vectors made by stacking the columns of X and B on top of each other. This leads us to introduce the *vec* operator.

We can take an m-by-n matrix A and form an mn-by-1 column vector by stacking the columns of A over each other.

DEFINITION 12.7 *(vec)*

Let A be an m-by-n matrix. Then $vec(A) = \begin{bmatrix} Col_1(A) \\ Col_2(A) \\ \vdots \\ Col_n(A) \end{bmatrix} \in \mathbb{C}^{mn+1}$. *Now if A is m-by-n, B n-by-p, and X n-by-p, we can reformulate the matrix equation* $AX = B$ *as* $(I_p \otimes A)\boldsymbol{x} = \mathbf{b}$, *where* $\boldsymbol{x} = vec(X)$ *and* $\mathbf{b} = vec(B)$. *If we can solve one system, we can solve the other. Let's look at the basic properties of vec. Note the* (i, j) *entry of A is the* $[(j-1)m + i]^{th}$ *element of* $vec(A)$.

THEOREM 12.17 (basic properties of vec)

1. *If* $\mathbf{x}$ *and* $\mathbf{y}$ *are columns vectors, not necessarily the same size, then* $vec(\mathbf{x}\mathbf{y}^T) = \mathbf{y} \otimes \mathbf{x}$.

2. *For any scalar c and matrix A,* $vec(cA) = c\, vec(A)$.

3. *For matrices A and B of the same size,* $vec(A + B) = vec(A) + vec(B)$.

4. *Let A and B have the same size, then $tr(A^T B) = [vec(A)]^T\ vec(B)$.*

5. *If A is m-by-n and B n-by-p, then*

$$vec(AB) = \begin{bmatrix} Acol_1 B \\ Acol_2(B) \\ \vdots \\ Acol_p(B) \end{bmatrix} = diag(A, A, \ldots, A)vec(B) = (I_p \otimes A)vec(B)$$
$$= (B^{\mathsf{T}} \otimes I_m)vec(A).$$

6. *Let A be m-by-n, B n-by-p, C p-by-q; then $vec(ABC) = (C^{\mathsf{T}} \otimes A)vec(B)$.*

PROOF The proofs are left as exercises. ▯

We have seen that $AX = B$ can be rewritten as $(I_p \otimes A)vec(X) = vec(B)$. More generally, if A is m-by-n, B m-by-q, C p-by-q, and X n-by-p, then $AXC = B$ can be reformulated as $(C^T \otimes A)vec(X) = vec(B)$. Another interesting matrix equation is $AX + XB = C$, where A is m-by-m, B is n-by-n, C is m-by-n, and X, m-by-n. This equation can be reformulated as $(A \otimes I_n + I_m \otimes B^{\mathsf{T}})vec(X) = vec(C)$. Thus, the matrix equation admits a unique solution iff $A \otimes I_n + I_m \otimes B^{\mathsf{T}}$ is invertible of size mn-by-mn.

Exercise Set 57

1. Prove the various claims of Theorem 12.16.

2. Prove the various claims of Theorem 12.17.

3. Is it true that $(A \otimes B)^+ = A^+ \otimes B^+$?

4. Work out $vec(\begin{bmatrix} a & b & c \\ d & e & f \\ g & h & i \end{bmatrix})$ explicitly.

5. Define $rvec(A) = [vec(A^T)]^T$. Work out $rvec(\begin{bmatrix} a & b & c \\ d & e & f \\ g & h & i \end{bmatrix})$ explicitly. Explain in words what $rvec$ does to a matrix.

6. Prove that $vec(AB) = (I_p \otimes A)vec(B) = (B^T \otimes I_m)vec(A) = (B^T \otimes A)vec(I_n)$.

7. Show that $vec(ABC) = (C^T \otimes A)vec(B)$.

8. Argue that $tr(AB) = vec(A^T)^T vec(B) = vec(B^T)^T vec(A)$.

9. Prove that $tr(ABCD) = vec(D^T)^T(C^T \otimes A)vec(B) = vec(D)^T(A \otimes C^T)vec(B^T)$.

10. Show that $\alpha A \otimes \beta B = \alpha\beta(A \otimes B) = (\alpha\beta A) \otimes B = A \otimes (\alpha\beta B)$.

11. What is $(A \otimes B \otimes C)(D \otimes E \otimes F) =?$

12. If $\lambda \in \lambda(A)$ and $\mu \in \lambda(B)$ with eigenvectors $\mathbf{v}$ and $\mathbf{w}$, respectively, argue that $\lambda\mu$ is an eigenvalue of $A \otimes B$ with eigenvector $\mathbf{v} \otimes \mathbf{w}$.

13. Suppose $\mathbf{v}$ and $\mathbf{w}$ are m-by-1. Show that $\mathbf{v}^T \otimes \mathbf{w} = \mathbf{w}\mathbf{v}^T = \mathbf{w} \otimes \mathbf{v}^T$ and $vec(\mathbf{v}\mathbf{w}^T) = \mathbf{w} \otimes \mathbf{v}$.

14. Prove that $vec(\overline{A}) = \overline{vec(A)}$.

15. Prove that $vec(A^*) = \overline{vec(A^T)}$.

16. Suppose $G \in A\{1\}$ and $H \in B\{1\}$. Prove that $G \otimes H \in A \otimes B\{1\}$.

17. Prove that the tensor product of diagonal matrices is a diagonal matrix.

18. Prove that the tensor product of idempotent matrices is an idempotent matrix. Is the same true for projections?

19. Is the tensor product of nilpotent matrices also nilpotent?

12.7.1 MATLAB Moment

12.7.1.1 Tensor Product of Matrices

MATLAB has a built-in command to compute the tensor (or Kronecker) product of two matrices. If you tensor an m-by-n matrix with a p-by-q matrix, you get an mp-by-nq matrix. You can appreciate the computer doing this kind of operation for you. The command is

kron(A,B).

Just for fun, let's illustrate with matrices filled with prime numbers.

$$>>A = zeros(2); A(:) = primes(2)$$

$$A = \begin{matrix} 2 & 5 \\ 3 & 7 \end{matrix}$$

$>>B = zeros(4); B(:) = primes(53)$

$$B = \begin{matrix} 2 & 11 & 23 & 41 \\ 3 & 13 & 29 & 43 \\ 5 & 17 & 31 & 47 \\ 7 & 19 & 37 & 53 \end{matrix}$$

$>>kron(A, B)$ ans =

4	22	46	82	10	55	115	205
6	26	58	86	15	65	145	215
10	34	62	94	25	85	155	235
14	38	74	106	35	95	185	265
6	33	69	123	14	77	161	287
9	39	87	129	21	91	203	301
15	51	93	141	35	119	217	329
21	57	111	159	49	133	259	371

Try kron(eye(2),A) and kron(A,eye(2)) and explain what is going on.

This is a good opportunity to create a function of our own that MATLAB does not have built in. We do it with an M-file. To create an .m file in a Windows environment, go to **File**, choose **New**, and then choose **M-File**. In simplest form, all you need is to type

$$B = A(:) function\ B = vec(A)$$

Then do a **Save.**

Try out your new function on the matrix B created above.

$$>>vec(B)$$

You should get a long column consisting of the first 16 primes.

Appendix A

Complex Numbers

A.1 What is a Scalar?

A scalar is just a number. In fact, when you ask the famous "person on the street" what mathematics is all about, he or she will probably say that it is about numbers. All right then, but what is a number? If we write "5" on a piece of paper, it might be tempting to say that it is the number five, but that is not quite right. In fact, this scratching of ink "5" is a *symbol* or *name* for the number and is not the number five itself. We got this symbol from the Hindus and Arabs. In ancient times, the Romans would have written "V" for this number. In fact, there are many numeral systems that represent numbers. Okay, so where does the number five, independent of how you symbolize it, exist? Well, this is getting pretty heavy, so before you think you are in a philosophy course, let's proceed as if we know what we are talking about. Do you begin to see why mathematics is called "abstract"?

We learned to count at our mother's knee: 1, 2, 3, 4, ... (Remember that " . . . " means "keep on going in this manner," or "etc." to a mathematician.)[1] These numbers are the *counting numbers,* or *whole numbers,* and we officially name them the "natural numbers." They are symbolized by $\mathbb{N} = \{1, 2, 3 \ldots\}$.

After learning to count, we learned that there are some other things you can do with natural numbers. We learned the *operations* of *addition, multiplication, subtraction,* and *division.* (Remember "gozinta" ? 3 gozinta 45 fifteen times.) It turns out that the first two operations can always be done, but that is not true of the last two. (What natural number is 3 minus 5 or 3 divided by 5?) We also learned that addition and multiplication satisfy certain *rules of computation.* For example, $m+n = n+m$ for all natural numbers m and n. (Do you know the name of this rule?) Can you name some other rules? Even though the natural numbers seem to have some failures, they turn out to be quite fundamental to all of mathematics. Indeed, a famous German mathematician, **Leopold Kronecker** (7 December 1823–29 December 1891), is supposed to have said that God

[1]Do you know the official name of the symbol " . . . "?

created the natural numbers and that all the rest is the work of humans. It is not clear how serious he was when he said that.

Anyway, we can identify a clear weakness in the system of natural numbers by looking at the equation:

$$m + x = n. \tag{A.1}$$

This equation cannot always be solved for x in $\mathbb{N}$ (sometimes yes, but not *always*). Therefore, mathematicians invented more numbers. They constructed the system of *integers,* $\mathbb{Z}$, consisting of all the natural numbers together with their opposites (the opposite of 3 is -3) and the wonderful and often underrated number zero. Thus, we get the enlarged number system:

$$\mathbb{Z} = \{\dots, -2, -1, 0, 1, 2, \dots\}.$$

We have made progress. In the system, $\mathbb{Z}$, equation (1) can always be solved for x. Note that $\mathbb{N}$ is contained in $\mathbb{Z}$, in symbols, $\mathbb{N} \subseteq \mathbb{Z}$. The construction of $\mathbb{Z}$ is made so that all the rules of addition and multiplication of $\mathbb{N}$ still hold in $\mathbb{Z}$.

We are still not out of the woods! What integer is 2 divided by 3? In terms of an equation,

$$ax = b \tag{A.2}$$

cannot always be solved in $\mathbb{Z}$, so $\mathbb{Z}$ also is lacking in some sense.

Well all right, let's build some more numbers, the ratios of integers. We call these numbers *fractions,* or *rational numbers,* and they are symbolized by:

$$\mathbb{Q} = \left\{\frac{a}{b} \mid a, b \in \mathbb{Z}, b \neq 0\right\}.$$

(Remember, we cannot divide by zero. Can you think of a good reason why not?) Since a fraction $\frac{a}{1}$ is as good as a itself, we may view the integers as contained in $\mathbb{Q}$, so that our ever expanding universe of numbers can be viewed as an inclusion chain:

$$\mathbb{N} \subseteq \mathbb{Z} \subseteq \mathbb{Q}.$$

Once again, the familiar operations are defined, so nothing will be lost from the earlier number systems, making $\mathbb{Q}$ a rather impressive number system.

In some sense, all scientific measurements give an answer in $\mathbb{Q}$. You can always add, multiply, subtract, and divide (not by zero), and solve equations

(A.1) and (A.2) in $\mathbb{Q}$. Remember the rules you learned about operating with fractions?

$$\frac{a}{b} + \frac{c}{d} = \frac{ad + bc}{bd} \tag{A.2}$$[1]

$$\left(\frac{a}{b}\right)\left(\frac{c}{d}\right) = \frac{ac}{bd} \tag{A.4}$$

$$\frac{a}{b} - \frac{c}{d} = \frac{ad - bc}{bd} \tag{A.5}$$

$$\left(\frac{a}{b}\right) \div \left(\frac{c}{d}\right) = \frac{ad}{bc} \tag{A.6}$$

("Invert and multiply rule")

It seems we have reached the end of the road. What more can we do that $\mathbb{Q}$ does not provide? Oh, if life were only so simple! Several thousand years ago, Greek mathematicians knew there was trouble. I think they tried to hush it up at the time, but truth has a way of eventually working its way out. Consider a square with unit side lengths (1 yard, 1 meter—it does not matter).

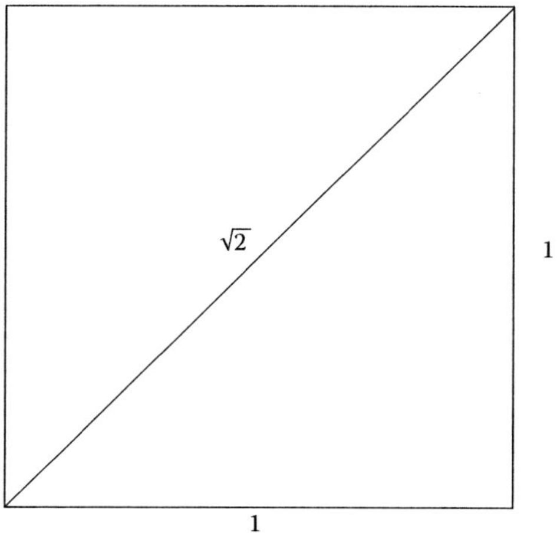

Figure A1.1: Existence of nonrational numbers.

[1]Would you believe we have seen some college students who thought $\frac{a}{b} + \frac{c}{d} = \frac{a+c}{b+d}$!? What were they thinking? It makes you shiver all over!

The diagonal of that square has a length whose square is two. (What famous theorem of geometry led to that conclusion?) The trouble is there are no ratios of integers $\frac{a}{b}$ whose square is two; that is, $x^2 = 2$ has no solution in $\mathbb{Q}$. Even $\mathbb{Q}$ is lacking in some sense, namely the equation:

$$x^2 = 2 \tag{A.7}$$

cannot be solved for x in $\mathbb{Q}$, although we can get really close to a solution in $\mathbb{Q}$. It took mathematicians quite awhile to get the next system built. This is the system $\mathbb{R}$ of real numbers that is so crucial to calculus. This system consists of all rational numbers as well as all irrational numbers, such as $\sqrt{2}, \sqrt{3}, \sqrt{5}, \pi, e, \ldots$. Our chain of number systems continues to grow:

$$\mathbb{N} \subseteq \mathbb{Z} \subseteq \mathbb{Q} \subseteq \mathbb{R}.$$

Once again, we extend our operations to $\mathbb{R}$ so we do not lose all the wonderful features of $\mathbb{Q}$. But now, we can solve $x^2 = 2$ in $\mathbb{R}$. There are exactly two solutions, which we can write as $\sqrt{2}$ and $-\sqrt{2}$. Indeed we can solve $x^2 = p$ for any nonnegative real number p. Aye, there's the rub. You can prove squares cannot be negative in $\mathbb{R}$. Did you think we were done constructing number systems? Wrong, square root breath! How about the equation

$$x^2 = -3? \tag{A.8}$$

No, the real numbers, a really big number system, is still lacking. You guessed it, though, we can build an even larger number system called the system of *complex numbers* $\mathbb{C}$ in which (8) can be solved.

Around 2000 B.C., the Babylonians were aware of how to solve certain cases of the quadratic equation (famous from our high school days):

$$ax^2 + bx + c = 0.$$

In the sixteenth century, **Girolamo Cardano** (24 September 1501–21 September 1576) knew that there were quadratic equations that could not be solved over the reals. If we simply "push symbols" using rules we trust, we get the roots of $ax^2 + bx + c = 0$ to be

$$r = \frac{b \pm \sqrt{b^2 - 4ac}}{2a} = \frac{-b}{2a} \pm \frac{\sqrt{b^2 - 4ac}}{2a}.$$

But what if $b^2 - 4ac$ is negative? The square root does not make any sense then in $\mathbb{R}$. For example, consider $x^2 + x + 1 = 0$. Then $r = \dfrac{-1 \pm \sqrt{1-4}}{2} = -\dfrac{1}{2} \pm \dfrac{\sqrt{-3}}{2}$. If you plug these "imaginary" numbers into the equation, you do get zero if you agree that $\sqrt{-1}\cdot\sqrt{-1} = -1$. What can I say? It works! You can tell by the name "imaginary" that these "numbers" were held in some mistrust. It took time but, by the eighteenth century, the remarkable work of **Leonard Euler** (15 April 1707–18 September 1783) brought the complex number system into good repute. Indeed, it was Euler who suggested using the letter i for $\sqrt{-1}$. (Yes, I can hear all you electrical engineers saying "i is current; we use j for $\sqrt{-1}$." I know, but this is a mathematics book so we will use i for $\sqrt{-1}$.) Our roots above can be written as

$$r = -\frac{1}{2} \pm \frac{\sqrt{3}}{2}i.$$

We are now in a position to give a complete classification of the roots of any quadratic equation $ax^2 + bx + c = 0$ with real number coefficients.

Case 1: $b^2 - 4ac$ is positive.
Then $r = \frac{-b}{2a} \pm \frac{\sqrt{b^2 - 4ac}}{2a}$ gives two distinct real roots of the equation, one for the + and one for the –.

Case 2 : $b^2 - 4ac$ is zero.
Then $r = -\frac{b}{2a}$ is a repeated real root of the equation.

Case 3 : $b^2 - 4ac$ is negative.

Then $r = \frac{-b}{2a} \pm \frac{\sqrt{(-1)(4ac - b^2)}}{2a} = \frac{-b}{2a} \pm \frac{\sqrt{4ac - b^2}}{2a}(\sqrt{-1}) = \left(\frac{-b}{2a}\right) \pm \left(\frac{\sqrt{4ac - b^2}}{2a}\right) i = \alpha \pm \beta i$ gives the so-called "complex conjugate pair" of roots of the quadratic. This is how complex numbers came to be viewed as numbers of the form $\alpha + \beta i$ or $\alpha + i\beta$.

It took much more work after Euler to make it plain that complex numbers were just as meaningful as any other numbers. We mention the work of **Casper Wessel** (8 June 1745–25 March 1818), **Jean Robert Argand** (18 July 1768–13 August 1822), **Karl Friedrich Gauss** (30 April 1777–23 February 1855) (his picture was on the German 10 Mark bill) and **Sir William Rowen Hamilton** (4 August 1805–2 September 1865) among others. We are about to have a good look at complex numbers, since they are the crucial scalars used in this book. One of our main themes is how the system of complex numbers and the system of (square) complex matrices share many features in common. We will use the approach of Hamilton to work out the main properties of the complex number system.

Further Reading

[Dehaene, 1997] S. Dehaene, *The Number Sense,* Oxford University Press, New York, (1997).

[Keedy, 1965] M. L. Keedy, *Number Systems: A Modern Introduction,* Addison-Wesley, Reading, MA, (1965).

[Nahin, 1998] P. J. Nahin, *An Imaginary Tale: The Story of* $\sqrt{-1}$, Princeton University Press, Princeton, NJ, (1998).

[Roberts, 1962] J. B. Roberts, *The Real Number System in an Algebraic Setting,* W. H. Freeman and Co., San Francisco, (1962).

A.2 The System of Complex Numbers

We will follow Hamilton's idea of defining a complex number z to be an ordered pair of real numbers. So $z = (a, b)$, where a and b are real, is a complex number; the first coordinate of z is called the *real part* of z, $Re(z)$, and the second is called the *imaginary part*, $Im(z)$. (There is really nothing imaginary about it, but that is what it is called.) Thus, complex numbers look suspiciously like points in the Cartesian coordinate plane $\mathbb{R}^2$. In fact, they can be viewed as vectors in the plane. This representation will help us draw pictures of complex numbers. It was Argand who suggested this and so this is sometimes called an *Argand diagram.*

The operations on complex numbers are defined as follows: (Let $z_1 = (x_1, y_1)$ and $z_2 = (x_2, y_2)$ be complex numbers):

1.1 **addition**: $z_1 \oplus z_2 = (x_1 + x_2, y_1 + y_2)$

1.2 **subtraction**: $z_1 \ominus z_2 = (x_1 - x_2, y_1 - y_2)$

1.3 **multiplication**: $z_1 \odot z_2 = (x_1x_2 - y_1y_2, x_1y_2 + y_1x_2)$

There are two special complex numbers **0** (zero) and **1** (one). We define:

1.4 **the zero** $\mathbf{0} = (0, 0)$

1.5 **the multiplicative identity** $\mathbf{1} = (1, 0)$

1.6 **division**: $\dfrac{z_1}{z_2} = \left(\dfrac{x_2x_1 + y_2y_1}{x_2^2 + y_2^2}, \dfrac{x_1y_2 - x_2y_1}{x_2^2 + y_2^2}\right)$ provided $z_2 \neq \mathbf{0}$.

Some remarks are in order. First, there are two kinds of operations going on here: those in the complex numbers $\oplus$, $\odot$, etc. and those in $\mathbb{R}$, $+$, $\cdot$ etc. We are going to play as if we know all about the operations in $\mathbb{R}$. Then, we will derive all the properties of the operations for complex numbers. Next, the definitions of multiplication and division may look a little weird, but they are exactly what they have to be to make things work.

Let's agree to write $\mathbb{C} = \mathbb{R}^2$ for the set of complex numbers equipped with the operations defined above. We can take a hint from viewing $\mathbb{C}$ as $\mathbb{R}^2$ and identify the real number a with the complex number $(a, 0)$. This allows us to view $\mathbb{R}$ as contained within $\mathbb{C}$. So our inclusion chain now reads

$$\mathbb{N} \subseteq \mathbb{Z} \subseteq \mathbb{Q} \subseteq \mathbb{R} \subseteq \mathbb{C}.$$

Also, we can recapture the usual way of writing complex numbers by

$$(a, b) = (a, 0) \oplus (0, b) = a \oplus (b, 0) \odot (0, 1) = a \oplus b \odot i$$

Since the operations in $\mathbb{R}$ and $\mathbb{C}$ agree for real numbers, there is no longer a need to be so formal with the circles around plus and times, so we now drop that. Let's just agree

$$(a, b) = a + bi.$$

APA Exercise Set 1

1. Use the definitions of addition, subtraction, multiplication, and division to compute $z_1 + z_2$, $z_1 - z_2$, $z_1 z_2$, z_1/z_2, and z_2^2, z_1^3, where $z_1 = 3 + 4i$, and $z_2 = 2 - i$. Express you answers in standard form $a + bi$.

2. If $i = (0, 1)$, use the definition of multiplication to compute i^2.

3. Find the real and imaginary parts of:
   (a) $3 - 4i$
   (b) $6i$
   (c) $(2 + 3i)^2$
   (d) $(\sqrt{3} - i)^3$
   (e) $(1 + i)^{50}$.

4. Find all complex numbers with $z^2 = i$.

5. Consider complex numbers of the form $(a, 0)$, $(b, 0)$. Form their sum and product. What conclusion can you draw?

6. Solve $z + (4 - 3i) = 6 + 5i$ for z.

7. Solve $z(4 - 3i) = 6 + 5i$ for z.

8. Let $z \neq 0$, $z = a + bi$. Compute $\frac{1}{z}$ in terms of a and b. What is $\frac{1}{i}, \frac{1}{2+i}$?

9. Prove $Re(iz) = -Im(z)$ for any complex number z.

10. Prove $Im(iz) = Re(z)$ for any complex number z.

11. Solve $(1 + i)z^2 + (3i)z + (5 - 4i) = 0$.

12. What is i^{1999}?

13. Draw a vector picture of $z = 3 + 4i$ in $\mathbb{R}^2$. Also draw $3 - 4i$ and $-4i$.

A.3 The Rules of Arithmetic in $\mathbb{C}$

Next, we will list the basic rules of computing with complex numbers. Once you have these, all other rules can be derived.

A.3.1 Basic Rules of Arithmetic in $\mathbb{C}$

A.3.1.1 Associative Law of Addition

$z_1 + (z_2 + z_3) = (z_1 + z_2) + z_3$ for all z_1, z_2, z_3 in $\mathbb{C}$.
(This is the famous "move the parentheses" law.)

A.3.1.2 Existence of a Zero

The element $\mathbf{0} = (0, 0)$ is neutral for addition; that is, $\mathbf{0} + z = z + \mathbf{0} = z$ for all complex numbers z.

A.3.1.3 Existence of Opposites

Each complex number z has an opposite $-z$ such that $z + (-z) = \mathbf{0} = (-z) + z$. More specifically, if $z = (x, y)$, then $-z = (-x, -y)$.

A.3.1.4 Commutative Law of Addition

$z_1 + z_2 = z_2 + z_1$ for all z_1, z_2 in $\mathbb{C}$.

A.3.1.5 Associative Law of Multiplication

$z_1(z_2z_3) = (z_1z_2)z_3$ for all z_1, z_2, z_3 in $\mathbb{C}$.

A.3.1.6 Distributive Laws

Multiplication distributes over addition; that is,

$z_1(z_2 + z_3) = z_1z_2 + z_1z_3$ for all z_1, z_2, z_3 in $\mathbb{C}$ and also $(z_1 + z_2)z_3 = z_1z_3 + z_2z_3$.

A.3.1.7 Commutative Law for Multiplication

$z_1z_2 = z_2z_1$ for all z_1, z_2 in $\mathbb{C}$.

A.3.1.8 Existence of Identity

One is neutral for multiplication; that is, $\mathbf{1} \cdot z = z \cdot \mathbf{1} = z$ for all complex numbers z in $\mathbb{C}$.

A.3.1.9 Existence of Inverses

For each nonzero complex number z, there is a complex number z^{-1} so that $zz^{-1} = z^{-1}z = \mathbf{1}$. In fact, if $z = (x, y) \neq \mathbf{0}$, then $z^{-1} = \left(\dfrac{x}{x^2 + y^2}, \dfrac{-y}{x^2 + y^2}\right)$.

We will illustrate how these rules are established and leave most of them to you. Let us set a style and approach to proving these rules. It amounts to "steps and reasons."

PROOF of 2.1

Let $z_1 = (a_1, b_1)$, $z_2 = (a_2, b_2)$, and $z_3 = (a_{3,} b_3)$.

Then $z_1 + (z_2 + z_3) = (a_1, b_1) + ((a_2, b_2) + (a_{3,} b_3)$ by substitution,

$= (a_1 b_1) + (a_2 + a_3, b_2 + b_3)$ by definition of addition in $\mathbb{C}$,

$= (a_1 + (a_2 + a_3), b_1 + (b_2 + b_3))$ by definition of addition in $\mathbb{C}$,

$= ((a_1 + a_2) + a_3, (b_1 + b_2) + b_3)$ by associative law of addition in $\mathbb{R}$,

$= (a_1 + a_2, b_1 + b_2) + (a_3 b_3)$ by definition of addition in $\mathbb{C}$,

$= ((a_1, b_1) + (a_2, b_2)) + (a_3, b_3)$ by definition of addition in $\mathbb{C}$,

$= (z_1 + z_2) + z_3$ by substitution. □

You see how this game is played? The idea is to push everything (using definitions and basic logic) back to the crucial step where you use the given properties of the reals $\mathbb{R}$. Now see if you can do the rest. We can summarize our discussion so far. Mathematicians say that we have established that $\mathbb{C}$ is a *field*.

APA Exercise Set 2

1. Prove all the other basic rules of arithmetic in $\mathbb{C}$ in a manner similar to the one illustrated above.

2. Prove $z_1 \cdot z_2 = \mathbf{0}$ iff $z_1 = \mathbf{0}$ or $z_2 = \mathbf{0}$.

3. Prove the cancellation law: $z_1 z = z_2 z$ and $z \neq 0$ implies $z_1 = z_2$.

4. Note subtraction was not defined in the basic rules. That is because we can define subtraction by $z_1 - z_2 = z_1 + (-z_2)$. Does the associative law work for subtraction?

5. Note division was not defined in the basic rules. We can define $z_1 \div z_2 = \frac{z_1}{z_2}$ by $z_1 \div z_2 = z_1 z_2^{-1}$, if $z_2 \neq 0$. Can you discover some rules that work for division?

6. Suppose $z = a + bi \neq 0$. Argue that $\sqrt{z} = \sqrt{\frac{\sqrt{a^2+b^2}}{2}}(\sqrt{1+\frac{a}{\sqrt{a^2+b^2}}} + i\sqrt{1-\frac{a}{\sqrt{a^2+b^2}}})$ if $b \geq 0$ and $\sqrt{z} = \sqrt{\frac{\sqrt{a^2+b^2}}{2}}(-\sqrt{1+\frac{a}{\sqrt{a^2+b^2}}} + i\sqrt{1-\frac{a}{\sqrt{a^2+b^2}}})$ if $b \leq 0$. Use this information to find $\sqrt{1+i}$.

7. Argue that $ax^2+bx+c = 0$ has solutions $z = -\left(\frac{b}{2a}\right)+\sqrt{\left(\frac{b}{2a}\right)^2 - \frac{c}{a}}$ and $z = -\left(\frac{b}{2a}\right) - \sqrt{\left(\frac{b}{2a}\right)^2 - \frac{c}{a}}$ even if a, b, and c are complex numbers.

A.4 Complex Conjugation, Modulus, and Distance

There is something new in $\mathbb{C}$ that really is not present in $\mathbb{R}$. We can form the *complex conjugate* of a complex number. The official definition of complex conjugate is: if $z = (a, b) = a + bi$, then $\overline{z} := (a, -b) = a - bi$. Next we collect what is important to know about complex conjugation.

A.4.1 Basic Facts About Complex Conjugation

(3.1) $\overline{\overline{z}} = z$ for all z in $\mathbb{C}$.
(3.2) $\overline{z_1 + z_2} = \overline{z}_1 + \overline{z}_2$ for all z_1, z_2 in $\mathbb{C}$.
(3.3) $\overline{z_1 - z_2} = \overline{z}_1 - \overline{z}_2$ for all z_1, z_2 in $\mathbb{C}$.
(3.4) $\overline{z_1 z_2} = \overline{z}_1 \overline{z}_2$ for all z_1, z_2 in $\mathbb{C}$.
(3.5) $\left(\frac{\overline{z_1}}{z_2}\right) = \frac{\overline{z}_1}{\overline{z}_2}$ for all z_1, z_2 in $\mathbb{C}$ with $z_2 \neq 0$.
(3.6) $\overline{\left(\frac{1}{z}\right)} = \frac{1}{\overline{z}}$ for all nonzero z in $\mathbb{C}$.

As illustration again, we will prove one of these and leave the other proofs where they belong, with the reader.

PROOF of 3.4 Let $z_1 = (a_1, b_1)$ and $z_2 = (a_2, b_2)$. Then,
$\overline{z_1 z_2} = \overline{(a_1, b_1)(a_2, b_2)}$ by substitution,
$= \overline{(a_1 a_2 - b_1 b_2, a_1 b_2 + b_1 a_2)}$ by definition of multiplication in $\mathbb{C}$,
$= (a_1 a_2 - b_1 b_2, -(a_1 b_2 + b_1 a_2))$ by definition of complex conjugate,
$= (a_1 a_2 - b_1 b_2, -a_1 b_2 - b_1 a_2)$ by properties of additive inverses in $\mathbb{R}$,
$= (a_1 a_2 - (-b_1)(-b_2), a_1(-b_2) + (-b_1)a_2)$ by properties of additive inverses in $\mathbb{R}$,
$= (a_1, -b_1)(a_2, -b_2)$ by definition of multiplication in $\mathbb{C}$,
$= \overline{z}_1 \overline{z}_2$ by substitution. □

Again, see how everything hinges on definitions and pushing back to where you can use properties of the reals.

Next, we extend the idea of *absolute value* from $\mathbb{R}$. Given a complex number $z = (a, b) = a + bi$, we define the *magnitude* or *modulus* of z by $|z| := \sqrt{a^2 + b^2}$. We note that if z is real, $z = (a, 0)$; then $|z| = \sqrt{a^2} = |a|$ the absolute value of a. Did you notice we are using $|\ |$ in two ways here? Anyway, taking the modulus of a complex number produces a real number. We collect the basics on magnitude.

A.4.2 Basic Facts About Magnitude

(3.7) $\overline{z}z = |z|^2$ for all complex numbers z,
(3.8) $|z| \geq 0$ for all complex numbers z,
(3.9) $|z| = 0$ iff $z = 0$,
(3.10) $|z_1 z_2| = |z_1| |z_2|$ for all z_1, z_2 in $\mathbb{C}$,
(3.11) $\left|\frac{z_1}{z_2}\right| = \frac{|z_1|}{|z_2|}$ for all z_1, z_2 in $\mathbb{C}$ with $z_2 \neq 0$,

(3.12) $|z_1 + z_2| \le |z_1| + |z_2|$ for all z_1, z_2 in $\mathbb{C}$ (the triangle inequality),

(3.13) $\dfrac{1}{z} = \dfrac{\overline{z}}{|z|^2}$ for all $z \neq 0$ in $\mathbb{C}$,

(3.14) $||z_1| - |z_2|| \le |z_1 - z_2|$ for all z_1, z_2 in $\mathbb{C}$,

(3.15) $|z| = |-z|$ for all z in $\mathbb{C}$,

(3.16) $|\overline{z}| = |z|$.

Again, we leave the verifications to the reader. We point out that you do not always have to go back to definitions. We have developed a body of theory (facts) that can now be used to derive new facts. Take the following proof for example.

PROOF of 3.10

$$
\begin{aligned}
|z_1 z_2|^2 &= (z_1 z_2)(\overline{z_1 z_2}) && \text{by 3.7,}\\
&= (z_1 z_2)(\overline{z}_1 \overline{z}_2) && \text{by 3.4,}\\
&= ((z_1 z_2)\overline{z}_1)\overline{z}_2 && \text{by 2.5,}\\
&= (z_1(z_2\overline{z}_1))\overline{z}_2 && \text{by 2.5,}\\
&= ((z_1\overline{z}_1)z_2)\overline{z}_2 && \text{by 2.5 and 2.7,}\\
&= (z_1\overline{z}_1)(z_2\overline{z}_2) && \text{by 2.5,}\\
&= |z_1|^2\,|z_2|^2 && \text{by 3.7,}\\
&= (|z_1|\,|z_2|)^2 && \text{by the commutative law of multiplication in } \mathbb{R}.
\end{aligned}
$$

Now, by taking square roots, 3.10 is proved. □

One beautiful fact is that when you have a notion of absolute value (magnitude), you automatically have a way to measure *distance,* in this case, the distance between two complex numbers. If z_1 and z_2 are complex numbers, we define the *distance* between them by $d(z_1, z_2) := |z_1 - z_2|$. Note that if $z_1 = (a_1, b_1)$ and $z_2 = (a_2, b_2)$, then $d(z_1, z_2) = |z_1 - z_2| = |(a_1 - a_2, b_1 - b_2)| = \sqrt{(a_1 - a_2)^2 + (b_1 - b_2)^2}$. This is just the usual distance formula between points in $\mathbb{R}^2$! Let's now collect the basics on distance. These follow easily from the properties of magnitude.

A.4.3 Basic Properties of Distance

(3.17) $d(z_1, z_2) \ge 0$ for all z_1, z_2 in $\mathbb{C}$.

(3.18) $d(z_1, z_2) = 0$ iff $z_1 = z_2$ for z_1, z_2 in $\mathbb{C}$.

(3.19) $d(z_1, z_2) = d(z_2, z_1)$ for all z_1, z_2 in $\mathbb{C}$.

(3.20) $d(z_1, z_3) \le d(z_1, z_2) + d(z_2, z_3)$ for all z_1, z_2, z_3 in $\mathbb{C}$ (triangle inequality for distance).

3.20 $d(z_1 + w, z_2 + w) = d(z_1, z_2)$ for all z_1, z_2, w in $\mathbb{C}$
(translation invariance of distance).

We leave all these verifications as exercises. Note that once we have distance we can introduce calculus since we can say how close two complex numbers are. We can talk of convergence of complex sequences and define continuous functions. But don't panic, we won't.

We end this section with a little geometry to help out our intuition.

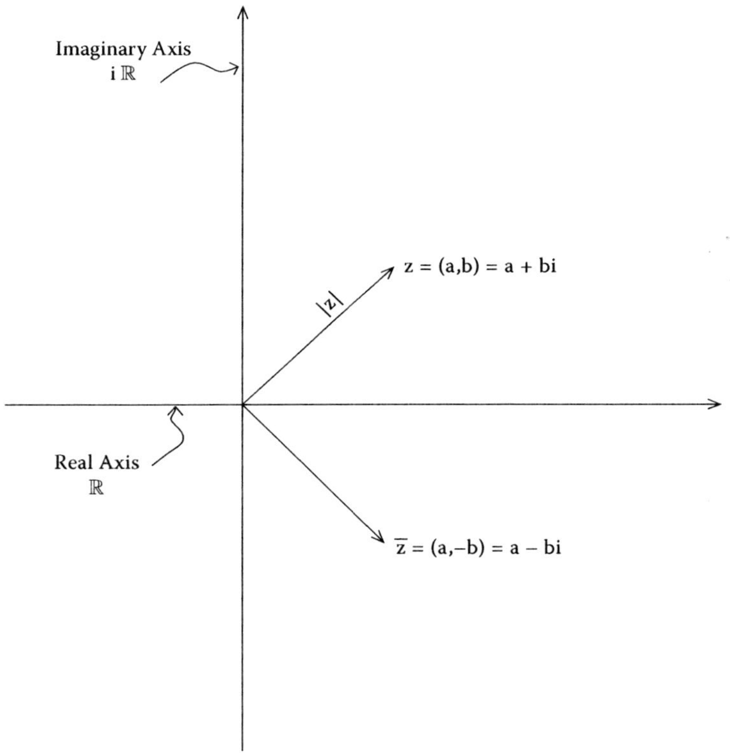

Figure A1.2: Magnitude and complex conjugate of $z = a + bi$.

Note that $|z|$ is just the distance from z to the origin and $\overline{z}$ is just the reflection of z about the real axis.

The addition of complex numbers has the interesting interpretation of addition as vectors according to the parallelogram law. (Remember from physics?)

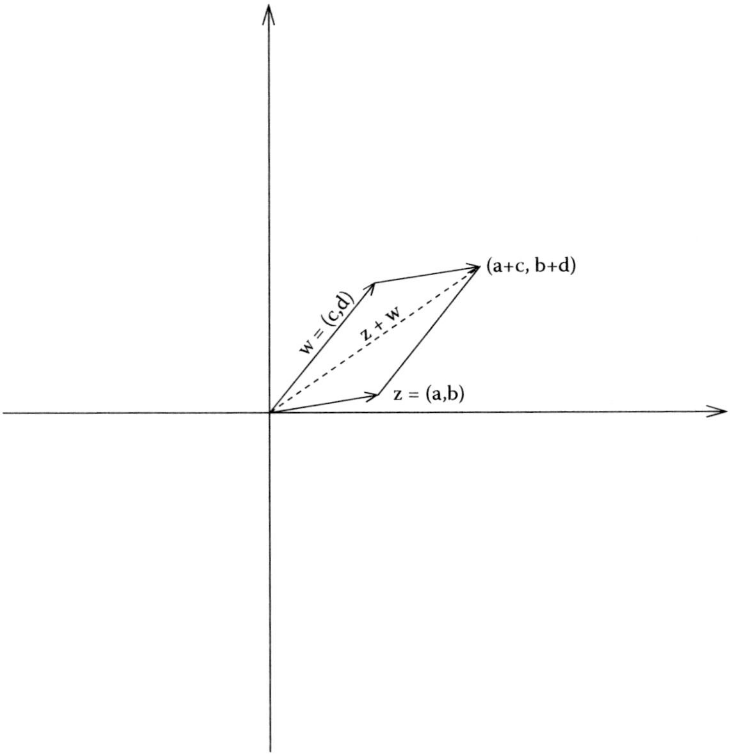

Figure A1.3: Vector view of addition of complex numbers.

(You can check the slopes of the sides to verify we have a parallelogram.) The multiplication of complex numbers also has some interesting geometry behind it. But, for this, we must develop another representation.

APA Exercise Set 3

1. Prove the basic facts about complex conjugation left unproved in the text.

2. Prove the basic facts about modulus left unproved in the text.

3. Prove the basic facts about distance left unproved in the text.

4. Find the modulus of $7 - 3i$, $4i$, $10 - i$.

5. Find the distance between $3 - 2i$ and $4 + 7i$.

6. Prove $\frac{z_1}{z_2} = \frac{z_1 \overline{z_2}}{|z_2|^2}$, if $z_2 \neq 0$.

7. Calculate $\frac{6 - 3i}{2 - i}$ using the formula in problem 6 above.

8. Find z if $(5 + 7i)\overline{z} = 2 - i$.

9. Compute $\overline{\frac{i + \overline{z}}{i - z}}$.

10. Prove $Re(z) = \frac{1}{2}(z + \overline{z})$, $Im(z) = \frac{1}{2i}(z - \overline{z})$ for any z in $\mathbb{C}$.

11. Prove $|\text{Re}(z)| \leq |z|$ for any z in $\mathbb{C}$.

12. Prove $|Re(z)| + |Im(z)| \leq \sqrt{2}\,|z|$.

13. Prove z is real iff $z = \overline{z}$; z is pure imaginary iff $\overline{z} = -z$.

14. Prove that for all z, w in $\mathbb{C}$, $|z + w|^2 + |z - w|^2 = 2\,|z|^2 + 2\,|w|^2$.

15. Prove that for all z, w in $\mathbb{C}$, $|zw| \leq \frac{1}{2}\left(|z|^2 + |w|^2\right)$.

16. For complex numbers a,b,c,d, prove that $|a - b|\,|c - d| + |a - d|\,|b - c| \geq |a - c|\,|b - d|$. Is there any geometry going on here? Can you characterize when equality occurs?

A.5 The Polar Form of Complex Numbers

You may recall from calculus that points in the plane have *polar coordinates* as well as rectangular ones. Thus, we can represent complex numbers in two ways. First, let's have a quick review. An angle in the plane is in *standard position* if it is measured counterclockwise from the positive real axis. Do not forget, these angles are measured in radians. Any angle θ (in effect, any real number) determines a unique point on the unit circle, so we take *cos* θ to be the first coordinate of that point and sin θ to be the second. Now, any complex number $z = (a, b) = a + bi \neq 0$ determines a *magnitude* $r = |z| = \sqrt{a^2 + b^2}$ and an *angle* $arg(z)$ in standard position. This angle θ is not unique since, if $\theta = \arg(z)$, then $\theta + 2\pi k$, $k \in \mathbb{Z}$ works just as well. We take the *principal argument* $\text{Arg}(z)$ to be the θ such that $-\pi < \theta \leq \pi$.

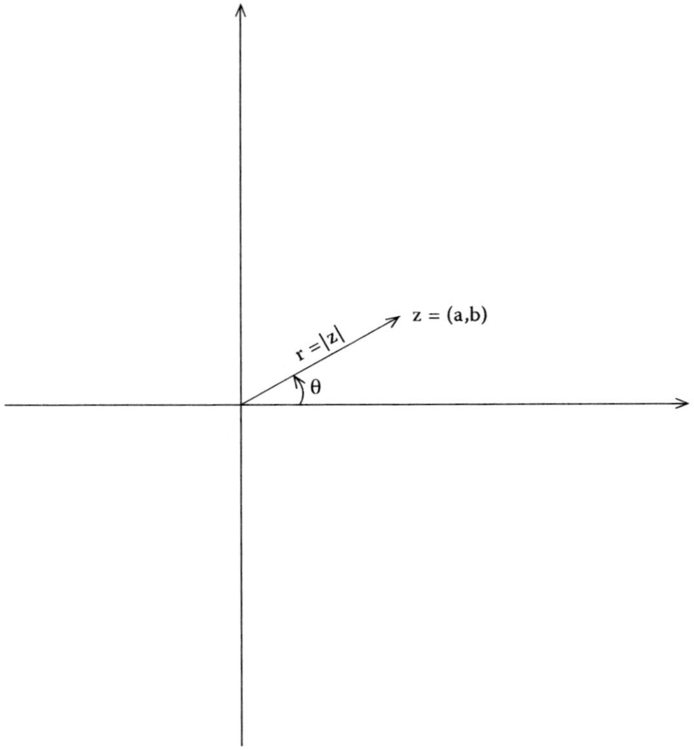

Figure A1.4: Polar form of a complex number.

From basic trigonometry we see $cos\ \theta = \frac{a}{|z|} = \frac{a}{r}$ and $sin\ \theta = \frac{b}{|z|} = \frac{b}{r}$. Thus $z = a + bi = r(\cos\theta + i\sin\theta)$.

There is an interesting connection here with the exponential function base e. Recall from calculus that

$$e^x = 1 + x + \frac{x^2}{2!} + \frac{x^3}{3!} + \frac{x^4}{4!} + \cdots + \frac{x^n}{n!} + \cdots .$$

Let's be bold and plug in a pure imaginary number $i\theta$ for x.

$$\begin{aligned} e^{i\theta} &= 1 + (i\theta) + \frac{(i\theta)^2}{2!} + \frac{(i\theta)^3}{3!} + \frac{(i\theta)^4}{4!} + \cdots \\ &= 1 + \frac{i\theta}{1!} - \frac{\theta^2}{2!} - i\frac{\theta^3}{3!} + \frac{\theta^4}{4!} + \cdots \\ &= (1 - \frac{\theta^2}{2!} + \frac{\theta^4}{4!} - \frac{\theta^6}{6!} + \frac{\theta^8}{8!} \cdots) + i(\theta - \frac{\theta^3}{3!} + \frac{\theta^5}{5!} - \frac{\theta^7}{7!} + \frac{\theta^9}{9!} \cdots) \\ &= \cos\theta + i\sin\theta. \end{aligned}$$

This is the famous **Euler formula**

$$e^{i\theta} = \cos\theta + i\sin\theta.$$

We can now write

$$z = a + bi = re^{i\theta},$$

where $r = |z|$ and $\theta = \arg(z)$ for any nonzero complex number z.
This leads to a very nice way to multiply complex numbers.

THEOREM A.1
Let $z_1 = r_1(\cos\theta_1 + i\sin\theta_1) = r_1e^{i\theta_1}$ *and* $z_2 = r_2(\cos\theta_2 + i\sin\theta_2) = r_2e^{i\theta_2}$. *Then* $z_1z_2 = r_1r_2\,[\cos(\theta_1 + \theta_2) + i\sin(\theta_1 + \theta_2)] = r_1r_2e^{i(\theta_1+\theta_2)}$.

PROOF We compute $z_1z_2 = r_1(\cos\theta_1 + i\sin\theta_1)r_2(\cos\theta_2 + i\sin\theta_2) = r_1r_2[(\cos\theta_1\cos\theta_2 - \sin\theta_1\sin\theta_2) + i(\sin\theta_1\cos\theta_2 + \cos\theta_1\sin\theta_2)] = r_1r_2[\cos(\theta_1 + \theta_2) + i\sin(\theta_1 + \theta_2)]$ using some very famous trigonometric identities. ▯

In words, our theorem says that, to multiply two complex numbers, you just multiply their magnitudes and add their arguments (any arguments will do). Many nice results fall out of this theorem. Notice that we can square a complex number $z = re^{i\theta}$ and get $z^2 = zz = r^2e^{2i\theta}$. Who can stop us now? $z^3 = r^3e^{3i\theta}$, $z^4 = r^4e^{4i\theta}$, ... Thus, we see that the theorem above leads to the famous theorem of **Abraham DeMoivre** (26 May 1667–27 November 1754).

THEOREM A.2 (DeMoivre's theorem)
If θ *is any angle,* n *is any integer, and* $z = re^{i\theta}$, *then* $z^n = r^ne^{in\theta} = r^n(\cos(n\theta) + i\sin(n\theta))$.

PROOF An inductive argument can be given for positive integers. The case $n = 0$ is easy to see. The case for negative integers has a little sticky detail. First, if $z = re^{i\theta} \neq 0$, then $(re^{i\theta})(\frac{1}{r}e^{-i\theta}) = e^0 = 1$. This says $z^{-1} = \frac{1}{r}e^{-i\theta}$. Now suppose $n = -m$, where m is a positive integer. Then $z^n = (re^{i\theta})^n = (re^{i\theta})^{-m} = ((re^{i\theta})^{-1})^m = (\frac{1}{r}e^{-i\theta})^m = r^{-m}e^{i(-m)\theta} = r^ne^{in\theta}$. ▯

One clever use of DeMoivre's theorem is to turn the equation around and find the nth roots of complex numbers. Consider $z^n = c$. If $c = 0$, this equation has only $z = 0$ as a solution. Suppose $c \neq 0$ and $c = \rho(\cos\varphi + i\sin\varphi) = z^n = r^n(\cos(n\theta) + i\sin(n\theta))$. Then we see $\rho = r^n$ and $\varphi = n\theta$ or, the other way around, $r = \rho^{1/n}$ and $\varphi = \varphi_0 + 2k\pi$, where φ_0 is the smallest nonnegative angle for c. That is, $\theta = \dfrac{\varphi_0 + 2k\pi}{n}$, where k is any integer. To get n distinct roots, we simply need $k = 0, 1, 2, \ldots, n-1$.

THEOREM A.3 (on Nth roots)
For $c = \rho(\cos\varphi_0 + i\sin\varphi_0)$, $\rho \neq 0$, the equation $z^n = c$, n a positive integer, has exactly n distinct roots, namely $z = \rho^{1/n}\left(\cos\left(\dfrac{\varphi_0 + 2k\pi}{n}\right) + i\sin\left(\dfrac{\varphi_0 + 2k\pi}{n}\right)\right)$, where $k = 0, 1, 2 \ldots, n-1$.

For example, let's find the fifth roots of $c = -2 + 2i$. First, $r = |-2 + 2i| = \sqrt{8} = 2^{3/2}$, so $\varphi_0 = 135^\circ = 3\pi/4$. Thus, $c = 2^{3/2}(\cos\dfrac{3\pi}{4} + i\sin\dfrac{3\pi}{4}) = 2^{3/2}e^{i3\pi/4}$. So the fifth roots are

$$2^{3/10}e^{3\pi/20} = 2^{3/10}(\cos\frac{3\pi}{20} + i\sin\frac{3\pi}{20}) = 2^{3/10}(\cos 27^\circ + i\sin 27^\circ),$$
$$2^{3/10}e^{10\pi/20} = 2^{3/10}(\cos\frac{11\pi}{20} + i\sin\frac{11\pi}{20}) = 2^{3/10}(\cos 99^\circ + i\sin 99^\circ),$$
$$2^{3/10}e^{i19\pi/20} = 2^{3/10}(\cos\frac{19\pi}{20} + i\sin\frac{19\pi}{20}) = 2^{3/10}(\cos 171^\circ + i\sin 171^\circ),$$
$$2^{3/10}e^{i27\pi/20} = 2^{3/10}(\cos\frac{27\pi}{20} + i\sin\frac{27\pi}{20}) = 2^{3/10}(\cos 243^\circ + i\sin 243^\circ),$$
$$2^{3/10}e^{i35\pi/20} = 2^{3/10}(\cos\frac{35\pi}{20} + i\sin\frac{35\pi}{20}) = 2^{3/10}(\cos 315^\circ + i\sin 315^\circ).$$

A special case occurs when $c = 1$. We then get what are called the *roots of unity.* In other words, they are the solutions to $z^n = 1$.

THEOREM A.4 (on roots of unity)
The n distinct nth roots of unity are

$$e^{\frac{2\pi k}{n}} = \cos\left(\frac{2\pi k}{n}\right) + i\sin\left(\frac{2\pi k}{n}\right), \text{ where } k = 0, 1, 2 \ldots, n-1.$$

Moreover, if we unite $\omega = \cos\left(\dfrac{2\pi}{n}\right) + i\sin\left(\dfrac{2\pi}{n}\right)$, then ω is the nth root of unity having the smallest positive angle and the nth roots of unity are $\omega, \omega^2, \omega^3, \ldots, \omega^{n-1}$. Note that all roots of unity are magnitude one complex numbers, so they all lie on the unit circle in $\mathbb{R}^2$. There is some nice geometry here. The roots of unity lie at the vertices of a regular polygon of n sides *inscribed in the unit circle with one vertex at the real number 1. In other words, the unit circle is cut into n equal sectors.*

Example A.1
Find the four fourth roots of 1 and plot them

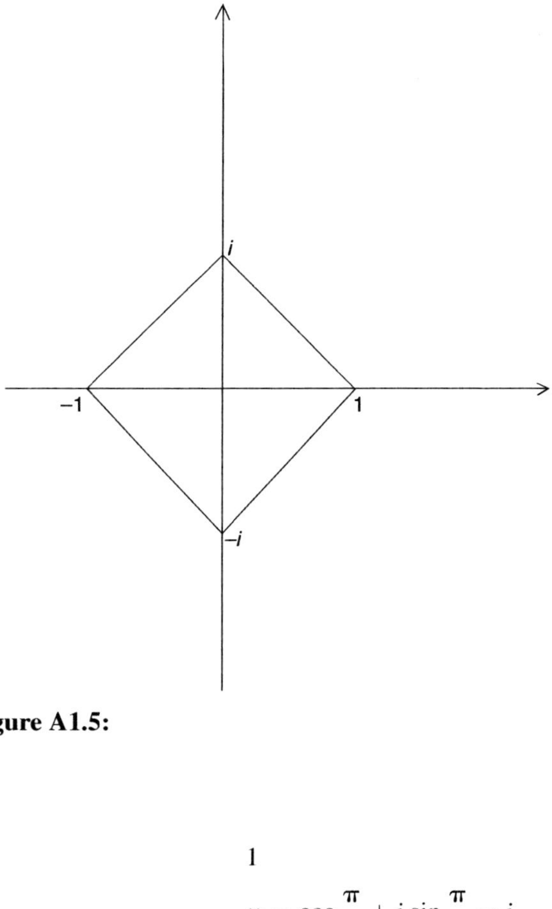

Figure A1.5:

$$1$$

$$\omega = \cos\frac{\pi}{2} + i\sin\frac{\pi}{2} = i,$$

$$\omega^2 = \cos\pi + i\sin\pi = -1,$$

$$\omega^3 = \cos\frac{3\pi}{2} + i\sin\frac{3\pi}{2} = -i.$$

APA Exercise Set 4

1. Express $10(\cos 225^\circ + i \sin 225^\circ)$ in rectangular form. Do the same for

a) $6(\cos 60^\circ + i \sin 60^\circ)$	h) $18(\cos 135^\circ + i \sin 135^\circ)$
b) $17(cos(270^\circ + i \sin 270^\circ)$	i) $41(\cos 90^\circ + i \sin 90^\circ)$
c) $2\sqrt{3}(\cos 150^\circ + i \sin 150^\circ)$	j) $5\sqrt{2}(\cos 225^\circ + i \sin 225^\circ)$
d) $6\sqrt{2}(\cos 45^\circ + i \sin 45^\circ)$	k) $10(cos 300^\circ + i \sin 300^\circ)$
e) $12(\cos 330^\circ + i \sin 330^\circ)$	l) $22(\cos 240^\circ + i \sin 240^\circ)$
f) $40(\cos 70^\circ + i \sin 70^\circ)$	m) $8(\cos 140^\circ + i \sin 140^\circ)$
g) $6(\cos 345^\circ + i \sin 345^\circ)$	n) $15(\cos 235^\circ + i \sin 235^\circ)$

2. Write the following complex numbers in polar form.

a) 8	j) 14	s) $18i$
b) $-19i$	k) $3\sqrt{2} - i3\sqrt{2}$	t) $-11\sqrt{3} + 11i$
c) $\sqrt{15} - i\sqrt{5}$	l) $-4 - 4i$	u) $10 - i10\sqrt{3}$
d) $-13 - i13\sqrt{3}$	m) $\sqrt{5} + i\sqrt{15}$	v) $2\sqrt{6} - i2\sqrt{2}$
e) $4 - 3i$	n) $12 + 5i$	w) $-6 + 12i$
f) $6 + 8i$	o) $-13 - 10i$	x) $7 - 11i$
g) $5\sqrt{2} + i5\sqrt{2}$	p) $-7 - 7i$	y) -23
h) $31i$	q) $-8\sqrt{3} + 8i$	z) $3 - i3\sqrt{3}$
i) $-4\sqrt{3} - 4i$	r) $15\sqrt{2} - i15\sqrt{2}$	aa) $10 + 20i$

3. Perform the indicated operations and express the results in rectangular form.

   a) $[2(\cos 18^\circ + i \sin 18^\circ)]\,[4(\cos 12^\circ + i \sin 12^\circ)]$
   b) $[10(\cos 34^\circ + 34^\circ)]\,[3(\cos 26^\circ + i \sin 26^\circ)]$
   c) $[7(\cos 112^\circ + i \sin 112^\circ)]\,[2(\cos 68^\circ + i \sin 68^\circ)]$
   d) $[6(\cos 223^\circ + i \sin 223^\circ)]\,[\cos 227^\circ + i \sin 227^\circ)]$
   e) $\dfrac{12(\cos 72^\circ + \sin 72^\circ)}{3(\cos 42^\circ + \sin 42^\circ)}$
   f) $\dfrac{24(\cos 154^\circ + \sin 154^\circ)}{6(\cos 64^\circ + i \sin 64^\circ)}$
   g) $\dfrac{42(\cos 8^\circ + i \sin 8^\circ)}{7(\cos 68^\circ + i \sin 68^\circ)}$
   h) $\dfrac{6\sqrt{2}(\cos 171^\circ + i \sin 171^\circ)}{2(\cos 216^\circ + i \sin 216^\circ)}$

4. Express in rectangular form.

   a) $[2(\cos 30^\circ + i \sin 30^\circ)]^4$
   b) $[4(\cos 10^\circ + i \sin 10^\circ)]^6$
   c) $[3(\cos 144^\circ + i \sin 144^\circ)]^5$
   d) $[2(\cos 210^\circ + i \sin 210^\circ)]^7$

e) $\left[\frac{1}{2}(\cos 18^\circ + i \sin 18^\circ)\right]^{-5}$
f) $\left[\frac{1}{3}(\cos 30^\circ + i \sin 30^\circ)\right]^{-6}$
g) $[(\cos 15^\circ + i \sin 15^\circ)]^{100}$
h) $(\cos 60^\circ + i \sin 60^\circ)^{50}$
i) $(\frac{\sqrt{2}}{2} - \frac{\sqrt{2}}{2}i)^{30}$
j) $(-\frac{1}{2} + \frac{\sqrt{3}}{2}i)^{40}$
k) $(\sqrt{3} + i)^5$
l) $(\sqrt{2} - i\sqrt{2})^9$

5. Find the fourth roots of $-8 + i8\sqrt{3}$.

6. Find in polar form the roots of the following. Draw the roots graphically.
a) $x^2 = 36(\cos 80^\circ + i \sin 80^\circ)$
b) $x^2 = 4(\cos 140^\circ + i \sin 140^\circ)$
c) $x^3 = 27(\cos 72^\circ + i \sin 72^\circ)$
d) $x^3 = 8(\cos 105^\circ + i \sin 105^\circ)$
e) $x^4 = 81(\cos 64^\circ + i \sin 64^\circ)$
f) $x^4 = 16(\cos 200^\circ + i \sin 200^\circ)$
g) $x^5 = (\cos 150^\circ + i \sin 150^\circ)$
h) $x^6 = 27(\cos 120^\circ + i \sin 120^\circ)$
i) $x^2 = 1 + i\sqrt{3}$
j) $x^2 = 8 - i8\sqrt{3}$
k) $x^3 + 4 + i\sqrt{3} = 0$
l) $x^3 - 8\sqrt{2} - i8\sqrt{6} = 0$
m) $x^4 = 2 - i\sqrt{3}$
n) $x^4 = -8 + i8\sqrt{3}$
o) $x^5 - 32 = 0$
p) $x^5 + 16\sqrt{3} + 16i = 0$

7. Find all roots and express the roots in both polar and rectangular form.
a) $x^2 + 36i = 0$
b) $x^2 = 32 + i32\sqrt{3}$
c) $x^3 - 27 = 0$
d) $x^3 - 8i = 0$
e) $x^3 + 216 = 0$
f) $x^3 + 27i = 0$
g) $x^3 - 64i = 0$
h) $x^4 + 4 = 0$
i) $x^4 + 81 = 0$
j) $x^4 = -32 - i32\sqrt{3}$
k) $x^6 - 64 = 0$

l) $x^6 + 8 = 0$
m) $x^5 - 1 = 0$
n) $x^5 - 243i = 0$

8. Find necessary and sufficient conditions on the complex numbers z and w so that $\overline{z}z = \overline{w}w$.

A.6 Polynomials over $\mathbb{C}$

It turns out that properties of polynomials are closely linked to properties of matrices. We will take a few moments to look at some basics about polynomials. But the polynomials we have in mind here have complex numbers as coefficients, such as $(4+2i)x^2+3ix+(\sqrt{\pi}+e^2i)$. We will use $\mathbb{C}[x]$ to denote the collection of all polynomials in the indeterminate x with coefficients from $\mathbb{C}$. You can add and multiply polynomials in $\mathbb{C}[x]$ just like you learned in high school. First, we focus on quadratic equations, but with complex coefficients. This, of course, includes the real coefficient case. Let $ax^2+bx+c=0$, where a, b, and c are in $\mathbb{C}$. We will show that the famous quadratic formula for the roots of the quadratic still works. The proof uses a very old idea, *completing the square.* This goes back at least to the Arab mathematician **Al-Khowarizmi** (circa 780–850) in the ninth century A.D. Now a is assumed nonzero, so we can write

$$x^2 + \frac{b}{a}x = -\frac{c}{a}.$$

Do you remember what to do? Sure, add $\left(\frac{b}{2a}\right)^2$ to each side (half the coefficient of the first degree term squared) and get

$$x^2 + \frac{b}{a}x + \left(\frac{b}{2a}\right)^2 = -\frac{c}{a} + \left(\frac{b}{2a}\right)^2.$$

The left-hand side is a perfect square!

$$\left(x + \frac{b}{2a}\right)^2 = x^2 + \frac{b}{a}x + \left(\frac{b}{2a}\right)^2 = -\frac{c}{a} + \left(\frac{b}{2a}\right)^2.$$

Now form a common denominator on the right

$$\left(x + \frac{b}{2a}\right)^2 = \frac{b^2 - 4ac}{4a^2}.$$

We have seen how to take square roots in $\mathbb{C}$, so we can do it here.

$$x + \frac{b}{2a} = \pm\frac{\sqrt{b^2 - 4ac}}{2a},$$

$$x = -\frac{b}{2a} \pm \frac{\sqrt{b^2 - 4ac}}{2a}.$$

In 1797, Gauss, when he was only 20 years old, proved probably the most crucial result about $\mathbb{C}[x]$. You can tell it must be good by what we call the theorem today; it is called the Fundamental Theorem of Algebra.

THEOREM A.5 (the fundamental theorem of algebra)
Every polynomial of degree $n \geq 1$ in $\mathbb{C}[x]$ has a root in $\mathbb{C}$.

We are in no position to prove this theorem right here. But that will not stop us from using it. The first consequence is the factor theorem, every polynomial in $\mathbb{C}[x]$ factors completely as a product of linear factors.

THEOREM A.6 (the factor theorem)
Let $p(x)$ be in $\mathbb{C}[x]$ and have degree $n \geq 1$. Then $p(x) = a(x - r_1)(x - r_2)\ldots(x - r_n)$, where $a, r_1, r_2, \ldots, r_n$ are complex numbers with $a \neq 0$. The numbers $r_1, r_2 \ldots r_n$ (possibly not distinct) are the roots of $p(x)$ and a is the coefficient of the highest power of x in $p(x)$.

PROOF By the fundamental theorem, $p(x)$ has a root, r_1. But then we can factor $p(x) = (x - r_1)q(x)$, where $q(x) \in \mathbb{C}[x]$ and the degree of $q(x)$ is one less than the degree of $p(x)$. An inductive argument then gets us down to a last factor of the form $a(x - r_n)$. □

In the case of real polynomials (i.e., $\mathbb{R}[x]$), the story is not quite as clean, but we still know the answer.

THEOREM A.7
Every polynomial $p(x)$ in $\mathbb{R}[x]$ of degree at least one can be factored as a product of linear factors and irreducible quadratic factors.

PROOF Let $p(x) \in \mathbb{R}[x]$ with $p(x) = a_n x^n + a_{n-1}x^{n-1} + \ldots + a_1 x + a_0$. Here the coefficients a_i are all real numbers. If r is a root of $p(x)$, possibly complex, then so is $\overline{r}$. (Verify $p(r) = p(\overline{r})$). Thus, nonreal roots occur in complex conjugate pairs. Therefore, if $s_1, \ldots s_k$ are real roots and $r_1, \overline{r}_1, r_2, \overline{r}_2 \ldots$ are the complex ones we get in $\mathbb{C}[x]$, $p(x) = a_n(x - s_1)\ldots(x - s_k)(x - r_1)(x - \overline{r}_1)\ldots$.

But $(x - r_j)(x - \overline{r}_j) = x^2 - (r_j + \overline{r}_j)x + r_j\overline{r}_j$ is an irreducible quadratic in $\mathbb{R}[x]$, so the theorem is proved. □

APA Exercise Set 4

1. Wouldn't life be simpler if fractions were easier to add? Wouldn't it be great if $\frac{1}{a} + \frac{1}{b} = \frac{1}{a+b}$? Are there any real numbers a and b that actually work in this formula? Are there any complex numbers that actually work in this formula?

Further Reading

[B&L, 1981] J. L. Brenner and R. C. Lyndon, Proof of the Fundamental Theorem of Algebra, The American Mathematical Monthly, Vol. 88, (1981), 253–256.

[Derksen, 2003] Harm Derksen, The Fundamental Theorem of Algebra and Linear Algebra, The American Mathematical Monthly, Vol. 110, No. 7, Aug-Sept (2003), 620–623.

[Fine, 1997] B. Fine and G. Rosenberger, *The Fundamental Theorem of Algebra*, Springer-Verlag, New York, (1997).

[H&H, 2004] Anthony A. Harkin and Joseph B. Harkin, Geometry of Generalized Complex Numbers, Mathematics Magazine, Vol. 77, No. 2, April, (2004), 118–129.

[Ngo, 1998] Viet Ngo, Who Cares if $x^2 + 1 = 0$ Has a Solution?, The College Mathematics Journal, Vol. 29, No. 2., March, (1998), 141–144.

A.7 Postscript

Surely by now we have come to the end of our construction of number systems. Surely $\mathbb{C}$ is the ultimate number system and you can do just about anything you want in $\mathbb{C}$. Well, mathematicians are never satisfied. Actually, the story goes on. If you can make ordered pairs in $\mathbb{R}^2$ into a nice number system, why not ordered triples in $\mathbb{R}^3$. Well, it turns out, as William Rowan Hamilton discovered about 150 years ago, you cannot. But this Irish mathematician discovered a number system you can make with four-tuples in $\mathbb{R}^4$. Today we call this system $\mathbb{H}$, the *quaternions.* If $q_1 = (a, b, c, d)$ and $q_2 = (x, y, u, v)$, then $q_1 \oplus q_2 = (a + x, b + y, c + u, d + r)$ defines an addition that you probably would have guessed. But take a look at the multiplication! $q_1 \odot q_2 = (ax - by - cu - dv, ay + bx + cu - du, au + cx + dy - bv, av + dx + bu - cy)$. How many of the properties of $\mathbb{C}$ also work for $\mathbb{H}$? There is a big one that does not. Can you find it? Now $\mathbb{C}$ can be viewed as being contained in $\mathbb{H}$ by taking four-tuples with the last two entries zero. Our chain now reads

$$\mathbb{N} \subseteq \mathbb{Z} \subseteq \mathbb{Q} \subseteq \mathbb{R} \subseteq \mathbb{C} \subseteq \mathbb{H}.$$

Mathematicians have invented even weirder number systems beyond the quaternions (see Baez [2002]). To close this section, let me tempt you to do some arithmetic in the finite system {0, 1, 2, 3} where the operations are defined by the following tables:

+	0	1	2	3
0	0	1	2	3
1	1	0	3	2
2	2	3	0	1
3	3	2	1	0

·	0	1	2	3
0	0	0	0	0
1	0	1	2	3
2	0	2	3	1
3	0	3	1	2

Can you solve $2x = 1$ in this system? Can you figure out which of the rules of $\mathbb{C}$ work in this system? Have fun!

Further Reading

[Baez, 2002] J. C. Baez, The Octonions, Bulletin of the American Mathematical Society, Vol. 39, No. 2, April (2002), 145–205. Available at <http://math.ucr.edu/home/baez/octonions.html>.

[Kleiner, 1988] I. Kleiner, Thinking the Unthinkable: The Story of Complex Numbers (With a Moral), Mathematics Teacher, Vol. 81, (1988), 583–592.

[Niev, 1997] Yves Nievergelt, History and Uses of Complex Numbers, UMAP Module 743, UMAP Modules 1997, 1–66.

Appendix B

Basic Matrix Operations

B.1 Introduction

In Appendix A, we looked at scalars and the operations you can apply to them. In this appendix, we take the same approach to matrices. A *matrix* is just a rectangular array of scalars. Apparently, it was **James Joseph Sylvester** (3 September 1814–15 March 1897) who first coined the term "matrix" in 1850. The scalars in the array are called the *entries* of the matrix. We speak of a matrix having *rows* and *columns* (like double-entry bookkeeping). Remember, columns hold up buildings, so they go up and down. Rows run from left to right. So, an m-by-n matrix A (we usually use uppercase Roman letters to denote matrices so as not to confuse them with scalars) is a rectangular array of scalars arranged in m horizontal rows and n vertical columns. In general then, an m-by-n matrix A looks like

$$A = \begin{bmatrix} a_{11} & a_{12} & \dots & a_{1n} \\ a_{21} & a_{22} & \dots & a_{2n} \\ \vdots & & & \vdots \\ a_{m1} & a_{m2} & \dots & a_{mn} \end{bmatrix} = \left[a_{ij}\right]_{m\times n} = \left[a_{i,j}\right]_{i,j=1}^{m,n} = \left[a_{ij}\right],$$

where each a_{ij} is in $\mathbb{C}$ for $1 \leq i \leq m$ and $1 \leq j \leq n$. In a more formal treatment, we would define a matrix as a function $A : [m] \times [n] \rightarrow \mathbb{C}$ where $[m] = \{1, 2, \dots, m\}$ and $[n] = \{1, 2, \dots, n\}$ and $A(i, j) = a_{ij}$, but we see no advantage in doing that here.

Note that we can locate any entry in the matrix by two subscripts; that is, the entry a_{ij} lives at the intersection of the ith row and jth column. For example, a_{34} denotes the scalar in the third row and fourth column of A. When convenient, we use the notation $ent_{ij}(A) = a_{ij}$; this is read "the (i, j) entry of A." The index i is called the *row index* and j the *column index*. By the ith row of A we mean the 1-by-n matrix

$$row_i(A) = [a_{i1}\ a_{i2}\ \dots a_{in}]$$

and by the jth column of A we mean the m-by-1 matrix

$$col_j(A) = \begin{bmatrix} a_{1j} \\ a_{2j} \\ \vdots \\ a_{mj} \end{bmatrix}.$$

We say two matrices are the *same size* if they have the same number of rows and the same number of columns. A matrix is called *square* if the number of rows equals the number of columns. Let $\mathbb{C}^{m\times n}$ denote the collection of all possible m-by-n matrices with entries from $\mathbb{C}$. Then, of course, $\mathbb{C}^{n\times n}$ would represent the collection of all square matrices of size n-by-n with entries from $\mathbb{C}$.

Let's have a little more language before we finish this section. Any entry in a matrix A with equal row and column indices is called a *diagonal element* of A; $diag(A) = [a_{11}\, a_{22}\, a_{33} \ldots]$. The entry a_{ij} is called *off-diagonal* iff $i \neq j$, *subdiagonal* if $i > j$, and *superdiagonal* if $i < j$. Look at an example and see if these names make any sense.

We now finish this section with the somewhat straightforward idea of matrix equality. We say two matrices are *equal* iff they have the same size and equal entries; that is, if $A, B \in \mathbb{C}^{m\times n}$, we say $A = B$ iff $ent_{ij}(A) = ent_{ij}(B)$ for all i, j with $1 \leq i \leq m$ and $1 \leq j \leq n$. This definition may seem pretty obvious, but it is crucial in arguments where we want to "prove" that two matrices that may not look the same on the surface really are. Notice we only speak of matrices being equal when they are the same size.

APB Exercise Set 1

1. Let $A = \begin{bmatrix} 1 & 2 & 3 \\ 4 & 5 & 6 \end{bmatrix}$. What is $ent_{23}(A)$? How about $ent_{32}(A)$? What are the diagonal elements of A? The subdiagonal elements? The superdiagonal elements? How many columns does A have? How many rows?

2. Suppose $\begin{bmatrix} x - y & y + z \\ 2x - 3w & 3w + 2z \end{bmatrix} = \begin{bmatrix} 9 & 2 \\ 6 & 5 \end{bmatrix}$. What are x, y, z, and w?

B.2 Matrix Addition

We now begin the process of introducing the basic operations on matrices. We can add any two matrices of the same size. We do this in what seems a perfectly reasonable way, namely entrywise. Let A and B be in $\mathbb{C}^{m \times n}$ with $A = \left[a_{ij}\right]$, $B = \left[b_{ij}\right]$. The matrix sum $A \oplus B$ is the m-by-n matrix whose (i, j) entry is $a_{ij} + b_{ij}$. Note that we have two notions of addition running around here, $\oplus$ used for matrix addition and + used for adding scalars in $\mathbb{C}$. We have defined one addition in terms of the other; that is $A \oplus B = \left[a_{ij} + b_{ij}\right]$. In other words,

$$ent_{ij}(A \oplus B) = ent_{ij}(A) + ent_{ij}(B).$$

For example, in $\mathbb{C}^{2\times 3}$,

$$\begin{bmatrix} 1 & 2 & 3 \\ 4 & 5 & 6 \end{bmatrix} \oplus \begin{bmatrix} 2 & 4 & 6 \\ 1 & 3 & 5 \end{bmatrix} = \begin{bmatrix} 3 & 6 & 9 \\ 5 & 8 & 11 \end{bmatrix}.$$

Now the question is: what rules of addition does this definition enjoy? Are they the same as those for the addition of scalars in $\mathbb{C}$? Let's see!

THEOREM B.1 (basic rules of addition in $\mathbb{C}^{m\times n}$)
Suppose A, B, and C are in $\mathbb{C}^{m\times n}$. Then

1. *Associative law of matrix addition*
   $A \oplus (B \oplus C) = (A \oplus B) \oplus C$.

2. *Existence of a zero*
   Let $\mathbb{O} = [0]_{m\times n}$ *(i.e.,* $ent_{ij}(\mathbb{O}) = 0$*). Then for any matrix A at all,* $\mathbb{O} \oplus A = A = A \oplus \mathbb{O}$.

3. *Existence of opposites*
   Given any matrix A in $\mathbb{C}^{m\times n}$, we can find another matrix B with $A \oplus B = \mathbb{O} = B \oplus A$. *In fact take B with* $ent_{ij}(B) = -ent_{ij}(A)$. *Denote* $B = -A$ *Then* $A \oplus (-A) = \mathbb{O} = (-A) \oplus A$.

4. *Commutative law of matrix addition*
   For any two matrices A, B in $\mathbb{C}^{m\times n}$, $A \oplus B = B \oplus A$.

PROOF We will illustrate the proofs here and leave most as exercises. We appeal to the definition of matrix equality and show that the (i, j) entry of the matrices on both sides of the equation are the same. The trick is to push the argument back to something you know is true about $\mathbb{C}$. Let's compute:

$$\begin{aligned}
&ent_{ij}[A \oplus (B \oplus C)] && \\
&= ent_{ij}(A) + ent_{ij}(B \oplus C) && \text{definition of matrix addition,} \\
&= ent_{ij}(A) + (ent_{ij}(B) + ent_{ij}(C)) && \text{definition of matrix addition,} \\
&= (ent_{ij}(A) + ent_{ij}(B)) + ent_{ij}(C) && \text{associative law of addition in } \mathbb{C}, \\
&= ent_{ij}(A \oplus B) + ent_{ij}(C) && \text{definition of matrix addition,} \\
&= ent_{ij}((A \oplus B) \oplus C) && \text{definition of matrix addition.} \quad \square
\end{aligned}$$

That was easy, wasn't it? All we did was use the definition of matrix addition and the fact that addition in $\mathbb{C}$ is associative to justify each of our steps. The big blow of course was the associative law of addition in $\mathbb{C}$. Now you prove the rest!

So far, things are going great. The algebraic systems $(\mathbb{C}, +)$ and $(\mathbb{C}^{m\times n}, \oplus)$ have the same additive arithmetic. The basic laws are exactly the same. We can even introduce matrix subtraction, just like we did in $\mathbb{C}$. Namely, for A, B in $\mathbb{C}^{m\times n}$, define $A \ominus B = A \oplus (-B)$. In other words,

$$ent_{ij}(A \ominus B) = ent_{ij}(A) - ent_{ij}(B) = a_{ij} - b_{ij}.$$

Now that the basic points have been made, we will drop the special notation of circles around addition and subtraction of matrices. The context should make clear whether we are manipulating scalars or matrices.

APB Exercise Set 2

1. Let $A = \begin{bmatrix} 3 & 1 & -1 \\ 2 & 0 & 1 \end{bmatrix}$, $B = \begin{bmatrix} 6 & 1 & 1 \\ -1 & 1 & 4 \end{bmatrix}$.
   Find $A + B$. Now find $B + A$. What do you notice? Find $A - B$. Find $B - A$. What do you notice now?

2. Suppose $A = \begin{bmatrix} 1 & 2 \\ 3 & 4 \end{bmatrix}$, $X = \begin{bmatrix} x & y \\ u & v \end{bmatrix}$, and $B = \begin{bmatrix} 5 & 10 \\ 14 & 20 \end{bmatrix}$.
   Solve for X in $A + X = B$.

3. What matrix would you have to add to $\begin{bmatrix} 7 & 3+4i \\ 5i & 7-2i \end{bmatrix}$ in $\mathbb{C}^{2\times 2}$ to get the zero matrix?

4. Prove the remaining basis laws of addition in $\mathbb{C}^{m\times n}$ using the format previously illustrated.

B.3 Scalar Multiplication

The first kind of multiplication we consider is multiplying a matrix by a scalar. We can do this on the left or on the right. So let A be a matrix in $\mathbb{C}^{m\times n}$, and let $a \in \mathbb{C}$ be a scalar. We define a new matrix aA as the matrix obtained by multiplying all the entries of A on the left by a. In other words, $ent_{ij}(aA) = a\ ent_{ij}(A)$; that is, $aA = a\left[a_{ij}\right] = \left[aa_{ij}\right]$. For example, in $\mathbb{C}^{2\times 2}$, we have $3\begin{bmatrix} 1 & 2 \\ 4 & 6 \end{bmatrix} = \begin{bmatrix} 3 & 6 \\ 12 & 18 \end{bmatrix}$.

Let's look at basic properties.

THEOREM B.2 (basic properties of scalar multiplication)

Let A, B be in $\mathbb{C}^{m\times n}$ and a, b be in $\mathbb{C}$. Then

1. $a(A + B) = aA + aB$.
2. $(a + b)A = aA + bA$.
3. $(ab)A = a(bA)$.
4. $1A = A$.
5. $a(Ab) = (aA)b$.

PROOF As usual, we illustrate only one of the arguments and leave the rest as exercises. The procedure is to compare (i, j) entries. Let's prove (1) together:

$ent_{ij}(a(A + B)) = a\ ent_{ij}(A + B)$	definition of scalar multiplication,
$= a\left[ent_{ij}(A) + ent_{ij}(B)\right]$	definition of matrix addition,
$= a(ent_{ij}(A)) + a(ent_{ij}(B))$	left distributive law in $\mathbb{C}$,
$= ent_{ij}(aA) + ent_{ij}(aB)$	definition of scalar multiplication,
$= ent_{ij}(aA + aB)$	definition of matrix addition,

Once again, our attack on this proof has been to appeal to definitions and then use a crucial property in $\mathbb{C}$. □

Our game plan is still going well, although it is a little unsettling that we are mixing apples and oranges, scalars and matrices. Here we multiply a matrix by a scalar and get another matrix. Can we multiply a matrix by a matrix and get another matrix? Yes, we can, and that is what we address next.

APB Exercise Set 3

1. Let $A = \begin{bmatrix} 3 & 1 & -1 \\ 2 & 0 & 1 \end{bmatrix}$, $B = \begin{bmatrix} 1 & 2 & 4 \\ 5 & 6 & 3 \end{bmatrix}$. Compute $5A - 3B$, $-4(A - B)$.

2. Prove the remaining basic properties in the manner illustrated. Formulate and prove similar properties for right scalar multiplication.

3. Let $C \in \mathbb{C}^{m\times n}$. Argue $C = A + iB$ where the entries of A and B are real. Argue A and B are unique.

B.4 Matrix Multiplication

The issue of matrix multiplication is not quite so straightforward. Suppose we have two matrices A and B of the same size. Since we defined addition entrywise, it might seem natural to define multiplication of matrices the same way. Indeed you can do that, and we will assign exercises to investigate properties of this definition, of multiplication. However, it turns out this is not the "right" definition. There is a sophisticated way to motivate the upcoming definition, but we will attempt one using systems of linear equations. Consider the problem of substituting one system of linear equations into another. Begin with $\begin{cases} a_{11}x_1 + a_{12}x_2 = c_1 \\ a_{21}x_1 + a_{22}x_2 = c_2 \end{cases}$. The coefficient matrix is $A = \begin{bmatrix} a_{11} & a_{12} \\ a_{21} & a_{22} \end{bmatrix}$.

Now consider a linear substitution for the xs : $\begin{cases} x_1 = b_{11}y_1 + b_{12}y_2 \\ x_2 = b_{21}y_1 + b_{22}y_2 \end{cases}$. The coefficient matrix here is $B = \begin{bmatrix} b_{11} & b_{12} \\ b_{21} & b_{22} \end{bmatrix}$. Now, putting the xs back into the original system, we get

$$\begin{aligned} a_{11}x_1 + a_{12}x_2 &= a_{11}(b_{11}y_1 + b_{12}y_2) + a_{12}(b_{21}y_1 + b_{22}y_2) \\ &= a_{11}b_{11}y_1 + a_{11}b_{12}y_2 + a_{12}b_{21}y_1 + a_{12}b_{22}y_2 \\ &= (a_{11}b_{11} + a_{22}b_{21})y_1 + (a_{11}b_{12} + a_{12}b_{22})y_2 \quad \text{and} \end{aligned}$$

$$\begin{aligned} a_{21}x_1 + a_{22}x_2 &= a_{21}(b_{11}y_1 + b_{12}y_2) + a_{22}(b_{21}y_1 + b_{22}y_2) \\ &= a_{21}b_{11}y_1 + a_{21}b_{12}y_2 + a_{22}b_{21}y_1 + a_{22}b_{22}y_2 \\ &= (a_{21}b_{11} + a_{22}b_{21})y_1 + (a_{21}b_{12} + a_{22}b_{22})y_2. \end{aligned}$$

If we use matrix notation for the original system, we get $AX = C$, where $X = \begin{bmatrix} x_1 \\ x_2 \end{bmatrix}$ and $C = \begin{bmatrix} c_1 \\ c_2 \end{bmatrix}$. We write the substitution as $X = BY$, where $Y = \begin{bmatrix} y_1 \\ y_2 \end{bmatrix}$. Then the new system is $AX = A(BY) = (AB)Y = C$ if you cherish the associative law. Thus, the coefficient matrix $AB = \begin{bmatrix} a_{11}b_{11} + a_{12}b_{21} & a_{11}b_{12} + a_{12}b_{22} \\ \\ a_{21}b_{11} + a_{21}b_{21} & a_{21}b_{12} + a_{22}b_{22} \end{bmatrix}$. So, if we make this our definition of matrix multiplication, we see the connection between the rows of A and the columns of B. Using this row by column multiplication produces the resulting matrix AB. This is what leads us to make the general definition of how to multiply two matrices together.

Suppose A is in $\mathbb{C}^{m\times n}$ and B is in $\mathbb{C}^{n\times p}$. Then the product matrix AB is the matrix whose (i, j) entry is obtained by multiplying each entry from the ith row of A by the corresponding entry from the jth column of B and adding the results. Notice that the product matrix has size m-by-p. Let's spell this out a bit more. The product matrix AB is the m-by-p matrix whose (i, j) entry is

$$ent_{ij}(AB) = \sum_{k=1}^{n} ent_{ik}(A)ent_{kj}(B).$$

Okay, still clear as mud? Let $A = [a_{ij}]$ and $B = [b_{ij}]$. Then

$$ent_{ij}(AB) = a_{i1}b_{1j} + a_{i2}b_{2j} + \cdots + a_{in}b_{nj}.$$

Now, let's look at an example. Let $A = \begin{bmatrix} 1 & 2 & 4 \\ 2 & 6 & 0 \end{bmatrix}$ and $B = \begin{bmatrix} 4 & 1 & 4 & 3 \\ 0 & -1 & 3 & 1 \\ 2 & 7 & 5 & 2 \end{bmatrix}$.

Let's compute the (2,3) entry of AB. We use the second row of A and third column of B:

$$2 \cdot 4 + 6 \cdot 3 + 0 \cdot 5 = 26. \text{ So } \begin{bmatrix} 1 & 2 & 4 \\ 2 & 6 & 0 \end{bmatrix} \begin{bmatrix} 4 & 1 & 4 & 3 \\ 0 & -1 & 3 & 1 \\ 2 & 7 & 5 & 2 \end{bmatrix} = \begin{bmatrix} * & * & * & * \\ * & * & 26 & * \end{bmatrix}.$$

Does that help? See if you can fill out the rest of AB. As you can see, multiplying matrices requires quite a bit of work. But that is why we have calculators and computers. Notice that to multiply matrix A by matrix B in the order AB, the matrices do not have to be the same size, but it is absolutely necessary that the number of columns of A is the same as the number of rows of B. If you remember the concept of dot product from your earlier studies, it might be helpful to view matrix multiplication in the following way:

$$ent_{ij}(AB) = row_i(A) \bullet col_j(B).$$

Before we move to basic properties, let's have a quick review of the useful sigma notation we have already used above. You may recall this notation from calculus when you were studying Riemann sums and integrals. Recall that the capital Greek letter Σ stands for summation. It tells us to add something up. It is a great shorthand for lengthy sums. So $a_1 + a_2 + \ldots + a_n = \sum_{i=1}^{n} a_i$. Now the (i, j) entry of AB can be expressed as $[a_{i1}a_{i2}\ldots a_{in}]\begin{bmatrix} b_{1j} \\ b_{2j} \\ \vdots \\ b_{nj} \end{bmatrix} = \left[\sum_{k=1}^{n} a_{ik}b_{kj}\right]$ using Σ-notation. Note the "outside" indices i and j are fixed and the "running index" is k, which matches on the "inside."

There are three basic rules when we do computations using the Σ-notation.

THEOREM B.3 (basic rules of sigma)

1. $\sum_{j=1}^{n}(a_j + b_j) = \sum_{j=1}^{n} a_j + \sum_{j=1}^{n} b_j$.

2. $\sum_{j=1}^{n} a\, b_j = a\sum_{j=1}^{n} b_j$.

3. $\sum_{j=1}^{m}\sum_{k=1}^{n} a_{jk} = \sum_{k=1}^{n}\sum_{j=1}^{m} a_{jk}$.

The proofs here are unenlightening computations and we will not even ask you to do the arguments as exercises. However, we would like to consider rule (3) about double summations. This may remind you of the Fubinni theorem in several variable calculus about interchanging the order of integration (then again, maybe not). Anyway, we can understand (3) by using matrices. Consider the matrix $\begin{bmatrix} a_{11} & a_{12} \\ a_{21} & a_{22} \\ a_{31} & a_{32} \end{bmatrix}$. Let T stand for the summation total of all the matrix entries. Well, there are (at least) two different ways we could go about computing T in steps. First, we could compute row sums and add these or we could first compute column sums and then add these. Evidently, we would get the same answer either way, namely T.

$$\begin{array}{ccccc} \begin{bmatrix} a_{11} & a_{12} \\ a_{21} & a_{22} \\ a_{31} & a_{32} \end{bmatrix} & \begin{array}{c} \rightarrow \\ \rightarrow \\ \rightarrow \end{array} & \begin{array}{c} \sum_{k=1}^{2} a_{1k} \\ \sum_{k=1}^{2} a_{2k} \\ \sum_{k=1}^{2} a_{3k} \end{array} \\ \begin{array}{cc} \downarrow & \downarrow \\ \sum_{j=1}^{3} a_{j1} & \sum_{j=1}^{3} a_{j2} \end{array} & & \end{array}$$

Thus, $T = \sum_{k=1}^{2} a_{1k} + \sum_{k=1}^{2} a_{2k} + \sum_{k=1}^{2} a_{3k} = \sum_{j=1}^{3}\sum_{k=1}^{2} a_{ik}$ or $T = \sum_{j=1}^{3} a_{j1} + \sum_{j=1}^{3} a_{j2} = \sum_{k=1}^{2}\sum_{j=1}^{3} a_{jk}$ or $T = \sum_{j=1}^{3} a_{j1} + \sum_{j=1}^{3} a_{j2} = \sum_{k=1}^{2}\sum_{k=1}^{3} a_{jk}$.

So you see, matrices are good for something already! Now let's look at the basic properties of matrix multiplication.

THEOREM B.4 (basic properties of matrix multiplication)

Let A, B, C be matrices and $a \in \mathbb{C}$. Then

1. *If A is m-by-n, B n-by-p, and C p-by-q, then* $A(BC) = (AB)C$ *in* $\mathbb{C}^{m\times q}$.
2. *If A is m-by-n, B n-by-p, and C n-by-p, then* $A(B + C) = AB + AC$ *in* $\mathbb{C}^{m\times p}$.
3. *If A is m-by-n, B is m-by-n, and C is n-by-p, then* $(A + B)C = AC + BC$ *in* $\mathbb{C}^{m\times p}$.
4. *If A is m-by-n, B n-by-p, and* $a \in \mathbb{C}$, *then* $a(AB) = (aA)B = A(aB)$.
5. $row_i(AB) = (row_i(A))B$.
6. $col_j(AB) = A(col_j(B))$.
7. *Let* I_n *denote the n-by-n matrix whose diagonal elements all equal one and all off-diagonal entries are zero. Let A be m-by-n. Then* $I_m A = A = A I_n$.

PROOF The proofs are left as exercises in entry verification in the manner already illustrated. ∎

What are some of the consequences of these basic properties? That is, $(\mathbb{C}, +, \cdot, 0,)$ and $(\mathbb{C}^{n\times n}, +, \cdot, \mathbb{O}_n)$ share many common algebraic features. Thus, the basic arithmetic of both structures is the same. However, now the cookies start to crumble. Let's look at some examples. Let $A = \begin{bmatrix} 6 & 9 \\ -4 & 6 \end{bmatrix}$ and $B = \begin{bmatrix} 1 & 2 \\ -1 & 0 \end{bmatrix}$ in $\mathbb{C}^{2\times 2}$. Note $AB = \begin{bmatrix} -3 & 12 \\ 2 & -8 \end{bmatrix} \neq \begin{bmatrix} -2 & -3 \\ -6 & -9 \end{bmatrix} = BA$.

Thus, the commutative law of multiplication breaks down for matrices. Therefore, great care must be exercised when doing computations with matrices. Changing the order of a product could really mess things up. Even more can go wrong; $A \neq \mathbb{O}_2$ but $A^2 = \begin{bmatrix} 0 & 0 \\ 0 & 0 \end{bmatrix} = \mathbb{O}_2$, so $\mathbb{C}^{2\times 2}$ has "zero divisors" even though $\mathbb{C}$ does not. These two facts bring in some big differences in the arithmetic of scalars versus matrices.

Understanding the nature of matrix multiplication is crucial to understanding matrix theory, so we will dwell on the concept a bit more. There are four ways to look at multiplying matrices and each gives insights in the appropriate context. We will illustrate with small matrices to bring out ideas.

View 1 ***(the row column view [or dot product view])***

Let $A = \begin{bmatrix} a & b \\ c & d \end{bmatrix}$ and $B = \begin{bmatrix} x & y \\ u & v \end{bmatrix}$. Recall the dot product notation $\begin{bmatrix} a & b \end{bmatrix}\begin{bmatrix} x \\ u \end{bmatrix} = [ax + bu] = [(a, b) \cdot (x, u)]$. Then $AB = \begin{bmatrix} a & b \\ \cdots & \cdots \\ c & d \end{bmatrix}$ $\begin{bmatrix} x & \vdots & y \\ u & \vdots & v \end{bmatrix} = \begin{bmatrix} (a, b) \cdot (x, u) & (a, b) \cdot (y, v) \\ (c, d) \cdot (x, u) & (c, d) \cdot (y, v) \end{bmatrix}$. This is the *row-column view*.

View 2 ***(the column view)***

Here we leave A alone but think of B as a collection of columns. Then $AB = \begin{bmatrix} a & b \\ c & d \end{bmatrix}\begin{bmatrix} x & \vdots & y \\ u & \vdots & v \end{bmatrix} = \left[\begin{bmatrix} a & b \\ c & d \end{bmatrix}\begin{bmatrix} x \\ u \end{bmatrix} \vdots \begin{bmatrix} a & b \\ c & d \end{bmatrix}\begin{bmatrix} y \\ v \end{bmatrix}\right] = \begin{bmatrix} ax + bu & \vdots & ay + bv \\ ex + du & \vdots & cu + du \end{bmatrix}$. We find it very handy in matrix theory to do things like $AX = A[x_1 \mid x_2 \mid x_3] = [Ax_1 \mid Ax_2 \mid Ax_3]$, which moves the matrix on the left in to operate on the columns of the matrix on the right one by one.

View 3 ***(the row view)***

Here we leave B alone and partition A into rows. Then $AB = \begin{bmatrix} a & b \\ \cdots & \cdots \\ c & d \end{bmatrix}$

$$\begin{bmatrix} x & y \\ u & v \end{bmatrix} = \begin{bmatrix} \begin{bmatrix} a & b \end{bmatrix} & \begin{bmatrix} x & y \\ u & v \end{bmatrix} \\ \cdots\cdots\cdots\cdots & \cdots\cdots\cdots\cdots \\ \begin{bmatrix} c & d \end{bmatrix} & \begin{bmatrix} x & y \\ u & v \end{bmatrix} \end{bmatrix} = \begin{bmatrix} ax+bu & ay+bv \\ cx+du & cy+dv \end{bmatrix}.$$

View 4 ***(the column-row view [or outer product view])***

Here we partition A into columns and B into rows. Then $AB = \begin{bmatrix} a & \vdots & c \\ b & \vdots & d \end{bmatrix}$

$\begin{bmatrix} x & y \\ \cdots & \cdots \\ u & v \end{bmatrix} = \begin{bmatrix} a \\ c \end{bmatrix} \begin{bmatrix} x & y \end{bmatrix} + \begin{bmatrix} b \\ d \end{bmatrix} \begin{bmatrix} u & v \end{bmatrix} = \begin{bmatrix} ax & ay \\ cx & cy \end{bmatrix} +$ $\begin{bmatrix} bu & bv \\ du & dv \end{bmatrix} = \begin{bmatrix} ax+bu & ay+bv \\ cx+du & cy+dv \end{bmatrix}$. Also notice that $\begin{bmatrix} a & b \\ c & d \end{bmatrix}$ $\begin{bmatrix} x \\ u \end{bmatrix} = \begin{bmatrix} ax+bu \\ cx+du \end{bmatrix} = \begin{bmatrix} ax \\ cx \end{bmatrix} + \begin{bmatrix} bu \\ du \end{bmatrix} = \begin{bmatrix} a \\ c \end{bmatrix} x + \begin{bmatrix} b \\ d \end{bmatrix} u$. We see that the matrix product AB is a right linear combination of the columns of A, with the coefficients coming from B. Indeed, $\begin{bmatrix} a & b \end{bmatrix} \begin{bmatrix} x & y \\ u & v \end{bmatrix} =$ $a\,[x\ y] + b\,[u\ v]$, so the product can also be viewed as a left linear combination of the rows of B, with coefficients coming from the matrix on the left.

B.5 Transpose

Given a matrix A in $\mathbb{C}^{m \times n}$ we can always form another matrix A^T, *A-transpose*, in $\mathbb{C}^{n \times m}$. In other words,

$$ent_{ij}(A^T) = ent_{ji}(A).$$

Thus, the transpose of A is the matrix obtained by simply interchanging the rows and columns of A. If $A = \begin{bmatrix} 1 & 2 & 3 \\ 4 & 5 & 6 \end{bmatrix}$, then $A^T = \begin{bmatrix} 1 & 4 \\ 2 & 5 \\ 3 & 6 \end{bmatrix}$.

THEOREM B.5 (basic facts about transpose)

1. *If* $A \in \mathbb{C}^{m\times n}$, *then* $A^{TT} = A$.
2. *If* $A, B \in \mathbb{C}^{m\times n}$, *then* $(A+B)^T = A^T + B^T$.
3. *If* $A \in \mathbb{C}^{m\times n}$ *and* $B \in \mathbb{C}^{n\times p}$, *then* $(AB)^T = B^T A^T$.
4. *If* A *is invertible in* $\mathbb{C}^{n\times n}$, *then* $(A^T)^{-1} = (A^{-1})^T$.
5. $(\alpha A)^T = \alpha A^T$.

PROOF We illustrate with (3). Compute

$$\begin{array}{ll} ent_{ij}((AB)^T) = ent_{ji}(AB) & \text{definition of transpose,} \\ = \sum_{k=1}^{n} a_{jk}b_{ki} & \text{definition of matrix multiplication,} \\ = \sum_{k=1}^{n} b_{ki}a_{jk} & \text{commutativity of } \mathbb{C}, \\ = \sum_{k=1}^{n} ent_{ik}(B^T)ent_{ki}(A^T) & \text{definition of transpose,} \\ = ent_{ij}((B^T)(A^T)) & \text{definition of matrix multiplication.} \end{array}$$

Since the (i, j) entries agree, the matrices are equal. ☐

We can use the notion of transpose to distinguish important families of matrices. For example, a matrix A is called *symmetric* iff $A = A^T$. A matrix A is called *skew-symmetric* iff $A^T = -A$.

The reader may notice that there seems to be a pretty strong analogy between complex conjugation $z \longmapsto \overline{z}$ and matrix transposition $A \longmapsto A^T$. Indeed, the analogy is quite strong. However, there is a problem. For complex numbers, $z\overline{z} = |z|^2 = 0$ implies $z = 0$. If we let

$A = \begin{vmatrix} 1 & i \\ i & -1 \end{vmatrix} \in \mathbb{C}^{2\times 2}$, then $AA^T = \begin{bmatrix} 1 & i \\ i & -1 \end{bmatrix}\begin{bmatrix} 1 & i \\ i & -1 \end{bmatrix} = \begin{bmatrix} 0 & 0 \\ 0 & 0 \end{bmatrix}$
but $A \neq \mathbb{O}$. Note that if we had restricted A to real entries, then $AA^T = \mathbb{O}$ implies $A = \mathbb{O}$ for $A \in \mathbb{R}^{n\times n}$. It boils down to the fact that, in $\mathbb{R}$, the sum of squares equaling zero implies each entry in the sum is zero. We can fix this problem for complex matrices. If $A \in \mathbb{C}^{m\times n}$, define the *conjugate transpose* A^* by

$$ent_{ij}(A^*) = ent_{ji}(\overline{A}) = \overline{ent_{ji}(A)};$$

that is,

$$A^* = \overline{(A^T)} = \overline{A}^T.$$

But first we define the *conjugate* of a matrix and list the basic properties. Define $\overline{A}$ by

$$ent_{ij}(\overline{A}) = \overline{ent_{ij}(A)}.$$

THEOREM B.6 (basic fact about the conjugate matrix)
Let $A, B \in \mathbb{C}^{n\times n}$ and $\alpha \in \mathbb{C}$. Then

1. $\overline{\overline{A}} = A$.
2. $\overline{(A + B)} = \overline{A} + \overline{B}$.
3. $\overline{AB} = \overline{A}\,\overline{B}$.
4. $A = \overline{A}$ *iff all entries of A are real.*
5. $\overline{\alpha A} = \overline{\alpha}\,\overline{A}$.
6. $(\overline{A})^T = \overline{A^T}$.

THEOREM B.7 (basic facts about conjugate transpose)
Let $A, B \in \mathbb{C}^{n\times n}$, and $\alpha \in \mathbb{C}$. Then

1. $(A^*)^* = A$.
2. $(A + B)^* = A^* + B^*$.
3. $(AB)^* = B^*A^*$.
4. $(\alpha A)^* = \bar{\alpha}A^*$.
5. $(AA^*)^* = AA^*$ *and* $(A^*A)^* = A^*A$.
6. $AA^* = \mathbb{O}$ *implies* $A = \mathbb{O}$.
7. $\overline{A}^* = A^T = \overline{(A^*)}$.
8. $(A \otimes B)^* = A^* \otimes B^*$.

Now the real numbers sit inside of the complex numbers, $\mathbb{R} \subseteq \mathbb{C}$. We can identify complex numbers as being real exactly when they are equal to their complex conjugates. This suggests that matrices of the form $A = A^*$ should play the role of real numbers in $\mathbb{C}^{n\times n}$. Such matrices are called *self-adjoint*, or *Hermitian*, in honor of the French mathematician **Charles Hermite** (24 December 1822–14 January 1901). Constructing Hermitian matrices is easy. Start with any matrix A and form AA^* or A^*A. Of course, Hermitian matrices are always square. So, in $\mathbb{C}^{n\times n}$, we can view the Hermitian matrices as playing the role of real numbers in $\mathbb{C}$. This suggests an important representation of matrices in $\mathbb{C}^{n\times n}$. Recall that if z is a complex number, then $z = a + bi$, where a and b are real numbers. In fact, we defined $\mathrm{Re}(z) = a$ and $\mathrm{Im}(z) = b$. We saw $\mathrm{Re}(z) = \dfrac{z+\overline{z}}{2}$ and $\mathrm{Im}(z) = \dfrac{z-\overline{z}}{2i}$, so the real and imaginary part of a complex number are uniquely determined. We can reason by analogy with matrices in $\mathbb{C}^{n\times n}$. Let $A \in \mathbb{C}^{n\times n}$. Define $\mathrm{Re}(A) = \dfrac{A + A^*}{2}$ and $\mathrm{Im}(A) = \dfrac{A - A^*}{2i}$. This leads to the so-called *Cartesian decomposition*. Let $A \in \mathbb{C}^{n\times n}$ be arbitrary. Then $\mathrm{Re}(A)$ and $\mathrm{Im}(A)$ are self-adjoint and $A = \mathrm{Re}(A) + \mathrm{Im}(A)i$. This has important philosophical impact. This representation suggests that if you know everything there is to know about Hermitian matrices, then you know everything there is to know about all (square) matrices.

Let's push this analogy one step forward. A positive real number is always a square; that is, $a \geq 0$ implies $a = b^2$ for some real number b. Thus, if we view a as a complex number, $a = bb = b\overline{b}$. This suggests defining a Hermitian matrix H to be positive if $H = SS^*$ for some matrix $S \neq \mathbb{O}$. We shall have more to say about this elsewhere.

APB Exercise Set 4

1. Establish all the unproved claims of this section.

2. Say everything you can that is interesting about the matrix
$$\begin{bmatrix} \frac{1}{3} & \frac{2}{3} & \frac{2}{3} \\ -\frac{2}{3} & -\frac{1}{3} & \frac{2}{3} \\ \frac{2}{3} & -\frac{2}{3} & \frac{1}{3} \end{bmatrix}.$$

3. Let $A = \begin{bmatrix} 1 & 1 \\ 1 & 1 \end{bmatrix}$. Find $A^2, A^3, \ldots, A^n$. What if $A = \begin{bmatrix} 1 & 1 & 1 \\ 1 & 1 & 1 \\ 1 & 1 & 1 \end{bmatrix}$?
Find $A^2, A^3, \ldots, A^n$. Can you generalize?

4. Argue that $(A + B)^2 = A^2 + 2AB + B^2$ iff $AB = BA$.

5. Argue that $(A + B)^2 = A^2 + B^2$ iff $AB = -BA$.

6. Suppose $AB = \alpha BA$. Can you find necessary and sufficient conditions so that $(A + B)^3 = A^3 + B^3$?

7. Suppose $AB = C$. Argue that the columns of C are linear combinations of the columns of A.

8. Let $a \in \mathbb{C}$. Argue that $\begin{bmatrix} 1 & \mathbb{O}_{1\times n-1} \\ \mathbb{O}_{n-1\times 1} & A \end{bmatrix} \begin{bmatrix} a & \mathbb{O} \\ \mathbb{O} & B \end{bmatrix} \begin{bmatrix} 1 & \mathbb{O} \\ \mathbb{O} & C \end{bmatrix} = \begin{bmatrix} a & \mathbb{O} \\ \mathbb{O} & ABC \end{bmatrix}$.

9. Argue that AB is a symmetric matrix iff $AB = BA$.

10. Suppose A and B are 1-by-n matrices. Is it true that $AB^T = BA^T$?

11. Give several nontrivial examples of 3-by-3 symmetric matrices. Do the same for skew-symmetric matrices.

12. Find several nontrivial pairs of matrices A and B such that $AB \neq BA$. Do the same with $AB = \mathbb{O}$, where neither A nor B is zero. Also find examples of $AC = BC$, $C \neq \mathbb{O}$ but $A \neq B$.

13. Define $[A, B] = AB - BA$. Let $A = \begin{bmatrix} 0 & \frac{\sqrt{2}}{2} & 0 \\ \frac{\sqrt{2}}{2} & 0 & \frac{\sqrt{2}}{2} \\ 0 & \frac{\sqrt{2}}{2} & 0 \end{bmatrix}$, $B = \begin{bmatrix} 0 & i\frac{\sqrt{2}}{2} & 0 \\ -i\frac{\sqrt{2}}{2} & 0 & i\frac{\sqrt{2}}{2} \\ 0 & i\frac{\sqrt{2}}{2} & 0 \end{bmatrix}$, and $C = \begin{bmatrix} -1 & 0 & 0 \\ 0 & 0 & 0 \\ 0 & 0 & 1 \end{bmatrix}$. Prove that $[A, B] = iC$, $[B, C] = iA$, $[C, A] = iB$, $A^2 + B^2 + C^2 = 2I_3$.

14. Determine all matrices that commute with $\begin{bmatrix} 0 & 1 \\ 0 & 0 \end{bmatrix}$. Do the same for $\begin{bmatrix} 0 & 1 & 0 \\ 0 & 0 & 1 \\ 0 & 0 & 0 \end{bmatrix}$. Can you generalize?

15. Find all matrices that commute with $\begin{bmatrix} \lambda & 1 & 0 \\ 0 & \lambda & 1 \\ 0 & 0 & \lambda \end{bmatrix}$. Now find all matrices that commute with $\begin{bmatrix} 0 & 0 & 1 \\ 0 & 1 & 0 \\ 1 & 0 & 0 \end{bmatrix}$.

16. Suppose $A = \begin{bmatrix} 0 & 0 & 1 \\ 0 & 1 & 0 \\ 1 & 0 & 1 \end{bmatrix}$ and $p(x) = x^2 - 2x + 3$. What is $p(A)$?

17. Suppose A is a square matrix and $p(x)$ is a polynomial. Argue that A commutes with $p(A)$.

18. Suppose A is a square matrix over the complex numbers that commutes with all matrices of the same size. Argue that there must exist a scalar λ such that $A = \lambda I$.

19. Argue that if A commutes with a diagonal matrix with distinct diagonal elements, then A itself must be diagonal.

20. Suppose P is an idempotent matrix ($P = P^2$). Argue that $P = P^k$ for all k in $\mathbb{N}$.

21. Suppose U is an upper triangular matrix with zeros down the main diagonal. Argue that U is nilpotent.

22. Determine all 2-by-2 matrices A such that $A^2 = I$.

23. Compute $\begin{bmatrix} 1 & i \\ i & -1 \end{bmatrix}^2$. Determine all 2-by-2 nilpotent matrices.

24. Argue that any matrix can be written uniquely as the sum of a symmetric matrix plus a skew-symmetric matrix.

25. Argue that for any matrix A, AA^T, A^TA, and $A + A^T$ are always symmetric while $A - A^T$ is skew-symmetric.

26. Suppose $A^2 = I$. Argue that $P = \frac{1}{2}(I + A)$ is idempotent.

27. Argue that for any complex matrix A, AA^*, A^*A, and $i(A - A^*)$ are always Hermitian.

28. Argue that every complex matrix A can be written as $A = B + iC$, where B and C are Hermitian.

29. Suppose A and B are square matrices. Argue that $A^k - B^k = \sum_{j=0}^{k-1} B^j(A - B)A^{k-1-j}$.

30. Suppose P is idempotent. Argue that $tr(PA) = tr(PAP)$.

31. Suppose $A = \begin{bmatrix} B & C \\ \mathbb{O} & I \end{bmatrix}$. What is A^2? A^3? Can you guess A^n?

32. For complex matrices A and B, argue that $\overline{AB} = \overline{A}\,\overline{B}$, assuming A and B can be multiplied.

33. Argue that $(I - A)(I + A + A^2 + \cdots + A^m) = I - A^{m+1}$ for any positive integer m.

34. Argue that $2AA^* - 2BB^* = (A+B)^*(A-B) + (A-B)^*(A+B)$ for complex matrices A and B.

35. This exercise involves the commutator, or "Lie bracket," of exercise 13. Recall $[A, B] = AB - BA$. First, argue that the trace of any commutator is the zero matrix. Next, work out the 2-by-2 case explicitly. That is, calculate $[\begin{bmatrix} a & b \\ c & d \end{bmatrix}, \begin{bmatrix} e & f \\ g & h \end{bmatrix}]$ explicitly as a 2-by-2 matrix.

36. There is an interesting mapping associated with the commutator. This is for a fixed matrix A, Ad_A by $Ad_A(X) = [A, X] = AX - XA$. Verify the following:

    (a) $Ad_A(A) = \mathbb{O}$.
    (b) $Ad_A(BC) = (Ad_A(B))C + B(Ad_A(C))$. (Does this remind you of anything from calculus?)
    (c) $Ad_{Ad_A}(X) = ([Ad_A, Ad_B])(X)$.
    (d) $Ad_A(\alpha X) = \alpha Ad_A(X)$, where α is a scalar.
    (e) $Ad_A(X+Y) = Ad_A(X) + Ad_A(Y)$.

37. Since we add matrices componentwise, it is tempting to multiply them componentwise as well. Of course, this is not the way we are taught to do it. However, there is no reason we cannot define a slotwise multiplication of matrices and investigate its properties. Take two m-by-n matrices, $A = \left[a_{ij}\right]$ and $B = \left[b_{ij}\right]$, and define $A \odot B = [a_{ij}b_{ij}]$, which will also be m-by-n. Investigate what rules hold for this kind of multiplication. For example, is this product commutative? Does it satisfy the associative law? Does it distribute over addition? What can you say about the rank of $A \odot B$? How about $vec(A \odot B)$?

38. We could define the absolute value or magnitude of a matrix $A = [a_{ij}] \in \mathbb{C}^{m \times n}$ by taking the magnitude of each entry: $|A| = [\left|a_{ij}\right|]$. (Do not confuse this notation with the determinant of A.) What can you say about $|A + B|$? How about $|\alpha A|$ where α is any scalar? What about $|AB|$, $|A^*|$, and $|A \otimes B|$?

39. This exercise is about the conjugate of a matrix. Define $ent_{ij}(\overline{A}) = \overline{ent_{ij}(A)}$. That is, $\overline{A} = [\overline{a_{ij}}]$. Prove the following:

    (a) $\overline{A+B} = \overline{A} + \overline{B}$
    (b) $\overline{AB} = \overline{AB}$
    (c) $\overline{\alpha A} = \bar{\alpha}\overline{A}$
    (d) $(\overline{A})^k = \overline{(A^k)}$.

40. Prove that $BAA^* = CAA^*$ implies $BA = CA$.

41. Suppose $A^2 = I$. Prove that $(I + A)^n = 2^{n-1}(I + A)$. What can you say about $(I - A)^n$?

42. Prove that if $H = H^*$, then B^*HB is Hermitian for any conformable B.

43. Suppose A and B are n-by-n. Argue that A and B commute iff $A - \lambda I$ and $B - \lambda I$ commute for every scalar λ.

44. Suppose A and B commute. Prove that $A^p B^q = B^q A^p$ for all positive integers p and q.

45. Consider the Pauli spin matrices $\sigma_x = \begin{bmatrix} 0 & 1 \\ 1 & 0 \end{bmatrix}$, $\sigma_y = \begin{bmatrix} 0 & -i \\ i & 0 \end{bmatrix}$, and $\sigma_z = \begin{bmatrix} 1 & 0 \\ 0 & -1 \end{bmatrix}$. Argue that these matrices anticommute ($AB = -BA$) pairwise. Also, compute all possible commutators.

B.5.1 MATLAB Moment

B.5.1.1 Matrix Manipulations

The usual matrix operations are built in to MATLAB. If A and B are compatible matrices,

$A + B$	matrix addition
$A * B$	matrix multiplication
$A - B$	matrix subtraction
$A\hat{}n$	matrix to the nth power
ctranspose(A)	conjugate transpose
transpose(A)	transpose of A
A'	conjugate transpose
$A.'$	transpose without conjugation
$A.\hat{}n$	raise individual entries in A to the nth power

Note that the dot returns entrywise operations. So, for example, $A * B$ is ordinary matrix multiplication but $A. * B$ returns entrywise multiplication.

B.6 Submatrices

In this section, we describe what a submatrix of a matrix is and establish some notation that will be useful in discussing determinants of certain square submatrices. If A is an m-by-n matrix, then a submatrix of A is obtained by deleting rows and/or columns from this matrix. For example, if

$$A = \begin{bmatrix} a_{11} & a_{12} & a_{13} & a_{14} \\ a_{21} & a_{22} & a_{23} & a_{24} \\ a_{31} & a_{32} & a_{33} & a_{34} \end{bmatrix},$$

we could delete the first row and second column to obtain the submatrix

$$\begin{bmatrix} a_{12} & a_{13} & a_{14} \\ a_{32} & a_{33} & a_{34} \end{bmatrix}.$$

From the three rows of A, there are $\binom{3}{r}$ choices to delete, where $r = 0, 1, 2, 3$, and from the four columns, there are $\binom{4}{c}$ columns to delete, where $c = 0, 1, 2, 3$. Thus, there are $\binom{3}{r}\binom{4}{c}$ submatrices of A of size (3-r)-by-(4-c).

A submatrix of an n-by-n matrix is called a *principal submatrix* if it is obtained by deleting the same rows and columns. That is, if the ith row is deleted, the ith column is deleted as well. The result is, of course, a square matrix. For example, if

$$A = \begin{bmatrix} a_{11} & a_{12} & a_{13} \\ a_{21} & a_{22} & a_{23} \\ a_{31} & a_{32} & a_{33} \end{bmatrix},$$

and we delete the second row and second column, we get the principal submatrix

$$\begin{bmatrix} a_{11} & a_{13} \\ a_{31} & a_{33} \end{bmatrix}.$$

The r-by-r principal submatrix of an n-by-n matrix obtained by striking out its last n-r rows and columns is called a *leading principal submatrix*. For the matrix A above, the leading principal submatrices are

$$[a_{11}], \begin{bmatrix} a_{11} & a_{12} \\ a_{21} & a_{22} \end{bmatrix}, \begin{bmatrix} a_{11} & a_{12} & a_{13} \\ a_{21} & a_{22} & a_{23} \\ a_{31} & a_{32} & a_{33} \end{bmatrix}.$$

It is handy to have a notation to be able to specify submatrices more precisely. Suppose k and m are positive integers with $k \leq m$. Define $Q_{k,m}$ to be the set of all sequences of integers $(i_1, i_2, \ldots, i_k)$ of length k chosen from the first m positive integers $[m] = \{1, 2, \ldots, m\}$:

$$Q_{k,m} = \{(i_1, i_2, \ldots, i_k) \mid 1 \leq i_1 < i_2 < \cdots < i_k \leq m\}.$$

Note that $Q_{k,m}$ can be linearly ordered lexicographically. That is, if α and β are in $Q_{k,m}$, we define $\alpha \leq \beta$ iff $\alpha = (i_1, i_2, \ldots, i_k)$, $\beta = (j_1, j_2, \ldots, j_k)$ and $i_1 < j_1$ or $i_1 = j_1$ and $i_2 < k_2$, or $i_1 = j_1$ and $i_2 = j_2$ and $i_3 < j_3, \ldots, i_1 = j_1$ and, $\ldots, i_{k-1} = j_{k-1}$ but $i_k < j_k$. This is just the way words are ordered in the dictionary. For example, in lexicographic order:

$$\begin{aligned}
Q_{2,3} &= \{(1,2), (1,3), (2,3)\},\\
Q_{2,4} &= \{(1,2), (1,3), (1,4), (2,3), (2,4), (3,4)\},\\
Q_{3,4} &= \{(1,2,3), (1,2,4), (1,3,4), (2,3,4)\},\\
Q_{1,4} &= \{(1), (2), (3), (4)\}.
\end{aligned}$$

It is an exercise that the cardinality of the set $Q_{k,m}$ is the binomial coefficient m choose k:

$$\left|Q_{k,m}\right| = \binom{m}{k}.$$

There is a simple function on $Q_{k,m}$ that will prove useful. It is the *sum function* $s : Q_{k,m} \to \mathbb{N}$, defined by $s(i_1, i_2, \ldots, i_k) = i_1 + i_2 + \cdots + i_k$. Now if A is m-by-n, $\alpha \in Q_{k,m}$, $\beta \in Q_{j,n}$ and the notation $A[\alpha \mid \beta]$ stands for the k-by-j submatrix of A consisting of elements whose row index comes from α and whose column index comes from β. For example, if

$$A = \begin{bmatrix} a_{11} & a_{12} & a_{13} & a_{14} & a_{15} \\ a_{21} & a_{22} & a_{23} & a_{24} & a_{25} \\ a_{31} & a_{32} & a_{33} & a_{34} & a_{35} \\ a_{41} & a_{42} & a_{43} & a_{44} & a_{45} \end{bmatrix}$$

and $\alpha \in Q_{2,4}$, $\alpha = (1,3)$, $\beta \in Q_{3,5}$ $\beta = (2,4,5)$, then

$$A[\alpha \mid \beta] = \begin{bmatrix} a_{12} & a_{14} & a_{15} \\ a_{32} & a_{34} & a_{35} \end{bmatrix},$$

$$A[\alpha \mid \alpha] = \begin{bmatrix} a_{11} & a_{13} \\ a_{31} & a_{33} \end{bmatrix},$$

and so on. We adopt the shorthand, $A[\alpha \mid \alpha] \equiv A[\alpha]$ for square matrices A. Note that $A[\alpha]$ is a principal submatrix of A. Also, we abbreviate $A[(i_1, i_2, \ldots, i_k)]$

by $A[i_1, i_2, \ldots, i_k]$. Thus, $A[1, \ldots, r]$ is the r-by-r leading principal submatrix of A. Using A above, $A[1, 2] = \begin{bmatrix} a_{11} & a_{12} \\ a_{21} & a_{22} \end{bmatrix}$.

Each $\alpha \in Q_{k,m}$ has a complementary sequence $\widehat{\alpha}$ in $Q_{k-m,m}$ consisting of the integers in $\{1, \ldots, m\}$ not included in α, but listed in increasing order. This allows us to define several more submatrices:

$$A[\alpha \mid \beta) \equiv A[\alpha \mid \widehat{\beta}],$$
$$A(\alpha \mid \beta] \equiv A[\widehat{\alpha} \mid \beta],$$
$$A(\alpha \mid \beta) \equiv A[\widehat{\alpha} \mid \widehat{\beta}].$$

For square matrices, $A(\alpha) \equiv A(\alpha \mid \alpha)$. Note if $A[\alpha]$ is a nonsingular principal submatrix of A, we can define its Schur complement as

$$A/A[\alpha] = A(\alpha](A[\alpha])^{-1}A[\alpha).$$

We sometimes abuse notation and think of a sequence as a set with a preferred ordering. For example, we write $A[\alpha \cup \{i_p\} \mid \alpha \cup \{j_p\}]$, where i_p and j_p are not in α but they are put in the appropriate order. This is called "bordering." Using A and $\alpha = (1, 3)$ from above, we see

$$A[\alpha \cup \{4\} \mid \alpha \cup \{5\}] = \begin{bmatrix} a_{11} & a_{13} & a_{15} \\ a_{31} & a_{33} & a_{35} \\ a_{41} & a_{43} & a_{45} \end{bmatrix}.$$

In this notation, we can generalize the notion of a Schur complement even further:

$$A/A[\alpha, \beta] = A[\widehat{\alpha} \mid \widehat{\beta}] - A[\widehat{\alpha} \mid \beta]\,(A[\alpha \mid \beta])^{-1}\,A[\alpha \mid \widehat{\beta}].$$

APB Exercise Set 5

1. What can you say about the principal submatrices of a symmetric matrix? A diagonal matrix? An upper triangular matrix? A lower triangular matrix? An Hermitian matrix?

2. How do the submatrices of A compare with the submatrices of A^T? Specifically, if $\widehat{A}$ is the r-by-s submatrix of the m-by-n matrix A obtained by deleting rows $i_1, i_2, \ldots, i_{m-r}$ and columns $j_1, j_2, \ldots, j_{n-s}$, and B is the s-by-r submatrix obtained from A^T by deleting rows $j_1, j_2, \ldots, j_{n-s}$ and columns, $i_1, i_2, \ldots, i_{m-r}$, how does B compare to $\widehat{A}$?

3. Prove that $\left|Q_{k,m}\right| = \binom{m}{k}$.

Further Reading

[B&R, 1986(2)] T. S. Blyth and E. F. Robertson, *Matrices and Vector Spaces*, Vol. 2, Chapman & Hall, New York, (1986).

[Butson, 1962] A. T. Butson, Generalized Hadamard Matrices, Proceedings of the American Mathematical Society, Vol. 13, (1962), 894–898.

[Davis, 1965] Philip J. Davis, *The Mathematics of Matrices: A First Book of Matrix Theory and Linear Algebra*, Blaisdell Publishing Company, New York, (1965).

[G&B, 1963] S. W. Golomb and L. D. Baumert, The Search for Hadamard Matrices, The American Mathematical Monthly, Vol. 70, (1963), 12–17.

[J&S, 1996] C. R. Johnson and E. Schreiner, The Relationship Between AB and BA, The American Mathematical Monthly, Vol. 103, (1996), 578–582.

[Kolo, 1964] Ignace I. Kolodner, A Note on Matrix Notation, The American Mathematical Monthly, Vol. 71, No. 9, November, (1964), 1031–1032.

[Leslie, 1945] P. H. Leslie, On the Use of Matrices in Certain Population Mathematics, Biometrika, Vol. 33, (1945).

[Lütkepohl, 1996] Helmut Lütkepohl, *Handbook of Matrices*, John Wiley & Sons, New York, (1996).

B.6.1 MATLAB Moment

B.6.1.1 Getting at Pieces of Matrices

Suppose A is a matrix. It is easy to extract portions of A using MATLAB.

$A(i, j)$ returns the i, j-entry of A.

$A(\ :\ , j)$ returns the jth column of A.

$A(i,\ :\ )$ returns the ith row of A.

$A(end,\ :\ )$ returns the last row of A.

$A(\ :\ , end)$ returns the last column of A.

$A(\ :\ )$ returns a long column obtained by stacking the columns of A.

[] is the empty 0-by-0 matrix.

Submatrices can be specified in various ways. For example,

$$A([i\ j], [p\ q\ r])$$

returns the submatrix of A consisting of the intersection of rows i and j and columns p, q, and r. More generally,

$$A(i : j, k : m)$$

returns the submatrix that is the intersection of rows i to j and columns k to m.

The size command can be useful; size(A), for an m-by-n matrix A, returns the two-element row vector $[m, n]$ containing the number of rows and columns in the matrix. $[m, n]$ = size(A) for matrix A returns the number of rows and columns in A as separate output variables.

Appendix C

Determinants

C.1 Motivation

Today, determinants are out of favor with some people (see Axler [1995]). Even so, they can be quite useful in certain theoretical situations. The concept goes back to **Seki Kowa** (March 1642–24 October 1708), a famous Japanese mathematician. In the west, **Gottfried Leibniz** (1 July 1646–14 November 1716) developed the idea of what we call Cramer's rule today. However, the idea lay dormant until 1750, when determinants became a major tool in the theory of systems of linear equations. According to Aitken [1962], Vandermonde may be regarded the founder of a notation and of rules for computing with determinants. These days, matrix theory and linear algebra play the dominant role, with some texts barely mentioning determinants. We gather the salient facts about determinants in this appendix; but first we consider some geometric motivation.

Consider the problem of finding the area of a parallelogram formed by two independent vectors $\overrightarrow{A} = (a, b)$ and $\overrightarrow{B} = (c, d)$ in $\mathbb{R}^2$.

The area of the parallelogram is base times height, that is $\left\|\overrightarrow{A}\right\| \left\|\overrightarrow{B}\right\| \sin(\theta)$. Thus, the squared area is $\left\|\overrightarrow{A}\right\|^2 \left\|\overrightarrow{B}\right\|^2 (\sin(\theta))^2 = \left\|\overrightarrow{A}\right\|^2 \left\|\overrightarrow{B}\right\|^2 (1-\cos^2(\theta))$ $= \left\|\overrightarrow{A}\right\|^2 \left\|\overrightarrow{B}\right\|^2 - \left\|\overrightarrow{A}\right\|^2 \left\|\overrightarrow{B}\right\|^2 \cos^2(\theta) = \left\|\overrightarrow{A}\right\|^2 \left\|\overrightarrow{B}\right\|^2 - (\overrightarrow{A} \cdot \overrightarrow{B})^2 = (a^2 + b^2)(c^2+d^2) - (ac+bd)^2 = a^2c^2 + a^2d^2 + b^2c^2 + b^2d^2 - a^2c^2 - 2acbd - b^2d^2 = a^2d^2 - 2adbc + b^2c^2 = (ad-bc)^2$. So the area of the parallelogram determined by independent vectors $\overrightarrow{A} = (a, b)$ and $\overrightarrow{B} = (c, d)$ is the absolute value of $ad - bc$. If we make a matrix $\begin{bmatrix} a & c \\ b & d \end{bmatrix}$, then we can designate this important number as $det \begin{bmatrix} a & c \\ b & d \end{bmatrix} = ad - bc$.

More generally, $det \begin{bmatrix} a_{11} & a_{12} \\ a_{21} & a_{22} \end{bmatrix} = a_{11}a_{22} - a_{12}a_{21}$. Notice the "crisscross" way we compute the determinant of a 2-by-2 matrix? Notice the minus sign? In two dimensions, then, the determinant is very closely related to the geometric idea of area.

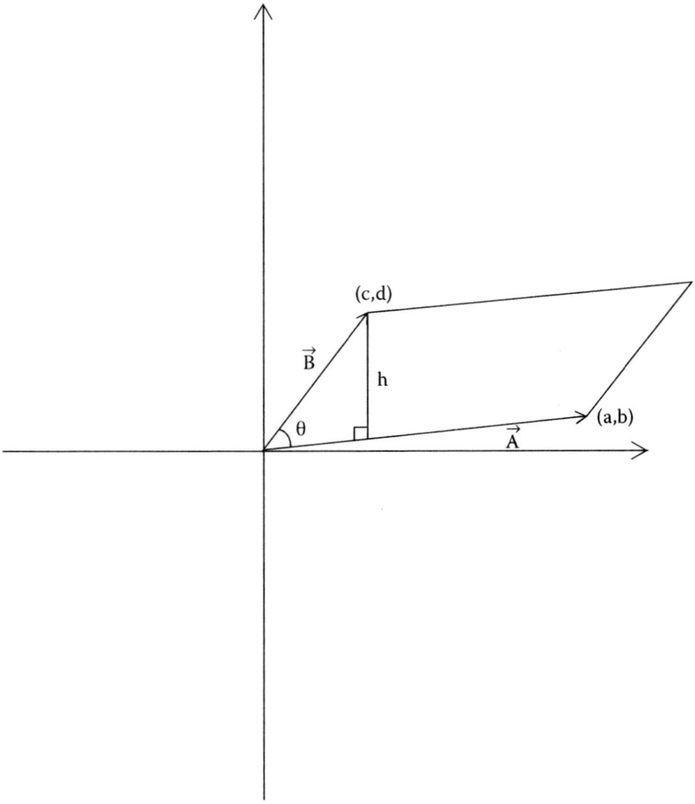

Figure C1.1: Area of a parallelogram.

Consider a parallelepiped in $\mathbb{R}^3$ determined by three independent vectors, $\overrightarrow{A} = (a_1, a_2, a_3)$, $\overrightarrow{B} = (b_1, b_2, b_3)$, and $\overrightarrow{C} = (c_1, c_2, c_3)$. The volume of this "box" is the area of its base times its altitude; that is,

$$\begin{aligned}
\text{volume} &= (\text{area of base})(\text{altitude}) \\
&= (\left\|\overrightarrow{B}\right\| \left\|\overrightarrow{C}\right\| \sin(\theta))(\left\|\overrightarrow{A}\right\| \cos(\theta)) = \left|\left(\overrightarrow{B} \times \overrightarrow{C}\right) \cdot \overrightarrow{A}\right| \\
&= |((b_2c_3 - c_2b_3), (b_3c_1 - b_1c_3), (b_1c_2 - c_1b_2)) \cdot (a_1, a_2, a_3)| \\
&= |a_1(b_2c_3 - c_2b_3) + a_2(b_3c_1 - b_1c_3) + a_3(b_1c_2 - c_1b_2)| \\
&= |a_1b_2c_3 - a_1c_2b_3 + a_2b_3c_1 - a_2b_1c_3 + a_3b_1c_2 - a_3c_1b_2| .
\end{aligned}$$

This is getting more complicated. Again we create a matrix whose columns are the vectors and define a 3-by-3 determinant: $det \begin{bmatrix} a_1 & b_1 & c_1 \\ a_2 & b_2 & c_2 \\ a_3 & b_3 & c_3 \end{bmatrix} = a_1b_2c_3 - a_1c_2b_3 + a_2b_3c_1 - a_2b_1c_3 + a_3b_1c_2 - a_3c_1b_2$.

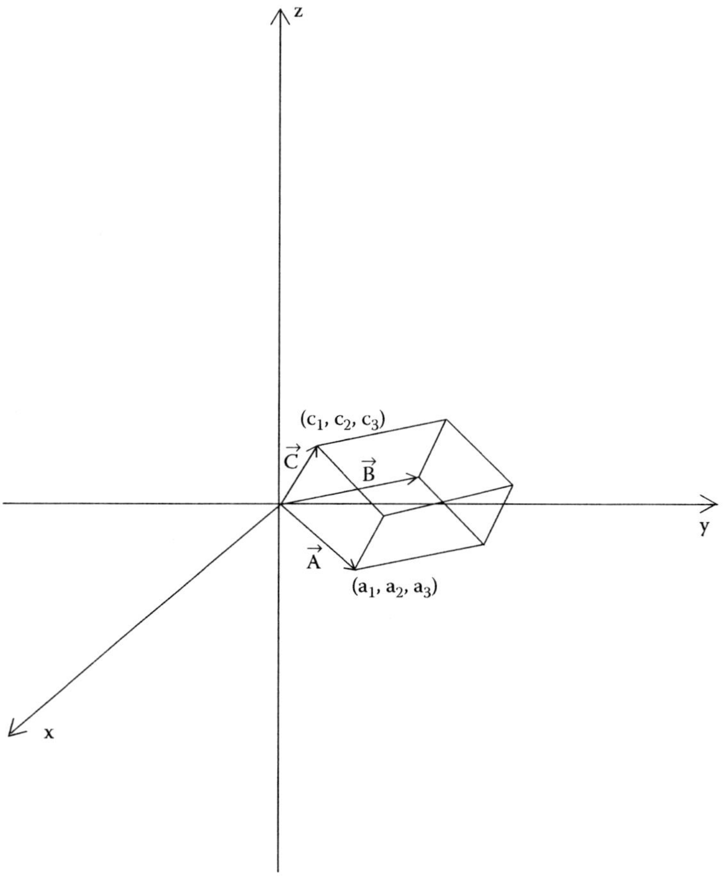

Figure C1.2: Volume of a parallelepiped.

Actually, there is a little trick to help us compute a 3-by-3 determinant by hand. Write the first two columns over again. Then go down diagonals with plus products and up diagonals with minus products:

$$det \begin{bmatrix} a_{11} & a_{12} & a_{13} \\ a_{21} & a_{22} & a_{23} \\ a_{31} & a_{32} & a_{33} \end{bmatrix} \begin{matrix} a_{11} & a_{12} \\ a_{21} & a_{22} \\ a_{31} & a_{32} \end{matrix}$$

$$= \begin{matrix} a_{11} & a_{22} & a_{33} & + & a_{12} & a_{23} & a_{31} & + & a_{13} & a_{21} & a_{32} \\ -a_{31} & a_{22} & a_{13} & - & a_{32} & a_{23} & a_{11} & - & a_{33} & a_{21} & a_{12} \end{matrix}.$$

Thus, in three dimensions, the idea of determinant is closely associated with the geometric idea of volume.

Though we cannot draw the pictures, we can continue this idea into $\mathbb{R}^n$. A "solid" in $\mathbb{R}^n$ whose adjacent sides are defined by a linearly independent set of vectors $\overrightarrow{\mathbf{x}}_1, \overrightarrow{\mathbf{x}}_2, \ldots, \overrightarrow{\mathbf{x}}_n$ is called an *n-dimensional parallelepiped.* Can we get hold of its volume? We can motivate our generalization by looking at the two-dimensional example again.

Form the matrix A whose columns are the vectors (a, b) and (c, d):

$$A = \begin{bmatrix} a & c \\ b & d \end{bmatrix}.$$

We note that the height of the parallelogram is the length of the orthogonal projection of $\overrightarrow{B}$ onto $\overrightarrow{A}$, which is $(I - P_{\overrightarrow{A}})(\overrightarrow{B})$, where we recall $P_{\overrightarrow{A}}(\overrightarrow{B}) = \left(\frac{\overrightarrow{A} \cdot \overrightarrow{B}}{\overrightarrow{A} \cdot \overrightarrow{A}}\right)\overrightarrow{A}$. Thus, the area of the parallelogram is $\left\|\overrightarrow{A}\right\| \left\|(I - P_{\overrightarrow{A}})(\overrightarrow{B})\right\|$. Now here is something you might not think of: Write the matrix A in its QR factorization

$$A = QR = \begin{bmatrix} q_{11} & q_{12} \\ q_{21} & q_{22} \end{bmatrix} \begin{bmatrix} r_{11} & r_{12} \\ 0 & r_{22} \end{bmatrix}.$$

Then, by direct computation using the orthonormality of the columns of Q, we find $r_{11} = \left\|\overrightarrow{A}\right\|$ and $r_{22} = \left\|(I - P_{\overrightarrow{A}})(\overrightarrow{B})\right\|$ and so the area of the parallogram is just the determinant of R. This idea generalizes.

Suppose we make a matrix A whose columns are these independent vectors, $A = [\mathbf{x}_1 \mid \mathbf{x}_2 \mid \ldots \mid \mathbf{x}_n]$. Then the volume is $\|\mathbf{x}_1\| \|(I - P_2)\mathbf{x}_2\| \cdots \|(I - P_n)\mathbf{x}_n\|$, where P_k is the orthogonal projection onto $span\,\{\mathbf{x}_1, \mathbf{x}_2, \cdots, \mathbf{x}_{k-1}\}$. Write A in its QR factorization. Then the volume is the product of the diagonal elements of R. Now $(det(A))^2 = det(A^T A) = det(R^T Q^T Q R) = det(R^T R) = (det(R))^2 = (\text{volume})^2$. But then, volume equals the absolute value of $det(A)$. Even in higher dimensions, the determinant is closely associated with the geometric idea of volume. Well, all this presumes we know some facts about determinants. It is time to go back to the drawing board and start from scratch.

C.2 Defining Determinants

We could just give a formula to define a determinant, but we prefer to begin more abstractly. This approach goes back to Kronecker. We begin by defining a function of the columns of a square matrix. (Rows could be used just as well.) The first step is to dismantle the matrix into its columns. Define

$$\text{col} : \mathbb{C}^{n \times n} \to \mathbb{C}^{n\times 1} \times \mathbb{C}^{n\times 1} \times \cdots \times \mathbb{C}^{n\times 1} \quad (n \text{ copies})$$

by

$$\text{col}(A) = (\text{col}_1(A), \text{col}_2(A), \ldots, \text{col}_n(A)).$$

For example,

$$\text{col}(\begin{bmatrix} 1 & 2 \\ 3 & 4 \end{bmatrix}) = (\begin{bmatrix} 1 \\ 3 \end{bmatrix}, \begin{bmatrix} 2 \\ 4 \end{bmatrix}) \in \mathbb{C}^{2\times 1} \times \mathbb{C}^{2\times 1}.$$

Clearly, col is a one-to-one and onto function. Then we introduce a "determinant function" on the n-tuple space with certain crucial properties. Define $D : \mathbb{C}^{n\times 1} \times \mathbb{C}^{n\times 1} \times \cdots \times \mathbb{C}^{n\times 1} \rightarrow \mathbb{C}$ to be *n-linear* if

1. $D(\cdots, a[x_i], \cdots) = a\ D(\cdots, [x_i], \cdots)$ for $i = 1, 2, \ldots, n$
2. $D(\cdots, [x_i] + [y_i], \cdots) = D(\cdots, [x_i], \cdots) + D(\cdots, [y_i], \cdots)$ $i = 1, 2, \ldots, n$.

We say D is *alternating* iff $D = 0$ whenever two adjacent columns are equal. Finally, $D : \mathbb{C}^{n\times 1} \times \mathbb{C}^{n\times 1} \times \cdots \times \mathbb{C}^{n\times 1} \rightarrow \mathbb{C}$ is a *determinant function* iff it is n-linear and alternating. We are saying that, as a function of each column, D is linear, hence the name n-linear. Then we have the associated *determinant*

$$\det(A) = D(\text{col}(A)).$$

Note that (2) above does *not* say $\det(A + B) = \det(A) + \det(B)$, which is false as soon as $n > 1$. Also note

$$\det : \mathbb{C}^{n\times n} \rightarrow \mathbb{C}.$$

To see some examples, fix a number b and consider $D_b(\begin{bmatrix} a_{11} \\ a_{21} \end{bmatrix}, \begin{bmatrix} a_{12} \\ a_{22} \end{bmatrix}) = b(a_{11}a_{22} - a_{12}a_{21})$. D_b is a determinant function. Define $D(\begin{bmatrix} a_{11} \\ a_{21} \\ \vdots \\ a_{n1} \end{bmatrix}, \begin{bmatrix} a_{12} \\ a_{22} \\ \vdots \\ a_{n2} \end{bmatrix}, \ldots, \begin{bmatrix} a_{1n} \\ a_{2n} \\ \vdots \\ a_{nn} \end{bmatrix}) = a_{11}a_{22}\cdots a_{nn}$. This is an n-linear function. Is it a determinant function? Let $\sigma \in S_n$ and define $D\sigma\left(\begin{bmatrix} a_{11} \\ a_{21} \\ \vdots \\ a_{n1} \end{bmatrix}, \begin{bmatrix} a_{12} \\ a_{22} \\ \vdots \\ a_{n2} \end{bmatrix}, \dots, \begin{bmatrix} a_{1n} \\ a_{2n} \\ \vdots \\ a_{nn} \end{bmatrix}\right) = a_{1\sigma(1)}a_{2\sigma(2)}\cdots a_{n\sigma(n)}$. $D\sigma$ is n-linear. Is it a determinant function?

Let $f : S_n \to \mathbb{C}$ be any function. Define D_f by $D_f\left(\begin{bmatrix} a_{11} \\ a_{21} \\ \vdots \\ a_{n1} \end{bmatrix}, \begin{bmatrix} a_{12} \\ a_{22} \\ \vdots \\ a_{n2} \end{bmatrix}, \dots, \begin{bmatrix} a_{1n} \\ a_{2n} \\ \vdots \\ a_{nn} \end{bmatrix}\right) = \sum_{\sigma \in S_n} f(\sigma)a_{1\sigma(1)}a_{2\sigma(2)}\dots a_{n\sigma(n)}$. D_f is an n-linear function. Is it a determinant function?

Our goal now is to show that there is a determinant function for every $n \in \mathbb{N}$ and this determinant function is uniquely determined by its value on the identity matrix I_n. If $n = 1$ this is easy: $D([a]) = D(a[1]) = aD([1]) = aD(I_1)$. First, we assemble some facts.

THEOREM C.1

Let $D : \mathbb{C}^{n\times 1} \times \mathbb{C}^{n\times 1} \times \cdots \times \mathbb{C}^{n\times 1} \to \mathbb{C}$ be a determinant function. Then

1. *$D = 0$ if there is a zero column.*
2. *If two adjacent columns are interchanged, D changes sign.*
3. *If any two columns are interchanged, then D changes sign.*
4. *If two columns are equal, $D = 0$.*
5. *If the jth column is multiplied by the scalar a and the result is added to the ith column, the value of D does not change.*
6. $D\left(\begin{bmatrix} a_{11} \\ a_{21} \\ \vdots \\ a_{n1} \end{bmatrix}, \begin{bmatrix} a_{12} \\ a_{22} \\ \vdots \\ a_{n2} \end{bmatrix}, \dots, \begin{bmatrix} a_{1n} \\ a_{2n} \\ \vdots \\ a_{nn} \end{bmatrix}\right) = \left[\sum_{\sigma \in S_n} sgn(\sigma)a_{1\sigma(1)}\cdots a_{n\sigma(n)}\right] D(I_n).$

PROOF In view of (1) in the definition of n-linear, (1) of the theorem is clear. To prove (2), let $[x]$ and $[y]$ be the ith and $(i+1)$st columns. Then

$$\begin{aligned}
0 &= D(\cdots, [x]+[y], [x]+[y], \cdots) \\
&= D(\cdots, [x], [x]+[y], \cdots) + D(\cdots, [y], [x]+[y], \cdots) \\
&= D(\cdots, [x], [x], \cdots) + D(\cdots, [x], [y], \cdots) + D(\cdots, [y], [x], \cdots) \\
&\quad + D(\cdots, [y], [y], \cdots) \\
&= 0 + D(\cdots, [x], [y], \cdots) + D(\cdots, [y], [x], \cdots) + 0.
\end{aligned}$$

Therefore,

$$D(\cdots, [x], [y], \cdots) = -D(\cdots, [y], [x], \cdots).$$

The proof of (3) uses (2). Suppose the ith and jth columns are swapped. Say $i < j$. Then by $j - i$ swaps of adjacent columns, the jth column can be put into the ith position. Next, the ith column is put in the jth place by $j - i - 1$ swaps of adjacent columns. The sign change is $(-1)^{2j-2i-1} = -1$. Now (4) is clear. For (5), we use linearity in the ith column. We compute

$$\begin{aligned}
&D([x_1], \ldots, [x_i] + a[x_j] \ldots, \\
&\qquad [x_j], \ldots, [x_n]) = D([x_1], \ldots, [x_i], \ldots, [x_j], \ldots, [x_n]) \\
&\qquad\qquad + aD([x_1], \ldots, [x_j] \ldots, [x_j], \ldots, [x_n]) \\
&\qquad = D([x_1], \ldots, [x_i], \ldots, [x_j], \ldots, [x_n]) + 0 \\
&\qquad = D([x_1], \ldots, [x_i], \ldots, [x_j], \ldots, [x_n]).
\end{aligned}$$

The proof of (6) is somewhat lengthy and the reader is referred to A&W [1992] for the details. □

In view of (6), we see that a determinant function is completely determined by its action on the identity matrix. Let D_a be such that $D_a(I_n) = a$. Indeed, $D_a = aD_1$ so we take det as the associated determinant to D_1, our uniquely defined determinant function that takes the value 1 on the identity matrix. Note, we could have just started with this formula

$$\det(A) = \sum_{\sigma \in S_n} sgn(\sigma) a_{1\sigma(1)} \cdots a_{n\sigma(n)}$$

and proved properties of this function but we like this abstract approach. You may already be familiar with the following results, which follow readily from this formula for $det(A)$.

COROLLARY C.1

1. *If $A \in \mathbb{C}^{n\times n}$ is upper or lower triangular, then $det(A) = a_{11}a_{22}\dots a_{nn}$.*
2. *$det(A) = det(A^T)$.*
3. *$det(AB) = det(A)det(B)$.*
4. *If S is invertible, $det(S^{-1}AS) = det(A)$.*
5. *If $i \neq j$, then $det(P_{ij}A) = -det(A)$.*
6. *If $i \neq j$, then $det(T_{ij}(a)A) = det(A)$.*
7. *If $\sigma \in S_n$, then $det(P(\sigma)) = sgn(\sigma)$.*
8. *$det(cA) = c^n det(A)$ if A is n-by-n.*

PROOF We will offer only two proofs as illustration and leave the rest to the reader. First, we illustrate the use of the formula inproving (2).

$$\begin{aligned}
\det(A^T) &= \sum_{\sigma\in S_n} sgn(\sigma)a_{\sigma(1)1}\cdots a_{\sigma(n)n} \\
&= \sum_{\tau\in S_n} sgn(\tau^{-1})a_{\tau^{-1}(1)1}\cdots a_{\tau^{-1}(n)n} \\
&= \sum_{\tau\in S_n} sgn(\tau)a_{\tau^{-1}(1)1}\cdots a_{\tau^{-1}(n)n} \\
&= \sum_{\tau\in S_n} sgn(\tau)a_{\tau\tau^{-1}(1)\tau(1)}\cdots a_{\tau\tau^{-1}(n)\tau(n)} \\
&= \sum_{\tau\in S_n} sgn(\tau)a_{1\tau(1)}\cdots a_{n\tau(n)} \\
&= \det(A).
\end{aligned}$$

Next, we illustrate a proof based on the abstract characterization of determinant. Look at (3). The proof from the formula is rather messy, but consider a function $D_B(A) = \det(AB)$. One easily checks that D_B is n-linear and alternating on the columns of A. Thus, $D_B(A) = b\det(A)$, but $b = D_B(I) = det(B)$, so the theorem follows. ▯

C.3 Some Theorems about Determinants

In this section, we present some of the more important facts about determinants. We need some additional concepts. Suppose $A = [a_{ij}] \in \mathbb{C}^{n \times n}$. We shall use the notation developed for submatrices in the previous appendix. The main theorems we develop are the Laplace expansion theorem and the Cauchy-Binet theorem. These are very useful theorems.

C.3.1 Minors

If $A \in \mathbb{C}^{m \times n}$, the determinant of any submatrix $A[\alpha \mid \beta]$ where $\alpha \in Q_{k,m}$ and $\beta \in Q_{k,n}$ is called a k-by-k *minor* of A or the (α, β)-minor of A. Here, $0 \leq k \leq \min\{m, n\}$. The *complementary minor* is the determinant of $A[\widehat{\alpha} \mid \widehat{\beta}]$ if this makes sense. If $k = m = n$, then $Q_{m,m}$ has only one element so there is only one minor of order m, namely $det(A)$. If $k = 1$, there are m^2 minors of order one, which we identify with the elements of the matrix A. There are $\binom{m}{k}$ elements in $Q_{k,m}$ and so there are $\binom{m}{k}^2$ minors of order k. The determinant of a principal submatrix $A[\alpha]$ is called a *principal minor* and the determinant of a leading principal submatrix is called a *leading principal minor*. For example, if $A = [a_{ij}]$ is 5-by-5, $\alpha = (2, 3, 5)$, and $\beta = (1, 2, 4)$, then the (α, β)-minor is

$$\det \begin{bmatrix} a_{21} & a_{22} & a_{24} \\ a_{31} & a_{32} & a_{34} \\ a_{51} & a_{52} & a_{54} \end{bmatrix}$$

and the complementary minor is

$$\det \begin{bmatrix} a_{13} & a_{15} \\ a_{43} & a_{45} \end{bmatrix}.$$

There are $\binom{5}{3} = 10$ minors of A of order three altogether.

C.3.2 The Cauchy-Binet Theorem

You probably recall the well-known theorem that "the determinant of a product of two square matrices is the product of their determinants." The generalization of this result has a history that can be read in Muir, [1906 pages 123–130]. We present a vast generalization of this theorem next. We follow the treatment in A&W [1992].

THEOREM C.2 (Cauchy-Binet, 1812)

Suppose $A \in \mathbb{C}^{m\times n}$, $B \in \mathbb{C}^{n\times r}$, *and* $C = AB \in \mathbb{C}^{m\times r}$. *Suppose* $1 \leq t \leq \min\{m, n, r\}$ *and suppose* $\alpha \in Q_{t,m}$, $\beta \in Q_{t,r}$. *Then*

$$\det(C[\alpha \mid \beta]) = \sum_{\gamma\in Q_{t,n}} \det(A[\alpha \mid \gamma])\det(B[\gamma \mid \beta]).$$

PROOF Suppose $\alpha = (\alpha_1, \ldots, \alpha_t) \in Q_{t,m}$ and $\beta = (\beta_1, \ldots, \beta_t) \in Q_{t,r}$. We compute the (i, j)-entry of $C[\alpha \mid \beta]$:

$$ent_{ij}(AB[\alpha \mid \beta]) = \text{row}_{\alpha_i}(A) \cdot \text{col}_{\beta_j}(B) = \sum_{k=1}^{n} a_{\alpha_i k} b_{k\beta_j}$$

so

$$AB[\alpha \mid \beta] = \begin{bmatrix} \sum_{k=1}^{n} a_{\alpha_1 k} b_{k\beta_1} & \cdots & \sum_{k=1}^{n} a_{\alpha_1 k} b_{k\beta_t} \\ \vdots & & \vdots \\ \sum_{k=1}^{n} a_{\alpha_t k} b_{k\beta_1} & & \sum_{k=1}^{n} a_{\alpha_t k} b_{k\beta_t} \end{bmatrix}.$$

Now, using the n-linearity of the determinant function, we get

$$\det(AB[\alpha \mid \beta]) = \sum_{k_1=1}^{n} \cdots \sum_{k_t=1}^{n} a_{\alpha_1 k_1} \cdots a_{\alpha_t k_t} \det \begin{bmatrix} b_{k_1\beta_1} & \cdots & b_{k_1\beta_t} \\ \vdots & & \vdots \\ b_{k_t\beta_1} & \cdots & b_{k_t\beta_{5t}} \end{bmatrix}.$$

If $k_i = k_j$ for $i \neq j$, then the ith and jth rows of the matrix on the right are equal. Thus, the determinant is zero in this case. The only nonzero determinants that appear on the right occur when the sequence $(k_1, k_2, \ldots, k_t)$ is a permutation of a sequence $\gamma = (\gamma_1, \ldots, \gamma_n) \in Q_{t,n}$. Let σ be a permutation in the symmetric group S_t such that $\gamma_i = k_{\sigma(i)}$ for $1 \leq i \leq t$. Then

$$\det \begin{bmatrix} b_{k_1\beta_1} & \cdots & b_{k_1\beta_t} \\ \vdots & & \vdots \\ b_{k_t\beta_1} & \cdots & b_{k_t\beta_{5t}} \end{bmatrix} = sgn(\sigma)\det B[\gamma \mid \beta].$$

Given $\gamma \in Q_{t,n}$, all possible permutations of γ are included in the summation above. Therefore,

$$\begin{aligned} \det(AB[\alpha \mid \beta]) &= \sum_{\gamma\in Q_{t,n}} (\sum_{\sigma\in S_t} (sgn(\sigma) a_{\alpha_1\gamma_{\sigma(1)}} \cdots a_{\alpha_t\gamma_{\sigma(t)}}) \det B[\gamma \mid \beta] \\ &= \sum_{\gamma\in Q_{t,n}} \det(A[\alpha \mid \gamma])\det(B[\gamma \mid \beta]). \end{aligned}$$

□

Numerous corollaries follow from this result. First, as notation, let $1 : n = (1, 2, 3, \ldots, n)$.

COROLLARY C.2

Suppose A is m-by-n and B is n-by-m where $m \leq n$. Then

$$\det(AB) = \sum_{\gamma \in Q_{m,n}} \det(A[1:m \mid \gamma]) \det(B[\gamma \mid 1:m]).$$

In other words, the determinant of the product AB is the sum of the products of all possible minors of the maximal order m of A with the corresponding minors of B of the same order. Also note that if $m > n$, $det(AB) = 0$.

COROLLARY C.3

For conformable square matrices A, B, $det(AB) = det(A)det(B)$.

PROOF If $\gamma \in Q_{n,n}$, then $\gamma = 1 : n$, so $A[\gamma \mid \gamma] = A$, $B[\gamma \mid \gamma]$ and $AB[\gamma \mid \gamma] = AB$. Thus $det(AB) = det(AB[\gamma \mid \gamma]) = \sum_{\gamma \in Q_{n,n}} \det(A[\gamma \mid \gamma]) \det(B[\gamma \mid \gamma]) = \det(A)\det(B)$. □

COROLLARY C.4

If A is k-by-n where $k \leq n$, then $det(AA^*) = \sum_{\gamma \in Q_{k,n}} |\det A[1:k \mid \gamma]|^2 \geq 0$.

COROLLARY C.5 (the Cauchy identity)

$$det \begin{bmatrix} \sum_{j=1}^{n} a_i c_i & \sum_{i=1}^{n} a_i d_i \\ \sum_{i=1}^{n} b_i c_i & \sum_{i=1}^{n} b_i d_i \end{bmatrix} = \sum_{1 \leq j < k \leq n} det \begin{bmatrix} a_j & a_k \\ b_j & b_k \end{bmatrix} det \begin{bmatrix} c_j & c_k \\ d_j & d_k \end{bmatrix}.$$

In other words,

$$\left(\sum_{i=1}^{n} a_i c_i\right)\left(\sum_{i=1}^{n} b_i d_i\right) - \left(\sum_{i=1}^{n} a_i d_i\right)\left(\sum_{i=1}^{n} b_i c_i\right) = \sum_{1 \leq j < k \leq n} (a_j b_k - a_k b_j) \times \left(c_j d_k - c_k d_j\right).$$

As a special case, we get the following corollary.

COROLLARY C.6 (Cauchy inequality)
Over $\mathbb{R}$,

$$\left(\sum_{i=1}^{n} a_i^2\right)\left(\sum_{i=1}^{n} b_i^2\right) - \left(\sum_{i=1}^{n} a_i b_i\right)^2 = \sum_{1\le i<k\le n} det \begin{bmatrix} a_i & a_k \\ b_i & b_k \end{bmatrix}^2$$

so over $\mathbb{R}$,

$$\left(\sum_{i=1}^{n} a_i b_i\right)^2 \le \left(\sum_{i=1}^{n} a_i^2\right)\left(\sum_{i=1}^{n} b_i^2\right).$$

C.3.3 The Laplace Expansion Theorem

Another classical theorem deals with expressing the determinant of a square matrix in terms of rows or columns and smaller order determinants. Recall the *sum map* on $Q_{t,n}$; $s : Q_{t,n} \to \mathbb{N}$ defined by $s(i_1, i_2, \ldots, i_t) = i_1 + i_2 + \cdots + i_t$.

THEOREM C.3 (Laplace expansion theorem)
Suppose $A \in \mathbb{C}^{n\times n}$ *and* $\alpha \in Q_{t,n}$ *for* $1 \le t \le n$. *Then*

1. *(fix* α*)* $det(A) = \sum_{\beta\in Q_{t,n}} (-1)^{s(\alpha)+s(\beta)} \det(A[\alpha \mid \beta]) \det(A[\widehat{\alpha} \mid \widehat{\beta}])$
   (expansion of det(A) by the rows in α*)*
2. *(fix* β*)* $det(A) = \sum_{\alpha\in Q_{t,n}} (-1)^{s(\alpha)+s(\beta)} \det(A[\alpha \mid \beta]) \det(A[\widehat{\alpha} \mid \widehat{\beta}])$
   (expansion of det(A) by the columns in β*)*

PROOF Fix α in $Q_{t,n}$ and define $D_\alpha(A) = \sum_{\beta\in Q_{t,n}} (-1)^{s(\alpha)+s(\beta)} \det(A[\alpha \mid \beta]) \det(A[\widehat{\alpha} \mid \widehat{\beta}])$. Then $D_\alpha : \mathbb{C}^{n\times n} \to \mathbb{C}$ is n-linear as a function of the columns of A. We need D_α to be alternating and $D_\alpha(I) = 1$ to prove the result. Then, by uniqueness, $D_\alpha = \det$. Suppose two columns of A are equal; say $col_p(A) = col_q(A)$ with $p < q$. If p and q are both in $\beta \in Q_{t,n}$, then $A[\alpha \mid \beta]$ will have two columns equal so $\det(A[\alpha \mid \beta]) = 0$. Similarly, if both p and q are in $\widehat{\beta} \in Q_{n-t,n}$, then $\det\left(A[\widehat{\alpha} \mid \widehat{\beta}]\right) = 0$. Thus, in evaluating $D_\alpha(A)$, it is only necessary to consider those $\beta \in Q_{t,n}$ such that $p \in \beta$ and $q \in \widehat{\beta}$ or vice versa. Suppose then that $p \in \beta$ and $q \in \widehat{\beta}$. Define a new sequence β' in $Q_{t,n}$ by replacing p in β by q. Then $\widehat{\beta'}$ agrees with $\widehat{\beta}$ except that q has been replaced by p. Thus, $s(\beta') - s(\beta) = q - p$. Now consider $(-1)^{s(\beta)} \det(A[\alpha \mid \beta]) \det(A[\widehat{\alpha} \mid \widehat{\beta}]) + (-1)^{s(\beta')} \det(A[\alpha \mid \beta']) \det(A[\widehat{\alpha} \mid \widehat{\beta'}])$. We claim this sum is zero. If we can show this, then $D_\alpha(A) = 0$ since β and β' occur in pairs in $Q_{t,n}$. It will follow that $D_\alpha(A) = 0$ whenever two columns of A agree, making D_α alternating. Suppose $p = \beta_k$ and $q = \widehat{\beta}_l$. Then β and β' agree except in the range from p to q, as do $\widehat{\beta}$ and $\widehat{\beta}'$. This includes a total of $q - p + 1$ entries. We have $\beta_1 < \cdots < \beta_k = p < \beta_{k+1} < \cdots < \beta_{k+r-1} < q < \beta_{k+r} < \cdots < \beta_t$ and $A[\alpha \mid \beta'] = A[\alpha \mid \beta]P(\omega^{-1})$ where ω is the r-cycle $(k + r - 1, k + r - 2, .., k)$. Similarly, $A[\widehat{\alpha} \mid \widehat{\beta}']) P(\omega')$ where ω' is a $(q - p + 1 - r)$-cycle. Thus $(-1)^{s(\beta')} \det(A[\alpha \mid \beta']) \det(A[\widehat{\alpha} \mid \widehat{\beta}']) = (-1)^{s(\beta')+(r-1)+(q-p)-r} \det(A[\alpha \mid \beta]) \det(A[\widehat{\alpha} \mid \widehat{\beta}])$. Since $s(\beta') + (r - 1) + (q - p) - r - s(\beta) = 2(q - p) - 1$ is odd, we conclude the sum above is zero. We leave as an exercise the fact that $D_\alpha(I) = 1$. So, $D_\alpha(A) = det(A)$ by the uniqueness characterization of determinants, and we are done. ▯

(2) Apply the result above to A^T.

We get the classical Laplace expansion theorem as a special case.

THEOREM C.4 (Laplace cofactor expansion theorem)
Let $A \in \mathbb{C}^{n\times n}$. Then

1. $\sum_{k=1}^{n}(-1)^{k+j} a_{ki} det(A[\widehat{k} \mid \widehat{j}]) = \delta_{ij} det(A)$.

2. $\sum_{i=1}^{n}(-1)^{i+j} a_{ij} det(A[\widehat{i} \mid \widehat{j}]) = det(A)$, *(expansion by the jth column).*

3. $\sum_{k=1}^{n}(-1)^{k+j} a_{ik} det(A[\widehat{j} \mid \widehat{k}]) = \delta_{ij} det(A)$.

4. $\sum_{j=1}^{n}(-1)^{i+j} a_{ij} det(A[\widehat{i} \mid \widehat{j}]) = det(A)$, *(expansion by the ith row).*

So, in theory, the determinant of an n-by-n matrix is reduced by the theorem above to the computation of n, $(n - 1)$-by-$(n - 1)$ determinants. This can be continued down to 3-by-3 or even 2-by-2 matrices but is not practical for large n. There is an interesting connection with inverses.

Let $A \in \mathbb{C}^{n\times n}$. We note that $A[i \mid j] = a_{ij} = ent_{ij}(A)$ and $A[\widehat{i} \mid \widehat{j}]$ is the matrix obtained from A by deleting the ith row and jth column. The (i,j)-*cofactor* of A is defined by $cof_{ij}(A) = (-1)^{i+j} det(A[\widehat{i} \mid \widehat{j}])$. Define the *cofactor matrix* $cof(A)$ by $ent_{ij}(cof(A)) = cof_{ij}(A) = (-1)^{i+j} det(A[\widehat{i} \mid \widehat{j}])$. The *adjugate* of A is then $adj(A) = cof(A)^T$ (i.e., the transpose of the matrix obtained from A by replacing each element of A by its cofactor). For example, let $A = \begin{bmatrix} 3 & -2 & 1 \\ 5 & 6 & 2 \\ 1 & 0 & -3 \end{bmatrix}$. Then $adj(A) = \begin{bmatrix} -18 & -6 & -10 \\ 17 & -10 & -1 \\ -6 & -2 & 28 \end{bmatrix}$.

In this notation, the Laplace expansions become

1. $\sum_{k=1}^{n} a_{ki}cof_{kj}(A) = \delta_{ij}det(A)$.

2. $\sum_{i=1}^{n} a_{ij}cof_{ij}(A) = det(A)$.

3. $\sum_{k=1}^{n} a_{ik}cof_{jk}(A) = \delta_{ij}det(A)$.

4. $\sum_{j=1}^{n} a_{ij}cof_{ij}(A) = det(A)$.

The main reason for taking the transpose of the cofactor matrix above to define the adjugate matrix is to get the following theorem to be true.

THEOREM C.5
Let $A \in \mathbb{C}^{n\times n}$.

1. $Aadj(A) = adj(A)A = det(A)I_n$.

2. *A is invertible iff $det(A) \neq 0$.*

In this case, $A^{-1} = det(A)^{-1}adj(A)$.

COROLLARY C.7
$A \in \mathbb{C}^{n\times n}$ is invertible iff there exists $B \in \mathbb{C}^{n\times n}$ with $AB = I_n$.

There is a connection with "square" systems of linear equations. Let $A\mathbf{x} = \mathbf{b}$, where A is n-by-n. Then $adj(A)A\mathbf{x} = adj(A)\mathbf{b}$ whence $det(A)\mathbf{x} = adj(A)\mathbf{b}$. Thus, at the element level, $det(A)x_i = \sum_{i=1}^{n}(adj(A))_{ji}b_i = \sum_{i=1}^{n}(-1)^{i+j}b_i det(A[\widehat{i} \mid \widehat{j}])$. This last expression is just the determinant of the matrix A with the jth column replaced by $\mathbf{b}$. Call this matrix B_j. What we have is the familiar *Cramer's rule*.

THEOREM C.6 (Cramer's rule)
Let $A\mathbf{x} = \mathbf{b}$, where A is n-by-n. Then this system has a solution iff $det(A) \neq 0$, in which case $x_i = \dfrac{\det(B_i)}{\det(A)}$ for $i = 1, \ldots, n$.

In the body of the text, we have used some results about the determinants of partitioned matrices. We fill in some details here.

THEOREM C.7
$det\begin{bmatrix} A & B \\ \mathbb{O} & C \end{bmatrix} = \det(A)\det(C)$ *for matrices of appropriate size.*

PROOF Define a function $D(A, B, C) = det\begin{bmatrix} A & B \\ \mathbb{O} & C \end{bmatrix}$, where A and B are fixed. Suppose C is n-by-n. Then D is n-linear as a function of the columns of C. Thus, by the uniqueness of determinants, $D(A, B, C) = (\det(C))D(A, B, I_n)$. Using transvections, we can zero out B and not change the value of D. Thus, $D(A, B, I_n) = D(A, \mathbb{O}, I_n)$. Now suppose A is m-by-m. Then $D(A, \mathbb{O}, I_n)$ is m-linear and alternating as a function of the columns of A. Thus $D(A, \mathbb{O}, I_n) = (\det(A))D(I_m, \mathbb{O}, I_n)$. But note that $D(I_m, \mathbb{O}, I_n) = 1$. $D(A, B, C) = (\det(C))$ $D(A, B, I_n) = D(A, B, C) = (\det(C))D(A, \mathbb{O}, I_n) = (\det(C))(\det(A))$. ▯

This theorem has many nice consequences.

COROLLARY C.8

1. $det\begin{bmatrix} A & \mathbb{O} \\ B & C \end{bmatrix} = \det(A)\det(C)$ *for matrices of appropriate size.*

2. $det\begin{bmatrix} A & \mathbb{O} \\ \mathbb{O} & C \end{bmatrix} = \det(A)\det(C)$ *for matrices of appropriate size.*

3. *Let* $M = \begin{bmatrix} A & B \\ C & D \end{bmatrix}$ *where A is nonsingular. Then* $det(M) = det(A)det(D - CA^{-1}B)$.

4. $det\begin{bmatrix} I & B \\ C & D \end{bmatrix} = \det(D - CB)$ *for matrices of appropriate size.*

5. $det\begin{bmatrix} I & B \\ C & I \end{bmatrix} = \det(I - CB)$ *for matrices of appropriate size.*

PROOF For the most part, these are easy. For three, note $\begin{bmatrix} A & B \\ C & D \end{bmatrix} =$
$\begin{bmatrix} I & \mathbb{O} \\ CA^{-1} & I \end{bmatrix}\begin{bmatrix} A & B \\ \mathbb{O} & D - CA^{-1}B \end{bmatrix}$. ▯

APC Exercise Set 1

1. Prove that if $A \in \mathbb{C}^{m\times m}$, $B \in \mathbb{C}^{n\times n}$, then $det(A\otimes B) = det(A)^n det(B)^m$.

2. Fill in the proofs for Theorem C.1 and Corollary C.1.

3. Fill in the proofs for Corollary C.2, Theorem C.4, and Corollary C.7.

4. Argue that A is invertible iff $adj(A)$ is invertible, in which case $adj(A)^{-1} = det(A)^{-1}A$.

5. Prove that $det(adj(A)) = det(A)^{n-1}$.

6. Prove that if $det(A) = 1$, then $adj(adj(A)) = A$.

7. Argue that $adj(AB) = adj(B)adj(A)$.

8. Prove that B invertible implies $adj(B^{-1}AB) = B^{-1}adj(A)B$.

9. Prove that the determinant of a triangular matrix is the product of the diagonal elements.

10. Prove that the inverse of a lower (upper) triangular matrix is lower (upper) triangular when there are no zero elements on the diagonal.

11. Argue that $adj(A^T) = (adj(A))^T$.

12. Argue that $det(adj(A)) = (det(A))^{n-1}$ for $n \geq 2$ and A n-by-n.

13. Give an example to show $det(A+B) = det(A)+det(B)$ does not always hold.

14. If A is n-by-n, argue that $det(aA) = a^n det(A)$. In particular, this shows that $det(det(B)A) = det(B)^n \det(A)$.

15. It is important to know how the elementary matrices affect determinants. Argue the following:
    (a) $det(D_i(a)A) = adet(A)$.
    (b) $det(T_{ij}(a)A) = det(A)$.
    (c) $det(P_{ij}A) = -det(A)$.
    (d) $det(AD_i(a)) = adet(A)$.
    (e) $det(AT_{ij}(a)) = det(A)$.
    (f) $det(AP_{ij}) = -det(A)$.

16. Argue that over $\mathbb{C}$, $det(AA^*) \geq 0$.

17. Argue that over $\mathbb{C}$, $det(\overline{A}) = \overline{det(A)}$. Conclude that if $A = A^*$, $det(A)$ is real.

18. If one row of a matrix is a multiple of another row of a matrix, what can you say about its determinant?

19. What is $\det \begin{bmatrix} 0 & a \\ b & 0 \end{bmatrix}$? How about $\det \begin{bmatrix} 0 & 0 & a \\ 0 & b & 0 \\ c & 0 & 0 \end{bmatrix}$? Can you generalize?

20. Suppose A is nonsingular. Argue that $(adj(A))^{-1} = det(A^{-1})A = adj(A^{-1})$.

21. Find a method for constructing integer matrices that are invertible and the inverse matrices also have integer entries.

22. If A is skew-symmetric, what can you say about $det(A)$?

23. What is $det(C(h))$, where $C(h)$ is the companion matrix of the polynomial $h(x)$.

24. Prove $det \begin{bmatrix} A & \mathbb{O} \\ -I & B \end{bmatrix} = det \begin{bmatrix} A & AB \\ -I & \mathbb{O} \end{bmatrix}$.

25. Suppose $A = \begin{bmatrix} 1 & 1 \\ -1 & 1 \end{bmatrix}, \begin{bmatrix} 1 & 1 & 0 \\ -1 & 1 & 1 \\ 0 & -1 & 1 \end{bmatrix}, \begin{bmatrix} 1 & 1 & 0 & 0 \\ -1 & 1 & 1 & 0 \\ 0 & -1 & 1 & 1 \\ 0 & 0 & -1 & 1 \end{bmatrix}, \ldots.$ Calculate $det(A)$ for each. Do you recognize these famous numbers?

26. Prove that $det \begin{bmatrix} 1 & 1 & \cdots & 1 \\ x_1 & x_2 & \cdots & x_n \\ x_1^2 & x_2^2 & \cdots & x_n^2 \\ \vdots & \vdots & & \vdots \\ x_1^{n-1} & x_2^{n-1} & \cdots & x_n^{n-1} \end{bmatrix} = \prod_{1 \leq j < i \leq n} (x_i - x_j)$. This is a famous determinant known as the *Vandermonde determinant.*

27. There is a famous sequence of numbers called the *Fibonacci sequence.* There is an enormous literature out there on this sequence. It starts out $\{1, 1, 2, 3, 5, \ldots\}$. Do you see the pattern?

Anyway, let F_n be the n-by-n matrix $F_n = \begin{bmatrix} 1 & i & 0 & 0 & \cdots & 0 \\ i & 1 & i & 0 & \cdots & 0 \\ 0 & i & 1 & i & \cdots & 0 \\ 0 & 0 & i & 1 & \ddots & \vdots \\ \vdots & \vdots & \vdots & \ddots & \ddots & i \\ 0 & 0 & 0 & \cdots & i & 1 \end{bmatrix}$.
Compute the determinants of F_1, F_2, F_3, F_4, and F_5, and decide if there is a connection with the Fibonacci sequence.

28. Find $det \begin{bmatrix} \frac{b+c}{a} & \frac{a}{b+c} & \frac{a}{b+c} \\ \frac{b}{c+a} & \frac{c+a}{b} & \frac{c}{c+a} \\ \frac{c}{a+b} & \frac{c}{a+b} & \frac{a+b}{c} \end{bmatrix}$. Notice anything interesting?

29. (F. Zhang) For any n-by-n matrices A and B, argue that $det \begin{bmatrix} A & B \\ -\overline{B} & \overline{A} \end{bmatrix} \geq 0$.

30. (R. Bacher) What is the determinant of a matrix of size n-by-n if its (i, j) entry is $a^{(i-j)^2}$ for $1 \leq i, j \leq n$?

31. Find $det \begin{bmatrix} 1+a & 1 & 1 \\ 1 & 1+b & 1 \\ 1 & 1 & 1+c \end{bmatrix}$. Can you generalize your findings?

32. Here is a slick proof of Cramer's rule. Consider the linear system $A\mathbf{x} = \mathbf{b}$, where A is n-by-n, where $\mathbf{x} = \begin{bmatrix} x_1 \\ x_2 \\ x_3 \\ \vdots \\ x_n \end{bmatrix}$. Replace the first column of the identity matrix by $\mathbf{b}$ and consider $A[\mathbf{x} \mid \mathbf{e}_2 \mid \mathbf{e}_3 \mid \cdots \mid \mathbf{e}_n] = [A\mathbf{x} \mid A\mathbf{e}_2 \mid A\mathbf{e}_3 \mid \cdots \mid A\mathbf{e}_n] = [\mathbf{b} \mid \text{col}_2(A) \mid \text{col}_3(A) \mid \cdots \mid \text{col}_n(A)]$. Taking determinants and using that the determinant of a product is the product of the determinants, we find

$$\det(A)\det([\mathbf{x} \mid \mathbf{e}_2 \mid \mathbf{e}_3 \mid \cdots \mid \mathbf{e}_n]) = \det([\mathbf{b} \mid \text{col}_2(A) \mid \text{col}_3(A) \mid \cdots \mid \text{col}_n(A)]).$$

But $\det([\mathbf{x} \mid \mathbf{e}_2 \mid \mathbf{e}_3 \mid \cdots \mid \mathbf{e}_n]) = x_1$, as we see by a Laplace expansion, so

$$x_1 = \frac{\det([\mathbf{b} \mid \text{col}_2(A) \mid \text{col}_3(A) \mid \cdots \mid \text{col}_n(A)])}{\det(A)}.$$

The same argument applies if $\mathbf{x}$ is placed in any column of the identity matrix. Make the general argument. Illustrate the argument in the 2-by-2 and 3-by-3 case.

33. Suppose all r-by-r submatrices of a matrix have determinant zero. Argue that all submatrices $(r+1)$-by-$(r+1)$ have determinant zero.

34. Prove that any minor of order r in the product matrix AB is a sum of products of minors of order r in A with minors of order r in B.

Further Reading

[A&W, 1992] William A. Adkins and Steven H. Weintraub, *Algebra: An Approach via Module Theory*, Springer-Verlag, New York, (1992).

[Aitken, 1962] A. C. Aitken, *Determinants and Matrices*, Oliver and Boyd, New York: Interscience Publishers, Inc., (1962).

[Axler, 1995] Sheldon Axler, Down with Determinants, The American Mathematical Monthly, Vol. 102, No. 2, February, (1995), 139–154.

[Axler, 1996] Sheldon Axler, *Linear Algebra Done Right*, Springer, New York, (1996).

[B&R, 1986] T. S. Blyth and E. F. Robertson, *Matrices and Vector Spaces*, Vol. 2, Chapman & Hall, New York, (1986).

[Bress, 1999] David M. Bressoud, *Proofs and Confirmations: The Story of the Alternating Sign Matrix Conjecture*, Cambridge University Press, Cambridge, (1999).

[B&S, 1983/84] R. A. Brualdi and H. Schneider, Determinantal Identities: Gauss, Schur, Cauchy, Sylvester, Krone, Linear Algebra and Its Applications, 52/53, (1983), 769–791, and 59, (1984), 203–207.

[C,D'E et al., 2002] Nathan D. Cahill, John R. D'Errico, Darren A. Narayan and Jack Y. Harayan, Fibonacci Determinants, The College Mathematics Journal, Vol. 33, No. 3, May, (2002), 221–225.

[Des, 1819] P. Desnanot, Complément de la théorie des équations du premier degré, Paris, (1819).

[Dodg, 1866] Charles L. Dodgson, Condensation of Determinants, Proceedings of the Royal Society, London, 15, (1866), 150–155.

[Garibaldi, 2004] Skip Garibaldi, The Characteristic Polynomial and Determinant are Not Ad Hoc Constructions, The American Mathematical Monthly, Vol. 111, No. 9, November, (2004), 761–778.

[Muir, 1882] Thomas Muir, *A Treatise on the Theory of Determinants*, Macmillan and Co., London, (1882).

[Muir, 1906–1923] Thomas Muir, *The Theory of Determinants in the Historical Order of Development*, 4 volumes, Macmillan and Co., London, (1906–1923).

[Muir, 1930] Thomas Muir, *Contributions to the History of Determinants, 1900–1920*, Blackie & Sons, London, (1930).

[Rob&Rum, 1986] David P. Robbins and Howard Rumsey, Determinants and Alternating Sign Matrices, Advances in Mathematics, Vol. 62, (1986), 169–184.

[Skala, 1971], Helen Skala, An Application of Determinants, The Americal Mathematical Monthly, Vol. 78, (1971), 889–990.

C.4 The Trace of a Square Matrix

There is another scalar that can be assigned to a square matrix that is very useful. The *trace* of a square matrix is just the sum of the diagonal elements.

DEFINITION C.1 *(trace)*
Let A be in $\mathbb{C}^{n\times n}$. *We define the trace of A as the sum of the diagonal elements of A. In symbols,* $tr(A) = \sum_{i=1}^{n} ent_{ii}(A) = a_{11} + a_{22} + \cdots + a_{nn}$. *We view tr as a function from* $\mathbb{C}^{n\times n}$ *to* $\mathbb{C}$.

Next, we develop the important properties of the trace of a matrix. The first is that it is a linear map.

THEOREM C.8
Let A, B be matrices in $\mathbb{C}^{n\times n}$. Then

1. $tr(A+B) = tr(A) + tr(B)$.
2. $tr(\alpha A) = \alpha tr(A)$.
3. $tr(AB) = tr(BA)$.
4. *If S is invertible, then* $tr(S^{-1}AS) = tr(A)$.
5. $tr(\alpha I_n) = n\alpha$.
6. $tr(ABC) = tr(BCA) = tr(CAB)$.
7. $tr(A^T B) = tr(AB^T)$.
8. $tr(A^T) = tr(A)$.
9. $tr(\overline{A}) = \overline{tr(A)}$.
10. $tr(A^*) = \overline{tr(A)}$.

The trace can be used to define an inner product on the space of matrices $\mathbb{C}^{n\times n}$.

THEOREM C.9
The function $\langle A \mid B\rangle = tr(B^*A)$ defines an inner product on $\mathbb{C}^{n\times n}$. In particular,

1. $tr(A^*A) = tr(AA^*) \geq 0$ *and* $= 0$ *iff* $A = \mathbb{O}$.
2. $tr(AX) = 0$ *for all X implies* $A = \mathbb{O}$.
3. $|tr(AB)| \leq \sqrt{tr(A^*A)tr(B^*B)} \leq \frac{1}{2}(tr(A^*A) + tr(B^*B))$.
4. $tr(A^2) + tr(B^2) = tr((A+B)^2) - 2tr(AB)$.

Can you see how to generalize (3) above?
There is an interesting connection with the trace and orthonormal bases of $\mathbb{C}^n$.

THEOREM C.10
Suppose $\mathbf{e}_1, \mathbf{e}_2, \cdots, \mathbf{e}_n$ is an orthonormal basis of $\mathbb{C}^n$ with respect to the usual inner product $\langle | \rangle$. Then $tr(A) = \sum_{i=1}^{n} \langle A\mathbf{e}_i \mid \mathbf{e}_i\rangle$.

There is also an interesting connection to the eigenvalues of a complex matrix.

THEOREM C.11
The trace of $A \in \mathbb{C}^{n \times n}$ *is the sum of the eigenvalues of* A.

APC Exercise Set 2

1. Fill in the proofs of the theorems in this section.
2. (G. Trenkler) Argue that $A^2 = -A$ iff $rank(A) = -tr(A)$ and $rank(A+I) = n + tr(A)$, where A is n-by-n.

Appendix D

A Review of Basics

D.1 Spanning

Suppose $\mathbf{v}_1, \mathbf{v}_2, \ldots, \mathbf{v}_p$ are vectors in $\mathbb{C}^n$. Suppose $\alpha_1, \alpha_2, \ldots, \alpha_p$ are scalars in $\mathbb{C}$. Then the vector $\mathbf{v} = \alpha_1\mathbf{v}_1 + \alpha_2\mathbf{v}_2 + \cdots + \alpha_p\mathbf{v}_p$ is called a *linear combination* of $\mathbf{v}_1, \mathbf{v}_2, \ldots, \mathbf{v}_p$. For example, $\begin{bmatrix} 2 \\ 9i \end{bmatrix}$ is a linear combination of $\begin{bmatrix} 1 \\ 3i \end{bmatrix}$ and $\begin{bmatrix} 0 \\ 1 \end{bmatrix}$ since $2\begin{bmatrix} 1 \\ 3i \end{bmatrix} + 3i\begin{bmatrix} 0 \\ 1 \end{bmatrix} = \begin{bmatrix} 2 \\ 9i \end{bmatrix}$. However, there is no way $\begin{bmatrix} 10 \\ 3 \\ 2 \end{bmatrix}$ could be a linear combination of $\begin{bmatrix} 0 \\ 1 \\ 1 \end{bmatrix}$ and $\begin{bmatrix} 0 \\ 3 \\ 1 \end{bmatrix}$ (why not?). Recall that the system of linear equations $A\mathbf{x} = \mathbf{b}$ has a solution iff $\mathbf{b}$ can be expressed as a linear combination of the columns of A. Indeed, if $\mathbf{b} = c_1 col_1(A) + c_2 col_2(A) + \cdots + c_n col_n(A)$, then $\mathbf{x} = \begin{bmatrix} c_1 \\ c_2 \\ \vdots \\ c_n \end{bmatrix}$ solves the system.

Consider a subset S of vectors from $\mathbb{C}^n$. Define the *span of* S (in symbols, $sp(S)$) as the set of all possible (finite) linear combinations that can be formed using the vectors in S. Let's agree the span of the empty set is the set having only the zero vector in it (i.e., $sp(\varnothing) = \{\overrightarrow{0}\}$). For example, if $S = \{(1, 0, 0)\}$, then $sp(S) = \{(\alpha, 0, 0) \mid \alpha \in \mathbb{C}\}$). Note how spanning tends to make sets bigger. In this example, we went from one vector to an infinite number. As another example, note $sp\{(1, 0), (0, 1)\} = \mathbb{C}^2$. We now summarize the basic facts about spanning.

THEOREM D.1 (basic facts about span)
Let S, S_1, S_2 be subsets of vectors in $\mathbb{C}^n$.

1. *For any subset S, $S \subseteq sp(S)$. (increasing)*

2. *For any subset S, $sp(S)$ is a subspace of $\mathbb{C}^n$. In fact, $sp(S)$ is the smallest*

subspace of $\mathbb{C}^n$ *containing* S.

3. *If* $S_1 \subseteq S_2$, *then* $sp(S_1) \subseteq sp(S_2)$. *(monotone)*

4. *For any subset* S, $sp(sp(S)) = sp(S)$. *(idempotent)*

5. M *is a subspace of* $\mathbb{C}^n$ *iff* $M = sp(M)$.

6. $sp(S_1 \cap S_2) \subseteq sp(S_1) \cap sp(S_2)$.

7. $sp(S_1 \cup S_2) = sp(S_1) + sp(S_2)$.

8. $sp(S_1) = sp(S_2)$ *iff each vector in* S_1 *is a linear combination of vectors in* S_2 *and conversely.*

PROOF The proofs are left as exercises. □

We can view *sp* as a function with certain properties from the set of all subsets of $\mathbb{C}^n$ to the set of all subspaces of $\mathbb{C}^n$, $sp : \mathcal{P}(\mathbb{C}^n) \to Lat(\mathbb{C}^n)$. The fixed points of *sp* are exactly the subspaces of $\mathbb{C}^n$.

If a subspace M of $\mathbb{C}^n$ is such that $M = sp(S)$, we call the vectors in S *generators* of M. If S is a finite set, we say M is *finitely generated*. It would be nice to have a very efficient spanning set in the sense that none of the vectors in the set are redundant (i.e., can be generated as a linear combination of other vectors in the set).

THEOREM D.2

Let $S = \{\mathbf{v}_1, \mathbf{v}_2, \dots, \mathbf{v}_p\}$ *where* $p \geq 2$. *Then the following are equivalent.*

1. $sp(\{\mathbf{v}_1, \mathbf{v}_2, \dots, \mathbf{v}_p\}) = sp(\{\mathbf{v}_1, \mathbf{v}_2, \dots, \mathbf{v}_{k-1}, \mathbf{v}_{k+1}, \dots, \mathbf{v}_p\})$.

2. $\mathbf{v}_k$ *is a linear combination of* $\mathbf{v}_1, \mathbf{v}_2, \dots, \mathbf{v}_{k-1}, \mathbf{v}_{k+1}, \dots, \mathbf{v}_p$.

3. *There exist scalars* $\alpha_1, \alpha_2, \dots, \alpha_p$, *not all zero, such that* $\alpha_1\mathbf{v}_1 + \alpha_2\mathbf{v}_2 + \dots + \alpha_p\mathbf{v}_p = \overrightarrow{0}$.

PROOF The proof is left as an exercise. □

The last condition of the theorem above leads us to the concept developed in the next section.

D.2 Linear Independence

A set of vectors $\{\mathbf{v}_1, \mathbf{v}_2, \ldots, \mathbf{v}_p\}$ is called *linearly dependent* iff there exist scalars $\alpha_1, \alpha_2, \ldots, \alpha_p$, not all zero, such that $\alpha_1\mathbf{v}_1 + \alpha_2\mathbf{v}_2 + \cdots + \alpha_p\mathbf{v}_p = \overrightarrow{0}$. Such an equation is referred to as a *dependency relation*. These are, evidently, not unique when they exist. If a set of vectors $\{\mathbf{v}_1, \mathbf{v}_2, \ldots, \mathbf{v}_p\}$ is not linearly dependent, it is called *linearly independent.* Thus, the set $\{\mathbf{v}_1, \mathbf{v}_2, \ldots, \mathbf{v}_p\}$ is linearly independent iff the equation $\alpha_1\mathbf{v}_1 + \alpha_2\mathbf{v}_2 + \cdots + \alpha_p\mathbf{v}_p = \overrightarrow{0}$ implies $\alpha_1 = \alpha_2 = \cdots = \alpha_p = 0$. For example, any set of vectors that has the zero vector in it must be linearly dependent (why?). A set with just two distinct vectors is dependent iff one of the vectors is a scalar multiple of the other. The set $\{(1,0), (0,1)\}$ is independent in $\mathbb{C}^2$.

THEOREM D.3
Let S be a set of two or more vectors.

1. *S is linearly dependent iff at least one vector in S is a linear combination of other vectors in S.*

2. *S is linearly independent iff no vector in S is expressible as a linear combination of vectors in S.*

3. *Any subset of a linearly independent set is linearly independent.*

4. *Any set containing a linearly dependent set is linearly dependent.*

5. *(Extension theorem) Let* $S = \{\mathbf{v}_1, \mathbf{v}_2, \ldots, \mathbf{v}_p\}$ *be a linearly independent set and* $\mathbf{v} \notin sp(S)$*. Then* $S^\sharp = \{\mathbf{v}_1, \mathbf{v}_2, \ldots, \mathbf{v}_p, \mathbf{v}\}$ *is also an independent set.*

6. *Let* $S = \{\mathbf{v}_1, \mathbf{v}_2, \ldots, \mathbf{v}_p\}$ *be a set of two or more nonzero vectors. Then S is dependent iff at least one vector in S is a linear combination of the vectors preceding it in S.*

7. *If S is a linearly independent set and* $\mathbf{v} \in sp(S)$*, then* $\mathbf{v}$ *is uniquely expressible as a linear combination of vectors from S.*

PROOF As usual, the proofs are left as exercises. □

D.3 Basis and Dimension

There is a very important result about the size of linearly independent sets in finitely generated subspaces that allows us to introduce the idea of dimension. We begin by developing some language. Let $\mathcal{B}$ be a set of vectors in a subspace M of $\mathbb{C}^n$. We say $\mathcal{B}$ is a *basis* of M iff (1) $\mathcal{B}$ is an independent set and (2) $sp(\mathcal{B}) = M$. For example, let $\mathcal{B} = \{(1,0),(0,1)\}$ in $\mathbb{C}^2$. Then clearly, $\mathcal{B}$ is a basis of $\mathbb{C}^2$. A subspace M of $\mathbb{C}^n$ is called *finitely generated* iff it contains a finite subset S with $sp(S) = M$. We see from above, $\mathbb{C}^2$ is finitely generated. Next is a fundamental result about finitely generated subspaces. It is such a crucial fact, we offer a proof.

THEOREM D.4 (Steinitz exchange theorem)
Let M be a finitely generated subspace of $\mathbb{C}^n$. Specifically, let $M = sp(\{\mathbf{v}_1, \mathbf{v}_2, \ldots, \mathbf{v}_p\})$. Let T be an independent set of vectors in M, say $T = \{\mathbf{w}_1, \mathbf{w}_2, \ldots, \mathbf{w}_m\}$. Then $m \leq p$. In other words, in a finitely generated subspace, you cannot have more independent vectors than you have generators.

PROOF First, note $\mathbf{w}_1 \in M$, so $\mathbf{w}_1$ is a linear combination of the $\mathbf{v}$s. Consider the set $T_1 = \{\mathbf{w}_1, \mathbf{v}_1, \ldots, \mathbf{v}_p\}$. Clearly $sp(T_1) = M$. Now T_1 is a dependent set since at least one vector in it, namely $\mathbf{w}_1$, is a linear combination of the others. By Theorem D.3, some vector in T_1 is a linear combination of vectors preceding it in the list, say $\mathbf{v}_j$. Throw it out and consider $S_1 = \{\mathbf{w}_1, \mathbf{v}_1, \ldots, \mathbf{v}_{j-1}, \mathbf{v}_{j+1}, \ldots, \mathbf{v}_p\}$. Note $sp(S_1) = M$ since $\mathbf{v}_j$ was redundant. Now we go again. Note $\mathbf{w}_2 \in M$ so $\mathbf{w}_2 \in sp(S_1) = M$. Consider $T_2 = \{\mathbf{w}_2, \mathbf{w}_1, \mathbf{v}_1, \ldots, \mathbf{v}_{j-1}, \mathbf{v}_{j+1}, \ldots, \mathbf{v}_p\}$. Clearly T_2 is a linearly dependent set since $\mathbf{w}_2$ is a linear combination of the elements in S_1. Again by Theorem D.3, some vector in T_2 is a linear combination of vectors previous to it. Could this vector be $\mathbf{w}_2$? No, since $\mathbf{w}_1$ and $\mathbf{w}_2$ are independent, so it must be one of the $\mathbf{v}$s. Throw it out and call the resulting set S_2. Note $sp(S_2) = M$ since the vector we threw out is redundant. Continue in this manner exchanging $\mathbf{v}$s for $\mathbf{w}$s. If we eliminate all the $\mathbf{v}$s and still have some $\mathbf{w}$s left over, then some $\mathbf{w}$ will be in the span of the other $\mathbf{w}$s preceding it, contradicting the independence of the $\mathbf{w}$s. So there must be more $\mathbf{v}$s than $\mathbf{w}$s or perhaps the same number. That is $m \leq p$. □

With such a beautiful theorem, we can reap a harvest of corollaries.

COROLLARY D.1
Any $n + 1$ vectors in a subspace generated by n vectors must be dependent.

COROLLARY D.2
Any $n+1$ vectors in $\mathbb{C}^n$ are necessarily dependent.

COROLLARY D.3
Any two bases of a finitely generated subspace have the same number of vectors in them.

PROOF (*Hint:* View one basis as an independent set and the other as a set of generators. Then change these roles.) ▯

This last result allows us to define the concept of *dimension*. A subspace M of $\mathbb{C}^n$ is *m-dimensional* iff M has a basis of m vectors. In view of the previous corollary, this is a uniquely defined number. The notation is $dim(M) = m$. Note $dim(\mathbb{C}^n) = n$.

COROLLARY D.4
If M is generated by n vectors and $S = \{\mathbf{v}_1, \mathbf{v}_2, \ldots, \mathbf{v}_n\}$ is an independent set of vectors in M, then S must be a basis for M.

COROLLARY D.5
If M has dimension n and $S = \{\mathbf{v}_1, \mathbf{v}_2, \ldots, \mathbf{v}_n\}$ spans M, then S must be a basis for M.

COROLLARY D.6
Suppose $M \neq \{\overrightarrow{0}\}$ is a finitely generated subspace of $\mathbb{C}^m$. Then

1. *M has a finite basis.*
2. *Any set of generators of M contains a basis.*
3. *Any independent subset of M can be extended to a basis of M.*

COROLLARY D.7
Suppose M and N are subspaces of $\mathbb{C}^m$ and $dim(N) = n$ and $M \subseteq N$. Then $dim(M) \leq \dim(N)$. Moreover, if in addition, $dim(M) = dim(N)$, then $M = N$.

These are wonderful and useful corollaries to the Steinitz theorem. Some results we want to review depend on making new subspaces from old ones. Recall that when M_1 and M_2 are subspaces of $\mathbb{C}^n$, we can form their *intersection* $M_1 \cap M_2$ and their *sum* $M_1 + M_2 = \{\mathbf{u} + \mathbf{v} \mid \mathbf{u} \in M_1 \text{ and } \mathbf{v} \in M_2\}$. It is easy to show that these constructions lead to subspaces. A sum is called a *direct sum* when $M_1 \cap M_2 = \{\overrightarrow{0}\}$. The notation is $M_1 \oplus M_2$ for a direct sum. If $\mathbb{C}^n = M_1 \oplus M_2$, we say M_2 is a *complement* of M_1 or that M_1 and M_2 are *complementary subspaces*. Typically, a given subspace has many complements.

THEOREM D.5
Suppose M_1 and M_2 are subspaces of $\mathbb{C}^n$ with bases $\mathcal{B}_1$ and $\mathcal{B}_2$ respectively. Then T.A.E.:

1. $\mathbb{C}^n = M_1 \oplus M_2$.
2. *For each vector* $\mathbf{w}$ *in* $\mathbb{C}^n$, *there exist unique vector* $\mathbf{v}$ *in* M_1 *and* $\mathbf{u}$ *in* M_2 *with* $\mathbf{w} = \mathbf{v} + \mathbf{u}$.
3. $\mathcal{B}_1 \cap \mathcal{B}_2 = \varnothing$ *and* $\mathcal{B}_1 \cup \mathcal{B}_2$ *is a basis for* $\mathbb{C}^n$.

PROOF The proof is left as an exercise. ▯

If $M \subseteq N$ and there exists a subspace K with $M \oplus K = N$, then K is called a *relative complement* of M in N.

THEOREM D.6
Suppose N is finitely generated and $M \subseteq N$. Then M has a relative complement in N.

PROOF As usual, the proof is left as an exercise. ▯

COROLLARY D.8
Any subspace of $\mathbb{C}^n$ has a complement.

We end with a famous formula relating dimensions of two subspaces.

THEOREM D.7 (the dimension formula)
Suppose M_1 and M_2 are subspaces of a finitely generated subspace M. Then M_1, M_2, M_1+M_2, and $M_1 \cap M_2$ are all finite dimensional and $\dim(M_1 + M_2) = \dim(M_1) + \dim(M_2) - \dim(M_1 \cap M_2)$.

PROOF Start with a basis of $M_1 \cap M_2$, say $\mathbf{u}_1, \mathbf{u}_2, \ldots, \mathbf{u}_a$. Then extend this basis to one of M_1 and one of M_2, say $\mathbf{u}_1, \ldots, \mathbf{u}_a, \mathbf{v}_1, \ldots, \mathbf{v}_b$ and $\mathbf{u}_1, \mathbf{u}_2, \ldots, \mathbf{u}_a, \mathbf{w}_1, \ldots, \mathbf{w}_c$, respectively. Now argue that $\mathbf{u}_1, \ldots, \mathbf{u}_a, \mathbf{v}_1, \ldots, \mathbf{v}_b, \mathbf{w}_1, \ldots, \mathbf{w}_c$ is a basis for $M_1 + M_2$. ▯

Does the dimension formula remind you of anything from probability theory?

APD Exercise Set 1

1. How would the dimension formula read if three subspaces were involved? Can you generalize to a finite number of subspaces?

2. Fill in the arguments that have been omitted in the discussion above.

3. Suppose $\{\mathbf{u}_1, \mathbf{u}_2, \dots, \mathbf{u}_p\}$ and $\{\mathbf{v}_1, \mathbf{v}_2, \dots, \mathbf{v}_q\}$ are two sets of vectors with $p > q$. Suppose each $\mathbf{u}_i$ lies in the span of $\{\mathbf{v}_1, \mathbf{v}_2, \dots, \mathbf{v}_q\}$. Argue that $\{\mathbf{u}_1, \mathbf{u}_2, \dots, \mathbf{u}_p\}$ is necessarily a linearly dependent set.

4. Suppose M_1 and M_2 are subspaces of $\mathbb{C}^n$ with $dim(M_1+M_2) = dim(M_1\cap M_2) + 1$. Prove that either $M_1 \subseteq M_2$ or $M_2 \subseteq M_1$.

5. Suppose M_1, M_2, and M_3 are subspaces of $\mathbb{C}^n$. Argue that
   $dim(M_1 \cap M_2 \cap M_3) + 2n \geq \dim(M_1) + \dim(M_2) + \dim(M_3)$.

6. Suppose $\{\mathbf{u}_1, \mathbf{u}_2, \dots, \mathbf{u}_n\}$ and $\{\mathbf{v}_1, \mathbf{v}_2, \dots, \mathbf{v}_n\}$ are two bases of $\mathbb{C}^n$. Form the matrix U where the columns are the $\mathbf{u}_i$s and the matrix V whose columns are the $\mathbf{v}_i$s. Argue that there is an invertible matrix S such that $SU = V$.

7. Suppose $\{\mathbf{u}_1, \mathbf{u}_2, \dots, \mathbf{u}_p\}$ span a subspace M. Argue that $\{A\mathbf{u}_1, A\mathbf{u}_2, \dots, A\mathbf{u}_n\}$ spans $A(M)$.

8. Suppose $M_1, \dots, M_k$ are subspaces of $\mathbb{C}^n$. Argue that $dim(M_1 \cap \dots \cap M_k) = n - \sum_{i=1}^{k}(n - \dim(M_i)) + \sum_{j=1}^{k-1}\{n - \dim((M_1 \cap \dots \cap M_j) + M_{j+1})\}$.

   Deduce that $dim(M_1 \cap \dots \cap M_k) \geq n - \sum_{i=1}^{k}(n - \dim(M_i))$ and $dim(M_1 \cap \dots \cap M_k) = n - \sum_{i=1}^{k}(n - \dim(M_i))$ iff for all $i = 1, \dots, k$, $M_i + (\bigcap_{j \neq i} M_j) = \mathbb{C}^n$.

9. Select some columns from a matrix B and suppose they are dependent. Prove that the same columns in AB are also dependent for any matrix A that can multiply B.

D.4 Change of Basis

You have no doubt seen in your linear algebra class how to associate a matrix to a linear transformation between vector spaces. In this section, we review how that went. Of course, we will need to have bases to effect this connection. First, we consider the *change of basis problem.*

Let V be a complex vector space and let $\mathcal{B} = \{\mathbf{b}_1, \mathbf{b}_2, \dots, \mathbf{b}_n\}$ be a basis for V. Let $\mathbf{x}$ be a vector in V. Then $\mathbf{x}$ can be written uniquely as $\mathbf{x} = \mathbf{b}_1\beta_1 + \mathbf{b}_2\beta_2 + \cdots + \mathbf{b}_n\beta_n$. That is, the scalars $\beta_1, \beta_2, \dots, \beta_n$ are uniquely determined by $\mathbf{x}$ and the basis $\mathcal{B}$. We call these scalars in $\mathbb{C}$ the "coordinates" of $\mathbf{x}$ relative to the basis $\mathcal{B}$. Thus, we have the correspondence

$$\mathbf{x} \longleftrightarrow \begin{bmatrix} \beta_1 \\ \beta_2 \\ \vdots \\ \beta_n \end{bmatrix} = Mat(\mathbf{x}; \mathcal{B}) \equiv [\mathbf{x}]_{\mathcal{B}}.$$

Now suppose $\mathcal{C} = \{\mathbf{c}_1, \mathbf{c}_2, \dots, \mathbf{c}_n\}$ is another basis of V. The same vector $\mathbf{x}$ can be expressed (again uniquely) relative to this basis as $\mathbf{x} = \mathbf{c}_1\gamma_1 + \mathbf{c}_2\gamma_2 + \cdots + \mathbf{c}_n\gamma_n$. Thus, we have the correspondence

$$\mathbf{x} \longleftrightarrow \begin{bmatrix} \gamma_1 \\ \gamma_2 \\ \vdots \\ \gamma_n \end{bmatrix} = Mat(\mathbf{x}; \mathcal{C}) \equiv [\mathbf{x}]_{\mathcal{C}}.$$

The question we seek to resolve is, what is the connection between these sets of coordinates determined by $\mathbf{x}$? First we set an exercise:

1. Show $Mat(\mathbf{x} + \mathbf{y}; \mathcal{B}) = Mat(\mathbf{x}; \mathcal{B}) + Mat(\mathbf{y}; \mathcal{B})$.

2. Show $Mat(\mathbf{x}\alpha; \mathcal{B}) = Mat(\mathbf{x}; \mathcal{B})\alpha$.

For simplicity to fix ideas, suppose $\mathcal{B} = \{\mathbf{b}_1, \mathbf{b}_2\}$ and $\mathcal{C} = \{\mathbf{c}_1, \mathbf{c}_2\}$. Let $\mathbf{x}$ be a vector in V. Now $\mathbf{c}_1$ and $\mathbf{c}_2$ are vectors in V and so are uniquely expressible in the basis $\mathcal{B}$. Say $\mathbf{c}_1 = \mathbf{b}_1\alpha + \mathbf{b}_2\beta$ so that $[\mathbf{c}_1]_{\mathcal{B}} = \begin{bmatrix} \alpha \\ \beta \end{bmatrix}$ and $[\mathbf{c}_2]_{\mathcal{B}} = \begin{bmatrix} \gamma \\ \delta \end{bmatrix}$, reflecting that $\mathbf{c}_2 = \mathbf{b}_1\gamma + \mathbf{b}_2\delta$. Let $[\mathbf{x}]_{\mathcal{C}} = \begin{bmatrix} \mu \\ \sigma \end{bmatrix}$. What is $[x]_{\mathcal{B}}$? Well, $\mathbf{x} = \mathbf{c}_1\mu + \mathbf{c}_2\sigma = (\mathbf{b}_1\alpha + \mathbf{b}_2\beta)\,\mu + (\mathbf{b}_1\gamma + \mathbf{b}_2\delta)\,\sigma = \mathbf{b}_1\,(\alpha\mu + \gamma\sigma) + \mathbf{b}_2\,(\beta\mu + \delta\sigma)$. Thus,

$$[\mathbf{x}]_{\mathcal{B}} = \begin{bmatrix} \alpha\mu + \gamma\sigma \\ \beta\mu + \delta\sigma \end{bmatrix} = \begin{bmatrix} \alpha & \gamma \\ \beta & \delta \end{bmatrix} \begin{bmatrix} \mu \\ \sigma \end{bmatrix} = \begin{bmatrix} \alpha & \gamma \\ \beta & \delta \end{bmatrix} [\mathbf{x}]_{\mathcal{C}}.$$

The matrix $\begin{bmatrix} \alpha & \gamma \\ \beta & \delta \end{bmatrix}$ gives us a computational way to go from the coordinates of $\mathbf{x}$ in the $\mathcal{C}$-basis to the coordinates of $\mathbf{x}$ in the $\mathcal{B}$-basis. Thus, we call the matrix $\begin{bmatrix} \alpha & \gamma \\ \beta & \delta \end{bmatrix}$ the *transition matrix* or *change of basis matrix* from the $\mathcal{C}$ to the $\mathcal{B}$ basis. We write

$$R_{\mathcal{B}\leftarrow\mathcal{C}} = \begin{bmatrix} \alpha & \gamma \\ \beta & \delta \end{bmatrix} = [[\mathbf{c}_1]_{\mathcal{B}} \mid [\mathbf{c}_2]_{\mathcal{B}}]$$

and note

$$[\mathbf{x}]_{\mathcal{B}} = R_{\mathcal{B}\leftarrow\mathcal{C}}\,[\mathbf{x}]_{\mathcal{C}}.$$

The general case is the same. Only the notation becomes more obscure. Let $\mathcal{B} = \{\mathbf{b}_1, \mathbf{b}_2, \ldots, \mathbf{b}_n\}$ and $\mathcal{C} = \{\mathbf{c}_1, \mathbf{c}_2, \ldots, \mathbf{c}_n\}$ be bases of V. Define $R_{\mathcal{B}\leftarrow\mathcal{C}} = [[\mathbf{c}_1]_{\mathcal{B}} \mid [\mathbf{c}_2]_{\mathcal{B}} \mid \cdots \mid [\mathbf{c}_n]_{\mathcal{B}}]$. Then for any $\mathbf{x}$ in V, $[\mathbf{x}]_{\mathcal{B}} = R_{\mathcal{B}\leftarrow\mathcal{C}}\,[\mathbf{x}]_{\mathcal{C}}$. There is a clever trick that can be used to compute transition matrices given two bases. We illustrate with an example. Let $\mathcal{B} = \{(2,0,1), (1,2,0), (1,1,1)\}$ and $\mathcal{C} = \{(6,3,3), (4,-1,3), (5,5,2)\}$. Form the augmented matrix $[\mathcal{B} \mid \mathcal{C}]$ and use elementary matrices on the left to produce the identity matrix:

$$\left[\begin{array}{ccc|ccc} 2 & 1 & 1 & 6 & 4 & 5 \\ 0 & 2 & 1 & 3 & -1 & 5 \\ 1 & 0 & 1 & 3 & 3 & 2 \end{array}\right] \to \left[\begin{array}{ccc|ccc} 1 & 0 & 0 & 2 & 2 & 1 \\ 0 & 1 & 0 & 1 & -1 & 2 \\ 0 & 0 & 1 & 1 & 1 & 1 \end{array}\right].$$

Then $R_{\mathcal{B}\leftarrow\mathcal{C}} = \begin{bmatrix} 2 & 2 & 1 \\ 1 & -1 & 2 \\ 1 & 1 & 1 \end{bmatrix}$. We finally make the connection with invertible matrices.

THEOREM D.8
With notation as above

1. $R_{\mathcal{A}\leftarrow\mathcal{B}}R_{\mathcal{B}\leftarrow\mathcal{C}} = R_{\mathcal{A}\leftarrow\mathcal{C}}$.

2. $R_{\mathcal{B}\leftarrow\mathcal{B}} = I$.

3. $(R_{\mathcal{B}\leftarrow\mathcal{C}})^{-1}$ *exists and equals* $R_{\mathcal{C}\leftarrow\mathcal{B}}$. *Moreover,* $[\mathbf{x}]_{\mathcal{B}} = R_{\mathcal{B}\leftarrow\mathcal{C}}\,[\mathbf{x}]_{\mathcal{C}}$ *iff* $(R_{\mathcal{B}\leftarrow\mathcal{C}})^{-1}\,[\mathbf{x}]_{\mathcal{B}} = [\mathbf{x}]_{\mathcal{C}}$.

PROOF The proof is left to the reader. ∎

Next, we tackle the problem of attaching a matrix to any linear transformation between two vector spaces that have finite bases. First, recall that a linear transformation is a function that preserves all the vector space structure, namely, addition and scalar multiplication. So let $T : V \to W$. T is a *linear transformation* iff (1) $T(\mathbf{x}+\mathbf{y}) = T(\mathbf{x}) + T(\mathbf{y})$ for all $\mathbf{x}$, $\mathbf{y}$ in V and (2) $T(\mathbf{x}r) = T(\mathbf{x})r$ for all $\mathbf{x}$ in V and r in $\mathbb{C}$. We shall now see how to assign a matrix to T relative to a pair of bases.

As above, we start with a simple illustration and then generalize. Let $\mathcal{B} = \{\mathbf{b}_1, \mathbf{b}_2\}$ be a basis of V and $\mathcal{C} = \{\mathbf{c}_1, \mathbf{c}_2, \mathbf{c}_3\}$ be a basis for W. Now $T(\mathbf{b}_1)$ is a vector in W and so must be uniquely expressible in the $\mathcal{C}$ basis, say $T(\mathbf{b}_1) = \mathbf{c}_1\alpha + \mathbf{c}_2\beta + \mathbf{c}_3\gamma$. The same is true for $T(\mathbf{b}_2)$, say $T(\mathbf{b}_2) = \mathbf{c}_1\delta + \mathbf{c}_2\varepsilon + \mathbf{c}_3\rho$. Define the *matrix* of T relative to the bases $\mathcal{B}$ and $\mathcal{C}$ by

$$Mat(T;\mathcal{B},\mathcal{C}) = \begin{bmatrix} \alpha & \delta \\ \beta & \varepsilon \\ \gamma & \rho \end{bmatrix} = \left[[T(\mathbf{b}_1)]_{\mathcal{C}} \mid [T(\mathbf{b}_2)]_{\mathcal{C}}\right].$$

Then a remarkable thing happens. Let $\mathbf{x}$ be a vector in V. Then say $\mathbf{x} = \mathbf{b}_1 r_1 + \mathbf{b}_2 r_2$. Then $[\mathbf{x}]_{\mathcal{B}} = \begin{bmatrix} r_1 \\ r_2 \end{bmatrix}$. Also, $T(\mathbf{x}) = T(\mathbf{b}_1 r_1 + \mathbf{b}_2 r_2) = T(\mathbf{b}_1) r_1 + T(\mathbf{b}_2) r_2 = (\mathbf{c}_1\alpha + \mathbf{c}_2\beta + \mathbf{c}_3\gamma) r_1 + (\mathbf{c}_1\delta + \mathbf{c}_2\varepsilon + \mathbf{c}_3\rho) r_2 = \mathbf{c}_1(\alpha r_1 + \delta r_2) + \mathbf{c}_2(\beta r_1 + \varepsilon r_2) + \mathbf{c}_3(\gamma r_1 + \rho r_2)$. Thus $Mat(T(\mathbf{x});\mathcal{C}) = \begin{bmatrix} \alpha r_1 + \delta r_2 \\ \beta r_1 + \varepsilon r_2 \\ \gamma r_1 + \rho r_2 \end{bmatrix} = \begin{bmatrix} \alpha & \delta \\ \beta & \varepsilon \\ \gamma & \rho \end{bmatrix} \begin{bmatrix} r_1 \\ r_2 \end{bmatrix} = Mat(T;\mathcal{B},\mathcal{C}) Mat(x;\mathcal{B})$. In other words,

$$[T(\mathbf{x})]_{\mathcal{C}} = Mat(T;\mathcal{B},\mathcal{C})[\mathbf{x}]_{\mathcal{B}}.$$

The reader may verify that this same formula persists no matter what the finite cardinalities of $\mathcal{B}$ and $\mathcal{C}$ are.

We end with a result that justifies the weird way we multiply matrices. The composition of linear transformations is again a linear transformation. It turns out that the matrix of a composition is the product of the individual matrices. More precisely, we have the following theorem.

THEOREM D.9

Suppose V, W, and P are complex vector spaces with bases $\mathcal{B}$, $\mathcal{C}$, $\mathcal{A}$, respectively. Suppose $T : V \to W$ and $S : W \to P$ are linear transformations. Then $Mat(S \circ T;\mathcal{B},\mathcal{A}) = Mat(S;\mathcal{C},\mathcal{A}) Mat(T;\mathcal{B},\mathcal{C})$.

PROOF Take any $\mathbf{x}$ in V. Then, using our formulas above,

$Mat(S\circ T;\mathcal{B},\mathcal{A})[\mathbf{x}]_{\mathcal{B}} = [(S\circ T)(\mathbf{x})]_{\mathcal{A}} = [S(T(\mathbf{x}))]_{\mathcal{A}} = Mat(S;\mathcal{C},\mathcal{A})[T(\mathbf{x})]_{\mathcal{C}} = Mat(S;\mathcal{C},\mathcal{A})(Mat(T;\mathcal{B},\mathcal{C})[\mathbf{x}]_{\mathcal{B}})$. Now this holds for all vectors $\mathbf{x}$ so it holds when we choose basis vectors for $\mathbf{x}$. But $[\mathbf{b}_1]_{\mathcal{B}} = \begin{bmatrix} 1 \\ 0 \\ \vdots \\ 0 \end{bmatrix}$, and generally,

$[\mathbf{b}_i]_{\mathcal{B}} = \mathbf{e}_i$, the standard basis vector. By arguments we have seen before, $Mat(S \circ T; \mathcal{B}, \mathcal{A})$ must equal the product $Mat(S;\mathcal{C},\mathcal{A})Mat(T;\mathcal{B},\mathcal{C})$ column by column. Thus, the theorem follows. □

If we had not yet defined matrix multiplication, this theorem would be our guide since we surely want this wonderful formula relating composition to matrix multiplication to be true.

APD Exercise Set 2

1. Find the coordinate matrix $Mat(\mathbf{x}; \mathcal{B})$ of $\mathbf{x} = (3, -1, 4)$ relative to the basis $\mathcal{B} = \{(1, 0, 0), (1, 1, 0), (1, 1, 1)\}$. Now find the coordinate matrix of $\mathbf{x}$ relative to $\mathcal{C} = \{(2, 0, 0), (3, 3, 0), (4, 4, 4)\}$. Suppose $Mat(\mathbf{v};\mathcal{B}) = \begin{bmatrix} 4 \\ 7 \\ 10 \end{bmatrix}$. Find v. Suppose $Mat(v;\mathcal{C}) = \begin{bmatrix} 4 \\ 7 \\ 10 \end{bmatrix}$. Find $\mathbf{v}$.

2. Let $\mathcal{B} = \{(1,\ 0,\ 1), (6,\ -4,\ 2), (-1,\ -6,\ 1)\}$ and $\mathcal{C} = \{(-1, -1, 0), (-1, -3, 2), (2, 3, -7)\}$. Verify that these are bases and find the transition matrix $R_{\mathcal{C}\leftarrow\mathcal{B}}$ and $R_{\mathcal{B}\leftarrow\mathcal{C}}$. If $Mat(\mathbf{v}; \mathcal{B}) = \begin{bmatrix} 2 \\ 4 \\ 6 \end{bmatrix}$, use the transition matrix to find $Mat(v; \mathcal{C})$.

3. Argue that $(R_{\mathcal{C}\longleftarrow\mathcal{B}})^{-1} = R_{\mathcal{B}\leftarrow\mathcal{C}}$, $R_{\mathcal{B}\leftarrow\mathcal{B}} = I$, and $R_{\mathcal{A}\leftarrow\mathcal{B}}R_{\mathcal{B}\leftarrow\mathcal{C}} = R_{\mathcal{A}\leftarrow\mathcal{C}}$.

4. Argue that if $\mathcal{C}$ is any basis of V and S is any invertible matrix over $\mathbb{C}$, then there is a basis $\mathcal{B}$ of V so that $S = R_{\mathcal{C}\longleftarrow\mathcal{B}}$.

5. Investigate the map $Mat(\_; \mathcal{B},\mathcal{C}) : Hom(V, W) \rightarrow \mathbb{C}^{m\times n}$ that assigns a $\mathbb{C}$-linear map between V and W and the matrix in $\mathbb{C}^{m\times n}$ relative to the bases $\mathcal{B}$ of V and $\mathcal{C}$ of W.

6. Suppose $T : V \rightarrow V$ is invertible. Is $Mat(T; \mathcal{B},\mathcal{B})$ an invertible matrix for any basis $\mathcal{B}$ of V?

7. How would you formulate the idea of the kernel and image of a $\mathbb{C}$-linear map?

8. Is $T_r : V \rightarrow V$ by $T_r(\mathbf{x}) = \mathbf{x}r$ a $\mathbb{C}$-linear map on the V? If so, what is its image and what is its kernel?

9. Let T be any $\mathbb{C}$-linear map between two vector spaces. Argue that
   (a) $T(\vec{0}) = \vec{0}$.
   (b) $T(-\mathbf{x}) = -T(\mathbf{x})$ for all $\mathbf{x}$.
   (c) $T(\mathbf{x} - \mathbf{y}) = T(\mathbf{x}) - T(\mathbf{y})$ for all $\mathbf{x}$ and $\mathbf{y}$.

10. Argue that a linear map is completely determined by its action on a basis.

Further Reading

[B&R, 1986(2)] T. S. Blyth and E. F. Robertson, *Matrices and Vector Spaces*, Vol. 2, Chapman & Hall, New York, (1986).

[B&B, 2003] Karim Boulabiar and Gerard Buskes, After the Determinants Are Down: A Criterion for Invertibility, The American Mathematical Monthly, Vol. 110, No. 8, October, (2003).

[Brown, 1988] William C. Brown, *A Second Course in Linear Algebra*, John Wiley & Sons, New York, (1988).

[Max&Mey, 2001] C. J. Maxon and J. H. Meyer, How Many Subspaces Force Linearity?, The American Mathematical Monthly, Vol. 108, No. 6, June–July, (2001), 531–536.

Index

D

E

F

G

H

I

J

K

L

M

S

T

U

V

W

Z